# Zoological Monographs

Volume 5

The book series *Zoological Monographs* is devoted to the study of the animal kingdom and covers all aspects of animals, both living and extinct, including taxonomy, systematics and evolution, biodiversity and biogeography, ecology, morphology and anatomy, physiology and behavior. The volumes in the series either discuss the biology of a single animal group, approach a topic from an interspecies level, or present recent methodologies. The series is intended for graduate students, university teachers, scientists and professionals.

More information about this series at http://www.springer.com/series/15188

Harald W. Krenn
Editor

# Insect Mouthparts

## Form, Function, Development and Performance

Springer

*Editor*
Harald W. Krenn
Department of Integrative Zoology
University of Vienna
Vienna, Austria

ISSN 2523-3904 ISSN 2523-3912 (electronic)
Zoological Monographs
ISBN 978-3-030-29653-7 ISBN 978-3-030-29654-4 (eBook)
https://doi.org/10.1007/978-3-030-29654-4



This Springer imprint is published by the registered company Springer Nature Switzerland AG.
The registered company address is: Gewerbestrasse 11, 6330 Cham, Switzerland

*Natura maxima miranda in minimis*
Carl von Linné

# Preface

Insects comprise the largest group of animals on Earth. The high complexity and functional morphology of insect mouthparts attracted my attention since my early years of study at the University of Vienna. I was puzzled by the perplexing diversity and performance of these minute feeding tools. The mouthparts of any insect—dead or alive—reflect the diet and provide insights into the ecology and life history of the respective species. Insect feeding behavior and therefore also insect mouthparts have a major impact on terrestrial ecosystems and influence much of human activity and health. As a consequence, entomologists have been studying various aspects of the morphology and function of feeding organs over a long period, and all entomological textbooks deal with morphology and evolution of insect mouthparts. However, to date there has been no comprehensive book on insect mouthparts. This compendium hopes to fill this gap and provides information on various contemporary aspects of insect mouthparts. Certainly, the book does not claim to contain all available information on insect mouthparts, but it includes several overviews and presents recent scientific advances that have contributed to an increasing understanding of the form, function, development, and performance of these organs.

This book is aimed at fostering and encouraging students and academics to pursue research on mouthparts. It is an attempt at summarizing the large body of knowledge about insect mouthparts and delineating new perspectives of investigation. This volume covers various aspects and presents intriguing results on the complexity of the feeding organs of insects. The topics start with overviews on functional morphology of mouthparts and the developmental genetic patterning of mouthparts in model species and further range from biomechanics to morphology and evolution of mouthparts in the phylogenetic context of particular taxa, ecomorphological aspects of flower-visiting groups, as well as morphology and feeding ecology in extinct fossil insects.

This book includes 17 chapters which were written by various experts who summarized the current knowledge or who presented novel avenues of research and new perspectives on form, function, and evolution of insect mouthparts.

Chapter 1 summarizes facts about the impact of insect feeding and the ecological roles of insects. It stresses that all feeding activities are associated with mouthparts and evaluates the ecological and commercial impact of insect feeding.

Chapter 2 provides an overview on the functional morphology of mouthparts and should be regarded as an introduction for students and all those who are new to insect mouthparts. In addition to the description on the general morphology and function, information about mouthpart morphology of the various insect orders is summarized and examples are selected to illustrate current research in functional morphology of mouthparts.

Chapter 3 deals with the huge diversity of sucking mouthparts in a comparative approach using functional types. Many examples of parallel evolution and convergent adaptations of feeding behavior in relation to particular food sources are presented.

Chapter 4 reviews the filter-feeding techniques of aquatic insects. The immature stages of various groups are highly important for maintaining clear water bodies, recycling of organic material, and transport of organic substances out of freshwater habitats back to the surrounding terrestrial habitats.

Chapter 5 is devoted to the developmental biology of insect mouthparts. The current knowledge of the genetic patterning and the modification of serially homologous appendages to different morphology and feeding techniques are presented. Concepts of evolution of serial homologous organs are discussed to understand organ evolution in principle.

Chapter 6 summarizes the current knowledge of the mouthpart morphology of basal entognathous and ectognathous taxa of Hexapoda. In addition, this chapter provides a quantitative analysis of the mandibular lever mechanics across this diverse taxon together with an exploration of phylogenetic signal.

Chapter 7 presents the feeding apparatus of cockroaches as a biomechanical model of biting–chewing mouthparts in the Polyneoptera. This comparative contribution combines information on microanatomy, morphology, kinematics, mechanics, and the muscular control of the various mouthpart components.

Chapter 8 introduces the biophysics of fluid-feeding in insects and explains the underlying parameters and physical laws that govern fluid uptake with a proboscis. This chapter discusses the associated wetting phenomena, the mechanisms of fluid uptake by proboscises, and the biomechanics of proboscis/sucking pump interactions.

Chapter 9 focuses on the cuticle surface properties of the proboscis of Lepidoptera. Recent studies are summarized demonstrating that the hydrophilic and hydrophobic properties of the cuticle are crucial in explaining the performance of fluid-feeding insects in general.

Chapter 10 presents an overview on the mouthparts of ants and focuses on a little studied feature: the fluid-feeding capabilities of ant workers that collect liquids as a main source of carbohydrate for the colony. A new method is presented to compare the adhesive surface areas of mouthparts.

Chapter 11 outlines the evolution of mouthparts in Neuroptera. This chapter compares larval and adult mouthparts in context with feeding strategies in different lineages, as well as the developmental reorganization in the pupal stage.

Chapter 12 compiles the current knowledge of the mouthparts of beetles. This includes recent findings on the form and function in larva and adults and a reconstruction of the evolution of the character system in the megadiverse Coleoptera.

Chapter 13 is devoted to the mouthparts of anthophilous Coleoptera that have been rarely studied in detail. This chapter presents an overview of previously studied examples and compiles data on the floral associations of these beetles.

Chapter 14 summarizes the available information on the evolutionary morphology of the mouthparts of Vespidae. These insects illustrate the adaptations of mouthparts to various tasks such as nest construction, hunting, and feeding on various food sources including nectar feeding with greatly elongated proboscises that are placed in context with the evolution of various lifestyles of wasps, such as eusociality.

Chapter 15 emphasizes nectar-feeding insects that evolved extreme proboscis lengths as adaptations to deep-tubed flowers. This chapter compares the morphology and evolution of the feeding apparatus in orchid bees, hawk moths and butterflies, and some brachyceran flies with proboscis length that exceeds the body length by far. This contribution lists the recorded nectar host plants of insects with very long proboscises.

Chapter 16 deals with the peculiar development of stylets inside the head of Hemiptera during intermolt periods and assembly into a functional piercing–sucking bundle during ecdysis. These important but poorly known phenomena are discussed based on literature and new observations.

Chapter 17 summarizes the fossil record of insect mouthparts. It presents the evolution of feeding strategies and functional mouthpart classes and provides inferences on the origin of major evolutionary events in mouthpart evolution.

On behalf of all the authors, I hope the contributions of this book will stimulate further research related to the various subjects explored in these chapters.

The editor is thankful to all the authors who have contributed excellent articles. This book relied on the contributions of many scientists with expert knowledge from a range of disciplines in addition to entomology, such as evolutionary morphology, hydrobiology, developmental genetics, biophysics, systematics, functional morphology, insect ecology, insect physiology, and paleobiology. I am grateful and honored by the positive collaboration with all authors over the past year. The interest and enthusiasm they have shown in working on this book has encouraged me in this undertaking.

I acknowledge all those who have made photos available, and all publishers who have provided the copyrights for figures from already published articles. Thanks to Julia Bauder, Florian Karolyi, and Norbert Milasowszki who helped edit manuscripts and to all the colleagues who contributed during the review process by reading draft manuscripts and providing valuable comments. Special thanks go to my wife Barbara-Amina Gereben-Krenn, who is also an entomologist, has provided me with loving encouragement and support, and has helped edit many of the articles.

I would also like to acknowledge the rector, of the University of Vienna, Heinz W. Engl who reduced my teaching duties over 6 months, which enabled me to focus on editing this book.

Especially, I wish to thank Springer Publisher for their endorsement to produce a book on mouthparts of insects. I hope that this volume will be an important reference in the future and will foster research on insect mouthparts and feeding.

Vienna, Austria
May 2019

Harald W. Krenn

# Contents

# Chapter 1
# Introduction: Ecological Importance of Insect Feeding

**Harald W. Krenn**

**Abstract** Insects are extremely diverse arthropods with highly diverse lifestyles. All kinds of organic material may be used by insects as food. Their feeding activities have an enormous ecological impact on all terrestrial and freshwater ecosystems. Insect feeding contributes to pollination, nutrient recycling, pest control and water purification, whereas it can be destructive to wild and cultivated plants and stored products and may transmit pathogens to plants and animals including humans. In this context, form and function of mouthparts are crucial to understand the feeding behaviour as well as the ecological and economic importance of insects.

## 1.1 Insect Feeding

Insects are the most diverse and abundant group of arthropods. They are the dominant group of invertebrates in most terrestrial and freshwater habitats of the world; and their activity, for instance, feeding, provides many important ecosystem functions. Insects have evolved diverse lifestyles comprising a great number of feeding preferences, feeding modes and specialized adaptations to various food sources. The nutrition of different groups ranges from phytophagy (i.e. feeding on various plant tissues; plant sap, nectar and pollen; and seeds, as well as aquatic grazing and consumption of tissue from induced plant galls), carnivory (i.e. feeding on, within or off other animals), fungivory (i.e. feeding on spores, hyphae and fungal bodies) and detritivory (i.e. feeding on decaying organic matter including saprophagy, coprophagy and xylophagy) to filter-feeding on suspended particles. Phytophagous insects make up to 25% of all living species of animals and have an enormous impact on plants and terrestrial food webs (Bernays 2009). At least a quarter of all insects are estimated to be parasites, parasitoids or predators of other arthropod species (Chapman et al. 2016). Detritivorous insects comprise a significant component of soil arthropods and are an integral part of subterranean food webs (Bagyaraj et al. 2016). Many insect species are highly specialized on particular food sources, while others are generalized and omnivorous showing various combinations of the major feeding types.

H. W. Krenn (✉)
Department of Integrative Zoology, University of Vienna, Vienna, Austria
e-mail: harald.krenn@univie.ac.at

 
H. W. Krenn (ed.), *Insect Mouthparts*, Zoological Monographs 5,
https://doi.org/10.1007/978-3-030-29654-4_1

During their life cycle, many insects switch their feeding style, which is often combined with a change in habitat, for example, a shift from aquatic immatures to terrestrial imagines or from soil-dwelling larvae to epigaeic or phytophilous adults.

## 1.2 Importance of Insect Feeding

Through their feeding activity, insects perform essential ecological roles in terrestrial and freshwater ecosystems. The estimated number of 5.5 million species (Stork 2018), each with innumerable individuals, makes it impossible to gain a realistic view of how much food insects consume in their habitats per year. However, the enormous impact of insect feeding on natural habitats and human-dominated environments including indoor habitats has been demonstrated (e.g. Scudder 2009; Chakravarthy and Sridhara 2016; Leong et al. 2017). This includes mutualistic interactions like pollination, the dispersal of seeds and decomposition but also antagonistic interactions, such as phytophagy, predation and parasitism. In this way, insects significantly contribute to vital ecosystem functions that are directly or indirectly associated with their feeding activities and the performance of their mouthparts (Fig. 1.1). The ecosystem services provided by insects may be beneficial to humans with a high economic value (Losey and Vaughan 2006; Chakravarthy and Sridhara 2016). However, insect feeding activities can also have a significant adverse impact on human beings, agricultural crops and livestock.

### *1.2.1 Pollination*

The most important beneficial activity that is directly related with insect feeding is the pollination of angiosperms. When insects search for floral food resources, i.e. mainly pollen and/or nectar, they transfer pollen onto the stigma of conspecific flowers. Through this activity, insects pollinate about 90% of the flowering plants on Earth (Ollerton et al. 2011). This mutualistic interaction between plants and insects was the starting point in the evolution of flowering plants in the Cretaceous (Grimaldi and Engel 2005; Hu et al. 2007). Still in the present, insect pollinators play a significant role in diversification and evolution of angiosperms (Van der Niet and Johnson 2012; Gervasi and Schiestl 2017).

Insects provide pollination services for a range of crop plants in agricultural landscapes. For the 100 most economically important crops used for human food, the economic value of insect pollination was estimated at 153 billion euros per year (Potts et al. 2010). Considering that only a small fraction of plants are used by humans, the feeding activities of flower-visiting insects are invaluable in the reproduction of many angiosperm plants.

Honey bees (*Apis mellifera*) may provide an estimate of the number of flowers visited by a pollinating insect and the mouthpart performance during food gathering. Mouthparts can be seen as minute tools for collecting of even very small amounts of nectar from

**Fig. 1.1** Insects feeding activities; (**a**) herbivorous locust (*Schistocerca gregaria*) takes up grass; (**b**) carnivorous lady beetle (*Harmonia axyridis*) feeds on plant sap-sucking aphids (photo by courtesy of H. May); (**c**) carnivorous ground beetle larva (*Nebria* sp.) eats a soil-living springtail; (**d**) nectar-feeding honey bee (*Apis mellifera*) with pollen grains on the body; (**e**) blood-feeding tsetse fly (*Glossina morsitans*) may transmit trypanosomes, the pathogen of sleeping sickness (photo by courtesy of R. Pospischil); (**f**) aquatic larva of *Simulium* sp. (Simuliidae) feeds on suspended particles in freshwater

flowers. Depending on the available nectar in flowers, approximately 1 million apple blossoms, about 3.3 million rapeseed flowers, or more than 10 million red clover flowers need to be visited by workers of the western honey bees (*Apis mellifera*) to produce 1 kg of honey (Farkas and Zajacz 2007; Morawetz pers. comm.). When considering these numbers in the context of global honey production, estimated to be 1,003,627 tons (FAO 2018), an inordinately large number of flower visits would be required.

In addition to honey bees and other Apidae, the significant role of non-bee pollinators in the global agricultural production is well supported (Rader et al. 2016; Ssymank et al. 2017). The activity of all flower-visiting insects would significantly add more flower visits by countless numbers of species in the global annual pollination service, without which there would be only inconspicuous flowers and less fruits or vegetables available on Earth. Considering that only a small fraction of plants are used by humans, the feeding activities of all flower-visiting insects are invaluable in the reproduction of angiosperm plants. In this way, pollination by insects crucially supports the primary production of biomass and plant biodiversity in terrestrial ecosystems by their feeding activities linked to the action of their mouthparts.

### 1.2.2 Nutrient Recycling

Many insects and other arthropods are responsible for litter breakdown, dung burial, turnover of soil through burrowing activities and accelerating the return of nutrients to the soil through their feeding behaviour and defecation. Detritus-feeding activities of myriads of soil-dwelling insects and non-insect hexapods achieve decomposition and nutrient recycling in terrestrial ecosystems that represent an enormous economic value (Scudder 2009). For example, the leaf-cutter ants are indirectly responsible for improving soil fecundity in Neotropical forests. Billions of worker ants cut off leaves by using their mouthparts; they carry the plant material into their underground nests and feed it to a symbiotic fungus which serves as food for these ants (Hölldobler and Wilson 1990). The waste generated by the ant colony and the metabolic products of the fungi are essential for soil fertility. In the same sense, termites are regarded as eco-engineers that are perhaps the most impressive decomposers of dead wood and plant material in the subtropical and tropical regions of the world (Scudder 2009; Bagyaraj et al. 2016). Furthermore, in freshwater habitats, many insects perform an essential ecosystem function by filter-feeding suspended organic microparticles. Vast numbers of insects help maintain clear water bodies through their feeding activities and provide purification of freshwater systems (Gullan and Cranston 2014). In this way, aquatic immatures and larvae are important contributors to link food webs through the uptake of nutrients in water and returning them to terrestrial ecosystems by the winged adult insects via their dispersal flights.

### 1.2.3 Pest Control

Numerous insect predators, parasitoids and parasites are invaluable for biological control and the balance of ecosystems. The activities associated with the feeding behaviour of these insects provide natural population control of other insects, arthropods, invertebrates and even some vertebrates. There is no estimate available for the number of insects that are eaten by other insects or are parasitized by larvae. However, biological pest control by wild or native insects is valued to be approximately USD$13.6 billion per year in the USA (Losey and Vaughan 2006). This estimate does not include mass-reared insect species, like parasitoid wasps or lady beetles that are regularly used in large numbers to protect crops grown in open fields or greenhouses against pest insects, without the use of insecticides.

### 1.2.4 Adverse Effects of Insect Feeding

Insects can destroy crops, parasitize livestock and humans or become a nuisance and health hazard to other organisms (McGavin 2016). In terms of the significant negative impact on the human population, feeding activities of many insects can have tremendous adverse effects on agriculture and horticulture and may be harmful to domestic animals and humans by transmitting pathogens. Defoliation caused by insects can alter ecosystems, and the feeding activities of keystone species cause a strong top-down effect on habitats (Carson et al. 2004), such as the feeding of a swarm of desert locusts with devastating effects on dryland habitats (Baron 1972). It was estimated that approximately one-fifth of all crops grown worldwide and stored products are lost to herbivorous insects annually (Sallam 1999), and further damage is caused by plant diseases transmitted by feeding insects. Destructive feeding activities of insects can also cause damage to wooden structures and a wide range of natural materials and fabrics. Despite this, less than 2% of phytophagous insects are potential pests of crops and agricultural products (Scudder 2009). Yet more severe, over 15% of human beings are affected by an insect-borne illness such as sleeping sickness, river blindness, yellow fever, malaria, etc., predominantly in tropical and subtropical regions of the world. More than 500 million people are at risk of exposure, and more than 1 million die from such diseases every year (Murray et al. 2012). The transmission of pathogens to plants, animals and humans occurs during insect feeding; thus, these activities have great importance for other organisms. Therefore, the knowledge on life history and feeding ecology of insects is significant and helps to recognize dangers from insects.

## 1.3 Why Study Insect Mouthparts?

All feeding activities are of crucial importance in the various ecological roles of insects, and feeding performance is always associated with form and function of the mouthparts. Most entomological textbooks provide descriptions of insect mouthparts and discuss various examples of functional anatomy, feeding behaviour and techniques of food uptake. The study of insect feeding organs and their function was recognized in the classical textbook of Berlese (1909) and in the benchmarking volumes of Weber (1933) and Snodgrass (1935). They illustrated many examples in detail and discussed how homologous structures developed according to functional demands in context with feeding ecology and particular food sources. Similarly, this applies to recent textbooks such as Gullan and Cranston (2014), Beutel et al. (2014), Chapman et al. (2016) and Grimaldi and Engel (2005)—the latter focuses on fossil insects. All devote chapters to these complex organs and give much emphasis to morphology, feeding ecology, physiology and evolution or the fossil records of mouthparts. Furthermore, a number of comparative studies emphasized the eco-morphology of insect mouthparts (e.g. Smith 1985; Krenn et al. 2005; Krenn and Aspöck 2012).

Studying mouthparts provides information about the feeding ecology of a particular insect species. The morphology of mouthparts allows for inference as to what type of food is consumed, such as plants or animals, solid or liquid food or dead or living organisms, even when only dead museum specimens or fossils are available.

Mouthparts of fossil insects may give information about past feeding preferences and can shed light on the evolution of food webs and the ecological roles of extinct insects millions of years ago (Labandeira 1997; Nel et al. 2018; see also Chap. 17). Furthermore, mouthparts provide fascinating examples of organ evolution that can be easily used to argue in favour of evolutionary theory since feeding organs impressively demonstrate the power of natural selection. Recently, biophysicists discovered the amazing functional performance of insect mouthparts and have started investigating feeding organs using principles of biomechanics, material science and biomimetics (e.g. Kornev et al. 2016; Li et al. 2017; Zhang et al. 2018; see also Chaps. 8 and 9). These new perspectives likewise give explanations for feeding performance under various ecological aspects. Thus, studying mouthparts provides a variety of information that can be used to understand different aspects of insect feeding and can help to better evaluate the significance of insects in the ecosystems on Earth.

**Acknowledgements** I thank Annalie Melin for linguistic help and Barbara-Amina Gereben-Krenn and Florian Karolyi for comments and suggestions to improve the manuscript. I am grateful to Reiner Pospischil and Helge May for providing excellent photos of feeding insects.

## References

Bagyaraj DJ, Nethravathi CJ, Nitin KS (2016) Soil biodiversity and arthropods: role in soil fertility. In: Chakravarthy AK, Sridhara S (eds) Economic and ecological significance of arthropds in diversified ecosystems. Springer Science+Business Media, Singapore, pp 17–51

Baron S (1972) The desert locust. Charles Scribner's Sons, New York

Berlese A (1909) Gli insetti, vol 1. Milan

Bernays EA (2009) Chapter 201 – Phytophageous insects. In: Resh VH, Cardé RT (eds) Encyclopedia of insects, 2nd edn. Academic, London, pp 798–800

Beutel RG, Friedrich F, Ge S-Q, Yang X-K (2014) Insect morphology and phylogeny – a textbook for students of entomology. De Gruyter Graduate, Berlin

Carson WP, Cronin JP, Long ZT (2004) A general rule for predicting when insects will have strong top-down effects on plant communities: on the relationship between insect outbreaks and host concentration. In: Weisser WW, Sieman E (eds) Insects and ecosystem function. Springer, Berlin, pp 193–211

Chakravarthy AK, Sridhara S (eds) (2016) Economic and ecological significance of arthropods in diversified ecosystems. Springer Science+Business Media, Singapore

Chapman RF, Simpson SJ, Douglas AE (eds) (2016) The insects structure and function, 5th edn. Cambridge University Press, Cambridge

FAO (2018) Food and agriculture organization of the United Nations. www.fao.org/faostat. Accessed 25 Oct 2018

Farkas Á, Zajácz (2007) Nectar production for the Hungarian honey industry. Eur J Plant Sci Biotechnol 1:125–151

Gervasi DDL, Schiestl FP (2017) Real-time divergent evolution in plants driven by pollinators. Nat Commun 8:14691

Grimaldi DA, Engel MS (2005) Evolution of the insects. Cambridge University Press, New York

Gullan PJ, Cranston PS (2014) The insects: an outline of entomology, 5th edn. Wiley, New York

Hölldobler B, Wilson OE (1990) The ants. Spinger, Berlin

Hu S, Dilcher DL, Jarzen DM, Taylor WD (2007) Early steps of angiosperm-pollinator coevolution. Proc Natl Acad Sci U S A 105:240–245

Kornev KG, Monaenkova D, Adler PH, Beard CE, Lee W-K (2016) The butterfly proboscis as a fiber-based, self-cleaning, micro-fluidic system. In: Proceedings of SPIE 9797, Bioinspiration, Biomimetics Bioreplication 2016, 979705

Krenn HW, Aspöck H (2012) Form, function and evolution of the mouthparts of blood-feeding Arthropoda. Arthropod Struct Dev 41:101–118

Krenn HW, Plant J, Szucsich NU (2005) Mouthparts of flower-visiting insects. Arthropod Struct Dev 34:1–40

Labandeira CC (1997) Insect mouthparts: ascertaining the paleobiology of insect feeding strategies. Annu Rev Ecol Syst 28:153–193

Leong M, Bertone MA, Savage AM, Bayless KM, Dunn RR, Trautwein MD (2017) The habitats humans provide: factors affecting the diversity and composition of arthropods in houses. Sci Rep 7:15347

Li L, Guo C, Xu S, Li X, Han C (2017) Morphology and nanoindentation properties of mouthparts in *Cyrtotrachelus longimanus* (Coleoptera: Curculionidae). Microsc Res Tech 80:704–711

Losey JE, Vaughan M (2006) The economic value of ecological services provided by insects. Bioscience 56:311–323

McGavin GC (2016) Prologue. In: Chapman RF (Edited by Simpson SJ, Douglas AE) The insects structure and function, 5th edn. Cambridge University Press, Cambridge, pp xii–xxxi

Murray CJL, Rosenfeld LC, Lim SS, Andrews KG, Foreman KJ (2012) Global malaria mortality between 1980 and 2010: a systematic analysis. Lancet 379(9814):413–431

Nel P, Bertrand S, Nel A (2018) Diversification of insects since the Devonian: a new approach based on morphological disparity of mouthparts. Sci Rep 8:3516

Ollerton J, Winfree R, Tarrent S (2011) How many flowering plants are pollinated by animals? Oikos 120:321–326

Potts SG, Biesmeijer JC, Kremen C, Neumann P, Schweiger O, Kunin WE (2010) Global pollinator declines: trends, impacts and driver. TREE 25:345–353

Rader R, Bartomeus I, Garibaldi LA, Garratte MPD, Howlett BG, Winfree R, Cunningham SA, Mayfield MM, Arthur AD, Andersson GKS, Bommarco R, Brittain C, Carvalheiro LG, Chacoff NP, Entling MH, Foully B, Freitas BM, Gemmill-Herren B, Ghazoul J, Griffin SR, Gross CL, Herbertsson L, Herzog F, Hipólito J, Jaggar S, Jauker F, Klein A-M, Kleijn D, Krishnan S, Lemos CQ, Lindström SAM, Mandelik Y, Monteiro VM, Nelson W, Nilsson L, Pattemore DE, Pereira N, Pisanty G, Potts SG, Reemer M, Rundlöf M, Sheffield CS, Scheper J, Schüepp C, Smith HG, Stanley DA, Stout JC, Szentgyörgyi H, Taki H, Vergara CH, Viana BF, Woyciechowski M (2016) Non-bee insects are important contributors to global crop pollination. Proc Natl Aacd Sci U S A 113:146–151

Sallam MN (1999) Insect damage: damage on post-harvest. FAO, International Centre of Insect Physiology and Ecology (ICIPE), Nairobi, pp 1–37

Scudder GGE (2009) The importance of insects. In: Foottit RG, Adler PH (eds) Insect biodiversity: science and society. Blackwell, London, pp 7–32

Smith JJB (1985) Feeding mechanisms. In: Kerkut GA, Gilbert LI (eds) Comprehensive insect physiology, biochemistry and pharmacology, vol 4. Pergamon Press, pp 33–85

Snodgrass RE (1935) Principles of insect morphology. McGraw-Hill Book, New York (reprint with a new foreword by George C Eickwort, Cornell University Press 1993)

Ssymank A, Kearns CA, Pape T, Thompson CF (2017) Pollinating flies (Diptera): a major contribution to plant diversity and agricultural production. Biodiversity 9:86–89

Stork NE (2018) How many species of insects and other terrestrial arthropods are there on earth? Annu Rev Entomol 63:31–45

Van der Niet T, Johnson SD (2012) Phylogenetic evidence for pollinator-driven diversification of angiosperms. TREE 27:353–361

Weber H (1933) Lehrbuch der Entomologie. Gustav Fischer Verlag, Stuttgart (reprint Verlag von Otto Koeltz, Koenigstein – Taunus 1968)

Zhang C, Beard CE, Adler PH, Kornev KG (2018) Effect of curvature on wetting and dewetting of proboscises of butterflies and moths. R Soc Open Sci 5:171241

# Chapter 2
# Form and Function of Insect Mouthparts

**Harald W. Krenn**

**Abstract** Insect mouthparts are modified appendages of head segments that are adapted to exploit different food sources. This chapter describes the general mouthpart morphology of Hexapoda, introduces basic feeding types in insects, and illustrates mouthpart function. Insect mouthparts include three appendages, the paired mandibles, the paired maxillae, and the unpaired labium as well as additional head structures, the labrum, and the hypopharynx. The noninsect lineages of Hexapoda possess entognathous mouthparts, which are concealed inside the head, while ectognathous mouthparts of Insecta articulate externally on the head capsule. Especially in winged insects, characteristic adaptations of mouthparts evolved in context with various food sources resulting in feeding specialization and enhanced functional performance. Insect mouthparts can be categorized in three principal functional types: (1) mandibulate biting and chewing mouthparts, (2) haustellate mouthparts forming variously composed proboscises, and (3) filter-feeding mouthparts of aquatic immature stages. The diversity of functional types and remarkable modifications are presented in various examples; characteristic patterns of mouthpart evolution are discussed. The composition of mouthparts in the various hexapod orders is summarized in a table. Additional functions, like defense, brood care, and male-male competition, modified the mouthparts in some insects. Rudimentary mouthparts are found in some nonfeeding adults of various insect taxa.

H. W. Krenn (✉)
Department of Integrative Zoology, University of Vienna, Vienna, Austria
e-mail: harald.krenn@univie.ac.at

 
H. W. Krenn (ed.), *Insect Mouthparts*, Zoological Monographs 5,
https://doi.org/10.1007/978-3-030-29654-4_2

## 2.1 Introduction

The feeding organs of Hexapoda[1] and Insecta[2] comprise modified arthropod appendages of the head segments. As in all arthropods, the mouthparts are a defined set of serially homologous limb structures (e.g., Snodgrass 1935; Matsuda 1965; Angelini and Kaufman 2005; Jockusch and Smith 2015), which serve as the main organs for the uptake and processing of food. The mouthparts of Insecta are highly diverse and have complex forms and functions. Their disparity in various taxa is the result of natural selection on different diets and optimized feeding techniques.

All entomological textbooks (e.g., Snodgrass 1935; Gullan and Cranston 2014; Beutel et al. 2014; Chapman et al. 2016) provide excellent general descriptions of mouthparts and give examples of characteristic modifications and functional types in various hexapods. The classical textbooks of Snodgrass (1935) and Weber (1933) precisely illustrate mouthpart morphology of many insects in previously unknown quality and still are valuable sources of information. The modern standard textbooks of entomology likewise include chapters which are devoted to the form and function of these complex organs. For example, Gullan and Cranston (2014) treat insect mouthparts as examples for understanding the functional types of various taxa in context with their lifestyle. Beyond this, Beutel et al. (2014) discuss the ground patterns in insect lineages and give additional information about the morphology of mouthparts, including their musculature in immature stages and adults of all hexapod orders in a phylogenetic context; this has been summarized in Table 2.1. The comprehensive textbook of Chapman et al. (2016) focuses on physiological and nutritional aspects of mouthparts and feeding. Although focused on fossilized insects, Grimaldi and Engel (2005) devote several chapters to mouthparts given that their morphology provides inference about the type of food that was consumed by extinct insects.

The aim of this chapter is to present a general description of hexapod mouthparts and the functional types in recent Insecta and to provide examples of characteristic modifications that illustrate their performance in the context of insect biology.

## 2.2 Entognathous and Ectognathous Mouthparts

The head of Hexapoda seems to comprise six fused segments. The labrum lies anterior to the mouth opening, while the paired mandibles and maxillae as well as the unpaired labium extend postorally from the last three head segments, called the gnathal segments (e.g., Posnien and Bucher 2010; Hartenstein and Chipman 2015; Jockusch and Smith 2015). Hexapoda, Insecta sensu lato, exhibit entognathous or

[1]The taxon Hexapoda (= insects sensu lato) comprises Protura, Collembola, Diplura, and Insecta sensu stricto (Kristensen 1975; Beutel et al. 2017).

[2]The taxon Insecta sensu stricto (= Ectognatha) comprises the Archaeognatha, Zygentoma (both wingless in all stages), and Pterygota (winged insects) (Kristensen 1975; Beutel et al. 2017).

**Table 2.1** Characteristic functional types and composition of mouthparts are summarized for major representatives in all orders of Hexapoda with remarks on life styles and feeding behavior. Particularly high diversity of mouthpart morphology occurs in Psocodea and in the species-rich lineages of Holometabola that also show differences in larvae and adults. Biting-chewing mouthparts refer to orthopteriod morphology; biting mouthparts possess mandibles without mola; sucking mouthparts are composed of different functional types of proboscises (see Sect. 2.5.2); filter-feeding mouthparts occur in aquatic insects. If fundamentally different, mouthpart compositions are listed in instars and adults (from Beutel et al. (2014) and further references); not all exceptions are mentioned within the taxa. Orientation of head: prognathous, mouthparts in horizontal position; orthognathous, mouthparts in vertical position; hypognathous, mouthparts directed caudally. Systematic arrangement was done according to Misof et al. (2014); common names are derived from Chapman et al. (2016). *lr* labrum; *ma* mandible; *mx* maxilla with components *c* cardo, *st* stipes, *lc* lacinia, *ga* galea, and *mxp* maxillary palpus and number of palpomers; *la* labium with components *smt* submentum, *mt* mentum, *pom* postmentum, *pm* prementum, *li* ligula, *gl* glossa, *pgl* paraglossa, and *lp* labial palpus with number of palpomers; *hy* hypopharynx. + indicates fusion of components, ma-mx functional coupling of mandibles and maxillae, and mx-la functional coupling forming the labiomaxillary complex. [] rudimentary; ? unclear

| Taxon | Mouthpart orientation and functional type | Mouthpart composition | Predominant feeding style, remarks on life history, diets, and additional mouthpart use | References in addition to Beutel et al. (2014) |
|---|---|---|---|---|
| **Hexapoda** | | | | |
| Protura (coneheads) | Entognathous piercing | lr, ma stylet shaped, mx (c, st, lc stylet shaped, mxp 1–5) la, lp 2 | Mycophagous, soil hexapods | François et al. (1992), Pass and Szucsich (2011) |
| Collembola (springtails) | Entognathous biting-chewing or piercing | lr, ma-mx (c, st, ga + lc, mxp 1) (ma, mx slender), la (lp), hy | Omnivorous, detritivorous, microphytophagous, mycophagous; soil hexapods | Eisenbeis and Wichard (1985), Blanke et al. (2015b) |
| Diplura (two-pronged bristletails) | Entognathous biting or piercing | lr, ma-mx (c, st, ga, lc, mxp 1–3), la, lp 1, plus labial admentum | Predaceous, detritivorous; soil hexapods | Koch (2000), Blanke and Machida (2016) |
| **Insecta (Ectognatha)** | Ectognathous | | | |
| Archaeognatha (bristletails) | Orthognathous biting-chewing | lr, ma-mx (c, st, lc, ga, mxp 7), la (pom, pm, gl, pgl, lp 3), hy | Detritivorous, microphytophagous; soil insects | Blanke et al. (2015a, b) |
| Zygentoma (silverfish) | Orthognathous biting-chewing | lr, ma, mx (c, st, lc, ga, mxp 5–6), la (pom, pm, gl, pgl, lp 4), hy | Detritivorous, microphytophagous; soil insects, may consume cellulose | Staniczek (2000), Blanke et al. (2015a) |

(continued)

**Table 2.1** (continued)

| Taxon | Mouthpart orientation and functional type | Mouthpart composition | Predominant feeding style, remarks on life history, diets, and additional mouthpart use | References in addition to Beutel et al. (2014) |
|---|---|---|---|---|
| **Pterygota Palaeoptera** | | | | |
| Ephemeroptera (mayflies) | Immatures: biting-chewing | Immatures: lr, ma, mx (c, st, lc + ga, mxp 2), la (lp 2), hy<br>Adults: reduced mouthparts | Phytophagous, detritivorous, predaceous; aquatic immature stages, nonfeeding adults | Wichard et al. (2002) |
| Odonata (damsel-flies and dragonflies) | Orthognathous biting-chewing; immatures with labial mask | lr, ma, mx (c, st, lc, mxp 1), la (pom, pm, gl + pgl, lp 1)<br>Immatures: prehensile elongated prm with grasping lp, hy | Predaceous; aquatic immature stages and aerial hunting adults | Büsse et al. (2017), Büsse and Gorb (2018) |
| **Neoptera Polyneoptera** | | | | |
| Zoraptera (ground lice, angel insects) | Orthognathous biting-chewing | lr, ma, mx (c, st, lc, ga, mxp 5), la (smt, mt, pm, gl, pgl, lp 3), hy | Microphytophagous, detritivorous, mycophagous; gregarious soil insects | Beutel and Weide (2005) |
| Dermaptera (earwigs) | Prognathous biting-chewing | lr, ma, mx (c, st, lc, ga, mxp 5), la (smt, mt, pm, gl + pgl, lp 3), hy | Omnivorous in all stages; soil insects, brood caring imagines | Neubert et al. (2017) |
| Plecoptera (stoneflies) | Prognathous biting and chewing | lr, ma, mx (c, st, lc, ga, mxp 5), la (smt, mt, pm, gl, pgl, lp 3), hy | Phytophagous, carnivorous; aquatic immatures, rarely feeding adults, in some +/− reduced mouthparts | Wichard et al. (2002), Zwick (2003) |
| Orthoptera (crickets and grasshoppers) | Orthognathous biting-chewing | lr, ma, mx (c, st, lc, ga, mxp 5), la (smt, mt, pm, gl, pgl, lp 3), hy | Phytophagous, omnivorous phytophilous, terricolous in all stages | Groll and Günther (2003) |
| Mantophasmatodea (gladiators) | Orthognathous biting | lr, ma (mola missing), mx (c, st, lc, ga, mxp 5), la (smt, mt, pm, gl, pgl, lp 3), hy | Predaceous; terricolous in all stages | Klass et al. (2003) |

| | | | | |
|---|---|---|---|---|
| Grylloblattodea (Notoptera) (ice crawlers) | Prognathous biting | lr, ma (mola missing), mx (c, st, lc, ga, mxp 5), la (smt, mt, pm, gl, pgl, lp 3), hy | Predaceous, sometimes scavenging; terricolous in all stages | Wipfler et al. (2011) |
| Embioptera (webspinners) | Prognathous biting-chewing | lr, ma (elongate in males), mx (c, st, lc, ga, mxp 5), la (pom, pm, gl, pgl, lp 3), hy | Detritivorous subsocial soil insects in silk tunnels; sexual dimorphic mandible in nonfeeding males | Rähle (1970) |
| Phasmatodea (stick and leaf insects) | Prognathous biting-chewing | lr, ma, mx (c, st, lc, ga, mxp 5), la (smt, mt, pm, gl, pgl, lp 3), hy | Herbivorous; phytophilous in all stages | Friedemann et al. (2011) |
| Mantodea (mantids) | Orthognathous biting | lr, ma (mola missing), mx (c, st, lc, ga, mxp 5), la (pom, pm, gl, pgl, lp 3), hy? | Predaceous; phytophilous in all stages | Klass and Ehrmann (2003) |
| Blattodea (cockroaches) | Ortho- or hypognathous biting-chewing | lr, ma, mx (c, st, lc, ga, mxp 5), la (smt, mt/smt + mt, pm, gl, pgl, lp 3), hy | Omnivorous; terricolous in all stages | Zhuzhikov (2007) |
| Isoptera (termites) | Prognathous biting-chewing | lr, ma (partly different shapes or reduced in soldiers), mx (c, st, lc, ga, mxp 5), la (pom, pm, gl, pgl, lp 3), hy | Detritivorous, xylophagous; eusocial insects, trophallaxis, use cellulose with help of symbiotic microorganisms | Watson and Gay (1991), Krishna et al. (2013) |
| **Condylognatha** | | | | |
| Thysanoptera (thrips) | Hypognathous, asymmetric piercing-sucking | lr, left ma, mx (c + st, stylet-shaped lc, mxp 3–8), la (pom, pm, gl, pgl, lp 1–5), hy | Plant sap feeding, pollen feeding; phytophilous, flower-visiting, gall-forming, small | Mickoleit (1963) |
| Hemiptera (true bugs, cicadas, and plant lice) | Pro-, ortho-, hypognathous piercing-sucking | Proboscis: lr, ma (long, stylet shaped), mx (lc? long, stylet shaped), long la, hy | Plant sap feeding, frugivorous, predaceous; phytophilous, gall-forming, terricolous, blood sucking, ectoparasitic | Weber (1933) |
| Psocodea Psocoptera (bark lice) | Ortho- or prognathous modified biting-chewing | lr, asymmetrical ma, mx (c + st, ga, chisel-shaped lc, mxp 4), la (pom, pm, [gl], lp 1–2), hy | Detritivorous, microphytophagous, mycophagous in all stages, silk | Dathe (2003) |

(continued)

**Table 2.1** (continued)

| Taxon | Mouthpart orientation and functional type | Mouthpart composition | Predominant feeding style, remarks on life history, diets, and additional mouthpart use | References in addition to Beutel et al. (2014) |
|---|---|---|---|---|
| Phthiraptera: Amplycera, Ischnocera (chewing lice) | Prognathous modified biting-chewing | lr, ma, mxp 4–5, la (pom, pm, lp 1), hy | Feather-, hair-, skin-feeding wingless ectoparasites on birds and mammals in all stages | Mey (2003) |
| Rhynchophthirina (elephant lice) | Biting (on elongated head) | ma (exodont), long pom (elongated clypeus and gena) | Blood-sucking wingless ectoparasites of elephants in all stages | Weber (1969) |
| Anoplura (sucking lice) | Piercing-sucking proboscis | Proboscis: ma + mx + la rudiments fused with head, stylet-shaped la, and hy form retractable proboscis | Blood-sucking ectoparasites on mammals in all stages | Tröster (1990) |
| **Holometabola** | | | | |
| Hymenoptera (sawflies, wasps, bees, and ants) | Orthognathous larva: biting-chewing<br>Adult: biting-chewing plus lapping-sucking or siphoning | Larva: lr, ma, mx-la (c, st, lc, ga, mxp); la (pom/smt, me, pm, lp 1–3)<br>Adult: lr, ma, mx-la∗ (c, st, lc, ga, mxp 2–6; pom/mt, lorum, pm, lp 3–4, gl, pgl), hy; variously composed proboscis in many nectar-feeding lineages<br>∗Lorum connects mx and la basally | Larvae: phytophagous, gall-forming, parasitoid and eusocial larvae with simplified mouthparts<br>Adults: flower-visiting, pollen- and nectar-feeding, predaceous, rarely nonfeeding; mouthpart used in nest construction in many; brood care and trophallaxis in social taxa | Weber (1933), Jervis (1998), Jervis and Vilhelmsen (2000), Krenn et al. (2005) |
| Raphidioptera (snakeflies) | Prognathous biting | lr, ma, mx (c, st, lc, ga, mxp 5), la (smt/pom, pm, li, lp 3), hy | Predaceous in all stages; under tree bark | Achtelig (1967), Beutel and Ge (2008) |
| Megaloptera (alderflies and dobsonflies) | Prognathous biting | lr, ma, mx (c, st, lc, ga, mxp 3–5), la (smt, me/pom, pm, li, lp 3–4), hy | Predaceous aquatic larvae; rarely feeding adults, huge mandibles in males of one taxon | Beutel and Friederich (2008) |

| | | | | |
|---|---|---|---|---|
| Neuroptera (antlions and lacewings) | Larva: prognathous sucking jaws<br>Adult: biting-chewing, fluid-feeding | Larva: lr, ma-mx (c, st, lc), la (smt, mt, pm, lp 2–7), hy?<br>Adult: lr, ma, mx (c, st, lc, ga, mpx 5), la (smt, mt, pm, li, lp 2–3) | Predaceous terrestrial or aquatic larvae perform extraoral digestion;<br>Predaceous, nectar- and pollen-feeing imagines | Aspöck et al. (2001), Krenn et al. (2008); see Chap. 11 |
| Strepsiptera (twisted-wing parasites) | Larva: simplified biting<br>Adult: reduced | Larva: ma, fused mx [mxp], la + hy (lp)<br>Adult: [ma], [mx], mxp 1 | Larvae: endoparasites of insect females simplified;<br>only adult males winged and free living, nonfeeding | Kinzelbach (1971), Pohl and Beutel (2005) |
| Coleoptera (beetles) | Prognathous biting, biting-chewing, few: sucking, filter-feeding, rarely reduced | Larva: lr, ma, mx (c, st, lc, ga, mxp 2), la + hy (smt, mt, pm, li, lp 3)<br>Adult: lr, ma, mx (c, st subdivided, lc, ga subdivided, mxp 4), la + hy (smt, mt, pm, li, lp 3) components rarely elongated | Larvae and imagines: detritivorous, xylophagous, fungivorous, predaceous (some with extraoral digestion), phytophagous in all terrestrial and fresh water habitats; some flower-visiting imagines pollen- and nectar-feeding; some adults nonfeeding | Beutel and Leschen (2005); see Chap. 12 |
| Trichoptera (caddis flies) | Few: sucking, filterfeeding, rarely reduced<br>Larva: prognathous biting-chewing<br>Adult: sponging | Larva: lr, ma (mola reduced), mx + la (mxp 4, smt, mt, pm, lp 1), la + hy (spinneret)<br>Adult: lr, [ma], mx + la (mxp 5–7, lp 3), hy | Aquatic larvae: phytophagous, predaceous, some net-building and filter-feeding<br>Adults: rarely fluid-feeding | Wichard et al. (2002), Holzenthal et al. (2015) |
| Lepidoptera (moths and butterflies) | Larva: orthognathous biting-chewing<br>Adults of basal lineage: biting-chewing<br>Most Glossata: siphoning<br>Some: reduced<br>Few: piercing | Larva: lr, ma, mx (c + st, ga, mxp 1), la + hy (pom, pm, lp 2)<br>Adults of basal lineage: lr, ma, mx (c, st, lc, ga, mxp 5) la (pom?, pm, gl, pgl, lp 1)<br>Adult of Glossata with coilable | Larvae: phytophagous, few predaceous, myrmecophilous<br>Adults of basal lineage: pollen-feeding<br>Adults of Glossata: fluid-feeding on nectar, sap, sweat, tears, fruits, | Kristensen (1998), Krenn (2010) |

(continued)

**Table 2.1** (continued)

| Taxon | Mouthpart orientation and functional type | Mouthpart composition | Predominant feeding style, remarks on life history, diets, and additional mouthpart use | References in addition to Beutel et al. (2014) |
|---|---|---|---|---|
| | | proboscis: lr, [ma] mx (c, st, ga, mxp 0–5), la (pm, lp 3), [hy] | few blood, rarely pollen-feeding; nonfeeding species in many taxa | |
| Siphonaptera (fleas) | Larva: prognathous, biting-chewing<br>Adult: piercing-sucking | Larva: lr, ma, mx? (mxp 2), la?<br>Adult with proboscis: lr, stylet-shaped lc, mxp 5, long la, lp, hy | Larvae: detritivorous in soil and nests<br>Adults: blood-feeding ectoparasites | Michelsen (1997), Kluge (2002) |
| Mecoptera (scorpionflies) | Larva: pro- or orthognathous biting-chewing<br>Adult: orthognathous biting on elongated head | Larva: lr, ma, mx, mxp 1–2, la (pom, pm, lp 2)<br>Adult: lr (fused with clypeus), ma, mx (c, st, lc, ga, mxp 5), la (mt, pm, lp 2), hy | Larvae: detritivorous, few aquatic<br>Adults: omnivorous, some kleptoparasites | Friedrich et al. (2013) |
| Diptera (true flies) | Larva: prognathous filter-feeding, biting-chewing biting, slashing mouth hooks<br>Adult: sponging, piercing-sucking, siphoning, reduced | Larva: lr, ma, [mx], mxp 1, lp 1<br>Cyclorrhapha: mouth hooks formed by modified ma + mx<br>Adult with proboscis: long lr, stylet-shaped ma, mx (c, st, stylet-shaped lc, mxp 1–5), la (pom, long prm, labellum), long hy<br>Brachycera: lr, lc, la (pom, long pm, labellum), long hy<br>Cyclorrhapha: long lr, [lc], mxp, long la (pm, labellum), long hy | Larvae: aquatic filter-feeding, predaceous, gall-forming, detritivorous, phytophagous, mycophagous, microphagous, parasitoid<br>Adults: fluid-feeding on open sources (nectar, sap, sweat, tears, fruits, dung, blood, etc.), flower-visitors on concealed nectar, ectoparasitic blood-feeding (mostly) females<br>Some taxa: nonfeeding | Gilbert and Jervis (1998), Nagatomi and Soroida (1985), Krenn et al. (2005), Grimaldi and Engel (2005), Krenn and Aspöck (2012) |

ectognathous mouthparts. In entognathous hexapod taxa, i.e., Protura, Collembola, and Diplura, the paired mouthparts are concealed inside the head by duplicatures of the head capsule. The mandibles and maxillae reside in the gnathal pouches (Fig. 2.1a). Thus, they are located between a small labrum, forming the frontal closure of the preoral cavity, and the labium which closes the preoral cavity ventro-posteriorly. Only the distal tips of mandibles and maxillae are visible from the ventral side, when the labrum opens the preoral cavity (Fig. 2.1b). The mandibles and maxillae are slender in shape with a single articulation point to the head capsule at the base of the gnathal pouch. Both paired appendages are protrusible and retractable; they can be closed and rotate to some degree. The labium is medially divided and sometimes bears short labial palpi (Grimaldi and Engel 2005; Blanke and Machida 2016). Functionally, the labium is connected to the hypopharynx, where the salivary glands open into the preoral cavity. Interacting mouthpart structures, such as apodemes and maxillary structures embracing the mandibles, evolved in entognathous Hexapoda that probably improve the functional performance and food uptake abilities of the mandibles and maxillae (Blanke et al. 2015b).

Ectognathous mouthparts occur in Insecta sensu stricto (= Ectognatha or true insects that comprise Archaeognatha, Zygentoma, and Pterygota) that are characterized by external articulations of all mouthpart components. The mandibles and maxillae articulate ventrally to the gena on the head capsule (Fig. 2.1c–f). Traditionally, it was considered that the mandibles were connected to the head either by one or two joints, described as a monocondylic or a dicondylic mandible articulation. However, recent studies have shown that all mandibles of Ectognatha are dicondylic or multicondylic. Only the posterior articulation is shaped like a ball and socket joint. Archaeognatha possess nonpermanent mandibular articulations, two joints with the head and one with the base of the maxillary palpus (Blanke et al. 2015a). The mandibles of Archaeognatha are equipped with eight muscles, which originate from the cranium, the anterior tentorium, a median ligamentous structure, and the hypopharynx. This arrangement allows additional rolling movements of the mandible. Additional structures of the maxillary palpus support mandible movements, similar to the structural interaction of mandibles and maxillae of entognathous hexapods (Blanke et al. 2015a).

Two distinct mandibular articulations of the head occur in Zygentoma and Pterygota, which together represent the majority of insects. However, the anterior articulation is flexible to some degree and allows gliding movements in Zygentoma and the immatures of Ephemeroptera (Staniczek 2000). The posterior condyle is formed by the mandible, while the anterior condyle is formed by the head capsule (Blanke et al. 2015a). The permanent dicondylic mandible articulation results in a hinged joint that reduces the degree of movability of the mandibles to a transverse plane but allows for more powerful biting movements in Neoptera (Fig. 2.1e, f). Closing movements of the mandibles are achieved by large adductor muscles that are attached at the median sides of large apodemes in Neoptera, whereas smaller abductor musculature is attached to the lateral sides (Wipfler et al. 2011; Beutel et al. 2014). A detailed contribution on the functional morphology of these fundamental types of mandible articulation is given in Chap. 6.

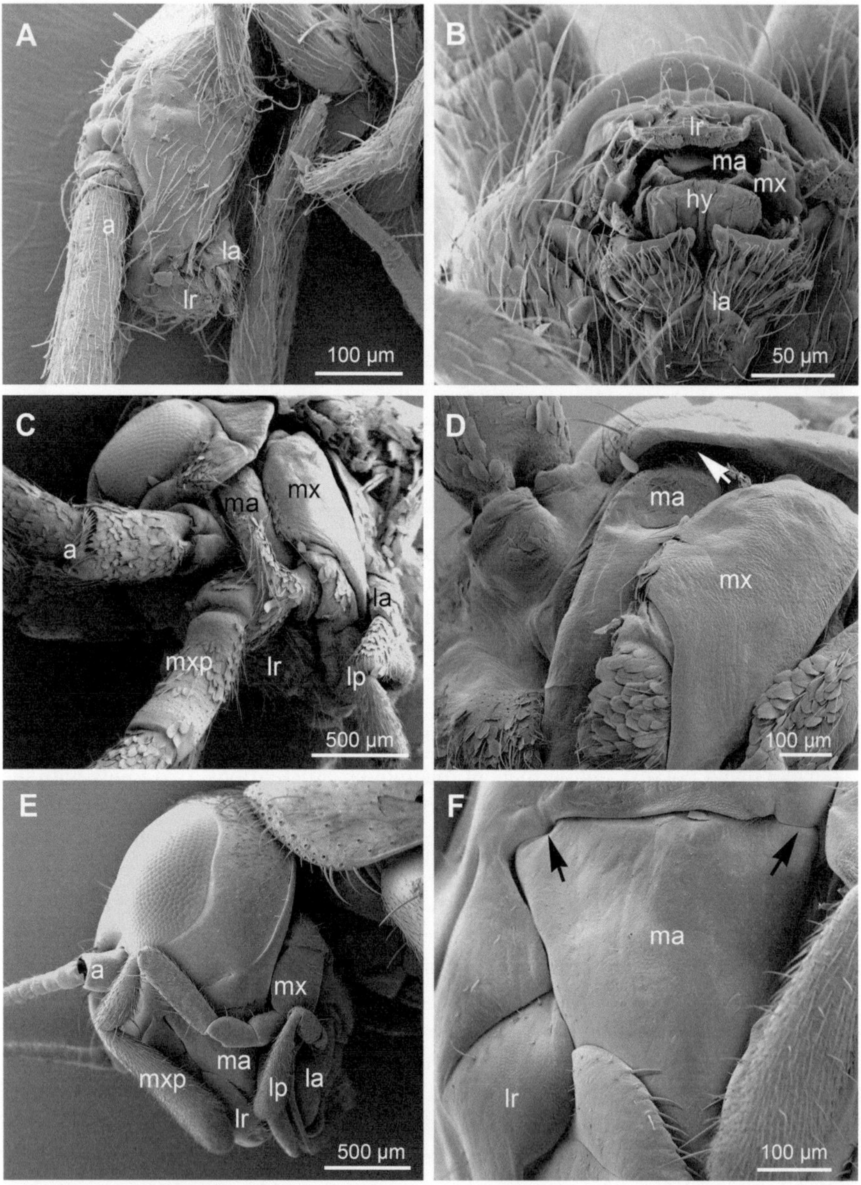


**Fig. 2.1** Principal morphology of mouthparts in Hexapoda (scanning electron microscope images). (**a**) Entognathous head of noninsect Hexapoda *Entomobrya* sp. (Entomobryidae, Collembola), lateral view. (**b**) Mouthparts of *Entomobrya* sp. are concealed in the head, ventral view. (**c**) Ectognathous mouthparts and head of *Machilis* sp. (Machilidae, Archaeognatha), fronto-lateral view. (**d**) Mandible has one external condyle (arrow) to the head (*Machilis* sp.), lateral view. (**e**) Head of pterygote insect *Nemobius sylvestris* (Gryllidae, Orthoptera), mouthparts in orthognathous position, lateral view. (**f**) Dicondylic mandible has two external condyles (arrows) to the head (*Nemobius sylvestris*), lateral view. *a* antenna, *hy* hypopharynx, *la* labium, *lp* labial palpus, *lr* labrum, *ma* mandible, *mx* maxilla, *mxp* maxillary palpus

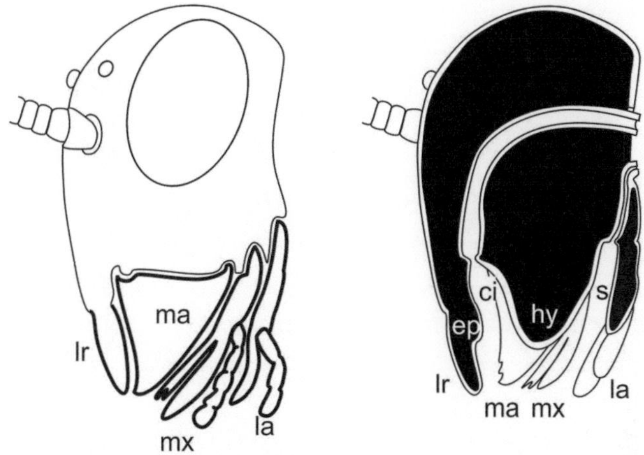


**Fig. 2.2** General morphology of head and mouthparts of a pterygote insect (schematic drawings). Head and mouthparts in lateral view and sagittal section of head illustrating preoral cavity. *ci* cibarium, *ep* epipharynx, *hy* hypopharynx, *la* labium, *lr* labrum, *ma* mandible, *mx* maxilla, *s* salivarium

## 2.3 Form and Function of Mouthpart Components

The generalized mouthparts include the anterior unpaired labrum, paired mandibles, paired maxillae bearing a pair of leglike maxillary palpi, and the posterior unpaired labium, bearing a pair of labial palpi as well as the hypopharynx in the preoral cavity (Figs. 2.2 and 2.3). The following description of the mouthpart components of Insecta (Ectognatha) focuses on their form and function and is based on Weber (1933), Snodgrass (1935), Smith (1985), Chapman (1995), Gullan and Cranston (2014), and Beutel et al. (2014). This type of insect mouthparts is often termed as mandibulate mouthparts.

### *2.3.1 Labrum*

The labrum is an unpaired, rounded sclerite, which is connected to the clypeus at the frontal head capsule by a hinged articulation (Figs. 2.2 and 2.4). The developmental origin of the labrum is still under debate, as it has either been regarded as a remnant of appendages of the first head segment (Kimm and Prpic 2006) or as a separate sclerite of the head capsule that is not derived from a head appendage at all (Rogers and Kaufman 1996; Posnien and Bucher 2010). The function of the labrum is a frontal covering of the mandibles and thus a closure of the frontal preoral cavity (Fig. 2.2). It can be moved by several extrinsic muscles from the head capsule; one pair is inserted at a small sclerotized attachment site, the tormae. The buccal side toward the preoral cavity is called the epipharynx. Its cuticle structures and intrinsic

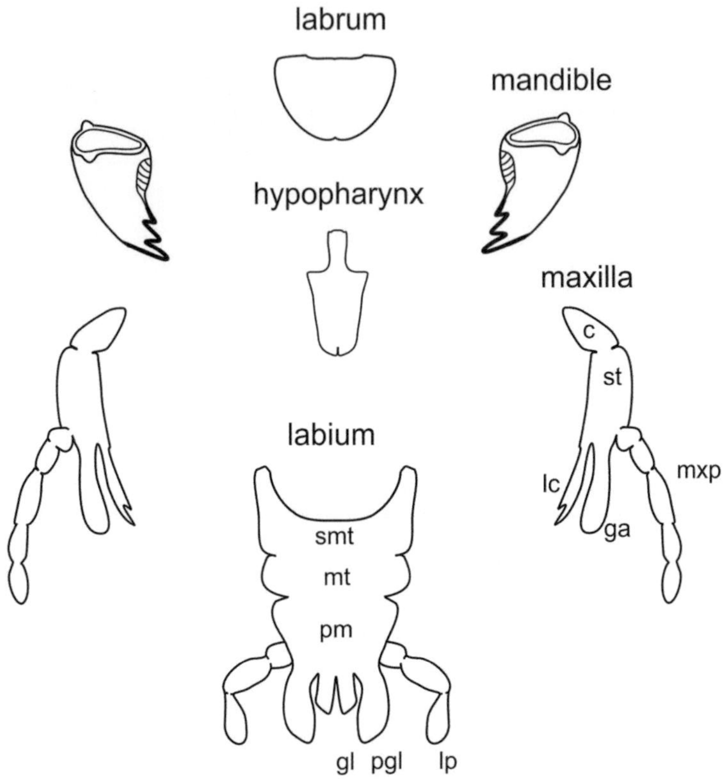


**Fig. 2.3** General morphology of the mouthpart components of a pterygote insect (schematic drawings). *c* cardo, *ga* galea, *gl* glossa, *lc* lacinia, *lp* labial palpus, *mt* mentum, *mxp* maxillary palpus, *pgl* paraglossa, *pm* prementum, *st* stipes, *smt* submentum

musculature assist in handling food inside the preoral cavity. The intrinsic labral muscles extend between the epipharynx and the frontal cuticle of this structure (for a detailed description of the musculature, see Chap. 7).

## 2.3.2 *Mandible*

The paired mandibles represent the main biting and chewing organs (Figs. 2.3 and 2.4). The primary function of mandibles is to bite off pieces from solid food items with the apical cusps and teeth forming the incisive region as well as to grind food particles with the mola at the mandibular base. The broad molar region is equipped with small teeth and rows of variously shaped structures, which are adapted to grind particles and crush food items within the preoral cavity before they are transported to the anatomical mouth and swallowed. A detailed description of the muscles and their functions is given in Chap. 7.

**Fig. 2.4** Head and biting-chewing mouthparts of a cockroach *Periplaneta americana* (Blattidae, Blattodea) [reproduced with permission from Oberösterreichisches Landsmuseum, Krenn and Aspöck (2010)]. *c* cardo, *ga* galea, *gl* glossa, *lc* lacinia, *lp* labial palpus, *mxp* maxillary palpus, *pgl* paraglossa, *pm* prementum, *st* stipes

### *2.3.3 Maxilla*

The paired maxillae articulate to the head capsule via their basal sclerites, the cardo and the stipes. These components are oriented to each other in an obtuse angle. The cardo and stipes are open on their median sides, where extrinsic muscles from the head enter and attach. The stipes bears a leglike maxillary palpus in addition to distal components (endits), the lacinia and the galea. In general, the medial lacinia possesses a sclerotized tip and functions as a small pincer to manipulate food particles. The lateral, rather membranous galea usually bears sensilla and covers the mouthparts laterally (Figs. 2.3 and 2.4). Extrinsic maxillary muscles originate at the head capsule or the tentorium and project through the whole maxilla; another extrinsic head muscle flexes the lacinia upon contraction. Additional small intrinsic flexor muscles of the lacinia and the galea arise inside the stipes. However, no separate extensor muscle of the lacinia and galea exists. Contractions of the extrinsic stipes muscles are responsible for the protrusions of the maxilla. The intrinsic muscles that arise inside the stipes move the maxillary palpus; its palpomers may contain single flexor muscles (for details, see Chap. 7) and bear numerous sensilla (e.g., Prakash et al. 1995; Hao et al. 2019).

### 2.3.4 *Labium*

The labium is composed of several sclerites of the last head segment (Fig. 2.2), which are homologous to the maxillary components (Angelini and Kaufman 2005). The basal component is equivalent to the fused maxillary cardines, probably including parts of the sternum of the last head segment. This fused sclerite is called the postmentum that in some taxa can be further divided into the submentum and mentum. The distal component, the prementum, is homologous to the fused maxillary stipites, since the prementum bears two pairs of distal endits, the median glossae and the lateral paraglossae, as well as a pair of labial palpi (Figs. 2.3 and 2.4). The hinged articulations to the head and the joints between the postmentum and prementum permit opening and closing movements of the preoral cavity. External muscles originating from the tentorium are attached to the prementum. Intrinsic flexor muscles extend between the submentum and prementum as well as between the prementum and the base of the glossa and paraglossa. In addition, muscles originating from the prementum can move the palpi, which may include a pair of antagonistic muscles within the proximal palpomer (for details of the muscles, see Chap. 7). The labial palpi are typically composed of three parts and bear sensilla (e.g., Hao et al. 2019) which often form apical sensory organs which are described for example in cockroaches (Prakash et al. 1995).

### 2.3.5 *Hypopharynx*

The unpaired hypopharynx lies between the paired mouthparts (Figs. 2.2 and 2.3). It is not serially homologous to head appendages; instead, it is part of the pharynx and is formed by the ventral part of the intercalary segment (Hartenstein and Chipman 2015). The hypopharynx often functions like a tongue. It is rather soft and bears cuticular structures and sensilla; it is suspended by small sclerites and moved by numerous small muscles. Swinging movements in anterior and posterior directions are achieved by a set of muscles, which originate at the frons of the head, the tentorium, and the labium, and insert at the small suspensory sclerites. In addition, intrinsic hypopharyngeal musculature may form part of the salivarium (Beutel et al. 2014). A detailed description of the hypopharynx is given for cockroaches in Chap. 7.

### 2.3.6 *Position of Mouthparts*

The principal orientation of the head and its mouthparts can be categorized in relation to the body axis. In the "orthognathous" condition, the head is directed perpendicular to the body axis, and the mouthparts are directed downward in a vertical plane (Fig. 2.1e), while a "prognathous" head is in a horizontal position (in the same axis as the body), and thus the mouthparts project forward (Fig. 2.5a). A "hypognathous" head condition is present when the head and mouthparts point in a posterior direction (Weber 1933; Beutel et al. 2014). In English textbooks, the term "hypognathous" is traditionally used when the head is directed vertically and "opisthognathous" or "opisthorhynchous" when it is oriented in a posterior direction and the mouthparts are positioned under the body (Grimaldi and Engel 2005; Gullan and Cranston 2014; Chapman et al. 2016). In this chapter, the term "orthognathous" is used to describe the vertical orientation of the head. It was shown to be the ancestral condition in the Pterygota, the winged insects (Wipfler et al. 2019).

## 2.4 Organs of the Preoral Cavity

The mouthparts enclose a preoral cavity of the head where food particles are mixed with saliva and transported into the pharynx. The epipharynx on the buccal side of the labrum constitutes the muscular fronto-anterior wall of the preoral cavity, whereas the posterior-ventral closure is formed by the proximal labial sclerites (Fig. 2.2).

The preoral cavity contains the hypopharynx, which aids in swallowing particles and in mixing saliva with food. This organ is positioned between the mouthparts and divides the preoral cavity into an anterior cibarium leading into the pharynx and a posterior salivarium where the salivary duct opens (Fig. 2.2). The cibarium forms a space in front of the mouth opening that can be enlarged by contraction of the epipharyngeal muscles. The salivarium is formed by basal parts of the hypopharynx and the buccal side of the labium. Here, saliva is delivered into the cibarium from the labial glands, which normally function as salivary glands in insects.

## 2.5 Functional Types of Mouthparts of Insecta

Insect mouthparts show a multitude of functional mechanisms for food uptake that can be distinguished into various functional categories (Fig. 2.5). Based on Metcalf (1929), Weber (1933), and Snodgrass (1935), three main functional types can be differentiated: (1) biting and chewing mouthparts; (2) elongated, often called haustellate, mouthparts comprising various types of proboscises; and (3) filter-feeding mouthparts. The mandibles of biting and chewing mouthparts are adapted to feeding on solid food. Haustellate mouthparts are derived and adapted to fluid-feeding. These more or less elongated structures enclose a food canal and form an unpaired proboscis. Filter-feeding mouthparts are equipped with specialized mandibles and use structures equipped with rows of hairs and bristles for collecting suspended material out of water. Labandeira (1997) distinguished 35 functional classes of mouthparts in recent and extinct Hexapoda. A comprehensive review on this categorization and the evolution of mouthparts is presented in Chap. 17.

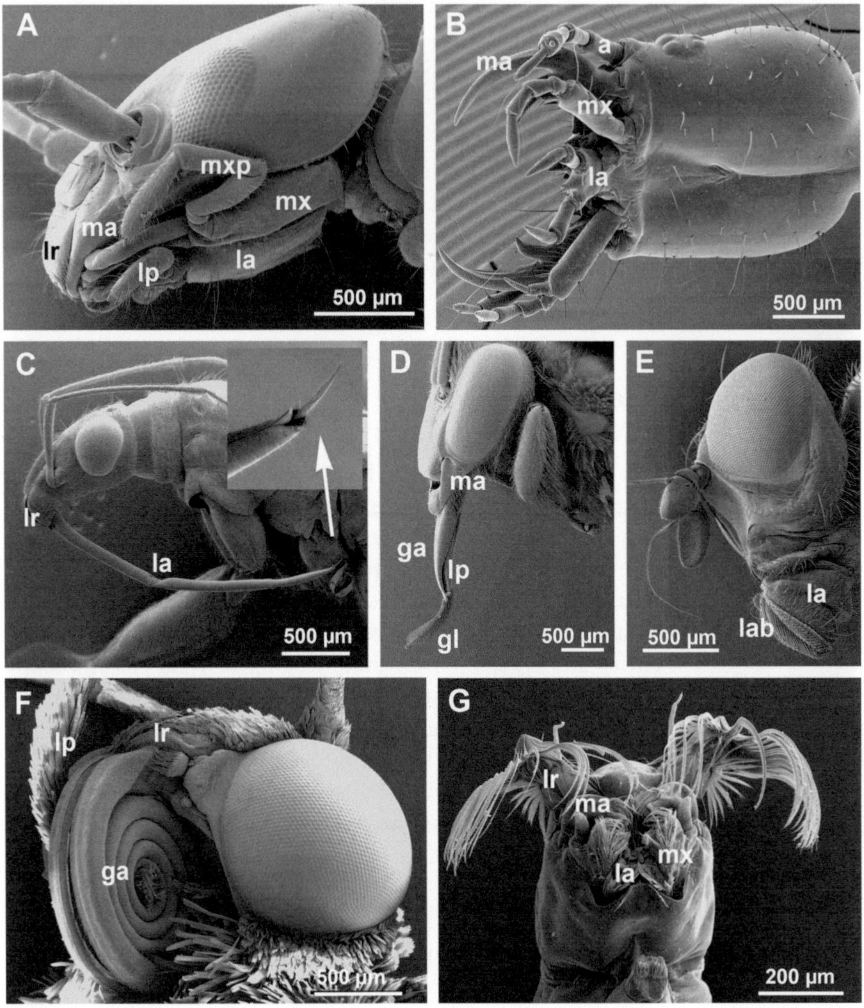


**Fig. 2.5** Functional types of mouthparts in insects (scanning electron microscope images). (**a**) Prognathous biting and chewing mouthparts of the omnivorous earwig *Forficula auricularia* (Forficulidae, Dermaptera). (**b**) Biting mouthparts of a predaceous ground beetle larva *Nebria* sp. (Carabidae, Coleoptera); mandibles and maxillae in opened position. (**c**) Piercing-sucking proboscis of a predaceous true bug *Nabis* sp. (Nabidae, Hemiptera); proboscis tip magnified; protruded piercing stylets (arrow). (**d**) Lapping-sucking proboscis of a stingless bee *Melipona* sp. (Apidae, Hymenoptera); hairy glossa extended. (**e**) Sponging mouthparts of the fly *Ceratitis capitata* (Tephritidae, Diptera) with apical labellum. (**f**) Siphoning proboscis of a nectar-feeding moth in coiled resting position; *Euplagia quadripunctaria* (Erebidae, Lepidoptera); left labial palpus removed. (**g**) Head and mouthparts of a filter-feeding black fly larva (*Simulium* sp., Simuliidae, Diptera). *a* antenna, *ga* galea, *gl* glossa, *la* labium, *lab* labellum, *lp* labial palpus, *lr* labrum, *ma* mandible, *mx* maxilla, *mxp* maxillary palpus

## 2.5.1 *Mandibulate Mouthparts for Biting and Chewing*

### 2.5.1.1 Occurrence of Biting and Chewing Mouthparts

Biting and chewing mouthparts are plesiomorphic in Insecta (Ectognatha). They occur in Archaeognatha, Zygentoma, Palaeoptera, Polyneoptera, as well as many Holometabola (Table 2.1). In Ephemeroptera, only the aquatic larvae take up food using variously modified biting-chewing mouthparts, while the imagines possess reduced mouthparts and do not feed at all. In Odonata, the mandibulate mouthparts are modified in the aquatic larvae which catch their prey with the uniquely shaped prehensile labium. The last common ancestor of the Polyneoptera had unspecialized biting-chewing mouthparts in orthognathous position that often are termed as orthopteriod (Wipfler et al. 2019). Both the immature stages and adults of the Polyneoptera consume similar nutritional resources from the same habitats. Whereas in many holometabolan taxa imagines of various lineages evolved derived haustellate mouthparts, which are specialized for the uptake of liquid food, the larvae predominantly use biting and chewing mouthparts to feed from solid food sources. Especially in Hymenoptera, Lepidoptera, Trichoptera, Diptera, and Siphonaptera, the biting-chewing feeding organs of larvae are fundamentally different from the mouthparts of adults in the same lineages. In contrast in non-holometabolan taxa, the same mouthpart types (i.e., biting-chewing or piercing-sucking) occur in immature and adult stages (Table 2.1). Details of the functional morphology of biting-chewing mouthparts are presented in Chap. 7.

### 2.5.1.2 Mechanism of Biting-Chewing Mandibles

The combined biting and chewing functions are performed by the closing movements of the mandibles supported by the two fixed points of articulation. The conspicuous apical cusps, teeth, and cutting edges perform the biting function; and the basal molar region chews and grinds the food particles (Fig. 2.4). In most insects, the two mandibles are asymmetrically shaped and fit into each other. The cuticle of the whole mandible is heavily sclerotized, and the apical region is usually darkly colored (Chapman 1995). The incisivi perforate the outer surface of a food item, while the proximal edges perform the cutting. Herbivorous and omnivorous insects typically possess biting-chewing mandibles with strong cusps and sharp cutting edges in the incisive parts as well as small cuticular structures at the mola. One mandible overlaps the other so that during movement, the inner surface shears along the outer cuticle of the second incisor region. The outer surface of the left incisor region and the inner cuticle of the right incisor region have been found to be much harder than other regions of the mandible. In this way, the cutting edges that bite off leaf fragments are self-sharpening, for example, in locusts (reviewed in Chapman 1995; Weihmann et al. 2015a, b).

The cutting edges of the mandibles are often reinforced against wear and tear by impregnating the cuticle with heavy metals, such as zinc, manganese, or iron (reviewed in Chapman 1995; Vincent and Wegst 2004). For example, in termites (Cribb et al. 2008) or leaf-cutter ants (Schofield et al. 2002), the metal content of the cuticle enhances the hardness of the incisive region. The degree of wear on the mandibles has been documented in several taxa and seems to be a widespread phenomenon in long-lived insects. It influences feeding in various aspects, such as reduction of feeding rate and prolonged feeding time (reviewed in Chapman 1995). Wear of the mandibles in particular is a critical problem in grazing immatures, but new fully functional mouthparts are produced anew after each molt. The frequent molts of aquatic immature insects which feed on epilithic algae have been interpreted as an adaptation to offset mandibular wear (Arens 1990).

The sensillar equipment of the mandibles includes scolopidia, sensilla campaniformia, and trichoid sensilla that detect mechanosensitive information. The mandible cusps are innervated by scolopidia, which probably monitor the hardness of the food, modulate the power output of the adductor muscle, and transmit information about mandibular wear. However, chemoreceptive sensilla are absent (reviewed in Chapman 1995).

#### 2.5.1.3 Modifications of the Mandible

The size, shape, and musculature of the mandible reflect feeding preferences; and these parameters are regarded as functional adaptations to specialized diets and performances. For example, in grass-feeding caterpillars and grasshoppers, the mandibular morphology and associated musculature differ between taxa, which prefer dicotyledonous plants. The biomechanical properties of these different diets and the degree to which plant material can be fragmented influence the rate and amount of nutrients assimilated. Mandible morphology and associated muscles determine the initiation of leaf fractures as well as the size of the bites taken and the degree of chewing required. The physical properties of the plant material determine the speed of feeding and modification of chewing behavior. It was concluded that the biomechanics of the mandibles and food processing must be considered to fully understand nutrition and feeding behavior of herbivorous insects (Clissold 2007). Similarly, the morphology of the mandibles of specialized phytophagous Coleoptera is a key characteristic. Many xylophage beetles possess short, stout, and wedge-shaped mandibles, while detritus- and dung-feeding species are equipped with spoon-shaped mandibles (Weber 1933; Snodgrass 1935; Nel and Scholtz 1990). Pollen-feeding Scarabaeidae are characterized by lamella-shaped mandibles with reduced incisive parts. Unable to bite, anthophilous beetle mandibles bear numerous bristles and a membranous lacinia mobilis (or prostheca) at their basal region, which is probably used for pollen manipulation (Karolyi et al. 2016).

Compared to herbivorous insects, the mandibles of raptorial insects often lack a molar region at their bases. This type of predominantly biting mouthpart occurs in predatory Mantodea, Mantophasmatodea, and Grylloblattodea (Table 2.1). Similar biting mouthpart morphology, but in prognathous orientation, evolved in larvae and adults of many adephageous Coleoptera (Fig. 2.5b), Raphidioptera, and larvae of

Megaloptera that all possess dagger-shaped mandibles characterized by reduced molar regions (Table 2.1). Most larvae of Neuroptera are equipped with unique grasping-biting mouthparts (Snodgrass 1935): the long, flat mandibles are interlocked with similarly shaped maxillae. Together, they form a food canal between these mouthpart structures, which is used to suck body fluids from insect prey (Aspöck and Aspöck 2003; see Chap. 11).

Most predatory Coleoptera have a prognathous head and forward-facing mandibles (Fig. 2.5b). They are armed with strong teeth and pointed structures and may form impressive mandibles, for example, in tiger beetles (Fig. 2.6b) (Ball et al. 2011), many ground beetles (Evans and Forsythe 1985; Acorn and Ball 1991), or some Polyphaga (e.g., Hao et al. 2019). Predaceous Carabidae dismember or macerate their prey using the mandibles involving proventricular teeth and various degrees of extraoral digestion. Comparative studies of carabid mandibles indicate an evolutionary transformation from an elongated and flat shape to a more robust shape with molar projections, in addition to premolar teeth and medial curvature. Remarkably, herbivorous carabid beetles tend to have shorter mandibles equipped with characteristic and special features (Acorn and Ball 1991). Predaceous water beetles, for example, Dytiscidae, possess sharp toothless mandibles in adult stages. Likewise, their larvae are equipped with sickle-shaped fangs which possess a channel for extraoral digestion and ingestion (Kehl 2014). The larvae of Lampyridae and Lycidae evolved functionally similar but morphologically different cannulated mandibles for sucking-feeding of prey (Cicero 1994).

Many social insects show an extraordinary diversity of mandible shapes, as seen in Isoptera (Krishna et al. 2013) and Formicidae (Hölldobler and Wilson 1990; Grimaldi and Engel 2005; Schmidt 2017). In a comparative study on various ant species, Paul (2001) reported that the closing speed and power as well as overall force generation of mandible musculature depend on the physiology of the various muscle fibers and the angle of attachment to the mandibles. Huge high-speed mandibles have evolved multiple times in various subfamilies of Formicidae that are primarily used for prey capture and defense (Larabee and Suarez 2014). These ants use various mechanisms for the extremely fast closing of their long and slender mandibles that involve different head structures as a latch, spring, and trigger to amplify their closing speed. In some lineages, the specialized labrum blocks and releases the mandibles, while in others a modified mandible insertion locks the jaws open during muscle loading where a specialized fast trigger muscle unlocks the mandibles from the joint and closes them with extremely high velocity (Gronenberg 1995a, b, 1996). *Myrmoteras* ants latch their mandibles open at an angle of 280°, while, prior to strike, the large adductor muscles load potential energy by deformation of the occipital head capsule that serves as a spring. Upon stimulation of the labral trigger hairs, the mandibles shut in approximately 0.5 ms with high peak velocity of nearly 50 $ms^{-1}$. In this way, the prey is smashed with the peak acceleration in the order of $10^4$ g. The adductor muscle fibers of the high-speed mandibles in *Myrmoteras* ants have particularly short sarcomeres, characteristic of fast-moving muscles. Their extremely fast closing movements are powered by a spring-loading mechanism to amplify the muscle power output (Larabee et al. 2017). Even higher speeds (over 60 $ms^{-1}$) of closing mandibles were reported from *Odontomachus* ants.

**Fig. 2.6** Mouthparts used in addition to feeding. (**a**) Body grooming, *Leptophyes punctatissima* (Tettigoniidae, Orthoptera). (**b**) Biting in defense, *Cicindela campestris* (Cicindelidae, Coleoptera). (**c**) Sleeping; the bee *Epeolus alpinus* (Apidae, Hymenoptera) clung to a leaf by the mandibles (photo by courtesy of Heinz Wiesbauer). (**d**) Nest construction, harvest termite (*Hodotermes* sp., Hodotermitidae, Isoterma) carrying an object. (**e**) Sexual dimorphic mandibles in the stag beetle *Lucanus cervus* (Lucanidae, Coleoptera). (**f**) Pollen accumulation for pollination by a female yucca moth *Tegeticula carnerosanella* (Prodoxidae, Lepidoptera); mouthparts with a pollen load; proboscis (white arrow); tentacle organ extending from maxillary palpus (black arrow) used for collection and deposition of pollen onto *Yucca* flowers (Asparagaceae, Asparagales) [reproduced with permission from National Academy of Sciences, U.S.A. from Pellmyr and Krenn (2002)]. (**g**) Silk nest-building, *Yponomeuta* sp. caterpillar (Yponomeutidae, Lepidoptera). (**h**) Leaf nest; *Pissodogryllacris* sp. (Gryllacrididae, Orthoptera) cricket builds leaf shelter using silk (photo by courtesy of Sylvain Hugel)

In addition to prey capture, they use jaw strikes for locomotion: by striking hard surfaces, the ant catapults itself into the air whereby the closing movements of left and right mandibles, over a short time lag, direct the jumps (Spanga et al. 2008).

Furthermore, differences in length and shape of mandibles within a species and polymorphic heads can be found in castes of ants and termites. Convergent in these taxa, the mandibles differ fundamentally in workers and soldiers according to their main tasks, such as nest construction, defense, etc. (Hölldobler and Wilson 1990; Mehdiabadi and Schultz 2010; Krishna et al. 2013).

#### 2.5.1.4 Movements of the Maxillae

The maxillae manipulate food particles by characteristic protrusion and retraction movements that are combined with a rapid closing motion of the strongly sclerotized laciniae, while the soft galeae on the lateral sides have mainly sensory functions (Fig. 2.4). The maxillary palpus functions as a sensory organ bearing various tactile and chemosensory sensilla which are concentrated at the apical palpomer (Chapman 1995; Prakash et al. 1995). The coordinated movements of the mandibles and maxillae were studied in vivo in cockroaches (Schmitt et al. 2014). X-ray synchrotron cineradiology revealed symmetrical and synchronous movements of the mouthparts during feeding whereupon the maxillae move in antiphase with respect to the mandibles. In comparison to the simple closing movements of the mandibles, the maxillae exhibit complex motions including protraction and retraction as well as opening and closing movements of the distal parts. The maxilla is protracted by the adduction of the cardo. During the backward rotation of the cardo, the stipes is flexed, and the galea and lacinia close to keep contact with the food. Retraction of the maxilla is enabled by the flexion of the stipes against the cardo and moves the tips apart. This functional model is explained by the antagonistic activity of the maxillary musculature supported by elastic articulation membranes. Food transport is achieved in several phases in which the maxillae grasp food particles, draw them backward, and prevent their loss during mastication of the closing mandibles. Some degree of modulation of the rhythmic pattern of mouthpart movements helps the insect to adjust to different mechanical properties of the food. The maxillary palpi are regularly moved back and forth but show only loose coupling with the activities of the other components (Schmitt et al. 2014).

#### 2.5.1.5 Specializations of Maxillae in Biting-Chewing Mouthparts

Modifications of the maxillae are rather inconspicuous in orthopteriod mouthparts apart from subdivision of stipes, galea, and lacinia in some taxa as well as the variation in the number of palpomers and length of the maxillary palpus. The maxillae of biting and chewing mouthparts are well developed and composed of the basal components bearing the multi-segmented palpus as well as variously shaped distal components. Studies on locusts highlight the importance of the maxillae in biting and chewing insects and emphasis that they are engaged in food inspection (Chapman et al. 2016). In predaceous insects, the maxillae are often equipped with spines which are strongly sclerotized (Fig. 2.5b). In Carabidae, the movements of the maxillae are in antiphase with the mandibles, for example, closing as the mandibles open. The structures of the laciniae serve to rake the food toward the mouth. When the mandibles close, the basal maxillary components push the food further into the cibarium (Forsythe 1982).

Inconspicuous, however functionally remarkable, modifications of the maxillae are found in the orthopteriod mouthparts of tropical crickets of the genus *Glomeremus* (Gryllacrididae) that evolved adaptations to take up fluids, for example, nectar. Despite their mouthparts not forming a proboscis, liquid can be taken up by a special patch of microtrichia at the tips of the galeae and is probably transported through a groove between each maxilla and the mandible into the preoral cavity (Krenn et al. 2016).

#### 2.5.1.6 Modified Labium in Biting-Chewing Mouthparts

The labium is composed of a varying number of sclerites and a pair of three segmented palpi. In many prognathous insects, the proximal sclerite of the labium attaches to the ventral head capsule via an additional plate of the head, called the gula. Labial modifications with regard to feeding mainly concern the glossae and paraglossae that are variously shaped, sometimes fused to form a median ligula, or they are absent. These lobe-shaped components often are densely covered with cuticle structures which form an enlarged surface for fluid uptake in many Hymenoptera (e.g., Jervis 1998; Jervis and Vilhelmsen 2000; Paul et al. 2002; Baranek et al. 2018). For example, Formicidae use the glossa for intake of carbohydrate liquids from easily accessible resources, such as honeydew which is an important energy source (Hölldobler and Wilson 1990; Josens et al. 1998). The protracted glossa functions like a tongue: it is equipped with numerous microtrichia to which fluid adheres that is then transported into the preoral cavity by retraction. Chapter 10 presents the fluid-feeding performance of the glossa in ecological context of various Formicidae.

In many Coleoptera, the glossa and paraglossa are fused to form a median ligula which is involved in fluid uptake, for example, in anthophilous Scarabaeidae (Nel and Scholtz 1990; Karolyi et al. 2016) and Lucanidae (Krenn et al. 2002). In carabid beetles, cuticle structures of the basal labium and bristles at the labial palpi help to keep food fragments in the preoral cavity (Forsythe 1982).

Remarkable labium modifications evolved in aquatic, predaceous instars of Odonata which have biting-chewing mouthparts that are equipped with a prehensile organ formed by the modified labium. This so-called labial mask can be moved forward very rapidly to catch prey with the modified labial palpi which form a pair of movable hooks or spoons (Olesen 1972; Tanaka and Hisada 1980). The prey strike of the labial mask involves a click mechanism at the prementum-postmentum joint which is composed of various cuticular elements including resilin pads (Büsse and Gorb 2018). The process of capturing prey includes propelling the labial mask toward the prey and grasping it using pointed apical structures. The retracted prey is processed by the maxillae and the biting mandibles. The laciniae grasp the prey and deliver it to the mandibles. The functions of single muscles during the predatory strike and the retraction of the prey have been studied in detail. The biomechanics of the feeding apparatus has been analyzed using musculoskeletal modelling (Büsse et al. 2017). A study of the cuticle properties shows resilin-containing cuticle pads that support mobility and provide shock absorption as well as detect high degrees of sclerotization in the incisivi of the mandibles, where high forces occur when crushing the prey (Büsse and Gorb 2018). Büsse et al. (2018) suggest a catapult system as the driving force, which includes resilin-dominated sclerites for energy storage in a spring-loaded system. The hydraulic mechanism of the striking process—as suggested in Tanaka and Hisada (1980)—could be further investigated using high-speed videography. Instead, hydraulic pressure is most likely used for recoil-preventing jet propulsion during the prey capturing process (Büsse et al. 2018).

Another extensible labial prey capture apparatus evolved in beetles, for example, in the genus *Stenus* (Staphylinidae). The prey capture strike is fast enough to catch springtails that stick to adhesive pads at the tip of the labium; and, after retraction, the prey is squashed using the mandibles (Betz 1998; Koerner et al. 2017). This specific labial prey capture apparatus has been considered a key innovation that has initiated the evolutionary success of *Stenus* beetles, one of the most species-rich animal genera comprising more than 3000 species worldwide (reviewed in Betz et al. 2018). Comparative experimental measurements of strike performance indicate that the adhesive force depends on the size and micromorphology of the paraglossae (i.e., numbers of adhesive outgrowths and adhesive contact points) as well as the generated force during strikes assessed using force sensors (Koerner et al. 2017). The pad area—dependent on micromorphology of the adhesive pads—scaled with positive allometry in relation to body size, whereas the generated forces had a negative relationship relative to body size between different *Stenus* species. Hunting behavior and prey capture success of species from various microhabitats could be further explained experimentally by assessing performance of the variously shaped adhesive capture apparatus and the potential relationship with prey size.

### 2.5.2 Haustellate Mouthparts Form Proboscises of Fluid-Feeding Insects

Various types of fluid-feeding mouthparts (Fig. 2.5c–f) evolved for the uptake of plant sap, nectar, various types of exposed fluids, and body fluids. Haustellate or elongated mouthparts are called “proboscis”, if the interacting components form an unpaired fluid canal in which nutritional fluids are transported into the preoral cavity. The morphology of fluid-feeding mouthparts is reviewed in detail in Chap. 3.

#### 2.5.2.1 Occurrence of Proboscises

In non-holometabolan insects, piercing-sucking mouthparts occur in Hemiptera, which are used to feed from various food sources. They evolved independently in ectoparasitic anopluran Psocodea (Phthiraptera). In both taxa, the respective composition of the proboscis is similar in immature and adult stages. In contrast, in Holometabola, haustellate mouthparts are present only in imagines, while larval stages possess biting and chewing or filter-feeding mouthparts (Table 2.1). In many adult Hymenoptera, the mandibulate mouthparts evolved lapping proboscises (Fig. 2.5d) multiple times independently in several lineages as an adaptation to nectar-feeding. They function in a similar manner, but vary in composition and length of individual components, which interact in complex ways, but always include protractible parts for fluid adhesion (reviewed in Krenn et al. 2005; Borrell and Krenn 2006). A high disparity of piercing-sucking proboscises occurs in blood-feeding Holometabola such as Siphonaptera, many nematoceran Diptera, and some brachyceran lineages of Diptera, which are distinguished by different morphologies of the proboscis (Table 2.1) (reviewed in Krenn and Aspöck 2012). Most other fluid-feeding Diptera are characterized by a flexible, sponging proboscis which lacks mandibles. It is intricately composed of several parts, but always includes the labrum and labium with modified labial palpi. The latter are called labella and form soft apical organs for fluid uptake by capillarity and salivary distribution (Fig. 2.5e). A siphoning (or sucking) type of proboscis evolved once in Lepidoptera and is an autapomorphy of the Glossata. It is unique in its spirally coiled resting position (Fig. 2.5f). The proboscis of glossatan Lepidoptera is used for ingestion of nectar, fruit juice, and honeydew. In addition, some Lepidoptera take up decaying substances, and a few species are specialized on tears or use their proboscis for piercing and even for blood-feeding (reviewed in Krenn 2010). In context with nectar-feeding from long-spurred flowers, the siphoning or sucking feeding technique also occurs in long-tongued Apidae and Vespidae as well as in some specialized, long-tongued flies (e.g., some Nemestrinidae, Bombyliidae, Conopidae, etc.) (Table 2.1) (reviewed in Krenn et al. 2005). The sucking feeding technique mainly relies on a pressure gradient created by a sucking pump in the head (Kingsolver and Daniel 1995) that allows nectar uptake with proboscises reaching more than 250 mm in length in tropical hawk moths (reviewed in Borrell and Krenn 2006). Chapter 15 gives comprehensive information about form and function of extremely long proboscises of insects.

#### 2.5.2.2 Characteristic Modifications of Mouthpart Components Forming a Proboscis

The slender and elongate labrum forms major parts of the proboscis in Diptera and Siphonaptera. In these taxa, the labrum-epipharynx complex forms the anterior part of the food canal. In many blood-sucking Diptera, it functions as a piercing organ. In contrast, the labrum is found to be short and inconspicuous in the proboscises of Hemiptera (Fig. 2.5c), Hymenoptera, and Lepidoptera (Fig. 2.5f) where it covers the basal region of the food canal and/or forms part of the proboscis' articulation to the head.

The mandibles, if present, are modified to piercing stylets in the proboscises of plant tissue-piercing, sap-sucking, or blood-feeding insects, such as Hemiptera (Beutel et al. 2014) or nematoceran Diptera, like mosquitoes and horseflies. Thysanoptera possess unique asymmetrical mouthparts in which only the left mandible is present which is composed of a short piercing device together with the maxillae (Mickoleit 1963). Not all piercing proboscises include mandibles, as in brachyceran flies, e.g., tsetse flies, true lice, or Anoplura (reviewed in Krenn and Aspöck 2012). Mandibles are rudimentary in the siphoning proboscis of glossatan Lepidoptera, long-tongued Brachycera, and male Culicidae and are not part of the proboscis of Hymenoptera. Similarly, the sponging proboscises of non-piercing Diptera mostly lack mandibles (Table 2.1).

Characteristic modifications of the maxillae include elongation of the distal components. They may form piercing stylets that evolved independently several times in Hemiptera (Fig. 2.5c), Diptera, and Siphonaptera. Extreme elongations of the maxillae are found in some plant sap-sucking aphids, where the piercing stylets may exceed the body length. In these insects, the proboscis is coiled, like a watch spring, in resting position within a head pocket (Brożek et al. 2015; Rakitov et al. 2018). They form both the food canal and the salivary duct and function as the piercing stylets which can be deeply inserted due to the protraction-retraction motions of the maxillae. The maxillary palpi envelop the piercing structures in Siphonaptera and Glossinidae (Diptera) (reviewed in Krenn and Aspöck 2012).

In many nectar-feeding insects, such as long-tongued bees (some taxa of Apoidea), the galeae are elongated but flat and form parts of the food canal (Fig. 2.5d). Similarly, they form the coilable proboscis of glossatan Lepidoptera (Fig. 2.5f) where the galeae alone enclose the central food canal. Less conspicuous elongations of various parts of maxillae are present in specialized nectar-feeding meliod beetles (Wilhelmi and Krenn 2012), few Trichoptera (Ulmer 1905), and some Neuroptera (reviewed in Krenn et al. 2005). These fluid-imbibing maxillary structures are characterized by hairy surfaces, which allow liquid uptake by capillary forces.

Remarkable elongations of the labium mainly concern the prementum in various types of proboscises. In the piercing-sucking mouthparts of Hemiptera (Fig. 2.5c), Siphonaptera, and blood-sucking Diptera, the labium evolved into a sheath, which encompasses the other mouthpart structures (reviewed in Krenn et al. 2005; Krenn and Aspöck 2012). In all these cases, the labium is rather soft and not inserted into the host's body or food source during feeding. However, in the blood-feeding anopluran Psocodea and cyclorrhaphan Diptera, the labium or the labella are modified into a piercing organ. In most cases, the labium is retractable in some way or can be flexed under the body.

In various nectar-feeding Hymenoptera and some anthophilous beetles (see Chap. 13), the more or less elongated labial components form a hairy organ, which is adapted to take up nectar by adhesion between the cuticular structures (Krenn et al. 2005). Elongated and flat labial palpi compose parts of the food canal in many long-tongued bees (Fig. 2.5d) as well as some other taxa of Hymenoptera (reviewed in Krenn et al. 2005). In Diptera, rudiments of the labial palpi form the apical labellum of the proboscis (Fig. 2.5e). In Brachycera, they are equipped with cuticular grooves, the pseudotracheae, which are used for fluid ingestion into the food canal and discharge saliva over the food source to liquefy food in a sponging feeding technique (Table 2.1). The food canal of the remarkably long proboscises of some flower-visiting nemestrinid and tabanid flies is formed by the enormously elongated prementum alone (Karolyi et al. 2012, 2014).

An elongated hypopharynx can be found in many proboscises of piercing, blood-sucking Diptera, such as mosquitoes, horseflies, and stable flies, but also in true lice (reviewed in Krenn and Aspöck 2012). The thin and pointed hypopharynx often functions as an additional piercing device to discharge salivary fluid at the tip of the proboscis directly into the host.

### *2.5.3 Filter-Feeding Mouthparts*

Several groups of aquatic insect larvae consume suspended material using filtering devices (Fig. 2.5g). They ingest fine detritus and small particles along with micro-organisms, algae, protozoans, and tiny invertebrates. Specific modifications of the mandibles are found in many aquatic nematoceran larvae, where hair brushes on the labrum and mandibles filter suspended food material. For example, the mouthparts of mosquito larvae (Culicidae) or black fly larvae (Simuliidae) possess an array of brushes or fanlike arrangements of bristles on different mouthpart components (Fig. 2.5g) that generate water currents or trap small particles and transport them to the mouth (Craig 1977; Merritt and Wallace 1981; Gullan and Cranston 2014). Other examples of filter-feeders are found in immature stages of Ephemeroptera and Trichoptera, including species that spin sticky nets to trap food particles from the water or use mouthpart brushes to remove captured particles from filtering structures on their legs. In addition, one group of aquatic beetle larvae (Scirtidae) are characterized by a complex arrangement of interacting and movable filtering bristles of the epipharynx, mandibles, maxillae, and the hypopharynx that collect suspended particles (Hannappel and Paulus 1987). A summary of form, function, and ecological importance of these filter-feeding aquatic insects is included in this book in Chap. 4.

## 2.6 Mouthpart Functions in Addition to Feeding

### *2.6.1 Various Tasks*

In addition to gathering and manipulating food or taking up water, the mouthparts can perform other important functions in the life history of insects (Fig. 2.6). Such additional functions include grooming, defense and fighting, carrying objects, brood care in social insects (e.g., honeybees), construction of shelters and nests (e.g., wasps, Mauss 2007; see Chap. 14), cocoon formation (e.g., silk moths), as well as emergence from the pupae or subterraneous hides (e.g., ground beetles). Not all these additional functions are correlated with obvious modifications of the mouthparts. For example, body grooming is a very common behavior of insects that can be observed in numerous species with mandibulate mouthparts (Valentine 1973, 1989; Zhukovskaya et al. 2013) (Fig. 2.6a). Usually legs and antennae are cleaned to remove small particles. Cleaning may improve olfactory reception (Böröczky et al. 2013), increase adhesive abilities necessary for walking over plants (Hosoda and Gorb 2011), or remove pathogens (Zhukovskaya et al. 2013). Many insects use their mandibles to bite in defense (Fig. 2.6b). Male solitary bees clasp objects with their mandibles to secure their body while resting overnight (Westrich 1990; Alcock 1998) (Fig. 2.6c). In exceptional cases, such as weevils, biting and chewing mouthparts at the apex of particularly elongated heads of females are involved in oviposition by boring tunnels to lay eggs into seeds (e.g., Wilhelm et al. 2011; Moon 2015). Although reduced in complexity, the mouth hooks of cyclorrhaphan larvae aid in body movement in addition to feeding (Roberts 1971).

Some insects are even able to produce sound with their feeding organs or evolved auditive organs on modified mouthparts. For example, in some catantopin grasshoppers, the movements of the medial grooved mandibles can produce a creaking noise for acoustic signaling during courtship (Roesti and Keist 2009). Reduviid bugs produce sounds by rasping the apex of the proboscis over cuticle grooves on the ventral side of the body between the coxae of the forelegs (Wachmann et al. 2006). The death's-head hawk moth of the genus *Acherontia* produces squeaking defense sounds by expelling air out of the pharynx that causes vibration of a sclerotized lobe of the epipharynx (Brehm et al. 2015). Other hawk moths (Sphingidae) possess auditive organs on their modified labial palpi, which interact with the bristle-bearing pilifers on the lateral labrum (Roeder et al. 1970; Göpfert and Wasserthal 1999).

### *2.6.2 Mouthpart Functions in Social Insects*

In some social insects, fundamental morphological modifications of mouthparts evolved in context with division of labor. Formicidae and Isoptera perform numerous tasks with their mouthparts that are not directly related to feeding, e.g., defense, nest construction, fungus gardening, carrying objects, etc. (Fig. 2.6d). In many species, the members of the defense caste of termites and ants, the soldiers, possess huge polymorphic heads and big mandibles related to defense of the colony. Their curved mandibles are equipped with teeth and cusps that are specialized to bite other insects or to fight off intruders by cutting them into pieces (Hölldobler and Wilson 1990; Traniello and Leuthold 2000; Krishna et al. 2013). For example, *Eciton* ant soldiers have huge mandibles on massive heads in contrast to the inconspicuous mandibles of the smaller workers. In leaf-cutter ants (genus *Atta*), considerable size differences and polymorphism of the mandibles are related to caste polyethism and division of labor in a colony (Wilson 1980; Hölldobler and Wilson 1990;

Mehdiabadi and Schultz 2010). Individual tasks of workers can be related to mouthpart performance in leaf-cutter ants since initially, mandible teeth are razor-sharp but older foragers end up with worn mandibles and these individuals change from cutting leaves to carrying leaves to the nest (Schofield et al. 2011). Likewise, in many termites, the mandibles of soldiers extend beyond the big heads and are particularly armored with teeth, or they are sickle shaped and toothless compared to the stout biting-chewing mandibles of imago-workers (Watson and Gay 1991). Special snapping mandibles are shaped like distorted blades that close like scissors upon sudden release of energy. The snapping action of a species of genus *Termes* is reported to be one of the fastest movements ever recorded in animals (Seid et al. 2008). In contrast, the mandibles of some nasute soldiers (Nasutitermitinae) are vestigial and nonfunctional; these soldiers cannot feed themselves and rely on trophallaxis (Watson and Gay 1991; Krishna et al. 2013).

### 2.6.3 Mouthparts in Reproductive Behavior

Likewise, enlarged, dimorphic, and exceptionally modified mandibles evolved in context with male-male competition in some stag beetles (e.g., Liu et al. 2015; Goyens et al. 2016) (Fig. 2.6e), some staphylinid beetles (Forsyth and Alcock 1990), orthopteran tree wetas (Field and Deans 2001; Kelly 2006; Gwynne and Kelly 2018), tusked males of *Synagris* wasps (Longair 2004), and male dobsonflies (Simonsen et al. 2008). Particularly well studied are Lucanidae where sexual selection led to enormously large and complex mandibles of male stag beetles, which are unsuitable for feeding solid diets (Shine 1989; Krenn et al. 2002). Depending on morphology and muscle volume, the mandibles of stag beetles may transfer extraordinarily high biting forces in fighting contests. The biomechanics of the mandibles under stress has been studied across various species, using finite element analyses; optimized morphology for stress distribution has been found. By including the large head musculature into the functional analysis, it is underlined that the entire head of a male stag beetle should be regarded as a weapon, providing an advantage in male-male competitions (Goyens et al. 2014, 2015, 2016). Likewise, males of other taxa use their enlarged mandibles to clasp females during copulation, for example, in Embioptera (Richards and Davies 1977) or tiger beetles (Ball et al. 2011; Jones and Conner 2018).

A remarkable new organ of the mouthparts evolved as a result of an obligate pollination mutualism between prodoxid moths and agave plants (Pellmyr 2003). Female yucca moths (*Tegeticula* and *Parategeticula*) actively pollinate *Yucca* flowers with an additional tentacle-shaped organ of the maxillary palpus which is used for pollen collection (Fig. 2.5f) and subsequent pollen placement on the stigma of the flower of the larval host plant species (Pellmyr and Krenn 2002).

### 2.6.4 *Application of Silk*

In many instars but few adult insects, labial glands produce silk (Sehnal and Sutherland 2008; Sehnal and Craig 2009). Silk is applied with the mouthparts to line nests, form brood cells and pupal cases, or build shelters. A spinneret is formed by parts of the hypopharynx and distal components of the labium that is used to discharge and apply silk (Akai et al. 2003). The most prominent example concerns silk production by lepidopteran larvae (Sehnal and Craig 2009). Caterpillars form nests, ropes, and patches of silk to roll up leaves for shelter (Fig. 2.6g). In addition, silk is used to form protective cases, fix pupae onto structures, and build cocoons as a protection for their pupae (Stehr 2009). Similarly, silk production and its application by mouthpart structures also occur in Psocoptera, larvae of sawflies, ants and wasps, and trichopteran and many nematoceran larvae (Sutherland et al. 2010). Some aquatic dipteran larvae use silk to catch food particles out of water currents. In particular, trichopteran larvae build protective cases, which are often strengthened with gravel, sand, twigs, or bitten-off pieces of plants (e.g., Holzenthal et al. 2015). Some caddis flies use their mouthparts to build nets and sweep up food particles from the net's surface with a brushlike labrum (Wallace and Merritt 1980; Holzenthal et al. 2015). Another remarkable example of silk use was reported in both the immatures and adults of raspy crickets (Gryllacrididae), which construct nests for resting. These nests are constructed using the labium and silk produced by the labial glands to sew together leaves (Fig. 2.6h) (Walker et al. 2012). Australian gryllacridid crickets have been reported to construct underground nests, which are lined with silk that is produced by the labial glands to coat the substrate (Rentz 1991).

Brood cell lining has been reported in various Apidae which use petals or other substances which are applied with the mouthparts during the construction of nests. For example, colletid bees imbibe Dufour's gland secretions from the partly exserted sting and regurgitate the secretion that is probably mixed with other substances for a cellophane-like, waterproof lining of their subterraneous brood nests. The female bees use their broadly bifid brushlike glossa for application of this polymerizing fluid onto the inner brood cell wall (Michener 2000; Almeira 2007).

## 2.7 Reduced Mouthparts

Mouthparts can be completely reduced and nonfunctional in adult stages of Ephemeroptera and a few other taxa of Holometabola (Table 2.1). Reduction of mouthparts occurs more frequently in nonfeeding males than in conspecific females, which possess mouthparts to consume food. Such dimorphism occurs in Coccoidea and some holometabolan insects, such as several taxa of Lepidoptera and few Diptera (Dathe 2003). Nonfunctional mouthparts are usually correlated with a short adult life span (Gullan and Cranston 2014). Mouthpart atrophy, however, never occurs in immature stages although the mouthparts can be simplified consisting of only few and short components, for example, in endophagous parasitoids (e.g., Strepsiptera) and cyclorrhaphan larvae. Other examples are known from endoparasitoid larvae of Aculeata where the mandibles are only simple spines and other components are not differentiated (Snodgrass 1956; Chapman et al. 2016). However, mouthpart reduction does not necessarily mean that all structures are entirely missing. In some Lepidoptera, even rudimentary mouthparts have been demonstrated to be very efficient in water uptake (Smedley and Eisner 1995). Even vestigial mouthparts may still have particular functions in otherwise nonfeeding insects, e.g., opening the puparium in male Strepsiptera or clasping females during mating in Embioptera (Beutel et al. 2014).

## 2.8 Conclusions

Convergent evolution has led to the development of various similar feeding mechanisms, which are adapted to particular modes of food consumption. The concept of functional types of mouthparts allows inferences on diets and feeding behavior in insects, even if feeding observations in the field are not possible, for example, in museum specimens or fossil insects. In particular, mouthparts of extinct insects document their feeding strategies and shed light on the evolution of insect diets and their impact on food webs over the past 400 million years (Labandeira 1997; Grimaldi and Engel 2005; Nel et al. 2018).

The plesiomorphic condition of mandibulate biting and chewing mouthparts has been retained in most non-holometabolan insect lineages, in both the instars and adults (Table 2.1). A striking exception occurs in the instars of Odonata that evolved for prey capture in aquatic environments. High mouthpart disparity evolved multiple times independently in ectoparasites—on plants or animal hosts—as well as in flower-visiting Holometabola (Table 2.1). Remarkable functional diversity occurs in the Psocodea ranging from biting-chewing mouthparts of Psocoptera and the modified mandibulate mouthparts of ectoparasitic Amplycera and Ischnocera to the highly derived piercing-sucking proboscis of Anoplura (Table 2.1). In many Holometabola, larval mouthparts are rather simple in morphology but distinctly differ in form and function from those of the imagines of the same species. The high disparity of mouthparts between larvae and adults in many lineages of Holometabola probably is linked to the fundamental reorganization of the body in the pupal stage. An incredible morphological diversity of proboscises has evolved in nectar-feeding holometabolan insects as a result of co-adaptations between flowers and mouthparts. This evolutionary interaction occurs only in imagines since nectar is the main diet only in adult holometabolan insects. Likewise, blood-feeding holometabolan insects belong to adult stages (with only few exceptions in Diptera), which feature numerous functional types of piercing-sucking proboscises that probably developed as adaptation to preferred hosts. The filter-feeding mouthparts are found in aquatic instars mainly in Holometabola. Extraordinary modifications of mandibles evolved in context of division of labor in specialized castes of ants and termites. Sexual dimorphic mandibles mainly developed for male-male competition. Major reductions and rudimentary mouthparts are found in holometabolan larvae which live (e.g., as parasitoids) inside soft food sources and in only facultative-feeding or nonfeeding adult insects. The complex morphology of mouthparts reflects their optimized performance to exploit all imaginable food sources including various additional tasks in the life cycle of insects.

**Acknowledgements** I thank Annalie Melin for linguistic help and Andy Sombke, Günther Pass, and Barbara-Amina Gereben-Krenn for comments and suggestions to improve the manuscript as well as to Florian Karolyi who helped with the line drawings. I am grateful to Heinz Wiesbauer and Sylvain Hugel for providing photos and to Daniela Gruber (Core Facility of Cell Imaging and Ultrastructural Research, University of Vienna, Austria).

## References

Achtelig M (1967) Über die Anatomie des Kopfes von *Rhaphidia flavipes* STEIN und die Verwandtschaftsbeziehungen der Raphidiidae zu den Megaloptera. Zool Jahrb Abt Anat Ontog Tiere 84:249–312

Acorn JH, Ball GE (1991) The mandibles of some adult ground beetles: structure, function, and the evolution of herbivory (Coleoptera: Carabidae). Can J Zool 69:638–650

Akai H, Hakim RS, Kristensen NP (2003) Labial glands, silk and saliva. In: Kristensen NP (ed) Lepidoptera: moths and butterflies, vol 2: Morphology, physiology, and development. Handbook of zoology, vol IV, part 36. Walter de Gruyter, Berlin

Alcock J (1998) Sleeping aggregations of the bee *Idiomelissodes duplocincta* (Cockerell) (Hymenoptera: Anthophorini) and their possible function. J Kansas Entomol Soc 71:74–84

Almeira EAB (2007) Colletidae nesting biology (Hymenoptera: Apoidea). Apidologie 39:16–29

Angelini DR, Kaufman TC (2005) Insect appendages and comparative ontogenetics. Dev Biol 286:57–77

Arens W (1990) Wear and tear of mouthparts: a critical problem in stream animals feeding on epilithic algae. Can J Zool 68:1896–1914

Aspöck U, Aspöck H (2003) Ordnung Neuroptera (Planipennia), Netzflügler. In Dathe HH (ed) Lehrbuch der Speziellen Zoologie, begründet von A Kaestner, 2. Auflage Bd 1: Wirbellose Tiere 5. Teil, Spektrum, Gustav Fischer, Heidelberg, pp 564–584

Aspöck U, Plant JD, Nemeschkal HL (2001) Cladistic analysis of Neuroptera and their systematic position within Neuropterida (Insecta: Holometabola: Neuropterida: Neuroptera). Syst Entomol 26:73–86

Ball GE, Acorn JH, Shpeley A (2011) Mandibles and labrum-epipharynx of tiger beetles: basic structure and evolution (Coleoptera, Carabidae, Cicindelitae). Zookeys 147:39–83

Baranek B, Kuba K, Bauder JA-S, Krenn HW (2018) Mouthpart dimorphism in male and female wasps *Vespula vulgaris* and *Vespula germanica* (Vespidae, Hymenoptera). Dtsch Entomol Z 65:65–74

Betz O (1998) Comparative studies on the predatory behaviour of *Stenus* spp. (Coleoptera: Staphylinidae): the significance of its specialized labial apparatus. J Zool 244:527–544

Betz O, Koerner L, Dettner K (2018) The biology of Steninae. In: Betz O, Irmler U, Klimaszewski J (eds) Biology of rove beetles (Staphylinidae): life history, evolution, ecology and distribution. Springer, New York, pp 229–283

Beutel RG, Friederich F (2008) Comparative study of larval head structures of Megaloptera (Hexapoda). Eur J Entomol 105:917–938

Beutel RG, Ge S-Q (2008) The larval head of *Rhaphidia* (Rphidioptera, Insecta) and its phylogenetic significance. Zoology 11:89–113

Beutel RG, Leschen RAB (eds) (2005) Handbook of zoology, vol IV Arthropoda: Insecta. Part 38. Coleoptera, vol 1: Morphology and systematics (Archostemata, Adephaga, Myxophaga, Polyphaga partim). Walter de Gruyter, Berlin

Beutel RG, Weide D (2005) Cephalic anatomy of *Zorotypus hubbardi* (Hexapoda: Zoraptera): new evidence for a relationship with Acercaria. Zoomorphology 124:121–136

Beutel RG, Friedrich F, Ge S-Q, Yang X-K (2014) Insect morphology and phylogeny – a textbook for students of entomology. De Gruyter Graduate, Berlin

Beutel RG, Yavorskaya MI, Mashimo Y, Fukui M, Meusemann K (2017) The phylogeny of Hexapoda (Arthopoda) and the evolution of megadiversity. Proc Arthopod Embryol Soc Jpn 51:1–15

Blanke A, Machida R (2016) The homology of cephalic muscles and endoskeletal elements between Diplura and Ectognatha (Insecta). Org Div Evol 16:241–257

Blanke A, Machida R, Szucsich NU, Wilde F, Misof B (2015a) Mandibles with two joints evolved much earlier in the history of insects: dicondyly is a synapomorphy of bristletails, silverfish and winged insects. Syst Entomol 40:357–364

Blanke A, Rühr PT, Mokso R, Villanueva P, Wilde F, Stampanoni M, Uesugi K, Machida R, Misof B (2015b) Structural mouthpart interaction evolved already in the earliest lineages of insects. Proc R Soc B 282:20151033

Böröczky K, Wada-Katsumata A, Batchelor D, Zhukovskaya M, Schal C (2013) Insects groom their antennae to enhance olfactory acuity. Proc Natl Acad Sci U S A 110:3615–3620

Borrell BJ, Krenn HW (2006) Nectar feeding in long-proboscis insects. In Herrel A, Speck T, Rowe NP (eds) Ecology and biomechanics: a mechanical approach to the ecology of animals and plants, CRC Press LLC, pp 185–212

Brehm G, Fischer M, Gorb S, Kleinteich T, Kühn B, Neubert D, Pohl H, Wipfler B, Wurdinger S (2015) The unique sound production of the Death's-head hawkmoth (*Acherontia atropos* (Linnaeus, 1758)) revisited. Sci Nat 102:43–56

Brożek J, Mróz E, Wylężek D, Depa Ł, Węgierek P (2015) The structure of extremely long mouthparts in the aphid genus *Stomaphis* Walker (Hemiptera: Sternorrhyncha: Aphididae). Zoomorphology 134:431–445

Büsse S, Gorb SN (2018) Material composition of the mouthpart cuticle in a damselfly larva (Insecta: Odonata) and its biomechanical significance. R Soc Open Sci 5:172117

Büsse S, Hörnschemeyer T, Gorb S (2017) The head morphology of *Pyrrhosoma nymphula* larvae (Odonata: Zygoptera) focusing on functional aspects of the mouthparts. Front Zool 14:25–38

Büsse S, Köhnsen A, Gorb SN (2018) Biomechanics of the predatory strike in dragonfly larvae (Insecta: Odonata). In: 111th Annual meeting of the German Zoological Society, Greifswald, 10–15 Sept 2018, abstracts, p 29

Chapman RF (1995) Mechanics of food handling by chewing insects. In: Chapman RF, de Boer G (eds) Regulatory mechanisms in insect feeding. Chapman and Hall, New York

Chapman RF, Simpson SJ, Douglas AE (eds) (2016) The insects structure and function, 5th edn. Cambridge University Press, Cambridge

Cicero JM (1994) Composite, haustellate mouthparts in netwinged beetle and firefly larvae (Coleoptera, Cantharoidea: Lycidae, Lampyridae). J Morphol 219:183–192

Clissold FJ (2007) The biomechanics of chewing and plant fracture: mechanisms and implications. Adv Insect Physiol 34:317–372

Craig DA (1977) Mouthparts and feeding behaviour of Tahitian larval Simuliidae (Diptera: Nematocera). Quaest Entomol 13:195–218

Cribb BW, Stewart A, Huang H, Truss R, Noller B, Rasch R, Zalucki M (2008) Insect mandibles – comparative mechanical properties and links with metal incorporation. Naturwissenschaften 95:17–23

Dathe HH (ed) (2003) Lehrbuch der Speziellen Zoologie, begründet von A Kaestner, 2. Auflage Bd 1: Wirbellose Tiere 5. Teil, Spektrum, Gustav Fischer, Heidelberg

Eisenbeis G, Wichard W (1985) Atlas zur Biologie der Bodenarthropoden. Gustav Fischer Verlag, Stuttgart, New York

Evans MEG, Forsythe TG (1985) Feeding mechanisms, and their variation in form, of some ground-beetles (Coleoptera: Caraboidea). J Zool Lond A 206:113–143

Field LH, Deans NA (2001) Sexual selection and secondary sexual characters of wetas and king crickets. In: Field LH (ed) The biology of wetas, king crickets and their allies. CABI Publishing, Wallingford, pp 179–204

Forsyth A, Alcock JD (1990) Female mimicry and resource defense polygyny by males of a tropical rove beetle, *Leistophorus versicolor* (Coleoptera: Staphylinidae). Behav Ecol Sociobiol 26:325–330

Forsythe TG (1982) Feeding mechanism of certain ground beetles (Coleoptera: Carabidae). Coleopt Bull 36:26–73

François J, Dallai R, Yin WY (1992) Cephalic anatomy of *Sinentomon erythranum* Yin (Protura: Sinentomidae). Int J Insect Morphol Embryol 21:199–121

Friedemann K, Wipfler B, Bradler S, Beutel RG (2011) On the head morphology of *Phyllium* and the phylogenetic relationships of Phasmatodea (Insecta). Acta Zool (Stockholm) 93:184–199

Friedrich F, Beckmann F, Beutel RG (2013) The head of *Merope tuber* (Meropidae) and the phylogeny of Mecoptera (Hexapoda). Arthropod Struct Dev 37:29–54

Gilbert F, Jervis M (1998) Functional, evolutionary and ecological aspects of feeding-related mouthpart specializations in parasitoid flies. Biol J Linn Soc 63:495–535

Göpfert MC, Wasserthal LT (1999) Hearing with the mouthparts: behavioural responses and structural basis of ultrasound perception in acherontiine hawkmoths. J Exp Biol 202:909–918

Goyens J, Soons J, Aerts P, Dierick M, Dirckx J (2014) Unravelling the biomechanics of stag beetle armature using microcomputered X-ray tomography. Microsc Anal 28:16–18

Goyens J, Dirckx J, Piessen M, Aerts P (2015) Role of stag beetle jaw bending and torsion in grip on rivals. J R Soc Interface 13:20150768

Goyens J, Dirckx J, Aets P (2016) Jaw morphology and fighting forces in stag beetles. J Exp Biol 219:2955–2961

Grimaldi DA, Engel MS (2005) Evolution of the insects. Cambridge University Press, New York

Groll EK, Günther KK (2003) Ordnung Saltatoria (Orthoptera), Heuschrecken, Springschrecken. In: Dathe HH (ed) Lehrbuch der Speziellen Zoologie, begründet von A Kaestner, 2 Auflage Bd 1: Wirbellose Tiere 5. Teil, Spektrum, Gustav Fischer, Heidelberg, pp 261–290

Gronenberg W (1995a) The fast mandible strike in the trap-jaw ant *Odontomachus*. I. Temporal properties and morphological characteristics. J Comp Physiol A 176:391–398

Gronenberg W (1995b) The fast mandible strike in the trap-jaw ant *Odontomachus*. II. Motor control. J Comp Physiol A 176:399–408

Gronenberg W (1996) The trap-jaw mechanism in the dacetine ants *Daceton armigerum* and *Strumigenys* sp. J Exp Biol 199:2021–2033

Gullan PJ, Cranston PS (2014) The insects: an outline of entomology, 5th edn. Wiley, Chichester

Gwynne DT, Kelly CD (2018) Successful use of radiotransmitters in tracking male tree weta *Hemideina crassidens* (Orthoptera: Tettigonioidea: Anostostomatidae). New Zeal Entomol 41:25–28

Hannappel U, Paulus HF (1987) Arbeiten zu einem phylogenetischen System der Helodidae (Coleoptera) – Feinstrukturuntersuchungen an europäischen Larven. Zool Beitr NF 31:77–150

Hao Y-N, Sun Y-X, Liu C-Z (2019) The functional morphology of the mouthparts of *Coccinella transversoguttata*, with reference to their feeding mechanism. J Morphol 280:701–711

Hartenstein V, Chipman AD (2015) Hexapoda: a *Drosophila*'s view of development. In: Wanninger A (ed) Evolutionary developmental biology of invertebrates 5: Ecdysozoa III: Hexapoda, Springer, Wien

Hölldobler B, Wilson OE (1990) The ants. Spinger, Berlin

Holzenthal RW, Thomsen RE, Rios-Touma B (2015) Order Trichoptera. In: Thorp JH, Rogers DC (eds) Thorp and Covich's fresh water invertebrates: ecology and general biology, Elsevier, Amsterdam, pp. 965–1003

Hosoda N, Gorb SN (2011) Friction force reduction triggers feet grooming behaviour in beetles. Proc R Soc B 278:1748–1752

Jervis M (1998) Functional and evolutionary aspects of mouthpart structures in parasitoid wasps. Biol J Linn Soc 63:461–493

Jervis M, Vilhelmsen L (2000) Mouthpart evolution in adults of the basal, 'symphytan', hymenopteran lineages. Biol J Linn Soc 70:121–146

Jockusch EL, Smith FW (2015) Hexapoda: comparative aspects of later embryogenesis and metamorphosis. In: Wanninger A (ed) Evolutionary developmental biology of invertebrates 5: Ecdysozoa III: Hexapoda, Springer, Wien

Jones TK, Conner WE (2018) Pre-mating reproductive isolation in tiger beetles (Carabidae: Cicindelinae): an examination of the role of visual and morphological feedback. J Insect Behav 31:672–688

Josens RB, Farina WM, Roces F (1998) Nectar feeding by the ant *Camponotus mus*: intake rate and crop filling as a function of sucrose concentration. J Insect Physiol 44:579–585

Karolyi F, Szucsich NU, Colville JF, Krenn HW (2012) Adaptations for nectar-feeding in the mouthparts of long-proboscid flies (Nemestrinidae: *Prosoeca*). Biol J Linn Soc 107:414–424

Karolyi F, Colville JF, Handschuh S, Metscher BD, Krenn HW (2014) One proboscis, two tasks: Adaptations to blood-feeding and nectar-extracting in long-proboscid horse flies (Tabanidae, *Philoliche*). Arthropod Struct Dev 43:403–413

Karolyi F, Hansal T, Krenn HW, Colville JF (2016) Comparative morphology of the mouthparts of the megadiverse South African monkey beetles (Scarabaeidae: Hopliini): feeding adaptations and guild structure. PeerJ 4:e1597

Kehl S (2014) Morphology, anatomy, and physiological aspects of dytiscids. In: Yee DA (ed) Ecology, systematics, and the natural history of predaceous diving beetles (Coleoptera: Dytiscidae). Springer, Berlin, pp 173–197

Kelly CD (2006) Fighting for harems: assessment strategies during male-male contests in the sexually dimorphic Wellington tree weta. Anim Behav 72:727–736

Kimm MA, Prpic N-M (2006) Formation of the arthropod labrum by fusion of paired and rotated limb-bud-like primordia. Zoomorphology 125:147–155

Kingsolver JG, Daniel TL (1995) Mechanics of food handling by fluid-feeding insects. In: Chapman RF, DeBoer G (eds) Regulatory mechanisms in insect feeding. Chapman & Hall, New York, pp 32–73

Kinzelbach RK (1971) Morphologische Befunde an Fächerflüglern und ihre phylogenetische Bedeutung (Insecta: Strepsiptera). In: Schaller F (ed) Zoologica Originalabhandlungen aus dem Gesamtgebiet der Zoologie, vol 119. E. Schweizerbart'sche Verlagsbuchhandlung, Stuttgart

Klass K-D, Ehrmann R (2003) Ordnung Mantodea, Fangschrecken, Gottesanbeterinnen. In: Dathe HH (ed) Lehrbuch der Speziellen Zoologie, begründet von A Kaestner, 2. Auflage Bd 1: Wirbellose Tiere 5. Teil, Spektrum, Gustav Fischer, Heidelberg, pp 182–197

Klass K-D, Zompro O, Adis JU (2003) Ordnung Mantophasmatodea. In: Dathe HH (ed) Lehrbuch der Speziellen Zoologie, begründet von A Kaestner, 2. Auflage Bd 1: Wirbellose Tiere 5. Teil, Spektrum, Gustav Fischer, Heidelberg, pp 161–166

Kluge NJ (2002) The homology of mouthparts in fleas (Insecta, Aphaniptera). Entomol Rev 82:808–816

Koch M (2000) The cuticular cephalic endoskeleton of primarily wingless hexapods: ancestral state and evolutionary changes. Pedobiology 44:374–385

Koerner L, Garamszegi LZ, Heethoff M, Betz O (2017) Divergent morphologies of adhesive predatory mouthparts of *Stenus* species (Coleoptera: Staphylinidae) explain differences in adhesive performance and resulting prey-capture success. Zool J Linnean Soc 181:500–518

Krenn HW (2010) Feeding mechanisms of adult Lepidoptera: structure, function, and evolution of the mouthparts. Annu Rev Entomol 55:307–327

Krenn HW, Aspöck H (2012) Form, function and evolution of the mouthparts of blood-feeding Arthropoda. Arthropod Struct Dev 41:101–118

Krenn HW, Pernstich A, Messner T, Hannappel U, Paulus HF (2002) Kirschen als Nahrung des männlichen Hirschkäfers, *Lucanus cervus* (LINNAEUS 1758) (Lucanidae: Coleoptera). Entomol Zeitsch 112:165–169

Krenn HW, Plant J, Szucsich NU (2005) Mouthparts of flower-visiting insects. Arthropod Struct Dev 34:1–40

Krenn HW, Gereben-Krenn B-A, Steinwender BM, Popov A (2008) Flower visiting Neuroptera: mouthparts and feeding behaviour of *Nemoptera sinuata* (Nemopteridae). Eur J Entomol 105:267–277

Krenn HW, Fournel J, Bauder JA-S, Hugel S (2016) Mouthparts and nectar feeding of the flower visiting cricket *Glomeremus orchidophilus* (Gryllacrididae). Arthropod Struct Dev 45:221–229

Krishna K, Grimaldi DA, Krishna V, Engel MS (2013) Treatise on the Isoptera of the world. Amnah Bull 377:1–202

Kristensen NP (1975) The phylogeny of hexapod "orders". A critical review of recent accounts. Zeitsch Zool Syst Evol 13:1–44

Kristensen NP (ed) (1998) Lepidoptera: moths and butterflies 1. Handbook of zoology IV, vol 35, De Gruyter, Berlin

Labandeira CC (1997) Insect mouthparts: ascertaining the paleobiology of insect feeding strategies. Annu Rev Ecol Syst 28:153–193

Larabee FJ, Suarez AV (2014) The evolution and functional morphology of trap-jaw ants (Hymenoptera: Formicidae). Myrmeco News 20:25–36

Larabee FJ, Gronenberg W, Suarez AV (2017) Performance, morphology and control of power-amplified mandibles in the trap-jaw ant *Myrmoteras* (Hymenoptera: Formicidae). J Exp Biol 220:3062–3071

Liu X, Hayashi F, Lavine LC, Yang D (2015) Is diversification in male reproductive traits driven by evolutionary trade-offs between weapons and nuptial gifts? Proc R Soc B 282:20150247

Longair RW (2004) Tusked males, male dimorphism and nesting behavior in a subsocial afrotropical wasp, *Synagris cornuta*, and weapons and dimorphism in the genus (Hymenoptera: Vespidae: Eumeninae). J Kansas Entomol Soc 77:528–557

Matsuda R (1965) Morphology and evolution of the insect head. Mem Am Ent Inst 4:1–334

Mauss V (2007) Evolution verschiedener Lebensformtypen innerhalb basaler Teilgruppen der Faltenwespen (Hymenoptera, Vespidae). Denisia 20:701–722

Mehdiabadi NJ, Schultz TR (2010) Natural history and phylogeny of the fungus-farming ants (Hymenoptera: Fromicidae: Myrmicinae: Attini). Myrmecol News 13:37–55

Merritt RW, Wallace JB (1981) Filter-feeding insects. Sci Am 244:132–147

Metcalf CL (1929) The mouthparts of insects. Trans Ill Acad Sci 21:109–135

Mey E (2003) Ordnung Phthiraptera, Tierläuse, Lauskerfe. In Dathe HH (ed) Lehrbuch der Speziellen Zoologie, begründet von A Kaestner, 2. Auflage Bd 1: Wirbellose Tiere 5. Teil, Spektrum, Gustav Fischer, Heidelberg, pp 308–330

Michelsen V (1997) A revised interpretation of the mouthparts in adult fleas (Insecta, Siphonaptera). Zool Anz 235:217–223

Michener CD (2000) The bees of the world. The John Hopkins University Press, Baltimore, MD

Mickoleit E (1963) Untersuchungen zur Kopfmorphologie der Thysanoptera. Zool Jb Anat 81:101–150

Misof B, Liu S, Meusemann K, Peters RS, Donath A, Mayer C, Frandsen PB, Ware J, Flouri T, Beutel RG, Niehuis O, Petersen M, Izquierdo-Carrasco F, Wappler T, Rust J, Aberer AJ, Aspöck U, Aspöck H, Bartel D, Blanke A, Berger S, Böhm A, Buckley T, Calcott B, Chen J, Friedrich F, Fukui M, Fujita M, Greve C, Grobe P, Gu S, Huang Y, Jermiin LS, Kawahara AY, Krogmann L, Kubiak M, Lanfear R, Letsch H, Li Y, Li Z, Li J, Lu H, Machida R, Mashimo Y, Kapli P, McKenna DD, Meng G, Nakagaki Y, Navarrete-Heredia JL, Ott M, Ou Y, Pass G, Podsiadlowski L, Pohl H, von Reumont BM, Schütte K, Sekiya K, Shimizu S, Slipinski A, Stamatakis A, Song W, Su X, Szucsich NU, Tan M, Tan X, Tang M, Tang J, Timelthaler G, Tomizuka S, Trautwein M, Tong X, Uchifune T, Walzl MG, Wiegmann BM, Wilbrandt J, Wipfler B, Wong TK, Wu Q, Wu G, Xie Y, Yang S, Yang Q, Yeates DK, Yoshizawa K, Zhang Q, Zhang R, Zhang W, Zhang Y, Zhao J, Zhou C, Zhou L, Ziesmann T, Zou S, Li V, Xu X, Zhang Y, Yang H, Wang J, Wang J, Kjer KM, Zhou X (2014) Phylogenomics resolves the timing and pattern of insect evolution. Science 346:763–767

Moon M-J (2015) Microstructure of mandibulate mouthparts in the greater rice weevil, *Sitophilus zeamais* (Coleoptera: Curculionidae). Entomol Res 45:9–15

Nagatomi A, Soroida K (1985) The structure of the mouthparts of the orthorrhaphous Brachycera (Diptera) with special reference to blood-sucking. Beitr Entomol 35:263–368

Nel A, Scholtz CH (1990) Comparative morphology of the mouthparts of adult Scarbaeoidea (Coleoptera). Entomol Mem Dep Agric Dev Repub S Afr 80:1–84

Nel P, Bertrand S, Nel A (2018) Diversification of insects since the Devonian: a new approach based on morphological disparity of mouthparts. Sci Rep 8:3516

Neubert D, Simon S, Beutel RG, Wipfler B (2017) The head of the earwig *Forficula auricularia* (Dermaptera) and its evolutionary implications. Arthropod Syst Phylo 75:99–124

Olesen J (1972) The hydraulic mechanism of labial extension and jet propulsion in dragonfly nymphs. J Comp Physiol 81:53–55

Pass G, Szucsich NU (2011) 100 years of research on the Protura: many secrets still retained. Soil Organ 83:309–334

Paul J (2001) Mandible movements in ants. Comp Biochem Physiol A 131:7–20

Paul J, Roces F, Hölldobler B (2002) How do ants stick out their tongues? J Morphol 254:39–52

Pellmyr O (2003) Yuccas, Yucca moths, and coevolution: a review. Ann Missouri Bot Gard 90:35–55

Pellmyr O, Krenn HW (2002) Origin of a complex key innovation in an obligate insect-plant mutualism. Proc Natl Acad Sci U S A 99:5498–5502

Pohl H, Beutel RG (2005) The phylogeny of Strepsiptera (Hexapoda). Cladistics 21:328–374

Posnien N, Bucher G (2010) Formation of the insect head involves lateral contribution of the intercalary segment, which depends on *Tc-labial* function. Dev Biol 338:107–116

Prakash S, Mendki MJ, Rao KM, Singh K, Sing RN (1995) Sensilla on the maxillary palps of the cockroach *Supella longipalpa* Fabricius (Dictyoptera: Blattellidae). Int J Insect Morphol Embryol 24:13–34

Rähle W (1970) Untersuchungen an Kopf und Prothorax von *Embia ramburi* Rimski-Korsakow 1906 (Embioptera: Embiidae). Zool Jahrb Abt Anat Ontog Tiere 87:248–330

Rakitov R, Karengin S, Karengina O (2018) How *Stomaphis* aphids use their giant proboscides. J Nat Hist 52:299–304

Rentz DCF (1991) Orthoptera (grasshoppers, locusts, katydids, crickets). In: CSIRO The insects of Australia, vol 1. Melbourne University Press, pp 369–393

Richards OW, Davies RG (1977) Embioptera. In: Imms' general textbook of entomology, vol 2. Springer, Dordrecht, pp 583–591

Roberts MJ (1971) On the locomotion of cyclorrhaphan maggots (Diptera). J Nat Hist 5:583–590

Roeder KD, Treat AE, Van de Berg JS (1970) Distal lobe of the pilifer: an ultrasonic receptor in choerocampine hawkmoths. Science 170:1098–1099

Roesti C, Keist B (2009) Die Stimmen der Heuschrecken. Haupt Verlag, Bern

Rogers TB, Kaufman TC (1996) Structure of the insect head as revealed by the EN protein pattern in developing embryos. Development 112:3419–3432

Schmidt C (2017) Morphological and functional diversity of ant mandibles. Tree of life web project: http://tolweb.org, accesssed 12.06.2018

Schmitt C, Rack A, Betz O (2014) Anasylses of the mouthpart kinematics in *Periplaneta americana* (Blattodea, Blattidae) using synchotron-based X-ray cinradiography. J Exp Biol 217:3095–3107

Schofield RM, Nesson MH, Richardson KA (2002) Tooth hardness increases with zinc-content in mandibles of young adult leaf-cutter ants. Naturwissenschaften 89(12):579–583

Schofield RMS, Emmett KD, Niedbala JC, Nesson MH (2011) Leaf-cutter ants with worn mandibles cut half as fast, spend twice the energy, and tend to carry instead of cut. Behav Ecol Sociobiol 65:969–982

Sehnal F, Craig C (2009) Silk production. In: Resh VH, Cardé RT (eds) Encyclopedia of insects, 2nd edn. Elsevier, Amsterdam, pp 921–924

Sehnal F, Sutherland T (2008) Silk production by insect labial glands. In: Scheibel T (ed) Fibrous proteins. Landes Biosciences, Austin, TX, pp 1–15

Seid MA, Scheffrahn RH, Niven JE (2008) The rapid mandible strike of a termite soldier. Curr Biol 18:PR1049–PR1050
Shine R (1989) Ecological causes for the evolution of sexual dimorphism: a review of the evidence. Q Rev Biol 64:419–461
Simonsen TJ, Dombroskie JJ, Lawrie DD (2008) Behavioral observations on the dobsonfly, *Corydalus cornutus* (Megaloptera: Corydalidae) with photographic evidence of the use of the elongate mandibles in the male. Am Entomol 64:167–169
Smedley SR, Eisner T (1995) Sodium uptake by puddling in a moth. Science 270:1816–1817
Smith JJB (1985) Feeding mechanisms. In: Kerkut GA, Gilbert LI (eds) Comprehensive insect physiology, biochemistry and pharmacology 4. Pergmon, Oxford, pp 33–85
Snodgrass RE (1935) Principles of insect morphology. McGraw-Hill Book, New York (reprint with a new foreword by Goerge C Eickwort, Cornell University Press 1993)
Snodgrass RE (1956) Anatomy of the honey bee. Comstock, Ithaca, NY
Spanga JC, Vakis AI, Schmidt CA, Patek SN, Zhang X, Tsutsui ND, Suarez A (2008) Phylogeny, scaling, and the generation of extreme forces in trap-jaw ants. J Exp Biol 211:2358–2368
Staniczek AH (2000) The mandible of silverfish (Insecta: Zygentoma) and mayflies (Ephemeroptera): its morphology and phylogenetic significance. Zool Anz 239:315–337
Stehr FW (2009) Caterpillars. In: Resh VH, Cardé RT (eds) Encyclopedia of insects, 2nd edn. Elsevier, Amsterdam, pp 135–137
Sutherland TD, Young JH, Weisman S, Hayashi CY, Merritt DJ (2010) Insect silk: one name, many materials. Annu Rev Entomol 55:171–188
Tanaka Y, Hisada M (1980) The hydraulic mechanism of the predatory strike in dragonfly larvae. J Exp Biol 88:1–19
Traniello JFA, Leuthold RH (2000) Behavior and ecology of foraging in termites. In: Abe T et al (eds) Termites: evolution, sociality, symbioses, ecology. Kluwer Academic, Dordrecht, pp 141–168
Tröster G (1990) Der Kopf von *Hybophthirus notophallus* (Neumann) (Phthiraptera:Anoplura). Eine funktionsmorphologische und konsequent-phylogenetische Analyse. Stuttg Beitr Naturkunde A (Biologie) 442:1–89
Ulmer G (1905) Über einige Trichopteren mit rüsselförmigen Kopfanhängen. Zool Anz 28:56–59
Valentine BD (1973) Grooming behavior in Coleoptera. Coleopt Bull 27:63–73
Valentine BD (1989) Grooming behavior in Embioptera and Zoraptera (Insecta). Ohio J Sci 86:150–152
Vincent JFV, Wegst UGK (2004) Design and mechanical properties of insect cuticle. Arthropod Struct Dev 33:187–199
Wachmann E, Melber A, Deckert J (2006) Wanzen 1 – Tierwelt Deutschlands, vol 77. Goecke & Evers, Kelterns
Walker AA, Weisman S, Church JS, Merritt DJ, Mudie ST, Sutherland TD (2012) Silk from crickets: a new twist on spinning. PLoS One 7:e30408
Wallace JB, Merritt RW (1980) Filter-feeding ecology of aquatic insects. Annu Rev Entomol 25:103–132
Watson JAL, Gay FJ (1991) Isoptera (termites). In: CSIRO The insects of Australia, vol 1. Melbourne University Press, Melbourne, pp 330–347
Weber H (1933) Lehrbuch der Entomologie. Gustav Fischer Verlag, Stuttgart (reprint Verlag von Otto Koeltz, Koenigstein – Taunus 1968)
Weber H (1969) Die Elefantenlaus *Haematomyzus elefantis* Piaget 1869. Versuch einer konstruktionsmorphologischen Analyse. Zoologica 116:1–154
Weihmann T, Reinhard L, Weissing K, Siebert T, Wipfler B (2015a) Fast and powerful: biomechanics and bite forces of the mandibles in the American cockroach *Periplaneta americana*. PLoS One 10:e0141226
Weihmann T, Kleinteich T, Grob SN, Wipfler B (2015b) Functional morphology of the mandibular apparatus in the cockroach *Periplaneta americana* (Blattodea: Blattidae) – a model species for omnivore insects. Arthopod Syst Phylo 73:477–488

Westrich P (1990) Die Wildbienen Baden-Württembergs Allgemeiner Teil: Lebensräume, Verhalten, Ökologie und Schutz, 2. Ulmer Verlag, Auflage
Wichard W, Arens W, Eisenbeis G (2002) Biological atlas of aquatic insects: with a preface of Vincent Resh. Apollo Verlag, new edition 2013
Wilhelm G, Handschuh S, Plant J, Nemeschkal HL (2011) Sexual dimorphism in head structures of the weevil *Rhopalapion longirostre* (Olivier 1807) (Coleoptera: Curculionoidea): a response to ecological demands of egg deposition. Biol J Linn Soc 104:642–660
Wilhelmi A, Krenn HW (2012) Elongated mouthparts of nectar-feeding Meloidae (Coleoptera). Zoomorphology 131:325–337
Wilson EO (1980) Caste and division of labor in leaf-cutter ants (Hymenoptera: Formicidae: *Atta*): I. The overall pattern in *A. sexdens*. Behav Ecol Sociobiol 7:143–156
Wipfler B, Machida R, Müller B, Beutel RG (2011) On the head morphology of Grylloblattodea (Insecta) and the systematic position of the order, with a new nomenclature for the head muscles of Dicondylia. Syst Entomol 36:241–266
Wipfler B, Letsch H, Frandsen PB, Kapli P, Mayer C, Bartel D, Buckley TR, Donath A, Edgerly-Rooks J, Fujita M, Liu S, Machida R, Mashimo Y, Misof B, Niehuis O, Peters RS, Peterson M, Podsiadlowski L, Schütte K, Shimizu S, Uchifune T, Wilbrandt J, Yan E, Zhou X, Simon S (2019) Evolutionary history of Polyneoptera and its implications for our understanding of early winged isects. Proc Natl Acad Sci U S A. www.pnas.org/cgi/doi/10.1073/pnas.1817794116
Zhukovskaya M, Yanagawa A, Forschler BT (2013) Grooming behavior as a mechanism of insect disease defense. Insects 2013:609–630
Zhuzhikov DP (2007) The mouthparts of cockroaches (Blattodea). Entomol Rev 87:191–203
Zwick P (2003) Ordnung Plecoptera, Steinfliegen, Frühlingsfliegen, Uferbolde. In Dathe HH (ed) Lehrbuch der Speziellen Zoologie, begründet von A Kaestner, 2. Auflage Bd 1: Wirbellose Tiere 5. Teil, Spektrum, Gustav Fischer, Heidelberg, pp 144–154

# Chapter 3
# Fluid-Feeding Mouthparts

**Harald W. Krenn**

**Abstract** The mouthparts of most specialized fluid-feeding insects consist of more or less elongated components forming a proboscis. Functional types of mouthparts evolved as adaptations to particular food sources. Characteristic feeding techniques are used which are based on a combination of capillarity and a pressure gradient created by sucking pumps. Biting-sucking mandibles occur in some predaceous insect larvae and are used for extraoral digestion. Lapping mouthparts evolved in nectar-feeding insects; such proboscises are characterized by a loose food canal and setose, pro- and retractable structures at the tip which take up fluids mainly by capillarity. In contrast, piercing-sucking proboscises have pointed components to penetrate the host's epidermis. The elongated components are firmly interlocked to form a tight food canal and a salivary duct. Piercing-sucking proboscises ingest fluid along a pressure gradient and occur in plant sap-sucking insects, blood feeders, and predaceous species. Sponging proboscises are rather short and retractable. Their soft and cushion-shaped apical components take up liquids from open fluid sources. Siphoning proboscises are particularly long and primarily adapted for nectar drinking. Such sucking mouthparts ingest fluid into the food canal predominantly by a pressure gradient.

## 3.1 Fluid-Feeding Without a Proboscis

### *3.1.1 Unspecialized Fluid-Feeding*

Many insects regularly ingest liquids or water in addition to solid food. Many insects use an adhesive fluid-feeding technique without haustellate mouthparts. It can be assumed that certain areas of biting and chewing mouthparts are hydrophilic and fluids adhere on these structures by capillary force. Rarely studied in detail, most insects are probably able to supply themselves at least with water. Representatives of many insect taxa occasionally feed on nectar using short unspecialized mouthparts (Proctor et al. 1996). Conspicuous

H. W. Krenn (✉)
Department of Integrative Zoology, University of Vienna, Vienna, Austria
e-mail: harald.krenn@univie.ac.at



H. W. Krenn (ed.), *Insect Mouthparts*, Zoological Monographs 5,
https://doi.org/10.1007/978-3-030-29654-4_3

adaptations to ingest liquids are scarce in non-holometabolan insects with orthopteriod mouthparts but have been reported in nectar-feeding crickets from genus *Glomeremus* (Gryllacrididae) that show specialized tips of the maxillae which can be dipped into sucrose fluids like nectar that is subsequently ingested (Krenn et al. 2016).

Most adult Hymenoptera feed on liquids using a labiomaxillary complex. This functional unit of maxillae and labium forms an extensible structure to lick up fluids (Jervis 1998; Jervis and Vilhelmsen 2000). In particular, ants regularly feed on extrafloral nectaries and honeydew. The workers do not possess elongated mouthparts, but are able to suck in fluids, or they use licking motions of the labial glossa which is equipped with rows of small microtrichia for fluid uptake (Gotwald 1969; Paul 2001). The glossa may work as an up-and-down moving shovel, supported by synchronous movements of the entire labium. This allows uptake of various concentrated sucrose fluids with an adhesive fluid-feeding mechanism (Josens et al. 1998). Some ants transport droplets of liquid between the mouthpart components for supplying the colony with carbohydrate-rich fluids (Paul et al. 2002). The functional mechanics of fluid-feeding in ants is summarized in Chap. 10 of this book.

### *3.1.2 Nectar-Feeding Mouthparts of Anthophilous Coleoptera*

In addition to pollen, many flower-visiting Coleoptera feed on nectar by using setaceous mouthpart components for fluid uptake (reviewed in Krenn et al. 2005). Many anthophilous Cetonidae lack biting mandibles, and South African monkey beetles (Hopliini) show more or less elongated maxillary structures for nectar uptake in addition to pollen-feeding (Karolyi et al. 2016). By applying an adhesive feeding technique, Scarabaeidae load nectar onto setose maxillary structures by sweeping movements (Johnson and Nicolson 2001). Particularly elongated maxillary components form a proboscis-like organ for fluid ingestion from concealed nectar sources in some anthophilous Meloidae (Handschin 1929; Chaudonneret 1990). In the genus *Leptopalpus* (Nemognathinae), the proboscis is composed by the elongated four-segmented maxillary palpi which are flexed under the body in resting position. The distal palpomers are covered with bristles on their median sides which form a fluid pathway during nectar drinking. In related genera of Nemognathinae, the elongated, hairy galeae are loosely connected and form the moveable fluid-feeding organ, but a true food tube is missing. The galeae bear long bristles on the median side that form a pointed apical brushlike structure (Wilhemi and Krenn 2012). Capillary forces probably draw nectar between the hairy mouthparts. However, details of mouthpart movements during flower visiting and biophysics of fluid transport remain unexamined in beetles. Morphological mouthpart adaptations of anthophilous beetles are reviewed in Chap. 13.

### 3.1.3 *Nectar-Feeding Mouthparts of Anthophilous Neuroptera*

An adhesive nectar-feeding technique occurs in some flower-visiting adult Nemopteridae (Neuroptera) which possess an elongated head with derived orthopteriod mouthparts. The elongated and brush-shaped laciniae, galeae, and maxillary palpi form a flexible functional unit that can be extended into a flower, from which nectar is taken up by the structures of the laciniae. Upon retraction, the adhered nectar is transported into the preoral cavity and ingested probably by the expansion of the muscular pharynx (Krenn et al. 2008). Chapter 11 examines the morphology, feeding preferences, and evolution of the mouthparts of Neuroptera.

### 3.1.4 *Biting-Sucking Mouthparts*

Predaceous larvae of Neuroptera and some Coleoptera catch their prey with strong prognathous mandibles and digest it extraorally. Curved biting mandibles convergently evolved which are equipped with a canal through which digestive fluid is discharged directly into the prey and through which liquefied tissue is ingested (Fig. 3.1).

The long, fanglike mandibles of larval Neuroptera are deeply grooved on the ventral surface. The longitudinal groove is closed by the similarly shaped bladelike maxilla which perfectly fits into the mandibular groove. Thus, a pair of tubular channels is formed between each mandible and maxillary that is used to suck out prey (Aspöck and Aspöck 2003) (Figs. 3.1a, b).

A functionally similar feeding apparatus evolved in the predaceous larvae of aquatic beetles, i.e., Dytiscidae, Gyrinidae, and Haliplidae, which possess long, curved fanglike mandibles, equipped with a median groove that forms a nearly closed canal from the cibarium to the apex of the mandible (Weber 1933; Kehl 2014) (Fig. 3.1c, d). Similarly, the snail-feeding larvae of Lampyridae (Coleoptera) possess biting-sucking mouthparts with each mandible forming a closed tube which leads from the base to a subterminal opening near the tip of the mandible (Fig. 3.1e, f). In the Lycidae, labral and mandibular parts compose the sucking jaws during embryogenesis that together form a closed sucking tube (Cicero 1994).

### 3.1.5 *Biting Blood-Feeding Mouthparts of Elephant Lice (Rhynchophthirina, Psocodea)*

A unique type of biting mouthparts occurs in ectoparasitic elephant lice (Rhynchophthirina). These blood feeders possess a greatly elongated rostrum-shaped head which apically bears small outward turned mandibles. The biting mandibles are equipped with large incisivi and can move in lateral directions. By these motions, the

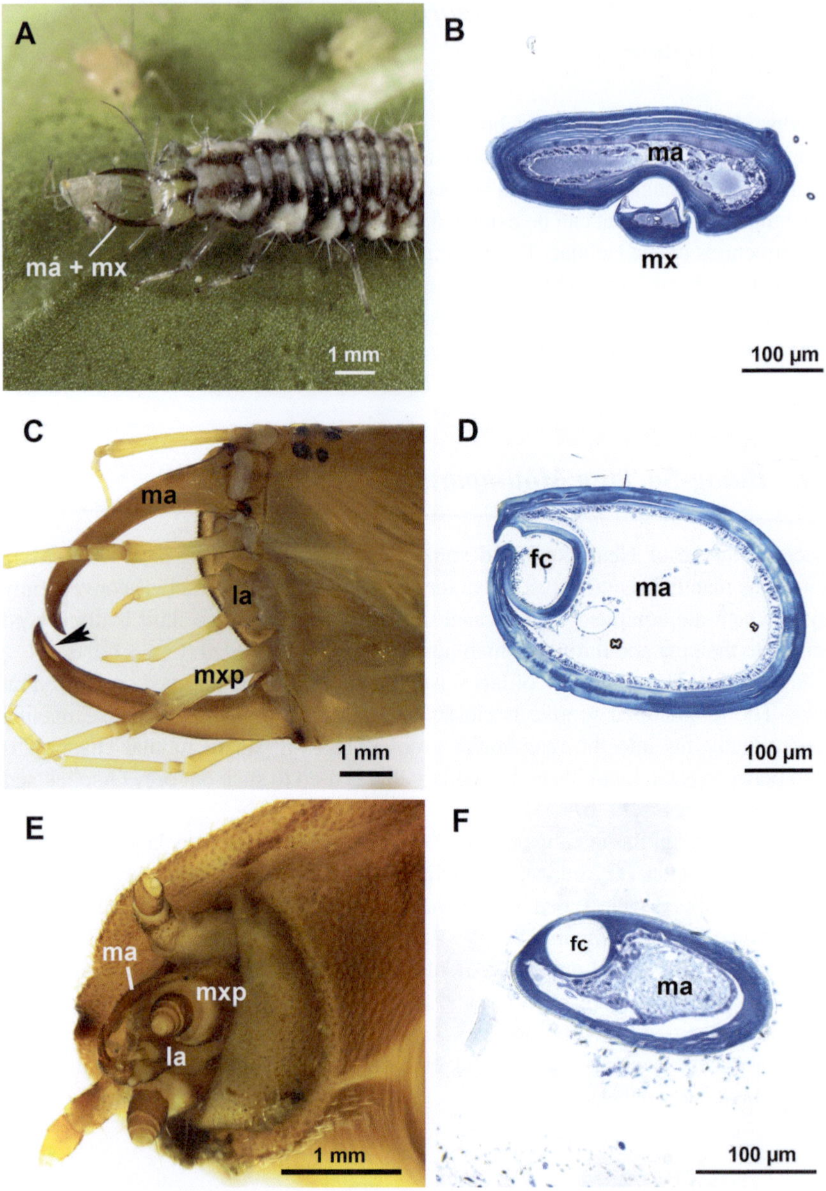


**Fig. 3.1** Larval biting-sucking mouthparts of Neuroptera and of some Coleoptera (microphotos and light microscope photos of histological sections). (**a**) Larva of *Chrysoperla* sp. (Chrysopidae, Neuroptera) feeds on an aphid (photo by courtesy of Kerry S. Matz). (**b**) Cross-section of the biting jaws of an ant lion (*Euroleon nostras*, Myrmeleontidae, Neuroptera); mandibles and maxillae form the food canal. (**c**) Head of a water beetle larva (*Cybister* sp., Dytiscidae, Coleoptera); arrow head indicates the subterminal opening of the food canal of the mandible. (**d**) Cross-section of the mandible of *Cybister* sp. (Dytiscidae). (**e**) Head of a glow worm larva (*Lampyris noctiluca*, Lampyridae, Coleoptera). (**f**) Sucking mandible with closed food canal in cross-section of *L. noctiluca* (Lampyridae). *fc* food canal, *ma* mandible, *mx* maxilla, *mxp* maxillary palpus, *la* labium

mandibles dig into the host's skin from which blood is ingested directly into the muscular preoral cavity (Weber 1969; Lehane 2005).

## 3.2 Fluid-Feeding Using a Proboscis

### 3.2.1 *Functional Components of a Proboscis*

A haustellate feeding apparatus comprises a proboscis and a sucking pump within the head. Elongated insect mouthparts are referred to as a proboscis if they are combined to form an unpaired functional unit for fluid-feeding. Fundamental disparity in functional morphology depends on whether the fluid sources are freely accessible or concealed. Consequently, the proboscis morphology of different taxa shows various adaptations in length, form, and function according to the preferred source; that may comprise plant sap, nectar, fruit juices, liquefied plant tissue (including pollen and seeds), body fluids of insects or vertebrates, excrements, honeydew, rotting semiliquid material, or water (Fig. 3.2).

Most proboscises can be functionally divided into (1) a basal articulation with the head capsule, (2) an elongated food canal with or without a separate salivary duct, and (3) a particularly shaped apical region.

### 3.2.2 *Basal Articulation Moves the Proboscis*

The base of the proboscis usually forms a movable connection to the head that, in most cases, allows a space-saving resting position. In addition, the proboscis base ensures a continuous fluid pathway from the food canal through the preoral cavity into the mouth and alimentary tract. Various characteristic protective resting positions can be observed in different taxa (Borrell and Krenn 2006). The proboscis can be (1) flexed back under the body (e.g., long-tongued Apidae, Hymenoptera), (2) retracted partly or entirely into the head (e.g., Bombyliidae, Diptera), (3) flexed in front of the head (e.g., Conopidae, Diptera), or (4) coiled under the head (e.g., Glossata, Lepidoptera) or inside the head (e.g., some Aphididae, Hemiptera). Movements from the resting position to a feeding position are usually initiated by the cranial muscles or by muscles of the basal mouthpart components. The most distal regions of a proboscis usually lack muscles. In addition, elastic cuticle properties and hydraulic mechanisms enable the full repertoire of movements (Snodgrass 1935; Van der Starre and Ruigrok 1980; Wenk 1980; Smith 1985; Krenn 1990, 2010). Sensory equipment at the basal proboscis regions and on the head detects the positions of these structures against each other. Mechanosensilla and proprioceptors monitor the correct posture for feeding and the resting position, but these aspects of proboscis movements have only been studied in few examples (Krenn 1998).

**Fig. 3.2** Food sources of various fluid-feeding insects. (**a**) Plant sap-feeding aphid *Uroleucon* cf. *grossum* (Aphididae, Hemiptera). (**b**) Flower-visiting "short-tongued" bee cf. *Lasioglossum* sp. (Halictidae, Hymenoptera) ingests nectar. (**c**) Blood-sucking mosquito *Aedes* sp. (Culicidae) (photo by courtesy of R. Pospischil). (**d**) Predatory robber fly (Asilidae, Diptera) feeding on a beetle. (**e**) A housefly *Musca domestica* (Muscidae, Diptera) feeding from freely accessible liquid. (**f**) Mud-puddling butterfly *Marpesia* sp. (Nymphalidae, Lepidoptera) feeds on damp soil

### 3.2.3 *Food Canal Conducts Fluid*

Elongated structures of the mouthparts are combined to form a food canal that can be regarded as a defined pathway to transport ingested fluids. In various taxa, this canal is composed of differently modified mouthpart components, often including the hypopharynx. The components are interlocked more or less firmly. Relatively short food canals are formed by rather loosely combined structures, while long and narrow feeding tubes are composed of firmly interlocked structures which close the food canal tightly (Fig. 3.3). A pressure gradient created by the cranial sucking pump, assisted by hydrophilic/hydrophobic properties of the mouthpart surface, transports liquids through the food canal. The pressure gradient and the hydrophilic properties contribute to a different degree to fluid transport depending on the morphology of the mouthparts (Kingsolver and Daniel 1995; Monaenkova et al. 2011; Lehnert et al. 2013; Kornev et al. 2017). Saliva is either expelled through the food canal, or a separate salivary duct is present. Sensilla in the food canal probably monitor the fluid transport. For example, in the food canal of butterflies, a row of chemoreceptive sensilla react to sugary fluids (Inoue et al. 2009). Similar sensilla can be found in the food canal forming galea of orchid bees (Düster et al. 2018). Mechanosensilla on the lateral side of the proboscis might provide information about the insertion depth into flowers (Krenn 1998; Düster et al. 2018).

### 3.2.4 *Proboscis Tip*

The apical region of a proboscis may comprise structures for capillary fluid uptake, anchoring the mouthparts, puncturing and penetration of hosts, and guiding liquid into the food canal as well as for salivary discharge. Depending on the proboscis type, the tip can be soft or heavily sclerotized; it may possess pointed structures or has structures that enlarge the surface (Fig. 3.3). In addition, the apical region may have specific structures forming openings into the food canal. Sensilla are responsible for food detection and for positioning the proboscis tip. For example, in Cicadellidae (Hemiptera), Apidae (Hymenoptera), Lepidoptera, and Diptera, the tip of the proboscises is equipped with chemo- and/or mechanosensilla (Fig. 3.3) which allow detection of nutritive substances and give the insect information on the location of the tip (Krenn 1998; Szucsich and Krenn 2000, 2002; Leopold et al. 2003; Lee and Craig 2009; Karolyi et al. 2012; Düster et al. 2018).

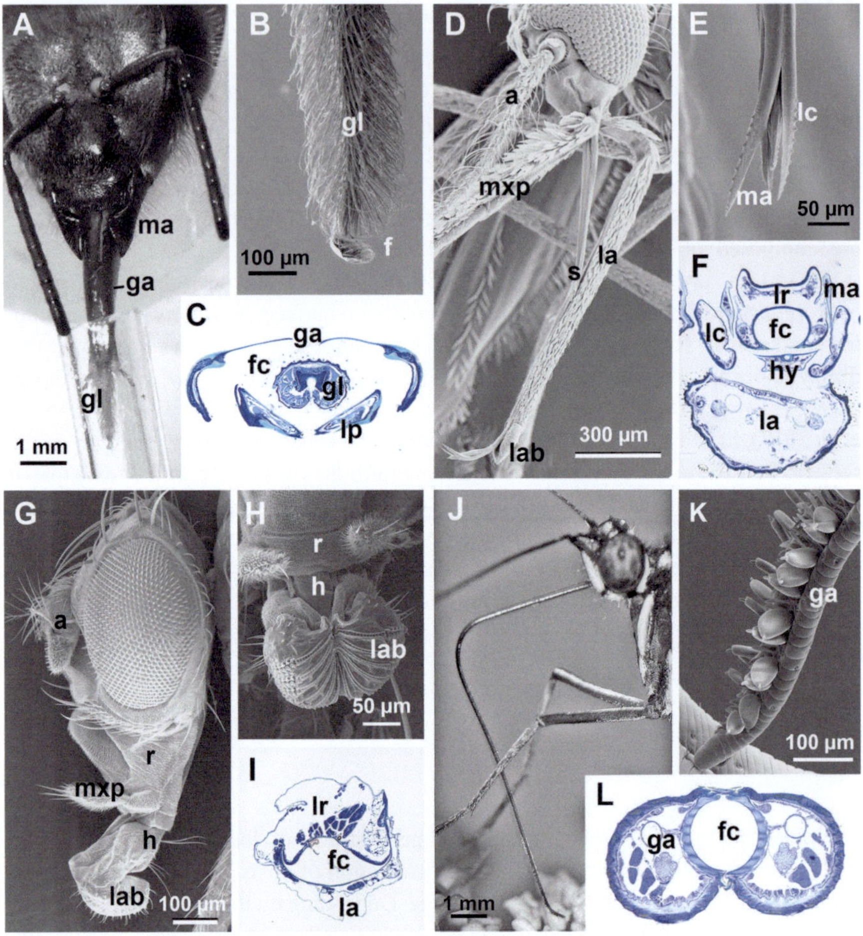


**Fig. 3.3** Functional types of proboscises in various insects (scanning electron microscope images and light microscope photos of histological sections). (**a**) Lapping proboscis of the western honey bee *Apis mellifera* (Apidae, Hymenoptera); head with extended proboscis, glossa protracted for fluid uptake out of a capillary. (**b**) Hairy glossa tip with apical flabellum. (**c**) Galeae and labial palpi form a loose food canal by enclosing the central glossa (cross-section through the proboscis). (**d**) Piercing-sucking proboscis of a female mosquito *Anopheles stephensi* (Culicidae, Diptera); head and mouthparts; labium encloses the piercing stylets. (**e**) Proboscis tip with extended mandibular and maxillary stylets. (**f**) Food canal composed of labrum and hypopharynx (cross-section through the proboscis). (**g**) Short sponging proboscis of *Drosophila melanogaster* (Drosophilidae, Diptera). (**h**) Proboscis tip with labellum characterized by pseudotracheae. (**i**) Food canal in mid-region of proboscis composed of labrum and labium (cross-section through the proboscis). (**j**) Siphoning proboscis of the butterfly *Heliconius melpomene* (Nymphalidae, Lepidoptera) composed of the elongated maxillary galeae. (**k**) Tip of galeae with sensilla in the drinking region of the proboscis. (**l**) Food canal composed of the galeae (cross-section through the mid-region of the proboscis). *a* antenna, *f* flabellum, *fc* food canal, *ga* galea, *gl* glossa, *h* haustellum, *hy* hypopharynx, *la* labium, *lab* labellum, *lc* lacinia, *lp* labial palpus, *lr* labrum, *ma* mandible, *mxp* maxillary palpus, *r* rostrum, *s* stylets

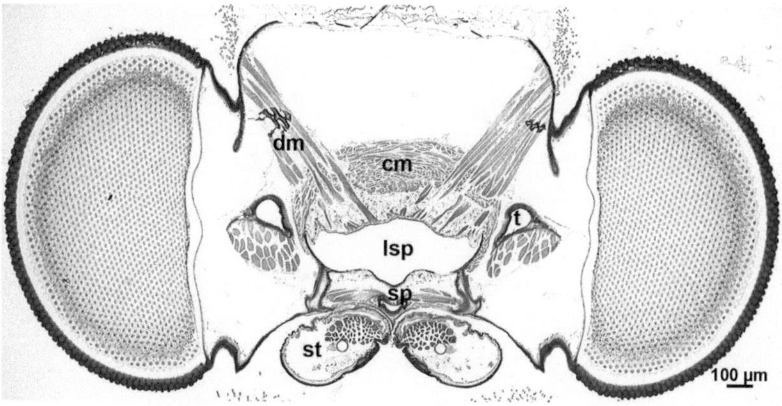


**Fig. 3.4** Sucking pump inside the head of *Heliconius melpomene* (Nymphalidae, Lepidoptera) (light microscope image of a histological section); big dilator muscles extend between the dorsal wall of the pump and the head capsule; compressor muscles surround the dorsal and lateral sides of the pump chamber; *cm* compressor muscles, *dm* dilator muscle, *lsp* lumen of sucking pump, *sp* salivary pump, *st* stipes, *t* tentorium

## 3.2.5 *Functional Morphology of Sucking Pumps*

### 3.2.5.1 Single-Chamber Pump

Fluid-feeding insects transfer liquids from the source through the food canal of their proboscis into the alimentary tract by action of pumps within the head capsule. The food canal is connected with the preoral cavity where muscular sucking pumps are formed by the cibarium, the stomodeum, and associated parts. In principle, all sucking pumps or suction pumps create a pressure gradient by expansion of the pump's lumen due to contractions of big dilator muscles (Fig. 3.4). Compression of ring muscles of the pumping chamber (Fig. 3.4), closure of a backflow valve, and opening of the mouth transport fluid further into the anterior parts of the alimentary canal where contractions of ring muscles transport fluid into the pharynx (Smith 1985).

Single-chamber pumps have been studied in detail in Hemiptera and Lepidoptera (Figs. 3.4 and 3.5a). In the Hemiptera, the cibarium forms the pump chamber. The expandable lumen is associated with various muscles extending between the head capsule and the dorso-anterior side of the sucking pump. The ventral side of the cibarium consists of a U-shaped, rigid component. A precibarial valve ensures the direction of flow through the sucking pump. The functional morphology of the cibarial sucking pump was studied in detail in the blood-feeding bug *Rhodnius* (Hemiptera) (Bennet-Clark 1963; Smith and Friend 1970; reviewed in Smith 1985) and can serve as a functional model for other Hemiptera regardless of whether they are blood-feeding or take up other liquid substances. The dorso-anterior component of the cibarial pump is connected with the ventro-posterior part by elastic ligaments and functions like a piston. The pump chamber is expanded and filled by contraction of the musculature attached to the dorso-anterior side. The sucking pump is emptied by elastic retraction of the

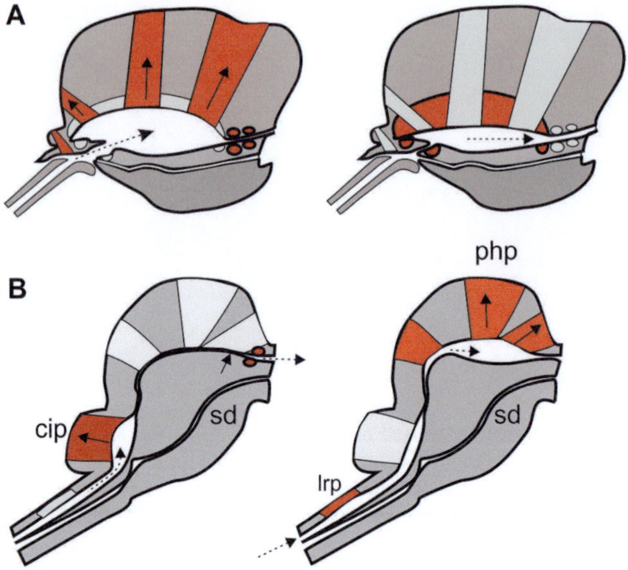


**Fig. 3.5** Functional mechanism of sucking pumps in insects, schematic drawings of a sagittal view of the head (contracted muscle is in red, arrows indicate direction of expansion, arrows with dotted lines show resulting liquid flow direction). (**a**) One-chamber pump of Lepidoptera. Left drawing: expansion of pump chamber by contraction of dilator muscles (arrows); posterior valve closed; fluid enters from the food canal. Right drawing: compressor muscles reduce volume of pump chamber and propel fluid into the esophagus. (**b**) Multichamber pump of Diptera shows phase-shifted expansion and compression of cibarial pump (cip) and pharyngeal pump (php). Left drawing: dilator muscles expand cibarial pump; fluid enters from food canal; other pump chambers closed and propel fluid into the esophagus. Right drawing: pharyngeal pump expands and fluid enters; labral pump (lrp) (present in some taxa) transports fluid through the food canal; salivary duct (sd) is indicated below the sucking pump; it opens into food canal proximally or distally. *cip* cibarial pump, *lrp* labral pump, *php* pharyngeal pump, *sd* salivary duct

ligaments of the dorso-anterior wall. The direction of flow toward the mouth is regulated by the precibarial valve and its associated muscles (Bennet-Clark 1963). Electromyograms could differentiate between filled and empty pump conditions and detect muscle contraction patterns between probing and drinking phases that indicated regular sequences of muscle action during blood-feeding (Arajo et al. 2011). The action of the cibarial pump supplies the force to move blood from the host's blood vessel into the alimentary tract. A fifth instar may consume approximately 300 μL of blood in 5 min that suggested that a high negative pressure is produced (Bennet-Clark 1963).

The sucking pump of Lepidoptera was studied from morphological, functional, and neurological points of view in representatives of butterflies and sphingid moths (Eastham and Eassa 1955; Miles and Brooker 1998; Eberhard and Krenn 2005; Davis and Hildebrand 2006; Krenn and Bauder 2017; see Chap. 15). The sucking pump of the Lepidoptera has been shown to consist of cibarial and buccopharyngeal components and comprises a cibarial valve, a buccal pump, and an esophageal sphincter valve. The cibarium forms an anterior valve at the connection to the food canal, but most of the pump is derived from the buccal chamber (Davis and Hildebrand 2006). Several pairs of large dilator muscles expand the pump lumen by pulling on its muscular dorso-anterior side (Fig. 3.5a). This opens the oral valve, and fluid is drawn into the pump chamber from the food canal. Transverse compressor muscles attached to either sides of the sclerotized ventro-posterior wall of the sucking pump reduce the volume of the lumen and passively close the oral valve. In this way, fluid is pressed through the relaxed posterior sphincter into the esophagus. By means of X-ray imaging, the discontinuous flow pattern from the food canal through the sucking pump into the foregut was visualized in real time in butterflies (Socha et al. 2007). Hungry nymphalid butterflies were found to ingest up to 23.6% of body weight in 2 min and up to 51.45% in ad libitum feeding trials (Knopp and Krenn 2003).

Miles and Brooker (1998) demonstrated that the sucking pump musculature is innervated from both the tritocerebrum and the frontal ganglion. The alternating contractions of cibarial dilator and compressor muscles are driven by a rhythmic pattern of the motor neurons of the frontal ganglion. This pattern of innervation indicates that large muscular parts of the sucking pump in hawk moths probably derived from the stomodeum and that the cibarial musculature is less prominent in Lepidoptera (Davis and Hildebrand 2006).

The functional coupling of the sucking pump and the proboscis is important for understanding the feeding habits of fluid-feeding insects in general. For various Lepidoptera, a biophysical model was created to describe the energy dissipation from liquid acquisition by the proboscis and the action of the attached sucking pump. Length and diameter of the food canal as well as the maximal expansion of the sucking pump and the chamber size turn out to be crucial parameters. This model categorizes fluid-feeding insects into a group with a long proboscis where most energy is spent on moving fluid through the food tube and insects with a short proboscis and a wide food canal where more energy is used to move the fluid from the pump into the alimentary tract. According to this model, and supported by flow analyses inside the food canal and pump, insects with a small pump to proboscis ratio use the majority of energy to overcome the viscous drag of liquid moving through the proboscis. In contrast, in insects with a larger chamber size to proboscis length ratio, energy is mostly spent to overcome the viscous drag of liquid on the moving pump plunger (Kornev et al. 2017). Chapter 8 proposes a biophysical model of sucking pumps in general.

In long-proboscid insects, like butterflies that can feed from wet surfaces (Fig. 3.2f), the action of sucking pumps cannot overcome the capillary pressures necessary to pull fluids from the substrate pores into the food canal. However, capillary action builds liquid bridges into and within the food canal to facilitate fluid transport (Monaenkova et al. 2011; Tsai et al. 2014; Lehnert et al. 2017).

### 3.2.5.2 Multichamber Pump

Two successive suction pumps can be found in the heads of Siphonaptera (Wenk 1953) and Diptera, like Culicidae, Simuliidae, and Tabanidae (Bonhag 1951; Lehane 2005; Karolyi et al. 2014). Syrphidae (Schuhmacher and Hoffmann 1982), Bombyliidae (Szucsich and Krenn 2000), and Nemestrinidae (Karolyi et al. 2012) possess additional pumps located in the labrum. Their action draws up fluid by expansion of the food canal in addition to the sucking pumps within the head capsule (Fig. 3.5b). Although comparative studies are missing, not all lineages of Diptera seem to have all pump types formed as distinct functional muscular units. In Syrphidae, a prelabral pump draws liquid over the labellar surface into the food canal by contraction of labellar muscles (Schuhmacher and Hoffmann 1982). In Bombyliidae, the contractions of labro-epipharyngeal muscles increase the food canal diameter and transport fluid toward the mouth (Szucsich and Krenn 2000). In most Diptera, the actions of the powerful cibarial pump, as well as the precerebral and postpharyngeal pump, are synchronized to ingest fluid into the alimentary tract (Gouin 1950; Bonhag 1951; Schiemenz 1957; Szucsich and Krenn 2000). Since the dimension of the epipharyngeal pumps and thus the sucking performance increase with proboscis length, they may represent an advantage for nectar-feeding insects with longer proboscises as they are evolutionarily adapted to longer nectar spurs (Szucsich and Krenn 2000). Chapter 15 presents the morphology of the sucking pumps in extremely long-proboscid nemestrinid flies in detail.

Mosquitos (Culicidae) are equipped with a cibarial and a pharyngeal dilator pump in a serial arrangement in the head (Schiemenz 1957; Snodgrass 1959). The functional morphology and fluid mechanics were recently studied in mosquitos in detail (Kim et al. 2011b; Kikuchi et al. 2018). The anterior pump formed by the preoral cavity has a solid floor and a flexible roof that can be lifted and compressed by muscle activity. The posterior pumping organ is located in the pharynx and can be extended by muscles. X-ray imaging of living mosquitos demonstrated improved pumping performance by exactly coordinated phase-shifted expansion and contraction of the two pumping chambers (Kim et al. 2011b). The cibarial pump starts to expand, and the pharyngeal pump follows with a time shift before the cibarial pump contracts (Fig. 3.5b). Likewise, the compression of the pumps is coordinated, whereby maximal flow rates are achieved. The pumping performance of nectar-drinking males was found to be lower than that of blood-feeding females. Female mosquitos are advantageous in taking up larger amounts of fluid into the pumping chamber. Flow rates are maximized by reduced relaxation times of the cibarial pump and increased pumping frequency of the sucking organs (Kim et al. 2011b). In mosquitos, two modes of pumping were revealed during drinking. The pattern of continuous pumping activity is interrupted by single large-volume strokes. This "burst mode" of pumping greatly increases the volume flow rate, but requires a massive increase in power, while continuous pumping seems to be more economical for drinking. This suggests that the advantage of a two-pump system lies in its flexibility and enables insects to feed either efficiently or fast and powerfully when demanded by environmental circumstances like irritation by the host (Kikuchi et al. 2018).

Adult fleas (Siphonaptera) possess a cibarial pump and separate pre- and postpharyngeal pumps in the head. The cibarial pump is separated by a valve from the pre- and postpharyngeal pumps that contain distinct groups of dilator muscles. These sucking pumps ensure efficient blood uptake and continuous direction of fluid transport from the food canal into the alimentary tract (Wenk 1953).

### 3.2.6 *Salivary Pump*

The musculature of the hypopharynx contributes to the formation of the salivarium where saliva is delivered. The salivary pump is situated ventro-posterior of the sucking pump and is prominent in many taxa. It is mainly composed of the hypopharynx and produces pressure to propel fluid into the salivary canal which is as long as the proboscis in many insects but has smaller diameter than the food canal. In many examples, the salivary canal is formed by the piercing structures which penetrate the host during feeding. In this way, saliva can be directly applied into the wound by the powerful salivary pump (Wenk 1953, 1962; Wenk et al. 2010; reviewed in Krenn and Aspöck 2012; Brożek and Bourgoin 2013). In the Reduviidae (Hemiptera), complex structures of the hypopharynx tightly connect the salivary pump to the maxillary salivary canal and ensure separate flows of blood and saliva while allowing the proboscis to be moved during feeding. Saliva may also flow between the mouthparts in addition to the salivary duct (Wenk et al. 2010). Basal valve-like structures were found likewise in butterflies (Lepidoptera) which separate the pathways of the nutritive fluid from the saliva when the food canal temporarily acts as a salivary duct (Eberhard and Krenn 2005). According to the functional model of the cibarial pump, saliva can be discharged out of the salivarium into the food canal during the compression phase of the sucking pump.

Saliva may play various roles in diluting fluids, liquefying food, as well as paralyzing prey and anesthetizing the puncture site at the host. The saliva of blood-feeding insects contains bioactive substances and inhibitors of platelet aggregation that prevent blood clotting and reduce the sensation of pain in the host (reviewed in Ribeiro 1987; Lehane 2005). Only minute quantities of anticoagulate substances sustain the blood flow during feeding. Since most of animal pathogens are injected with saliva, pathogens access the body of host during feeding (e.g., plasmodium transmitted by anopheline mosquitos or *Trypanosoma* transmitted by tsetse flies). In addition, the discharge of insect saliva may lead to allergic reactions in animal hosts. An overview of human-related pathogens, infections, and diseases transmitted by blood-feeding arthropods is given in Lehane (2005), Krenn and Aspöck (2012), and VectorBase (Giraldo-Calderón et al. 2015). Similarly, plant sap-feeding insects, like aphids, cicadas, or leafhoppers, are the important vectors of viruses or bacteria which can severely impair plant health (reviewed in Weintraub and Beanland 2006; Redak et al. 2004). Vascular-feeding Cicadellidae (Hemiptera) secrete a salivary sheath around the piercing stylets that has been identified as lipoproteinacious material (Leopold et al. 2003).

## 3.3 Functional Proboscis Types

### 3.3.1 Principal Fluid-Feeding Techniques

Characteristic form and function of the proboscises evolved in context with various feeding techniques applied in different insect taxa (Fig. 3.3). Fluid-feeding involves a combination of passive and active liquid transport (Kim et al. 2011a). Recent studies indicated that capillary forces and effects of hydrophilic cuticle have a significant influence on uptake mechanisms (Lehnert et al. 2017); see Chap. 9. Hence, the traditional categorization of adhesive feeding techniques, like lapping and sponging, or techniques dominated by a pressure gradient, like sucking and siphoning, is questionable. However, they may be still useful to discuss form and function of various commonly categorized functional types, namely, a lapping proboscis, a piercing-sucking proboscis, a sponging proboscis, and a siphoning proboscis that are presented in the following paragraphs.

### 3.3.2 Lapping Proboscises of Hymenoptera

Lapping was defined as a mode of feeding in which fluids are transferred from the food source to the mouth by adhering to a protractible organ (Smith 1985). Lapping proboscises combine extensile mouthpart structures with pump organs which transport fluid along a pressure gradient through a food canal (Kingsolver and Daniels 1995). Liquid or semiliquid food adheres to this functional "tongue" where fluid is loaded by capillarity onto, presumably, hydrophilic cuticle structures. The retraction of these mouthpart components transfers liquid between more proximal mouthpart structures from which it is transported further to the cibarium. Lapping mouthparts of various lengths and variously composed lapping-sucking proboscises occur in many flower-visiting nectar-feeding Hymenoptera (Table 3.1).

The mouthparts of the adult Hymenoptera are particularly diverse and complex in form and function. In addition to the biting mandibles, the maxillae and the labium are proximally connected and form the labiomaxillary complex. It operates as a functional unit and is the major organ for fluid uptake. In many taxa, elongations of the labiomaxillary complex occur (Krenn et al. 2005).They may form a proboscis which is adapted to ingest fluids, mainly nectar, while the mandibles can be used for additional functions, like catching prey, gathering pollen, nest construction, brood care, and defense. Despite some morphological differences, hymenopteran proboscises functioning according to a lapping and sucking feeding technique include (1) moveable labial structures where fluid first adheres and (2) a food canal composed by elongated components (Fig. 3.3a–c). The basal maxillary structures, the proximal parts of the labium, and the hypopharynx can be elongated and may increase the functional length of the proboscis even if they do not take part in the composition of the food canal (Krenn et al. 2005). Basally the fluid pathway is closed by the soft

**Table 3.1** Representatives of Hymenoptera with a long proboscis which functions according to a presumably lapping feeding mode; proboscis components are particularly elongated (+) or greatly elongated (++); listed taxa may contain exceptions or may not include all species of the taxon (modified from Krenn et al. 2005)

| Taxon | | Components of the labiomaxillary complex | | | | | | | References |
|---|---|---|---|---|---|---|---|---|---|
| | | Glossa | Para-glossa | Galea | Labial palpi | Lacinia | Stipes | Prementum | |
| **Tenthredinoidea** | | | | | | | | | |
| Pergidae: Euryinae | *Eurys laetus, E. rutilans* | + | + | + | | | + | + | Jervis and Vilhelmsen (2000) |
| **Ichneumonoidea** | | | | | | | | | |
| Braconidae: Agathidinae | *Agathis nixoni* | ++ | | ++ | ++ | | | | Jervis (1998) |
| Cardiochilinae | *Cardiochiles minutus* | ++ | | | | | | | Jervis (1998) |
| Ichneumonidae: Banchinae | *Agathilla bradleyi* | ++ | | | | | | | Jervis (1998) |
| Labeninae | *Cerionotus monticola* | ++ | | | | | | | Jervis (1998) |
| Ophioninae | *Agathophiona fulvicornis* | ++ | | | | | | | Jervis (1998) |
| **Aculeata** | | | | | | | | | |
| Chrysididae | *Parnopes grandior, P. fischer* | ++ | | ++ | | | | | Plant unpublished |
| Sphecidae: Sphecinae | *Prionyx* | + | + | + | | | | | Plant unpublished |
| Bembicinae | *Bembix, Bicyrtes, Stizus,* | + | + | + | | | | | Plant unpublished |
| | *Steniolia, Zyzzyx chilensis* | ++ | | ++ | | | | | Plant unpublished |

(continued)

**Table 3.1** (continued)

| Taxon | | Components of the labiomaxillary complex | | | | | | | References |
|---|---|---|---|---|---|---|---|---|---|
| | | Glossa | Para-glossa | Galea | Labial palpi | Lacinia | Stipes | Prementum | |
| Ammophilina | *Ammophila, Podalonia,* | + | | + | | | | | Plant unpublished |
| | *Eremnophila* | + | | + | | | | | Plant unpublished |
| Vespidae: Masarinae | Various species | + | | | | | | | Carpenter (1996), Krenn et al. (2002) |
| | *Metaparagia* (Paragiina) | ++ | | | | | | | Carpenter (1996), Krenn et al. (2002) |
| Eumeninae (sensu lato) | *Raphiglossa, Psiloglossa,* | ++ | | ++ | | | | | Plant unpublished |
| | *Zetha* | + | + | + | | | | | Plant unpublished |
| | *Pterocheilus phaleratus* | ++ | + | + | | | | | Haeseler (1975) |
| Scoliidae: Scoliinae | For example, *Scolia* | + | + | | | + | + | + | Osten (1982) |
| Campsomerinae | | + | + | | | + | + | + | Plant unpublished |
| **Apoidea** *"Short-tongued bees"* | | | | | | | | | |
| Colletidae: Hylaeinae | *Palaeorhiza papuana* | + | | | | | | | Michener (1965) |
| Xeromelissinae | *Chilimelissa, Xeromelissa* | | | | + | | + | + | Plant unpublished |
| Colletinae | *Leioproctus filamentosus* | | | ++[a] | ++ | | | | Laroca et al. (1989) |
| Melittidae | *Pseudophilanthus* | + | | | | | | | Michener (1981) |
| Halictidae: Rophitinae | Many species | + | | + | + | | | | Michener (1965) |

| | | | | | | | | | |
|---|---|---|---|---|---|---|---|---|---|
| | *Dufourea longiglossa* | + | | | | | | | Ebmer (1993) |
| Nomiinae | *Lipotriches testacea* | + | | | | | | | Pauly (1984) |
| Andrenidae, Andreninae | Species of several tribes | + | | + | + | | | | Michener and Brooks (1984) |
| | *Andrena violae* | + | | | | | | | LaBerge (1986) |
| Panurginae | Species of several tribes | + | | + | + | | | | Michener and Brooks (1984) |
| | *Perdita hurdi* | ++ | | + | | | + | + | Hurd and Linsley (1963) |
| | *Neffapis, Nolanomelissa* | ++ | | | ++ | | | | Rozen and Ruz (1995), Rozen (2003) |
| *"Long-tongued bees"* | | | | | | | | | |
| Megachilidae | Most species | + | | + | + | | | | E.g., Michener (2007), Plant and Paulus (2015) |
| Megachilidae | Lithurgini | ++ | | ++ | ++ | | | | Plant unpublished |
| Apidae | Most species | + | | + | + | | | | E.g., Michener (2007), Plant and Paulus (2015) |
| Apidae | Euglossini, Anthophorini | ++ | | ++ | ++ | | | | E.g., Alves dos Santos and Wittmann (1999), Michener (2007) |

[a]Filaments of galea

epipharynx; the hypopharynx extends to the prementum at the base of the glossa (Jervis and Vilhelmsen 2000). In the preoral cavity, an infrabuccal pouch is present in many Hymenoptera that is formed by an invagination of the basal hypopharynx which serves for food compression (Vilhelmsen 1996).

The labiomaxillary complex is shorter than the head in unspecialized fluid-feeding adult Hymenoptera but forms a fluid pathway along the labial components. This short proboscis permits only consumption of liquid from easily accessible food sources such as honeydew or freely accessible floral nectar (Jervis 1998; Jervis and Vilhelmsen 2000). It is composed of the extensible prementum bearing short glossa and paraglossae which possess transverse rows of cuticle structures covering their anterior/dorsal surface (Jervis and Vilhelmsen 2000). Nectar is loaded onto the exposed glossa and passes the proximal mouthparts after retraction (Jervis 1998).

In more specialized anthophilous Hymenoptera, elongated maxillary and/or labial components form a long proboscis equipped with a food canal to feed from concealed floral nectar. Lengthening of the food canal necessitates elongation of opposing structures. The lengthening of the labiomaxillary complex often includes elongation of the glossa simultaneously with enlargement of the prementum which contains the muscles needed to produce the licking movements. In the distal proboscis region, the composition of a more or less closed food canal varies in different lineages but normally consists of elongated and flat maxillary components on the dorso-anterior side and a glossa longer than the prementum on the ventro-posterior side. In some taxa, this functional unit is encompassed by the flat labial palpi. A proboscis longer than the head evolved many times independently in various lineages of Hymenoptera. Termed as long proboscises (Krenn et al. 2005), they evolved independently among the representatives of a few basal hymenopteran lineages (Jervis and Vilhelmsen 2000), some Apocrita (Osten 1982, 1988, 1991; Jervis 1998) and, for instance, Apoidea (Michener 2000; reviewed in Krenn et al. 2005; Plant and Paulus 2015). Eight types of elongated mouthparts ("concealed nectar extraction apparatus") have been distinguished in Hymenoptera (Jervis 1998; Jervis and Vilhelmsen 2000); however, the fluid-feeding technique of these proboscises is not known in all species. Table 3.1 lists various examples of a long proboscis among representatives of various lineages of Hymenoptera that probably primarily function according to a lapping technique of fluid uptake.

Only few representatives of these taxa are studied in greater detail. Compared to other Aculeata, a long proboscis occurs in several representatives of the Sphecidae and Scoliidae that feed on nectar from open flowers. The mouthpart morphology and the feeding movement have been described for various Scoliidae (Osten 1982, 1988, 1991). The glossa and paraglossae rapidly protract and retract for nectar uptake during flower visits. At extension, the paraglossae diverge off laterally, whereas on retraction they converge, sweeping nectar and pollen toward the median glossa where liquid adheres to special hairs. Glossa and paraglossae are about as long as the prementum and are pulled back into a pocket of the prementum in the state of retraction.

Although flower-visiting behavior is common in adult Vespidae, only few species have a proboscis (Table 3.1). Elongated mouthparts occur in some Eumeninae sensu lato and many Masarinae (see Chap. 14). The latter, the pollen wasps, possess variously long protractible glossa and paraglossae which take up nectar between the numerous lamella-shaped cuticle structures that form a food canal over the bifid glossa into the preoral cavity. In the resting position, the glossa and paraglossae are folded in front of the prementum under the head. Some lineages evolved a particularly long glossa which forms a closed food tube where nectar is probably transported mainly by suction forces (Krenn et al. 2002). Chapter 14 is devoted to the evolution of the mouthparts of Vespidae and discusses their functional mechanism in pollen wasps.

Two principal proboscis types can be distinguished in Apoidea (see Michener 2000): in "short-tongued" bees, like Andrenidae, Colletidae, Melittidae, and Halictidae, the glossa is shorter than the prementum (Harder 1983), while in "long-tongued" bees, like Megachilidae and Apidae, the glossa is longer than the prementum and a distinct food canal is formed by the flat, elongated galeae and labial palpi that surround the glossa (e.g., Snodgrass 1956; Michener 2000; Plant and Paulus 2015) (Table 3.1). Body size measured as intertegular distance was found to be a good predictor in an allometric model to estimate the proboscis length and foraging distances in the various families of the Apoidea (Cariveau et al. 2016; Melin et al. 2019). Among both the "short-tongued" and "long-tongued" Apoidea, a high disparity of proboscis constructions occurs in some species that results in even greater proboscis length, which can be regarded as convergent adaptations to feed from concealed nectar of differently shaped flowers (e.g., Laroca et al. 1989; Alves dos Santos and Wittmann 1999; summarized in Krenn et al. 2005). In long-tongued bees, the moveable, greatly elongated glossa outreaches the head in most taxa. It is equipped with a flat round structure at the apex of the glossa, the flabellum (Fig. 3.3b), which bears variously shaped sensilla (Galić 1971; Düster et al. 2018). The longest bee proboscis is reported from orchid bees (*Euglossa* species) that may exceed even the body length in folded resting position (Borrell and Krenn 2006). Its principal composition is similar to other long-tongued bees. However, some *Euglossa* species use a primarily sucking nectar-feeding technique without glossa movements (Borrell 2003). Feeding experiments showed that the extended glossa stays more or less stationary during ingesting large fluid quantities and lapping glossa movements occur to take up small amounts of fluid (Düster et al. 2018). The mouthpart movements are similar to other long-tongued bees, for example, bumblebees (Harder 1982). However, the prementum can be extended further and greatly enhances the protraction of the glossa resulting in a functional proboscis length of orchid bees that is about 50% longer than the galeae and labial palpi which build up the food canal (Düster et al. 2018).

The biomechanics of the feeding technique has been extensively investigated in honey bees (*Apis mellifera*, Apidae) that can serve as a model to understand the principal lapping fluid-feeding mechanics in long-tongued bees (Snodgrass 1956; Kingsolver and Daniel 1995; Yang et al. 2014; Wu et al. 2015). The proboscis lies flexed under the head in the resting position. In a first step, the labiomaxillary complex unfolds by turning at the cardines (the basal sclerites of the maxillae) at their articulation with the cranium. The basal components of maxillae unfold the labial structures passively. In a second step, the distal labial components, the prementum and glossa, are extended. By unfolding the proboscis to the initial feeding position, the galeae and labial palpi assemble themselves to form a food tube surrounding the glossa (Fig. 3.3a, c). During feeding, the maxillae remain relatively motionless, while movements of the postmentum slide the prementum back and forth whereby the glossa retracts and protracts ("licking" or "lapping"). The flexible glossa is composed of transverse cuticle rings or annuli, each bearing long stiff microtrichia ("hairs"). The flexibility is achieved by a soft cuticle membrane which separates the annuli. The glossa is internally reinforced along its entire length by the glossal rod. On the ventro-posterior side, a hair-fringed groove of the glossal rod forms the salivary channel. The glossa movements consist of three-phased licking cycles (Kingsolver and Daniel 1995): (1) extension, where nectar is loaded onto the exposed surface of the glossa between the hairs; (2) retraction, where nectar is drawn into the food tube; and (3) unloading of nectar inside the tube to be transported into the mouth by action of the sucking pump and/or by squeezing the food canal due to the elasticity of the galeae and labial palpi which surround the glossa (Smith 1985). When the glossa is fully extended, the cuticle membrane between each annulus is stretched, and the microtrichia of the exposed glossa spread out where nectar is "loaded" by adhesion (Fig. 3.6). The erection of the glossal hairs opens a space between them contributing to nectar loading between the microtrichia. When the glossa is retracted and the tension is relaxed, the microtrichia flatten, and liquid is squeezed off ("unloaded") into the food canal by the pressure of surrounding mouthpart components (Yang et al. 2014; Wu et al. 2015). In this way, the glossa retraction transports the fluid into a food canal where it is further conveyed by sucking pump action toward the mouth (Snodgrass 1956; Kingsolver and Daniels 1995).

## 3.3.3 *Piercing-Sucking Proboscis*

### 3.3.3.1 Various Piercing Techniques

Piercing-sucking mouthparts evolved as adaptation to hemolymph- or blood-feeding as well as to access plant sap from various tissues including fruits and seeds. The key features of a piercing-sucking proboscis are strongly sclerotized piercing structures. In many cases, piercing stylets are present that are enveloped by a soft guiding structure that is not inserted into the host's body. However, these characteristic functional components are derived from different organs of the basic set of the mouthparts and are not homologous in the various taxa (Lehane 2005; Krenn and Aspöck 2012; Garzo et al. 2012; Zhao et al. 2010). Table 3.2 lists examples of the functional morphology of piercing-sucking proboscises from Thysanoptera, Hemiptera, Anoplura, Diptera, Siphonaptera, and Lepidoptera.

A piercing-sucking proboscis is first anchored, so that forces can be transmitted efficiently by the piercing structures. In the second step, the thin piercing stylets puncture and penetrate the host's body (Smith 1985). Krenn and Aspöck (2012)

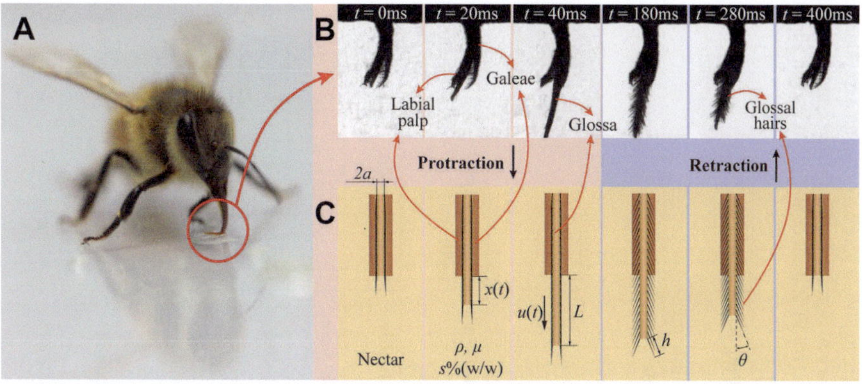


**Fig. 3.6** Lapping fluid-feeding technique of the western honey bee *Apis mellifera* (Apidae, Hymenoptera). (**a**) Proboscis in feeding position. (**b**) Selected frames showing the licking cycle of the proboscis (upper row of micrographs): protraction extends the glossa beyond the food canal ($t = 0$ ms, 20 ms, 40 ms), retraction ($t = 180$ ms, 280 ms, 400 ms); glossal hairs erect at the beginning of glossa retraction. (**c**) Physical model illustrating the drinking process (lower row): the glossa is simplified into a cylinder with dense hairs (length $h$, erection angle $\theta$); the labial palpi and galeae are treated as a sucking tube that surrounds the cylinder. The cylinder has a diameter 2a, extended glossa length $L$, movement velocity $u(t)$, and length $x(t)$ while protracting glossa; the nectar has a density $\rho$, mass concentration s% (percent of sugar), and viscosity $\mu$ [reprinted with permission from Wu et al. (2015)]

distinguish several principal piercing mechanisms: (1) the jigsaw principle, where the piercing components alternately protract and retract whereby apical teeth alternately hold and thrust into the host's tissue, (2) the scissor principle that is characterized by medially directed movements of paired cutting organs, and (3) the drilling principle where rasping structures perform rotating movements to carve into the skin (Fig. 3.7). (4) Another technique occurs in thrips and fleas that can be compared to the function of a percussion drill where punching movements push in the piercing organs. In many taxa, it was shown that the mechanical forces and acceleration of the piercing components are enhanced by torsion, bending of elastic structures, and/or click-release of compressed resilin pads (Wenk 1980). When the penetrating structures have reached an appropriate area inside the host or food source, liquids are imbibed directly into the food canal through the apical opening of the food canal by the action of sucking pumps in the head.

#### 3.3.3.2 Thysanoptera

Juvenile and adult Thysanoptera possess short piercing and sucking mouthparts which are retracted into a mouthcone in resting position. Most species are very small and suck from single plant cells, pollen grains, detritus, and fungal hyphae or

**Table 3.2** Morphology of piercing-sucking proboscises in various insects and function of structures during piercing; listed taxa may contain exceptions or may not include all species of the taxon; compiled and updated from Bletchly (1954, 1955), Smith (1985), Nagatomi and Soroida (1985), Elzinga and Broce (1986), McKeever et al. (1988), Chaudonneret (1990), Tröster (1990), Lehane (2005), Molleman et al. (2005), Dikow (2009), Zaspel et al. (2011), Krenn and Aspöck (2012); abbreviations: – absent, ? unclear

| Taxon | Proboscis composition | | | Function during feeding | | |
|---|---|---|---|---|---|---|
| | Sheath and guidance | Food canal | Salivary canal | Puncturing structure | Anchorage | Penetrating structure |
| **Thysanoptera** | Mouth cone? | Lacinia | Hypopharynx | Left mandible | – | Laciniae |
| **Hemiptera** | | | | | | |
| Heteroptera<br>Auchenorrhyncha<br>Sternorrhyncha | Labium | Maxilla | Maxilla | Mandible | Mandibular teeth | Maxilla, mandible |
| **Psocodea** | | | | | | |
| Phthiridae<br>Pediculidae | Evagination of labium | Hypopharynx | Hypopharynx | Epipharyngeal teeth | Epipharyngeal teeth | Hypopharynx, prementum |
| **Diptera** | | | | | | |
| Culicidae | Labium | Labrum/epipharynx, hypopharynx | Hypopharynx | Labrum, mandible, laciniae | Lacinial teeth | Lacinia |
| Corethrellidae | Labium? | Labrum/epipharynx? | Hypopharynx | Labrum? mandible, lacinia? | ? | ? |
| Simuliidae | Labium | Labrum/epipharynx, hypopharynx | Hypopharynx | Labrum, mandible | Lacinial teeth | Mandible, lacinia |
| Ceratopogonidae | Labium | Labrum/epipharynx, mandible, lacinia | Mandible, hypopharynx | Mandible | Lacinia | Mandible |
| Psychodidae: Phlebotominae | Labium | Labrum/epipharynx, mandible, lacinia | Mandible, hypopharynx | Mandible | Lacinia | Mandible |
| Tabanidae | Labium | Labrum/epipharynx, mandibles | Hypopharynx | Mandible, lacinia | Lacinial teeth | Mandible, lacinia |
| Rhagionidae | Labium | Labrum/epipharynx? | Hypopharynx? | Mandible, lacinia? | Lacinial teeth? | Mandible, lacinia? |

| | | | | | | |
|---|---|---|---|---|---|---|
| Asilidae | Labium? | Prementum | Hypopharynx | Lacinia, hypophar-ynx, labellum? | Hypopharyngeal bristles? | Lacinia? |
| Empididae | Labium | Labrum/epipharynx, hypopharynx | Hypopharynx | Labrum, hypopharynx? | ? | ? |
| Glossinidae, Hippoboscidae, Streblidae, Nycteribiidae | Maxillary palpi | Labrum | Hypopharynx | Prestomal teeth of labellum | – | Prestomal teeth of labellum |
| Muscidae: Stomoxinae | – | Labrum | Hypopharynx | Prestomal teeth of labellum | – | Prestomal teeth of labellum |
| **Siphonaptera** | Labium, labial palpi | Labrum/epipharynx, lacinia | Lacinia | Lacinia | Lacinial teeth | Lacinia |
| **Lepidoptera** | | | | | | |
| Erebidae: Calpini | – | Galea | Galea | Galeal apex | Galeal sessile | Galea |
| Nymphalidae: Charaxinae | – | Galea | Galea | Galeal apex | Galeal apex | Galea |

induce galls. Some larger species are predaceous on mites (Mound 2005). The mouthparts are asymmetrically developed. They are composed of only the left mandibular stylet and the paired lacinia which form interlocked stylets that enclose the food canal (Heming 1978; Chaudonneret 1990). Thrips feed with a punch-and-suck mechanism: the substrate is punctured by the mandible, and then food is extracted by the maxillary stylets. The food substrate is punctured by repeated thrusts of the mandible tip. It is forced out of the mouthcone by the indirect action of the large clypeo-labral muscle which presses a mandible guiding structure against the arched mandibular bases whereby the mandible tip is propelled out of the mouthcone four to five times per second (Heming 1978). The to-and-fro movements of the maxillary stylets replace the mandible, penetrate deeper, and ingest fluids into the feeding tube (Chrisholm and Lewis 1984). Action of the cibarial pump sucks liquid into the feeding canal formed by the maxillary laciniae (Heming 1993). The salivary pump releases saliva to lubricate the mouthparts and partly predigest the food. Saliva is propelled through the same canal formed by the laciniae (Heming 1978, 1993).

#### 3.3.3.3 Hemiptera

Piercing and sucking proboscises (sometimes called a rostrum) are present in juvenile as well as in adult stages of all species of Hemiptera. All Sternorrhyncha (plant lice, aphids, scale insects, psyllids, white flies) and Auchenorrhyncha (cicadas, leafhoppers, plant hoppers, and spittlebugs) suck saps from various plant tissues, for example, xylem or phloem sap; some are mycetophagous. Some Heteroptera (true bugs) likewise feed on plant saps, others feed mainly on fruits and seeds where the tissue is liquefied prior to ingestion, or they are predaceous and feed on insects; some representatives have a mixed diet (Dathe 2003). Few true bugs imbibe blood from vertebrate hosts, including humans (Lehane 2005). While many predaceous and blood-feeding Heteroptera suck from their host only for a short time, most plant sap-feeding Hemiptera ingest large quantities over relatively long periods of time and have a semi-sessile lifestyle. Because of their economic importance, detailed descriptions of proboscis morphology and function have existed for a long time (e.g., Weber 1930; Pollard 1970, 1972, 1973; reviewed in Smith 1985; Leopold et al. 2003; Reinhardt and Siva-Jothy 2007; Krenn and Aspöck 2012).

Regardless of the preferred food source, the principal composition of the proboscis is similar in all Hemiptera: it consists of the paired mandibles and maxillary structures which form the elongated, tightly interlocked piercing stylets that are completely enclosed by the segmented labium from the posterior-ventral side (Fig. 3.7a–c). The short labrum covers the base of the stylet bundles on the frontal side; palpi are missing. In the resting position, the whole proboscis, comprising mandibles, laciniae, and the labium, is flexed under the body and only the subdivided labium is visible from outside (Fig. 3.7a). The food canal is formed by the opposed laciniae which are interlocked by a tongue and groove junction (e.g., Cicadellidae; Leopold et al. 2003) which allows the stylets to slide on each other while maintaining the food canal.

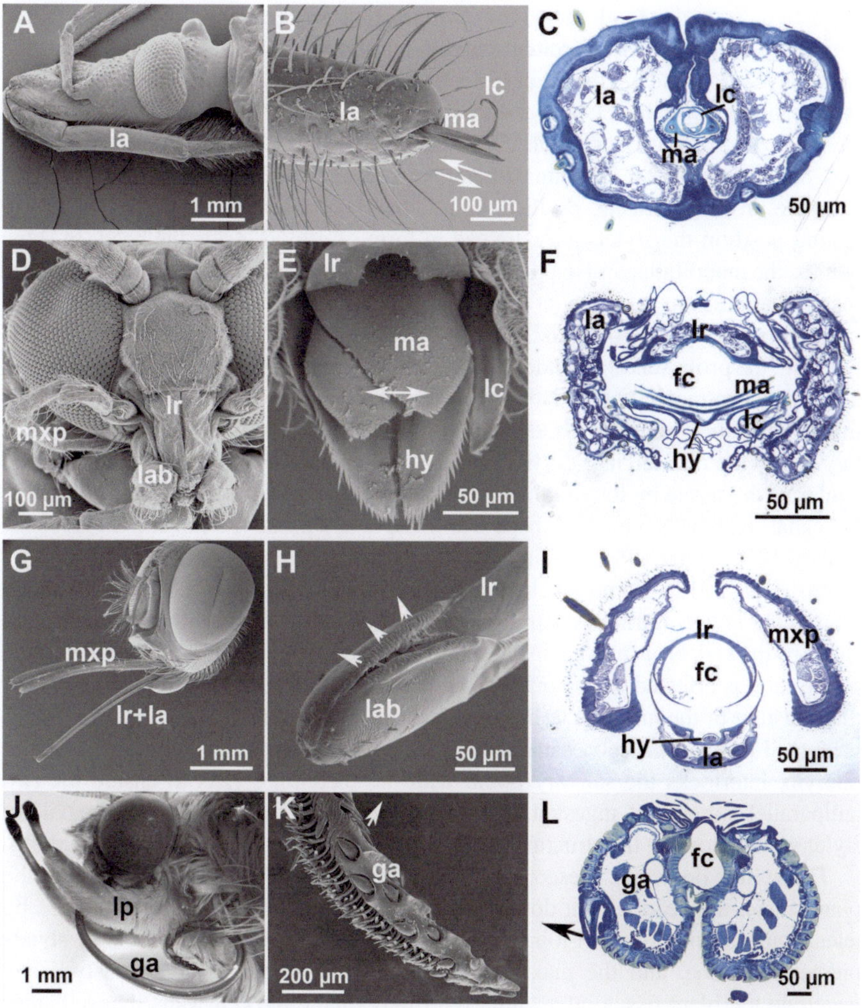


**Fig. 3.7** Piercing-sucking proboscises use different mechanisms (scanning electron microscope images and light microscope photos of histological cross-sections; arrows indicate movements of the structures during piercing). (**a–c**) Blood-feeding assassin bug *Triatoma infestans* (Reduviidae, Hemiptera): (**a**) head and proboscis in resting position; (**b**) proboscis tip, mandibular and maxillary piercing stylets protracted; antiparallel movements effect piercing like a jigsaw; (**c**) cross-section through the proboscis; labium tightly encloses the interlocked piercing stylets which form the food canal and salivary duct. (**d–f**) Black fly *Simulium* sp. (Simuliidae, Diptera): (**d**) Head in frontal view; (**e**) mandibles move like scissors during piercing; (**f**) composition of the proboscis (cross-section in mid-length). (**g–i**) Blood-feeding tsetse fly (*Glossina* sp., Glossinidae, Diptera): (**g**) proboscis lowered from the resting position between the maxillary palpi; (**h**) labella can evert the margins to drill into the host; (**i**) proboscis composition (cross-section); labrum and labium form food canal and enclose the thin hypopharynx which forms the salivary canal. (**j–l**) Fruit-piercing moth *Eudocima* sp. (Erebidae, Lepidoptera): (**j**) head and partly uncoiled proboscis; (**k**) proboscis tip equipped with erectile barbs and other piercing armatures of the galeae; (**l**) anatomical composition in proboscis cross-section; bulging outward of the galeal wall would erect cuticle barb (arrow). *fc* food canal, *ga* galea, *hy* hypopharynx, *la* labium, *lab* labellum, *lc* lacinia, *lp* labial palpus, *lr* labrum, *ma* mandible, *mxp* maxillary palpus

Separated from the food canal, the laciniae enclose the very thin salivary canal (Fig. 3.7c). The apically serrated mandibular stylets are in lateral position to the maxillary stylets and are the main piercing component. The stylets penetrate into the tissue and directly inject saliva into the tissue (e.g., Weber 1930; Smith 1985; Leopold et al. 2003; Uzest et al. 2010; Zhao et al. 2010; Garzo et al. 2012; Kumar and Sahayaraj 2012). Piercing structures of conspicuous lengths are found, for example, in scale insects, psyllids, and some Heteroptera like Aradidae where in resting position the stylets are retracted in form of a loop inside the head. In some aphids, the mandibular and maxillary stylets may be extremely long; when not in use, these structures are stored in an invaginated membranous sac between the head and the thorax (Weber 1930; Brożek et al. 2015; Rakitov et al. 2018). In the feeding position, the proboscis is extended in a ventral or frontal position; the piercing stylets alternately protract out of the labial sheath during puncturing of the food source. The mandibles puncture the host and anchor the proboscis superficially. Then the maxillary stylets are inserted into the tissue. The stylet bundle is protracted and retracted to penetrate the tissue by longitudinal antiparallel movements of the stylets. The piercing canal can have a branched shape within the host's tissue (Friend and Smith 1971; Pollard 1973; Wirtz 1987; Tjallingii and Esch 1993; Reinhardt and Siva-Jothy 2007; Garzo et al. 2012). Some Auchenorrhyncha produce a sheath of solidified saliva around the stylets during penetration of the plant host that possibly lubricates and supports the stylets or seals against leakage (Leopold et al. 2003). In Reduviidae, deep penetration is effected by the interlocked maxillae. When a blood vessel is reached, the thrusting of the maxillae terminates, and the left maxilla is drawn back and folds backward to open the food canal at the tip (Wirtz 1987). In Reduviidae, the hypopharynx interlocks the salivary pump with the piercing stylets to ensure the two antiparallel streams of ingested blood and discharged saliva within the maxillary stylet while allowing the structures to be moved back and forth (Wenk et al. 2010).

The aphids possess a telescope-like labium: as the stylet bundle advances in the plant tissue, the head is bent down toward the surface, and the labium shortens by telescoping at the proximal joints. Alternating protractions of the mandible stylets pierce the tissue, and the maxillary stylets are protracted simultaneously. They penetrate the plant tissue intercellularly or may pierce the cell walls by oscillatory or vibratory movements of the mouthparts (review in Pollard 1973; Smith 1985). The micromorphology of homopteran mouthparts, the interlocking structures, and the sensilla of the piercing stylets as well as the apical structures and the penetration of the mandibles and maxillae have been studied in some detail, for example, in Cicadellidae (Leopold et al. 2003). The connecting structures of the mandibular and maxillary stylets show an unexpectedly high morphological disparity in the representatives of Cixiidae and some other plant hoppers (Brożek and Bourgoin 2013).

Chapter 16 reviews the morphogenesis of hemipteran mouthparts in detail and discusses the assembly of the piercing stylets after molting.

#### 3.3.3.4 Anoplura

Anoplura are specialized wingless, ectoparasitic insects that are hematophagous throughout all stages. They usually remain on their mammal hosts including humans throughout their lives. Hooklike claws on their legs enable them to cling at hair, but mostly they remain attached to the skin by the mouthparts (Lehane 2005). The piercing apparatus of *Pediculus* consists of the elongated, thin, and half-pipe-shaped labium and hypopharynx that lie on top of each other. The ventral stylet is formed by the prementum, while the dorsal one is derived from the hypopharynx. In resting position, both components are fully retracted into a proboscis sheath inside the preoral cavity (Tröster 1990; Chaudonneret 1990). The frontal side of the head forms the buccal funnel which corresponds to the labrum. Its apical opening is equipped with backward-directed teeth that allow the louse to retain its grip in the skin during piercing (Wenk and Renz 2003). Contractions of the head muscles force the piercing stylets out of the proboscis sheath inside the head. The ventral stylet constitutes the actual piercing organ, while the dorsal one forms the food canal. The piercing components penetrate by alternating movement (Tröster 1990).

#### 3.3.3.5 Diptera

All imagines of Diptera possess sucking mouthparts, and many have piercing structures that enable these insects to puncture the hosts or prey to ingest their body fluid. A variety of piercing techniques are used in the different lineages of blood-sucking Diptera (Figs. 3.2c, 3.3d–e, and 3.7d–i). Likewise, predaceous flies use a piercing proboscis which is equipped with pointed components to feed on other insects (e.g., Asilidae; Fig. 3.2d). Lehane (2005) listed the taxa of Diptera which contain blood-sucking species. Based on the phylogeny of flies, it was concluded that hematophagy evolved at least 12 times independently in Diptera (Wiegmann et al. 2011). This is reflected by the morphological and functional disparity of the proboscises, indicated, for example, by the various piercing structures in Culicidae, Simuliidae, Tabanidae, Ceratopogonidae, and Glossinidae (Table 3.2). Rarely studied in detail, the short proboscises of various predaceous Diptera show considerable morphological disparity, for example, in Asilidae, Empididae, and Rhagionidae (Bletchly 1954, 1955; Nagatomi and Soroida 1985; Dikow 2009) (Table 3.2).

The blood- and hemolymph-feeding behavior is associated with the evolution of a more or less elongated proboscis consisting of two principal compositions (Krenn and Aspöck 2012). In nematoceran Diptera and other non-cyclorrhaphan flies, the piercing-sucking proboscises have a complex composition and include an arsenal of up to six piercing stylets which are more or less elongated, pointed, and apically serrated (Table 3.2). The prominent prementum with the small, apical labella encloses the piercing components, but does not penetrate the host (Figs. 3.2c and 3.3d). The food canal is composed by the channel-like labrum/epipharynx which is opposed by the hypopharynx. The gutter-shaped labium often embraces other likewise elongated components, such as mandibles and laciniae, as well as the blade-shaped hypopharynx which includes the salivary canal (Fig. 3.3d–f) (reviewed in Krenn and Aspöck 2012).

The feeding mechanism of female mosquitos can be taken as a model for how the skin is penetrated and the delicate piercing components are inserted into the skin as well as how blood is ingested. Like in other nematoceran Diptera, only adult females are blood feeders. Their proboscis is composed of the acute labrum forming the main parts of the food canal, the pairs of thin stylet-shaped mandibles and laciniae, as well as the thin pointed hypopharynx which forms the salivary canal. All components are enclosed by the gutter-shaped labium which bears the apical pair of labella (Snodgrass 1959; Kong and Wu 2009, 2010; Kim et al. 2011b). In males, the mandibles and maxillae are reduced or missing, since their proboscis is used for nectar-feeding only (Snodgrass 1959; Wahid et al. 2003). The mechanoreceptive bristles and variously shaped chemosensilla of the labella probably provide information on the suitability of the host (Lee and Craig 2009). On contact of the host's skin with the labella, the sharp-tipped and serrate laciniae subsequently cut into the skin by protraction and retraction like a jigsaw, while the other piercing components are pulled along. Backward-pointing teeth on the tips of these piercing components (Kong and Wu 2009) draw the other structures, like the hypopharynx, deeper into the host's tissue. When blood is detected, the mandibular retractor muscles contract and expose the opening into the food canal (Lee and Craig 2009). Mosquitos are exceptional in their ability to pierce human skin painlessly. Measurements of the insertion force showed that mosquitos use a very low penetrating force. This may be due to the vibratory movements of the piercing stylets and their minute tips having a radius in the order of nanometers. Analyses of the mechanical aspects of stylet insertion could predict deformation and structural failure of the human skin (Kong and Wu 2009).

The sensilla of the labella and the labium include long mechanoreceptive bristles and short chemosensilla that are probably involved in finding a host and a suitable place for piercing. Spreading and closing of the labella seems to be controlled by chordotonal organs at the base of the labella. Labial bristles are involved in monitoring the "bending status" of the labium while the stylets are inserted (Lee and Craig 2009). Sensory papillae in the cibarium probably monitor the chemistry of ingested fluid, while trichoid sensilla register the fluid flow. Apical and subapical sensilla of the labrum detect blood during piercing that may influence the action of the cibarial and pharyngeal pumps. Sensilla of the ventral side of the pump probably provide information to distinguish between sugar solutions which enter the ventral diverticulum of the alimentary tract and blood which goes into the midgut (Lee and Craig 1983). The labella lobes house numerous long gustatory sensilla and short thornlike chemosensilla which detect a large variety of olfactory cues (Kwon et al. 2006; Saveer et al. 2018). The sensilla of the labella detect a range of gustatory and volatile odorant substances (e.g., Tauxe et al. 2013). Saveer et al. (2018) reported on chemoreceptor transcriptome profiles together with electrophysiological studies of labial olfactory sensilla. Analyses of the behavioral responses of female mosquitos indicated that the olfactory information from the labellum contributes to host-seeking.

In many other blood-feeding fly taxa, e.g., Simuliidae (Fig. 3.7d–f), Ceratopogonidae, and Tabanidae, the serrate mandibles have an active role in opening the skin by scissor-like movements (Sutcliffe and McIver 1984; McKeever et al. 1988; Chaudonneret 1990, reviewed in Krenn and Aspöck 2012). The labella remain outside during piercing but have a guiding function for the piercing structures. In addition, sensilla on the labella provide information about an appropriate feeding site. In Simuliidae, torsion of the short mandibles occurs during sideward movement. In the abduction position, the mandible blades are bent inward, and the stored energy is released during adduction. In this way, the host's tissue is snipped by the rows of small teeth on the fronto-median side of the mandibles (Fig. 3.7e) (Wenk 1962, 1980). Ceratopogonidae are among the smallest hematophagous insects; the volume of a blood meal of *Culicoides imicola* was found to range from only 0.01 to 0.11 μl (De Beer et al. 2018).

Another type of a piercing-sucking proboscis occurs in blood-feeding Cyclorrhapha. They lack piercing mandibles as well as maxillary stylets and are derived from nectar-feeding ancestors. Their proboscises use a drilling mechanism for piercing. In the Glossinidae, Hippoboscidae, Strebilidae, and Nycteribiidae as well as *Stomoxys* (Muscidae), the labella evolved into a secondary piercing organ and functionally replace laciniae and mandibles (Ziegler 2003; Krenn and Aspöck 2012). In the Glossinidae, the proboscis is laterally embraced by the maxillary palpi in the frontal position at rest; it is flexed downward for feeding. The proboscis consists of the food canal formed by the labrum and the labium which together cover the thin central stylet-shaped hypopharynx (Fig. 3.7g–i) (Smith 1985). The inner margins of the labella are equipped with rows of small toothlike cuticle structures. Fast rotatory everting movements of the serrated labella ridges drill these rasping structures into the hosts (Wenk and Renz 2003). Repeated penetrations form large subdermal pools of blood from which the flies rapidly consume their meals (Wirtz 1989). In blood-feeding Muscidae, the proboscis is unfolded from the retracted position for feeding, and the labella are spread apart whereby the sharp ridges of their median sides are pressed against the skin. The modified prestomal teeth of the labella perform rotating movements that drill into the tissue, causing painful bleeding under the skin from which the stable flies feed (Lehane 2005).

#### 3.3.3.6 Siphonaptera

Male and female adult fleas pierce the skin of mammals or birds to drink their blood. The proboscis is about as long as the head. It consists of the acute labrum-epipharynx unit on the frontal side and the paired laciniae which all together form the food canal. These piercing components are enclosed by the labium and slender labial palpi on the posterior side; mandibles are missing (Wenk 1953; Michelsen 1996/1997; Kluge 2002). The sclerotized buccal side of the elongated labrum, often termed as the epipharynx, forms major parts of the food canal. The laciniae originate from a lever-like component of the maxillary base. Muscles of the maxillary base pull back the lacinia and press it against elastic cuticle. Upon muscle release of a click mechanism, the compressed resilin pad thrusts the lacinia downward assisting the penetration (Smith 1985; Chaudonneret 1990). The rapid sequence of this hammering mechanism forces the lacinia into the skin of the host and pulls the labrum with it. The apex of the laciniae bears backward-directed teeth, while the epipharynx is equipped with sensilla (Wenk 1953; Chaudonneret 1990; Wenk and Renz 2003). Each lacinia possesses a salivary canal extending from the short hypopharynx to the tip. The hypopharynx basally connects the piercing components in a way to allow longitudinal movements. The maxillary palpus that is situated lateral to the proboscis probably helps to find the right place for piercing (Wenk 1953). The enlarged basal segment ("maxillary lobe") serves as an attachment site of maxillary muscles responsible for the hammering mechanism (Michelsen 1996/1997). The labium does not penetrate the host but serves as a guiding structure for the piercing stylets (Wenk 1953).

#### 3.3.3.7 Lepidoptera

Two groups of Lepidoptera (i.e., Calpinae and Charaxinae) independently evolved piercing-sucking techniques of fluid-feeding that can be applied to different food sources. In comparison to most other piercing mouthparts, the proboscis of an adult Lepidoptera is simply composed of the two coilable galeae (Fig. 3.7j–l). They are firmly linked together and enclose the central food canal which also functions as the salivary canal. The labrum is small, the mandibles are reduced, and the labium forms a triangular plate on the ventral side of the head that bears the three segmented palpi (reviewed in Krenn 2010). Proboscis morphology and the hydraulic mechanism of movement are uniform in glossatan Lepidoptera, but cuticle structures and sensilla near the tip are modified for piercing (Bänziger 1970; Büttiker et al. 1996).

The Calpini (Erebidae) mainly use their proboscis for fruit-piercing. However, in several species, males occasionally suck blood from large mammals, including humans (Bänziger 1970, 1989, 2007; Zaspel et al. 2007, 2011). The acute proboscis of these fruit- or skin-piercing moths is armored with long, robust thornlike cuticle structures (Fig. 3.7k) for puncturing and anchoring the galeae during the piercing process. The biggest structures are derived from sensilla (Büttiker et al. 1996) and are erectile (Fig. 3.7l) due to an increase of hemolymph pressure within the galeae (Bänziger 1970, 1980; Zaspel et al. 2011). For feeding, the proboscis is uncoiled by increasing the internal hemolymph pressure (Bänziger 1971). The tip is pressed against the skin of fruits or mammals, and the galeae alternately move forward in longitudinal direction. This leads to lateral bending of the proboscis until the apical structures carve into the skin and anchor the proboscis. One galea pierces further, while the structures of the other galea serve as anchors resulting in a spindle movement of the proboscis. About 15 alternating protraction movements of the galea per second thrust the proboscis even into thick-skinned fruits (such as mandarins) or the skin of large mammals, like horse or tapir (Bänziger 1975, 1980, 1989, 2007; Zaspel et al. 2007, 2011). Withdrawing the proboscis is likewise achieved by alternate movements of the galeae. However, their thorns are folded backward due to decreased internal pressure (Bänziger 1970, 1975, 1980). Amazingly, a related moth of Malagasy Calpinae was found to drink tears from the eyes of sleeping birds using a similarly shaped proboscis, probably without irritation of the bird (Hilgartner et al. 2007).

In contrast to the impressive proboscis armature of Calpini moths, the proboscis morphology of charaxin butterflies (Nymphalidae) is inconspicuous (Krenn 2010). The representatives of the Charaxinae do not visit flowers, but only feed from soft rotting fruits or plunge their proboscis into excrements (DeVries 1987; Molleman et al. 2005). The proboscis of these butterflies is rather short, stout, thick walled, and blunt at the tip. Repeated downward movements of the whole body drive the extended proboscis into such rotting substances without visible antiparallel movements of the galeae. The proboscis tips were found to bear short sensilla styloconica which are arranged in rows along the slits leading into the food canal (Molleman et al. 2005).

## 3.3.4 *Sponging Mouthparts*

### 3.3.4.1 Proboscis of Short-Tongued Diptera

The mouthparts of many representatives of non-piercing Diptera function according to the sponging fluid-feeding technique. Proboscises using a sponging mode of fluid-feeding are rather short (Figs. 3.3g and 3.8). These flies mop up liquids and suspended particles without licking movements. Liquid food adheres to a large surface of the apical components of the labium, the labellum. This pair of soft lobes is homologous to the labial palpi. Sponging mouthparts are primarily used on open and easily accessible fluid sources (Fig. 3.2e), such as sweat, fresh dung, decaying substances, rotting fruit, or nectaries of open flowers, extrafloral nectaries, or honeydew. Some are able to liquefy food and suspend particles, like pollen, in discharged saliva (e.g., Graham-Smith 1930; Gilbert 1981; Vijaysegara et al. 1997; Krenn et al. 2005). The sponging mode of feeding is performed with the short haustellum, which is equipped with broad labella as the main organs of fluid uptake. All Diptera have the labial palpi fused at the apex of the labium. In the Brachycera, this structure is further modified into a broad pad with radiating gutters that lead into the food canal (Grimaldi and Engel 2005). These microscopic furrows on the labella surface which are called pseudotracheae form channels that are optimized to take up trace amounts of fluid from porous surfaces. Feeding trials with magnetic nanoparticles showed that the sponging proboscis of a cyclorrhaphan fly uses capillary forces for fluid uptake and ingestion into the food canal (Lehnert et al. 2017).

The sponging feeding technique probably represents the ancestral method of food uptake in adult Diptera, since the characteristic labellum can be found in representatives of Tipulimorpha, many non-piercing nematoceran Diptera, and Brachycera (Ziegler 2003; Wiegmann et al. 2011). In Tipulidae, the head capsule and the basal mouthparts form an immobile rostrum. The rather small prementum forms parts of the internal composition and bears the larger apical cushion-shaped labella which are covered with a system of microscopic furrows on the fronto-median side

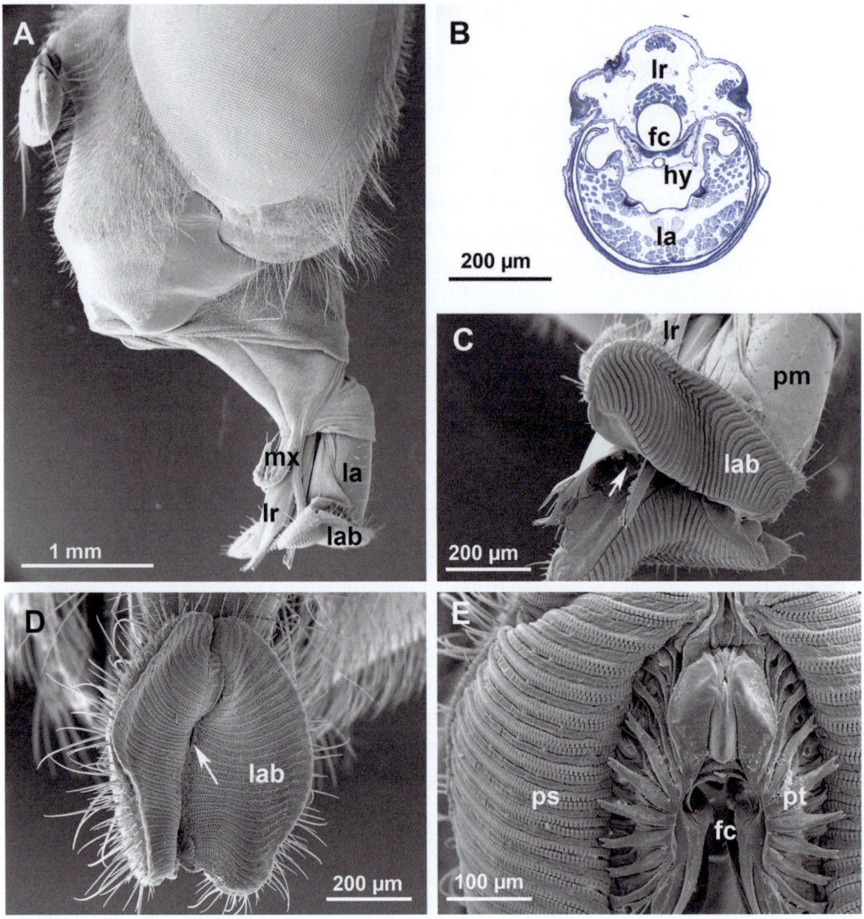


**Fig. 3.8** Sponging proboscis of Diptera (scanning electron microscope images and light microscope photos of histological cross-sections). (**a**) Head of the hoverfly *Eristalis* sp. (Syrphidae), short proboscis extended for feeding. (**b**) Proboscis of *Volucella bombylans* (Syrphidae) in cross-section, food canal formed by the buccal side of the labrum; the hypopharynx closes the food canal; labium forms posterior side of the haustellum. (**c**) Labella of *Eristalis* sp. are the main organs of fluid uptake; pseudotracheae and superimposed grooves lead fluid to the opening of the food canal (arrow). (**d**) Labellum of the blow fly *Calliphora vomitoria* (Calliphoridae) in cupping position; fluid can be ingested directly into the food canal (entrance indicated by an arrow). (**e**) Labellum of the flesh fly *Sarcophaga* sp.; pseudotracheae of the labella, prestomal teeth, surround the opening into the food canal. *fc* food canal, *hy* hypopharynx, *la* labium, *lab* labellum, *lr* labrum, *mx* maxilla, *pm* prementum, *ps* pseudotrachea, *pt* prestomal tooth

(Chaudonneret 1990). Similar to the labellum of Brachycera, these pseudotracheae take up fluid from a wet surface by capillarity (Proctor et al. 1996).

The typical brachyceran proboscis comprises (1) the proximal rostrum that forms a movable connection with the head, (2) the haustellum which composes the fluid canal, and (3) the apical labella equipped with the pseudotracheae (Figs. 3.3g–i and

3.8). The functional morphology was studied in several short-tongued nectar-feeding flies (Gilbert and Jervis 1998), such as Bombyliidae (Szucsich and Krenn 2000, 2002), Syrphidae (Gilbert 1981, 1985; Schuhmacher and Hoffmann 1982), as well as non-flower-visiting Tephritidae (Vijaysegara et al. 1997; Coronado-Gonzalez et al. 2008), Muscidae, and Calliphoridae (Graham-Smith 1930). In *Calliphora* (Calliphoridae), the mouthpart movements have been studied in detail and can be regarded as a model for other Cyclorrhapha (Graham-Smith 1930; Thomson 1977). Table 3.3 lists examples of various compositions of the proboscises in different lineages of Diptera.

The rostrum is derived from the clypeus, cibarium, and basal parts of the proboscis. It forms an articulation that increases the operational length and maneuverability of the proboscis. In Cyclorrhapha and some basal Brachycera, the rostrum allows retraction of the proboscis into a resting position in the preoral cavity of the head by folding the haustellum (Ziegler 2003). The muscles that are attached at sclerites of the cibarium (termed as fulcrum) fold and rotate the haustellum, sometimes in a Z-shape, against the head (Szucsich and Krenn 2000). The proboscis movements are probably supported by a hydraulic mechanism (Van der Starre and Ruigrok 1980).

The components of the haustellum form the food canal. The posterior side of the labrum, the epipharynx, closes the anterior side, while posteriorly the hypopharynx completes the canal. The hypopharynx also contains the salivary duct. The labium comprises the largest part of the mouthparts which enclose the other components by forming a gutter (Fig. 3.8). In sponging mouthparts, the mandibles are vestigial or lost; the maxillae are rudimentary, forming rodlike components that support the rostrum and function as lever arms to transmit movement during proboscis folding (Szucsich and Krenn 2000). The maxillary palpi probably have sensory function but are not part of the proboscis.

The prementum bears the labella, which vary in size, shape, and number of sensilla as well as in the presence of various additional structures. In most Brachycera, the labella consist of two more or less broad, moveable cushion-like lobes with pseudotracheae to perform a sponging fluid uptake. The surface is traversed by small channels (diameter 2–28 μm) which are strengthened by chitinous rings. These pseudotracheae merge into collecting channels which empty into the opening of the food canal (Figs. 3.3h and 3.8e). The pseudotracheae are particularly variable in number, width, microstructures, and shape of their openings and differ between feeding guilds and lineages (Zaitzev 1983; Broadhead 1984; Elzinga and Broce 1986; Gilbert and Jervis 1998; Ziegler 2003). In Calyptrata and Phoridae, additional cuticle structures called prestomal teeth are present near the apical opening of the food canal at the base of the labella (Fig. 3.8e). These teeth vary in number, length, and degree of sclerotization in different feeding guilds, e.g., weakly sclerotized in coprophagous flies, heavily sclerotized but small in hematophagous flies, or smooth and sharply pointed in predaceous flies. They serve as rasping structures and to mash substances or function as cutting and piercing structures (Elzinga and Broce 1986; Giangaspero and Broce 1993; Gilbert and Jervis 1998). Numerous sensilla of various types can be found on the labella, including bristle-shaped types at the margin of the

**Table 3.3** Sponging proboscises of Trichoptera and Diptera; haustellum forms food canal and bears apical structures for fluid uptake; listed taxa may contain exceptions or may not include all species of the taxon; compilation from Nagatomi and Soroida (1985), Elzinga and Broce (1986), Chaudonneret (1990), Gilbert and Jervis (1998), Krenn et al. (2005)

| Taxon | Haustellum composition | Tip |
|---|---|---|
| **Trichoptera** | | |
| Some species from various families | Labrum, labium, hypopharynx | Labium |
| **Diptera** | | |
| Tipulidae | Head, maxilla, labium, hypopharynx | Labellum |
| Stratiomyidae, Pantophthalmidae, Rachiceridae, Xylophagidae, Coenomyiidae, Heterostomidae, Exeretoneuridae, Therevidae, etc. | Labrum/epipharynx, lacinia, labium, hypopharynx | Labellum |
| Dolichopodidae[a] | Labrum/epipharynx, labium, hypopharynx | Labellum |
| Bombyliidae (short-tongued species) | Labrum/epipharynx, lacinia, prementum, hypopharynx | Labellum |
| Syrphidae | Labrum/epipharynx, lacinia, labium, hypopharynx | Labellum |
| Many orthorrhaphous Brachycera[b] | Labrum/epipharynx, mandible, lacinia, labium, hypopharynx | Labellum |
| Many Cyclorrhapha[b] | Labrum/epipharynx, labium, hypopharynx | Labellum |

[a]Mouthparts of Dolichopodidae form short predaceous haustellum
[b]Proboscises of species using piercing and sucking or siphoning feeding techniques are differently composed

labium and dome-shaped sensilla near the pseudotracheae. The sensory equipment most likely detects food substances and provides information about the position of the labella on food sources (e.g., Schuhmacher and Hoffmann 1982; Szucsich and Krenn 2002; Coronado-Gonzalez et al. 2008).

Labella movements are important for feeding from different food sources (Graham-Smith 1930; Gilbert 1981; Schuhmacher and Hoffmann 1982; Vijaysegara et al. 1997; Szucsich and Krenn 2000). In short-tongued Bombyliidae, the labella are folded together, and in resting position they are held in the longitudinal axis of the labium. They are flexed forward and opened by spreading the small sclerites of the labellar base (Szucsich and Krenn 2000). During feeding, the labella are placed on the fluid surface and gradually close as the fluid meniscus descends (Vijaysegara et al. 1997). In *Calliphora* flies, the labella can attain different positions (Graham-Smith 1930). At rest, the lobes are flexed posteriorly against the haustellum with the pseudotracheae in apposition. During feeding, the labella are spread out flat, and the pseudotracheae are applied to the substrate (Fig. 3.2e). Liquid is sucked through the pseudotracheal openings, but large particles are filtered out. In another position, the labella are turned downward; in this dome-shaped position, fluid is ingested between the lobes (Fig. 3.8d). The labella can also be turned upward with the prestomal teeth fully exposed for rasping over the food source, or the labella are turned up- and outward against the sides of the haustellum, and the aperture of the food canal directly ingests unfiltered fluid. Similar feeding positions were found in tephritid flies which are additionally able to feed from freely hanging droplet (Vijaysegara et al. 1997). Non-calyptran flies cannot evert the labella and are not able to scrape the surface of the food source (Elzinga and Broce 1986). A detailed study in hoverflies revealed additional interpseudotracheal folds of the labellar surface (Fig. 3.8c) that create channels overlying the pseudotracheae through which fluid is transported to the food canal. These fluid pathways are probably maintained by hemolymph pressure and are crucial for the uptake of suspended particles, like pollen grains (Schuhmacher and Hoffmann 1982). In this context, discharge of saliva is important to dissolve sugary fluids and suspend particles that are subsequently ingested (Vijaysegara et al. 1997; Gilbert and Jervis 1998).

#### 3.3.4.2 Haustellum of Trichoptera

The short haustellum of some adult Trichoptera is mainly composed of the labium which is fused with the hypopharynx and basal parts of the maxillae. The labrum forms the anterior side of the short food canal (Weber 1933; Chaudonneret 1990). The soft apex of the haustellum bears channels fringed with hairs that direct the fluid to the oral opening. The usually long maxillary and labial palpi do not take part in food uptake. The feeding technique is assumed to be similar to a sponging feeding mode. In Phryganeidae, a moveable haustellum was described that mainly consists of the elongated and soft hypopharynx which bears channels on its anterior surface (Chaudonneret 1990). Since adult Trichoptera are rarely observed feeding (Holzenthal et al. 2015), the functional morphology of their mouthparts was never studied in detail. Flower-visiting behavior and considerable elongation of the proboscis were noticed in some representatives of Plectrotarsidae, Kokiriidae, and Dipseudopsidae (Ulmer 1905); however, recent investigations on form and function of the mouthparts are missing.

### *3.3.5 Siphoning Proboscises*

#### 3.3.5.1 Fluid-Feeding Dominated by Pressure Gradient

Sucking or siphoning mouthparts mainly use a pressure gradient to ingest fluids (Kingsolver and Daniels 1995). The mouthparts form a long proboscis and do not perform licking movements. Fluid transport is primarily achieved by sucking pumps in the head (Kingsolver and Daniel 1995; Kornev et al. 2017). Characteristic features are a (1) fluid-tight food tube and (2) a soft apical drinking region (Figs. 3.3j-l and 3.9). Siphoning proboscises evolved in flower-visiting nectar-feeding holometabolan insects, like Apidae and Vespidae (Hymenoptera), in glossatan Lepidoptera, and in some Diptera (Table 3.4) (reviewed in Krenn et al. 2005; Borrell and Krenn 2006). Evolutionary scenarios have postulated co-adaptations of proboscis length with host flower morphology in some Lepidoptera and the particularly long-tongued Diptera (e.g., Nilsson 1988; Johnson and Steiner 1997; Whittall and Hodges 2007; Anderson and Johnson 2008; Pauw et al. 2009; Bauder et al. 2013; Khramov and Lukashevich 2019). Particularly long proboscises of these flower-visiting insects have the advantage of a privileged access to the nectar sources from deep flowers; however, these flower visitors are not necessarily pollinators of their nectar host plants (Bauder et al. 2011, 2013, 2015a, b).

#### 3.3.5.2 Siphoning Proboscises in Hymenoptera

Many Hymenoptera use a lapping nectar-feeding technique (see Sect. 3.3.2). A few lineages developed a proboscis longer than the body in which a predominantly sucking feeding technique can be assumed. One lineage of Masarinae (Vespidae) evolved a particularly long and thin proboscis which is composed of only the glossa (Krenn et al. 2002). Thin, curved cuticle structures of the glossa overlap to form a food canal (Fig. 3.9a–c). Slits are formed in the short bifid apical region (Fig. 3.9d), through which fluid is ingested. Modifications of the basal glossa and its articulation with the prementum are responsible for the protraction of the proboscis out of the resting position in which the glossa is flexed back in one or several loops inside the prementum that forms a special sac below the head (Schremmer 1961; Krenn et al. 2002). Chapter 14 discusses the functional morphology of these sucking mouthparts in Masarinae.

Orchid bees (Euglossini) may also use their proboscis for suction feeding without lapping movement of the glossa (Borrell 2003). In these particularly long-tongued bees, nectar can be directly imbibed into the food tube which is formed by the galeae, the labial palpi, and a motionless glossa. Observations showed that orchid bees are flexible in the applied method of nectar consumption. Small amounts of nectar are ingested with the lapping feeding technique (Düster et al. 2018). The proboscis composition and movements are similar to those in other long-tongued Apidae, except that the postmentum in *Euglossa* can be additionally extended. This leads to a further protraction of the elongated glossa, resulting in a maximal functional length that is about 50% longer than the length of the galeae and the labial palpi which compose the food canal. The extent of the proboscis and glossa extension depends on the available corolla length (Düster et al. 2018).

Other examples can be found in several lineages of Apoidea where distinctly elongated maxillary and labial palpi form a very long proboscis (e.g., Alves dos Santos and Wittmann 1999), but the glossa and paraglossae are short (Jervis 1998; Krenn et al. 2005). Jervis (1998) and Jervis and Vilhelmsen (2000) described modified proboscises ("concealed nectar-extracting apparatus types 7 and 8") in some representatives of parasitoid wasps, such as Ichneumonidae and Braconidae. They are composed by greatly elongated maxillary palpi and/or labial palpi where no setose structures are discernible that could be used for a lapping feeding technique. It can be concluded from the morphology that these mouthparts function according to a pure sucking technique of fluid-feeding (Table 3.4).

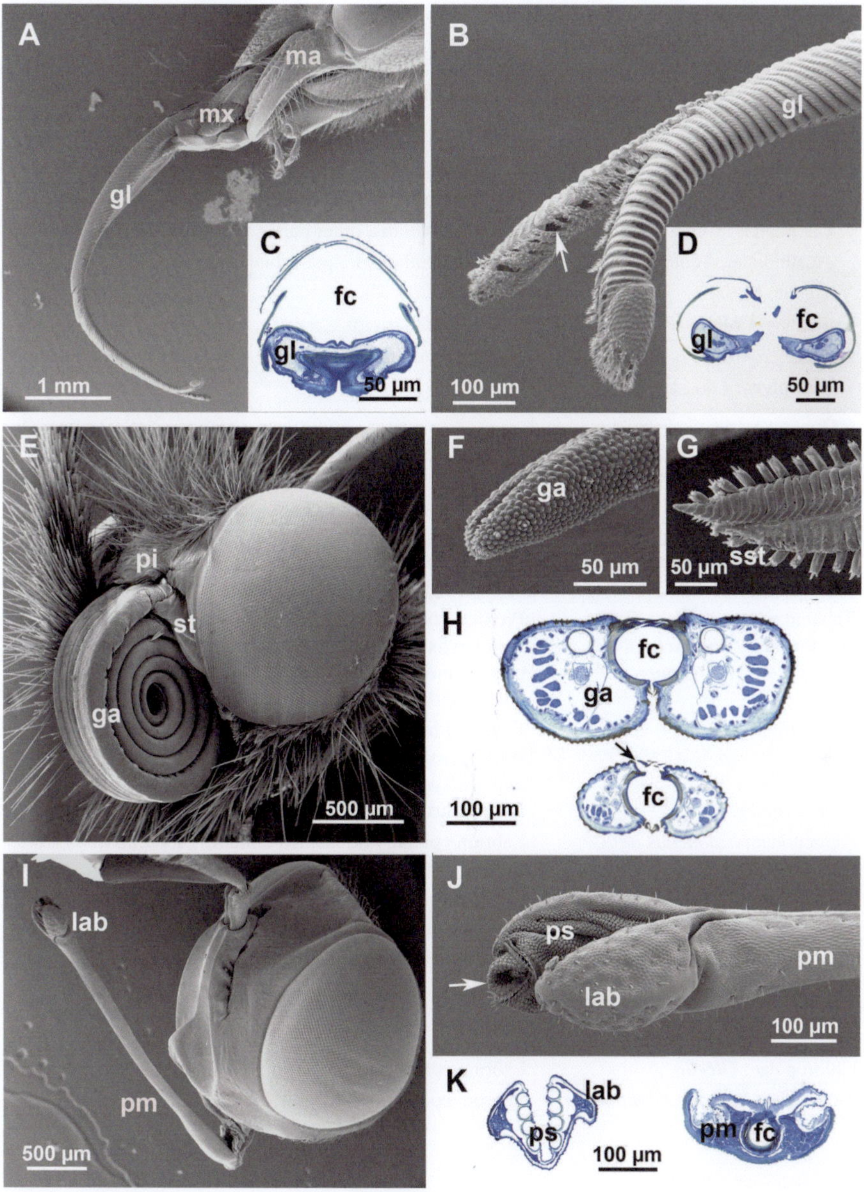


**Fig. 3.9** Siphoning proboscises (scanning electron microscope images and light microscope photos of histological cross-sections). (**a–d**) The pollen wasp *Ceramius hispanicus* (Vespidae, Hymenoptera): (**a**) head and extended proboscis; (**b**) bifid tip of the proboscis (arrow indicates opening into the food canal); (**c**) cross-sections of food canal at mid-length of the proboscis; (**d**) cross-section near the tip. (**e**) Head and coiled proboscis of *Zerynthia polyxena* (Papilionidae, Lepidoptera), one labial palpus removed. (**f**) Proboscis tip of *Z. polyxena* (Papilionidae) with inconspicuous sensilla. (**g**) Proboscis tip of *Vanessa cardui* (Nymphalidae, Lepidoptera) with prominent sensilla styloconica. (**h**) Cross-sections of the proboscis of *Z. polyxena* (Papilionidae) at mid-length (above) and through drinking region (below); arrow points to a drinking slit between the legulae. (**i–j**) Thick-headed fly *Physocephala* sp. (Conopidae, Diptera): (**i**) head with proboscis

**Table 3.4** Siphoning or sucking proboscises are much longer than the head; food canal is composed of various parts; tip region is variously long; compilation from Nagatomi and Soroida (1985), Elzinga and Broce (1986), Chaudonneret (1990), Gilbert and Jervis (1998), Jervis and Vilhelmsen (2000), Szucsich and Krenn (2000, 2002), Borrell (2003), Krenn et al. (2005), Karolyi et al. (2012, 2013)

| Taxon | Composition of food canal | Tip region |
|---|---|---|
| **Hymenoptera** | | |
| Braconidae: Braconidae: *Bracon* | Palpus maxillaris, other parts? | ? |
| Cheloninae: *Chelonus* | Palpus maxillaris, other parts? | ? |
| Agathidinae: *Agathis* | Palpus maxillaris, Palpus labialis | ? |
| Vespidae: Masarini | Glossa | Bifid glossa |
| Colletidae: Colletinae: *Euhesma* | Palpus maxillaris | ? |
| Apidae: *Euglossa* | Galea, labial palpi, glossa | Flabellum |
| **Lepidoptera:** Glossata | Galea | Galeal drinking region |
| **Diptera** | | |
| Culicidae: *Toxorhynchites* (males and females), in other Culicidae only males | Labrum/epipharynx, lacinia, labium, hypopharynx | Labellum |
| Vermileonidae: *Lampromyia intermedia* | Labrum, maxilla, prementum?, hypopharynx | Labellum |
| Tabanidae: Pangoniinae (males and females): *Philoliche*, *Corizoneura*, *Pangonius* | Proximal haustellum: Labrum/epipharynx, mandible, lacinia, prementum, hypopharynx<br>Distal haustellum: prementum | Labellum |
| Nemestrinidae (Tangle-veined flies): *Prosoeca*, *Moegistorhynchus*, *Stenobasipteron* | Proximal haustellum: Labrum/epipharynx, lacinia, prementum, hypopharynx<br>Distal haustellum: prementum | Labellum |
| Acroceridae: *Eulonchus halli* | Labrum/epipharynx, lacinia, labium, hypopharynx | Labellum |
| Bombyliidae (Beeflies): *Bombylius* | Labrum/epipharynx, lacinia, prementum, hypopharynx | Labellum |
| Conopidae: *Conops*, *Stylogaster*, etc. | Labrum/epipharynx, labium | Labellum (extremely elongated in some species) |
| Chloropidae: *Olcella* | Labrum/epipharynx?, labium | Labellum |
| Tachinidae: *Siphona*, *Sipholeskia* | Labrum/epipharynx, labium? | Labellum |

**Fig. 3.9** (continued) in resting position; (**j**) slender labella form the proboscis tip; arrow indicates opening into the pseudotrachea. (**k**) *Sicus* sp. (Conopidae), cross-sections of proboscis tip formed by the labella (left) and mid-region (right); food canal is formed by the labial prementum. *fc* food canal, *ga* galea, *gl* glossa, *lab* labellum, *ma* mandible, *mx* maxilla, *pi* pilifer, *pm* prementum, *ps* pseudotrachea, *sst* sensilla styloconica, *st* stipes

#### 3.3.5.3 Proboscis of Glossatan Lepidoptera

The feeding apparatus of adult glossatan Lepidoptera represents one of the best studied examples of a siphonate (or sucking) proboscis. The proboscis of butterflies may exceed twice the body length reaching up to 52.7 mm (Bauder et al. 2014), whereas the maximal reported proboscis length in hawk moths (Sphingidae) is 280 mm representing the longest proboscis in insects by far (Amsel 1938) (see Chap. 15). In comparison to sucking mouthparts of other insects, the proboscis of Lepidoptera has a relatively simple morphology, but it is unique in its spirally coiled resting position. The proboscis of glossatan Lepidoptera is composed of the two extremely elongated galeae which are interlocked by special cuticle structures (termed as legulae) on the dorsal and ventral side and enclose the central food canal. Near the tip (termed as tip region in Krenn 2010), the legulae form slits into the food canal on the dorsal side (Figs. 3.3k and 3.9h) (Krenn and Kristensen 2000). This drinking region is highly maneuverable allowing butterflies and moths to feed from variously shaped flowers in a very efficient way as well as from many other additional open fluid sources (Fig. 3.2f) (DeVries 1987; Krenn et al. 2001; Knopp and Krenn 2003; reviewed in Krenn et al. 2005; Krenn 2010). The drinking region is shorter in nectar feeders than in non-flower-visiting butterflies (Krenn et al. 2001; Lehnert et al. 2016). The latter feeding guild includes species which are specialized to ingest liquid from surfaces of rotting fruits, honeydew, or tree sap (see Chap. 9). Lehnert et al. (2013) discovered that the cuticle of the drinking region and the food canal is hydrophilic, whereas other regions of the proboscis are hydrophobic. These cuticle properties propagate fluid ingestion into the food canal (see Chaps. 8 and 9). Various feeding guilds are distinguished also in other ditrysian Lepidoptera, including specialized tear-feeding moths which are characterized by typical features of the proboscis (e.g., Büttiker et al. 1996; reviewed in Krenn 2010).

A hydraulic mechanism uncoils the proboscis from the spirally coiled resting position. Most likely, the compressions of stipital hemolymph pumps lead to stepwise increase of the hemolymph pressure inside each galea resulting in their uncoiling. When the internal pressure decreases, the elastic properties of the proboscis coil it into a loosely spiral. The intrinsic galeal musculature returns the proboscis to its resting position by tightening the coil and tucking it underneath the head (Schmitt 1938; Bänziger 1971; Krenn 1990, 2000; Wannenmacher and Wasserthal 2003; for a review of the mechanism of movements, see Krenn 2010). The proboscis wall is composed of various cuticle types with different elastic properties (Hepburn 1971). The presence of resilin was proven in the cuticle, but it remains unclear how this elastic cuticle protein contributes to the mechanical properties of the galeal wall and the proboscis movements (Hepburn 1971). The basal galeal articulation connects the proboscis halves with the stipites on each side. The basal galeal muscles are derived from extrinsic galeal muscles and allow up-and-down movements of the whole proboscis (Krenn 1990; Krenn and Kristensen 2004).

The evolution of the lepidopteran proboscis can be reconstructed based on the well-studied phylogeny of these insects (most recent Mitter et al. 2017). The proboscis evolved only once in the stem group of the Glossata. The earliest fossil of a glossatan moth, and hence the proboscis, is 125 million years old (Grimaldi and Engel 2005). Myoglossata possess novel musculature in the lumen of the galeae for tightly coiling the proboscis (Kristensen and Nielsen 1981a). The principal composition of the proboscis is similar in all Glossata except the species-poor Neopseustidae where each galea forms a functionally closed food tube. Both are linked by special structures extending from the ventral side and form a short double-tube proboscis. Each galea contains a single intrinsic galeal muscle in addition to the basal extrinsic muscle (Kristensen and Nielsen 1981b). The representatives of the Eulepidoptera have a long proboscis with particular adaptations for extracting nectar from concealed flowers. This includes the pilifers (bristle-bearing organs at the proboscis base), complex interlocking structures of the proboscis halves, and elaborate sensory equipment (Figs. 3.3k and 3.9e–h) (Krenn and Kristensen 2000). The proposed hypothetical evolutionary pathway of the intrinsic galeal musculature indicates the recruitment of muscle fibers from the basal galeal muscles by shifting their attachment sites beyond the basal joint. A hypothetical shift of the muscle origin resulted in the intrinsic galeal muscles of the Eulepidoptera. This was followed by the multiplication of the musculature in context with proboscis elongation in the lineage of Ditrysia and the split into two series of overlapping muscles resulting in a dense arrangement of intrinsic galeal musculature as found in of Macrolepidoptera, including butterflies (Figs. 3.3l, 3.7l, and 3.9h) (Krenn and Kristensen 2000, 2004; reviewed in Krenn 2010).

Adult Eulepidoptera use a variety of fluid sources in addition to floral nectar, such as rotting fruits, tree sap, mammal sweat and tears, fresh dung, muddy water, and pollen (reviewed in Krenn 2010). The morphological adaptations to special food sources are found in proboscis length and near the tip (called the "tip region" in Krenn 1990, 2010) where specially shaped sensilla and cuticle structures form specialized uptake regions for the various fluids (Figs. 3.3k, 3.7k and 3.9f, g) (e.g., Büttiker et al. 1996; Krenn and Penz 1998; Krenn et al. 2001; Knopp and Krenn 2003; Molleman et al. 2005; Hilgartner et al. 2007; Bauder et al. 2011, 2013; Zaspel et al. 2011; Zenker et al. 2011; Xue and Hua 2014). The conspicuous structures of the proboscis drinking region are the reason why the sensory equipment of the lepidopteran proboscis has been studied from ecomorphological, functional, and evolutionary points of view many times in detail (e.g., Faucheux 1999; Krenn et al. 2001; Petr and Stewart 2004; reviewed in Krenn 2010). Bristle-shaped mechanosensitive sensilla occur only on the external surface of the proboscis. They probably provide information on the depth of proboscis insertion during flower probing. In addition, the bristle-shaped sensilla monitor the coiling status in the resting position, since they are in contact with the surface of the adjacent coil (Krenn 1990, 1998). The bristles of the pilifers probably detect movements of the entire proboscis against the head (Krenn and Kristensen 2000). Dome-shaped sensilla basiconica are arranged in rows on the external proboscis and in the food canal. A terminal pore may be present or absent; in exceptional cases, the cone is multiporous. Deduced from their ultrastructure, these sensilla are regarded as chemo- and/or mechanoreceptors (e.g., Altner and Altner 1986; Krenn

1990, 1998; Faucheux 1999). In electrophysiological tests, the sensilla of the food canal respond to sucrose solutions and thus may provide information on ingested fluid (Inoue et al. 2009). Special sensilla styloconica are arranged in rows in the distal half of the proboscis (Figs 3.3k and 3.9g) where they occur close to the drinking slits which lead into the food canal (Fig. 3.9h). The sensilla styloconica consist of a variously shaped stylus and a shorter terminal sensory cone. A striking diversity of sensilla styloconica lengths and shapes are found in Papilionoidea, Noctuoidea, Geometridae, and Pyralidae (e.g., Paulus and Krenn 1996; Büttiker et al. 1996; Faucheux 1999; Petr and Stewart 2004; Bauder et al. 2013) that were interpreted to reflect various feeding preferences, like nectar, rotting fruits, tears, etc. They may help to extract additional substances in pollen-feeding and self-medicating butterflies (Krenn and Penz 1998; Krenn et al. 2001; reviewed in Krenn 2010; Zenker et al. 2011; Zaspel et al. 2013). The internal ultrastructure is similar in all examined taxa indicating that they serve as combined chemo-mechanosensilla (Altner and Altner 1986; Krenn 1998; Nagnan-Le Meillour et al. 2000; Kvello et al. 2006). Sensilla styloconica are sensitive to certain mono- and oligosaccharides and other substances, like ethanol (Salama et al. 1984; Blaney and Simmonds 1988; Ômura et al. 2008). Their stimulation with sucrose increases proboscis movements and food ingestion (Jørgensen et al. 2006). Chapter 15 deals with the proboscis morphology of particularly long-proboscid Lepidoptera and illustrates the anatomy of galeae and the stipes pumps.

The ability of Lepidoptera to drink liquids from various sources such as rotting fruit, floral tubes (Fig. 3.3j), and wet soil (Fig. 3.2f) raised the question whether the conventional view of the proboscis as a drinking straw can explain the ability to imbibe fluids from porous substances or of fluid films. X-ray imaging of living butterflies revealed that fluid is transported into the food canal through the drinking slits by capillary forces and the pressure gradient created by the sucking pump transports the fluid through the food tube (Monaenkova et al. 2011). Lehnert et al. (2013) proved hydrophilic cuticle surface in the drinking region where the slits actively lead fluid into the food canal. The hydrophilic/hydrophobic properties of the external proboscis enable butterflies to efficiently take up liquid from wet surfaces (Lehnert et al. 2013, 2016). Chapters 8 and 9 in this book discuss the biophysics of the proboscis of Lepidoptera in detail.

#### 3.3.5.4 Proboscis of Long-Proboscid Diptera

In long-proboscid Diptera (also called as long-tongued flies), the siphoning or sucking feeding technique is predominant. Several lineages evolved particularly long but variously composed proboscises (Table 3.4). The characteristic labella are slender and cannot be spread. Further, they are equipped with few pseudotracheae, but take up nectar directly into the pseudotracheae that open at the margins of the labella and directly lead into the food canal. Such proboscises independently evolved in representatives of the flower-visiting flies, such as Nemestrinidae, Acroceridae, some Bombyliidae, Syrphidae, Conopidae (Fig. 3.9i–k), and pangoniin Tabanidae

(Table 3.4) (Schremmer 1961; Dierl 1968; Nagatomi and Soroida 1985; Krenn et al. 2005; Karolyi et al. 2012, 2013, 2014). In addition to non-piercing *Toxorhynchites* mosquitos, nectar-feeding males of other blood-feeding long-proboscid nematoceran Diptera probably use this feeding technique for nectar-feeding (Bonnet and Hu 1951).

The functional morphology, proboscis movements, and possible adaptations to particular nectar resources have been studied in detail in Bombyliidae (Szucsich and Krenn 2002), Syrphidae (Gilbert 1981), Tabanidae (Karolyi et al. 2014), and Nemestrinidae (Karolyi et al. 2012, 2013). In Bombyliidae, the particularly long proboscis of some taxa is characterized by interlocked slender labella which bear only few apically open pseudotracheae. Together with the lengthening of ventral parts of the rostrum, these features increase the operational length of proboscis and allow nectar uptake with closed labella from laterally open flowers (Szucsich and Krenn 2002). Adaptations to narrow corolla tubes were also described in some Syrphidae, such as *Rhingia* (Gilbert 1981). This hoverfly has a long, foldable proboscis which is equipped with particularly slender labella that extend the functional length and allow feeding from narrow tubular flowers which physically prevent the labella from opening (Gilbert 1981; Krenn et al. 2005). Particularly elongated labella are recorded from some Conopidae (Proctor et al. 1996); however, detailed studies are missing.

The proboscises of Nemestrinidae from the genera *Prosoeca*, *Stenobasipteron*, and *Moegistorhinchus* are particularly variable in length but may exceed the body length by far. In *Moegistorhinchus longirostris*, the proboscis is up to 100 mm long and represents the longest proboscis of all known Diptera species worldwide (Johnson 2010; Barraclough and Slotow 2010; Barraclough 2017). The mouthpart morphology and flower-visiting behavior were studied in *Prosoeca* (Karolyi et al. 2012, 2013). The proximal proboscis is composed like in other brachyceran flies, but the distal two-thirds are formed only by the very slender labium (Table 3.4). The prementum is rolled up longitudinally and forms the tightly closed food canal (see Chap. 15 for more details). Apically it bears particularly short labella where nectar is ingested through the pseudotracheae which open frontally. The mean meal size is about 1 μl nectar per flower visit in the natural habitat (Krenn et al. 2018). Several large sucking pumps in the head transport the ingested fluid to the mouth (Karolyi et al. 2012).

The proboscis of female long-proboscid Pangoniinae (Tabanidae) is particularly amazing, since it is composed of two functional units: The short proximal part is adapted for piercing blood-sucking and a very long distal region for nectar-feeding. The proximal proboscis is equipped with mandibular and maxillary stylets like in other female horseflies. However, the greatly elongated distal region is composed of the labial prementum alone that forms a tightly sealed tube (Karolyi et al. 2014). Female representatives of the Pangoniinae are known to use the shorter piercing apparatus to penetrate the vertebrate skin and drink blood while the elongated prementum is flexed away (Dierl 1968; Morita 2008, 2011). In an unique way, this proboscis combines the piercing-sucking feeding type with the siphoning proboscis in one feeding organ demonstrating the characteristic functional features of both feeding types. Long-proboscid flies are considered among the first pollinators of early angiosperms and appear to have co-evolved with flowering plants since the Late Jurassic (Labandeira 1997; Ren 1998; Nel et al. 2018). Chapter 15 presents the ecomorphology and evolution of superlong proboscises of these Diptera.

## 3.4 Directions of Future Studies

Fluid-feeding insects show an enormous diversity of proboscis designs which permit ingestion and transport of various liquids with amazing efficiency. The various proboscis formations are striking examples of convergent evolution that shaped a set of homologous organs by natural selection. The morphological and functional disparity of mouthparts evolved as adaptations to different kinds of fluid sources and feeding behaviors. In context with blood-feeding and plant sap feeding, remarkably diverse piercing techniques evolved that are studied only in few examples in detail so far. Particularly long proboscises developed multiple times independently in many flower-visiting Holometabola as an adaptation to ingest concealed nectar from long-tubed flowers. Their functional performance is rather easy to measure, and comparative experimental studies gave evidence for the functional interpretation of morphological features from ecomorphological points of view in butterflies (e.g., Bauder et al. 2015a, b; Lehnert et al. 2016, 2017), but other flower-visiting insects—many of them potential pollinators—await detailed investigation.

Technological advances of the last decades (e.g., X-ray microtomography) permit exceptionally detailed investigation of insect mouthparts and their function. These new imaging techniques allow examination of living insects at the microscopic level during feeding. This resulted in new conclusions about feeding mechanisms based on biophysics and material sciences even in well-studied insects like honey bees or butterflies. Recent results indicated that micromorphology, capillarity, and pressure gradients created by sucking pumps interplay at all levels of fluid ingestion and transport (e.g., Lehnert et al. 2013; Yang et al. 2014; Wu et al. 2015; Kornev et al. 2017). As a result, the categorization into adhesion- or pressure-dominated feeding techniques has to be reconsidered after more examples have been studied from a biophysical point of view. Recently, biophysicists discovered the remarkable functional performance of insect mouthparts and started investigating these complex mechanical systems under aspects of biomechanics and material science (e.g., Kornev et al. 2016; Li et al. 2017; Zhang et al. 2018). Bioengineers showed that mouthparts have a potential as templates for bioinspired technology, for example, the piercing structures of mosquitos served as models for the invention of painless hypodermic microneedles (Izumi et al. 2011). Future studies should include examination of micro-biomechanics of mouthpart structures and the elastic properties of the cuticle. This could lead to modelling of, for example, foldable structures such as the proboscis of honey bees which is assembled anew each time it is extended.

The cuticle properties of interacting structures seem to be amazingly diverse at microscopic level. Small areas of wettability have enormous importance for the function of the whole proboscis, and knowledge of hydrophilic cuticle properties helps to explain the principles of fluid transport through small tubes (Lehnert et al. 2017; Zhang et al. 2018). Thus, insect proboscises could be the model for developing microtools to transfer minute amounts of fluids with high precision. Finally, uncovering the genetic basis of various cuticle properties and their exact localization on the components could be a challenging but fruitful avenue for future research to better understand the form and function of fluid-feeding mouthparts in insects.

**Acknowledgements** I thank Philipp Maier for linguistic help, as well as Barbara-Amina Gereben-Krenn, Roman Rakitov, and Florian Karolyi for comments and suggestions to improve the manuscript. I am grateful to Daniela Gruber (Core Facility of Cell Imaging and Ultrastructure Research, University of Vienna, Austria) for her assistance with scanning electron microscopy, Julia Bauder for preparing histological sections, Kerstin Stejskal for help with microphotography of honey bees, Florian Karolyi for help with the line drawings, and Kerry S. Matz and Reinhard Pospischil for the photos of feeding insects.

## References

Altner H, Altner I (1986) Sensilla with both terminal pore and wall pores on the proboscis of the moth, *Rhodogastria bubo* Walker (Lepidoptera: Arctiidae). Zool Anz 216:129–150

Alves dos Santos I, Wittmann D (1999) The proboscis of the long-tongued *Ancyloscelis* bees (Anthophoridae/Apoidea), with remarks on flower-visits and pollen collectiing with the mouthparts. J Kansas Entomol Soc 72:277–288

Amsel HG (1938) *Amphimoea walkeri* Bsd., der Schwärmer mit dem längsten Rüssel! Entomol Rundschau 55:165–167

Anderson B, Johnson SD (2008) The geographical mosaic of coevolution in a plant-pollinator mutualism. Evolution 62:220–225

Arajo RN, Gontijo NF, Guarneri AA, Gontijo AF, Soares AC, Pereira MH (2011) Electromyogram of the cibarial pump and the feeding process in hematophagous Hemiptera. In: J Mizrahi (ed) Advances in applied electromyography. IntechOpen. https://doi.org/10.5772/10647

Aspöck U, Aspöck H (2003) Ordnung Neuroptera (Planipennia), Netzflügler. In: Dathe HH (ed) Lehrbuch der Speziellen Zoologie Begründet von A Kaestner Band 1: Wirbellose Tiere, 5. Teil: Insecta, 2nd edn. Spektrum Akademischer Verlag, Heidelberg, pp 564–584

Bänziger H (1970) The piercing mechanism of the fruit-piercing moth *Calpe* [*Calyptra*] *thalictri* Bkh. (Noctuidae) with reference to the skin-piercing blood-sucking moth *C. eustrigata* Hmps. Acta Trop 27:53–88

Bänziger H (1971) Extension and coiling of the lepidopterous proboscis – a new interpretation of the blood-pressure theory. Mitt schweiz Entomol Ges 43:225–239

Bänziger H (1975) Skin-piercing bloodsucking moths I: Ecological and ethological studies on *Calpe eustrigata* (Lepid., Noctuidae). Acta Trop 32:125–144

Bänziger H (1980) Skin-piercing bloodsucking moths III: Feeding act and piercing mechanism of *Calyptra eustringata* (Hmps.) (Lep., Noctuidae). Mitt Schweiz Entomol Ges 53:127–142

Bänziger H (1989) Skin-piercing blood-sucking moths. V. Attacks on man by 5 *Calyptra* spp. (Lepidoptera, Noctuidae) in S and SE Asia. Mitt Schweiz Entomol Ges 62:215–233

Bänziger H (2007) Skin-piercing blood-sucking moths VI: fruit-piercing habits in *Calyptra* (Noctuidae) and notes on the feeding strategies of zoophilous and frugivorous adult Lepidoptera. Mitt Schweiz Entomol Ges 80:271–288

Barraclough DA (2017) Nemestrinidae (tangle-veined flies). In: Kirk-Spriggs AH, Sinclair BJ (eds) Manual of Afrotropical Diptera, vol 2. Nematocerous Diptera and lower Brachycera. Suricata 5. South African National Biodiversity Institute, Pretoria, pp 995–1003

Barraclough DA, Slotow R (2010) The South African keystone pollinator *Moegistorhynchus longirostris* (Wiedemann, 1819) (Diptera: Nemestrinidae): notes on biology, biogeography and proboscis length variation. Afr Invertebr 51:397–403

Bauder J, Lieskonig N, Krenn HW (2011) The extremely long-tongued neotropical butterfly *Eurybia lycisca* (Riodinidae): proboscis morphology and flower handling. Arthropod Struct Dev 40:122–127

Bauder JA-S, Handschuh S, Metscher BD, Krenn HW (2013) Functional morphology of the feeding apparatus and evolution of proboscis length in metalmark butterflies (Lepidoptera: Riodinidae). Biol J Linn Soc 110:291–304

Bauder JA-S, Warren AD, Krenn HW (2014) Evolution of extreme proboscis lengths in Neotropical Hesperiidae (Lepidoptera). J Res Lepidopt 47:65–71

Bauder JA-S, Morawetz L, Warren AD, Krenn HW (2015a) Functional constraints on the evolution of long butterfly proboscides: lessons from Neotropical skippers (Lepidoptera: Hesperiidae). J Evol Biol 28:678–687

Bauder JA-S, Warren AD, Krenn HW (2015b) The ecological role of extremely long-proboscid Neotropical butterflies (Lepidoptera: Hesperiidae) in plant-pollinator networks. Arthrop Plant Interact 9:415–424

Bennet-Clark HC (1963) Negative pressure produced in the pharyngeal pump of the blood-sucking bug, *Rhodnius prolixus*. J Exp Biol 40:223–229

Blaney WM, Simmonds SJ (1988) Food selection in adults and larvae of three species of Lepidoptera: a behavioural and electrophysiological study. Entomol Exp Appl 49:111–121

Bletchly JD (1954) The mouth-parts of the dance fly, *Empis livida* (Diptera, Empididae). Proc Zool Soc London 124:317–334

Bletchly JD (1955) The mouth-parts of the down-looker fly, *Rhagio* (= *Leptis*) *scolopacea* (L.) (Diptera, Rhagionidae). Proc Zool Soc London 125:779–794

Bonhag PF (1951) The skeleto-muscular mechanism of the head and abdomen of the adult horsefly (Diptera: Tabanidae). Trans Am Entomol Soc 77:131–202

Bonnet DD, Hu SMK (1951) The introduction of *Toxorhynchites brevipalpis* Theobald into the territory of Hawaii. Proc Hawaiian Entomol Soc 14:237–242

Borrell BJ (2003) Suction feeding in orchid bees (Apidae: Euglossini). Proc R Soc Lond 271:164–166

Borrell BJ, Krenn HW (2006) Nectar feeding in long-proboscis insects. In Herrel A, Speck T, Rowe NP (eds) Ecology and biomechanics: a mechanical approach to the ecology of animals and plants, CRC Press LLC, pp 185–212

Broadhead (1984) Adaptation to fungal grazing in lauxaniid flies. J Nat Hist 18:639–649

Brożek J, Bourgoin T (2013) The phylogenetic information carried by a new set of morphological characters in planthoppers: the internal mouthpart structures and test in the Cixiidae model (Hemiptera: Fulgomorpha). Zoomorphology 132:403–420

Brożek J, Mróz E, Wylężek D, Depa Ł, Węgierek P (2015) The structure of extremely long mouthparts in the aphid genus *Stomaphis* Walker (Hemiptera: Sternorrhyncha: Aphididae). Zoomorphology 134:431–445

Büttiker W, Krenn HW, Putterill J (1996) The proboscis of eye-frequenting and piercing Lepidoptera. Zoomorphology 116:77–83

Cariveau DP, Nayak GK, Bartomeus I, Zientek J, Ascher JS, Gibbs J, Winfree R (2016) The allometry of bee proboscis length and its uses in ecology. PLoS One 11(3):e0151482

Carpenter J (1996) Generic classification of the Australian pollen wasps (Hymenoptera: Vespidae; Masarinae). J Kansas Entomol Soc Suppl 69:384–400

Chaudonneret J (1990) Les pièces buccales des insectes. Thème et variations. Édition hors Série du Bulletin Scientifique de Bourgogne, Imprimerie Berthier & Cie, Dijon

Chrisholm IF, Lewis T (1984) A new look at thrips (Thysanoptera) mouthparts, their action and effects of feeding on plant tissue. Bull Entomol Res 74:663–675

Cicero JM (1994) Composite, haustellate mouthparts in netwinged beetle and firefly larvae (Coleoptera, Cantharoidea, Lycidae, Lampyridae). J Morphol 219:183–192

Coronado-Gonzalez PA, Vijaysegaran S, Robinson AS (2008) Functional morphology of the mouthparts of the adult Mediterranean fruit fly, *Ceratitis capitata*. J Insect Sci 8:73

Dathe HH (ed) (2003) Lehrbuch der Speziellen Zoologie Begründet von A Kaestner Band 1: Wirbellose Tiere, 5. Teil: Insecta, 2nd edn. Berlin, Spektrum Akademischer Verlag, Heidelberg

Davis NT, Hildebrand JG (2006) Neuroanatomy of the sucking pump of the moth, *Manduca sexta* (Sphingidae, Lepidoptera). Arthropod Struct Dev 35:15–33

De Beer CJ, Venter GJ, Mans B (2018) Comparing methods for the assessment of blood meal volume of *Culicoides imicola* Kieffer (Ceratopogonidae) fed on an in vitro feeding system. In: Kirk-Spriggs AH, Muller BS (eds) Abstract volume, 9th International Congress of Dipterology, 25–30 Nov 2018, Windhoek, p 62

DeVries PJ (1987) The butterflies of Costa Rica and their natural history, vol 1: Papilionidae, Pieridae, Nymphalidae. Princton University Press

Dierl W (1968) Zur Nahrungsaufnahme von *Corizoneura longirostris* (Hardwicke) (Diptera: Tabanidae). Khumbu Himal 3:76–81

Dikow T (2009) Phylogeny of Asilidae inferred from morphological characters of imagines (Inescta: Diptera: Brachycera: Asiloidea). Bull Am Mus Nat Hist 319:1–175

Düster JV, Gruber MH, Karolyi F, Plant JD, Krenn HW (2018) Drinking with a very long proboscis: functional morphology of orchid bee mouthparts (Euglossini, Apidae, Hymenoptera). Arthropod Struct Dev 47:25–35

Eastham LES, Eassa YEE (1955) The feeding mechanism of the butterfly *Pieris brassicae* L. Philos Trans R Soc Lond B Biol Sci 239:1–43

Eberhard SH, Krenn HW (2005) Anatomy of the oral valve in nymphalid butterflies and a functional model for fluid uptake in Lepidoptera. Zool Anz 243:305–312

Ebmer AW (1993) Die westpaläarktischen Arten der Gattung *Dufourea* Lepeletier 1841 mit illustrierten Bestimmungstabellen, dritter Nachtrag. Linzer Biol Beitr 25:15–42

Elzinga RJ, Broce AB (1986) Labellar modifications of muscomorpha flies (Diptera). Ann Entomol Soc Am 79:150–290

Faucheux MJ (1999) Biodiversité et unité des organes sensoriels des Insectes Lépidoptères. Bull Soc Sci Nat de l'ouest France, Suppl Hors Série, pp 1–296

Friend WG, Smith JJB (1971) Feeding in *Rhodnius prolixus*: mouthpart activity and salivation, and their correlation with changes of electrical resistance. J Insect Physiol 17:233–243

Galić M (1971) Die Sinnesorgane an der Glossa, dem Epipharynx und dem Hypopharynx der Arbeiterin von *Apis mellifica* L. (Insecta, Hymenoptera). Z Morph Tiere 70:201–228

Garzo E, Bonani JP, Lopes JRS, Fereres A (2012) Morphological description of the mouthparts of the Asian citrus psyllid, *Diaphorina citri* Kuwayama (Hemiptera: Psyllidae). Arthropod Struct Dev 41:78–86

Giangaspero A, Broce AB (1993) Micromorphology of the prestomal teeth and feeding behaviour of *Musca autumnalis*, *M. larvipara* and *M. osiris* (Diptera: Muscidae). Med Vet Entomol 7:398–400

Gilbert FS (1981) Foraging ecology of hoverflies: morphology of the mouthparts in relation to feeding on nectar and pollen. Ecol Entomol 6:245–262

Gilbert FS (1985) Ecomorphological relationships in hoverflies (Diptera, Syrphidae). Proc R Soc Lond B 224:91–105

Gilbert FS, Jervis M (1998) Functional, evolutionary and ecological aspects of feeding-related mouthparts specializations in parasitoid flies. Biol J Linn Soc 63:495–535

Giraldo-Calderón GI, Emrich SJ, MacCallum RM, Maslen G, Dialynas E, Topalis P, Ho N, Gesing S, VectorBase Consortium, Madey G, Collins FH, Lawson D (2015) VectorBase: an updated bioinformatics resource for invertebrate vectors and other organisms related with human diseases. Nucleic Acids Res 43:D707–D713

Gotwald WH (1969) Comparative morphological studies of the ants with particular reference to the mouthparts. Cornell Univ Agric Exp Station Memoir 408:1–150

Gouin F (1950) Évolution de la région clypéo-cibariale chez les Diptères Brachycères. Trans VIII Intern Congr Entomol, Stockholm, pp 554–556

Graham-Smith GS (1930) Further observations on the anatomy and function of the proboscis of the blow-fly, *Calliphora erythrocephala* L. Parasitology 22:47–115

Grimaldi D, Engel MS (2005) Evolution of the insects. Cambridge University Press, Cambridge

Haeseler V (1975) *Pterocheilus phaleratus* (Hymenoptera: Vespoidae), ein Nektardieb an den Blüten von *Lotus corniculatus* (Fabales: Fabaceae). Entomol German 1:213–221

Handschin E (1929) Ein neuer Rüsseltypus bei einem Käfer. Biologische und morphologische Beobachtungen an *Leptopalpus rostratus* F. Z Morphol Ökol Tiere 14:513–521
Harder LD (1982) Measurement and estimation of functional proboscis length in bumblebees (Hymenoptera: Apidae). Can J Zool 60:1073–1079
Harder LD (1983) Functional differences of the proboscides of short- and long-tongued bees (Hymenoptera, Apoidea). Can J Zool 61:1580–1586
Heming BS (1978) Structure and function of the mouthparts in larvae of *Haplothrips verbasci* (Osborn) (Thysanoptera, Tubulifera, Phlaeothripidae). J Morphol 156:1–38
Heming BS (1993) Structure, function, ontogeny, and evolution of feeding in *Thrios* (Thysanoptera). In: Schaefer CW, Leschen CW (eds) Functional morphology of insect feeding. Thomas Say Publications in Entomology. Entomological Society of America, Lanham, MD, pp 3–41
Hepburn HR (1971) Proboscis extension and recoil in Lepidoptera. J Insect Physiol 17:637–656
Hilgartner R, Raoilison M, Büttiker W, Lees DC, Krenn HW (2007) Malagasy birds as hosts for eye-frequenting moths. Biol Lett 3:117–120
Holzenthal RW, Thomsen RE, Rios-Touma B (2015) Order Trichoptera. In: Thorp JH, Rogers DC (eds) Ecology and general biology, Thorpand Covich's fresh water invertebrates. Elsevier, Amsterdam, pp 965–1003
Hurd PD, Linsley EG (1963) Pollination of the unicorn plant (Martyniaceae) by an oligolectic, corolla-cutting bee (Hymenoptera: Apidae). J Kan Entomol Soc 36:248–252
Inoue T, Asaoka K, Seta K, Imeda D, Ozaki M (2009) Sugar receptor response of the food-canal taste sensilla in a nectar-feeding swallowtail butterfly, *Papilio xuthus*. Naturwiss 96:355–363
Izumi H, Suzuki M, Aoyagi S, Kanzaki T (2011) Realistic imitation of mosquito' proboscis: electrochemically etched sharp and jagged needles and their cooperative inserting motion. Sensors Actuators A 165:115–123
Jervis M (1998) Functional and evolutionary aspects of mouthpart structures in parasitoid wasps. Biol J Linn Soc 63:461–493
Jervis M, Vilhelmsen L (2000) Mouthpart evolution in adults of the basal, "symphytan", hymenopteran lineages. Biol J Linn Soc 70:121–146
Johnson SD (2010) The pollination niche and its role in the diversification and maintenance of the southern African flora. Philos Trans R Soc Lond B Biol Sci 365(1539):499–516
Johnson SA, Nicolson SW (2001) Pollen digestion by flower-feeding Scarabaeidae: protea beetles (Cetoniini) and monkey beetles (Hopliini). J Insect Physiol 47:725–733
Johnson SD, Steiner KE (1997) Long-tongued fly pollination and evolution of floral spur length in the *Disa draconis* complex (Orchidaceae). Evolution 51:45–53
Jørgensen K, Kvello P, Almaas TJ, Mustaparta H (2006) Two closely located areas in the suboesophageal ganglion and the tritocerebrum receive projections of gustatory receptor neurons located on the antennae and the proboscis in the moth *Heliothis virescens*. J Comp Neurol 496:121–134
Josens RB, Farina WM, Roces F (1998) Nectar feeding by the ant *Camponotus mus*: intake rate and crop filling as a function of sucrose concentration. J Insect Physiol 44:579–585
Karolyi F, Szucsich NU, Colville JF, Krenn HW (2012) Adaptations for nectar-feeding in the mouthparts of long-proboscid flies (Nemestrinidae: *Prosoeca*). Biol J Linn Soc 107:414–424
Karolyi F, Morawetz L, Colville JF, Handschuh F, Metscher BD, Krenn HW (2013) Time management and nectar flow: flower handling and suction feeding in long-proboscid flies (Nemestrinidae: *Prosoeca*). Naturwissenschaften 100:1083–1093
Karolyi F, Colville JF, Handschuh S, Metscher BD, Krenn HW (2014) One proboscis, two tasks: adaptations to blood-feeding and nectar-extracting in long-proboscid horse flies (Tabanidae, *Philoliche*). Arthropod Struct Dev 43:403–413
Karolyi F, Hansal T, Krenn HW, Colville JF (2016) Comparative morphology of the mouthparts of the megadiverse South African monkey beetles (Scarabaeidae: Hopliini): feeding adaptations and guild structure. PeerJ 4:e1597
Kehl S (2014) Morphology, anatomy, and physiological aspects of dytiscids. In: Yee DA (ed) Ecology, systematics, and the natural history of predaceous diving beetles (Coleoptera: Dytiscidae). Springer Science+Business Media BV

Khramov AV, Lukashevich ED (2019) A Jurassic dipteran pollinator with an extremely long proboscis. Gondwana Res 71:210–215

Kikuchi K, Stremler MA, Chatterjee S, Lee W-K, Mochizuki O, Socha JJ (2018) Burst mode pumping: a new mechanism of drinking in mosquitos. Sci Rep 8:4885

Kim W, Gilet T, Bush JWM (2011a) Optimal concentrations in nectar feeding. PNAS 108(40):16618–16621

Kim BH, Kim HK, Lee SJ (2011b) Experimental analysis of the blood-sucking mechanism of female mosquitos. J Exp Biol 214:1163–1169

Kingsolver JG, Daniel TL (1995) Mechanics of food handling by fluid-feeding insects. In: Chapman RF, DeBoer G (eds) Regulatory mechanisms in insect feeding. Chapman & Hall, New York, pp 32–73

Kluge NJ (2002) The homology of mouthparts in fleas (Insecta, Aphaniptera). Entomol Rev 82:808–816

Knopp MCN, Krenn HW (2003) Efficiency of fruit juice feeding in *Morpho peleides* (Nymphalidae, Lepidoptera). J Insect Behav 16:67–77

Kong XQ, Wu CW (2009) Measurement and prediction of insertion force for the mosquito fascicle penetrating into human skin. J Bionic Eng 6:143–152

Kong XQ, Wu CW (2010) Mosquito proboscis: an elegant biomicroelectromechanical system. Phys Rev E82:011910

Kornev KG, Monaenkova D, Adler PH, Beard CE, Lee WK (2016) The butterfly proboscis as a fiber-based, self-cleaning, micro-fluidic system. Proc Int Soc Optic Photon 9797:979705

Kornev KG, Salamatin AA, Adler PH, Beard CE (2017) Structural and physical determinants of the proboscis-sucking pump complex in the evolution of fluid-feeding insects. Sci Rep 7

Krenn HW (1990) Functional morphology and movements of the proboscis of Lepidoptera (Insecta). Zoomorphology 110:105–114

Krenn HW (1998) Proboscis sensilla in *Vanessa cardui* (Nymphalidae, Lepidoptera) – functional morphology and significance in flower-probing. Zoomorphology 118:23–30

Krenn HW (2000) Proboscis musculature in the butterfly *Vanessa cardui* (Nymphalidae, Lepidoptera): settling the proboscis recoiling controversy. Acta Zool 81:259–266

Krenn HW (2010) Feeding mechanisms of adult Lepidoptera: structure, function, and evolution of the mouthparts. Annu Rev Entomol 55:307–327

Krenn HW, Aspöck H (2012) Form, function and evolution of the mouthparts of blood-sucking arthropods. Arthropod Struct Dev 41:101–118

Krenn HW, Bauder JA-S (2017) Morphological fine tuning of the feeding apparatus to proboscis length in Hesperiidae (Lepidoptera). J Morphol 279:396–408

Krenn HW, Kristensen NP (2000) Early evolution of the proboscis of Lepidoptera (Insecta): external morphology of the galea in basal glossatan moths lineages, with remarks on the origin of the pilifers. Zool Anz 239:179–196

Krenn HW, Kristensen NP (2004) Evolution of proboscis musculature in Lepidoptera. Eur J Entomol 101:565–575

Krenn HW, Penz CM (1998) Mouthparts of *Heliconius* butterflies (Lepidoptera: Nymphalidae): a search for anatomical adaptations to pollen-feeding behaviour. Int J Insect Morphol Embryol 27:301–309

Krenn HW, Zulka KP, Gatschnegg T (2001) Proboscis morphology and food preferences in Nymphalidae (Lepidoptera, Papilionoidea). J Zool Lond 253:17–26

Krenn HW, Mauss V, Plant J (2002) Evolution of the suctorial proboscis in pollen wasps (Masarinae, Vespidae). Arthropod Struct Dev 31:103–120

Krenn HW, Plant J, Szucsich NU (2005) Mouthparts of flower-visiting insects. Arthropod Struct Dev 34:1–40

Krenn HW, Gereben-Krenn B-A, Steinwender BM, Popov A (2008) Flower visiting Neuroptera: Mouthparts and feeding behavious of *Nemoptera sinuata* (Nemopteridae). Eur J Entomol 105:267–277

Krenn HW, Fournel J, Bauder JA-S, Hugel S (2016) Mouthparts and nectar feeding of the flower visiting cricket *Glomeremus orchidophilus* (Gryllacrididae). Arthropod Struct Dev 45:221–229

Krenn HW, Karolyi F, Melin A, Colville J (2018) Flowers fuel flies – supply of nectar for long-proboscid flies (*Prosoeca* sp., Nemestrinidae). In: Kirk-Spriggs AH, Muller BS (eds) Abstract volume, 9th International Congress of Dipterology, 25–30 Nov 2018, Windhoek, p 141

Kristensen NP, Nielsen ES (1981a) Intrinsic proboscis musculature in non-ditrysian Lepidoptera-Glossata: structure and phylogentic significance. Entomol Scand Suppl 15:299–304

Kristensen NP, Nielsen ES (1981b) Double-tube proboscis configuration in neopseustid moths (Lepidoptera: Neopseustidae). Intern J Insect Morphol Embryol 10:483–486

Kumar MS, Sahayaraj K (2012) Gross morphology and histology of head and salivary apparatus of the predatory bug, *Rhynocoris marginatus*. J Insect Sci 12:19

Kvello P, Almaas TJ, Mustaparta H (2006) A confined taste area in a lepidopteran brain. Arthropod Struct Dev 35:35–45

Kwon HW, Lu T, Rutzler M, Zwiebel LJ (2006) Olfactory responses in a gustatory organ of the malaria vector mosquito *Anopheles gambiae*. Proc Natl Acad Sci U S A 103:13526–13531

Labandeira CC (1997) Insect mouthparts: ascertaining the paleobiology of insect feeding strategies. Annu Rev Ecol Syst 28:153–193

LaBerge WE (1986) A revision of the bees of the genus *Andrena* of the western hemisphere. Part XI. Minor subgenera and subgeneric key. Trans Am Entomol Soc 111:441–567

Laroca S, Michener CD, Hofmeister RM (1989) Long mouthparts among "short-tongued" bees and the fine structure of the labium in *Niltonia* (Hymenoptera, Colletidae). J Kansas Entol Soc 62:400–410

Lee RMKW, Craig DA (1983) Cibarial sensilla and armature in mosquito adults (Diptera: Culicidae). Can J Zool 61:633–646

Lee RMKW, Craig DA (2009) Fine structure of the sense organs on the labella and labium of the mosquito *Aedes aegypti* (L.). Open Entomol J 3:7–17

Lehane M (2005) The biology of blood-sucking in insects, 2nd edn. Cambridge University Press, Cambridge

Lehnert MS, Monaenkova D, Andrukh T, Beard CE, Adler PH, Kornev KK (2013) Hydrophobic-hydrophilic dichotomy of the butterfly proboscis. J R Soc Interface 10:20130336

Lehnert MS, Beard CE, Gerand PD, Kornev KG, Adler PH (2016) Structure of the lepidopteran proboscis in relation to feeding guild. J Morphol 277:166–182

Lehnert MS, Bennett A, Reiter KE, Gerard PD, Wei Q-H, Byler M, Yan H, Lee W-K (2017) Mouthpart conduit sizes of fluid-feeding insects detemine the ability to feed from pores. Proc R Soc B 284:20162026

Leopold RA, Freeman TP, Buckner JS, Nelson DR (2003) Mouthpart morphology and stylet penetration of host plants by the glassy-winged sharpshooter, *Homalodisca coagulate*, (Homoptera: Cicadellidae). Arthrop Struct Dev 32:189–199

Li L, Guo C, Xu S, Li X, Han C (2017) Morphology and nanoindentation properties of mouthparts in *Cyrtotrachelus longimanus* (Coleoptera: Curculionidae). Microsc Res Tech 80:704–711

McKeever S, Wright MD, Hagan DV (1988) Mouthparts of females of four *Culicoides* species (Diptera: Ceratopogonidea). Entomol Soc Am 81(2):332–341

Melin A, Krenn HW, Beale CM, Manning JC, Colville JF (2019) The allometry of proboscis length in Melittidae (Hymenoptera: Apoidae) and an estimate of their foraging distance using museum collections. PLoS One 14(6):e0217839

Michelsen V (1996/1997) A revised interpretation of the mouthparts in adult fleas (Insecta, Siphonaptera). Zool Anz 235:217–223

Michener CD (1965) A classification of the bees of the Australian and South Pacific Regions. Bull Am Mus Nat Hist 130:1–362

Michener CD (1981) Classification of the bee family Melittidae with a review of the species of Meganomiinae. Contr Am Entomol Inst 18:1–135

Michener CD (2000) The bees of the world. John Hopkins University Press, Baltimore

Michener CD (2007) The bees of the world, 2nd ed. John Hopkins University Press, Baltimore

Michener CD, Brooks RW (1984) Comparative study of the glossae of bees. Contrib Am Entomol Institut 22:1–73

Miles CI, Brooker R (1998) The role of the frontal ganglion in the feeding and eclosion behavior of the moth *Manduca sexta*. J Comp Physiol A 174:755–767

Mitter C, Davis DR, Cummings MP (2017) Phylogeny and evolution of Lepidoptera. Annu Rev Entomol 62:265–283
Molleman F, Krenn HW, Van Alphen ME, Brakefield PM, DeVries PJ, Zwaan BJ (2005) Food intake of fruit-feeding butterflies: evidence for adaptive variation in proboscis morphology. Biol J Linn Soc 86:333–343
Monaenkova D, Lehnert MS, Andrukh T, Beard CE, Rubin B, Tokarev A, Lee WP, Adler H, Kornev KG (2011) Butterfly proboscis: combining a drinking straw with a nanosponge facilitated diversification of feeding habits. J Royal Soc Interface 9:720–726
Morita SI (2008) A revision of the *Philoliche aethiopica* species complex (Diptera: Tabanidae). Afr Invertebr 49:129–158
Morita SI (2011) Repeatability and precision in proboscis length measurements for long proboscid flies. Zootaxa 3112:49–58
Mound LA (2005) Thysanoptera: diversity and interactions. Annu Rev Entomol 50:247–269
Nagatomi A, Soroida K (1985) The structure of the mouthparts of the orthorrhaphous Brachycera (Diptera) with special reference to blood-sucking. Beitr Entomol 35:263–368
Nagnan-Le Meillour P, Cain AH, Jacquin-Joly E, Francois MC, Ramachandran S, Maida R, Steinbrecht RA (2000) Chemosensory proteins from the proboscis of *Mamesta brassicae*. Chem Senses 25:541–553
Nel P, Bertrand S, Nel A (2018) Diversification of insects since the Devonian: a new approach based on morphological disparity of mouthparts. Sci Rep 8:3516
Nilsson AL (1988) The evolution of flowers with deep corolla tubes. Nature 334:147–149
Ômura H, Honda K, Asaoka K, Inoue TA (2008) Tolerance to fermentation products in sugar reception: gustatory adaptation of adult butterfly proboscis for feeding on rotting foods. J Comp Physiol A 194:545–555
Osten T (1982) Vergleichend-funktionsmorphologische Untersuchungen der Kopfkapsel und der Mundwerkzeuge ausgewählter "Scolioidea" (Hymenoptera, Aculeata). Stuttg Beitr Naturk A (Biologie) 354:1–60
Osten T (1988) Die Mundwerkzeuge von *Proscolia spectator* Day (Hymenoptera, Aculeata). Ein Beitrag zur Phylogenie der Scolioidea. Stuttg Beitr Naturk A (Biologie) 414:1–30
Osten T (1991) Konvergente Entwicklung der Mundwerkzeuge von *Epomidiopteron* (Tiphiidae) und der Scoliidae (Hymenoptera). Stuttg Beitr Naturk A (Biologie) 466:1–7
Paul J (2001) The mouthparts of ants. Dissertation Bayerische Julius-Maximilian-Universität Würzburg, Würzbug, p 145
Paul J, Roces F, Hölldobler B (2002) How do ants stick out their tongues? J Morphol 254:39–52
Paulus HF, Krenn HW (1996) Vergleichende Morphologie des Schmetterlingsrüssels und seiner Sensillen – Ein Beitrag zur phylogenetischen Systematik der Papilionoidea (Insecta, Lepidoptera). J Zool Syst Evol Res 34:203–216
Pauly A (1984) Contribution a l'etude des genres afrotropicaux de Nomiinae (Hymenoptera Apoidea Halictidae). Rev Zool Afr 98:693–702
Pauw A, Stofberg J, Waterman RJ (2009) Flies and flowers in Darwin's race. Evolution 63:268–279
Petr D, Stewart KW (2004) Comparative morphology of sensilla styloconica on the proboscis of North American Nymphalidae and other selected taxa (Lepidoptera): systematic and ecological considerations. Trans Am Entomol Soc 130:293–409
Plant JD, Paulus HF (2015) Evolution and phylogeny of bees: a review and a cladistic analysis in light of morphological evidence (Hymenoptera, Apoidea). Zoologica 161:1–368
Pollard DG (1970) The mechanism of stylet movement in *Psylla mali* Schmidberger (Homoptera: Psyllidae). Zool J Linnean Soc 49:295–271
Pollard DG (1972) The stylet structure of leafhopper (*Eupteryx melissae* Curtis: Homoptera, Cicadellidae). J Nat Hist 6:261–271
Pollard DG (1973) Plant penetration by feeding aphids (Hemiptera, Aphidoidea): a review. Bull Entomol Res 62:631–714
Proctor M, Yeo P, Lack A (1996) The natural history of pollination. Harper Collins, London
Rakitov R, Karengin S, Kerengina O (2018) How *Stomaphis* aphids use their giant proboscides. J Nat Hist 52:299–304

Redak RA, Purcell AH, Lopes JRS, Blua MJ, Mizell RF III, Andersen PC (2004) The biology of xylem fluid-feeding insect vectors of *Xylella fastidiosa* and their relation to disease epidemiology. Annu Rev Entomol 49:243–270

Reinhardt K, Siva-Jothy MT (2007) Biology of the bed bugs (Cimicidae). Annu Rev Entomol 52:351–374

Ren D (1998) Flower-associated Brachycera flies as fossil evidence for Jurassic angiosperm origins. Science 280:85–87

Ribeiro JMC (1987) Role of saliva in blood-feeding by arthropods. Annu Rev Entomol 32:463–478

Rozen JG Jr (2003) A new tribe, genus, and species of South American panurgine bee (Andrenidae, Panurginae), oligolectic on *Nolana* (Nolanaceae). In: Melo GAR, Alves dos Santos I (eds) Apoidea Neotropica: homenagem aos 90 Anos de Jesus Santiago Moure. Editora UNESC, Cricioma

Rozen JG Jr, Ruz L (1995) South American panurgine bees (Andrenidae: Panurginae), part II adults, immature stages, and biology of *Neffapis longilingua*, a new genus and species with an elongate glossa. Am Mus Novitates 3136:1–15

Salama HS, Khalifa A, Azmy N, Sharaby A (1984) Gustation in the lepidopterous moth *Spodoptera littorialis* (Boid.). Zool Jb Physiol 88:165–178

Saveer AM, Pitts RJ, Ferguson ST, Zwiebel LJ (2018) Characterization of chemosensory responses on the labellum of the malaria vector mosquito, *Anopheles coluzzii*. Sci Rep 8:5656

Schiemenz H (1957) Vergleichende funktionell-anatomische Untersuchungen der Kopfmuskulatur von *Theobaldia* und *Eristalis* (Dipt. Culicidae und Syrphidae). Dtsche Entomol Zeitschr NF 4:268–331

Schmitt JB (1938) The feeding mechanism of adult Lepidoptera. Smiths Misc Collect 97:1–28

Schremmer F (1961) Morphologische Anpassungen von Tieren – insbesondere Insekten – an die Gewinnung von Blumennahrung. Verh Dtsch Zool Ges (Zool Anz Suppl) 25:375–401

Schuhmacher H, Hoffmann H (1982) Zur Funktion der Mundwerkzeuge von Schwebfliegen bei der Nahrungsaufnahme (Diptera: Syrphidae). Entomol Gen 7:327–342

Smith JJB (1985) Feeding mechanisms. In: Kerkut GA, Gilbert LI (eds) Comprehensive insect physiology, biochemistry and pharmacology 4. Pergmon, Oxford, pp 33–85

Smith JJB, Friend WG (1970) Feeding of *Rhodnus prolixus* responses to artificial diets as revealed by changes in electrical resistance. J Insect Physiol 16:1709–1720

Snodgrass RE (1935) Principles of insect morphology. McGraw-Hill Book Company, New York & London (reprint with a new foreword by Goerge C Eickwort, Cornell University Press 1993)

Snodgrass RE (1956) Anatomy of the honey bee, 1st publ. Comstock Publishing Associates, Ithaca, NY

Snodgrass RE (1959) The anatomical life of the mosquito. Smiths Misc Coll 139:1–87

Socha JJ, Westneat MW, Harrison JF, Waters JS, Lee W-K (2007) Real-time phase-contrast x-ray imaging: a new technique for the study of animal from and function. BMC Biol 5:6

Sutcliffe JB, McIver SB (1984) Mechanics of blood-feeding in black flies (Diptera: Simuliidae). J Morphol 180:125–144

Szucsich NU, Krenn HW (2000) Morphology and function of the proboscis in Bombyliidae (Diptera, Brachycera) and implications for proboscis evolution in Brachycera. Zoomorphology 120:79–90

Szucsich NU, Krenn HW (2002) Flies and concealed nectar sources: morphological innovations in the proboscis of Bombyliidae (Diptera). Acta Zool 83:183–192

Tauxe GM, MacWilliam D, Boyle SM, Guda T, Ray A (2013) Targeting a dual detector of skin and CO2 to modify mosquito host seeking. Cell 155:1365–1379

Thomson AJ (1977) The mechanics of proboscis extension by the blowfly *Phormia regina* Meigen. Can J Zool 55:602–606

Tjallingii WF, Esch TH (1993) Fine Structure of aphid stylet routes in plant tissues in correlation with EPG signals. Physiol Entomol 18:317–328

Tröster G (1990) Der Kopf von *Hybophthirus notophallus* (Neumann) (Phthiraptera: Anoplura). Eine funktionsmorphologische und konsequent-phylogenetische Analyse. Stuttg Beitr Naturkunde A (Biologie) 442:1–89

Tsai C-C, Monaenkova D, Beard CE, Adler PH, Kornev KG (2014) Paradox of the drinking-straw model of the butterfly proboscis. J Exp Biol 217:2130–2138
Ulmer G (1905) Über einige Trichopteren mit rüsselförmigen Kopfanhängen. Zool Anz 28:56–59
Uzest M, Gargani D, Dombrovsky A, Carevieille C, Cot D, Blanc S (2010) The "acrostyle": a new described anatomical structure in aphid stylets. Arthrop Struct Develop 39:221–229
Van der Starre H, Ruigrok T (1980) Proboscis extension and retraction in the blowfly, *Calliphora vicina*. Physiol Entomol 5:87–92
Vijaysegara S, Walter GH, Drew RAI (1997) Mouthpart structures, feeding mechanisms, and natural food sources of adult *Bactrocera* (Diptera: Tephritidae). Ann Entomol Soc Am 90:184–201
Vilhelmsen L (1996) The preoral cavity of lower Hymenoptera (Insecta): comparative morphology and phylogenetic significance. Zool Scripta 25:143–170
Wahid I, Sunahara T, Mogi M (2003) Maxillae and mandibles of male mosquitoes and female autogenous mosquitos (Diptera: Culicidae). Entomol Soc Am 40:150–158
Wannenmacher G, Wasserthal LT (2003) Contribution of the maxillary muscles to proboscis movement in hawkmoths (Lepidoptera: Sphingidae) – an electrophysiological study. J Insect Physiol 49:765–776
Weber H (1930) Biologie der Hemiptera. Eine Naturgeschichte der Schnabelkerfe. I-VII. Biologische Studienbücher XI. Juluis Springer, Berlin
Weber H (1933) Lehrbuch der Entomologie. Gustav Fischer Verlag, Stuttgart (reprint Verlag von Otto Koeltz, Koenigstein – Taunus 1968)
Weber H (1969) Die Elefantenlaus *Haematomyzus elefantis* Piaget 1869. Versuch einer konstruktionsmorphologischen Analyse. Zoologica 116:1–154
Weintraub PG, Beanland L (2006) Insect vectors of phytoplasmas. Annu Rev Entomol 51:91–111
Wenk P (1953) Der Kopf von *Ctenocephalus canis* (Curt.)(Aphaniptera). Zool Jb Anat 73:103–164
Wenk P (1962) Anatomie des Kopfes von *Wilhelmia equina* L. (Simuliidae syn. Melusinidae, Diptera). Zool Jb Anat 80:81–134
Wenk P (1980) How bloodsucking insects perforate the skin of their hosts. In: Traub R, Starcke H (eds) Fleas. AA Balkema, Rotterdam, pp 329–335
Wenk P, Renz A (2003) Parasitologie: Biologie der Humanparasiten. Georg Thieme Verlag, Stuttgart, New York
Wenk P, Lucic S, Betz O (2010) Functional anatomy of the hypopharynx and the salivary pump in the feeding apparatus of the assassin bug *Rhodnius prolixus* (Reduviidae, Heteroptera). Zoomorphology 129:225–234
Whittall JB, Hodges SA (2007) Pollinator shifts drive increasingly long nectar spurs in columbine flowers. Nature 447:706–710
Wiegmann BM, Trautwein MD, Winkler IS, Barr NB, Kim J-W, Lambkin C, Bertone MA, Cassel BK, Bayless KM, Heimberg AM, Wheeler BM, Peterson KJ, Pape T, Sinclair BJ, Skevington JH, Blagoderov V, Caravas J, Kutty SN, Schmidt-Ott U, Kampmeier GE, Thompson FC, Grimaldi DA, Beckenbach AT, Courtney GW, Friedrich M, Meier R, Yeates DK (2011) Episodic radiations in the fly tree of life. PNAS 108:5690–5695
Wilhemi A, Krenn HW (2012) Elongated mouthparts of nectar-feeding Meloidae (Coleoptera). Zoomorphology 131:325–337
Wirtz H-P (1987) Eindringen der Mundwerkzeuge von Raubwanzen durch eine Membran (Hemiptera: Reduviidae). Entomol Gen 12:147–153
Wirtz H-P (1989) Struktur, Funktion und Eindringweise der Mundwerkzeuge von *Glossina palpalis palpalis* (Diptera; Glossinidae). Entomol Mitt Zool Mus Hamburg 9:295–304
Wu J, Zhu R, Yan S, Yang Y (2015) Erection pattern and section-wise wettability of a honeybee's glossal hairs in nectar feeding. J Exp Biol 218:664–667
Xue S, Hua B-Z (2014) Proboscis sensilla of the black cutworm *Agrotis ypsilon* (Rottemberg) (Lepidoptera: Noctuidae). J Asia-Pacific Entomol 17:295–301
Yang H, Wu J, Yan S (2014) Effects of erectable glossal hairs on a honeybee's nectar-drinking strategy. Appl Phys Lett 104:263701.4
Zaitzev VF (1983) Anthophilie und Rüssellabellenstruktur der Dipteren. Verh 10th International Symposium Entomofaunistik Mitteleuropa. SIEEC X, Budapest, pp 169–171

Zaspel JM, Kononenko VS, Goldstein PZ (2007) Another blood feeder? Experimental feeding by a fruit-piercing moth on human blood in the Primorye region of Far Eastern Russia. J Insect Beh 20:437–451

Zaspel JM, Weller SJ, Branham MA (2011) A comparative survey of proboscis morphology and associated structures in fruit-piercing, tear-feeding, and blood-feeding moths in Calpinae (Lepidoptera: Erebidae). Zoomorphology 130:203–225

Zaspel JM, Coy S, Habanek K, Weller SJ (2013) Presence and distribution of sensory structures on the mouthparts of self-medicating moths. Zool Anz 253:6–10

Zenker MM, Penz C, de Paris M, Specht A (2011) Proboscis morphology and its relationship to feeding habits in noctuid moth. J Insect Sci 11/42:1–10

Zhang C, Beard CE, Adler PH, Kornev KG (2018) Effect of curvature on wetting and dewetting of proboscises of butterflies and moths. R Soc Open Sci 5:171241

Zhao L, Dai W, Zhang C, Zhang Y (2010) Morphological characterization of the mouthparts of the vector leafhopper *Psammotettix striatus* (L.) (Hemiptera: Cicadellidae). Micron 41:754–759

Ziegler J (2003) Ordnung Diptera, Zweiflügler (Fliegen und Mücken). In Dathe HH (ed) Lehrbuch der Speziellen Zoologie, begründet von A Kaestner, 2. Auflage Bd 1: Wirbellose Tiere, 5. Teil, Spektrum, Gustav Fischer, Heidelberg, pp 756–860

# Chapter 4
# Suspension and Filter Feeding in Aquatic Insects

**Donald A. Yee and Michael G. Kaufman**

**Abstract** Aquatic insect feeding occurs at the nexus of habitat, food source and size, and behavior and relies largely on the complexities of mouthpart morphology. This intersection has important consequences for tropic interactions, nutrient processing, and ecosystem function. In aquatic habitats, immature insects feed in a variety of ways; however, consumption of small suspended particles (seston) in the water column is a common mode for representatives of several insect groups. Ingestion of seston can occur via active or passive removal and broadly encompasses filter and suspension feeding. In this chapter, we explore the ways in which various aquatic insects acquire food particles. We focus on food sources and particle sizes, feeding behavior, morphology of mouthparts, and trophic importance. The major groups explored include Ephemeroptera (mayflies), Diptera (true flies), and Trichoptera (caddisflies), each of which have evolved unique strategies for obtaining particles from the water column. Members of this feeding group are critical as food sources for aquatic and terrestrial organisms, they play large roles in nutrient cycling, and some are vectors of important human and animal diseases.

## 4.1 Introduction

Insects feed on almost all available food sources, including live and dead plant material, animals, fungi, bacteria, and protozoans. However, within the insects there are some taxa that specialize on feeding on suspended particles of food within the water column. Also known as collector filters or suspension feeders (hereafter, filter feeders), these insects comprise an important group of animals in both flowing (lotic) and stagnant (lentic) waters. In moving waters, filter feeding is often passive,

D. A. Yee (✉)
School of Biological, Environmental, and Earth Sciences, University of Southern Mississippi, Hattiesburg, MS, USA
e-mail: donald.yee@usm.edu

M. G. Kaufman
Department of Entomology, Michigan State University, East Lansing, MI, USA
e-mail: kaufma15@msu.edu



H. W. Krenn (ed.), *Insect Mouthparts*, Zoological Monographs 5,
https://doi.org/10.1007/978-3-030-29654-4_4

inasmuch as insects rely on the current to deliver particles to them. In standing waters, an active filtering behavior is often observed, either using moving mouthparts to filter particles out of the water column or via the use of burrows that take advantage of currents produced in proximity to their tubes. Feeding styles are not mutually exclusive, as some inhabitants of moving water also use burrows to facilitate particle capture (e.g., some mayflies). Regardless of the mode, these animals contribute to energy processing and conversion of dead biomass into living tissue and are also important for the transport of resources across habitats (i.e., resource subsidies). Our objectives in this review are to examine the mechanisms of suspension feeding across aquatic insect taxa, the source and composition of their food sources, their importance within aquatic food webs, and research directions and challenges for the future. Earlier reviews of some of these topics can be found in Cummins and Klug (1979), Wallace and Merritt (1980), and Merritt et al. (2008).

## 4.2 Mechanisms for Gathering and Ingesting

Insects in the filter-feeding guild have evolved a myriad of modifications for particle capture, some of which do not necessarily involve specialized mouthparts (Wallace and Merritt 1980; Merritt and Wallace 1981; Huryn et al. 2008; Hershey et al. 2010). Contrasting mechanistic strategies that serve the same basic purpose can be seen in the diversity of stream dwelling species across several taxonomic groups that are generally considered to be passive filter feeders or filter collectors (Merritt et al. 2008). These insects take advantage of consistent water flow while primarily in a sedentary mode. Other members of the guild inhabit low flow or stagnant water and are more active filter feeders in that they employ a variety of strategies to move fluid and particles past trapping structures and mouthparts for collection and ingestion.

### *4.2.1 Passive Filter Feeders*

Perhaps the best examples of passive filter feeders with modified mouthparts are black fly larvae (Diptera: Simuliidae). They are widespread in lotic systems, both tropical and temperate, and have evolved elaborate cephalic (labral) fans that are deployed in the current to collect particles; they are then retracted toward the oral cavity where trapped particles are removed by other mouthparts and ingested. Cephalic fans represent a truly specialized feeding adaptation by this group, and the feeding action and hydrodynamic considerations have been extensively reviewed elsewhere (Currie and Craig 1987; Craig and Galloway 1987; Adler and Currie 2008). Particle entrapment efficiency is related to current velocity and density of setae on the fans. Species with large fans and densely packed setae are associated with lower velocity, whereas species with smaller fans and larger spaces between setae are found in faster flowing stream sections (Palmer and Craig 2000). However, there is much variability within this generalization as fan size and setal density change with instar and larval environment (Lucas and Hunter 1999). Spacing between setae and number of setae on labral fan rays is phenotypically plastic: it is affected by current velocity and food concentrations in which the larvae develop (Lucas and Hunter 1999).

Within the same lotic habitats, and often found adjacent to black fly larvae on a substrate, are species of Trichoptera (e.g., Hydropsychidae, Philopotamidae, Polycentropidae, Dipseudopsidae) and Diptera (Chironomidae: *Rheotanytarsus*) that construct nets out of silk to collect particles from the current. These organisms anchor themselves to a substrate and build nets of various mesh sizes and shapes (Wallace and Merritt 1980; Merritt and Wallace 1981). The nets are then periodically cleaned or consumed by the larvae using largely biting and chewing mouthparts (e.g., mandibles and maxillae). Larvae of net builders are otherwise sedentary often living in retreats built with silk and available materials. Similar to black fly larvae, net mesh size varies with species, instar, and current velocity (Plague and McArthur 2003; Wiggins 2005). Philopotamidae larvae construct very fine mesh tubular nets (mesh size in the $<1$ μm range) and use an extended membranous labrum to harvest trapped material (Wiggins 1996, 2005), often small detrital particles (Shapas and Hilsenhoff 1976). In contrast, some late instars of Hydropsychidae species generally build nets with mesh sizes in the 500 μm range and trap more intact invertebrates and large particles than fine particulate matter (Wallace and Merritt 1980; Wotton 1994). *Rheotanytarsus* (Chironomidae) larvae utilize both silken nets and sticky secretions from salivary glands on filaments at the opening of their tube dwellings to collect particles in flowing water (Merritt and Wallace 1981).

Additional means of collecting suspended particulates in flowing water is accomplished through setaceous limbs in some groups (e.g., prothoracic legs in Isonychia and meso- and meta-thoracic legs in *Brachycentrus*). In these cases, the insects face the current with legs extended laterally to collect passing particles, which are then harvested via setaceous mouthparts directly (*Isonychia*) or manipulated into a bolus by the forelegs and transferred to the mouth (*Brachycentrus*) (Merritt and Wallace 1981, but see Hershey et al. 2010).

### 4.2.2 Active Filter Feeders

Active filter feeders in non-flowing water utilize brush-like modifications of mouthparts to generate fluid movement on their own (e.g., mosquito larvae) and simultaneously collect and ingest particles brought toward the oral cavity with other mouthparts, or create fluid movements with body undulations and gill movements in constructed tubes or burrows (e.g., Chironomidae, Ephemeridae, Dipseudopsidae) to move water through silken nets and setaceous appendages and mouthparts. Larval mosquito feeding has been extensively studied and currents generated by mouthparts have been detailed (Clements 1999 and references therein). Currents generated by the feeding of this group are generally vertical and lateral, moving particles up and into to the oral cavity and expelling fluid downward and laterally.

Other active filterers include tube- or burrow-dwelling Chironomidae (*Chironomus*) and Ephemeroptera (Ephemeridae, Polymitarcyidae) found in sediments in lentic habitats or in depositional zones of lotic habitats. *Chironomous* species utilize silk for tube lining and to construct nets that capture particulates brought into the tube or burrow by body undulations. Ephemeridae (e.g., *Ephemera*) construct U-shaped burrows in sediment and also use body undulations and gill movements to bring particulates in and collect them on foreleg setae and mouthparts (Wallace and Merritt 1980). Similarly, Dipseudopsidae larvae construct silk lined tubes in sediment and use body undulations to move water through, collecting trapped material from the inner surfaces with setaceous mandibles (Wiggins 1996, 2005).

It is important to note that although filter feeding per se may be the primary mode of food acquisition in the groups discussed here, most species are flexible in modes of obtaining food. Both mosquito and black fly larvae, for example, regularly switch to browsing on surfaces to harvest available biofilms. Thus, the same mouthparts used in filter feeding are used to brush or scrape microorganisms and detritus from surfaces. In the case of mosquito larvae, this feeding mode can predominate in some species or under certain conditions (low suspended organic matter or presence of predators) (Merritt et al. 1992; Yee et al. 2004; Yee and Kehl 2014; Roberts 2014).

## 4.3 Relationship Among Filter-Feeding Taxa

Recent phylogenetic analyses suggest a long and complicated evolutionary history for insects, dating back approximately 479 million years (Early Ordovician) (Misof et al. 2014). Diversification has continued unabated, producing not only differences in morphology and feeding modes, but also in types of metamorphosis, emergence of flight, behavioral differences, and ecological diversity. Aquatic insects appear within 12 different insect orders, and invasions by terrestrial forms into freshwater have occurred at least 50 times (Klaas-Douwe et al. 2014). Given this, it is also likely that filter feeding evolved independently several times. Specifically, Ephemeroptera (mayflies), Trichoptera (caddisflies), and Diptera (true flies) contain a high number of filter-feeding taxa. However, these three groups are not closely related, with the true flies and caddisflies being the most closely related among the three, although these two groups are likely still separated by tens of millions of years (Misof et al. 2014). Within Diptera, there do seem to be strong relationships in the feeding apparatus of Culicidae, Chaoboridae, and Dixidae (Wagner et al. 2008), and more distantly with Simuliidae (Craig 1974); all these Diptera do appear in the same Infraorder (Culicomorpha, Wagner et al. 2008). We might speculate that given the diversity of filter-feeding modes outlined elsewhere in this review, the evolution of filter feeding likely was due to the availability of various niches within different aquatic systems at different times, and not due to a single instance of the evolution of mouthparts or behavior. However, at present there is no single review on the evolution of filter feeding across aquatic insect groups.

## 4.4 Suspension Feeding Across Insect Taxa

Of the approximately 1 million species of insects identified, three orders dominate those that use filter feeding to obtain food: Ephemeroptera (mayflies), Diptera (true flies), and Trichoptera (caddisflies). There are other orders that also exhibit filtering, including Coleoptera; however given that perhaps only one-fifth of all insect species have been described (Stork 2018), it is likely that more species that exhibit filter feeding are yet to be found. Also of note is that the vast majority of individuals that exhibit filter feeding are larvae, which likely is explained by the fact that adult forms of these groups are terrestrial and have different modes of obtaining food, or don't feed at all (e.g., Ephemeroptera). The key distinguishing features among species in this group are the morphology of specialized mouthparts or hairs, and in some instances, the use of silk for capture and sieving of particles. Moreover, filter-feeding insects can be divided into those that filter using their body parts (e.g., mouthparts or legs) or constructed habitats (e.g., burrows, tubes, nets) to collect particles.

### *4.4.1 Coleoptera*

Beetles are the most diverse group of animals on Earth, representing about 40% of all insect species (Stork 2018); however when examining their feeding behavior the vast majority rely on modes other than filter feeding. Nevertheless, at least two aquatic families of beetles are known to use filtering to obtain food particles: Scirtidae and Spercheidae. Although the mechanism and ecology of filter feeding is not well studied in either group, the fact that it exists in some beetles may hint at the use of this feeding mechanism in other less studied families.

Scirtidae (marsh beetles) are a widely distributed family of beetles with short-lived adults occurring in the terrestrial environment (reviewed in Yee and Kehl 2014). Larvae are aquatic, and often can be found in lotic or lentic waters, including marshes, swamps, and ground pools; they also reside in phytotelmata, including tree holes (Kitching 2000). In all habitats, larvae are shredders or detritivores, feeding on fungi, algae, and other organic matter. However, small particles are captured using a complex filtering structure, which is present on the hypopharynx, and various types of microorganisms are filtered by a dense maxillary or mandibular set of bristles (Fig. 4.1), or collected from the detrital or container surface (Lawrence 2016). These particles are subsequently sorted on a complex and greatly modified hypopharynx. Other aspects of feeding in this family can be found in Hannappel and Paulus (1987).

Spercheidae (filter-feeding water scavenger beetles) are represented by a single genus (*Spercheus*) containing about 20 species (Yee and Kehl 2014). Found in shallow lentic waters, these are unique among beetles as both larvae and adults use filter feeding to obtain food. Both life history stages stay on the underside of leaves, where they may sit and filter. In adults, the clypeal bristles rests above the surface of the water, whereas the complex clypeus itself remains below (Rothmeier

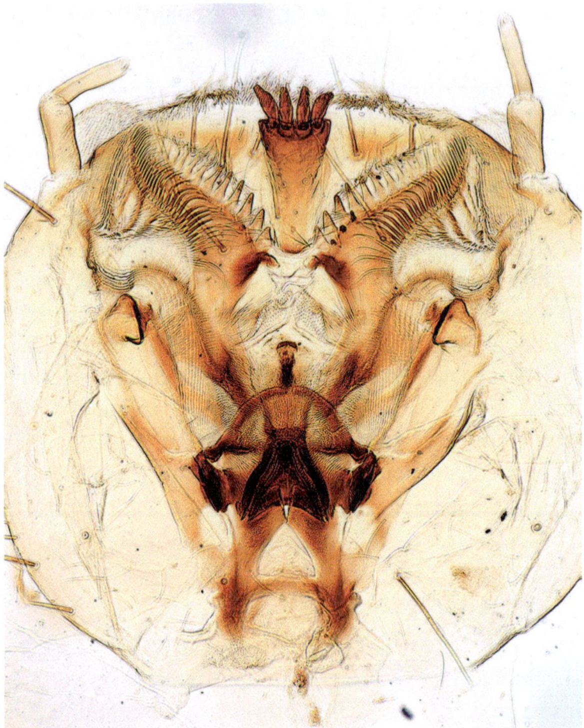

**Fig. 4.1** Head and mouthpart of larval *Prionocyphon* sp. (Coleoptera: Scirtidiae) noting dense maxillary or mandibular set of bristles. Phase contrast image by R. Ruta, University of Wroclaw, Poland

and Jäch 1986). Adults remove food particles from the bristles by the use of a galea comb. A more comprehensive evaluation of the feeding of adults can be found in Rothmeier and Jäch (1986). For larvae, food is often detritus or small invertebrates, whereas adults feed on algae or decaying plant material (Archangelsky 1997; Hansen 1997). The mouthparts of larvae and adult are both well adapted to filtering and contain several sections of bristles and setae (Fig. 4.2).

### *4.4.2 Ephemeroptera*

Mayflies comprise a well-studied group of filter-feeding insects with about 3200 species (Stork 2018). Most Ephemeroptera juveniles (nymphs) feed, whereas short-lived adults have vestigial mouthparts and do not. Beyond other forms of feeding, including scraping algae and predation, there are at least nine families where filter feeding occurs, including the Baetidae, Coloburiscidae, Ephemeridae, Heptageniidae, Isonychiidae, Leptophlebiidae, Oligoneuriidae, Polymitarcidae, and Siphlonuridae (Brittain 1982; Merritt et al. 2008; Ramírez and Gutiérrez-Fonseca 2014). Among these families, there is wide variation in how filtering ensues, and how it may be classified (e.g., passive versus active filtering). For instance, *Curotenetes*

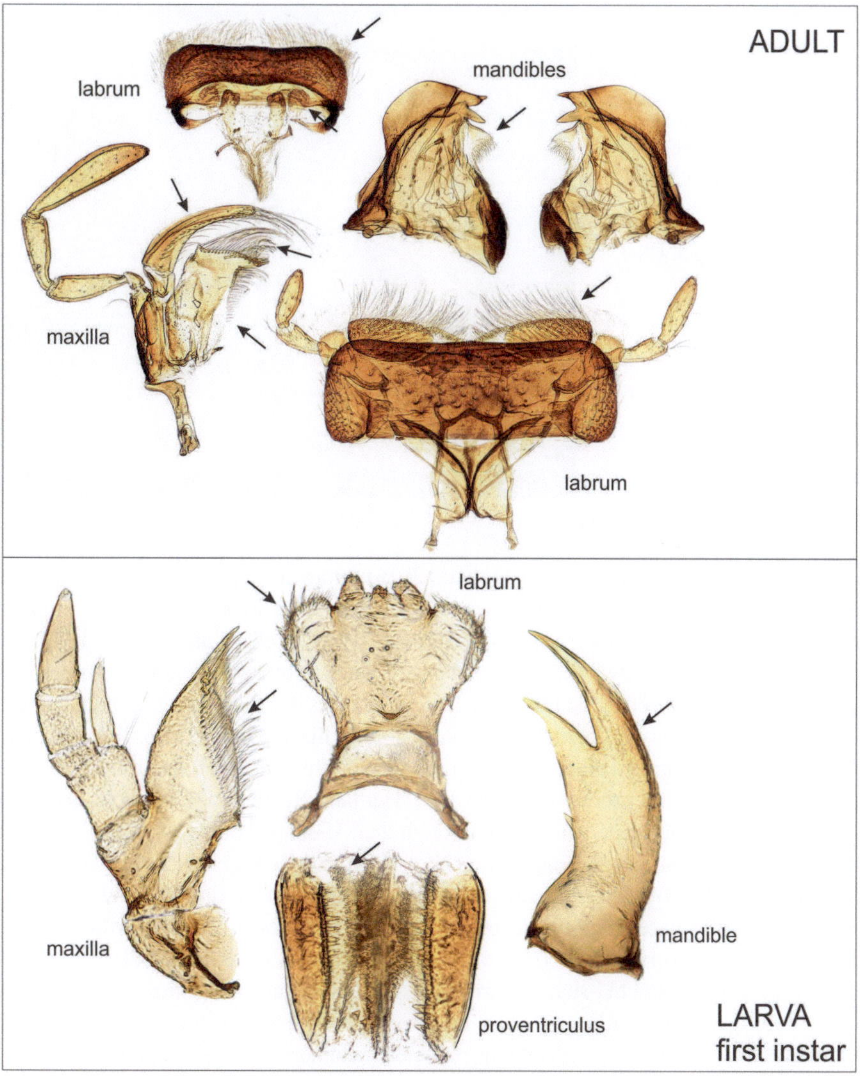


**Fig. 4.2** Mouthpart of *Spercheus emarginatus* (Coleoptera: Sphericidae) 1st instar larvae and adult. Arrows indicate setae associated with the mouthparts used in filtering. Images by M. Fikáček, National Museum, Czech Republic

*albomanicatus* extend their long seti-covered forelegs into the water current to capture particles before bringing them close to their mouthparts for feeding (Clemens 1917), thus filtering via morphological adaptations of the body (e.g., Fig. 4.3). A similar behavior is seen in *Oligoneuriella rhenana*, where labial and maxillary palps remove particles from foreleg setae (Elpers and Tomka 1995). Long fringes of setae on legs do not necessarily indicate a filter-feeding function, as hairs on the hind legs are more

**Fig. 4.3** Ventral view of head of mayfly larvae (Ephemeroptera: Isonychiidae) showing interlocking setae on legs. Inset: Close-up of the setae used for filtering on leg segments. Photo by D.A. Yee, University of Southern Mississippi, USA

likely for swimming (Lancaster and Downes 2013). More direct particle collection can be achieved by fringes of setae on the mouthparts, as seen in some genera including *Oligoneuriella* and *Isonychia* (Elpers and Tomka 1995; Wallace and O'Hop 1979). In the Coloburiscidae, Isonychiidae, and Oligoneuriidae, nymphs may also possess coxal gill tufts that can be used to aid filtering (Zhou 2010). Besides morphological adaptations, some mayflies construct burrows, especially among the Polymitarcidae. For instance, *Povilla* sp. dig a burrow, often in submerged wood, which are then lined with silk-like proteinaceous material produced via the anus (Hartland-Rowe 1953). Nymphs use their abdominal gills to increase water flow through their U-shaped burrows, where particles may then land on various portions of their body (Hartland-Rowe 1953, 1958). This effect is further enhanced by the presence of secondary hairs on the filtering setae, which when interlocked with adjacent hairs can capture very small particles (4–8 μm) (Hartland-Rowe 1953, 1958). For *Tortopus* sp. (Polymitarcyidae), nymphs have mandibular tusks that they also use to construct U-shaped burrows, especially in mud along stream banks (Scott et al. 1959), and have filtering setae on several body parts including the tibia and mandibles that are likely for gathering particles (Molineri et al. 2010). These particles are then removed via the palps (Scott et al. 1959). Rhythmic moving of gills in many burrowing mayflies (e.g., *Ephemera*) likely aid in current movement and enhance particle deposition (Eastham 1939).

### *4.4.3 Diptera*

The most speciose order of insects to contain filter-feeding members is the Diptera, or true flies, with about 155,000 species (Storks 2018). However, within the taxon, there are only five out of over 180 families of flies that contain a high proportion of filter-

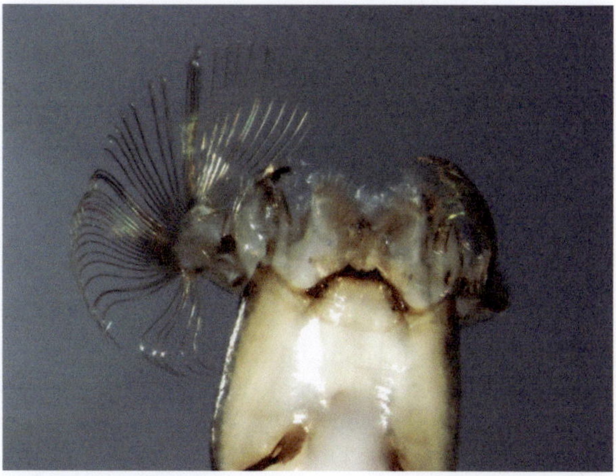

**Fig. 4.4** Ventral view of a black fly larval (Diptera: Simuliidae) head with cephalic fan (right side extended, left side collapsed). Photo by D.A. Yee, University of Southern Mississippi, USA

feeding species: Simuliidae, Culicidae, Dixidae, Chaoboridae, and Chironomidae. However flies do exist with filter-feeding attributes in other families, including Stratiomyiidae, Syrphidae, and Calliphoridae. However, as has been pointed out elsewhere, there is a great need for research into the feeding biology of other fly larvae (Wallace and Merritt 1980).

#### 4.4.3.1 Simuliidae

Black flies are represented by 2300 species worldwide (Adler and Crosskey 2018), and are most often found in lotic waters. Adults are high pestiferous for their biting behavior on humans. Blackfly larvae have a highly adapted filter-feeding structure known as a cephalic fan (Fig. 4.4), which individually are located between their antenna and mouths, and are capable of retracting and folding depending on feeding activity (Craig 1974; Wallace and Merritt 1980). Working in conjunction with current speed, a unique body position, and beating of the fans, black fly larvae are able to collect a variety of particles (summarized by Merritt et al. 1996). The fans trap particles of 0.09 to 350 μm in size, including bacteria, algae, diatoms, other insects, and detritus (reviewed in Wallace and Merritt 1980). However, the considerably smaller particle size found in their guts suggests that another mechanism for capture may be involved. Specifically, Ross and Craig (1980) identified a mucosubstance associated with the cephalic fan of several genera, which when applied to the fans acts in a way to retain particles smaller than the fan alone can capture. Entrapment of particles was initially presumed to be enhanced by this endogenous mucous secretion on the fans, but the source of mucous-like substances on fan setae surfaces appears to be derived from flocculation of dissolved organic matter (Ciborowski et al. 1997). Regardless of the source, this amorphous material may also aid in entrapment of organic matter and serve as a food source for larvae. Particles in general are removed by the larvae via sweeping their mouthparts over the surface of the fan. Finelli et al. (2002) showed that under experimental trials black fly larvae feeding behavior was more related to benthic water velocity than to food concentration. Not all simuliids are equipped with large fans (e.g., species in the genera *Gymnopais* and *Twinnia*, Craig 1974) but instead may scrape or browse on surfaces (Wallace and Merritt 1980). Although simuliids produce silk as a holdfast onto surfaces in fast-moving lotic habitats, they do not appear to use it for particle capture.

#### 4.4.3.2 Culicidae

Mosquitoes contain approximately 3500 species worldwide, and all species have larvae that occur in freshwater to brackish lentic or slow-moving lotic environments (Laird 1988). These habitats also include container systems, including both natural (e.g., phytotelmata like bromeliads, tree holes, bamboo stumps) and artificial (e.g., vehicle tires, cemetery vases) containers (Kitching 2000; Vezzani 2007; Yee 2008). Mosquitoes are insects best known for their association with disease, being responsible for millions of new infections by pathogens and hundreds of thousands of deaths in humans each year. Perhaps because of this, we know a good deal about mosquito feeding, especially among those genera most active in vectoring pathogens. Mosquitoes have an aquatic larval phase followed by a terrestrial adult phase, and although adults may still feed on plant nectar, growth is the sole purview of the larval phase. Outside of a few predatory taxa (e.g., all *Toxorhynchites*, some *Psorphora*) mosquito larvae have specific adaptations for filter feeding and obtain nourishment from heterotrophic microorganisms, algae, and detritus (e.g., Walker et al. 1988). The mouthparts of mosquitoes are adapted for straining particles from the water column and from surfaces. The entire apparatus, referred to as mouth brushes (Fig. 4.5), are primarily composed of well-developed mandibles and maxillae, with lateral palatal brushes located on a reduced labrum (reviewed by Pucat 1965). The action of these brushes, other mouth structures including setae, and pumping action of the pharynx combine to create strong water currents surrounding the mouth, bringing particles within reach (Merritt et al. 2008). In general, actively feeding larvae move suspended particles toward the mouth with modified mouth brushes, creating local currents that can extend into the surrounding water up to several centimeters. These brushes are not necessarily the primary mode of particle capture but serve as fluid movers: acting more like paddles than sieving mechanisms (Clements 1999). The mouth brushes do pick up some particles, but interestingly, they are cleaned not only by other mouthparts, but also by other structures (e.g., comb scales, pecten) located on the penultimate abdominal segment and respiratory siphon. The length of the setae influences the distance that a particle may be retrieved, and currents generated by brushes can move particles as far as from 40 mm away toward the mouth (Merritt et al. 1992; Clements 1999). Among genera, mosquitoes may utilize different feeding behaviors, so there are great differences in food type and performance in different environments. Categorization of mosquito larval feeding behavior has been based on particle size range and the general location of the food item (Merritt et al. 1992). Collector-filtering or filtering is found to be

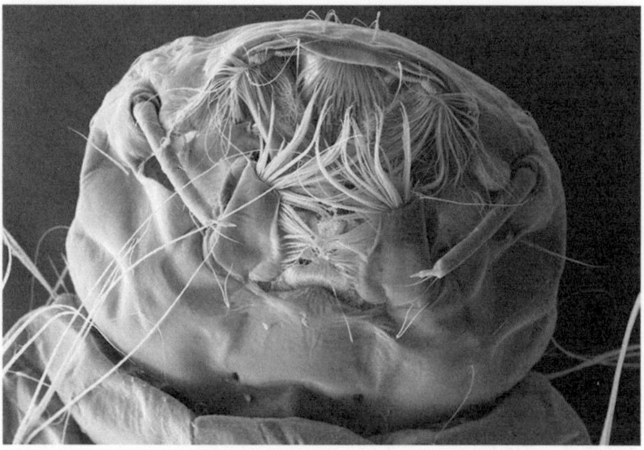

**Fig. 4.5** Scanning electron micrograph of a head and mouthparts of *Aedes albopictus* (Diptera: Culicidae) showing the mouth bristles. Photo by J. H. Deerman, University of Southern Mississippi, USA

dominant in the *Culex*, *Anopheles*, and *Culiseta*, wherein larvae remove particles that are suspended in the water column or floating on or at the water surface. Other genera, like *Aedes* and *Wyeomyia*, feed by removing particles on or loosely connected to surfaces like submerged rocks and vegetation ("browsers," Merritt et al. 1992; Clements 1999). Some Culicine larvae (e.g., *Aedes* and *Culex* spp.) filter feed primarily beneath the surface, suspended beneath their respiratory siphon. In contrast, *Anopheles* larvae feed at the air–water interface and primarily collect particles and material in the surface microlayer. This group generates currents that travel mainly parallel to the water surface and any expelled fluid moves downward and away from the head capsule (Clements 1999). Food sources and differences among larval assimilation and acquisition rates may also influence multiple measures of mosquito life history (e.g., Yee et al. 2015; Yee 2016). This can affect pathogen transmission by influencing the body size and nutritional reserves of emerging females, as well as the numbers of emerging females from any particular habitat (Juliano et al. 2014; Alto et al. 2015). The rich literature on mosquito larval nutrition has been reviewed by Clements (1999) and Dadd (1973).

#### 4.4.3.3 Dixidae

With 173 species, dixids are a small group of flies that are found in association with aquatic habitats and are closely related to mosquitoes (Wagner et al. 2008). Also called "meniscus midges," dixids are found in the surface tension around stones and organic substrates in slow-moving lotic or lentic habitats like ponds, lakes, and marshes (Wallace and Merritt 1980). Larvae also share similar morphological adaptations with Culicidae, including the anatomy of the labral brushes, but often consume just algae and detritus. Larvae hang in the water surface of leaves of macrophytes or riparian vegetation, and take on a curled "U-shape"; adults do not feed.

#### 4.4.3.4 Chaoboridae

With only 50 species, phantom midges are also found in similar habitats as both their closest relatives, the mosquitoes and dixids (Wagner et al. 2008). Although most species are predatory, *Australomochlonyx nitidus* filter feeds exclusively, using a large, conspicuous, fan-like set of hairs on the mandibles (Colless 1977). In addition, this species contains a unique morphological adaptation to feeding, the oral bullae. The oral bullae consist of a hair-covered membranous structure that lies on either side of the pharyngeal orifice but below the mandible, which are likely used to transfer food from the mouthparts to the mouth itself (Colless 1977).

#### 4.4.3.5 Chironomidae

Midges, or non-biting midges, are one of the most specious families of flies, with greater than 20,000 species (Merritt et al. 2008). Adults generally do not feed and are poor fliers, but are often more abundant than larvae. Larvae are small and occupy more aquatic and semi-aquatic habitats of any other aquatic insect, including permanent and temporary lotic and lentic habitats, and can be found across a vast range of temperatures, elevations, and environmental conditions. Many filter-feeding larvae employ silk as a means of particle capture. Often, a net is spun across the opening of a small burrow located within the substrate, and larvae move their body in small undulations to facilitate water movement across the net (Berg 1950). This net, along with associated particles, is consumed before the larvae spin another one to replace it. *Rheotanytarus muscicola* constructs small silk cases that are attached to the substrate in lotic environments. In later instars, the larvae add a salivary secretion to two to five small protuberances, which look like small arms, incorporated into the case (Kullberg 1988). These secretions are then periodically consumed along with any particles. Another chironomid, *Odontomesa fulva*, directly filter feeds using setae associated with their mouthparts (reviewed in Pinder 1986), which seems to be a rare condition in filter-feeding Chironomidae.

### *4.4.4 Trichoptera*

With over 14,300 species (Storks 2018), caddisflies are one of the more diverse filter-feeding insect orders. Although there are caddisflies that have evolved morphological adaptations to filter, most species use woven nets of silk for filtering particles from lotic water. This silk is produced via the labial glands of the mouth and is very similar in composition to Lepidopterans (Sehnal and Sutherland 2008). Nets of silk vary in size, pore diameter, and location, but are all produced via the salivary glands (Wallace and Merritt 1980; Merritt et al. 2008). For instance, pore size ranges from >200,000 $\mu m^2$ in Arctopsuchinae to <200 $\mu m^2$ in Macronematinae (Wallace and Merritt 1980). Moreover, the type of particles captured can vary with species, habitat, and net morphology, with diatoms, algae, and detritus being the most common types of food. Caddisflies are divided into three broad suborders, the Annulipalpia, Integripalpia, and the Spicipalpia (Holzenthal et al. 2007). The most common filter feeders belong to the family Hydropsychoidea (Annulipalpia), which often dominates freshwater streams in North America (Wallace and Merritt 1980). These caddisflies build shelters, or retreats, of silk nets that may also incorporate material from the surrounding area, including organic particles, detritus, and mineral fragments. Mesh size within this family is often based on environmental conditions, including current speed and temperature. For instance, net mesh size tends to be larger in species that reside in cold fast-moving upstream sites, but smaller mesh sizes are found in downstream sites with slower current speeds (Merritt et al. 2008). Other families that construct silken nets within the Annulipalpia include the Polycentropodidae, Dipseudopsidae, and Philopotamidae. The smallest mesh sizes are found in the Philopotamidae (0.4 $\mu m^2$) (Wallace and Malas 1976). Mesh can be produced rapidly with as many as 70 individual strands being excreted at a time, and nets are often found on the underside of rocks in slow-moving currents (Wallace and Malas 1976).

Silk is often used to construct caddisfly cases (caddis), wherein a variety of particles, including pieces of leaves, small rocks, snail shells, wood, or other debris, are adhered together. Construction particles are often specific to certain species, and cases come in a dazzling array of sizes and shapes (Ross 1964; Merritt et al. 2008). When constructing a caddis, a larva starts with an oval frame of silk, to which they attach a larger net in a set of complex behaviors, involving anchoring, resting, and weaving (Ross 1964). Similar to other complex behaviors in insects, the actual type and size of net or caddis is the result of several highly conserved steps, which when added together produce the variety of filter-feeding structures seen in Trichoptera. Besides the protection afforded by the net itself, the caddis can serve as a food-capturing device, wherein particles that enter through the larger upstream opening adhere to the silk lining of the tube. Once prey or particles are passively captured on the nets, they are removed in a number of distinct ways (Merritt and Wallace 1981). Elongated forelimbs are used by some hydropsychoids to remove small living prey that was captured in their nets. Other species that capture organic particles may remove them using specialized, densely arranged bristles, located either on their mouthparts (e.g., *Macronema* sp.) or on their forelegs (e.g., *Phylocentropus* sp.). Others sweep particles into their mouth using setae along the upper labrum (Merritt and Wallace 1981).

Some caddisflies use a combination of approaches for capturing food particles via direct filtering, often via net spinning and tube building. These include *Macronema* sp. mentioned above, which build a short, upward pointing tube in wood on bark in streams that contains a small section laced with a silk net (Wallace and Sherberger 1974). Particles that land on the netting are then removed via the legs or mouthparts. Members of the genus *Phylocentropus* (Dipseudopsidae) first build a long Y-shaped tunnel below the substrate of the stream. One side of the tunnel is longer than the other, and it normally extends upward and protrudes far above the bottom, whereas the other is shorter and often does not protrude as far. The larva also builds a bulge into the shorter tube, where it spins a silken net. By occupying the longer tube and moving its body in an undulating motion, the larva causes a current to be produced passing from the longer tube to the short tube across the net. In this way, the larvae achieves particle capture (Wiggins 2005). Caddisflies in the genus *Neureclipsis* (Polycentropodidae) dispense with a burrow or tube and instead build a large, cornucopia-shaped net, which can be as long as 20 cm, and is often attached to a underwater structure like a branch (Wallace and Merritt 1980).

Other species do not build a caddis, burrow, or use silk to filter feed. Some Brachycentridae use long setae on their middle and hind legs to aid in particle capture, whereas several *Drusus* sp. (Limnephilidae) possess spines or long hairs on the head and body that allow for prey capture (Bohle 1983; Graf et al. 2005).

## 4.5 Food Sources

Although populations of filter feeders are limited by many factors, the abundance and quality of food items available for capture is certainly a primary constraint. For most filter feeders, where you are (habitat) defines what you eat. Most immature filter feeders are either relatively sessile (attached to substrates directly or via constructed refugia) or restricted to small, defined habitats (e.g., container breeding mosquito larvae). Movement to higher quality habitats after hatching from the egg may be possible (e.g., stream dwelling insects drifting downstream), but such movements can increase predation risks and a new habitat is no guarantee of higher quality food resources. This general lack of choice is further constrained by body size and specific tools for particle capture.

### *4.5.1 Food Size Range*

As might be expected within a phylogenetically and morphologically diverse feeding guild, food items can vary considerably with taxon and size (instar) of the individual. Generally speaking, most filter-feeding insects collect and ingest a mixture of particles ranging from colloidal (nanometer size range) to coarse particulate organic matter (CPOM $>$ 1 mm) and whole macroinvertebrates (sometimes even younger conspecifics). FPOM (fine particulate organic matter, $>$0.45 μm, $<$1 mm) is considered the "sweet spot" in terms of ingestibility and food value (Bundschuh and McKie 2016). Indeed, most studies of insect filter feeders show this size range of particles to be predominant in the guts of filter feeders, with a majority of species or life stages feeding on sources between 1 and 300 μm (Huryn et al. 2008). Although there is evidence that DOM (dissolved organic matter)—defined as material passing through a glass fiber (GF/F) filter in many cases, but more precisely defined as being less than 0.45 μm in diameter (Nebbioso and Piccolo 2013)—is concentrated and assimilated, this reflects natural aggregation, adsorption, and consolidation of dissolved substances (Ciborowski et al. 1997; Kaplan and Cory 2016) more than any targeted mechanism of ingestion by filter feeders. Assimilation of smaller soluble compounds such as amino acids and sugars likely necessitates the presence of particles and colloidal compounds for adequate fluid ingestion and movement through the digestive tract (Merritt et al. 1992). Mosquito larvae, for example, primarily consume particles in the 1–50 μm size range, but this varies with species and instar (Merritt et al. 1992). Larger particles can also be ingested, depending on the shape (e.g., strands of algal filaments or nematodes), but cross-sectional diameter for these particles is usually within the size ranges noted above (Clements 1999). Other examples include black fly larvae, which consume a high percentage of large (>40 μm) particles (Kurtak 1978; Wotton 1994), yet still are capable of ingesting colloidal size material (Wotton 2009).

### 4.5.2 *Types and Selection of Ingested Particles*

The particulate components consumed by filter and suspension feeders include a wide range of detritus (animal and plant-based) and fecal material, microorganisms, small metazoans, and live invertebrates, in addition to mineral and inert materials. Gut content analysis of filter feeders has identified bacteria, algae, protozoans, micro-metazoans, fungi, small invertebrates, and often a dominance of amorphous detritus (Wallace and Merritt 1980; Merritt et al. 1992; Clements 1999; Wiggins 2005; Huryn et al. 2008). Detritus itself is an important substrate for microorganisms and ingestion of plant detritus in particular is considered a means of harvesting-associated microbial biomass rather than the generally refractory substrate (Cummins and Klug 1979; Cummins et al. 2008). This may not necessarily be the case for ingestion of animal-derived detritus, in which the substrate itself (soft tissues, small pieces of chitinous exoskeleton) can be digested and assimilated without microbial intervention. The higher relative food value of animal vs. plant detritus has been demonstrated in studies of larval mosquitoes (Yee and Juliano 2006; Yee et al. 2007, 2015; Winters and Yee 2012).

Apart from size range restrictions noted above, many filter feeders show little selection of ingested particles. Particles with no food value (e.g., clay, charcoal, plastic) are readily consumed and passed through the gut, even in the presence of particles or solutes with actual food value (Merritt et al. 1992; Clements 1999). Recent work has shown that not only are inert materials, such as small plastic particles, ingested by mosquito larvae, they may be transferred to the adult stage and presumably to higher trophic levels (Al-Jaibachi et al. 2018). Any “choice” of food source is largely made by females selecting oviposition sites, because larval or nymphal habitat determines the type and quality of ingestible material. Some direct selection of food may occur with large passive filter feeders, such as net spinning caddisflies in the Hydropyschidae subfamily Arctopsychinae, where larvae appear to harvest high-quality food items (e.g., insects) from nets while ignoring or discarding other large detritus particles (Wallace and Merritt 1980; Wiggins 2005; Huryn et al. 2008).

**Bacteria** For almost all filter-feeding insects bacteria represent a constant and important food source. Their contribution to the growth of many filter-feeding species is well documented, including important indirect roles in nutrition and initiation of development (Strand 2017; Valzania et al. 2018). Bacteria are ubiquitous in aquatic habitats, are in the size range (generally 0.5–5 μm) of particles collected by the majority of filter feeders, and are associated with the surfaces of bigger particles ingested by some larger filter-feeding species as well. Bacterial carbon was estimated to contribute up to 67% to black fly larval growth in some river systems (Meyer and Edwards 1987) and black flies have reportedly been reared to pupation on bacterial suspensions alone (Adler and Currie 2008). However, bacteria appear to be inadequate for complete mosquito development (Valzania et al. 2018), and in many cases, it is not clear how much bacteria contribute to filter feeder production relative to other food sources. Further, bacterial diversity is such that considering the category to be homogenous in food value to each filter feeder is untenable. Digestibility of bacterial species by aquatic insects can vary considerably (e.g., Austin and Baker 1988), and it has been suggested that filter-feeding organisms in non-flowing systems ultimately select for indigestible forms in their immediate environment (e.g., Kaufman et al. 2002). The recent research emphasis on gut bacterial communities in insects, and especially mosquitoes, suggests that many forms of ingested bacteria survive digestion and become residents, even passing transtadially into adults (Strand 2017; Guégan et al. 2018). Nonetheless, bacteria in the mosquito gut originate from and reflect the larval environment (Strand 2017; Guégan et al. 2018). This exploration of gut bacterial communities, therefore, can also give a picture of what types of bacteria are ingested and subsequently digested, providing information about the food value of specific particles harvested by this filter-feeding group. Additionally, studies of bacterial communities with and without filter feeders present can be useful in identifying food sources. For example, *Flavobacterium* was abundant in habitats without *Aedes triseriatus* larvae, but was greatly reduced in their presence (Xu et al. 2008). Subsequent studies showed this bacterium to be readily digested and assimilated, and capable of supporting larval growth (Chen et al. 2014).

**Algae** Many types of algae (single cells and small colonies across a range of taxa) are also important as food for many filter feeders. This is especially true for *Anopheles* mosquitoes, which feed primarily at the air–water interface in many permanent and semi-permanent aquatic habitats (Kaufman and Walker 2006; Tuno et al. 2018). Black fly larvae, mainly those below lake impoundments, harvest suspended algal cells at high rates in lotic habitats and grow better when they do (Wotton 1994). The same is true for net-spinning caddisflies exploiting lake outlet habitats (Wiggins 2005). It has been shown that black fly larvae can measurably reduce algae particles in stream water, but do not have the same effect on bacteria (Parkes et al. 2004). Like bacterial food sources, algal species vary greatly in their digestibility and food value to filter feeders to the extent that use of certain resistant algal types has been proposed as a means of larval mosquito control (Marten 1987, 2007; Garros et al. 2008a, b), and particular algal species have been identified as keys to mosquito production from larval habitats (Tuno et al. 2018).

**Protozoans and Meiofauna** Non-photosynthetic protists (e.g., ciliates, flagellates, amoebas) are thought to be important components of the food resources for many filter-feeding diptera, such as mosquitoes. Several studies have shown their decline after larval feeding, presumably reflecting ingestion and digestion (e.g., Kaufman et al. 2002; Walker et al. 2010). However, recent studies suggest that their consistent role as food sources for mosquitoes is questionable and that they may even compete with insect filter feeders for bacteria and micro-eukaryotes (Skiff and Yee 2015; Duguma et al. 2017). Similarly, meiofauna (e.g., rotifers, nematodes) may be nutritious food for many filter feeders, but their relative contribution to growth is unknown.

**Fungi** Fungi (usually yeasts and other Ascomycota) are generally less recognized as food items for filter feeders than they are for other functional feeding groups such as shredders (Cummins and Klug 1979; Cummins et al. 2008), but their presence in fragmented CPOM derived from leaf material and colonization of FPOM are potentially sources of valuable nutrition in many stream systems (Cummins and Klug 1979; Cummins et al. 2008; Webster et al. 2016). The value of fungi to mosquito nutrition has long been recognized and recent studies identify yeasts as an important food for complete development in larval mosquitoes (Díaz-Nieto et al. 2016; Souza et al. 2016; Steyn et al. 2016; Valzania et al. 2018).

**Detritus and Other Material** Although microorganisms are key components in the diets of most filter feeders, the bulk of ingested material is in the ill-defined detritus category. Origins of ingested detritus include senescent leaf material, fecal material from other arthropods, and carcasses of insects and arthropods. The relative importance of allochthonous vs. autochthonous detrital food sources in filter feeder diets will vary with habitat, but allochthonous inputs of terrestrial plant material are key to many stream systems and larval mosquito habitats (Cummins and Klug 1979; Cummins et al. 2008). Terrestrial leaf input into headwater stream systems is well documented for its effect on insect communities, and the cascade of FPOM and fecal material made available after initial processing of the leaf inputs is harvested by many groups of filter feeders (Wotton and Malmqvist 2001; Cummins et al. 2008; Bundschuh and McKie 2016). Fecal material produced by filter feeders is also a food source for members of the same functional group and sometimes the same species of filter feeder (Wotton et al. 1998). Even though the food value of fecal material to filter feeders is initially low due to prior digestion, colonization by microorganisms on the high surface area material greatly increases the nutritional content (Wotton and Malmqvist 2001). Other plant-derived allochthonous inputs include flower parts and pollen, which have been shown to contribute to the growth of mosquito larvae (Kaufman et al. 2010; Wondwosen et al. 2018). Pollen from grasses, maize, and sugar cane deposited on the surface of *Anopheles* habitats can be directly consumed and digested by the larvae (Asmare et al. 2017; Wondwosen et al. 2017, 2018), circumventing the usual plant detritus to microbial biomass transformation pathway usually necessitated by more refractory plant parts.

As mentioned above, insect and animal detritus inputs also can significantly increase filter feeder production. Some of these inputs are of terrestrial (allochthonous) origin, but others can be autochthonous (e.g., consumption of arthropod cadavers originating within the system—Tsurim and Silberbush 2016). The filter-feeding caddisfly larvae, *Brachycentrus*, has been observed feeding on moth larvae that had fallen into a stream (Ohkawa and Ito 2001), although the distinction between this being a detritus consuming process or predation event is vague. Aquatic insects in general are thought to benefit from animal carcass inputs (e.g., anadromous salmon—Claeson et al. 2006), but specific effects on filter feeders are not well known.

## 4.6 Trophic Importance

Filter-feeding insects are often the most abundant functional group within aquatic habitats. Black fly larvae and net-spinning caddisflies, for example, dominate the benthic invertebrate community in many stream and river sections, particularly those below lake outlets or impoundments (Huryn and Wallace 2000). Production estimates for the groups have sometimes approached 1000 g dry weight per $m^2$ in certain locations (Wotton 1988). In lentic systems, sediment dwelling filter feeders can also dominate the benthos, and large filter-feeding mayfly emergence swarms from parts of the Great Lakes and Mississippi basin (Brittain 1982) are even detectable by weather radar. Mosquito emergence from arctic habitats is also legendary, if not yet completely quantified in terms of biomass (Culler et al. 2018). Some work has been done to estimate production estimates for medically important mosquitoes, like *Aedes aegypti* (Focks and Chadee 1997; Morrison et al. 2006), including work in New Orleans, Louisiana, which found city blocks to produce 362–558 adult females per day (Focks et al. 1981).

Although biomass and production of filter feeders varies greatly within aquatic ecosystems, the group has important roles in transformation, retention, and export of organic matter from systems (Cummins and Klug 1979; Cummins et al. 2008). As discussed above, the filter-feeding group functions as primary and secondary consumers, detritivores, and even predators. They can also be classified as decomposers in that they process decaying organic matter, primarily in the FPOM category. Their consumption of detritus and microorganisms, and production and consumption of fecal material, contributes to nutrient spiraling within stream systems, and via adult emergence, they transfer organic carbon and other nutrients upstream and back to the terrestrial environment (Newbold et al. 1982; Wallace and Hutchens 2000; Cummins et al. 2008; Webster et al. 2016). Perhaps more importantly, filter feeders serve as prey items for a variety of aquatic predators including other aquatic insects and fish, linking microbial biomass, primary production, and detrital dynamics to higher trophic levels in the system (e.g., Curtis et al. 2018). Black fly larvae, for example, are often the dominant prey item for predaceous stoneflies (Malmqvist 1994). In some systems, such as phytotelmata habitats of mosquitoes, filter feeders themselves may represent the highest trophic level (Kitching 2001).

## 4.7 Future Directions

Like any subject related to our knowledge of insects, filter feeding is an underexplored area with vast gaps in data and understanding. This is particularly true with respect to quantifying the diversity of species that use this feeding mode, and the degree to which those taxa use filter feeding with respect to other forms of feeding. Most, if not all, filter-feeding aquatic insects are capable of obtaining food in other ways. For instance, mosquito larvae can switch between filtering in the water column and browsing surfaces given the concentration of food available in the environment (Merritt et al. 1992; Yee et al. 2004). Thus, although the adaptations for filter feeding, like net building or tunnel building in caddisflies, may appear to limit food choices, this flexibility in food gathering likely means that the contribution of these species to detrital processing and their trophic position are much more complicated than currently known.

Perhaps the two largest future challenges for filter-feeding aquatic taxa (and for those that study them) will be climate change and invasions by non-native species, with the former likely exacerbating the latter. Among filter-feeding taxa, mosquitoes contain the most invasive species, some of which are of immense importance due to their proclivity to spread human and animal disease. *Aedes aegypti* (yellow fever mosquito) and *Aedes albopictus* (Asian tiger mosquito) are the best studied of these invasives, as they are not only found throughout the world due to the actions of human actions but also are important as vectors for several major arboviruses (dengue, chikungunya, Zika) (Lounibos 2002). Other species, like *Culex quinquefasciatus/pipiens*, is also found distributed across the globe, and is a main vector of West Nile virus (Lounibos 2002). When any of these species comes in contact with native fauna, it has the potential to disrupt ecological interactions, detrital processing, and disease dynamics. Invasive species may also affect taxa important as filter feeding in aquatic systems. For example, invasive dreissenid mussels, which have invaded many lentic water ways throughout North America, can negatively affect resident mayfly larvae (*Hexagenia* sp.) via the bioturbation activities of the mussels (Osterling et al. 2007).

Future climate change, especially increasing temperatures, is going to have significant effects on ectotherms, including insects (Deutsch et al. 2008). These effects include modifying distributions and influencing population sizes via alternations in thermal limits and habitat suitability. For instance, Hering et al. (2009) investigated the potential for a changing climate to affect 1134 species and subspecies of Trichoptera and found that many endemic species were likely sensitive to a changing climate; however, this study did not specifically separate species by feeding type. Although investigations of climate change have been conducted for steam insects (Bonada et al. 2007; Durance and Ormerod 2007), there are almost no investigations of filter-feeding insects specifically, or how filter feeders as a group may be affected by increasing temperatures. Understanding how filter-feeding insects will be affected by changes in climate will be important, given their role in processing detritus as well as their importance as prey for many other insect and non-insect predators alike.

## References

Adler PH, Crosskey RW (2018) World blackflies (Diptera: Simuliidae): a comprehensive revision of the taxonomic and geographical inventory. Available from: https://biomia.sites.clemson.edu/pdfs/blackflyinventory.pdf. Accessed 28 Sep 2018

Adler PH, Currie DC (2008) Simuliidae. In: Merritt RW, Cummins KW, Berg MB (eds) Aquatic insects of North America, 4th edn. Kendall/Hunt, Dubuque, pp 825–846

Al-Jaibachi R, Cuthbert RN, Callaghan A (2018) Up and away: ontogenic transference as a pathway for aerial dispersal of microplastics. Biol Lett 14:20180479

Alto BW, Bettinardi DJ, Ortiz S (2015) Interspecific larval competition differentially impacts adult survival in dengue vectors. J Med Entomol 52:163–170

Archangelsky M (1997) Studies on the biology ecology and systematics of the immature stages of New World Hydrophiloidea (Coleoptera: Staphyliniformia). Bull Ohio Biol Sur New Ser 12:1–207

Asmare YR, Hopkins J, Tekie H, Hill SR, Ignell R (2017) Grass pollen affects survival and development of larval *Anopheles arabiensis* (Diptera: Culicidae). J Ins Sci 17:93. https://doi.org/10.1093/jisesa/iex067

Austin DA, Baker JH (1988) Fate of bacteria ingested by larvae of the freshwater mayfly, *Ephemera danica*. Microb Ecol 15:323–332

Berg CO (1950) Biology of certain Chironomidae reared from Potamogeton. Ecol Monogr 20:83–10

Bohle HW (1983) Drift-catching and feeding behaviour of the larvae of *Drusus discolor* (Trichoptera: Limnephilidae). Archiv für Hydrobiologie 97:455–470

Bonada N, Dolédec S, Statzner B (2007) Taxonomic and biological trait differences of stream macroinvertebrate communities between mediterranean and temperate regions: implications for future climatic scenarios. Global Change Biol 13:1658–1671

Brittain JE (1982) Biology of Mayflies. Annu Rev Entomol 27:119–147

Bundschuh M, McKie BG (2016) An ecological and ecotoxicological perspective on fine particulate organic matter in streams. Fresh Biol 61:2063–2074

Chen S, Kaufman MG, Korir ML, Walker ED (2014) Ingestibility digestibility and engineered biological control potential of *Flavobacterium hibernum* isolated from larval mosquito habitats. Appl Environ Microbiol 80:1150–1158

Ciborowski JJH, Craig DA, Fry KM (1997) Dissolved organic matter as food for black fly larvae (Diptera: Simuliidae). J N Am Benthol Soc 16:771–780

Claeson SM, Li JL, Compton JE, Bisson PA (2006) Response of nutrients biofilm and benthic insects to salmon carcass addition. Can J Fish Aquat Sci 63:1230–1241

Clemens WA (1917) An ecological study of the mayfly *Chirotenetes*. Univ Toronto Biol Ser 17:1–43

Clements AN (1999) The biology of mosquitoes, vol 2, Sensory reception and behavior. CAB International, Wallingford

Colless DH (1977) A possibly unique feeding mechanism in the dipterous larvae. J Aust Entomol Soc 16:335–339

Craig DA (1974) The labrum and cephalic fans of larval Simuliidae (Diptera: Nematocera). Can J Zool 52:133–159

Craig DA, Galloway MM (1987) Hydrodynamics of larval black flies. In: Kim KC, Merritt RW (eds) Black flies: ecology population management and annotated world list. Pennsylvania State University Press, University Park

Culler LE, Ayres MP, Virginia RA (2018) Spatial heterogeneity in the abundance and fecundity of Arctic mosquitoes. Ecosphere 9:e02345

Cummins KW, Klug MJ (1979) Feeding ecology of stream invertebrates. Annu Rev Ecol Syst 10:147–172

Cummins KW, Merritt RW, Berg MB (2008) Ecology and distribution of aquatic insects. In: Merritt RW, Cummins KW, Berg MB (eds) Aquatic insects of North America, 4th edn. Kendall/Hunt, Dubuque, pp 105–122

Currie DC, Craig DA (1987) Feeding strategies of larval black flies. In: Kim KC, Merritt RW (eds) Black flies: ecology population management and annotated world list. Penn State University Press, University Park, pp 155–170

Curtis WJ, Gebhard AE, Perkin JS (2018) The river continuum concept predicts prey assemblage structure for an insectivorous fish along a temperate riverscape. Freshw Sci 37:618–630

Dadd RH (1973) Insect nutrition: current developments and metabolic implications. Annu Rev Entomol 18:381–420

Deutsch CA, Tewksbury JJ, Huey RB, Sheldon KS, Ghalambor CK, Haak DC, Martin PR (2008) Impacts of climate warming on terrestrial ectotherms across latitude. Proc Natl Acad Sci U S A 105:6668–6672

Díaz-Nieto LM, D'Alessio C, Perotti CM, Beron CM (2016) *Culex pipiens* development is greatly influenced by native bacteria and exogenous yeast. PLoS One 11 doi.org/10.1371/journal.pone.0153133

Duguma D, Kaufman MG, Domingos ABS (2017) Aquatic microfauna alter larval food resources and affect development and biomass of West Nile and Saint Louis encephalitis vector *Culex nigripalpus* (Diptera: Culicidae). Ecol Evol 7:3507–3519

Durance I, Ormerod SJ (2007) Climate change effects on upland stream macroinvertebrates over a 25-year period. Global Change Biol 13:942–957

Eastham LES (1939) Gill movements of nymphal *Ephemera danica* and the water currents caused by them. J Exp Biol 16:18–33

Elpers C, Tomka I (1995) Food-filtering mechanism of the larvae of *Oligoneuriella rhenana* Imhoff (Ephemeroptera: Oligoneuriidae). In: Corkum LD, Ciborowski J (eds) Current directions in research on Ephemeroptera. Canadian Scholars' Press, Toronto, pp 283–293

Finelli SM, David DH, Merz RA (2002) Stream insects as passive suspension feeders: effects of velocity and food concentration on feeding performance. Behav Ecol 11:145–153

Focks DA, Chadee DD (1997) Pupal survey: an epidemiologically significant surveillance method for *Aedes aegypti*: an example using data from Trinidad. Am J Trop Med Hyg 56:159–167

Focks DA, Sackett SR, Bailey DL, Dame DA (1981) Observations on container-breeding mosquitoes in New Orleans, Louisiana, with an estimate of the population density of *Aedes aegypti* (L.). Am J Trop Med Hyg 30:1329–1335

Garros C, Ngungi N, Githeko AE, Tuno N, Yan G (2008a) Gut content identification of larvae of the *Anopheles gambiae* complex in western Kenya using a barcoding approach. Mol Ecol Res 8:512–518

Garros C, Van Nguyen C, Trung HD, Van Bortel W, Coosemans M, Manguin S (2008b) Distribution of *Anopheles* in Vietnam, with particular attention to malaria vectors of the *Anopheles minimus* complex. Malaria J 7:11

Graf W, Lubini V, Pauls SU (2005) Larval description of *Drusus muelleri* McLachlan 1868 (Trichoptera: Limnephilidae) with some notes on its ecology and systematic position within the genus *Drusus*. Ann Limnol Int J Lim 41:93–98

Guégan M, Zouache K, Démichel C, Minard G, Tran Van V, Potier P, Mavingui P, Valiente Moro C (2018) The mosquito holobiont: fresh insight into mosquito-microbiota interactions. Microbiome 6:49

Hannappel U, Paulus HF (1987) Arbeiten zu einem phylogenetischen System der Helodidae (Coleoptera) - Feinstrukturuntersuchungen an Larven. Zool Beitr NF 31:77–150

Hansen M (1997) Phylogeny and classification of the staphyliniform beetle families (Coleoptera). Biologiske Skrifter 48:1–339

Hartland-Rowe R (1953) Feeding mechanism of an ephemeropteran nymph. Nature 172:1109–1110

Hartland-Rowe R (1958) The biology of a tropical mayfly *Povilla adusta* Navas with special reference to the lunar rhythm of emergence. Rev Zool Bot Afr 58:185–202

Hering D, Schmidt-Kloiber A, Murphy J, Lücke S, Zamora-MuÇoz C, López-Rodríguez MJ, Huber T, Graf W (2009) Potential impact of climate change on aquatic insects: a sensitivity analysis for European caddisflies (Trichoptera) based on distribution patterns and ecological preferences. Aquat Sci 71:3–14

Hershey AE, Lamberti DT, Northington RM (2010) Aquatic insect ecology In: Thorp JH, Covich PA (eds) Ecology and classification of North American freshwater invertebrates. Elsevier Science, Amsterdam, pp 659–694

Holzenthal RW, Blahnik RJ, Prather AL, Kjer LM (2007) Order Trichoptera Kirby 1813 (Insecta) Caddisflies. In: Zhang Z–Q, Shear WA (eds) Linnaeus tercentenary: progress in invertebrate taxonomy. Zootaxa 1668:1–766

Huryn AD, Wallace JB (2000) Life history and production of stream insects. Annu Rev Entomol 45:83–110

Huryn AD, Wallace JB, Anderson NH (2008) Habitat life history secondary production and behavioral adaptations of aquatic insects. In: Merritt RW, Cummins KW, Berg MB (eds) Aquatic insects of North America, 4th edn. Kendall/Hunt, Dubuque, pp 55–104

Juliano SA, Ribeiro GS, Maciel-de-Freitas R, Castro MG, Codeco C, Lourenco-de-Oliveira R, Lounibos LP (2014) She's a femme fatale: low-density larval development produces good disease vectors. Mem Inst Oswaldo Cruz 109:1070–1077

Kaplan LA, Cory RM (2016) Dissolved organic matter in stream ecosystems: forms functions and fluxes of watershed tea. In: Jones JB, Stanley EH (eds) Stream ecosystems in a changing environment. Academic, Boston, pp 241–320

Kaufman MG, Walker ED (2006) Indirect effects of soluble nitrogen on growth of *Ochlerotatus triseriatus* larvae in container habitats. J Med Entomol 43:677–688

Kaufman MG, Goodfriend WA, Kohler-Garrigan A, Walker ED, Klug MJ (2002) Soluble nutrient effects on microbial communities and mosquito production in *Ochlerotatus triseriatus* habitats. Aquat Microb Ecol 29:73–88

Kaufman MG, Pelz-Stelinski KS, Yee DA, Juliano DA, Ostrom PH, Walker ED (2010) Stable isotope analysis reveals detrital resource base sources of the tree hole mosquito *Aedes triseriatus*. Ecol Entomol 35:586–593

Kitching RL (2000) Food webs and container habitats. The natural history and ecology of phytotelmata. Cambridge University Press, England

Kitching RL (2001) Food webs in phytotelmata: "bottom-up" and "top-down" explanations for community structure. Annu Rev Entomol 46:729–760

Klaas-Douwe BD, Monaghan MT, Pauls SU (2014) Freshwater biodiversity and diversification. Annu Rev Entomol 59:143–163

Kullberg A (1988) The case mouthparts silk and silk formation of *Rheotanytarsus muscicola* Kieffer (Chironomidae: Tanytarsini). Aqu Insect 10:249–255

Kurtak DC (1978) Efficiency of filter feeding of black fly larvae (Diptera: Simuliidae). Can J Zool 56:1608–1623

Laird M (1988) The natural history of larval mosquito habitats. Academic, London

Lancaster J, Downes NJ (2013) Aquatic entomology. Oxford University Press, Oxford

Lawrence JF (2016) Scirtidae. In: Beutel RG, Leschen RAB (eds) Handbook of zoology Volume IV Arthropoda Part 38 Coleoptera beetles, vol 1: Morphology and systematics (Archostemata Adephaga Myxophaga Polyphaga partim), 2nd edn. Walter de Gruyter, Berlin, pp 215–225

Lounibos LP (2002) Invasions by insect vectors of human disease. Annu Rev Entomol 47:233–266

Lucas P, Hunter FF (1999) Phenotypic plasticity in the labral fan of simuliid larvae (Diptera): effect of seston load on primary-ray number. Can J Zool 77:1843–1849
Malmqvist B (1994) Preimaginal blackflies (Diptera: Simuliideae) and their predators in a central Scandinavian lake outlet stream. Ann Zool Fennici 31:245–255
Marten GG (1987) The potential of mosquito-indigestible phytoplankton for mosquito control. J Am Mosq Cont Assoc 3:105
Marten GG (2007) Larvicidal algae. J Am Mosq Cont Assoc 23:177–183
Merritt RW, Wallace JB (1981) Filter-feeding insects. Sci Am 244:132–147
Merritt RW, Dadd RH, Walker ED (1992) Feeding behavior natural food and nutritional relationships of larval mosquitoes. Annu Rev Entomol 37:379–376
Merritt RW, Craig DA, Wotton RS, Walker ED (1996) Feeding behavior of aquatic insects: case studies on black fly and mosquito larvae. Invert Biol 3:206–217
Merritt RW, Cummins KW, Berg MB (eds) (2008) Aquatic insects of North America, 4th edn. Kendall/Hunt, Dubuque
Meyer JL, Edwards RT (1987) Bacteria as a food source for black fly larvae in a blackwater river. J N Am Benthol Soc 6:241–250
Misof B, Liu S, Meusemann K, Peters RS, Donath A, Mayer C, Frandsen PB, Ware J, Flouri T, Beutel RG, Niehuis O, Petersen M, Izquierdo-Carrasco F, Wappler T, Rust J, Aberer AJ, Aspöck U, Aspöck H, Bartel D, Blanke A, Berger S, Böhm A, Buckley T, Calcott B, Chen J, Friedrich F, Fukui M, Fujita M, Greve C, Grobe P, Gu S, Huang Y, Jermiin LS, Kawahara AY, Krogmann L, Kubiak M, Lanfear R, Letsch H, Li Y, Li Z, Li J, Lu H, Machida R, Mashimo Y, Kapli P, McKenna DD, Meng G, Nakagaki Y, Navarrete-Heredia JL, Ott M, Ou Y, Pass G, Podsiadlowski L, Pohl H, von Reumont BM, Schütte K, Sekiya K, Shimizu S, Slipinski A, Stamatakis A, Song W, Su X, Szucsich NU, Tan M, Tan X, Tang M, Tang J, Timelthaler G, Tomizuka S, Trautwein M, Tong X, Uchifune T, Walzl MG, Wiegmann BM, Wilbrandt J, Wipfler B, Wong TK, Wu Q, Wu G, Xie Y, Yang S, Yang Q, Yeates DK, Yoshizawa K, Zhang Q, Zhang R, Zhang W, Zhang Y, Zhao J, Zhou C, Zhou L, Ziesmann T, Zou S, Li Y, Xu X, Zhang Y, Yang H, Wang J, Wang J, Kjer KM, Zhou X (2014) Phylogenomics resolves the timing and pattern of insect evolution. Science 346:763–767
Molineri CA, Siegloch E, Righi-Cavallaro KO (2010) The nymph of *Tortopus harrisi* Traver (Ephemeroptera: Polymitarcyidae). Zootaxa 2436:65–68
Morrison AC, Sihuincha M, Stancil JD, Zamora E, Astete H, Olson JG, Vidal-Ore C, Scott TW (2006) *Aedes aegypti* (Diptera: Culicidae) production from non-residential sites in the Amazonian city of Iquitos. Peru Ann Trop Med Parasitol 100(Suppl 1):S73–S86
Nebbioso A, Piccolo A (2013) Molecular characterization of dissolved organic matter (DOM): a critical review. Anal Bioanal Chem 405(1):109–124. https://doi.org/10.1007/s00216-012-6363-2
Newbold JD, O'Neill RV, Elwood JW, Winkle WV (1982) Nutrient spiralling in streams: implications for nutrient limitation and invertebrate activity. Am Nat 120:628–652
Ohkawa A, Ito T (2001) Terrestrial insect ingestion by filter feeding caddisfly larvae Brachycentrus *Brachycentrus americanus* (Trichoptera). J Freshw Ecol 16:263–266
Osterling EM, Bergman E, Greenberg LA, Baldwin BS, Mills EL (2007) Turbidity-mediated interactions between invasive filter-feeding mussels and native bioturbating mayflies. Fresh Biol 52:1602–1610
Palmer RW, Craig DA (2000) An ecological classification of primary labral fans of filter-feeding black fly (Diptera: Simuliidae) larvae. Can J Zool 78:199–218
Parkes AH, Kalff J, Boisvert J, Cabana G (2004) Feeding by black fly (Diptera: Simuliidae) larvae causes downstream losses in phytoplankton but not bacteria. J N Am Benthol Soc 23:780–792
Pinder LCV (1986) Biology of freshwater Chironomidae. Annu Rev Entomol 3:1–23
Plague GR, McArthur JV (2003) Phenotypic plasticity of larval retreat design in a net-spinning caddisfly. Behav Ecol 14:221–226
Pucat AM (1965) The functional morphology of the mouthparts of some mosquito larvae. Quaestiones Entomologicae 1:41–86

Ramírez A, Gutiérrez-Fonseca PR (2014) Functional feeding groups of aquatic insect families in Latin America: a critical analysis and review of existing literature. Revista de Biología Tropical 62:155–167

Roberts D (2014) Mosquito larvae change their feeding behavior in response to kairomones from some predators. J Med Entomol 51:368–374

Ross DH (1964) Evolution of caddisworm cases and nets. Am Zool 4:209–220

Ross DH, Craig DA (1980) Mechanisms of fine particles capture by the larval black flies (Diptera: Simuliidae). Can J Zool 58:1186–1192

Rothmeier G, Jäch MA (1986) Spercheidae, the only filter-feeders among Coleoptera. Proceedings of the third European congress of entomology (Amsterdam) 1986:133–137

Scott DC, Berner L, Hirsch A (1959) The nymph of the mayfly genus *Tortopus*. Annu Entomol Soc 52:205–213

Sehnal F, Sutherland T (2008) Silks produced by insect labial glands. Prion 2:145–153

Shapas TJ, Hilsenhoff WL (1976) Feeding habits of Wisconsin's predominant lotic Plecoptera, Ephemeroptera, and Trichoptera. Great Lakes Entomol 9:175–188

Skiff JJ, Yee DA (2015) The effects of protozoans on larval container mosquito performance. Annu Entomol Soc Am 108:282–288

Souza RS, Diaz-Albiter HM, Dillon VM, Dillon RJ, Genta FA (2016) Digestion of yeasts and beta-1 3-glucanases in mosquito larvae: physiological and biochemical considerations. PLoS One 11: e0151403. https://doi.org/10.1371/journal.pone.0151403

Steyn A, Roets F, Botha A (2016) Yeasts associated with *Culex pipiens* and *Culex theileri* mosquito larvae and the effect of selected yeast strains on the ontogeny of *Culex pipiens*. Microb Ecol 71:747–760. https://doi.org/10.1007/s00248-015-0709-1

Stork NE (2018) How many species of insects and other terrestrial arthropods are there on Earth? Annu Rev Entomol 63:31–45

Strand M (2017) The gut microbiota of mosquitoes: diversity and function. In: Wikel S, Aksoy S, Dimopoulos G (eds) Arthropod vector: controller of disease transmission, vol 1, pp 185–199

Tsurim I, Silberbush A (2016) Detrivory competition and apparent predation by *Culiseta longiareolata* in a temporary pool ecosystem. Israel J Ecol Evol 62:138–142

Tuno N, Kohzu A, Tayasu I, Nakayama T, Githeko A, Yan G (2018) An algal diet accelerates larval growth of *Anopheles gambiae* (Diptera: Culicidae) and *Anopheles arabiensis* (Diptera: Culicidae). J Med Entomol 55:600–608

Valzania L, Martinson VG, Harrison RE, Boyd BM, Coon KL, Brown MR, Strand MR (2018) Both living bacteria and eukaryotes in the mosquito gut promote growth of larvae. PLoS Negl Trop Dis 12:e0006638

Vezzani D (2007) Review: artificial container-breeding mosquitoes and cemeteries: a perfect match. Trop Med Inter Health 12:299–313

Wagner R, Barták M, Borkent A, Courtney G, Goddeeris B, Haenni J-P, Knutson L, Pont A, Rotheray GE, Rozkošný R, Sinclair B, Woodley N, Zatwarnicki T, Zwick P (2008) Global diversity of dipteran families (Insecta Diptera) in freshwater (excluding Simulidae, Culicidae, Chironomidae, Tipulidae and Tabanidae). In: Balian EV, Lévêque C, Segers H, Martens K (eds) Freshwater animal diversity assessment. Springer, Dordrecht, pp 489–519

Walker ED, Olds EJ, Merritt RW (1988) Gut content analysis of mosquito larvae (Diptera: Culicidae) using DAPI stain and epifluorescence microscopy. J Med Entomol 25:551–554

Walker ED, Kaufman MG, Merritt RW (2010) An acute trophic cascade among microorganisms in the tree hole ecosystem following removal of omnivorous mosquito larvae. Comm Ecol 11:171–178

Wallace J, Hutchens JJ (2000) Effects of invertebrates on lotic ecosystem processes. In: Coleman DC, Hendrix PF (eds) Invertebrates as webmasters in ecosystems. CAB International, Wallingford, pp 73–96

Wallace JB, Malas D (1976) The fine structure of capture nets of larval Philopotamidae with special emphasis on *Dolophilodesd istinctus*. Can J Zool 54:1788–1802

Wallace JB, Merritt RW (1980) Filter-feeding ecology of aquatic insects. Annu Rev Entomol 25:103–132

Wallace JB, O'Hop J (1979) Fine particle suspension-feeding capabilities of *Isonychia* spp (Ephemeroptera: Siphlonuridae). Annu Entomol Soc Am 72:353–357
Wallace JB, Sherberger FF (1974) The larval retreat and feeding net of *Macronema carolina* Banks (Trichoptera: Hydropsychidae). Hydrobiology 45:177–184
Webster JR, Newbold JD, Lin L (2016) Nutrient spiraling and transport in streams: the importance of in-stream biological processes to nutrient dynamics in streams. In: Jones JB, Stanley E (eds) Stream ecosystems in a changing environment. Elsevier, pp 181-239
Wiggins GB (1996) Larvae of the North American caddisfly genera (Trichoptera). University of Toronto Press
Wiggins GB (2005) Caddisflies the underwater architects. NRC Press co-published with The University of Toronto Press and the Royal Ontario Museum, Ottawa
Winters AE, Yee DA (2012) Variation in performance of two co-occurring mosquito species across diverse resource environments: insights from nutrient and stable isotope analyses. Ecol Entomol 37:56–64
Wondwosen B, Hill SR, Birgersson G, Seyoum E, Tekie H, Ignell R (2017) A(maize)ing attraction: gravid *Anopheles arabiensis* are attracted and oviposit in response to maize pollen odours. Mallar J 16:39
Wondwosen B, Birgersson G, Tekie H, Torto B, Ignell R, Hill SR (2018) Sweet attraction: sugarcane pollen-associated volatiles attract gravid *Anopheles arabiensis*. Mal J 17:90
Wotton RS (1988) Very high secondary production at a lake outlet. Freshw Biol 20:341–346
Wotton RS (1994) The biology of particles in aquatic systems, 2nd edn. Taylor & Francis
Wotton RS (2009) Feeding in blackfly larvae (Diptera: Simuliidae) – the capture of colloids. Acta Zool Lituanica 19:64–67
Wotton RS, Malmqvist B (2001) Feces in aquatic ecosystems feeding animals transform organic matter into fecal pellets which sink or are transported horizontally by currents; these fluxes relocate organic matter in aquatic ecosystems. BioScience 51:537–544
Wotton R, Malmqvist B, Muotka T, Larsson K (1998) Fecal pellets from a dense aggregation of suspension-feeders in a stream: an example of ecosystem engineering. Limno Ocean 43:719–725
Xu Y, Chen S, Kaufman MG, Maknojia S, Bagdasarian M, Walker ED (2008) Bacterial community structure in tree hole habitats of *Ochlerotatus triseriatus*: influences of larval feeding. J Am Mosq Control Assoc 24:219–227
Yee DA (2008) Tires as habitats for mosquitoes: a review of studies within the eastern United States. J Med Entomol 45:581–593
Yee DA (2016) What can larval ecology tell us about the success of *Aedes albopictus* (Diptera: Culicidae) in the United States? J Med Entomol 53:1002–1012
Yee DA, Juliano SA (2006) Consequences of detritus type in an aquatic microsystem: assessing water quality, microorganisms, and the performance of the dominant consumer. Fresh Biol 51:448–459
Yee DA, Kehl S (2014) Order Coleoptera (Vol I Chapter 39) In: Thorp JH, Rogers C, Tockner K (eds) Vol I: Ecology and general biology. In: Thorp JH, Covich A (eds) Freshwater invertebrates, pp 1004–1043
Yee DA, Kesavaraju B, Juliano SA (2004) Interspecific differences in feeding behavior and survival under food-limited conditions for larval *Aedes albopictus* and *Aedes aegypti* (Diptera: Culicidae). Annu Entomol Soc Am 97:720–728
Yee DA, Kaufman MG, Juliano SA (2007) The significance of ratios of detritus types and microorganism productivity to competitive interactions between aquatic insect detritivores. J Anim Ecol 76:1105–1115
Yee DA, Kaufman MG, Ezeakacha NF (2015) How diverse detrital environments influence nutrient stoichiometry between males and females of the co-occurring container mosquitoes *Aedes albopictus Ae aegypti* and *Culex quinquefasciatus*. PLoS One. https://doi.org/10.1371/journal.pone.0133734
Zhou CF (2010) Accessory gills in mayflies (Ephemeroptera). Stuttgarter Beiträge zur Naturkunde A Neue Serie 3:79–84

# Chapter 5
# Theme and Variation in the Development of Insect Mouthparts

**David R. Angelini and Frank W. Smith**

**Abstract** Insect mouthparts are serially homologous appendages. As such, their development and evolution are nonindependent. Arthropod appendages share similarities in their developmental origins and underlying genetics. Here, we review the development, specification, and patterning of insect mouthparts, with comparisons to the legs of *Drosophila melanogaster*. The expression and function of genes in the arthropod head give clues as to the homology of the labrum. The activity of Hox genes establishes appendage-specific gene expression and interactions allowing for the development of unique appendage types. Many similarities exist in the patterning of gnathal appendages and legs; however, unique variations in gene function in each appendage type provide clues to the developmental origins of mouthpart morphologies. We examine what is known about mouthpart patterning in mandibulates, as exemplified from several beetle species, as well as in the proboscis of *Drosophila melanogaster* and in the hemipteran rostrum of *Oncopeltus fasciatus*. With these findings in mind, we reflect on the evolution of serially homologous structures.

## 5.1 Introduction

The mouthparts and other appendages of arthropods possess a versatile developmental program. The segmented body plan of these animals makes it possible for the redeployment of a conserved developmental system, which nevertheless admits variations enabling evolution and adaptation. Arthropods confront their environment with a varied array of tools for different lifestyles. Their success seems supreme in species diversity, if not also anatomical disparity. As far as we now understand it, this diversity arises from a shared set of developmental events and the genes that control them. Nevertheless, investigations of comparative developmental biology

D. R. Angelini (✉)
Department of Biology, Colby College, Waterville, ME, USA

F. W. Smith
Department of Biology, University of North Florida, Jacksonville, FL, USA
e-mail: frank.smith@unf.edu



H. W. Krenn (ed.), *Insect Mouthparts*, Zoological Monographs 5,
https://doi.org/10.1007/978-3-030-29654-4_5

and genetics have uncovered a mixture of conservation and divergence in insect appendage development.

Here, we will attempt to contextualize the patterns of the evolution in insect mouthpart and appendage development through analogy to the musical ideas of "theme and variation." Compared between species and between appendage types, mouthparts and other insect appendages are both special and serial homologs, respectively. These appendages share a great deal in their developmental origins and underlying genetics. This is the common "theme." But key differences exist and influence the generation of morphological variations. While consistent themes run throughout, individual variations enable novel life histories.

In this chapter, we will review thematic aspects of development common to arthropod species and appendage types, reflecting primarily on the mouthparts of insects. We will also explore variations that allow for unique appendage types and for the unique features of individual lineages.

**Theme and Variation** In classical western music, the compositional technique of theme and variations uses a theme as the central musical idea of the piece, usually a memorable melody or chord progression. As the piece progresses, the theme is repeated again and varied in a different way. This cycle continues several times, providing the structure for the piece of music. Often the conclusion returns more closely to the theme or has a dramatic or poignant variation.

## 5.2 Homology: Theme and Variation

The shared developmental features of insect appendages reflect their complex evolutionary history, and it is useful to distinguish between the different ways in which these structures are related to one another. An important issue is that morphology and developmental similarities reflect both a history of common descent (homology or, formally speaking, special homology) and the shared deployment of developmental programs at different positions in the body (serial homology).

The first appreciation of morphological similarity in western science was closer to our current notion of serial homology and explicitly implicated development. The poet, statesman, and botanist, Johann Wolfgang von Goethe, carefully observed the development of plants and noted the similarities between leaves and floral organs (1790). Goethe described that these different structures grew from a similar meristem but diverged as development proceeded. He described the differences in their structures as arising from differences in "expansion" or "contraction" (Pfau 2010), although it seems clear he meant more than simply allometric differences. Goethe's observation of this connection has direct historical continuity to our present idea of serial homology. Moreover, Goethe also contemplated the implications of his idea for species diversity. He considered that his model of development could, starting from the "Urpflanze" (the archetypal or primordial plant), "invent plants without limit." This concept could also be universal: "The same law will permit itself to be applied to everything that is living" (Goethe 1814; Pfau 2010).

It is perhaps ironic that the term "homology" was coined by Richard Owen (1843), who vocally opposed the idea of species evolution. Nevertheless, Owen clarified the ideas first expressed by Goethe, crediting him for his influential observations (1848). Owen explicitly defined what he called "serial homology" as the repeated appearance of structures, such as vertebrae, within the body of an animal. He distinguished this from "special homology," which he described as "correspondency of a part or organ, determined by its relative position and connections, with a part or organ in a different animal" (1848). Without recognizing the possibility of evolution, Owen drew the distinction to what he called "general homology," "... that in which a part or series of parts stands to the fundamental or general type, and its enunciation involves and implies a knowledge of the type on which a natural group of animals ... is constructed."

After Darwin, the concepts of special and general homology collapsed into one, as writers on the subject came to understand (special) homology as arising from shared ancestry. By the mid-twentieth century, Boyden (1943, 1947) argued that the literature had gone too far and confused serial and special homology, complicating the use of characters in taxonomy. In the 1980s, evolutionary biologists considering the implications of development (e.g., Van Valen 1982; Roth 1984) and developmental biologists considering the implications of evolution (e.g., Raff and Kaufman 1983; Wagner 1989) began to reconsider concepts of homology, arguing for a more mechanistic basis and drawing clear distinctions between special and serial homology.

In recent decades, detailed mechanistic studies of development in anatomically disparate organisms (e.g., Hinman et al. 2003; Davidson 2006) have meant that considerations of the evolution of characters often depend on consideration of their generative mechanisms. Günter Wagner (2007) has argued that the unit of homology should be considered to be the developmental genetic system responsible for the identity of a particular trait, what he terms the character identity network (ChIN).

We will return to the idea of homology in our conclusions and explore how insect appendage development reflects general principles in the evolution of homologous structures. The anatomy of insect mouthparts will be detailed elsewhere in this volume. So we will only briefly summarize their structure here, focusing on taxa relevant to studies of development.

## 5.3 Overview of Insect Mouthpart Anatomy

The ancestral and most common state of insect mouthparts is the mandibulate type (Grimaldi and Engel 2005; Misof et al. 2014), which is fixed in several prominent orders such as Odonata, Orthoptera, Coleoptera, and Hymenoptera (Marshall 2006). Mandibulate mouthparts are primarily used for chewing, and they appear in both generalist and specialist taxa. From anterior to posterior, the mouthpart appendages consist of the labrum, mandibles, maxillae, and labium (Snodgrass 1930, 1935). The labrum's status as an appendage remains controversial (e.g., Popadić et al. 1998; Haas et al. 2001; Kimm and Prpic 2006; Posnien et al. 2009), and this question is considered below. Anatomically, the labrum acts as an upper lip and roof to the oral cavity. The mandibles are unjointed appendages used in chewing, and they are typically robust and well-muscled. The maxillae are paired, jointed appendages, which branch distally. The basal-most segment of the maxilla, the cardo, is jointed to the ventral head. The next segment is the stipes, which articulates with two medial endites, the lacinia and galea, which are fringed with setae in many species. Laterally, the stipes is also jointed to the maxillary palps. The palps typically consist of multiple segments, although their number may vary between different taxa. The palps typically function in the recognition of food. Chemosensory receptors on the surface of the palps aid the insect in identifying its target food (Snodgrass 1930; Chapman 1998). The posterior mouthpart appendage is the labium. The proximal labial segments fuse medially, forming the mentum and prementum. These segments may be jointed, or the joint between them may fuse, as in *Tribolium* (Sokoloff 1972; Angelini et al. 2012a). Medially, the prementum articulates to a set of endites in most species. There may be as many as four labial endites, two medial glossae and two lateral paraglossae, although these are reduced or fused in some taxa (Snodgrass 1930, 1935). Lateral of the endites, the labium also articulates with a pair of palps, similar in their structure and function to the maxillary palps. The number of labial palpomeres also varies among taxa. The hypopharynx is a fleshy, non-appendicular structure that acts as a tongue or the bottom of the oral cavity in some taxa. While not prominent in many mandibulate insects, the hypopharynx is an essential component of some derived mouthpart morphologies.

Fossils and phylogenetic evidence establish mandibulate anatomy as the ancestral state for insects (Grimaldi and Engel 2005; Misof et al. 2014). Among extant orders, at least 24 of the 32 (as recognized by Misof et al. 2014) are characterized by mandibulate mouthparts. The development of mandibulate mouthparts has been examined in model species representing multiple orders, including the cricket *Gryllus bimaculatus* (reviewed by Liu and Popadić 2017) and the beetle species *Tribolium castaneum* (Angelini et al. 2012a), *Onthophagus taurus* (Simonnet and Moczek 2011), and *Cyclommatus metallifer* (Gotoh et al. 2017).

However, some of the most successful groups of insects have exploited variations on the mandibulate theme. Among these novel morphologies is the principle insect model of development and genetics, the fruit fly *Drosophila melanogaster*. Diptera are characterized by the modification of mouthparts to piercing or sponging functions. In Muscomorpha, such as *D. melanogaster*, this involves the reduction and fusion of mouthpart appendages and surrounding head structures into a proboscis. The labial palps are absent, and the labium ends in a modified area called the labellum that is used for collection of liquid or particulate food (Snodgrass 1944). Mosquitos have evolved bladelike mandibles and maxillary laciniae, with an elongated hypopharynx used to secrete saliva (Snodgrass 1959). Emerging models of vector biology, such as *Anopheles gambiae* (Adolfi and Lycett 2018), have the potential to serve as comparative models of mosquito mouthpart development in the future.

The milkweed bug *Oncopeltus fasciatus* has also emerged as an informative system for developmental genetics (Chipman 2017; Panfilio et al. 2018), and this species represents the diverse Hemiptera. In this order, the labium is modified into a medially fused rostrum with multiple joints and no endites, while the mandibles and maxillae form thin stylets used in piercing and fluid feeding. Lepidoptera are another lineage in which existing model species, such as the silk moth *Bombyx mori* (Tomita and Kikuchi 2009; Ando et al. 2018), may be amenable to developmental genetic studies of mouthparts. Lepidopteran larvae retain chewing mandibulate mouthparts. Except for the early-branching lineage of Micropterigidae, adult Lepidoptera have evolved mouthparts in which the maxillary galeae form a proboscis typically used for nectar feeding (Krenn 2010). Secondarily, adults of the ghost moths (Hepialoidea) have reduced or absent maxillary palps and galeae. The mouthparts of these moths are vestigial, and the adults do not feed (Powell and Opler 2009). A fascinating novelty exists in Prodoxidae, where female Yucca moths develop an enlarged maxillary palpomere that is used independently of the proboscis to pollinate their host plant (Davis 1967; Pellmyr and Krenn 2002).

Other groups present intriguing mouthpart modifications, but few models currently lend themselves to developmental genetic investigations. For example, Thysanoptera present an interest comparison to Hemiptera, their sister taxon. The mouthparts of thrips are asymmetrical, with a single left mandible modified to form a piercing stylet. The maxillae differ in size, but each possesses a medial stylet and a small lateral palp. The thysanopteran labium is much closer in morphology to that of mandibulates. It is symmetrical, with a medial mentum and prementum, ending distally in medial endites and lateral palps (Jones 1954; Hunter and Ullman 1992). Siphonaptera (fleas) are another insect group with independently derived piercing mouthpart morphologies (Snodgrass 1946). In fleas the mandibles are absent, but bladelike mouthparts are formed by elongation of the labrum and laciniae. The maxillae and labium retain palps. Unfortunately, despite their medical importance, developmental studies of Siphonaptera have lagged behind other groups.

## 5.4 Development of Insect Mouthparts

### *5.4.1 The Embryonic Origins of Insect Mouthparts*

In all hemimetabolous and many holometabolous insects, the mouthparts originate as ventral-lateral outgrowths from the embryo (Fig. 5.1; Snodgrass 1928; Butt 1949; Van Horn 1966). Limb buds appear soon after segment formation. Therefore, in species with short germ band development, the limb buds of the gnathal and thoracic segments appear before obvious external segmentation in the abdomen is completed. Initially, limb buds consist exclusively of ectoderm, but mesodermal cells from the body of each segment contribute to the appendages forming the muscles (Eastham 1931;

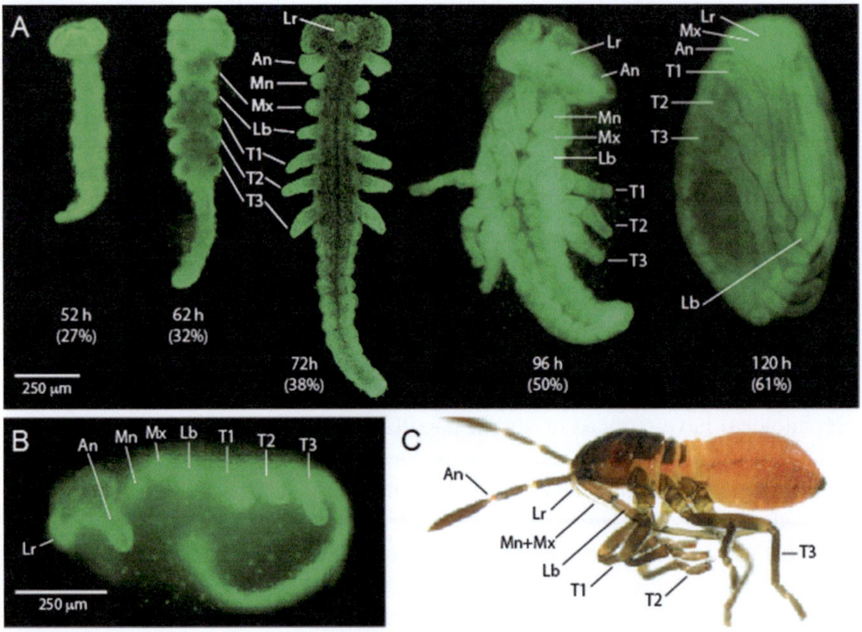


**Fig. 5.1** In most insect species, appendages develop from three-dimensional embryonic limb buds, such as in the milkweed bug *Oncopeltus fasciatus*. (**a**) *O. fasciatus* embryos of different ages are shown stained with Sytox, a fluorescent dye that binds to DNA, highlighting nuclei. Ages are given as hours post egg-laying and as a percentage of total average developmental time. Embryos have been dissected away from yolk for clarity. (**b**) A 72-h embryo stained with Sytox is shown from a lateral view with the yolk intact. (**c**) Lateral view of an *O. fasciatus* first-instar nymph. Notice that appendages are visible early, before abdominal segmentation is complete. The limb buds grow rapidly, and by 96 h, regionalization within the appendages is apparent. The labial appendages are initially separate but migrate ventrally, and by 120 h, they fuse together at the midline. *An* antenna, *Lr* labrum, *Mn* mandible, *Mx* maxilla, *Lb* labium, *T1–3* thoracic legs

Heming 1980). In Holometabola, muscle stem cells are associated with the imaginal discs and also give rise to the appendicular muscles at metamorphosis (Snodgrass 1935). During the germ band stage, specific gene expression establishes the components of differing character identity networks to define each appendage type.

### 5.4.2 Postembryonic Development of Appendages

Ancestrally, insects have a more-or-less direct development of the body plan. While adult structures such as wings and genitalia only appear after the adult molt (or in the subimago of Ephemeroptera; Edmunds and McCafferty 1988), among members of the hemimetabolous orders, which lack a complete metamorphosis, juveniles hatch with appendages similar in structure to those of the adult, differing only in relative size and cuticle or sensory features. Nevertheless, the number of segments in some distal appendage structures can vary by developmental stage. For example, in *Oncopeltus* juveniles, the legs have two tarsomeres on each leg, while adults have three, apparently due to the formation of a novel joint within the distitarsus.

In the Holometabola, species undergo a complete metamorphosis with a non-motile pupa. During this stage, appendages undergo a more dramatic repatterning. In most holometabolous orders, legs and mouthparts are present in juveniles but have a less complex morphology compared to adults. For example, the distal segments of *Tribolium* juvenile legs are much smaller than in the adult, and the tibiotarsus exists as a single segment that will become two in the adult (Angelini et al. 2012b). Adult structures are produced by cells from corresponding larval structures (Švácha 1992). An extreme "indirect" form appendage development exists in some Holometabola. *Drosophila* is a familiar example, in which larval appendages are visible externally only as small sensory Keilin's organs (Dambly-Chaudière and Ghysen 1986). In fruit flies and other Muscomorpha, most of the larval epidermis is polyploid (Smith and Orr-Weaver 1991) and must be replaced during metamorphosis. Imaginal discs give rise to the appendages and much of the surrounding body wall, while imaginal histoblasts produce to the remainder of the adult cuticle (Mandaravally Madhavan and Schneiderman 1977).

## 5.5 The Mystery of the Labrum

The labrum is an apical appendage-like structure on the insect head. It functions as the upper lip of insects (Snodgrass 1935); houses many sensory structures, such as setae, pressure receptors, trichoid sensilla, and coeloconic sensilla (Smith et al. 2014b); and serves a chemosensory function (Ortega-Hernández and Budd 2016). Several long-standing questions regarding the labrum have perplexed biologists (Budd 2002; Scholtz and Edgecombe 2006; Ortega-Hernández et al. 2017). Is the labrum a segmental structure, and if so, which segment is the labrum associated with? Is the labrum homologous to the paired ventral appendages that characterize insects and other arthropods? Lastly, what structure, if any, is the labrum homologous to in most distant relatives of Arthropoda? Several hypotheses have been proposed for each of these questions based on comparative studies of morphology and embryogenesis (Fig. 5.2). More recently, advances in developmental genetic techniques have provided an additional approach to testing hypotheses regarding the nature of the labrum. Here, we review the hypotheses for the nature of the insect labrum and summarize recent advances in our understanding of the labrum based on studies of developmental genes.

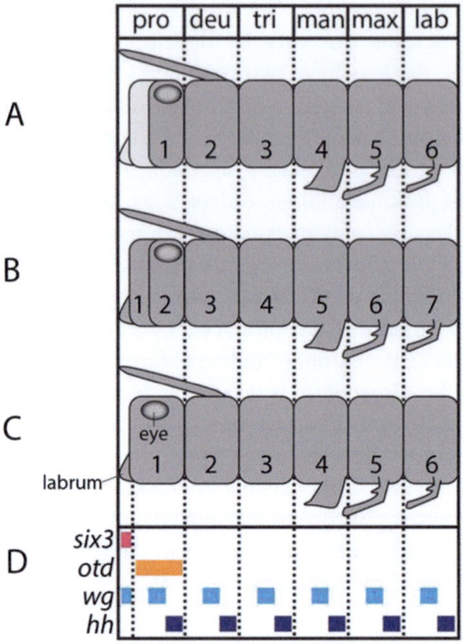


**Fig. 5.2** Gene expression and models of labrum identity in the insect head. Segments are numbered according to the different models. See main text for references. Segmental regions are shaded dark gray, and non-segmental regions are shaded light gray. (**a**) The protocerebral region is composed of a segmental and non-segmental region. (**b**) The protocerebral region is composed of two segments. (**c**) The protocerebral region is composed of a single segment and does not include non-segmental tissue. (**d**) Developmental gene expression patterns. The boundaries between *wg* and *hh* expression mark the parasegmental boundaries. The segment polarity gene *wg* is expressed in the labrum and in the ocular region of the developing protocerebrum. *deu* deutocerebral segment, *lab* labial segment, *man* mandibular segment, *max* maxillary segment, *pro* protocerebral region, *tri* tritocerebral segment

### *5.5.1 Where Is the Axial Origin of the Labrum?*

The labrum has been hypothesized to be a component of the intercalary segment—the segment that gives rise to the tritocerebral brain neuromere (Butt 1960; Haas et al. 2001), the acron—an unsegmental anterior-most region of the insect head (Brusca and Brusca 2003), or the first segment of the insect head (Budd 2002). The intercalary segment hypothesis is supported by several pieces of evidence, each of which has recently come under scrutiny in the literature. The first piece of evidence is based on the position of the labrum in the insect head. The stomodeum, which the labrum is closely associated with, sits somewhere between the intercalary segment and the antennal segment in models of insect head segmentation (Rempel 1975; Schmidt-Ott and Technau 1992; Rogers and Kaufman 1997; Haas et al. 2001). However, during embryogenesis, the stomodeum migrates posteriorly from an apical-most region (Khila and Grbić 2007). Furthermore, expression of the gene *six3*, which marks the apical-most region of the developing body axis of annelids, hemichordates, and onychophorans, also marks the labrum of insects (Fig. 5.2d; Steinmetz et al. 2010). These developmental studies suggest that the labrum originates in an apical position in the insect body axis, rather than in the intercalary segment, i.e., the ultimate position of the labrum does not reflect the position at which the labrum originates during embryogenesis. The second piece of evidence favoring an intercalary segment origin for the labrum is the fact that the labrum is innervated by the tritocerebral brain neuromere in the locusts *Schistocerca gregaria* and *Locusta migratoria* (Boyan et al. 2002). However, the labrum is innervated by the deutocerebrum in the horseshoe crab *Limulus polyphemus* (Mittmann and Scholtz 2003). The innervation of the labrum by either the deutocerebrum or tritocerebrum in euarthropods may represent derived conditions related to the ultimate position of the labrum, rather than its segmental origin (Scholtz and Edgecombe 2006; Bitsch and Bitsch 2010). Third, in the crustacean *Porcellio scaber* (Abzhanov and Kaufman 1999) and the centipede *Lithobius atkinsoni* (Hughes and Kaufman 2002b), the Hox gene *labial* (*lab*), which labels the intercalary/tritocerebral segment in all arthropods, is also expressed in the labrum (Haas et al. 2001). However, *lab* is not expressed in the labrum of other euarthropods investigated, including insects (Mlodzik et al. 1988; Rogers and Kaufman 1997; Peterson et al. 1999; Nie et al. 2001; Posnien and Bucher 2010), chelicerates (Damen et al. 1998; Sharma et al. 2012), and millipedes (Janssen and Damen 2006). This more comprehensive survey of *lab* expression suggests that its expression in the labrum of *P. scaber* and *L. atkinsoni* is likely a derived condition of the lineages leading to these species and is not indicative of the segmental origin of the labrum. In summary, most researchers now agree that the labrum originates in the insect protocerebral region. This hypothesis is supported by the expression of *six3* in the labrum (Fig. 5.2d; Steinmetz et al. 2010) and the fact that the labrum migrates posteriorly from an apical-most position during insect development (Khila and Grbić 2007).

### *5.5.2 Is the Labrum a Segmental Structure?*

While a consensus exists regarding the position of the labrum on the protocerebrum, there remains debate regarding the segmental nature of the protocerebrum. Current debates revolve around whether the protocerebrum represents a single segment, two fused segments, or a composite between a non-segmental and a segmental region. These debates have important implications for interpretations of the evolution of the labrum.

The existence of a non-segmental apical region in the insect head, and the heads of other euarthropods, originated with the Articulata hypothesis, which posits a sister-group relationship between Euarthropoda and Annelida, and a common origin of segmentation between these lineages (Scholtz 2002; Scholtz and Edgecombe 2006). The apical-most region of Annelida, referred to as the prostomium, lacks signatures of segmentation that are exhibited by body segments, such as nephridia and coelomic sacs, and unlike the body segments of Clitellata (earthworms and leaches), it does not develop from a posterior growth zone (Nielsen 2001; Ackermann et al. 2005; Scholtz and Edgecombe 2006). In polychaetes, distinct morphogenetic mechanisms underlie larval and juvenile segment development, but neither of these mechanisms is involved in development of the larval episphere, which gives rise to the prostomium (Ackermann et al. 2005; Scholtz and Edgecombe 2006). Therefore, this anterior-most region of the body axis may truly be regarded as non-segmental in nature (Scholtz and Edgecombe 2006). By extension, if segmentation is homologous between annelids and arthropods, then arthropods should exhibit an anterior-most non-segmental region.

In annelids, the prostomium is marked by *six3* expression during development, while the first segment—the peristomium—is marked by expression of the insect homolog of the gene *orthodenticle (otx)*. Likewise, *six3* marks the anterior-most region of the body axis of insects and other arthropods, while *otx* marks a slightly more posterior region (Fig. 5.2d; Steinmetz et al. 2010). These expression domains both lie within the protocerebral region in insects and other arthropods. Therefore, in accordance with the Articulata hypothesis, the protocerebral region would represent a composite between an anterior non-segmental region and a posterior segmental region (Fig. 5.2a), much as the annelid head is composed of the prostomium and the peristomium. The labrum lies within the expression domain of *six3*, in insects and other euarthropods (Steinmetz et al. 2010). Since this region is predicted to be homologous to the annelid prostomium—a non-segmental region, according to the Articulata hypothesis, the labrum would represent a non-segmental structure (Fig. 5.2a).

Molecular analyses have revealed that arthropods and annelids are not closely related (Aguinaldo et al. 1997; Dunn et al. 2008). Based on these analyses, the Articulata hypothesis has been replaced by the Ecdysozoa hypothesis, which posits that insects and other arthropods are more closely related to several unsegmented phyla than they are to annelids. The Ecdysozoa hypothesis suggests that segmentation evolved independently in Euarthropoda and Annelida. While the Ecdysozoa hypothesis has now reached a consensus in the field (Giribet and Edgecombe 2017), whether an apical unsegmented region exists in the head of insects and other euarthropods remains an open question (Budd 2002; Scholtz and Edgecombe 2006; Posnien et al. 2010). This possibility might be expected, if annelids and euarthropods evolved segmentation in parallel from shared ancestral developmental mechanisms that were reiterated along an unsegmented body axis, as has been proposed (Chipman 2010). Two observations based on studies of the red flour beetle *Tribolium castaneum* suggest that an apical non-segmental region does exist within the protocerebrum of insects (Posnien et al. 2010). First, the V-shaped median apical-most region that gives rise to the labrum lacks the parasegment-like gene expression patterns that reliably demarcate body segments along the rest of the insect body axis (Fig. 5.2d; Posnien et al. 2009, 2010). Second, the gene regulatory network that patterns the V-shaped region is not reiterated in segmental patterns (Li et al. 1996; Schroder et al. 2000; Economou and Telford 2009; Posnien et al. 2009; Steinmetz et al. 2010). Taken together, these observations suggest that the insect protocerebrum may be composed of a median apical non-segmental region *and* a posterolateral segmental region.

The remaining hypotheses regarding the segmental nature of the protocerebrum region agree that this region is segmental. By extension, these hypotheses argue that the labrum is a segmental structure. However, they disagree about the number of segments that compose the protocerebrum. In one hypothesis, the protocerebrum is composed of a fusion between two ancestrally independent segments (Fig. 5.2b; Strausfeld 2012; Cong et al. 2014). In insects and other arthropods, two regions can be recognized within the protocerebrum—the anterior region is referred to as the prosocerebrum and includes the labrum and the posterior region is referred to as the archicerebrum and includes the optic lobes and mushroom bodies of the brain (Urbach and Technau 2003). According to this hypothesis, the labrum represents a fused pair of segmental appendages of a protocerebral segment, while the stalked eyes of stem group euarthropods—homologs of insect compound eyes—represent the segmental appendages of an archicerebrum segment (Strausfeld 2012; Cong et al. 2014). In both insects and other euarthropods, segment polarity genes are typically expressed in a one-stripe per segment pattern but are expressed independently in both the labrum and ocular regions of the protocerebrum (Fig. 5.2d; Damen 2002; Farzana and Brown 2008; Posnien et al. 2009; Janssen 2012), which lends some developmental support to this hypothesis (Ortega-Hernández et al. 2017).

The lack of fossil evidence for the transition between a leg and a stalked eye, a prediction of the dual segment origin of the protocerebrum, challenges this hypothesis (Ortega-Hernández et al. 2017). Additionally, it now seems clear that the insect protocerebral region is homologous to the head of tardigrades (Smith et al. 2016, 2018) and the eye-bearing segment of onychophorans (Eriksson et al. 2010). Stalked eyes evolved in the euarthropod lineage, after this lineage diverged from Tardigrada and Onychophora (Park et al. 2018). Therefore, the dual segment origin predicts that two appendage pairs should be found in the protocerebral region of tardigrades and onychophorans, but a single appendage pair—the frontal appendages—is found in this region in onychophorans, and either no appendages or a single appendage pair is found in this region in tardigrades, depending on whether the teeth-like stylets of tardigrades are derived from legs or not (Nielsen 2001).

The remaining hypothesis argues that the protocerebrum represents a single segment, with the labrum representing a fused appendage pair of this segment (Budd 2002; Budd and Telford 2009; Ortega-Hernández et al. 2017). According to this hypothesis, the independent expression domains of segment polarity genes in the insect protocerebrum are the result of co-option of these genes for novel functions in the protocerebrum, possibly in development of the ocular lobes (Ortega-Hernández et al. 2017). In this hypothesis, each segment of ancient panarthropods housed a pair of appendages, and the labrum represents the appendage pair of a single protocerebral segment (Budd 2002; Budd and Telford 2009; Ortega-Hernández and Budd 2016). This hypothesis aligns well with recent conclusions about the homology of the protocerebral region across Panarthropoda based on developmental studies (Smith et al. 2016, 2018) and fossil evidence (Park et al. 2018). Yet it remains possible that the protocerebrum evolved from a fusion of two segments. If so, based on current evidence, this fusion must have happened in the stem group of Panarthropoda (Ortega-Hernández et al. 2017), rather than in the stem group of Euarthropoda (Strausfeld 2012; Cong et al. 2014).

### 5.5.3 Is the Labrum Serially Homologous to the Ventral Appendages?

Studies of labrum development have clear consequences for our interpretations of the homology of this structure to the ventral appendages of insects and other euarthropods—including the gnathal appendages. One way to gauge homology is to test whether similar mechanisms control the development of the labrum and the ventral appendages. Like the ventral appendages, the labrum originates as paired bud-like structures during insect development (Scholtz and Edgecombe 2006; Posnien et al. 2009). Furthermore, the distal appendage-patterning gene *Distal-less* (*Dll*) and other components of the appendage-patterning network are active in the developing labrum of several insect species investigated (Angelini and Kaufman 2004; Ronco et al. 2008; Ohde et al. 2009; Posnien et al. 2009; Simonnet and Moczek 2011; Smith et al. 2014b; Yoshiyama et al. 2013). These results support homology between the labrum and the ventral appendages.

Although similar mechanisms control development of the labrum and the ventral appendages, there are compelling differences. The ventral appendages develop at parasegmental boundaries. The Wnt signaling protein encoded by *wingless* (*wg*) is expressed on the anterior side of parasegmental boundaries, and *hedgehog* (*hh*) is expressed on the posterior side of the boundaries (Fig. 5.2d; Hidalgo 1991; Posnien et al. 2009). In *T. castaneum* and other insects, *wg* and *hh* are required for activation of *Dll* (Morata 2001; Posnien et al. 2009). Targeting *hh* or *wg* with RNAi during *T. castaneum* embryogenesis leads to loss of *Dll* expression where ventral appendages normally develop and, in the case of *hh*, complete deletion of all ventral appendages (Posnien et al. 2009). By contrast, the labrum does not develop at a parasegmental boundary, and RNAi targeting *hh* or *wg* treatments does not affect *Dll* expression in the labrum or lead to deletions of the labrum (Posnien et al. 2009). These results suggest that there are no parasegmental boundaries in the region where the labrum develops and that different mechanisms activate *Dll* expression in the labrum compared to ventral appendages. These conclusions are consistent with the hypothesis that the labrum develops in a non-segmental region of the insect head and suggest that the labrum is not a serial homolog of the ventral appendages (Posnien et al. 2009). Additionally, the Notch pathway activates *Dll* expression in the labrum, but not in the ventral appendages (Siemanowski et al. 2015).

If the labrum is not a serial homolog of the ventral appendages, then why are there are so many similarities between labrum development and ventral appendage development? One hypothesis is that the labrum is a novel structure that evolved by co-option of the ventral appendage-patterning network (Posnien et al. 2009;

Simonnet and Moczek 2011; Smith et al. 2014b). This hypothesis underpins a counterintuitive possibility. As an appendage, the labrum may not be homologous to the ventral appendages, while the developmental mechanisms that control development of the labrum and ventral appendages may be homologous.

### 5.5.4 How Does the Insect Labrum Relate to Structures in Other Animals?

A protocerebral appendage pair is predicted to be an ancient characteristic of Panarthropoda (Budd 2002). This ancestral appendage pair is thought to have given rise to the frontal appendages of onychophorans and possibly the teeth-like stylets of tardigrades (Nielsen 2001). This ancient appendage pair is exemplified by the "great appendages" of stem group euarthropods (Budd 2002). According to this hypothesis, the insect labrum—and the labra of other euarthropods—evolved from this ancient appendage pair. This hypothesis finds developmental support from expression of *six3*; *six3* is expressed in the developing antenna-like frontal appendages of onychophorans and the euarthropod labrum (Steinmetz et al. 2010; Eriksson et al. 2013). More recently, several genes that are expressed in the developing euarthropod labrum were found not to be expressed in the developing onychophoran frontal appendage, casting doubt on the significance of expression patterns of a single gene, *six3*, for inferring homology of the euarthropod labrum and onychophoran frontal appendage (Janssen 2017b). In other words, the fact that *six3* is expressed in both the labrum and frontal appendages may reflect the fact that they both develop in a homologous region of the body axis, rather than representing evidence that they share structural homology. On the other hand, differences in developmental patterning mechanisms should not be surprising, given how morphologically different the euarthropod labrum is compared to the onychophoran frontal appendages. Additional studies of labrum development and frontal appendage development need to be performed to better gauge the homology of these structures.

### 5.5.5 Current Outlook on Identity and Evolution of the Labrum

Although there is much to be determined regarding the origin of the labrum, the above discussion reveals three elements related to the evolution of the labrum that have reached a near consensus among zoologists. First, fossil evidence (Cong et al. 2014; Park et al. 2018) and developmental studies of Onychophora (Eriksson et al. 2010, 2013) strongly support a model in which an ancient ancestor of euarthropods had an appendage pair on the protocerebral region. Second, the labrum develops in the protocerebral region of the body axis (Steinmetz et al. 2010). Third, similar mechanisms control patterning of both the labrum and the ventral appendages (Smith et al. 2014b). However, determining whether the labrum is homologous to frontal appendages of onychophorans and ancient panarthropods and whether it is homologous to the ventral appendages requires additional studies. An important step toward addressing these questions will be to determine the segmental composition of the protocerebral region. New paleontological insights and developmental studies of a more diverse group of insects, additional euarthropods, and even onychophorans and tardigrades may be required to finally solve the mystery of the labrum.

## 5.6 Identity Specification of the Gnathal Appendages

The body plans of animals are established early in embryonic development. Anterior-to-posterior axial gradients activate a series of conserved transcription factors in adjacent and sometimes overlapping domains. Loss of function in these genes results in homeosis, the development of one anatomical structure in the position normally held by another. In many species these genes are linked in adjacent positions on the chromosomes. Their homeotic mutant phenotypes and linkage in a genetic complex gave them their name: Hox genes. In the 1990s and 2000s, evolutionary developmental biology (evo-devo) grew as a field in part by exploring the connections between Hox gene function and arthropod body plan variations (reviewed by Hughes and Kaufman 2002b; Angelini and Kaufman 2005). These genes are active during embryonic development, but the specification of appendage identity is an ongoing process, as evidenced by the transformation of appendages during metamorphic or juvenile-to-adult development following knockdown by RNA interference (e.g., Tomoyasu et al. 2005; Wasik et al. 2010; Aspiras et al. 2011).

### *5.6.1 The Mandible*

The mandible is the anterior-most head appendage that is not associated with a brain-housing segment. This appendage articulates with the head capsule but otherwise lacks joints. In zoological terms, it consists of the coxopodite (proximal) component, but not the telopodite (distal) component of the generalized insect appendicular appendage (Snodgrass 1935). In line with its coxopodite identity, the insect mandible lacks expression of the telopodite maker *Distal-less* during embryogenesis (Rogers et al. 2002). Additional gene expression studies suggest that the mandible is primarily composed of single endite of a single basal podomeres (Coulcher and Telford 2013).

While genetic screens of *Drosophila melanogaster* have laid the foundation for our understanding of how appendage identities are specified during development, fruit flies lack mandibles. For this appendage type, knowledge of mandible identity specification arose from studies of other insect species. As with other gnathal

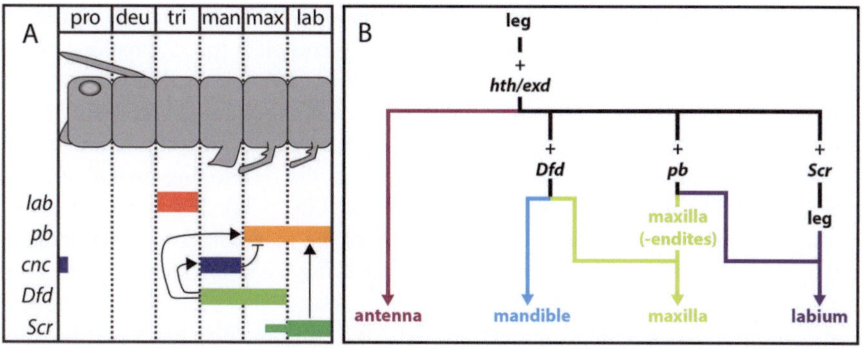


**Fig. 5.3** Head appendage identity specification based on studies of *T. castaneum* and other insects. See main text for references. (**a**) Expression domains of the Hox genes *labial* (*lab*), *proboscipedia* (*pb*), *Deformed* (*Dfd*), and *Sex combs reduced* (*Scr*) and the gene *cap'n'collar* (*cnc*) in the insect head. Known regulatory interactions are shown. Arrows indicate activation of expression. The horizontal bar indicates repression of expression. The thin line indicates a more restricted expression domain of *Scr* in the maxillary segment. (**b**) A model for appendage identity specification in insects. The default identity is leg (top). Expression of appendage identity selector genes in appendage anlagen (+ gene name) modifies the default leg state. Pathways leading to modified appendage identities are color-coded. *deu* deutocerebral segment, *lab* labial segment, *man* mandibular segment, *max* maxillary segment, *pro* protocerebral region, *tri* tritocerebral segment

appendages, the Hox genes play important roles in regulating mandible identity. In winged insects, the only Hox gene that is strongly expressed in the mandible is *Deformed* (*Dfd*) (Fig. 5.3a; Rogers and Kaufman 1997; Brown et al. 1999a; Hughes and Kaufman 2000; Rogers et al. 2002; Angelini et al. 2005). However, the insect ortholog of *Hox3*, *zerknüllt* (*zen*), which is typically expressed extraembryonically during insect development (Schmidt-Ott et al. 2010), is also expressed in a more typical Hox gene pattern in the apterygote insect *Thermobia domestica* (Hughes et al. 2004). In this species, *Hox3* is expressed in the mesoderm of the developing mandibles and maxillae. *Hox3* was most likely also expressed in the developing mandibles of the last common ancestor of insects, given that it is expressed in the mandibles of crustaceans (Papillon and Telford 2007) and centipedes (Hughes and Kaufman 2002a) and given that Zygentoma—the apterygote lineage that includes *T. domestica*—is an out-group of all winged insects that have been investigated (Yeates et al. 2016). Like in *T. domestica*, *Hox3* expression is restricted to the mesodermal layer of the developing mandibles of the crustacean *Daphnia pulex* (Papillon and Telford 2007), suggesting that this gene played a role in regulating development of mesodermal derivatives in the mandibles ancestrally in insects. Additionally, *Sex combs reduced* (*Scr*) is expressed at low levels in the mandibles of *T. domestica* (Passalacqua et al. 2010).

At this juncture, the function of *zen* and *Scr* in the developing mandibles of *T. domestica* is unknown. However, the function of *Dfd* during mandible development has been investigated in insects with generalized mandibulate mouthparts—the flour beetle *Tribolium castaneum*—and insects with highly derived mouthparts—the milkweed bug *Oncopeltus fasciatus*. In *T. castaneum*, null *Dfd* mutants and RNA interference (RNAi) targeting *Dfd* result in nearly complete transformations of the larval mandible to antenna (Fig. 5.3a; Brown et al. 1999b). In this species, Dfd activates the transcription factor-coding genes *cap'n'collar* (*cnc*) and *paired* (*prd*) during embryogenesis (Coulcher and Telford 2012). Dfd activates expression of *cnc* broadly across the mandible segment, including in the developing mandibles, and *prd* specifically in the endites of the mandibles (Fig. 5.3a). RNAi targeting *cnc* during embryogenesis results in transformation of the mandible to maxilla, indicating that this gene plays an important role in specifying mandible identity (Fig. 5.3b; Coulcher and Telford 2012). Expression of *cnc* is restricted to the mandible segment and labrum across mandibulate euarthropods. By contrast, it is expressed broadly across the developing embryo of chelicerates (Sharma et al. 2014) and onychophorans (Janssen 2017a). These results support a model in which the mandible characteristic of Mandibulata evolved by specialization of *cnc* function in this lineage.

In contrast to its function during embryogenesis, *Dfd* does not appear to be required for establishing mandible identity during metamorphosis in *T. castaneum* (Smith and Jockusch 2014). Instead, targeting *Dfd* during this period results in minor defects in mandible morphology but does not affect the identity of this appendage type. A similar result was recovered from studies of the postembryonic function of *Dfd* in a hemimetabolous insect species, the termite *Nasutitermes takasagoensis* (Toga et al. 2013). In this species, male minor workers can molt into either presoldiers or medium workers. The mandibles regress in size between the male minor worker and presoldier molt. When *Dfd* is targeted with RNAi, mandible regression is inhibited, i.e., presoldiers of *Dfd* RNAi treatments have larger mandibles than presoldiers of control treatments (Toga et al. 2013). This result suggests that *Dfd* functions to determine the size of presoldier mandibles postembryonically. As with postembryonic *Dfd* RNAi in *T. castaneum*, mandible identity is not affected by postembryonic *Dfd* RNAi in *N. takasagoensis* (Toga et al. 2013).

*Oncopeltus fasciatus* are true bugs (Hemiptera), and like other true bugs, they exhibit highly derived piercing-sucking mouthparts. In bugs, the mandibles and maxillae are modified into long thin stylets. The mandibles and maxillae form a piercing-sucking tube, with the mandible on the outside and the maxillae fused on the inside, with space between them for fluid to flow. The labial palps sheath and provide support to the feeding stylets. Of the Hox genes, only *Dfd* plays a role in establishing mandible identity in *O. fasciatus* (Hughes and Kaufman 2000). RNAi targeting this gene results in a transformation of the mandible to an antenna with multiple joints. The recognizable components of the ectopic antenna appear to exhibit distal antenna identity. Therefore, although bugs exhibit morphologically derived mandibles, *Dfd* functions to specify mandibular identity in the same manner as it does in insects with generalized mandible morphologies, by blocking antennal identity during embryogenesis.

## 5.6.2 The Maxilla

The Hox genes *pb* and *Dfd* are both expressed in the developing insect maxilla of most species that have been investigated (Fig. 5.3a; Brown et al. 1999a; Shippy et al. 2000; Curtis et al. 2001; Hughes and Kaufman 2002b; Angelini et al. 2005), and *Scr* is expressed in the maxillae of some insects that have been investigated (Passalacqua et al. 2010). Several null *pb* mutations cause nearly complete transformations of maxilla to leg in the homozygous state during embryogenesis in *T. castaneum* (Beeman et al. 1993; Shippy et al. 2000). Severely affected larvae of embryonic RNAi treatments targeting *pb* also exhibit nearly complete transformations of maxilla to leg (Shippy et al. 2000). Both loss-of-function *pb* mutations and larval RNAi targeting *pb* in *T. castaneum* also lead to transformations of the maxillae to leg during metamorphosis (Beeman et al. 1989; Smith and Jockusch 2014). In this case, only the palps are transformed, and they exhibit transformation to distal leg (femur, tibia, tarsus, pretarsus, claw). Together, these results suggest that *pb* played an ancient role in specifying maxilla identity in insects (Fig. 5.3b).

One might expect that the maxillae would develop into mandibles in the absence of *pb* function in *T. castaneum*. After all, in the absence of *pb* function, *Dfd* is the only Hox gene predicted to be expressed in the maxillae, and *Dfd* is required for specification of mandible identity (see above). Yet, the maxillae are transformed into legs when *pb* function is disrupted. This result can be explained by the fact that *cnc* is required for mandible development, and unlike *Dfd*, this gene is expressed in the developing mandibles, but not the maxillae (Fig. 5.3a; Coulcher and Telford 2012). However, *Dfd* does play an important role in maxillae development. In *T. castaneum*, *Dfd* loss-of-function embryos exhibit the telopodite component of the maxilla but lack the endite component (Fig. 5.3b; Brown et al. 2000). This suggests that *Dfd* is required for development of maxillary endites. When both *Dfd* and *pb* function are simultaneously disrupted, the maxilla develops into an antenna (Brown et al. 2002). Disrupting the function of *Dfd* and *Scr* simultaneously also results in maxilla to antenna transformations (Brown et al. 2002). The mechanism behind this result is unclear, but it most likely indicates that Dfd normally activates *pb* expression in the maxilla, but Scr can compensate for this function in the absence of *Dfd* function (Fig. 5.3a; Brown et al. 2002). In this model, when both *Dfd* and *Scr* function are compromised, *pb* is not expressed, resulting in transformation of the maxilla to antenna. There is some merit to this idea since Scr is required to activate *pb* expression in the labium of *T. castaneum* embryos (DeCamillis et al. 2001). This model of maxilla identity specification leaves open an interesting question. How does Scr affect expression of *pb* in the maxilla, since Scr is not expressed in the maxilla of *T. castaneum* embryos (Passalacqua et al. 2010)? It is possible that Scr is expressed in the maxilla when *Dfd* function is compromised, due to an inhibitory regulatory interaction between *Dfd* and *Scr*, but this possibility has not been tested in *T. castaneum*.

During *T. castaneum* metamorphosis, the roles that *pb* and *Dfd* play in maxilla identity specification are similar to their roles during embryogenesis (Smith and

Jockusch 2014). However, as with mandible development, it appears that slightly different mechanisms are active during metamorphosis. First, disrupting *Dfd* function with RNAi does not delete maxillary endites (Smith and Jockusch 2014), although this result is predicted based on studies of embryogenesis (Brown et al. 2000). Second, targeting *Dfd* and *Scr* simultaneously with RNAi does not cause homeotic transformations of the maxilla (Smith and Jockusch 2014), while the embryonic model predicts that this treatment should result in transformations of the maxillae to antenna (Brown et al. 2002). The simplest explanation for this difference is that, unlike during embryogenesis, *pb* expression does not require activation by Dfd or Scr in the maxilla during metamorphosis (Smith and Jockusch 2014).

Functional data and expression data make it clear that *pb* played a primary role in specifying maxilla identity in the last common ancestor of insects (Rogers et al. 2002). Intriguingly, however, *pb* is not expressed in the developing maxillae of the milkweed bug *O. fasciatus*, nor is this gene required for specification of maxilla identity in this species (Hughes and Kaufman 2000; Rogers et al. 2002; Angelini et al. 2005). In fact, the mechanisms that specify maxilla identity in *O. fasciatus* resemble those that specify mandible identity (Hughes and Kaufman 2000; Rogers et al. 2002). These similarities in specification resemble morphological similarities—both the mandible and maxilla are long unjointed appendages in *O. fasciatus* and other true bugs. By contrast, in other insect species, the maxilla is morphologically much more similar to the labium. Therefore, the loss of *pb* function in the maxilla of true bugs correlates with the evolution of the maxilla in this lineage toward a mandible-like morphology (Hughes and Kaufman 2000; Rogers et al. 2002). This change in morphology coupled with the loss of gene expression recalls the loss-of-function homeotic transformation of body segments that can be produced in Hox mutations in fruit flies and other animals. This correlation has led some authors (Rogers et al. 2002) to tentatively suggest that hemipteran mouthparts represent the success of a hopeful monster (Gould 1977; West-Eberhard 2003), the rare case in which a mutation of large phenotypic effect is favored and fixed by natural selection.

### 5.6.3 The Labium

The Hox genes *pb* and *Scr* are both expressed in the developing insect labium (Fig. 5.3a; Hughes and Kaufman 2000; Shippy et al. 2000; Curtis et al. 2001; DeCamillis et al. 2001; Hughes and Kaufman 2002b; Rogers et al. 2002; Angelini et al. 2005; Zhang et al. 2005; Hrycaj et al. 2010; Passalacqua et al. 2010). Structurally, the labium is very similar to the maxillae—consisting of basal podomeres with endites and terminal palps. However, unlike in the maxillae, the contralateral basal podomeres and endites are fused medially in the labium. Mirroring their morphological similarities, very similar mechanisms specify the maxillary and labial identities. For instance, as with the maxillae, disrupting *pb* function leads to transformations of the palps of the labium to distal leg in insect species that have been investigated (Pultz et al. 1988; Beeman et al. 1993; Hughes and Kaufman 2000; Smith and Jockusch 2014). These results indicate that *pb* plays a primary role in insects in promoting palp morphology during development. In contrast to the typical developing insect maxilla, *Scr* is typically strongly expressed in the developing labium (Fig. 5.3a; Hughes and Kaufman 2000; Curtis et al. 2001; DeCamillis et al. 2001; Rogers et al. 2002; Zhang et al. 2005; Hrycaj et al. 2010; Passalacqua et al. 2010). Therefore, *Scr* may be playing specific roles in distinguishing the labium from the maxillae. It is difficult to test this possibility during embryogenesis because *Scr* function is typically required for expression of *pb* in the labium (Fig. 5.3b; DeCamillis et al. 2001; Angelini et al. 2005). Loss of *Scr* function leads to loss of *pb* function, and the labium develops into antennae (Curtis et al. 2001; DeCamillis et al. 2001). Therefore, discriminating between *Scr* specific functions and functions of *Scr* that are mediated through its role in regulating *pb* expression are difficult in studies of insect embryogenesis. However, *Scr* does not appear to regulate *pb* expression during *T. castaneum* metamorphosis (see above). When *Scr* is targeted with RNAi during metamorphosis, the labial palps and endites develop characteristics that are typically restricted to the maxillae (Smith and Jockusch 2014). This result supports a role for *Scr* in promoting labium specific morphologies, while *pb* might play a more generic role in promoting the development of palp containing appendages.

### 5.6.4 The Role of Homothorax and Extradenticle in Specifying Mouthpart Identities

The protein products of genes *homothorax* (*hth*) and *extradenticle* (*exd*) must come together in the cytoplasm and form a heterodimer in order to be transported to the nucleus, where they function, in tandem, as transcription factors (Abu-Shaar and Mann 1998; Abu-Shaar et al. 1999; Kurant et al. 1998; Pai et al. 1998; Rieckhof et al. 1997). Therefore, the developmental functions of these genes perfectly overlap. Disrupting the function of either *hth* or *exd* results in homeotic transformations of gnathal appendage identities in *Gryllus bimaculatus* (Ronco et al. 2008), *O. fasciatus* (Angelini and Kaufman 2004), *Onthophagus taurus* (Simonnet and Moczek 2011), and *D. melanogaster* (Rauskolb et al. 1995; Inbal et al. 2001). These transformations most likely reflect the fact that Hth and Exd act as cofactors for Hox proteins and, as such, influence the specificity of Hox proteins for DNA regulatory elements (Chang et al. 1995; Chan et al. 1996; Johnson et al. 1995). In the absence of either Hth or Exd, Hox proteins are unable to properly regulate gene expression. This explains why the resulting phenotypes when *hth* or *exd* function is disrupted phenocopy the results of experiments in which Hox gene function is disrupted. Therefore, the roles that *hth* and *exd* play in specifying gnathal appendage identities are most likely mediated through direct interactions of their corresponding proteins with Hox proteins.

### 5.6.5 A General Model of Gnathal Appendage Identity Specification

Based on studies that began with *D. melanogaster* but have since expanded across diverse insects, it appears that highly conserved mechanisms control appendage identity specification in insects. The identities of most ventral appendages, including gnathal appendages, are determined by the Hox genes that are expressed in them (Hughes and Kaufman 2002a; see above). This is true for all ventral appendages except for the antennae. Hox genes are not expressed in the antennal segment (Fig. 5.3a; Hughes and Kaufman 2002a). In the absence of Hox gene function in the developing antennae, *hth* and *exd* promote antennal identity in insects (Fig. 5.3b; Struhl 1982a; Casares and Mann 1998, 2001; Mito et al. 2008; Ronco et al. 2008; Moczek and Rose 2009; Smith et al. 2014a; Setton et al. 2017). Antennal identity is specified by these genes, at least in part, by positively regulating the expression of the bHLH-PAS family transcription factor-coding gene *spineless* (Struhl 1982b; Duncan et al. 1998; Dong et al. 2002; Emmons et al. 2007; Shippy et al. 2009; Angelini et al. 2009; Toegel et al. 2009; Smith et al. 2014a; Setton et al. 2017). In developing legs, Hox genes repress *ss* expression (Duncan et al. 2010). In the absence of Hox gene activity, all ventral appendages develop as antennae (Struhl 1982a; Casares and Mann 1998, 2001; Brown et al. 2002; Smith and Jockusch 2014). While this might suggest that antennal identity is the default state of developing appendages, this is not the case. Disruption of *hth/exd* results in transformations of antenna to leg, even in the absence of Hox gene activity (Casares and Mann 2001; Dong et al. 2002; Ronco et al. 2008; Smith et al. 2014a). This suggests that leg identity is the default identity for ventral appendages (Fig. 5.3b; Casares and Mann 2001). To summarize the current model of ventral appendage identity specification, leg identity is most likely the default state, *hth/exd* promotes antennal identity in the absence of Hox gene activity, and Hox genes promote specific gnathal and leg identities combinatorially by suppressing antennal identity and the identities of other appendage types and/or by promoting particular ventral appendage identities (Fig. 5.3b).

Several features of the insect appendage identity specification mechanism predate the origin of insects. The Hox genes that pattern the gnathal appendages exhibit remarkably conserved expression patterns across Panarthropoda (Damen et al. 1998; Telford and Thomas 1998; Jager et al. 2006; Janssen and Damen 2006; Eriksson et al. 2010; Sharma et al. 2012; Janssen et al. 2014; Smith et al. 2016). Additionally, Hox genes are not expressed in the deutocerebral segment—the segment that houses antennae in insects—in Arthropoda or Onychophora (Damen et al. 1998; Telford and Thomas 1998; Jager et al. 2006; Janssen and Damen 2006; Eriksson et al. 2010; Sharma et al. 2012; Janssen et al. 2014). This suggests that specification of the appendage type that is associated with the deutocerebral segment without input from Hox genes is an ancient feature within Panarthropoda. Furthermore, RNAi targeting *hth* results in homeotic transformations of chelicerae—the deutocerebral appendages of Chelicerata—to leg in the harvestman *Phalangium opilio* (Sharma et al. 2015).

This indicates that *hth* was required for specification of deutocerebral appendage identity in the last common ancestor of Euarthropoda. Taken together, these results indicate that interactions among Hox genes and between Hox genes and *hth* were important for specifying appendage identities—including those of direct homologs of the insect gnathal appendages—in stem group Euarthropods and possibly earlier.

## 5.7 Developmental Genetic Patterning of Insect Appendages

While components of the core character identity network, such as Hox genes, establish the fate of different appendages, these genes activate a set of downstream genes and developmental events that direct the morphogenesis of the unique appendage types. Some of the genes involved have expression patterns and interactions that are similar across appendage types, while many are specific to the identity of the appendage. Most of our knowledge of this phase of appendage patterning comes from *D. melanogaster* and particularly from the leg imaginal disc. However, some studies in the fruit fly and other insects have examined patterning in diverse appendages, such as the mouthparts. Before considering the development of mouthparts, it will be useful to reflect on the thematic pattern demonstrated by development in the legs of insects. Several detailed reviews on the developmental genetics of insect appendages exist (Angelini and Kaufman 2005; Jockusch and Smith 2015; Jockusch 2017; Ruiz-Losada et al. 2018). Readers interested in an authoritative account of the developmental genetics of insect appendages should refer to Jockusch and Smith (2015).

### *5.7.1 Initiation of Appendage Primordia*

The cells that are competent to give rise to ventral appendages are specified at the anterior-posterior parasegment boundaries (Estella et al. 2003). In *D. melanogaster*, cells adjacent to the posterior of the boundary express the secreted protein Hedgehog (Hh) (Ingham 1993). To the anterior, Hh induces production of secreted Wingless (Wg), in ventral cells, and Decapentaplegic (Dpp), in dorsal cells (Basler and Struhl, 1994). The areas of *wg* and *dpp* expression maintain mutually repressive interactions, reinforcing their identities (Jiang and Struhl 1996; Theisen et al. 1996). The appendage primordia ultimately inherit cells from each compartment and the expression of these segment polarity genes marking their boundaries (Diaz-Benjumea et al. 1994; Theisen et al. 1996).

Outside of *Drosophila*, it is unclear whether these signaling pathways also initiate the expression of appendage development genes. The expression pattern of *wg* is known to extend laterally into the nascent appendages in diverse species, including the mayfly *Ephoron leukon* (O'Donnell and Jockusch 2010), the orthopterans *G. bimaculatus* (Niwa et al. 2000) and *Schistocerca americana* (Jockusch et al. 2000), the milkweed bug *O. fasciatus* (Angelini and Kaufman 2004), and the flour beetle *T. castaneum* (Bolognesi et al. 2008). However functional tests of *wg* in *G. bimaculatus* (Miyawaki et al. 2004) and *O. fasciatus* (Angelini and Kaufman 2004) appendage development do not produce defects in appendage growth or patterning. In the *T. castaneum* embryo, *wg* RNAi prevents appendage initiation (Ober and Jockusch 2006), suggesting that Wnt activation of appendage development may have evolved within Holometabola.

The transcription factor *Distal-less* (*Dll*) is one of first genes to be activated in the appendage primordia. In *D. melanogaster*, Wg promotes the expression of *Dll*, and its expression is restricted to a ventral-lateral domain in each embryonic body segment by inhibition from Dpp, dorsally, and epidermal growth factor (EGF), ventrally (Goto and Hayashi 1997). A subset of cells at the dorsal part of the Dll-expressing embryonic leg primordia contribute to the wing and haltere imaginal discs (Requena et al. 2017). In *Drosophila*, once the imaginal disc has formed, the initiation and maintenance of *Dll* expression is regulated by two separate enhancers. The first element is activated only by high levels of Wg and Dpp. Subsequently, an autoregulatory element is activated by Dll, independent of input from Wg or Dpp (Estella et al. 2008).

Dorsal-ventral specification within the leg imaginal disc is also controlled, independently, by Dpp and Wg (Estella and Mann 2008; Svendsen et al. 2009). These signaling molecules activate expression of transcription factors encoded by *optomotor blind* (*omb*) and *H15* in dorsal and ventral territories, respectively (Maves and Schubiger 2003; Wilder and Perrimon 1995). Orthologs of *omb* and *H15* are expressed in similar dorsal and ventral territories in the limb buds of the pill millipede *Glomeris marginata* (Prpic et al. 2005), but the expression of *H15* is reduced in the spider *Cupiennius salei* (Prpic et al. 2003) and actually appears in a dorsal area of the limb buds in the onychophoran *Euperipatoides kanangrensis* (Janssen et al. 2015). Moreover, patterns of *wg* and especially of *dpp* expression do not conform with the *Drosophila* model in most other arthropod species (Angelini and Kaufman 2005; Janssen et al. 2015). These results suggest that, while the specification of dorsal-ventral polarity may be conserved within insects, its establishment may rely on as yet unidentified factors.

### 5.7.2 Specification of Proximal-to-Distal Domains

By the late second instar, gene expression begins to differentiate discrete domains along the proximal-to-distal axis of the *Drosophila* leg imaginal disc (Lecuit and Cohen 1997), and similar patterns have been found in other insects (Fig. 5.4; Angelini and Kaufman 2005; Jockusch and Smith 2015). High levels of both Wg and Dpp occur only in the center of the *D. melanogaster* leg imaginal disc, where cells expressing the two signals are near each other spatially along the parasegment

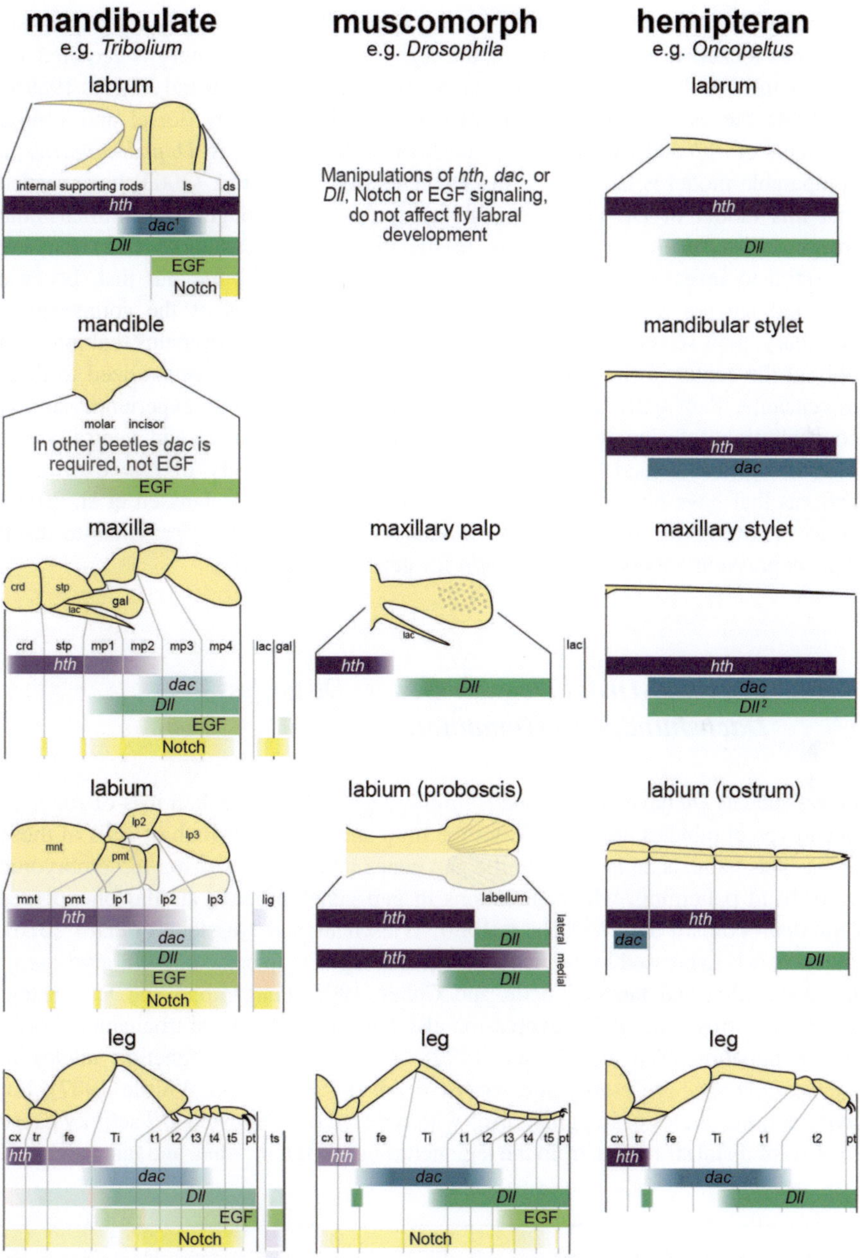


**Fig. 5.4** Summary of the requirement for appendage-patterning genes in the development of three insect species with different mouthpart morphologies. Distal structures are to the right in each panel, and lateral is up, except in the diagrams of legs where dorsal is up. Colored bars highlight structures affected by the manipulation of *hth*, *dac*, *Dll* and different components of the EGF and Notch signaling pathways. Notes: 1. While *dac* is expressed in an intermediate domain of the embryonic *T. castaneum* labrum (Prpic et al. 2001), embryonic *dac* RNAi has not been reported.

boundary (Lecuit and Cohen 1997; Wu and Cohen 1999). In this way, *Dll* expression becomes locked in at the center of the leg disc, where its activity is required for development of the telopodite, the distal region of the leg (Cohen and Jürgens 1989b).

While the establishment of the proximal-to-distal axis by dorsal and ventral gradients of *dpp* and *wg* expression has been well described in *D. melanogaster*, a comparable model is lacking for insects generally. Prpic et al. (2003) have argued that this model of *Dll* activation, in the context of a roughly two-dimensional imaginal disc, does not generalize to the three-dimensional limb buds that are ancestral to insects and other arthropods. These authors point out that, because *dpp* and *wg* are expressed along dorsal and ventral sides of the compartment boundary, their secreted products form two hyperbola-shaped domains that intersect only at the center of the disc. However, if the same model is generalized to three dimensions, then cells along the length of the limb would experience similar concentrations of signaling proteins produced from the dorsal and ventral sides. This theoretical consideration helps to explain the diversity of *dpp* expression patterns that have been found (Angelini and Kaufman 2005; Janssen et al. 2015). However, it is still not clear what mechanism produces unique proximal-to-distal gene expression outside of *Drosophila* for genes such as *Dll*.

### *5.7.3 Proximal-Distal Domain Genes: Distal-less, Dachshund, and Homothorax*

Genes such as *Dll* have been dubbed "limb gap genes" because their loss-of-function phenotype eliminates structures from the limb and reduces growth of cells in those areas. This name is an analogy to the gap genes involved in *Drosophila* embryonic germ band patterning, where mutations in gap genes produce similar phenotypes (Nüsslein-Volhard and Wieschaus 1980; Wieschaus and Nüsslein-Volhard 2016). *Distal-less* is expressed in the *D. melanogaster* leg disc in cells that will give rise to the distal tibia and tarsus (Lecuit and Cohen 1997; Panganiban and Rubenstein 2002). A narrow ring of *Dll* expression also appears in the distal trochanter shortly before pupation (Wu and Cohen 1999). Strong *Dll* loss-of-function alleles in *D. melanogaster* are embryonic recessive lethal (Sunkel and Whittle 1987), but hypomorphic alleles or imaginal discs with reduced or eliminated *Dll* activity cause the loss of distal structures from the leg, including the femur, tibia, and tarsus (Cohen

---

**Fig. 5.4** (continued) Metamorphic-stage *dac* RNAi does not produce noticeable defects in the labrum (Smith et al. 2014). 2. *Dll* is expressed in the embryonic maxillary appendages in *O. fasciatus*, but *Dll* RNAi has no noticeable effect on their development. *crd* cardo, *cx* coxa, *ds* distal sclerite of the labrum, *fe* femur, *gal* galea, *lac* lacinia, *lig* ligula (single labial endite), *lp1–3* labial palp segments 1–3, *ls* labral sclerite, *mnt* mentum, *mp1–4* maxillary palp segments 1–4, *pmt* prementum, *pt* pretarsus, *stp* stipes, *t1–5* tarsomeres 1–5, *Ti* tibia, *tr* trochanter, *ts* tibial spurs

and Jürgens 1989b). The expression pattern of *Dll* orthologs is well conserved in the distal legs of diverse insects and other animals (Jockusch and Smith 2015). Mutations or RNA interference reducing *Dll* activity has also produced deletion of the legs, distal to the trochanter, in several hemi- and holometabolous insect species (Fig. 5.4; Beermann et al. 2001; Angelini and Kaufman 2004; Ohde et al. 2009; Yoshiyama et al. 2013; Angelini et al. 2012b; Moczek and Rose 2009).

The proximal domain of the insect leg is marked by expression of the homeobox transcription factor *homothorax* (*hth*). Wg and Dpp act to inhibit the expression of *hth* in central parts of the leg imaginal disc, restricting its expression to the periphery (Abu-Shaar and Mann 1998; Wu and Cohen 1999). This pattern of *hth* expression in developing legs appears conserved in many insects (Prpic et al. 2003; Angelini and Kaufman 2004; Inoue et al. 2002) and in other arthropods (Prpic and Tautz 2003). In *D. melanogaster*, Hth functions by binding with its cofactor encoded by *extradenticle* (*exd*; Abu-Shaar and Mann 1998; Rieckhof et al. 1997). Leg imaginal discs that lack *hth* develop with a fusion of proximal leg structures, aberrant joint formation, or a proximal-to-distal transformation of podomeres (Casares and Mann 1998, 2001). A similar leg phenotype is found with *hth* or *exd* RNAi in *O. fasciatus* (Fig. 5.4; Angelini and Kaufman 2004), *G. bimaculatus* (Mito et al. 2008; Ronco et al. 2008), and *T. castaneum* (Smith and Jockusch 2014).

A unique intermediate domain becomes established later in the second instar leg imaginal disc with the expression of *dachshund* (*dac*) (Mardon et al. 1994; Giorgianni and Mann 2011). Over time, the area of *dac* expression expands to encompass cells that will give rise to the femur, tibia, and basitarsus. As with the activation of *Dll*, Wg and Dpp promote the expression of *dac* in the *D. melanogaster* leg imaginal disc (Lecuit and Cohen 1997). Its area of expression is refined through co-activation by Brinker (Brk), which is expressed in areas of the disc outside the influence of Dpp (Estella and Mann 2008). Dll also directly binds to a *dac* regulatory element to initiate its expression (Giorgianni and Mann 2011). Later in the third instar, *Dll* and *dac* distinguish the distal and intermediate domains of the leg through mutually antagonistic interactions (Dong et al. 2001). Orthologs of *dac* are expressed in similar patterns in the developing legs of diverse insects (Abzhanov and Kaufman, 2000; Schaeper et al. 2013; Inoue et al. 2002; Prpic et al. 2001; Angelini and Kaufman 2004; Tanaka and Truman 2007), although some differences exist among taxa in the dynamics and precise proximal or distal limits of *dac* expression (Jockusch and Smith 2015). Mutations eliminating *dac* activity in *D. melanogaster* reduce the length of the leg by eliminating the tibia, giving this gene its name in reference to the short-legged dog breed. Maternal RNAi in *O. fasciatus* produces embryos with similar deletion of the tibia (Fig. 5.4). Surprisingly, *dac* RNAi in *T. castaneum* embryos produces only minor leg defects (Lee et al. 2013), although RNAi during metamorphosis in the species results in deletion of the tibia (Angelini et al. 2012b), similar to the *D. melanogaster dac* mutant phenotype (Fig. 5.4).

Studies in diverse insects have largely supported the conservation of *Dll*, *dac*, and *hth* in establishing the pattern of proximal-to-distal domains in the leg. While small differences in the precise limits of expression and in timing exist (reviewed by Jockusch and Smith 2015), the homology of this network within leg development seems certain. In *Drosophila*, the interactions that define expression boundaries between the proximal-to-distal domain genes have been examined through elegant clonal analysis studies. Using methods for timed mosaic generation of cells with deletion alleles (Xu and Rubin 1993; Lee and Luo 1999), it is possible to see how cells lacking, for example, a distal gene change their expression of other genes or interact with neighboring wild-type cells. Using these methods, it has been found that the three principal proximal-distal domain genes, *Dll*, *dac*, and *hth*, interact antagonistically in a way that helps define each area (Dong et al. 2001; Wu and Cohen 1999).

The initial pattern established by *Dll*, *dac*, and *hth* is elaborated as other genes also become expressed in the leg, directing smaller aspects of local identity (reviewed by Angelini et al. 2012b; Jockusch and Smith 2015). The distal segmentation of the tarsus and development of the pretarsal structures are controlled by EGF signaling in *D. melanogaster* (Campbell 2002; Galindo et al. 2002). This terminal appendage-patterning role for EGF appears to be widely conserved. Knockdown of the EGF ligand during metamorphosis also eliminated the tarsus and tibial spurs in the legs of *T. castaneum* (Grossmann and Prpic 2012; Angelini et al. 2012b). Similarly, RNAi targeting the EGF receptor prevented regeneration of the distal tarsus and pretarsus in the legs of *G. bimaculatus* (Nakamura et al. 2008). Another well-conserved aspect of later appendage development is the requirement for Notch signaling in joint formation. In *D. melanogaster*, the Notch ligands Delta and Serrate are expressed adjacent to the locations of joint formation (de Celis et al. 1998; Bishop et al. 1999; Rauskolb and Irvine 1999; Tajiri et al. 2011), and the terminal EGF signal helps determine the position of joints in the leg by regulating the expression Notch pathway genes (Galindo et al. 2005). The role of Notch signaling in joint formation has been confirmed by RNAi in the insects *G. bimaculatus* (Mito et al. 2011) and *T. castaneum* (Angelini et al. 2012b). The spider *Cupiennius salei* also requires Notch signaling activity for leg growth and joint formation, leading to the suggestion that this function is an ancestral and defining feature of all euarthropods (Prpic and Damen 2009).

## 5.8 Developmental Genetic Patterning of Mandibulate Mouthparts

The developmental patterning of mouthparts is similar in many ways to the theme represented by legs. Unique morphologies are reflected by variations in the developmental system. Mandibulate mouthparts are the ancestral state for insects (Snodgrass 1935), but they also bear the closest resemblance to the theme established by leg development (Angelini et al. 2012a). The development of mandibulate mouthparts has been investigated through functional genetic tests in hemimetabolous and holometabolous species, including the primitively wingless insect *Thermobia domestica* (Schaeper et al. 2013), the cricket *G. bimaculatus* (Ronco et al. 2008), the beetles *Onthophagus taurus* (Simonnet and Moczek 2011) and *Tribolium castaneum* (Angelini et al. 2012a), and the stag beetle *Cyclommatus metallifer* (Gotoh et al. 2017).

### *5.8.1 The Mandible*

The mandible is the most anterior gnathal appendage, and it is unique in many ways. The insect mandible is unjointed, consisting of a single heavily muscled segment. The relative simplicity of its anatomy and its resemblance to the proximal-most segments of other appendages gave rise to the suggestion that the insect mandible is homologous only to other proximal appendage segments (Snodgrass 1935; Kukalová-Peck 1998). However it has also been suggested that the mandible evolved by reduction and elimination of joints, essentially retaining homology with the full proximal-to-distal extent of other appendages (Manton 1964). The gnathobasic hypothesis has been supported by developmental genetic studies of the distal appendage gene *Dll*, which is not expressed in the mandibles in insects (Panganiban et al. 1994; Scholtz et al. 1998; Popadić et al. 1998), and its suppression by RNAi does not affect mandible development (Niimi et al. 2005; Moczek and Rose 2009; Beermann et al. 2001; Angelini et al. 2012a; Gotoh et al. 2017; Yoshiyama et al. 2013). In *T. castaneum*, the Hox gene *Dfd* activates expression of *cnc* in the mandibular body segment, which inhibits expression of *Dll* (Coulcher and Telford 2012).

However, studies of different beetle species have revealed diverse roles for other genes in shaping the mandible. A functional study of 13 candidate appendage-patterning genes in the tenebrionid *T. castaneum* identified a role for EGF signaling in the mandible (Fig. 5.4; Angelini et al. 2012a). *EGF* RNAi significantly reduced mandible length, reducing the medial-distal incisor area in the flour beetle. This finding was unexpected, since EGF is required for formation of distal leg structures in diverse insects, including *G. bimaculatus* (Nakamura et al. 2008), *T. castaneum* (Grossmann and Prpic 2012; Angelini et al. 2012b), and *D. melanogaster* (Campbell 2002; Galindo et al. 2002). RNA interference targeting other appendage-patterning genes, including *dac* and *hth*, did not produce defects in the mandible of *T. castaneum*. In contrast, studies in scarabaeoid species *O. taurus* (Simonnet and Moczek 2011) and *C. metallifer* (Gotoh et al. 2017) found that RNAi suppression of *dac* caused reduction of mandibular teeth or incisors. Male *C. metallifer* have enlarged mandibles, and *dac* RNAi also significantly reduced their growth. Both studies also identified unique aspects of mandible patterning in these species. Depletion of *hth* modified a ridge between the molar and incisor regions in *O. taurus* (Simonnet and Moczek 2011) and eliminated the development of the medial mandibular teeth in *C. metallifer* (Gotoh et al. 2017). Other genes have not yet been examined in *O. taurus*, but RNAi targeting the distal leg gene *aristaless* also eliminated the mandibular teeth in *C. metallifer* (Gotoh et al. 2017). In contrast to its prominent role in the mandible of *T. castaneum*, *EGF* RNAi in *C. metallifer* did not cause noticeable defects.

Fully evaluating the gnathobasic hypothesis will require additional functional studies of mandibulate insects, especially among early-branching insect lineages. One possibility is that, while the ancestral state for insects may be gnathobasic, the existing interactions among appendage-patterning genes, necessary for the development of other appendage types, may have facilitated the evolutionary co-option of these genes into mandible development for roles in patterning novel structures, such as mandibular teeth.

### 5.8.2 The Maxilla and Labium

Patterning of the maxillae and labium is similar, reflecting similarities in their morphology. Their development also requires the same proximal-distal patterning genes known from leg development (Fig. 5.4). Exactly how these and other developmental regulatory genes direct appendage-specific anatomy is not completely clear. However, a striking difference is that the extent of overlap in gene expression is much greater for *Dll*, *dac*, and *hth* in the maxilla and labium than it is in the legs (reviewed by Angelini and Kaufman 2005; Jockusch and Smith 2015). It is likely that the combination of these transcription factors, along with regulatory proteins unique to these body segments, such as specific Hox protein combinations, directs specific target genes that ultimately leads to morphogenesis of these appendage types.

*Distal-less* is expressed in the palps and endites of both appendage types across diverse mandibulate insects (reviewed by Angelini and Kaufman 2005; Jockusch and Smith 2015). These structures are also eliminated by *Dll* mutation or RNAi in *T. castaneum* (Beermann et al. 2001; Angelini et al. 2012a). However, some variations in the requirement of *Dll* may exist among species, since metamorphic-stage *Dll* RNAi in the dung beetle *O. taurus* does not affect development of the maxillary endites (Simonnet and Moczek 2011), although other appendages have phenotypes resembling similar experiments in *T. castaneum* (Angelini et al. 2012a). Knockdown of *Dll* by RNAi in the sawfly *Athalia rosae* (Yoshiyama et al. 2013) and in the firebrat *Thermobia domestica* (Ohde et al. 2009) also reduced the maxillary and labial palps, causing fusion of palp segments. These studies did not report potential effects of *Dll* RNAi on the medial endites.

Expression of the intermediate domain gene, *dachshund*, occurs in the second maxillary and labial appendage segments (the stipes and prementum), in the maxillary and labial endites, and in a proximal to intermediate region of the palps (reviewed by Angelini and Kaufman 2005; Jockusch and Smith 2015). Tests for the functional requirement of *dac* in the mandibulate maxilla and labium have only been reported from the beetles *T. castaneum* (Angelini et al. 2012a) and *O. taurus* (Simonnet and Moczek 2011). In both species, *dac* RNAi reduces the length of and number of joints within the maxillary palps. However, in the labium, the two species have different *dac* RNAi phenotypes. In *O. taurus*, *dac* RNAi causes reduction of the prementum, while in *T. castaneum* the *dac* RNAi phenotype is similar in the maxilla and labium, with reductions in the length and joint number in the palps. Wider phylogenetic sampling is needed, but the serial homology among gnathal appendages suggests that a *dac* function in the palps may be ancestral.

Proximal appendage genes, *hth* and *exd*, are expressed across a much larger area of the maxilla and labium than they are in the legs (reviewed by Angelini and Kaufman 2005; Jockusch and Smith 2015). This creates a large degree of overlap between genes known for proximal and distal specification in the leg. While interactions among these genes have not been tested in the maxilla or labium of mandibulate insects, we predict that they should not exhibit the same antagonism seen during *D. melanogaster* leg development (e.g., Wu and Cohen 1999; Dong et al. 2001). RNA interference targeting *hth* in *T. castaneum* causes effects across the maxilla and labium (Angelini et al. 2012a). Shape changes occur in the proximal segments that may represent transformation toward more distal identities. Endites are present but reduced. Only the distal segments of the palps are unaffected. Depletion of *hth* in *O. taurus* produced similar phenotypes although the endites and palps of the maxilla appeared normal (Simonnet and Moczek 2011). The function of *hth* has also been tested in *G. bimaculatus*; however, its depletion by RNAi caused the transformation of mouthparts toward a mixed antenna-leg identity (Ronco et al. 2008). As discussed above, Hth also functions as a cofactor for Hox protein function.

Other aspects of gnathal appendage development follow the theme set by the legs (Fig. 5.4). Joints express and require locally restricted components of the Notch signaling pathway (Mito et al. 2011; Angelini et al. 2012a), and terminal patterning genes, such as *aristaless*, are required for development of the endites and the distal tip of the palps (Miyawaki et al. 2002; Angelini et al. 2012a).

## 5.9 Patterning Variations in Derived Mouthpart Morphologies

A number of successful insect lineages have evolved variations on the mandibulate theme. Two representatives of such taxa have been studied at the developmental genetic level: the fruit fly *Drosophila melanogaster* and the milkweed bug *Oncopeltus fasciatus*.

### *5.9.1 The Labellate Proboscis of* Drosophila

Muscomorpha have evolved an adult proboscis used to lap up liquid or semiliquid foods. In *Drosophila*, the proboscis is derived mostly from the labium, although labial palps or endites are absent. Maxillae are present on the lateral sides of the proboscis. A small maxillary base branches into a lacinia and maxillary palp. Adult mandibular structures are reduced and incorporated into the head capsule.

Signals that establish the proximal-distal axis of developing adult mouthparts are expressed in the late third instar larva, later than in the legs. Wnt and Dpp signaling is required for development of the maxillary field within the eye-antennal imaginal disc and for development of the labial imaginal discs (Joulia et al. 2005; Yasunaga et al. 2006; Doumpas et al. 2013), as they are in leg development. However, the timing of these signals is critical for the identity and patterning of both structures. If Wnt expression is activated early, the maxillary field develops as an ectopic antenna (Lebreton et al. 2008). In the labial disc, the Hox protein Pb represses *hh*, which results in reduced expression of *wg* and *dpp* (Joulia et al. 2005). In the absence of this repression, the labial disc develops as pair of ectopic legs.

*Distal-less* expression is activated in the developing adult maxillary field and labial imaginal disc by Wg and Dpp (Joulia et al. 2005; Yasunaga et al. 2006), although its expression is less intense than in the leg imaginal disc. *Dll* is expressed across the distal third of the labial disc (Fig. 5.4; Abzhanov et al. 2001; Joulia et al. 2005; Yasunaga et al. 2006). Mosaic mutant cells lacking *Dll* in the maxillary field fail to form maxillary palps, although the proximal base and lacinia remain (Cohen and Jürgens, 1989b). Distal structures of the labium are eliminated by loss of *Dll* mosaic clones in the labial disc (Cohen and Jürgens, 1989a; Yasunaga et al. 2006). Levels of *Dll* expression are also controlled by negative regulation from Scr (Abzhanov et al. 2001).

There appears to be no role for *dac* in the development of *Drosophila* mouthparts. The presence of Pb causes suppression of *dac*, and the maxillary field and labial disc in *D. melanogaster* show no expression of Dac (Abzhanov et al. 2001; Joulia et al. 2005). Abnormal phenotypes in the mouthparts have not been reported for *dac* loss of function.

The proximal leg patterning gene *hth* is expressed across the entire lateral layer of the labial imaginal disc, where it overlaps with *Dll*, but it is limited to the proximal two thirds as the medial layer (Yasunaga et al. 2006). Expression of *hth* occurs throughout the maxillary field (Pai et al. 1998). Exd, the cofactor required for Hth function, has very little expression in the maxillary field (Abzhanov et al. 2001). Nevertheless, loss of *hth* from the eye-antennal disc eliminates the maxillary palps (Stultz et al. 2012). The interactions among *Dll* and *hth* differ in the medial and lateral layers of the labial disc (Yasunaga et al. 2006). *Dll* represses *hth* only in the lateral layer, while the two genes are co-expressed medially. Since *Dll* and *hth* also are co-expressed in the antenna (reviewed by Angelini and Kaufman 2005; Jockusch and Smith 2015), the labium has been suggested as developing through an intermediate developmental program, rather than a completely unique one (Yasunaga et al. 2006).

### 5.9.2 *The Rostrum of Oncopeltus*

The mouthparts of Hemiptera are modified into a rostrum used for feeding by piercing and sucking liquid foods (Meek 1903; Snodgrass 1921). Different species have applied this strategy to predation and to phytophagy. One of the major differences in hemipteran mouthparts is that the difference in similarities among the gnathal appendages. While in mandibulate mouthparts the maxilla resembles the labium, in Hemiptera, it more closely resembles the mandibular appendages. However, both the mandibular and maxillary appendages of Hemiptera are modified into long, slender stylets. These interlock along their length, forming multiple channels for secretion of saliva and the uptake of liquefied food. The labial appendages are fused medially. The stylets are held in a midventral groove, and the rostrum is manipulated by the insect by means of joints between four labial segments.

The developmental genetic patterning of the hemipteran rostrum has been studied in the milkweed bug *Oncopeltus fasciatus* (Angelini and Kaufman 2004, 2005). The anatomical similarity in mandibular and maxillary appendages in *O. fasciatus* correlates with similar expression patterns in the gnathal Hox genes. In most insects, including *Drosophila* and *Tribolium*, the mandibular and maxillary body segments express the Hox gene *Dfd*, while *pb* is expressed in the maxillary and labial segments. The labial segment is also distinguished by expression of *Scr* (reviewed by Hughes and Kaufman 2002a). However, in *O. fasciatus*, the expression of *pb* is limited to the labial appendages (Rogers and Kaufman 1997; Angelini et al. 2005). As a result, the mandibular and maxillary segments develop with only Dfd for Hox regulation. RNA interference targeting the gnathal Hox genes produces transformations of the appendages in segments where these genes are normally expressed (Hughes and Kaufman 2000).

Proximal-distal domain genes also demonstrate the similarity of mandibular and maxillary development in Hemiptera (Angelini and Kaufman 2004). In embryos of *O. fasciatus*, the mandibular and maxillary limb buds both express *dac* and *hth* throughout their length. Both of these genes are also required for proper development of the stylets. Knockdown of *dac* by RNAi causes failure of the embryonic appendages to differentiate into stylets. Following *hth* RNAi, only the distal tips of the stylets differentiate, but the proximal majority of the appendage fails to and does not invaginate and coil into the head as normal (Dorn and Hoffmann 1983; Newcomer 1948). RNAi targeting *Dll* has no effect on stylet development in *O. fasciatus*, although the antennae, labium, and legs are all truncated (Angelini and Kaufman 2004). Juvenile-stage *Dll* RNAi also affects development of the male and female genitalia in *O. fasciatus* (Aspiras et al. 2011). The absence of a functional requirement for *Dll* in the hemipteran mandibular and maxillary and stylets is similar to what has been found in the development of mandibles in other insects. Unexpectedly, *Dll* mRNA and protein are expressed strongly in the maxillary limb buds of *O. fasciatus* embryos, although not in the mandibular limb buds (Rogers et al. 2002; Angelini and Kaufman 2004). This suggests two implications: First, a mechanism must exist to inhibit the function of the Dll transcription factor specifically in the maxillary body segment. Second, activation of *Dll* in the maxillary appendages is likely independent of Hox regulation, since both the mandibular and maxillary segments share the same Hox protein milieu. One possibility is activation of *Dll* by the gap gene *hunchback* (*hb*), which is expressed throughout the future head region in blastoderm-stage embryos, but *hb* expression is markedly more intense in the maxillary and labial body segments (Liu and Kaufman 2004).

The homology of hemipteran mouthpart structures to those of mandibulates has been uncertain (Meek 1903; Snodgrass 1921). Anatomists have proposed homology of the maxillary stylets to several components of the mandibulate maxilla, including the stipes (Cobben 1979), palpigers (Muir and Kershaw 1911a), lacinia (Crampton 1923; Hamilton 1981; Muir and Kershaw 1911b, 1912; Newcomer 1948; Snodgrass 1938, 1944), and to the entire maxilla (Bourgoin 1986; Parsons 1964, 1974). Perhaps this is asking the wrong question? The shift in *pb* expression and the functional similarities in *Dll*, *dac*, and *hth* in the mandibular and maxillary appendages suggests that, rather than modification of maxillary structures, the hemipteran maxillary stylets may have evolved by redeploying the mandibular developmental program within the maxillary appendages. Viewed in this way, both hemipteran stylets are homologous to the ancestral insect mandible. Moreover, the prominent functions of *dac* and *hth* in *O. fasciatus* stylet development fit well with developmental and anatomical evidence suggesting that the mandible is a proximal, gnathobasic structure.

## 5.10 Differences in Embryonic and Postembryonic Appendage Patterning

Appendages do not reach their final state in an individual insect until adulthood. Wings and genitalia are extreme in this regard, since they are not fully functional until adulthood. The subimago of Ephemeroptera is an exception, having functional wings in the last preadult stage (Edmunds and McCafferty 1988). However, other appendages, such as the antennae, mouthparts, and legs, appear in the juveniles of most insect groups and undergo subsequent development and repatterning during nymphal or pupal molts. Juvenile legs typically lack joints in distal structures, such as the tarsus or tibiotarsus. Experiments in beetle *Tenebrio molitor* using amputation (Huet and Lenoir-Rousseaux 1976) suggest that the entire larval leg contributes to the adult leg with cells maintaining their approximate relative position within the limb.

Once structures are formed during development, it is unclear to what extent their identity is irreversibly determined or whether they require continuous expression of genes to maintain their identity. Such a requirement may differ by species or between hemi- and holometabolous insects. The dramatic delay in appendage development in *Drosophila* has meant that the fruit fly has not provided its usual insights into development regarding differences in embryonic and postembryonic appendage patterning. Instead, other model species have provided these comparisons.

In the hemimetabolous hemipteran *O. fasciatus*, all juvenile stages have legs that closely resemble the adult in gross anatomy but have two tarsomeres to each leg. At the imaginal molt, the distal tarsomere is divided by a new joint, producing three tarsomeres in total. While the leg distal of the trochanter is lost if *Dll* is suppressed by RNAi during embryogenesis (Angelini and Kaufman 2004), only the distal tarsal joints of the adult are affected by *Dll* RNAi during the last juvenile instar (Aspiras et al. 2011). In contrast, the holometabolous species *T. castaneum* requires *Dll* activity continuously to maintain the growth and identity of leg structures. *Tribolium* adult legs have four to five tarsomeres, but larvae have a fused tibiotarsus. *Dll* mutations in *T. castaneum* affect larval and adult legs distal of the trochanter, causing reduced growth and an absence of distal identity affecting the femur, tibia, tarsus, and pretarsus (Beermann et al. 2001). When *Dll* is targeted by RNAi during the pupal stage, even structures such as the femur, which were properly formed in the same individuals as larvae, can be affected by the loss of *Dll* activity (Angelini et al. 2012b). Examination of in situ gene expression in embryonic, larval, and pupal legs of *Manduca sexta* (Lepidoptera), another holometabolous species with robust larval legs, has found a continuity of expression of *hth*, *dac*, and *Dll* at their respective proximal to distal levels in the leg as individuals undergo metamorphosis (Tanaka and Truman 2007).

Similarly, in the mouthparts of *T. castaneum*, gene activity is required to maintain the identity of specific regions during metamorphic development. Pupal-stage RNAi targeting *hth*, *dac*, or *Dll* produces defects in structures that were present in the larval maxilla and labium (Angelini et al. 2012a). While the requirement for these genes is limited to the tarsus during adult development of the legs in *O. fasciatus*, the mandibular and maxillary stylets continue to require *dac* activity for proper development during the imaginal molt (Aspiras et al. 2011), similar to its role during embryonic development (Angelini and Kaufman 2004). While the gross anatomy of the adult labium (rostrum) of *O. fasciatus* is not obviously altered by juvenile RNAi targeting proximal-distal domain genes, the length of the labium is reduced by *Dll* RNAi at this stage (Angelini and Kaufman 2005; Aspiras et al. 2011).

## 5.11 The Future of Research on Insect Appendage Development

Developmental genetics is still far from a detailed understanding of how genetic networks sculpt anatomy. However, we are beginning to appreciate how character identity networks initiate the development of specific structures. Mutant screens and functional analyses such as the production of mosaic discs have provided deep insights into appendage development in *D. melanogaster*. In contrast, studies of other species have relied heavily on a candidate-based approach, in which orthologs of genes from *D. melanogaster* developmental models are preferentially tested for roles in other species, highlighting instances of conservation or difference. While this path has been fruitful, it leads to a perspective of diversity that is likely biased toward conserved features of development. We often view other insects in terms of how they are "not like fruit flies" rather than how they uniquely generate their own morphologies.

Given the disparity of insect mouthparts, much remains to be learned about how genes direct this diversity of forms. Many model species are amendable to developmental and molecular genetic studies in ecologically and economically important insect groups with unique mouthparts. First among these are the Lepidoptera. A genetic model for the development of the galeate proboscis of moths or butterflies would provide important insights into this key innovation of the Lepidoptera.

Thankfully, the increasing accessibility of genomics and functional genetic manipulations is beginning to change the current situation. For example, a recent study of water striders used transcriptome comparisons among different legs to identify novel genes associated with the evolution and development of a unique, taxon-specific fan structure at the distal end of the midleg (Santos et al. 2017). RNA interference enabled tests of the gene's requirement in fan development, as well as the fan's function in the insect's locomotion. Similar applications of genomic methods should enable more sophisticated approaches that are not constrained by the assumptions of conservation with traditional genetic model species.

## 5.12 Returning to the Theme of Homology

Focusing on the genes responsible for development of traits underscores both special homology (inheritance from a common ancestor) and serial homology (deployment at multiple locations across the body plan). Importantly, this concept can also contextualize anatomical themes and variation seen across organisms. As we have recounted, some aspects of the developmental network may be conserved among serially homologous structures, such as the specification of appendage identity by Hox genes and the requirement of *Dll* for development of distal appendage structures, while other aspects, such as local interactions among patterning genes, may vary.

Importantly, serial homologs are not evolutionarily independent (Wagner 2007; Angelini et al. 2012b; Jockusch and Smith 2015). Serial homologs share developmental mechanisms and the genes that comprise their components. In this way, they have a shared evolutionary history, by virtual of their common genes, and a shared developmental history, via redeployment of those genes in different locations. Nevertheless, serial homologs can experience different selection pressures and may therefore evolve independently over time. Mutations in different lineages may affect development in ways that are general or specific with regard to serial homologs (Fig. 5.5). We will term evolutionary change causing uniform, similar changes

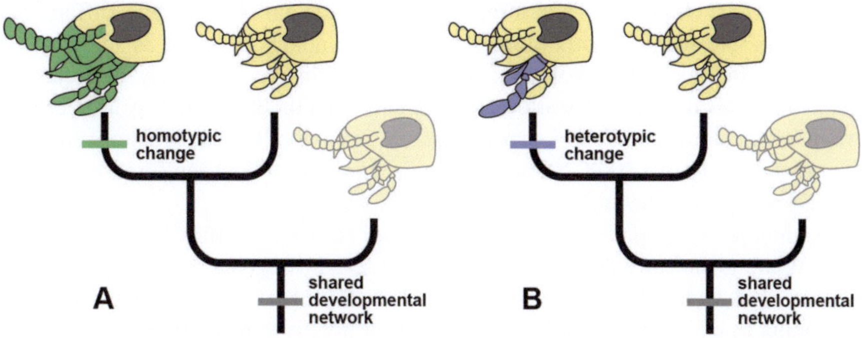


**Fig. 5.5** Serial homologs may evolve in concert, via homotypic change, or independently, via heterotypic change. These differing types of evolution are depicted here for mouthparts of a generic beetle-like insect. (**a**) In homotypic change, all serial homologs, such as the appendages, are affected pleiotropically and exhibit similar changes compared to the ancestral state. In this example, all appendages become enlarged and green. (**b**) In contrast, heterotypic change is limited to one serial homolog. In this example, the maxilla increases in size and changes color to blue. We predict mutations causing heterotypic change to be qualitatively different, such that their effects are limited in scope. The most likely mechanism for this specificity is change in a gene's regulatory elements, controlling expression in a given region

in all serial homologs homotypic change, while a change affecting a subset of serial homologs may be known as a heterotypic change.

With some understanding of the developmental system, we might begin to ask, what kinds of mutations are likely to result in homotypic or heterotypic change? An intuitive hypothesis might focus on the distinction between "core" genes, which function early and upstream in the developmental network in all serial homologs, and homolog-specific genes function later and downstream to effect unique morphology (Davidson 2006). This model would predict homotypic changes would result from mutations in the "core" genes and heterotypic changes would result from mutations in the more downstream, homolog-specific genes. However, our understanding of insect appendage development does not support this hypothesis (Fig. 5.4).

From what is known of appendage development, serial homologs differ from one another in their development at all levels. A comparison of *D. melanogaster* labial and leg imaginal discs helps illustrate this point. In both appendages, specification of the proximal-distal axis is accomplished by Wg and Dpp gradients, but this is indirectly modulated by *pb* activity in the labial disc in a way that is essential for labial development (Joulia et al. 2005; Yasunaga et al. 2006). Downstream of axis specification, the proximal-distal domain genes *Dll* and *hth* are co-expressed in the medial labial disc, but not laterally, as in the leg (Yasunaga et al. 2006). Similarly, *Dll*, *dac*, and *hth* are all expressed in antennae, mouthparts, and legs of diverse insects, but their areas of overlap differ (Fig. 5.4).

Rather than fixing mutations at different levels of the regulatory hierarchy, heterotypic evolution of serial homologs appears to have proceeded through changes in the regulatory interactions among genes that are part of a common theme in appendage development. Therefore, we predict that the key genetic differences underlying the unique morphologies of serial homologs, as well as the sites of mutation affecting their evolution, will be found in the regulatory elements of genes required for their formation (Rebeiz and Tsiantis 2017). If it is correct that morphological evolution among species proceeds more often via regulatory changes (Stern and Orgogozo 2008; Stern 2010), then similarities between serial homologs are likely to be retained to some extent by pleiotropy.

Additional developmental genetic studies are necessary to fully test this hypothesis. While technically demanding, functional tests of regulatory elements will be needed in a wider diversity of insect models. Genome editing technologies are beginning to make this possible. Ultimately, our goal is to read the musical genetic notation and understand how variations in morphological diversity emerge from the theme.

**Acknowledgments** Portions of this material are based upon work supported by US National Science Foundation grant IOS-1350207 to DRA. Special thanks to Joanna Smit Nelson for her perspective on theme and variation in musicology.

## References

Abu-Shaar M, Mann RS (1998) Generation of multiple antagonistic domains along the proximodistal axis during *Drosophila* leg development. Development 125:3821–3830

Abu-Shaar M, Ryoo HD, Mann RS (1999) Control of the nuclear localization of extradenticle by competing nuclear import and export signals. Genes Dev 13:935–945

Abzhanov A, Kaufman TC (1999) Homeotic genes and the arthropod head: expression patterns of the *labial*, *proboscipedia*, and *Deformed* genes in crustaceans and insects. Proc Natl Acad Sci U S A 96:10224–10229. https://doi.org/10.1073/pnas.96.18.10224

Abzhanov A, Kaufman TC (2000) Homologs of *Drosophila* appendage genes in the patterning of arthropod limbs. Dev Biol 227:673–689. https://doi.org/10.1006/dbio.2000.9904

Abzhanov A, Holtzman S, Kaufman TC (2001) The *Drosophila* proboscis is specified by two Hox genes, *proboscipedia* and *Sex combs reduced*, via repression of leg and antennal appendage genes. Development 128:2803–2814. https://doi.org/10.1016/S0925-4773(96)00649-1

Ackermann C, Dorresteijn A, Fischer A (2005) Clonal domains in postlarval *Platynereis dumerilii* (Annelida: Polychaeta). J Morphol 266:258–280

Adolfi A, Lycett GJ (2018) Opening the toolkit for genetic analysis and control of *Anopheles* mosquito vectors. Curr Opin Insect Sci 30:8–18. https://doi.org/10.1016/j.cois.2018.07.014

Aguinaldo AM, Turbeville JM, Linford LS et al (1997) Evidence for a clade of nematodes, arthropods and other moulting animals. Nature 387:489–493. https://doi.org/10.1038/387489a0

Ando T, Fujiwara H, Kojima T (2018) The pivotal role of *aristaless* in development and evolution of diverse antennal morphologies in moths and butterflies. BMC Evol Biol 18:8. https://doi.org/10.1186/s12862-018-1124-2

Angelini DR, Kaufman TC (2004) Functional analyses in the hemipteran *Oncopeltus fasciatus* reveal conserved and derived aspects of appendage patterning in insects. Dev Biol:271, 306–321. https://doi.org/10.1016/j.ydbio.2004.04.005

Angelini DR, Kaufman TC (2005) Insect appendages and comparative ontogenetics. Dev Biol 286:57–77. https://doi.org/10.1016/j.ydbio.2005.07.006

Angelini DR, Liu PZ, Hughes CL, Kaufman TC (2005) Hox gene function and interaction in the milkweed bug *Oncopeltus fasciatus* (Hemiptera). Dev Biol 287:440–455. https://doi.org/10.1016/j.ydbio.2005.08.010
Angelini DR, Kikuchi M, Jockusch EL (2009) Genetic patterning in the adult capitate antenna of the beetle *Tribolium castaneum*. Dev Biol 327:240–251. https://doi.org/10.1016/j.ydbio.2008.10.047
Angelini DR, Smith FW, Aspiras AC et al (2012a) Patterning of the adult mandibulate mouthparts in the red flour beetle, *Tribolium castaneum*. Genetics 190:639–654. https://doi.org/10.1534/genetics.111.134296
Angelini DR, Smith FW, Jockusch EL (2012b) Extent with modification: leg patterning in the beetle *Tribolium castaneum* and the evolution of serial homologs. G3 Genes Genomes Genet 2:235–248. https://doi.org/10.1534/g3.111.001537
Aspiras AC, Smith FW, Angelini DR (2011) Sex-specific gene interactions in the patterning of insect genitalia. Dev Biol 360:369–380. https://doi.org/10.1016/j.ydbio.2011.09.026
Basler K, Struhl G (1994) Compartment boundaries and the control of *Drosophila* limb pattern by hedgehog protein. Nature 368:208–214. https://doi.org/10.1038/368208a0
Beeman RW, Stuart JJ, Haas MS, Denell RE (1989) Genetic analysis of the homeotic gene complex (HOM-C) in the beetle *Tribolium castaneum*. Dev Biol 133:196–209
Beeman RW, Stuart JJ, Brown SJ, Denell RE (1993) Structure and function of the homeotic gene complex (HOM-C) in the beetle, *Tribolium castaneum*. BioEssays 15:439–444
Beermann A, Jay DG, Beeman RW et al (2001) The *short antennae* gene of *Tribolium* is required for limb development and encodes the orthologue of the *Drosophila* Distal-less protein. Development 128:287–297
Bishop SA, Klein T, Arias AM, Couso JP (1999) Composite signalling from Serrate and Delta establishes leg segments in *Drosophila* through Notch. Development 126:2993–3003
Bitsch J, Bitsch C (2010) The tritocerebrum and the clypeolabrum in mandibulate arthropods: segmental interpretations. Acta Zool 91:249–266. https://doi.org/10.1111/j.1463-6395.2009.00402.x
Bolognesi R, Farzana L, Fischer TD, Brown SJ (2008) Multiple Wnt genes are required for segmentation in the short-germ embryo of *Tribolium castaneum*. Curr Biol 18:1624–1629
Bourgoin T (1986) Valeur morphologique de la lame maxillaire chez les Hemiptera; remarques phylogénétiques. Ann La Soc Entomol France 22:413–422
Boyan GS, Williams JLD, Posser S, Bräunig P (2002) Morphological and molecular data argue for the labrum being non-apical, articulated, and the appendage of the intercalary segment in the locust. Arthropod Struct Dev 31:65–76
Boyden A (1943) Homology and analogy: a century after the definitions of “homologue” and “analogue” of Richard Owen. Q Rev Biol 18:228–241
Boyden A (1947) Homology and analogy. A critical review of the meanings and implications of these concepts in biology. Am Midl Nat 37:648–669. https://doi.org/10.2307/2421470
Brown SJ, Mahaffey JP, Lorenzen MD, Denell RE, Mahaffey JW (1999a) Using RNAi to investigate orthologous homeotic gene function during development of distantly related insects. Evol Dev 1:11–15. https://doi.org/10.1046/j.1525-142x.1999.99013.x
Brown S, Holtzman S, Kaufman T, Denell R (1999b) Characterization of the *Tribolium Deformed* ortholog and its ability to directly regulate Deformed target genes in the rescue of a *Drosophila Deformed* null mutant. Dev Genes Evol 209:389–398. https://doi.org/10.1007/s004270050269
Brown S, DeCamillis M, Gonzalez-Charneco K et al (2000) Implications of the *Tribolium Deformed* mutant phenotype for the evolution of Hox gene function. Proc Natl Acad Sci U S A 97:4510–4514. https://doi.org/10.1073/pnas.97.9.4510
Brown SJ, Shippy TD, Beeman RW, Denell RE (2002) *Tribolium* Hox genes repress antennal development in the gnathos and trunk. Mol Phylogenet Evol 24:384–387. https://doi.org/10.1016/S1055-7903(02)00205-1
Brusca RC, Brusca GJ (2003) Invertebrates, 2nd edn. Sinauer Associates, Inc, Sunderland, MA
Budd GE (2002) A palaeontological solution to the arthropod head problem. Nature 417:271–275. https://doi.org/10.1038/417271a

Budd GE, Telford MJ (2009) The origin and evolution of arthropods. Nature 457:812–817. https://doi.org/10.1038/nature07890

Butt FH (1949) Embryology of the milkweed bug, *Oncopeltus fasciatus* (Hemiptera). Mem Cornell Univ Agric Exp Stn 283:1–43

Butt FH (1960) Head development in the arthropods. Biol Rev 35:43–91

Campbell G (2002) Distalization of the *Drosophila* leg by graded EGF-receptor activity. Nature 418:781–785. https://doi.org/10.1038/nature00971

Casares F, Mann RS (1998) Control of antennal versus leg development in *Drosophila*. Nature 392:723–726. https://doi.org/10.1038/33706

Casares F, Mann RS (2001) The ground state of the ventral appendage in *Drosophila*. Science 293:1477–1480. https://doi.org/10.1126/science.1062542

Chan S-K, Pöpperl H, Krumlauf R, Mann RS (1996) An Extradenticle-induced conformational change in a HOX protein overcomes an inhibitory function of the conserved hexapeptide motif. EMBO J 15:2476–2487

Chang CP, Shen WF, Rozenfeld S, Lawrence HJ, Largman C, Cleary ML (1995) Pbx proteins display hexapeptide-dependent cooperative DNA binding with a subset of Hox proteins. Genes Dev 9:663–674

Chapman RF (1998) The insects: structure and function, 4th edn. Cambridge University Press, Cambridge

Chipman AD (2010) Parallel evolution of segmentation by co-option of ancestral gene regulatory networks. BioEssays 32:60–70

Chipman AD (2017) *Oncopeltus fasciatus* as an evo-devo research organism. Genesis:e23020. https://doi.org/10.1002/dvg.23020

Cobben RH (1979) On the original feeding habits of the Hemiptera (Insecta): a reply to Merrill Sweet. Ann Entomol Soc Am 72:711–715

Cohen SM, Jürgens G (1989a) Proximal-distal pattern formation in *Drosophila*: graded requirement for *Distal-less* gene activity during limb development. Rouxs Arch Dev Biol 198:157–169. https://doi.org/10.1007/BF02438941

Cohen SM, Jürgens G (1989b) Proximal-distal pattern formation in *Drosophila*: cell autonomous requirement for *Distal-less* gene activity in limb development. EMBO J 8:2045–2055

Cong P, Ma X, Hou X, Edgecombe GD, Strausfeld NJ (2014) Brain structure resolves the segmental affinity of anomalocaridid appendages. Nature 513:538–542

Coulcher JF, Telford MJ (2012) *Cap'n'collar* differentiates the mandible from the maxilla in the beetle *Tribolium castaneum*. EvoDevo 3:25. https://doi.org/10.1186/2041-9139-3-25

Coulcher JF, Telford MJ (2013) Comparative gene expression supports the origin of the incisor and molar process from a single endite in the mandible of the red flour beetle *Tribolium castaneum*. EvoDevo 4:1. https://doi.org/10.1186/2041-9139-4-1

Crampton G (1923) A phylogenetic comparison of the maxillae throughout the orders of insects. J New York Entomol Soc 31:77–107

Curtis CD, Brisson JA, DeCamillis MA, Shippy TD, Brown SJ, Denell RE (2001) Molecular characterization of *Cephalothorax*, the *Tribolium* ortholog of *Sex combs reduced*. Genes J Genet Dev 30:12–20

Dambly-Chaudière C, Ghysen A (1986) The sense organs in *the Drosophila* larva and their relation to the embryonic pattern of sensory neurons. Rouxs Arch Dev Biol 195:222–228. https://doi.org/10.1007/BF02438954

Damen WGM (2002) Parasegmental organization of the spider embryo implies that the parasegment is an evolutionary conserved entity in arthropod embryogenesis. Development 129:1239–1250

Damen WG, Hausdorf M, Seyfarth EA, Tautz D (1998) A conserved mode of head segmentation in arthropods revealed by the expression pattern of Hox genes in a spider. Proc Natl Acad Sci U S A 95:10665–10670. https://doi.org/10.1073/pnas.95.18.10665

Davidson EH (2006) The regulatory genome: gene regulatory networks in development and evolution. Academic, Cambridge, MA

Davis DR (1967) A revision of the moths of the subfamily prodoxinae (Lepidoptera: Incurvariidae). Bull U S Natl Mus 255:1–170. https://doi.org/10.5479/si.03629236.255.1
de Celis JF, Tyler DM, de Celis J, Bray SJ (1998) Notch signalling mediates segmentation of the *Drosophila* leg. Development 125:4617–4626
DeCamillis MA, Lewis DL, Brown SJ, Beeman RW, Denell RE (2001) Interactions of the *Tribolium Sex combs reduced* and *proboscipedia* orthologs in embryonic labial development. Genetics 159:1643–1648
Diaz-Benjumea FJ, Cohen B, Cohen SM (1994) Cell interaction between compartments establishes the proximal-distal axis of *Drosophila* legs. Nature 372:175–179. https://doi.org/10.1038/372175a0
Dong PD, Chu J, Panganiban G (2001) Proximodistal domain specification and interactions in developing *Drosophila* appendages. Development 128:2365–2372
Dong PDS, Dicks JS, Panganiban G (2002) *Distal-less* and *homothorax* regulate multiple targets to pattern the *Drosophila* antenna. Development 129:1967–1974
Dorn A, Hoffmann P (1983) Segmentation and differentiation of appendages during embryogenesis of the milkweed bug *Oncopeltus fasciatus*: a scanning electron microscopical study. Zool Jahrbücher 109:277–298
Doumpas N, Jékely G, Teleman AA (2013) Wnt6 is required for maxillary palp formation in *Drosophila*. BMC Biol 11:104. https://doi.org/10.1186/1741-7007-11-104
Duncan DM, Burgess EA, Duncan I (1998) Control of distal antennal identity and tarsal development in *Drosophila* by *spineless-aristapedia*, a homolog of the mammalian dioxin receptor. Genes Dev 12:1290–1303. https://doi.org/10.1101/gad.12.9.1290
Duncan D, Kiefel P, Duncan I (2010) Control of the *spineless* antennal enhancer: direct repression of antennal target genes by Antennapedia. Dev Biol 347:82–91. https://doi.org/10.1016/j.ydbio.2010.08.012
Dunn CW, Hejnol A, Matus DQ, Pang K, Browne WE, Smith SA, Seaver E, Rouse GW, Obst M, Edgecombe GD, Sørensen MV, Haddock SHD, Schmidt-Rhaesa A, Okusu A, Kristensen RM, Wheeler WC, Martindale MQ, Giribet G (2008) Broad phylogenomic sampling improves resolution of the animal tree of life. Nature 452:745–749. https://doi.org/10.1038/nature06614
Eastham LES (1931) The embryology of *Pieris rapae*: organogeny. Philos Trans Roy Soc B 219:1–50
Economou AD, Telford MJ (2009) Comparative gene expression in the heads of *Drosophila melanogaster* and *Tribolium castaneum* and the segmental affinity of the *Drosophila* hypopharyngeal lobes. Evol Dev 11:88–96
Edmunds GF, McCafferty WP (1988) The mayfly subimago. Annu Rev Entomol 33:509–538
Emmons RB, Duncan D, Duncan I (2007) Regulation of the *Drosophila* distal antennal determinant *spineless*. Dev Biol 302:412–426. https://doi.org/10.1016/j.ydbio.2006.09.044
Eriksson BJ, Tait NN, Budd GE, Janssen R, Akam M (2010) Head patterning and Hox gene expression in an onychophoran and its implications for the arthropod head problem. Dev Genes Evol 220:117–122. https://doi.org/10.1007/s00427-010-0329-1
Eriksson BJ, Samadi L, Schmid A (2013) The expression pattern of the genes *engrailed*, *pax6*, *otd* and *six3* with special respect to head and eye development in *Euperipatoides kanangrensis* Reid 1996 (Onychophora: Peripatopsidae). Dev Genes Evol 223:237–246
Estella C, Mann RS (2008) Logic of Wg and Dpp induction of distal and medial fates in the *Drosophila* leg. Development 135:627–636. https://doi.org/10.1242/dev.014670
Estella C, Rieckhof G, Calleja M, Morata G (2003) The role of *buttonhead* and *Sp1* in the development of the ventral imaginal discs of *Drosophila*. Development 130:5929–5941. https://doi.org/10.1242/dev.00832
Estella C, McKay DJ, Mann RS (2008) Molecular integration of wingless, decapentaplegic, and autoregulatory inputs into *Distalless* during *Drosophila* leg development. Dev Cell 14:86–96. https://doi.org/10.1016/j.devcel.2007.11.002
Farzana L, Brown SJ (2008) Hedgehog signaling pathway function conserved in *Tribolium* segmentation. Dev Genes Evol 218:181–192. https://doi.org/10.1007/s00427-008-0207-2

Galindo MI, Bishop SA, Greig S, Couso JP (2002) Leg patterning driven by proximal-distal interactions and EGFR signaling. Science 297:256–259. https://doi.org/10.1126/science.1072311

Galindo MI, Bishop SA, Couso JP (2005) Dynamic EGFR-Ras signalling in *Drosophila* leg development. Dev Dyn 233:1496–1508. https://doi.org/10.1002/dvdy.20452

Giorgianni MW, Mann RS (2011) Establishment of medial fates along the proximodistal axis of the *Drosophila* leg through direct activation of *dachshund* by distalless. Dev Cell 20:455–468. https://doi.org/10.1016/j.devcel.2011.03.017

Giribet G, Edgecombe GD (2017) Current understanding of Ecdysozoa and its internal phylogenetic relationships. Integr Comp Biol 57:455–466

Goto S, Hayashi S (1997) Specification of the embryonic limb primordium by graded activity of Decapentaplegic. Development 124:125–132

Gotoh H, Zinna RA, Ishikawa Y, Miyakawa H, Ishikawa A, Sugime Y, Emlen DJ, Lavine LC, Miura T (2017) The function of appendage patterning genes in mandible development of the sexually dimorphic stag beetle. Dev Biol 422:24–32. https://doi.org/10.1016/j.ydbio.2016.12.011

Gould SJ (1977) The return of hopeful monsters. Nat Hist 86:22–30

Grimaldi D, Engel MS (2005) Evolution of the insects. Cambridge University Press, New York

Grossmann D, Prpic NM (2012) Egfr signaling regulates distal as well as medial fate in the embryonic leg of *Tribolium castaneum*. Dev Biol 370:264–272. https://doi.org/10.1016/j.ydbio.2012.08.005

Haas MS, Brown SJ, Beeman RW (2001) Pondering the procephalon: the segmental origin of the labrum. Dev Genes Evol 211:89–95

Hamilton FGA (1981) Morphology and evolution of the rhychotan head (Insecta: Hemiptera, Homoptera). Can Ent 113:953–974

Heming BS (1980) Development of the mouthparts in embryos of *Haplothrips verbasci* (Osborn) (Insecta, Thysanoptera, Phlaeothripidae). J Morphol 164:235–263. https://doi.org/10.1002/jmor.1051640303

Hidalgo A (1991) Interactions between segment polarity genes and the generation of the segmental pattern in *Drosophila*. Mech Dev 35:77–87

Hinman VF, Nguyen AT, Cameron RA, Davidson EH (2003) Developmental gene regulatory network architecture across 500 million years of echinoderm evolution. Proc Natl Acad Sci U S A 100:13356–13361. https://doi.org/10.1073/pnas.2235868100

Hrycaj S, Chesebro J, Popadic A (2010) Functional analysis of *Scr* during embryonic and post-embryonic development in the cockroach, *Periplaneta americana*. Dev Biol 341:324–334. https://doi.org/10.1016/j.ydbio.2010.02.018

Huet C, Lenoir-Rousseaux JJ (1976) Etude de la mise en place de la patte imaginable de Tenebrio molitor. 1. Analyse expérimentale des processus de restauration au cours de la morphogenése. J Embryol Exp Morphol 35:303–321

Hughes CL, Kaufman TC (2000) RNAi analysis of *Deformed*, *proboscipedia* and *Sex combs reduced* in the milkweed bug On*copeltus fasciatus*: novel roles for Hox genes in the Hemipteran head. Development 127:3683–3694

Hughes CL, Kaufman TC (2002a) Hox genes and the evolution of the arthropod body plan. Evol Dev 499:459–499

Hughes CL, Kaufman TC (2002b) Exploring the myriapod body plan: expression patterns of the ten Hox genes in a centipede. Development 129:1225–1238. https://doi.org/10.1006/dbio.2002.0683

Hughes CL, Liu PZ, Kaufman TC (2004) Expression patterns of the rogue Hox genes *Hox3/zen* and *fushi tarazu* in the apterygote insect *Thermobia domestica*. Evol Dev 6:393–401. https://doi.org/10.1111/j.1525-142X.2004.04048.x

Hunter WB, Ullman DE (1992) Anatomy and ultrastructure of the piercing-sucking mouthparts and paraglossal sensilla of *Frankliniella occidentalis* (Pergande) (Thysanoptera: Thripidae). Int J Insect Morphol Embryol 21:17–35. https://doi.org/10.1016/0020-7322(92)90003-6

Inbal A, Halachmi N, Dibner C, Frank D, Salzberg A (2001) Genetic evidence for the transcriptional-activating function of Homothorax during adult fly development. Development 128:3405–3413

Ingham PW (1993) Localized *hedgehog* activity controls spatial limits of *wingless* transcription in the *Drosophila* embryo. Nature 366:560–562. https://doi.org/10.1038/366560a0

Inoue Y, Mito T, Miyawaki K, Matsushima K, Shinmyo Y, Heanue TA, Mardon G, Ohuchi H, Noji S (2002) Correlation of expression patterns of *homothorax*, *dachshund*, and *Distal-less* with the proximodistal segmentation of the cricket leg bud. Mech Dev 113:141–148

Jager M, Murienne J, Clabaut C, Deutsch J, Le Guyader H, Manuel M (2006) Homology of arthropod anterior appendages revealed by Hox gene expression in a sea spider. Nature 441:506–508

Janssen R (2012) Segment polarity gene expression in a myriapod reveals conserved and diverged aspects of early head patterning in arthropods. Dev Genes Evol 222:299–309

Janssen R (2017a) Comparative analysis of gene expression patterns in the arthropod labrum and the onychophoran frontal appendages, and its implications for the arthropod head problem. EvoDevo 8:1

Janssen R (2017b) A molecular view of onychophoran segmentation. Arthropod Struct Dev 46:341–353

Janssen R, Damen W (2006) The ten Hox genes of the millipede *Glomeris marginata*. Dev Genes Evol 216:451–465

Janssen R, Eriksson BJ, Tait NN, Budd GE (2014) Onychophoran Hox genes and the evolution of arthropod Hox gene expression. Front Zool 11:22

Janssen R, Jörgensen M, Prpic N-M, Budd GE (2015) Aspects of dorso-ventral and proximo-distal limb patterning in onychophorans. Evol Dev 17:21–33. https://doi.org/10.1111/ede.12107

Jiang J, Struhl G (1996) Complementary and mutually exclusive activities of *decapentaplegic* and *wingless* organize axial patterning during *Drosophila* leg development. Cell 86:401–409

Jockusch EL (2017) Developmental and evolutionary perspectives on the origin and diversification of arthropod appendages. Integr Comp Biol 57:533–545. https://doi.org/10.1093/icb/icx063

Jockusch EL, Smith FW (2015) Hexapoda: comparative aspects of later embryogenesis and metamorphosis. In: Wanninger A (ed) Evolutionary developmental biology of invertebrates 5. Springer, Vienna, pp 111–208

Jockusch EL, Nulsen C, Newfeld SJ, Nagy LM (2000) Leg development in flies versus grasshoppers: differences in *dpp* expression do not lead to differences in the expression of downstream components of the leg patterning pathway. Development 127:1617–1626

Johnson FB, Parker E, Krasnow MA (1995) Extradenticle protein is a selective cofactor for the *Drosophila* homeotics: role of the homeodomain and YPWM amino acid motif in the interaction. Proc Natl Acad Sci U S A 92:739–743

Jones T (1954) The external morphology of *Chirothrips hematus* (Trybom) (Thysanoptera). Trans R Entomol Soc London 105:163–187

Joulia L, Bourbon H-M, Cribbs DL (2005) Homeotic *proboscipedia* function modulates *hedgehog*-mediated organizer activity to pattern adult *Drosophila* mouthparts. Dev Biol 278:496–510. https://doi.org/10.1016/j.ydbio.2004.11.003

Khila A, Grbić M (2007) Gene silencing in the spider mite *Tetranychus urticae*: dsRNA and siRNA parental silencing of the *Distal-less* gene. Dev Genes Evol 217:241–251

Kimm M, Prpic N-M (2006) Formation of the arthropod labrum by fusion of paired and rotated limb-bud-like primordia. Zoomorphology 125:147–155

Krenn HW (2010) Feeding mechanisms of adult Lepidoptera: structure, function, and evolution of the mouthparts. Annu Rev Entomol 55:307–327. https://doi.org/10.1146/annurev-ento-112408-085338

Kukalová-Peck J (1998) Arthropod phylogeny and "basal" morphological structures. In: Fortey RA, Thomas RH (eds) Arthropod relationships. Chapman Hall, London, pp 249–268

Kurant E, Pai CY, Sharf R, Halachmi N, Sun YH, Salzberg A (1998) *Dorsotonals/homothorax*, the *Drosophila* homologue of *meis1*, interacts with *extradenticle* in patterning of the embryonic PNS. Development 125:1037–1048

Lebreton G, Faucher C, Cribbs DL, Benassayag C (2008) Timing of wingless signalling distinguishes maxillary and antennal identities in *Drosophila melanogaster*. Development 135:2301–2309. https://doi.org/10.1242/dev.017053

Lecuit T, Cohen SM (1997) Proximal-distal axis formation in the *Drosophila* leg. Nature 388:139–145

Lee T, Luo L (1999) Mosaic analysis with a repressible cell marker for studies of gene function in neuronal morphogenesis. Neuron 22:451–461

Lee AK, Sze CC, Kim ER, Suzuki Y (2013) Developmental coupling of larval and adult stages in a complex life cycle: insights from limb regeneration in the flour beetle, *Tribolium castaneum*. EvoDevo 4:20. https://doi.org/10.1186/2041-9139-4-20

Li Y, Brown SJ, Hausdorf B, Tautz D, Denell RE, Finkelstein R (1996) Two *orthodenticle*-related genes in the short-germ beetle *Tribolium castaneum*. Dev Genes Evol 206:35–45

Liu PZ, Kaufman TC (2004) *hunchback* is required for suppression of abdominal identity, and for proper germband growth and segmentation in the intermediate germband insect *Oncopeltus fasciatus*. Development 131:1515–1527

Liu J, Popadić A (2017) Early development and diversity of *Gryllus* appendages. In: Horch HW, Mito T, Popadić A, Ohuchi H, Noji S (eds) The cricket as a model organism. Springer, Tokyo, pp 17–30

Mandaravally Madhavan M, Schneiderman HA (1977) Histological analysis of the dynamics of growth of imaginal discs and histoblast nests during the larval development of *Drosophila melanogaster*. Wilhelm Roux's Arch Dev Biol 183:269–305. https://doi.org/10.1007/BF00848459

Manton SM (1964) Mandibular mechanisms and the evolution of arthropods. Philos Trans R Soc Lond Ser B Biol Sci 247:1–183. https://doi.org/10.1098/rstb.1964.0001

Mardon G, Solomon NM, Rubin GM (1994) *dachshund* encodes a nuclear protein required for normal eye and leg development in *Drosophila*. Development 120:3473–3486

Marshall SA (2006) Insects: their natural history and diversity: with a photographic guide to insects of eastern North America. Firefly Books, Buffalo

Maves L, Schubiger G (2003) Transdetermination in *Drosophila* imaginal discs: a model for understanding pluripotency and selector gene maintenance. Curr Opin Genet Dev 13:472–479

Meek WJ (1903) On the mouth-parts of the Hemiptera. Kansas Univ Sci Bull 2:256–277, Pl. 7–11

Misof B, Liu S, Meusemann K, Peters RS, Donath A, Mayer C, Frandsen PB, Ware J, Flouri T, Beutel RG, Niehuis O, Petersen M, Izquierdo-Carrasco F, Wappler T, Rust J, Aberer AJ, Aspöck U, Aspöck H, Bartel D, Blanke A, Berger S, Böhm A, Buckley T, Calcott B, Chen J, Friedrich F, Fukui M, Fujita M, Greve C, Grobe P, Gu S, Huang Y, Jermiin LS, Kawahara AY, Krogmann L, Kubiak M, Lanfear R, Letsch H, Li Y, Li Z, Li J, Lu H, Machida R, Mashimo Y, Kapli P, McKenna DD, Meng G, Nakagaki Y, Navarrete-Heredia JL, Ott M, Ou Y, Pass G, Podsiadlowski L, Pohl H, von Reumont BM, Schütte K, Sekiya K, Shimizu S, Slipinski A, Stamatakis A, Song W, Su X, Szucsich NU, Tan M, Tan X, Tang M, Tang J, Timelthaler G, Tomizuka S, Trautwein M, Tong X, Uchifune T, Walzl MG, Wiegmann BM, Wilbrandt J, Wipfler B, Wong TK, Wu Q, Wu G, Xie Y, Yang S, Yang Q, Yeates DK, Yoshizawa K, Zhang Q, Zhang R, Zhang W, Zhang Y, Zhao J, Zhou C, Zhou L, Ziesmann T, Zou S, Li Y, Xu X, Zhang Y, Yang H, Wang J, Wang J, Kjer KM, Zhou X (2014) Phylogenomics resolves the timing and pattern of insect evolution. Science 346:763–767. https://doi.org/10.1017/CBO9781107415324.004

Mito T, Ronco M, Uda T, Nakamura T, Ohuchi H, Noji S (2008) Divergent and conserved roles of *extradenticle* in body segmentation and appendage formation, respectively, in the cricket *Gryllus bimaculatus*. Dev Biol 313:67–79. https://doi.org/10.1016/j.ydbio.2007.09.060

Mito T, Shinmyo Y, Kurita K, Nakamura T, Ohuchi H, Noji S (2011) Ancestral functions of Delta/Notch signaling in the formation of body and leg segments in the cricket *Gryllus bimaculatus*. Development 138:3823–3833. https://doi.org/10.1242/dev.060681

Mittmann B, Scholtz G (2003) Development of the nervous system in the "head" of *Limulus polyphemus* (Chelicerata: Xiphosura): morphological evidence for a correspondence between the segments of the chelicerae and of the (first) antennae of Mandibulata. Dev Genes Evol 213:9–17. https://doi.org/10.1007/s00427-002-0285-5

Miyawaki K, Inoue Y, Mito T, Fujimoto T, Matsushima K, Shinmyo Y, Ohuchi H, Noji S (2002) Expression patterns of aristaless in developing appendages of *Gryllus bimaculatus* (cricket). Mech Dev 113:181–184

Miyawaki K, Mito T, Sarashina I, Zhang H, Shinmyo Y, Ohuchi H, Noji S (2004) Involvement of Wingless/Armadillo signaling in the posterior sequential segmentation in the cricket, *Gryllus bimaculatus* (Orthoptera), as revealed by RNAi analysis. Mech Dev 121:119–130

Mlodzik M, Fjose A, Gehring WJ (1988) Molecular structure and spatial expression of a homeobox gene from the labial region of the *Antennapedia*-complex. EMBO J 7:2569–2578

Moczek AP, Rose DJ (2009) Differential recruitment of limb patterning genes during development and diversification of beetle horns. Proc Natl Acad Sci U S A 106:8992–8997. https://doi.org/10.1073/pnas.0809668106

Morata G (2001) How *Drosophila* appendages develop. Nat Rev Mol Cell Biol 2:89–97

Muir F, Kershaw JC (1911a) On the later embryological stages of *Pristhesancus papuensis* (Reduviidae). Psyche 18:75–79

Muir F, Kershaw JC (1911b) On the homologies and mechanism of the mouth-parts of Hemiptera. Psyche (Stuttg) 18:1–12

Muir F, Kershaw JC (1912) The development of the mouthparts in the Homoptera, with observations on the embryo of *Siphanta*. Psyche 19:77–89

Nakamura T, Mito T, Miyawaki K, Ohuchi H, Noji S (2008) EGFR signaling is required for re-establishing the proximodistal axis during distal leg regeneration in the cricket *Gryllus bimaculatus* nymph. Dev Biol 319:46–55. https://doi.org/10.1016/j.ydbio.2008.04.002

Newcomer WS (1948) Embryological development of the mouthparts and related structures of the milkweed bug *Oncopeltus fasciatus*. J Morphol 82:365–411

Nie W, Stronach B, Panganiban G, Shippy T, Brown SJ, Denell RE (2001) Molecular characterization of *Tc-labial* and the 3′ end of the *Tribolium* homeotic complex. Dev Genes Evol 211:244–251

Nielsen C (2001) Animal evolution: interrelationships of the living phyla, 2nd edn. Oxford University Press, London

Niimi T, Kuwayama H, Yaginuma T (2005) Larval RNAi applied to the analysis of postembryonic development in the ladybird beetle, *Harmonia axyridis*. J Insect Biotechnol Sericology 74:95–102

Niwa N, Inoue Y, Nozawa A, Saito M, Misumi Y, Ohuchi H, Yoshioka H, Noji S (2000) Correlation of diversity of leg morphology in *Gryllus bimaculatus* (cricket) with divergence in *dpp* expression pattern during leg development. Development 127:4373–4381

Nüsslein-Volhard C, Wieschaus E (1980) Mutations affecting segment number and polarity in *Drosophila*. Nature 287:795–801

O'Donnell BC, Jockusch EL (2010) The expression of *wingless* and *Engrailed* in developing embryos of the mayfly *Ephoron leukon* (Ephemeroptera: Polymitarcyidae). Dev Genes Evol 220:11–24. https://doi.org/10.1007/s00427-010-0324-6

Ober KA, Jockusch EL (2006) The roles of *wingless* and *decapentaplegic* in axis and appendage development in the red flour beetle, *Tribolium castaneum*. Dev Biol 294:391–405

Ohde T, Masumoto M, Yaginuma T, Niimi T (2009) Embryonic RNAi analysis in the firebrat, *Thermobia domestica*: *Distal-less* is required to form caudal filament. J Insect Biotechnol Sericology 78:99–105

Ortega-Hernández J, Budd GE (2016) The nature of non-appendicular anterior paired projections in Palaeozoic total-group Euarthropoda. Arthropod Struct Dev 45:185–199. https://doi.org/10.1016/j.asd.2016.01.006

Ortega-Hernández J, Janssen R, Budd GE (2017) Origin and evolution of the panarthropod head – a palaeobiological and developmental perspective. Arthropod Struct Dev 46:354–379

Owen R (1843) Lectures on the comparative anatomy and physiology of the invertebrate animals, delivered at the Royal College of Surgeons in 1843. Longman, Brown, Green, and Longmans, London

Owen R (1848) On the archetype and homologies of the vertebrate skeleton. John Van Voorst, London

Pai CY, Kuo TS, Jaw TJ, Kurant E, Chen CT, Bessarab DA, Salzberg A, Sun YH (1998) The Homothorax homeoprotein activates the nuclear localization of another homeoprotein, extradenticle, and suppresses eye development in *Drosophila*. Genes Dev 12:435–446

Panfilio KA, Jentzsch IMV, Benoit JB, Erezyilmaz D, Suzuki Y, Colella S, Robertson HM, Poelchau MF, Waterhouse RM, Ioannidis P, Weirauch MT, Hughes DST, Murali SC, Werren JH, Jacobs CGC, Duncan EJ, Armisén D, Vreede BMI, Baa-Puyoulet P, Berger CS, Chang CC, Chao H, Chen MJM, Chen YT, Childers CP, Chipman AD, Cridge AG, Crumière AJJ, Dearden PK, Didion EM, Dinh H, Doddapaneni H, Dolan A, Dugan-Perez S, Extavour CG, Febvay G, Friedrich M, Ginzburg N, Han Y, Heger P, Holmes CJ, Horn T, Hsiao Y-M, Jennings EC, Johnston JS, Jones TE, Jones JW, Khila A, Koelzer S, Kovacova V, Leask M, Lee SL, Lee C-Y, Lovegrove MR, Lu H-L, Lu Y, Moore PJ, Munoz-Torres MC, Muzny DM, Palli SR, Parisot N, Pick L, Porter M, Qu J, Refki PN, Richter R, Rivera-Pomar R, Rosendale AJ, Roth S, Sachs L, Santos ME, Seibert J, Sghaier E, Shukla JN, Stancliffe RJ, Tidswell O, Traverso L, van der Zee M, Viala S, Worley KC, Zdobnov EM, Gibbs RA, Richards S (2018) Molecular evolutionary trends and feeding ecology diversification in the Hemiptera, anchored by the milkweed bug genome. bioRxiv:201731. https://doi.org/10.1101/201731

Panganiban G, Rubenstein JLR (2002) Developmental functions of the *Distal-less/Dlx* homeobox genes. Development 129:4371–4386

Panganiban G, Nagy L, Carroll SB (1994) The role of the *Distal-less* gene in the development and evolution of insect limbs. Curr Biol 4:671–675

Papillon D, Telford MJ (2007) Evolution of *Hox3* and *ftz* in arthropods: insights from the crustacean *Daphnia pulex*. Dev Genes Evol 217:315–322

Park T-YS, Kihm J-H, Woo J, Park C, Lee WY, Smith MP, Harper DAT, Young F, Nielsen AT, Vinther J (2018) Brain and eyes of *Kerygmachela* reveal protocerebral ancestry of the panarthropod head. Nat Commun 9:1019

Parsons MC (1964) The origin and development of the hemipteran cranium. Can J Zool 42:409–432

Parsons MC (1974) The morphology and possible origin of the hemipteran loral lobes. Can J Zool 52:189–202

Passalacqua KD, Hrycaj S, Mahfooz N, Popadic A (2010) Evolving expression patterns of the homeotic gene *Scr* in insects. Int J Dev Biol 54:897–904. https://doi.org/10.1387/ijdb.082839kp

Pellmyr O, Krenn HW (2002) Origin of a complex key innovation in an obligate insect-plant mutualism. Proc Natl Acad Sci U S A 99:5498–5502. https://doi.org/10.1073/pnas.072588699

Peterson MD, Rogers BT, Popadic A, Kaufman TC (1999) The embryonic expression pattern of labial, posterior homeotic complex genes and the teashirt homologue in an apterygote insect. Dev Genes Evol 209:77–90

Pfau T (2010) "All is Leaf": difference, metamorphosis, and Goethe's phenomenology of knowledge. Stud Romanticism 49:3–41

Popadić A, Panganiban G, Rusch D, Shear WA, Kaufman TC (1998) Molecular evidence for the gnathobasic derivation of arthropod mandibles and for the appendicular origin of the labrum and other structures. Dev Genes Evol 208:142–150

Posnien N, Bucher G (2010) Formation of the insect head involves lateral contribution of the intercalary segment, which depends on *Tc-labial* function. Dev Biol 338:107–116. https://doi.org/10.1016/j.ydbio.2009.11.010

Posnien N, Bashasab F, Bucher G (2009) The insect upper lip (labrum) is a nonsegmental appendage-like structure. Evol Dev 11:480–488. https://doi.org/10.1111/j.1525-142X.2009.00356.x

Posnien N, Schinko JB, Kittelmann S, Bucher G (2010) Genetics, development and composition of the insect head – a beetle's view. Arthropod Struct Dev 39:399–410. https://doi.org/10.1016/j.asd.2010.08.002

Powell JA, Opler PA (2009) Moths of Western North America. University of California Press, Berkeley, CA

Prpic NM, Damen WGM (2009) Notch-mediated segmentation of the appendages is a molecular phylotypic trait of the arthropods. Dev Biol 326:262–271. https://doi.org/10.1016/j.ydbio.2008.10.049

Prpic NM, Tautz D (2003) The expression of the proximodistal axis patterning genes *Distal-less* and *dachshund* in the appendages of *Glomeris marginata* (Myriapoda: Diplopoda) suggests a special role of these genes in patterning the head appendages. Dev Biol 260:97–112. https://doi.org/10.1016/S0012-1606(03)00217-3

Prpic NM, Wigand B, Damen WGM, Klingler M (2001) Expression of *dachshund* in wild-type and *Distal-less* mutant *Tribolium* corroborates serial homologies in insect appendages. Dev Genes Evol 211:467–477. https://doi.org/10.1007/s004270100178

Prpic NM, Janssen R, Wigand B, Klingler M, Damen WGM (2003) Gene expression in spider appendages reveals reversal of *exd/hth* spatial specificity, altered leg gap gene dynamics, and suggests divergent distal morphogen signaling. Dev Biol 264:119–140. https://doi.org/10.1016/j.ydbio.2003.08.002

Prpic NM, Janssen R, Damen WGM, Tautz D (2005) Evolution of dorsal-ventral axis formation in arthropod appendages: *H15* and *optomotor-blind/bifid*-type T-box genes in the millipede *Glomeris marginata* (Myriapoda: Diplopoda). Evol Dev 7:51–57. https://doi.org/10.1111/j.1525-142X.2005.05006.x

Pultz MA, Diederich RJ, Cribbs DL, Kaufman TC (1988) The *proboscipedia* locus of the *Antennapedia* complex: a molecular and genetic analysis. Genes Dev 2:901–920. https://doi.org/10.1101/gad.2.7.901

Raff RA, Kaufman TC (1983) Embryos, genes, and evolution: the developmental-genetic basis of evolutionary change. Macmillan, New York City, NY

Rauskolb C, Irvine KD (1999) Notch-mediated segmentation and growth control of the *Drosophila* leg. Dev Biol 210:339–350. https://doi.org/10.1006/dbio.1999.9273

Rauskolb C, Smith KM, Peifer M, Wieschaus E (1995) Extradenticle determines segmental identities throughout *Drosophila* development. Development 121:3663–3673

Rebeiz M, Tsiantis M (2017) Enhancer evolution and the origins of morphological novelty. Curr Opin Genet Dev 45:115–123. https://doi.org/10.1016/j.gde.2017.04.006

Rempel J (1975) Evolution of the insect head: the endless dispute. Quaest Entomol 11:7–25

Requena D, Álvarez JA, Gabilondo H, Loker R, Mann RS, Estella CS (2017) Origins and specification of the *Drosophila* wing. Curr Biol 27:3826–3836.e5. https://doi.org/10.1016/j.cub.2017.11.023

Rieckhof GE, Casares F, Ryoo HD, Abu-Shaar M, Mann RS (1997) Nuclear translocation of extradenticle requires *homothorax*, which encodes an *extradenticle*-related homeodomain protein. Cell 91:171–183. https://doi.org/10.1016/S0092-8674(00)80400-6

Rogers BT, Kaufman TC (1997) Structure of the insect head in ontogeny and phylogeny: a view from *Drosophila*. Int Rev Cytol 174:1–84. https://doi.org/10.1016/S0074-7696(08)62115-4

Rogers BT, Peterson MD, Kaufman TC (2002) The development and evolution of insect mouthparts as revealed by the expression patterns of gnathocephalic genes. Evol Dev 4:96–110. https://doi.org/10.1046/j.1525-142X.2002.01065.x

Ronco M, Uda T, Mito T, Minelli A, Noji S, Klingler M (2008) Antenna and all gnathal appendages are similarly transformed by *homothorax* knock-down in the cricket *Gryllus bimaculatus*. Dev Biol 313:80–92. https://doi.org/10.1016/j.ydbio.2007.09.059

Roth VL (1984) On homology. Biol J Linn Soc 22:13–29. https://doi.org/10.1111/j.1095-8312.1984.tb00796.x

Ruiz-Losada M, Blom-Dahl D, Córdoba S, Estella C (2018) Specification and patterning of *Drosophila* appendages. J Dev Biol 6:17. https://doi.org/10.3390/jdb6030017

Santos ME, Le Bouquin A, Crumière AJJ, Khila A (2017) Taxon-restricted genes at the origin of a novel trait allowing access to a new environment. Science 358:386–390. https://doi.org/10.1126/science.aan2748

Schaeper ND, Wimmer EA, Prpic N-M (2013) Appendage patterning in the primitively wingless hexapods *Thermobia domestica* (Zygentoma: Lepismatidae) and *Folsomia candida* (Collembola: Isotomidae). Dev Genes Evol 223:341–350. https://doi.org/10.1007/s00427-013-0449-5

Schmidt-Ott U, Technau GM (1992) Expression of *en* and *wg* in the embryonic head and brain of *Drosophila* indicates a refolded band of seven segment remnants. Development 116:111–125

Schmidt-Ott U, Rafiqi AM, Lemke S (2010) *Hox3/zen* and the evolution of extraembryonic epithelia in insects. Adv Exp Med Biol 689:133–144

Scholtz G (2002) The Articulata hypothesis–or what is a segment? Org Divers Evol 2:197–215

Scholtz G, Edgecombe GD (2006) The evolution of arthropod heads: reconciling morphological, developmental and palaeontological evidence. Dev Genes Evol 216:395–415. https://doi.org/10.1007/s00427-006-0085-4

Scholtz G, Mittmann B, Gerberding M (1998) The pattern of *Distal-less* expression in the mouthparts of crustaceans, myriapods and insects: new evidence for a gnathobasic mandible and the common origin of Mandibulata. Int J Dev Biol 42:801–810

Schroder R, Eckert C, Wolff C, Tautz D (2000) Conserved and divergent aspects of terminal patterning in the beetle *Tribolium castaneum*. Proc Natl Acad Sci U S A 97:6591–6596. https://doi.org/10.1073/pnas.100005497

Setton EVW, March LE, Nolan ED, Jones TE, Cho H, Wheeler WC, Extavour CG, Sharma PP (2017) Expression and function of *spineless* orthologs correlate with distal deutocerebral appendage morphology across Arthropoda. Dev Biol 430:224–236. https://doi.org/10.1016/j.ydbio.2017.07.016

Sharma PP, Schwager EE, Extavour CG, Giribet G (2012) Hox gene expression in the harvestman *Phalangium opilio* reveals divergent patterning of the chelicerate opisthosoma. Evol Dev 14:450–463. https://doi.org/10.1111/j.1525-142X.2012.00565.x

Sharma PP, Gupta T, Schwager EE, Wheeler WC, Extavour CG (2014) Subdivision of arthropod *cap-n-collar* expression domains is restricted to Mandibulata. EvoDevo 5:3. https://doi.org/10.1186/2041-9139-5-3

Sharma PP, Tarazona OA, Lopez DH, Schwager EE, Cohn MJ, Wheeler WC, Extavour CG (2015) A conserved genetic mechanism specifies deutocerebral appendage identity in insects and arachnids. ProcRSocB 282:20150698

Shippy TD, Guo J, Brown SJ, Beeman RW, Denell RE (2000) Analysis of *maxillopedia* expression pattern and larval cuticular phenotype in wild-type and mutant *Tribolium*. Genetics 155:721–731. https://doi.org/10.1002/jssc.201100117

Shippy TD, Yeager SJ, Denell RE (2009) The *Tribolium spineless* ortholog specifies both larval and adult antennal identity. Dev Genes Evol 219:45–51. https://doi.org/10.1007/s00427-008-0261-9

Siemanowski J, Richter T, Dao VA, Bucher G (2015) Notch signaling induces cell proliferation in the labrum in a regulatory network different from the thoracic legs. Dev Biol 408:164–177. https://doi.org/10.1016/J.YDBIO.2015.09.018

Simonnet F, Moczek AP (2011) Conservation and diversification of gene function during mouthpart development in *Onthophagus* beetles. Evol Dev 13:280–289. https://doi.org/10.1111/j.1525-142X.2011.00479.x

Smith FW, Jockusch EL (2014) Hox genes require *homothorax* and *extradenticle* for body wall identity specification but not for appendage identity specification during metamorphosis of *Tribolium castaneum*. Dev Biol 395:182–197. https://doi.org/10.1016/j.ydbio.2014.08.017

Smith AV, Orr-Weaver TL (1991) The regulation of the cell cycle during *Drosophila* embryogenesis: the transition to polyteny. Development 112:997–1008

Smith FW, Angelini DR, Jockusch EL (2014a) A functional genetic analysis in flour beetles (Tenebrionidae) reveals an antennal identity specification mechanism active during metamorphosis in Holometabola. Mech Dev 132:13–27. https://doi.org/10.1016/j.mod.2014.02.002

Smith FW, Angelini DR, Gaudio MS, Jockusch EL (2014b) Metamorphic labral axis patterning in the beetle *Tribolium castaneum* requires multiple upstream, but few downstream, genes in the appendage patterning network. Evol Dev 16:78–91

Smith FW, Boothby TC, Giovannini I, Rebecchi L, Jockusch EL, Goldstein B (2016) The compact body plan of tardigrades evolved by the loss of a large body region. Curr Biol 26:224–229

Smith FW, Cumming M, Goldstein B (2018) Analyses of nervous system patterning genes in the tardigrade *Hypsibius exemplaris* illuminate the evolution of panarthropod brains. EvoDevo 9:19

Snodgrass RE (1921) The mouthparts of the cicada. Proc Entomol Soc Washingt 23:1–15

Snodgrass RE (1928) Morphology and evolution of the insect head and its appendages. Smithson Misc Collect 81:1–158

Snodgrass RE (1930) Morphology and evolution of the insect head and its appendages. Smithson Misc Collect 81:1–158

Snodgrass RE (1935) Principles of insect morphology. Cornell University Press

Snodgrass RE (1938) The loral plates and the hypopharynx of Hemiptera. Proc Entomol Soc Washington 40:228–236

Snodgrass RE (1944) The feeding apperatus of biting and sucking insects affecting man and animals. Smithson Misc Collect 104:1–133

Snodgrass RE (1946) The skeletal anatomy of fleas (siphonaptera) (with 21 plates). Smithson Misc Collect 104:1–89

Snodgrass RE (1959) The anatomical life of the mosquito. Smithson Misc Collect 139:1–87

Sokoloff A (1972) The biology of *Tribolium*, vol 1. Clarendon Press, Oxford

Steinmetz PRH, Urbach R, Posnien N, Eriksson J, Kostyuchenko RP, Brena C, Guy K, Akam M, Bucher G, Arendt D (2010) *Six3* demarcates the anterior-most developing brain region in bilaterian animals. EvoDevo 1:1–9

Stern DL (2010) Evolution, development, and the predictable genome. Greenwood Village, Colorado, Roberts and Co.

Stern DL, Orgogozo V (2008) The loci of evolution: how predictable is genetic evolution? Evolution 62:2155–2177. https://doi.org/10.1111/j.1558-5646.2008.00450.x

Strausfeld NJ (2012) Arthropod brains: evolution, functional elegance, and historical significance. Belknap Press of Harvard University Press, Cambridge

Struhl G (1982a) Genes controlling segmental specification in the *Drosophila* thorax. Dev Biol 79:7380–7384. https://doi.org/10.1073/pnas.79.23.7380

Struhl G (1982b) *Spineless-aristapedia*: a homeotic gene that does not control the development of specific compartments in *Drosophila*. Genetics 102:737–749

Stultz BG, Park SY, Mortin MA, Kennison JA, Hursh DA (2012) Hox proteins coordinate peripodial decapentaplegic expression to direct adult head morphogenesis in *Drosophila*. Dev Biol 369:362–376. https://doi.org/10.1016/J.YDBIO.2012.07.012

Sunkel CE, Whittle JRS (1987) *Brista*: a gene involved in the specification and differentiation of distal cephalic and thoracic structures in *Drosophila melanogaster*. Rouxs Arch Dev Biol 196:124–132. https://doi.org/10.1007/BF00402034

Švácha P (1992) What are and what are not imaginal discs: reevaluation of some basic concepts (Insecta, Holometabola). Dev Biol 154:101–117. https://doi.org/10.1016/0012-1606(92)90052-I

Svendsen PC, Formaz-Preston A, Leal SM, Brook WJ (2009) The *Tbx20* homologs *midline* and *H15* specify ventral fate in the *Drosophila* melanogaster leg. Development 136:2689–2693. https://doi.org/10.1242/dev.037911

Tajiri R, Misaki K, Yonemura S, Hayashi S (2011) Joint morphology in the insect leg: evolutionary history inferred from *Notch* loss-of-function phenotypes in *Drosophila*. Development 138:4621–4626. https://doi.org/10.1242/dev.067330

Tanaka K, Truman JW (2007) Molecular patterning mechanism underlying metamorphosis of the thoracic leg in *Manduca sexta*. Dev Biol 305:539–550

Telford MJ, Thomas RH (1998) Expression of homeobox genes shows chelicerate arthropods retain their deutocerebral segment. Proc Natl Acad Sci U S A 95:10671–10675. https://doi.org/10.1073/pnas.95.18.10671

Theisen H, Haerry TE, O'Connor MB, Marsh JL (1996) Developmental territories created by mutual antagonism between Wingless and Decapentaplegic. Development 122:3939–3948

Toegel J, Wimmer E, Prpic N-M (2009) Loss of *spineless* function transforms the *Tribolium* antenna into a thoracic leg with pretarsal, tibiotarsal, and femoral identity. Dev Genes Evol 219:53–58

Toga K, Saiki R, Maekawa K (2013) Hox gene *Deformed* is likely involved in mandibular regression during presoldier differentiation in the nasute termite *Nasutitermes takasagoensis*. J Exp Zool Part B Mol Dev Evol 320:385–392

Tomita S, Kikuchi A (2009) Abd-B suppresses lepidopteran proleg development in posterior abdomen. Dev Biol 328:403–409. https://doi.org/10.1016/j.ydbio.2009.01.040

Tomoyasu Y, Wheeler SR, Denell RE (2005) *Ultrabithorax* is required for membranous wing identity in the beetle *Tribolium castaneum*. Nature 433:643–647

Urbach R, Technau GM (2003) Molecular markers for identified neuroblasts in the developing brain of *Drosophila*. Development 130:3621–3637

van Horn SN (1966) Studies on the embryogenesis of *Aulocara elliotti* (Thomas) (Orthoptera, Acrididae). I. External morphogenesis. J Morphol 120:83–113. https://doi.org/10.1002/jmor.1051200105

Van Valen LM (1982) Homology and causes. J Morphol 173:305–312. https://doi.org/10.1002/jmor.1051730307

von Goethe JW (1790) Versuch die Metamorphose der Pflanzen zu erklären. Carl Wilhelm Ettinger, Gotha

von Goethe JW (1814) Aus meinem Leben: Dichtung und Wahrheit. Cotta, Tübingen

Wagner GP (1989) The biological homology concept. Annu Rev Ecol Syst 20:51–69. https://doi.org/10.1146/annurev.es.20.110189.000411

Wagner GP (2007) The developmental genetics of homology. Nat Rev Genet 8:473–479. https://doi.org/10.1038/nrg2099

Wasik BR, Rose DJ, Moczek AP (2010) Beetle horns are regulated by the hox gene, *Sex combs reduced*, in a species- and sex-specific manner. Evol Dev 12:353–362. https://doi.org/10.1111/j.1525-142X.2010.00422.x

West-Eberhard MJ (2003) Developmental plasticity and evolution. Oxford University Press, Oxford

Wieschaus E, Nüsslein-Volhard C (2016) The Heidelberg screen for pattern mutants of *Drosophila*: a personal account. Annu Rev Cell Dev Biol 32:1–46. https://doi.org/10.1146/annurev-cellbio-113015-023138

Wilder EL, Perrimon N (1995) Dual functions of *wingless* in the *Drosophila* leg imaginal disc. Development 121:477–488

Wu J, Cohen SM (1999) Proximodistal axis formation in the *Drosophila* leg: subdivision into proximal and distal domains by *Homothorax* and *Distal-less*. Development 126:109–117

Xu T, Rubin GM (1993) Analysis of genetic mosaics in developing and adult *Drosophila* tissues. Development 117:1223–1237

Yasunaga K, Saigo K, Kojima T (2006) Fate map of the distal portion of *Drosophila* proboscis as inferred from the expression and mutations of basic patterning genes. Mech Dev 123:893–906. https://doi.org/10.1016/J.MOD.2006.08.008

Yeates DK, Meusemann K, Trautwein M, Wiegmann B, Zwick A (2016) Power, resolution and bias: recent advances in insect phylogeny driven by the genomic revolution. Curr Opin Insect Sci 13:16–23

Yoshiyama N, Tojo K, Hatakeyama M (2013) A survey of the effectiveness of non-cell autonomous RNAi throughout development in the sawfly, *Athalia rosae* (Hymenoptera). J Insect Physiol 59:400–407. https://doi.org/10.1016/j.jinsphys.2013.01.009

Zhang H, Shinmyo Y, Mito T, Miyawaki K, Sarashina I, Ohuchi H, Noji S (2005) Expression patterns of the homeotic genes *Scr*, *Antp*, *Ubx*, and *abd-A* during embryogenesis of the cricket *Gryllus bimaculatus*. Gene Expr Patterns 5:491–502. https://doi.org/10.1016/j.modgep.2004.12.006

# Chapter 6
# The Early Evolution of Biting–Chewing Performance in Hexapoda

**Alexander Blanke**

**Abstract** Insects show a plethora of different mandible shapes. It was advocated that these mandible shapes are mainly a function of different feeding habits. This hypothesis was tested on a larger sampling of non-holometabolan biting–chewing insects with additional tests to understand the interplay of mandible function, feeding guild, and phylogeny. The results show that at the studied systematic level, variation in mandible biting–chewing effectivity is regulated to a large extent by phylogenetic history and the configuration of the mandible joints rather than the food preference of a given taxon. Additionally, lineages with multiple mandibular joints such as primary wingless hexapods show a wider functional space occupation of mandibular effectivity than dicondylic insects (= silverfish + winged insects) at significantly different evolutionary rates. The evolution and occupation of a comparably narrow functional performance space of dicondylic insects is surprising given the low effectivity values of this food uptake solution. Possible reasons for this relative evolutionary "stasis" are discussed.

## 6.1 Introduction

Insecta *sensu lato* (= Hexapoda) display a high diversity of mouthpart shapes within the early evolved lineages which started to radiate approximately 479 million years ago (Misof et al. 2014). These shape changes were described qualitatively and were often stated to relate mainly to the type of food consumed (Yuasa 1920; Isely 1944; Evans and Forsythe 1985; Chapman and de Boer 1995). To the knowledge of the author, this and related statements regarding mouthpart mechanics being shaped by functional demands have never been tested in a quantitative framework.

Here, available evidence for mouthpart function and biomechanics in the early branched lineages of Hexapoda will be reviewed followed by an analysis of biomechanical performance changes of the mandibles across several non-holometabolan lineages. The analytical part of this chapter is restricted to the mandibles due to the

A. Blanke (✉)
Institute of Zoology, Biocenter, University of Cologne, Köln, Germany
e-mail: a.blanke@uni-koeln.de



H. W. Krenn (ed.), *Insect Mouthparts*, Zoological Monographs 5,
https://doi.org/10.1007/978-3-030-29654-4_6

paucity of knowledge regarding maxilla motion and its biomechanical parameters as will be outlined below. The biomechanical role of the labium during food uptake is even less known and difficult to compare between entognathous hexapods where the labium is fused with the pleural folds and Ectognatha, displaying a freely moveable labium.

## 6.2 Mouthpart and Muscle Configurations in Primary Wingless and Early Evolved Winged Hexapoda

As already stated in Chap. 2, a general difference in hexapod mouthpart configuration are the entognathous mouthparts of Protura, Collembola, and Diplura and the ectognathous mouthparts of Insecta *sensu stricto* (Chapman 1998; Grimaldi and Engel 2005).

Entognathous mandibles and maxillae are hidden within the head, lying in so-called gnathal pouches formed by a fusion of the enlarged subgenae and the labium (François 1970; François et al. 1992; Ikeda and Machida 1998). The gnathal pouches are mainly thin cuticular sheaths, enclosing mandibles and maxillae. In being sheathed into sack-like pouches, the movements of both the mandibles and the maxillae are mainly restricted to piercing motions (protraction) through the comparably narrow functional mouth opening, supplemented by a rotation and, to a minor degree, chewing motions (Koch 2001). Although the comparably narrow functional mouth opening in entognathous Hexapoda principally restricts larger mouthpart movements and wide gape angles, the diversity of different mouthpart articulations, muscle equipments, and thus movement types is remarkable and deserves further attention in order to understand the principal evolutionary transitions towards the prevailing (dicondylic) mouthpart configuration observable in more derived insects.

### 6.2.1 *Protura*

Protura (coneheads) have prognathous stylet-like mandibles and maxillae which are used for piercing into plant roots or fungal hypha by repeated protraction and retraction (Dunger 1983). The mandibles are approximately ten times longer than wide and have an elongated mandibular orifice half the length the mandible at their posterior part (Fig. 6.1). Although it is frequently stated that proturan mandibles articulate posteriorly with parts of the gnathal pouch and are thus non-permanent monocondylic (François et al. 1992), deduction of possible mandible movements based on the muscular equipment suggests that the gnathal pouch is rather used as a guiding structure in order to prevent lateral evasions of the mandibles during protraction and retraction. It remains to be tested whether the mandibles really articulate in a joint-like manner at their posterior end with parts of the gnathal pouch. The mandibular musculature is composed of several muscles, some of

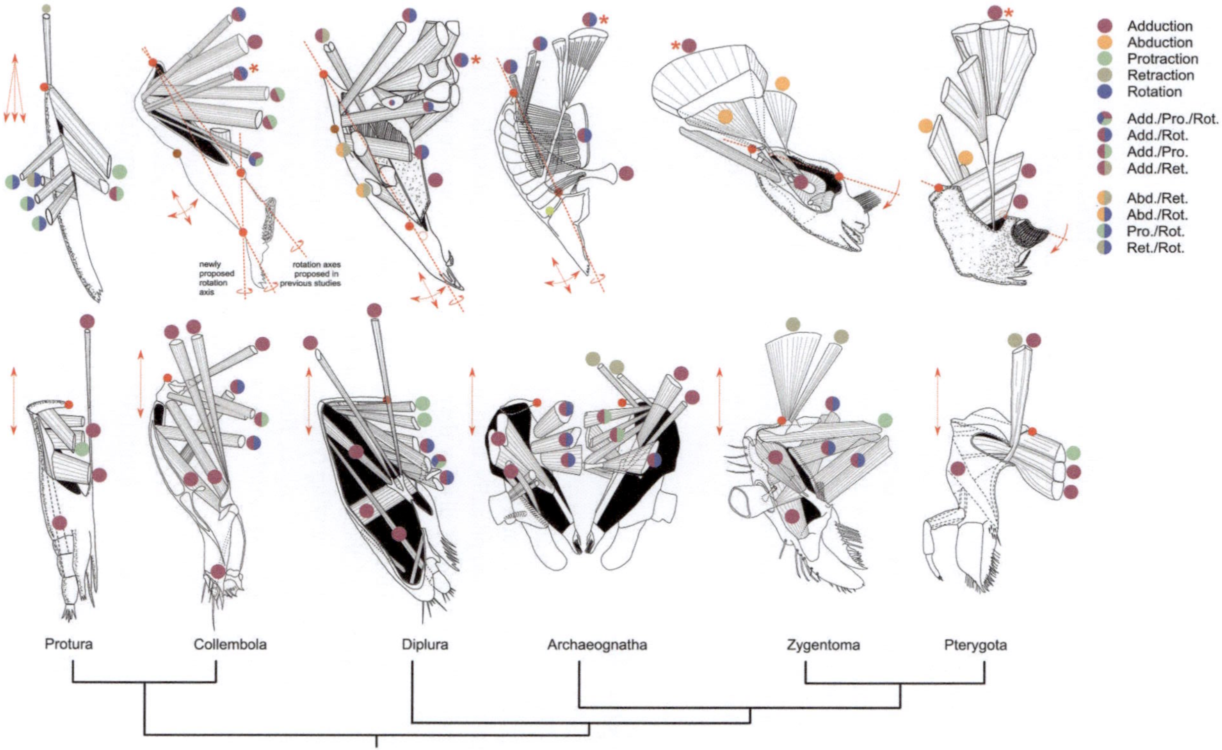


**Fig. 6.1** Muscular configuration of the mandibles (upper row) and maxillae (lower row), muscle functions (color coded), and principal mouthpart movement types in early branched Hexapoda together with their phylogenetic relationships (Misof et al. 2014). The red asterisk denotes M. craniomandibularis internus. The red dot denotes points of articulation, the brown dot in Collembola denotes an articulation point of the maxilla with the mandible. Mandibles of Protura and Diplura in dorsal view, all others in cranial view. Drawings based on specimens of *Sinentomon erythranum* (Protura), *Tomocerus catalanus* and *T. plumbeus* (Collembola), *Campodea chardardi* (Diplura), *Thermobia domestica* (Zygentoma), and *Oniscigaster wakefieldi* (Ephemeroptera). Redrawn and modified after François (1970, François et al. 1992), Denis (1928), Hofmann (1908), Bitsch (1963), Chaudonneret (1950), and Staniczek (2001)

them with a double function (Fig. 6.1): protractor-rotators originating at the cephalic wall between the pseudoculus and the pharyngeal muscles, a protractor-adductor muscle originating from the mandibular arm, a protractor originating from the central body of the fulcro-tentorium, and a retractor originating at the dorsal occiput (François et al. 1992). Note that the muscular equipment in Protura seems to be variable concerning the number of protractors and retractors (François 1968; François et al. 1992).

The proturan maxilla is composed of cardo, stipes, galea, lacinia, and maxillary palpus (Fig. 6.1). As in more derived Hexapoda, the proturan maxillary musculature principally allows for protraction/retraction and adduction/abduction of the distal maxilla parts via the lever relations of the cardo with the head and of the cardo with the stipes. The cardo articulates with the posterior arm of the fulcro-tentorium (François et al. 1992) and is connected to the stipes via a syndesmosis. Galea, lacinia, and palpus are broadly connected at their base to the stipes. All parts of the maxilla lie within the gnathal pouch; only the tips of galea, lacinia, and maxillary palpus protrude from it anteriorly together with the mandible tips (Eisenbeis and Wichard 1985). The musculature is composed of a protractor at the cardo which originates at the posterior arm of the endoskeleton and several adductors inserted at the stipes which originate at various parts of the endoskeleton (Fig. 6.1).

### 6.2.2 *Collembola*

The majority of Collembola (springtails) show orthognathous biting–chewing mandibles and maxillae. *Anurida sp.* and Pseudarchorutidae (Richards 1979) have stylet-like mouthparts which are supposedly used to penetrate the cell walls of fungi hyphen and roots. These lineages display derived features according to all current phylogenetic estimates based on morphological (D'Haese 2003; Schneider and D'Haese 2013) and molecular datasets (D'Haese 2002; Xiong et al. 2008; Schneider and D'Haese 2013). Therefore, the groundplan condition of collembolan mouthparts is assumed to be of the biting–chewing type and is described in the following. The diet of Collembola may consist of algae, fungi, pollen, and detritus (Paclt 1956; Eisenbeis and Wichard 1985). Mandibles show an elongated distal incisival part which is used for scraping off food particles from the substrate and a proximal molar area which is used for grinding the scraped off particles (Fig. 6.1). The collembolan mandible motion is mainly composed of a rotatory motion and adduction/abduction as well as protraction/retraction (Hofmann 1908; Manton and Harding 1964; Koch 2001).

There are multiple interactions of the mandibles with other head parts and the maxilla, some of which can be interpreted as articulations given the muscular equipment: At the proximal base of the mandible, there is a condyle-like structure which is moveably attached to the gnathal pouch via a ligament (Hofmann 1908; Denis 1928; Koch 2001). Further distal at the caudal side of the mandible, there is an impression into which a stud of the stipes fits. This stud can articulate with the mandible during the simultaneous mouthpart movement of the mandibles and maxillae (Blanke et al. 2015b). Further disto-medial is a mandibular hump which articulates with parts of the endoskeleton (Hofmann 1908; Koch 2001). Even further distal, opposite to the molar area, there is a depression which can articulate with a strongly thickened part of the head capsule in the region of the clypeus. All of the mandible muscles of Collembola insert between the two proximal mandibular "condyles." Several rotator-adductors insert on both sides of the mandibular orifice and several protractor-adductors within the orifice (Fig. 6.1). Due to the small size of Collembola and the entognathous mouthparts, it was not possible so far to generate video footage to visualize mouthpart motion. Statements made here and in the literature are therefore necessarily of qualitative nature and leave room for interpretation. Previous studies suggested that the mandibles mainly rotate around the axis generated by the proximal condyle and more distal articulations with the endoskeleton or the head capsule (Hofmann 1908; Denis 1928; Koch 2001) (Fig. 6.1). However, given the inclination of the molae, this would not allow for a grinding of particles between the molae during all stages of food processing. It appears likely that there exists another principal axis of rotation which is more aligned with the mola (Fig. 6.1). This axis of rotation is generated by the thickened part of the clypeus which articulates with a part of the mandible lateral of the mola and the mandibular hump which articulates with parts of the endoskeleton. Clearly, the mouthpart motions in Collembola deserve further study to understand the interaction of the musculature and the significance of condyle-like structures observed.

The collembolan maxilla is composed of cardo, stipes, a fused galea and lacinia, and a short palpus. The principal composition and spatial configuration of the cardo and stipes are the same as in Protura: the cardo articulates with a part of the endoskeleton and is moveably connected to the elongated stipes via a syndesmosis. Considerable disagreement still exists concerning the exact delimitations of galea, lacinia, and palpus to each other since these structures are fused to various degrees in Collembola which is why they were termed "terminal lobe" (Folsom 1900), "maxillar head" (Chen et al. 1997), or "claw" (Hofmann 1908). In any case, the maxilla as a whole is again mainly adapted to protraction/retraction given the muscular equipment so that food particles can be hauled toward the preoral cavity. Protraction is achieved through muscle bundles inserted at the inner side of the cardo; the various muscles inserting at the stipes and the claw allow for adduction and, to a minor degree, rotation of the whole maxilla (Fig. 6.1).

### *6.2.3 Diplura*

Diplura (two-pronged bristletails) have prognathous biting–chewing mouthparts. They are traditionally divided into Projapygoidea + Campodeoidea and Japygoidea based on the shape of the cerci and the presence of a prostheca, a moveable appendage on the mandibular gnathal edge (Bitsch and Jacques 2000; Richter et al. 2002; Koch 2009). While most Campodeoidea are omnivorous with a preference for saprophygy and microphytophagy, Japygoidea are mainly predatory. Despite these differences in general lifestyle, Diplura show a comparably uniform general mouthpart organization: the mandible lacks a distinct molar area; instead the entire gnathal edge is composed of sharp teeth whose left and right sides interlock, just like in neopteran insects (François 1970; Koch 2001). Apart from taxonomically relevant differences in the number of incisivi and similar characters (Allen 2002), more notable differences on the subordinal level concern the origin and insertions of certain muscles such as those responsible for the movement of the cardo or retractors of the mandible (François 1970; Blanke and Machida 2015). The mandibular muscle composition and observations of feeding in Japygidae (*Occasjapyx japonica*, Blanke pers. obs) suggest more forceful adduction capabilities in Diplura compared to Collembola. As in Collembola, the dipluran mandible shows several interactions with parts of the gnathal pouch, endoskeletal elements, or parts of the maxilla which are currently debated regarding their homology (Koch 2000, 2001; Blanke et al. 2015b; Blanke and Machida 2015; Koch 2016). The proximal part of the mandible is formed like a pointed tip and fits into a sclerotized part of the gnathal pouch. Most muscles of the mandible are rotator-adductors, namely those which insert at the dorsal and ventral sides of the thickened mandibular rim. A transverse mandibular tendon connects a large adductor muscle which spans from the left to the right mandible. Two abductor-retractors attach at the dorsal distal rim of the mandible, and further distal, near the end of the mandibular orifice, a retractor-adductor inserts.

The maxilla is composed of cardo, stipes, galea, lacinia, and a very short palpus. According to the muscle equipment, protraction is likely the principal movement of the maxilla so that larger food particles can be hauled toward the mandibles for further processing. The cardo articulates with posterior structures, termed as "lingual stalks" (Koch 2000) or "fulturae" (François 1970), so that activity of the muscles inserting at the cardo or the proximal parts of the stipes results in a protraction of the maxillae. The musculature furthermore consists of several distinct adductors of the lacinia (Fig. 6.1).

### *6.2.4 Archaeognatha*

Archaeognatha (bristletails) have orthognathous biting–chewing mouthparts which are used to consume lichen and detritus (Eisenbeis and Wichard 1985; Dettner and Peters 2011). Although the archaeognathan mandible shows several similarities to collembolan mandibles, such as an elongated incisival area and a pronounced molar area, mandibular articulations are clearly different and the muscle equipment is composed of different functional groups (Bitsch 1963; Koch 2001; Blanke et al. 2015a). Due to the larger size of Archaeognatha, observations of feeding movements were possible (Blanke et al. 2015a). Principal mandible movements consist in large parts of rotation and adduction/abduction. Only a minor portion of rhythmic protraction is realized. Archaeognatha possess three mandibular articulations, two with parts of the head and one with the maxillary palpus (Blanke et al. 2015a). These articulations are all non-permanent. The posterior mandibular articulation is located at the proximal most part of the mandibles. It is a distinct condyle which articulates with the ventral margin of the sclerotized head capsule at height of the gena just ventral of the eyes (Bitsch 1963). As in Diplura, this connection is non-permanent, because the condyle does not touch the infolded genal socket when the mandible is slightly protracted [see supplementary video material in Blanke et al. (2015a)]. The anterior mandibular articulation is located at the base of the molar area. Due to the strong curvature in this part of the mandible, this area serves as a depression which articulates with the region where the anterior tentorial arms connect to the clypeus during rotation of the mandible. Like the posterior articulation, the anterior one is non-permanent. During stronger protraction movements, the base of the mola is not in touch with the tentorio-clypeal area. Due to the striking fine structural similarities of this anterior articulation to the anterior articulation in Zygentoma (see below), it was suggested that these articulations are homologous (Blanke et al. 2015a). The third articulation is composed of a depression at the medio-frontal side of the mandible which serves as a socket for a knob located at the inner side of the first maxillary palpomere (Blanke et al. 2015a). Via this articulating coupling structure and the rest of the base of the maxillary palpus, an anterior and lateral evasion of the mandibles during the rhythmic rotatory movements is prevented. The mandibular musculature consists of a large adductor-rotator which inserts on the posterior mandibular rim near the base of the mola and several additional rotator-adductors which insert at different locations on the anterior side of the mandibular rim (Fig. 6.1). A large muscle bundle inserts within the mandibular orifice and connects to the dorsoventrally oriented parts of the anterior tentorium. In comparison to Diplura and Collembola, fewer muscles move the mandibles although the principal degrees of freedom of the mandibles are the same in all three taxa.

The archaeognathan maxilla is composed of cardo, stipes, galea, lacinia, and a seven-segmented palpus. Again, the principal movement of the maxilla is a protraction so that small food particles can be scraped off the substrate and particles are transported to the mandibles for further processing. There is one articulation of the cardo with the posterior tentorium and another articulation of the base of the palpus with the mandible as mentioned above. In line with its double function as a clamping structure for the mandibles and to reallocate food to the molae, Archaeognatha possess the highest number of distinct maxillary muscle bundles among early evolved Hexapoda (Fig. 6.1). Protractors insert at the cardo and the proximal part of the stipes, and rotator-adductors insert further distal on the ventral wall of the stipes. The lacinia is moved by four distinct adductors and the galea by one adductor and its antagonist, while the palpus is moved by another three muscle bundles (two extensors and one flexor) (Bitsch 1963).

### 6.2.5 Zygentoma

Zygentoma (silverfish) have orthognathous biting–chewing mouthparts which are used to consume algae, lichen, detritus, and cellulose. Zygentoma are the oldest extant lineage displaying a reduction of the degrees of freedom of mandible movement due to the joint configuration (Staniczek 2000; von Lieven 2000). Their bowl-shaped mandibles have an incisival area which reaches proximally the molar area without an intermediate area (Blanke et al. 2014; Koch 2001). Two principal articulations, a posterior and an anterior one, allow for an adduction and, to a minor extent, translation (Blanke pers. obs.). The posterior articulation is composed of a pointed tip which interacts with a pyramidal condylus on the head capsule (von Lieven 2000; Blanke et al. 2014). The anterior articulation is formed as a slight depression at the anterior dorsal margin lateral of the molar area. The head part is a caliper-like structure formed by parts of the anterior tentorial arms and the clypeus. This caliper fits around the dorsal rim of the mandible; its outer part interacts with the depression on the mandible during adduction/abduction (von Lieven 2000; Blanke et al. 2014, 2015a). The caliper furthermore allows translation of the mandible along the main axis of the caliper opening. The mandibular muscles in Zygentoma are characterized by a reduction in their double functions compared to non-dicondylic hexapods. Mandible adduction is accomplished by a cranial main adductor inserting at the posterior medial rim and a group of smaller tentorial muscles inserting within the mandibular orifice. Abduction is realized by two cranial muscles inserting at the lateral parts of the anterior mandibular rim.

The zygentoman maxilla shows the typical orthopteroid configuration. Unlike in Archaeognatha and entognathous Hexapoda, the bowl-shaped cardo articulates with the head capsule near the confluence of the posterior tentorial arms with the head. Cardo and stipes are connected via a broad syndesmosis. The stipes bears a five-segmented palpus, a pointed and sclerotized lacinia, and sickle-shaped galea. Compared to the mandibular musculature, protraction and adduction movements of the maxillae are still achieved by several muscles with double functions. Three protractors/retractors insert at the cardo, and several adductor-rotators insert at the ventral stipital wall or inside the stipes, while lacinia and galea are moved by a stipital adductor each (Fig. 6.1).

More detailed biomechanical information concerning the functional properties of the mouthparts, their piercing or biting strength, or the incurred strains on the head capsule is currently not available for the taxa mentioned above. The author is also not aware of any biomechanical studies using modern engineering methods which would cover the head mechanics of Protura, Collembola, Diplura, Archaeognatha, or Zygentoma in a quantitative framework.

### 6.2.6 *Pterygota*

Winged insects (Pterygota) show a reduction in mandibular degrees of freedom (DOFs) and mouthpart musculature (von Lieven 2000; Staniczek 2000, 2001). The mandibles of most pterygotes display a roughly quadratic dorsal opening toward the head lumen. Usually the tendon of the main mandibular adductor attaches at the medial corner, whereas the abductor attaches at the lateral side. The anterior and posterior mandibular articulations with the head are located at the remaining corners. The number of incisivi, shape, and presence of a mola, however, vary strongly among biting–chewing winged insect lineages, as do the areas between these regions (Chapman 1998; Grimaldi and Engel 2005; Beutel et al. 2014). The most severe deviation from the general setup of the mandibular articulations and muscle insertions is shown by mayflies (= Ephemeroptera), which have a posterior cylinder-like joint and an anterior articulation complex largely similar to the one of Zygentoma (Staniczek 2000, 2001), whereas Odonata and Neoptera uniformly show ball-and-socket joint configurations. These two fixed ball-and-socket joints generate a fixed rotation axis, restricting the DOFs of the mandible to a single plane. Note that although Zygentoma and Ephemeroptera also possess an anterior mandibular articulation, their joint configuration is not permanent, so that movement is not restricted to a single plane; this morphology was termed "facultative dicondyly" by Staniczek (2000, 2001). The ball-and-socket joint system of Odonata and Neoptera is maintained stable irrespective of the particular food preference of a given lineage although the rest of the mandible shape can be remarkably variable. Despite the reduced DOFs, the muscular equipment in Ephemeroptera and Odonata is the same compared to Zygentoma, with the exception of the absence of one abductor muscle (Blanke et al. 2012). In Neoptera, a further reduction of the mandibular musculature took place. The muscles inserting at the anterior and posterior mandibular rim are absent in most lineages; the tentorial muscles inserting within the mandibular orifice are strongly reduced and in many cases absent (Wipfler et al. 2011; Blanke et al. 2012). The maxilla of winged insects also shows a further reduction in musculature although the DOFs of the craniocardinal articulation are the same as for primary wingless insects. The cardo is moved by one retractor and one protractor, the stipes is adducted by two muscles which originate from the tentorium, and lacinia and galea are adducted by one muscle each (Fig. 6.1). Again, this muscle equipment can vary considerably in more derived lineages, a detailed account for each order is, however, out of scope of this contribution.

### 6.2.7 *Trends in Biting–Chewing Mouthpart Evolution*

The above-presented brief outline of the functional morphology of the mouthparts and their muscle configuration can be summarized as follows:

- The rotation axis of the mandible is nearly aligned with the distal incisivi in entognathous Hexapoda and is "lifted" anteriorly toward the cephalo-caudal axis in silverfish and winged insects (Fig. 6.1, red dotted line).
- The mandible muscle equipment becomes reduced from Protura to winged insects although exceptions exist for the maxillae (see Archaeognatha).
- The DOFs of the mandible become reduced. Adduction/abduction becomes the only movement type in winged insects.
- Articulation types with the head and endoskeletal elements range from multicondylic (Collembola, Diplura) to facultative dicondylic (Archaeognatha, Zygentoma, and Ephemeroptera) to an obligate dicondyly (Odonata, Neoptera). The mouthpart articulations in Protura need to be reinvestigated.

## 6.3 Comparative Biomechanics of Biting–Chewing Mouthparts

### *6.3.1 Experimental Assessments*

Due to the small size of most insect species (Chown and Gaston 2010), even simple biomechanical assessments such as bite force measurements involve complex experimental setups which are currently limited by the size of the force-sensing element which can be introduced between the mandibles. Weihmann et al. (2015a) used a bespoke strain gauge-based 2D force transducer with a tip element of 0.8 mm diameter; David et al. (2016a) used a piezoelectric 1D force sensor with 0.63 mm diameter. Therefore, insects with a gape no less than approximately 5 mm should be measured with currently available setups in order to ensure that the adductor muscles operate near their maximum force outputs (Blümel et al. 2012b, c). Due to the size/gape problematic, the bite forces of only 21 species from three insect "orders" (Odonata, Blattodea, Coleoptera) were measured so far. Head width (Fig. 6.2) and muscle size (Wheater and Evans 1989) were found to be reliable predictors of bite force, while other morphometric data such as body length or body weight are poor predictors of bite force (Wheater and Evans 1989). Note that head width is by far not a universal predictor of bite force (Senawi et al. 2015). Rather, this metric likely depends on the particular arrangement of the head muscles for which head width can be one proxy among many. Due to the abovementioned difficulties with regard to insect size, and the size of their mouthparts in particular, modeling approaches came into focus recently.

### *6.3.2 Functional Morphology of Biting–Chewing Mouthparts using Musculoskeletal Modeling*

An emerging technique to study aspects of the biomechanics, such as the kinematics and kinetics, of an arbitrarily sized organism is musculoskeletal modeling of the particular movement system (Curtis et al. 2008; Shi et al. 2012; Watson et al. 2014). Such simulations are subsumed under the umbrella term multibody dynamics analysis (MDA), placing an emphasis on the investigation of moving structures which are connected to each other by joints and/or muscles. Given accurate information about skeletal geometries and muscle characteristics, MDA allows to predict the resulting forces in a movement system, such as bite forces or joint reaction forces with acceptable accuracy (Fig. 6.2) (Curtis et al. 2008, 2010; Gröning et al. 2013; Blanke et al. 2017b) so that even fossils can be studied (Bates and Falkingham 2012; Snively et al. 2013). MDAs for insects so far used simple non-Hill-type muscle models in conjunction with experimentally measured bite forces. These studies showed that the measured bite forces can be simulated with MDA when approximately 80% of muscle activation are assumed (David et al. 2016a, b). This seems to be in accordance with measured muscle activation levels in relation to maximum force outputs in insect locomotion systems (Blümel et al. 2012a, b). However, a further refinement of the muscle models used for MDAs of insects, i.e., an investigation of mouthpart muscle properties, should have high priority to increase the accuracy of the predictions made.

Despite the obvious relevance of MDA to study forces and moments in arbitrarily sized biological systems, MDA requires a lot of raw data from different sources such as precise geometric data, movement data, muscle properties, and bite forces. Therefore, the study of biomechanical systems using MDA on an

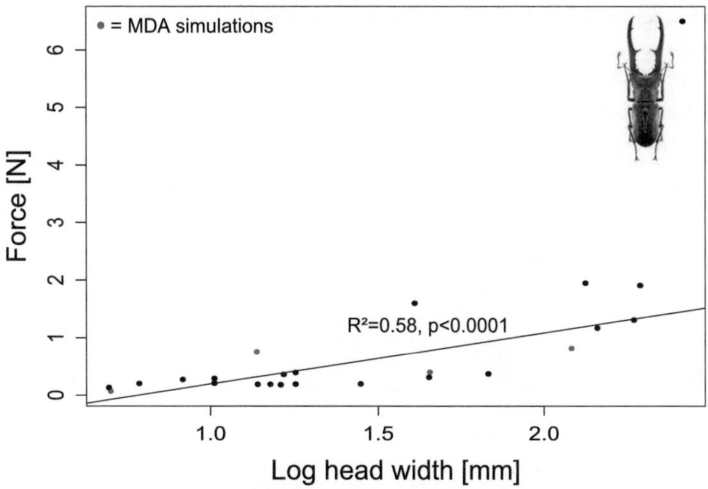


**Fig. 6.2** Relationship between bite force and head width in selected insects for which bite force measurements are currently available. Data points in red are results from multibody dynamics analyses (MDA), i.e., simulations of mandible biting. So far male stag beetles were shown to have unusually high bite forces given their head width (Goyens et al. 2014). The regression was calculated without the data for the male stag beetle. Image of *C. metallifer* courtesy of Udo Schmidt. Data based on Wheater and Evans (1989), Goyens et al. (2014), Weihmann et al. (2015b), and David et al. (2016a, b)

evolutionary scale including many taxa is currently not feasible. On a broader scale, biomechanical determinants such as the mechanical advantage (MA), which are directly measureable on a given specimen, could yield initial insights into the performance transitions in evolving movement systems such as insect mouthparts.

### *6.3.3 The Mechanical Advantage (MA) as a Performance Metric for Insect Mouthparts*

The MA is a straightforward biomechanical metric which in a biological context was first introduced for vertebrates (Westneat 1995, 2004) and was used since in many studies on vertebrate and arthropod jaw mechanics (Cooper and Westneat 2009; Sakamoto 2010; Habegger et al. 2011; Dumont et al. 2014; Senawi et al. 2015; Cox and Baverstock 2015; Weihmann et al. 2015b; Fujiwara and Kawai 2016; Blanke et al. 2017a; Fabre et al. 2017; Olsen 2017). The MA is defined as the inlever to outlever ratio. For dicondylic insect mandibles, the inlever is the distance between the application of the input force and the joint axis, while the outlever arm is the distance from the biting point to the joint axis (Fig. 6.3).

The MA thus indicates the percentage of force transmitted to the food item (i.e., the effectivity of the lever system). Although more detailed investigations concerning muscular insertion angles, muscle volumes, spatial arrangements, and muscle characteristics would be needed to quantify the forces applied to the food, the MA constitutes a useful mechanical performance index: it allows a size independent comparison of the relative efficiencies of force transmission within the mandibular lever system and it can be readily measured in a wide array of dried museum specimens as well as freshly collected ones.

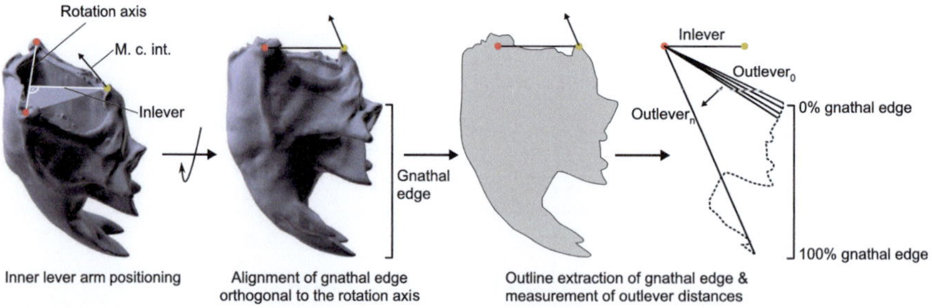


**Fig. 6.3** Workflow for extraction of the mandibular mechanical advantage illustrated on the mandible of *Neopetalia punctata* (Odonata: Anisoptera). Note that for a comparison across different insect lineages only the inlever component of the main mandibular adductor M. craniomandibularis internus (M. c. int.) was calculated (Inlever distance between red and yellow dot). This muscle solely has an adductive function in Dicondylia and a mixed adductive-rotatory function in entognathous Hexapoda. Dotted line = gnathal edge; Red = articulation points; Yellow = tendon insertion of M. craniomandibularis internus

The MA here is used on a phylogenetically diverse sample of Hexapoda ranging from Collembola to Psocodea. Given the morphologies and functional changes across early evolved Hexapoda summarized in the introduction, the extent to which such a simple lever measurement might serve as a proxy for the more complex changes observed with respect to joint configuration, main axes of rotation, DOFs, and musculoskeletal configurations was studied. Furthermore, the size and relative location of the performance space of each lineage expressed as the relative efficiency of force transmission of the mandibular lever system was investigated.

## 6.4 Studying the Mechanical Advantage in Early Evolved Hexapoda

Seventy-seven taxa ranging from Collembola to Psocodea were studied for the mechanical advantage (MA) of their mandibles. Species were investigated using micro-computed tomography (μCT) carried out at several synchrotron facilities: Beamline BW2 and IBL P05 of the outstation of the Helmholtz Zentrum Geesthacht at the Deutsches Elektronen Synchrotron (DESY), the beamline TOMCAT at the Paul Scherrer institute (PSI), the TOPO-TOMO beamline of the Karlsruhe Institute of Technology (KIT), and beamline BL47XU of the Super Photon Ring 8GeV (SPring-8, Table 6.1).

MA measurements were carried out on the segmentations of the left mandible for each specimen. Automatic segmentations were performed using the software ITK-snap (Yushkevich et al. 2006) after which STL files were imported into the software Blender (www.blender.org) for further processing (Fig. 6.3). The gnathal edge was defined *sensu* Richter et al. (2002) as the area from the *pars molaris* (proximal to the mouth opening) to the *pars incisivus* (distalmost tooth). Since the homology of subparts of the gnathal area is debated (Staniczek 2000; Richter et al. 2002; Fleck 2011), the gnathal outline, as seen when orienting the mandible in line with the rotation axis (Fig. 6.3), was scaled as a percentage of tooth row length. For this, ~800 points for each specimen were wrapped against the gnathal outline in Blender and the distance between each point orthogonal to the mandibular rotation axis (= outlever) was measured. Similarly, one point was placed at the insertion point of M. craniomandibularis internus on the mandible and the distance from this point orthogonal to the rotation axis was measured (= inlever). All measurements and calculations were carried out in the R software environment (v. 1.1.383) using custom scripting. The MAs for each specimen were computed and polynomial functions of the first–sixth order were fitted against each MA profile. The Akaike and Bayes information criteria (AIC and BIC) were used to determine the polynomial function with the best relative fit which was then used for further analysis.

In order to be able to compare MA values between taxa with different joint configurations (mainly entognathous vs. ectognathous lineages), and to prevent a violation of the homology hypotheses for muscles across Hexapoda (Rühr et al.

**Table 6.1** Taxon sampling studied (in alphabetical order) and coding of discrete character traits

| Order | Species | Coll. | Food | Cond. type | DOF | DTA | CT |
|---|---|---|---|---|---|---|---|
| Archaeognatha | *Machilis germanica* | BL | H | fac. dic | 3 | L | A |
| Archaeognatha | *Meinertellus cundinamarcensis* | BL | H | fac. dic | 3 | L | A |
| Archaeognatha | *Pedetontus unimaculatus* | BL | H | fac. dic | 3 | L | A |
| Archaeognatha | *Trigoniophthalmus alternatus* | BL | H | fac. dic | 3 | L | A |
| Collembola | *Pogonognathellus flavescens* | BL | O | multic. | 4 | L | A |
| Dermaptera | *Diplatys flavicollis* | MFN | O | dic. | 1 | F | P |
| Dermaptera | *Forficula auricularia* | BL | O | dic. | 1 | F | P |
| Dermaptera | *Labidura riparia* | MFN | C | dic. | 1 | F | P |
| Diplura | *Atlasjapyx cf atlas* | ZFMK | C | multic. | 3 | L | A |
| Diplura | *Campodea sp.* | BL | H | multic. | 3 | L | A |
| Diplura | *Catajapyx aquilionaris* | BL | C | multic. | 3 | L | A |
| Diplura | *Lepidocampa weberi* | BL | H | multic. | 3 | L | A |
| Diplura | *Metriocampa sp.* | BL | H | multic. | 3 | L | A |
| Diplura | *Occasjapyx japonicus* | BL | C | multic. | 3 | L | A |
| Embioptera | *Antipaluria urichi* | BÜ | H | dic. | 1 | L | P |
| Embioptera | *Aposthonia japonica* | SU | H | dic. | 1 | L | P |
| Embioptera | *Embia ramburi* | BL | H | dic. | 1 | L | P |
| Embioptera | *Metoligotoma* sp. | MFN | H | dic. | 1 | L | P |
| Ephemeroptera | *Ephemera danica* | BL | D | fac. dic | 2 | L | P |
| Ephemeroptera | *Epeorus* sp. | BL | H | fac. dic | 2 | L | P |
| Ephemeroptera | *Siphlonurus lacustris* | STAN | H | fac. dic | 2 | L | P |
| Grylloblattodea | *Grylloblatta bifratrilecta* | BL | O | dic. | 1 | L | P |
| Odonata | *Aeshna cyanea* | BL | C | dic. | 1 | F | P |
| Odonata | *Aeschnophlebia longistigma* | ZFMK | C | dic. | 1 | F | P |
| Odonata | *Anaciaeshna isoceles* | ZFMK | C | dic. | 1 | F | P |
| Odonata | *Anotogaster sieboldii* | ZFMK | C | dic. | 1 | F | P |
| Odonata | *Calopteryx virgo* | BL | C | dic. | 1 | F | P |
| Odonata | *Coenagrion puella* | BL | C | dic. | 1 | F | P |
| Odonata | *Cordulegaster bidentata* | BL | C | dic. | 1 | F | P |
| Odonata | *Cordulia aenea* | BL | C | dic. | 1 | F | P |
| Odonata | *Crocothemis erythraea* | BL | C | dic. | 1 | F | P |
| Odonata | *Epiophlebia superstes* | BL | C | dic. | 1 | F | P |
| Odonata | *Epophthalmia elegans* | ZFMK | C | dic. | 1 | F | P |
| Odonata | *Hagenius brevistylus* | ZFMK | C | dic. | 1 | F | P |
| Odonata | *Libellula depressa* | BL | C | dic. | 1 | F | P |
| Odonata | *Lestes virens* | BL | C | dic. | 1 | F | P |
| Odonata | *Macromia taeniolata* | ZFMK | C | dic. | 1 | F | P |
| Odonata | *Mecistogaster linearis* | ZFMK | C | dic. | 1 | F | P |
| Odonata | *Mnais* sp. | ZFMK | C | dic. | 1 | F | P |
| Odonata | *Neopetalia punctata* | ZFMK | C | dic. | 1 | F | P |

(continued)

**Table 6.1** (continued)

| Order | Species | Coll. | Food | Cond. type | DOF | DTA | CT |
|---|---|---|---|---|---|---|---|
| Odonata | *Oligoaeshna pryeri* | ZFMK | C | dic. | 1 | F | P |
| Odonata | *Onychogomphus forcipatus* | ZFMK | C | dic. | 1 | F | P |
| Odonata | *Phenes raptor* | ZFMK | C | dic. | 1 | F | P |
| Odonata | *Phyllopetalia apicalis* | ZFMK | C | dic. | 1 | F | P |
| Odonata | *Platycnemis pennipes* | BL | C | dic. | 1 | F | P |
| Odonata | *Pyrrhosoma nymphula* | BL | C | dic. | 1 | F | P |
| Odonata | *Sympetrum vulgatum* | BL | C | dic. | 1 | F | P |
| Odonata | *Tachopteryx thoreyi* | ZFMK | C | dic. | 1 | F | P |
| Odonata | *Zonophora baetesi* | ZFMK | C | dic. | 1 | F | P |
| Orthoptera | *Acheta domesticus* | BL | O | dic. | 1 | F | P |
| Orthoptera | *Comicus calcaris* | ZSM | O | dic. | 1 | F | P |
| Orthoptera | *Conocephalus dorsalis* | GÖ | O | dic. | 1 | F | P |
| Orthoptera | *Diaphanogryllacris laeta* | ZFMK | O | dic. | 1 | F | P |
| Orthoptera | *Gryllus bimaculatus* | BL | O | dic. | 1 | F | P |
| Orthoptera | *Hemideina crassidens* | GÖ | O | dic. | 1 | F | P |
| Orthoptera | *Meconema meridionale* | GÖ | C | dic. | 1 | F | P |
| Orthoptera | *Papuaistus* sp. | GÖ | O | dic. | 1 | F | P |
| Orthoptera | *Prosopogryllacris* sp. | MFN | O | dic. | 1 | F | P |
| Orthoptera | *Stenobothrus lineatus* | GÖ | H | dic. | 1 | F | P |
| Orthoptera | *Stenopelmatus* sp. | ZSM | O | dic. | 1 | F | P |
| Orthoptera | *Pholidoptera griseoaptera* | GÖ | O | dic. | 1 | F | P |
| Orthoptera | *Tettigonia viridissima* | GÖ | O | dic. | 1 | F | P |
| Orthoptera | *Tridactylus* sp. | ZSM | H | dic. | 1 | F | P |
| Orthoptera | *Troglophilus neglectus* | ZSM | O | dic. | 1 | F | P |
| Orthoptera | *Xya variegata* | NHM | D | dic. | 1 | F | P |
| Phasmatodea | *Agathemera crassa* | ZSM | H | dic. | 1 | F | P |
| Phasmatodea | *Peruphasma schultei* | BL | H | dic. | 1 | L | P |
| Plecoptera | *Oemopteryx* sp. | MFN | C | dic. | 1 | L | P |
| Plecoptera | *Perla marginata* | BL | C | dic. | 1 | L | P |
| Psocodea | *Caecilius* sp. | FF | H | dic. | 1 | F | P |
| Zoraptera | *Zorotypus caudelli* | BL | H | fac. dic | 2 | L | A |
| Zygentoma | *Atelura formicaria* | BL | H | dic. | 1 | L | P |
| Zygentoma | *Lepisma saccharina* | BL | H | fac. dic | 2 | L | A |
| Zygentoma | *Maindronia neotropicalis* | BL | C | fac. dic | 2 | L | A |
| Zygentoma | *Thermobia domestica* | BL | H | fac. dic | 2 | L | A |
| Zygentoma | *Tricholepidion gertschi* | BL | H | fac. dic | 2 | L | A |

*Coll.* collection, *cond. type* articulation type, *DOF* degrees of freedom, *DTA* connection of dorsal tentorial arms to head, *CT* corpotentorium, *BL* collection by the author, *BÜ* collection by Dr. Sebastian Büsse, *SU* collection of the Sugadaira Montane Research Center, Japan, *STAN* material provided by Dr. Arnold Staniczek, *GÖ* material provided by Dr. Fanny Leubner and Dr. Sven Bradler, Göttingen, *FF* material provided by Dr. Frank Friedrich, Hamburg; H, herbivore, *O* omnivore, *C* carnivore, *D* detrivore, *fac. dic.* facultative dicondyly, *multic.* multicondyly, *dic.* obligate dicondyly, *L* muscular/fibrillous connection, *F* fixed/sclerotized connection, *A* absent, *P* present, *MFN* Museum für Naturkunde Berlin, *ZFMK* Zoological Research Museum Alexander Koenig, *ZSM* Bavarian State Collection of Zoology, *NHM* Natural History Museum Vienna

in preparation; Wipfler et al. 2011; Blanke et al. 2012; Blanke and Machida 2015), several simplifications had to be made. Firstly, only the MA with respect to the main mandibular adductor (M. craniomandibularis internus fide Wipfler et al. 2011) was calculated. This adductor muscle delivers the main part of the bite force in dicondylic insects (David et al. 2016a, b), but it is acting mainly as a rotator in Collembola, and as a rotator-adductor in Diplura and Archaeognatha as explained in the introduction. As such, the MA in Collembola, Diplura, and Archaeognatha also is an index for the effectiveness of performing a rotatory motion rather than solely an adduction. Since rotatory mandible movements are mainly used by these lineages for food uptake, the biological meaning of the MA—the effectivity of the force transmission to the food item—is maintained in these cases. A further simplification had to be made with regard to the anterior points of contact between the mandible and the various head structures across Hexapoda. For Collembola and Diplura, the dorso-anterior interaction points of the mandibles with other head structures were interpreted as anterior articulation points. It has to be emphasized that this does not imply homology of these interaction points with the anterior mandibular articulations of Ectognatha although evidence exists in favor of such an interpretation (Koch 2001).

Phylognetic signal was assessed using the most recent phylogenetic estimate of the 1kite consortium (www.1kite.org) as a basis (pers. comm. B. Misof on behalf of 1kite). The phylogeny was pruned in order to contain only the taxa analyzed here. Phylogenetic signal was assessed using the *K* statistic as implemented in geomorph v.3.0.5 (Adams 2014a) with 10,000 random permutations. This test statistic was found to be the most efficient approach to test for phylogenetic signal (Pavoine and Ricotta 2013). Since significant phylogenetic signal was detected, a principal component analysis (PCA) as well as phylogenetic PCA as implemented in the phytools package v.0.6–44 (Revell 2009) was carried out. Tests for phylogenetic signal were also conducted for alternative (and debated) deep level relationships within Hexapoda, namely the potential sister-group relationship of Diplura with Ellipura [Protura + Collembola; (Dell'Ampio et al. 2009, 2014)] and the Metapterygota (Odonata + Neoptera) and Chiastomyaria (Ephemeroptera + Neoptera) hypotheses (Simon et al. 2018). Because the statistical significance of the phylogenetic signal was not influenced by these alternative topologies, only the results for the topology with Diplura as sister group to Ectognatha and Odonata + Ephemeroptera (= Palaeoptera), which represents the most recent phylogenetic estimate of the deep level relationships within Hexapoda, are presented in the following. The tempo of the MA variation was tested using the evolutionary rate parameter under a Brownian motion model of evolution as implemented in geomorph (Adams 2014b). It was tested whether the rates of evolution varied significantly depending on several group designations: the taxonomic rank of orders, the joint configuration (multicondylic, facultative dicondylic, obligate dicondylic), the connection type of the dorsal tentorial arms with the head (sclerotized or ligamentous), fusion of the anterior and posterior tentoria (corpotentorium absent/present), the degrees of freedom of the mandible (DOF), and the food preference (herbivorous, omnivorous, carnivorous, detrivorous). See Table 6.1 for group designations to each species.

## 6.5 Characteristics of Mechanical Advantage (MA) Evolution of Mouthparts and Correlation with Food Preference

Calculation of the MA along the entire gnathal edge revealed characteristic MA curve progressions for several systematic groups within Hexapoda (Fig. 6.4). The collembolan showed an exponential increase in the MA until approximately 80% of the tooth row followed by an almost linear decline. Diplura all showed a strongly parabolic decline from high MA values approximately 1.2–1.8 down to 0.05–0.8. Archaeognatha showed an almost linear increase in their MA; in Zygentoma curve progression is vice versa with an almost linear decrease toward the 100% tooth row position. In Ephemeroptera and Odonata, the otherwise linear decline in MA from 0 to 100% tooth row length is characterized by a peak in MA at the 40% and 70% tooth row position, respectively. All other investigated taxa show an almost linear decrease in MA from the 0 to 100% tooth row position.

A polynomial function of the fifth order resulted in the best relative fit on the MA curves according to the AIC value (−765.4). The first four principal components (PCs) accounted for 98.3% of the variation in MA (Table 6.2). Visual representation of the PCs of the polynomial coefficients showed a lineage-specific distinction between the above-described curve progressions (Fig. 6.5).

Taxa such as Collembola, Diplura, and Archaeognatha with mandible motions composed of rotatory and adduction-abduction movements (and corresponding MAs which reflect this diversity in mandible motion types) mostly scored at the extremes of the PCs. For example, the springtail scores at the positive extreme of PC1, while Diplura, although they occupy a comparably wide variance space, mostly cluster near the negative extreme of PC1. Archaeognatha mostly score at the positive extreme of PCs 1, 3, and 4; by comparison, Zygentoma (with the exception of *Maindronia neotropicalis*) score near the center of each PC, while Ephemeroptera occupy a comparably wide performance space along PC2. All other taxa cluster near the center of each PC.

Since significant phylogenetic signal ($K = 1.19$, $p = 0.0001$) was detected in the data, a phylogenetic PCA as implemented in the phytools package (Revell 2012) was carried out. This resulted in a similar lineage-dependent distribution of the data with most of the variance associated with PC1 (Fig. 6.5).

In both PCAs PC1 mainly codes for the inclination of the MA curve and its slope, the curve progression from high to low values, and the vertical position of the curve. PC2 mainly codes for whether there is a local maximum with sharp fall-offs on both sides. PC3 likely codes whether the curve progression goes from higher MA values to lower ones or vice versa, and PC4 likely codes for the slope of the MA curve and the general progression from high to lower MA values. However, due to the low variance associated with PCs3 + 4 further observations are necessary to elucidate whether the association of MA progressions remains stable for these PCs.

Analysis of the evolutionary rates of MA variation resulted in nonsignificant rate ratios when partitioning the data at the taxonomic rank of orders (observed rate ratio

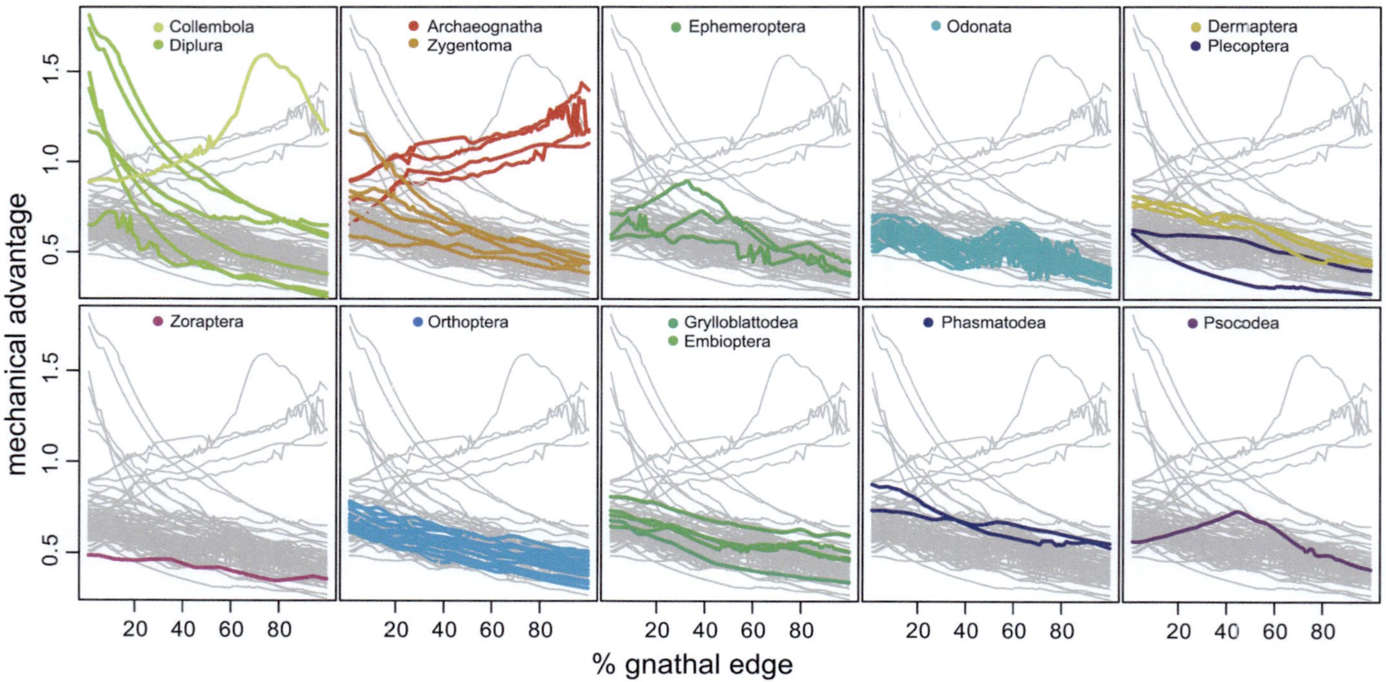


**Fig. 6.4** Progression of biting efficiency as expressed by the mechanical advantage against percent gnathal edge position (see Fig. 6.3 right) in selected non-holometabolous biting–chewing insects. For clarity biting efficiency of each lineage (colored) is shown with respect to the rest of the taxon sampling (gray)

**Table 6.2** Summary of the principal components of the mandibular mechanical advantage progression for the uncorrected (upper rows) and the phylogeny corrected data (lower rows)

| | PC1 | PC2 | PC3 | PC4 | PC5 | PC6 |
|---|---|---|---|---|---|---|
| Uncorrected data | | | | | | |
| Standard deviation | 0.80 | 0.27 | 0.20 | 0.12 | 0.09 | 0.07 |
| Proportion of variance | 0.82 | 0.09 | 0.05 | 0.02 | 0.01 | 0.01 |
| Cumulative proportion | 0.82 | 0.91 | 0.96 | 0.98 | 0.99 | 1.00 |
| Phylogeny taken into account | | | | | | |
| Standard deviation | 1.30 | 0.55 | 0.39 | 0.27 | 0.25 | 0.21 |
| Proportion of variance | 0.73 | 0.13 | 0.07 | 0.03 | 0.03 | 0.02 |
| Cumulative proportion | 0.73 | 0.86 | 0.92 | 0.95 | 0.98 | 1.00 |

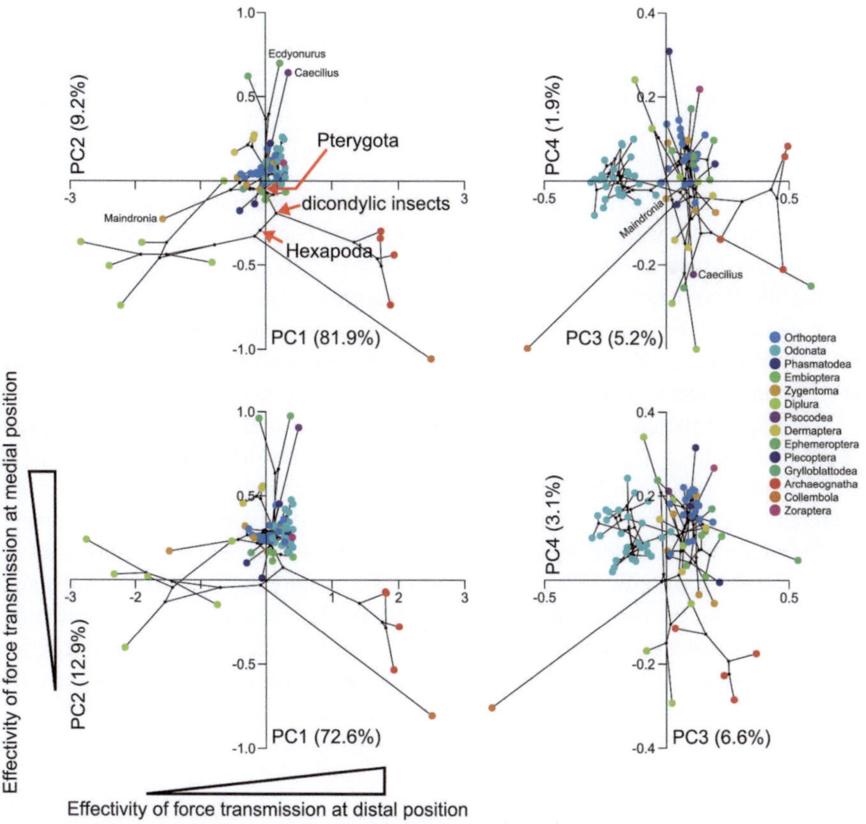


**Fig. 6.5** Principal component analysis (PCA) of MA progression with the first four PCs shown. 2D plots of the first four uncorrected (upper row) and phylogeny-corrected PCs (lower row). Same lineage-specific color code as in Fig. 6.4. Note that the indication of the functional interpretation of the PCs in the lower left plot is also applicable to the upper left plot. Functional interpretations for PCs3+4 are not shown due to the low variance associated with these PCs

**Table 6.3** Correlations of discrete morphological character states related to joint and endoskeleton morphology with MA variation across the taxon sampling for the uncorrected (upper rows) and the phylogeny corrected data (lower rows)

| | SS | MS | $R^2$ | $F$ | $Z$ | $p$ |
|---|---|---|---|---|---|---|
| Uncorrected data | | | | | | |
| DOF | 29.14 | 14.57 | 0.49 | 107.29 | 5.16 | 0.0001 |
| Joint config. | 13.49 | 6.75 | 0.23 | 49.67 | 4.48 | 0.0001 |
| CT present | 3.59 | 3.59 | 0.06 | 26.40 | 3.45 | 0.0001 |
| dicondyly:food | 2.48 | 0.83 | 0.04 | 6.10 | 5.08 | 0.0001 |
| Food pref. | 1.21 | 0.40 | 0.02 | 2.96 | 2.08 | 0.0199 |
| DTA connection | 0.48 | 0.48 | 0.01 | 3.54 | 1.62 | 0.0428 |
| Phylogeny taken into account | | | | | | |
| DOF | 45.21 | 22.61 | 0.26 | 15.06 | 5.39 | 0.0001 |
| Joint config. | 8.83 | 4.41 | 0.05 | 2.94 | 3.88 | 0.0001 |
| dicondyly:food | 17.09 | 5.70 | 0.10 | 3.80 | 5.38 | 0.0001 |
| Food pref. | 5.60 | 1.87 | 0.03 | 1.24 | 1.85 | 0.0420 |
| CT present | 1.55 | 1.55 | 0.01 | 1.03 | 2.77 | 0.0014 |
| DTA connection | 1.97 | 1.97 | 0.01 | 1.31 | 1.43 | 0.1103 |

(ORR): 26.4, $p = 0.661$) or according to food preference (ORR: 7.01, $p = 0.25$). Evolutionary rates were significantly different when partitioning the data according to the general type of articulation (ORR: 11.4, $p = 0.012$) with multicondylic taxa having the fastest rates (1.70), followed by facultative dicondylic taxa (0.74) and obligate dicondylic taxa (0.15). Evolutionary rates were also significantly different between taxa with different connection types of the dorsal tentorial arms and fused anterior and posterior tentoria (ORRs: 5.21/7.64, $p = 0.008/0.0001$).

Analyses of the correlation of the MA progressions with several morphological characters such as the degrees of freedom of the mandibles (DOFs), the joint configuration (multicondylic, facultatively dicondylic, obligate dicondylic), food preferences (herbivorous, omnivorous, carnivorous, detrivorous), and the general configuration of the endoskeleton resulted in significant correlations of almost all of these characters with the MA progressions of the mandibles (Table 6.3). DOFs and joint configurations explained 0.49 and 0.23% of the variance observed, while all other factors explained only a very minor proportion of the variance. Corrected for shared ancestry, coefficients of determination dropped considerably to 0.26 and 0.05 for the DOFs and the joint configuration, respectively.

## 6.6 Interpretation of the Mandibular Performance Space Occupation

Overall, the PCA of the MA progression reflects the principal lever arm differences resulting from the diverse set of mandible types analyzed here: taxa without obligate dicondylic mandibles such as Collembola, Diplura, Archaeognatha, Zygentoma, and

Ephemeroptera have the tendency to score at the extremes of the PCs due to their different MA progressions (Fig. 6.4). Closer inspection of the PCs reveals interesting functional associations between the different lineages. Collembola and Archaeognatha both scored at the positive extreme of PC1. Both lineages possess biting–chewing mandibles which perform a rolling motion around the dorsoventral axis with the rotation axis nearly aligned with the position of the distal incisivi (Fig. 6.1). Both taxa also possess a pronounced molar area to grind particles which is at a near orthogonal inclination to the rotation axis and therefore is advantageous regarding the effectiveness of the force transmission to the food item.

In contrast, Diplura scored at the negative extreme of PC1 and occupy a comparably wide lever arm performance space with no clear separation between Projapygoidea + Campodeoidea and the predatory Japygoidea. Diplura are characterized by prognathous mandibles whose main axis of rotation is not aligned with the distal row of incisivi and which are also capable of limited protraction and retraction movements of their mandibles. This is likely advantageous to rip off food particles from the substrate or prey. The comparably wide performance space occupation is mainly related to the vertical positioning of the slopes which shows that the distance of the gnathal edge to the rotation axis is more variable in Diplura compared to the other studied lineages.

In Zygentoma, the shift of the axis of rotation between the anterior and posterior joints leads to a shift of the biomechanical role of the M. craniomandibularis internus toward a purely adductive motion. This functional shift is also reflected in the PCs of the MA due to the clustering with the rest of Dicondylia. The silverfish *Maindronia neotropicalis* occupies a remarkable position within the dipluran performance space. The position of this species in PC space indicates that a considerable proportion of protraction and adduction might characterize the mandible motion. Living specimens should be investigated for their feeding habits, especially the amount of protraction/retraction, to corroborate this data. The rest of Zygentoma and winged insects agglomerated near the center of the first two PCs which account already for >90% of the variance associated with the mandibular mechanical advantage (Fig. 6.5). This clearly reflects the increased restriction of mandibular motion to adduction and abduction and the shift of the rotation axis toward the cephalo-caudal axis.

## 6.7 Phylogenetic Signal in the Mechanical Advantage (MA)

Significant phylogenetic signal in the MA data with a $K$ value greater than 1 means that there is a constraining phylogenetic effect on the variation in the effectivity of mandibular force transmissions: measured MAs are more similar than would be expected under a Brownian motion model of evolution, i.e., by a random unrestricted evolution of MA values (Adams 2014a; Blomberg et al. 2003).

However, this general pattern of phylogenetic signal in the MA data is not uniformly distributed. Analysis of the evolutionary rates (Adams 2014b) revealed that MAs associated with multicondyly and facultative dicondyly evolved at the fastest rates compared to the relative evolutionary stasis of obligate dicondylic taxa (Odonata + Neoptera). Associated with these significant rate changes in joint configuration are the connection types of the dorsal tentorial arms and the fusion of the anterior and posterior elements of the endoskeleton. Together with the phylogenetic signal in MAs, these data in summary suggest that the type of mandibular articulation (with its implications for rotation axis shifts) and the configuration of the endoskeleton lead to the comparatively static MA values in Odonata and Neoptera.

The detection of phylogenetic signal in this functional metric contrasts the common notion that specific functional selection pressures such as optimization toward more effective force transmission or, alternatively, toward faster biting movements (Westneat 1994, 2004; Sakamoto 2010) are the main drivers of the biomechanical performance space of food processing in the studied lineages at this taxonomic level. Phylogenetic signal in the MA with a $K$-value higher than 1 has also been detected for the mandibles of theropod dinosaurs (Sakamoto 2010), but it remained unclear whether this signal shows a uniform distribution or is dependent on certain lineages or functional types as is the case for the present data. Generally, a $K$-value higher than one is unusual for morphology-related data such as the MA (Blomberg et al. 2003). Based on a meta-analysis of literature data it appears that in most cases, $K$ is lower than one (Blomberg et al. 2003) which would mean that morphological traits are less similar among species than expected from a Brownian motion model. This is intuitive, as these traits should be more malleable as a reaction to environmental effects rather than being constrained by shared ancestry as indicated here.

## 6.8 Mouthpart Performance in Dicondylic and Winged Insects

According to the phylomorphospace reconstruction of the principal components of MA progression variation, the most recent common ancestors (MRCA) of dicondylic insects, Pterygota and Hexapoda, respectively, score near the center of the first PC (Fig. 6.5, red arrows). This suggests that all three MRCAs probably possessed mouthpart configurations which showed MA progressions similar to the ones of recent obligate dicondylic insects. Regarding PC2, the three MRCAs are located on the negative side with a gradual transition toward the center of PCs1 + 2 from Hexapoda to Pterygota. The investigated Phasmatodea show a comparable location, and, interestingly, they are the ones with the highest MA values at the distal incisivi among the studied Dicondylia (Fig. 6.4). Given the phylomorphospace reconstruction, it appears that higher mandibular force transmission effectivities can be postulated for the MRCA of Hexapoda with a decrease toward the MRCA of winged insects. The type of mandibular articulation which realized this higher mandibular effectivity in force transmission to the food items, however, remains unclear.

Early branched Hexapoda show a high disparity in mouthpart shapes and biomechanics. It is clear that these changes cannot be attributed to a single effect, e.g., miniaturization in Protura, Collembola, and Diplura as these taxa again show different modes of food uptake. The results from this study indicate that primary wingless taxa have mandible configurations which are partly, e.g., in the molar area, advantageous and effective with regard to the lever system. In a timeframe of ~twenty million years [MRCA Zygentoma → MRCA Pterygota (Misof et al. 2014)], a rather ineffective dicondylic mandible system with regard to the distal incisivi evolved. This dicondylic mandible system remains a stable configuration across all insects with biting–chewing mouthparts even beyond the ones studied here (such as beetles, bees, and ants). The fact that this type of "ineffective" mandible setup is an evolutionary stable solution might be related to several other factors besides phylogenetic inheritance. For example, with dicondylic mandibles wider gape angles can be realized compared to the other observed solutions and a fixed axis of rotation requires fewer muscles within the system to control mandible movement. Therefore, the larger gape angles of dicondylic mandibles principally should allow to take up more energy per time interval. Flight imposes high energetic costs, and this might have positively selected the intake of large food volumes in a rapid manner rather than a time-consuming preprocessing of the food through chewing–grinding motions seen in entognathous Hexapoda, bristletails, and silverfish. The fact that a dicondylic food uptake system occurred almost concomitantly with wings and other aerial locomotion types (Hasenfuss 2002; Yanoviak et al. 2009) is certainly worth considering in future studies regarding the evolutionary linkage of different body parts.

**Acknowledgements** Sven Bradler, Sebastian Büsse, Frank Friedrich, Fanny Leubner, Ryuichiro Machida, and Arnold Staniczek kindly donated specimens or provided μCT scans for this study. Additional specimens were loaned from several natural history museums: the Zoological Research Museum Alexander Koenig (ZFMK), the Museum für Naturkunde Berlin (MFN), the Natural History Museum Vienna (NHM), the Bavarian State Collection of Zoology (ZSM), and the Staatliches Museum für Naturkunde Stuttgart. Peter Rühr and Leif Moritz are thanked for their assistance during the beamtime experiments at the various synchrotron facilities. Christian Voss is thanked for collecting data of the dermapteran taxa. Members of the workgroup of Prof. Büschges at the University of Cologne provided further support in the lab.

**Funding Statement** The author was supported by an ERC starting grant (Project "MechEvoInsect"; ID: 754290). μCT scanning was supported under several facility specific grants allowed to the author (DESY: I-20080169, I-20090211, I-20120065, I-20130002, I-20140099 EC, I-20170190, I-20170896; PSI: 20091024, 20110069, 20140056, 20150464, 20171469; SPring-8: 2014B1046, 2016A1269).

## References

Adams DC (2014a) A generalized K statistic for estimating phylogenetic signal from shape and other high-dimensional multivariate data. Syst Biol 63:685–697. https://doi.org/10.1093/sysbio/syu030

Adams DC (2014b) Quantifying and comparing phylogenetic evolutionary rates for shape and other high-dimensional phenotypic data. Syst Biol 63:166–177. https://doi.org/10.1093/sysbio/syt105

Allen RT (2002) A synopsis of the Diplura of North America: keys to higher taxa, systematics, distributions and descriptions of new taxa (Arthropoda: Insecta). Trans Am Entomol Soc 128:403–466

Bates KT, Falkingham PL (2012) Estimating maximum bite performance in *Tyrannosaurus rex* using multi-body dynamics. Biol Lett 8:660–664. https://doi.org/10.1098/rsbl.2012.0056

Beutel RG, Friedrich F, Ge SQ, Yang XK (2014) Insect morphology and phylogeny. De Gruyter, Berlin

Bitsch J (1963) Morphologie céphalique des machilides (Insecta Thysaunura). Ann Sci Nat Zool 12:585–706

Bitsch C, Jacques B (2000) The phylogenetic interrelationships of the higher taxa of apterygote hexapods. Zool Scr 29:131–156. https://doi.org/10.1046/j.1463-6409.2000.00036.x

Blanke A, Machida R (2015) The homology of cephalic muscles and endoskeletal elements between Diplura and Ectognatha (Insecta). Org Divers Evol 16:241–257. https://doi.org/10.1007/s13127-015-0251-5

Blanke A, Wipfler B, Letsch H, Koch M, Beckmann F, Beutel R, Misof B (2012) Revival of Palaeoptera—head characters support a monophyletic origin of Odonata and Ephemeroptera (Insecta). Cladistics 28:560–581. https://doi.org/10.1111/j.1096-0031.2012.00405.x

Blanke A, Koch M, Wipfler B, Wilde F, Misof B (2014) Head morphology of *Tricholepidion gertschi* indicates monophyletic Zygentoma. Front Zool 11:16. https://doi.org/10.1186/1742-9994-11-16

Blanke A, Machida R, Szucsich NU, Wilde F, Misof B (2015a) Mandibles with two joints evolved much earlier in the history of insects: dicondyly is a synapomorphy of bristletails, silverfish and winged insects. Syst Entomol 40:357–364. https://doi.org/10.1111/syen.12107

Blanke A, Rühr PT, Mokso R, Villanueva P, Wilde F, Stampanoni M, Uesugi K, Machida R, Misof B (2015b) Structural mouthpart interaction evolved already in the earliest lineages of insects. Proc R Soc B 282:20151033. https://doi.org/10.1098/rspb.2015.1033

Blanke A, Schmitz H, Patera A, Dutel H, Fagan MJ (2017a) Form–function relationships in dragonfly mandibles under an evolutionary perspective. J R Soc Interface 14:20161038. https://doi.org/10.1098/rsif.2016.1038

Blanke A, Watson PJ, Holbrey R, Fagan MJ (2017b) Computational biomechanics changes our view on insect head evolution. Proc R Soc B Biol Sci 284:20162412. https://doi.org/10.1098/rspb.2016.2412

Blomberg SP, Garland T, Ives AR (2003) Testing for phylogenetic signal in comparative data: behavioral traits are more labile. Evolution 57:717–745. https://doi.org/10.1111/j.0014-3820.2003.tb00285.x

Blümel M, Guschlbauer C, Daun-Gruhn S, Hooper SL, Büschges A (2012a) Hill-type muscle model parameters determined from experiments on single muscles show large animal-to-animal variation. Biol Cybern 106:559–571. https://doi.org/10.1007/s00422-012-0530-6

Blümel M, Guschlbauer C, Hooper SL, Büschges A (2012b) Using individual-muscle specific instead of across-muscle mean data halves muscle simulation error. Biol Cybern 106:573–585. https://doi.org/10.1007/s00422-011-0460-8

Blümel M, Hooper SL, Guschlbauerc C, White WE, Büschges A (2012c) Determining all parameters necessary to build Hill-type muscle models from experiments on single muscles. Biol Cybern 106:543–558. https://doi.org/10.1007/s00422-012-0531-5

Chapman RF (1998) The insects: structure and function. Cambridge University Press, Cambridge

Chapman RF, de Boer G (1995) Regulatory mechanisms in insect feeding. Springer, Dordrecht

Chaudonneret J (1950) La morphologie céphalique de *Thermobia domestica* (Packard) (Insecte Aptérigota Thysannoure). Ann Sci Nat Zool 12:145–302

Chen B, Snider RJ, Snider RM (1997) Mouthparts of three collembolan species: comparative morphology. Invertebr Biol 116:355–362. https://doi.org/10.2307/3226868

Chown SL, Gaston KJ (2010) Body size variation in insects: a macroecological perspective. Biol Rev 85:139–169. https://doi.org/10.1111/j.1469-185X.2009.00097.x

Cooper WJ, Westneat MW (2009) Form and function of damselfish skulls: rapid and repeated evolution into a limited number of trophic niches. BMC Evol Biol 9:24. https://doi.org/10.1186/1471-2148-9-24

Cox PG, Baverstock H (2015) Masticatory muscle anatomy and feeding efficiency of the American Beaver, *Castor canadensis* (Rodentia, Castoridae). J Mamm Evol:1–10. https://doi.org/10.1007/s10914-015-9306-9

Curtis N, Kupczik K, O'higgins P, Moazen M, Fagan M (2008) Predicting skull loading: applying multibody dynamics analysis to a macaque skull. Anat Rec 291:491–501. https://doi.org/10.1002/ar.20689

Curtis N, Jones MEH, Lappin AK, O'Higgins P, Evans SE, Fagan MJ (2010) Comparison between in vivo and theoretical bite performance: using multi-body modelling to predict muscle and bite forces in a reptile skull. J Biomech 43:2804–2809. https://doi.org/10.1016/j.jbiomech.2010.05.037

D'Haese CA (2002) Were the first springtails semi-aquatic? A phylogenetic approach by means of 28S rDNA and optimization alignment. Proc R Soc B Biol Sci 269:1143–1151. https://doi.org/10.1098/rspb.2002.1981

D'Haese CA (2003) Morphological appraisal of collembola phylogeny with special emphasis on poduromorpha and a test of the aquatic origin hypothesis. Zool Scr 32:563–586. https://doi.org/10.1046/j.1463-6409.2003.00134.x

David S, Funken J, Potthast W, Blanke A (2016a) Musculoskeletal modeling of the dragonfly mandible system as an aid to understanding the role of single muscles in an evolutionary context. J Exp Biol 219(Pt 7):1041–1049. https://doi.org/10.1242/jeb.132399

David S, Funken J, Potthast W, Blanke A (2016b) Musculoskeletal modelling under an evolutionary perspective: deciphering the role of single muscle regions in closely related insects. J R Soc Interface 13:20160675. https://doi.org/10.1098/rsif.2016.0675

Dell'Ampio E, Szucsich NU, Carapelli A, Frati F, Steiner G, Steinacher A, Pass G (2009) Testing for misleading effects in the phylogenetic reconstruction of ancient lineages of hexapods: influence of character dependence and character choice in analyses of 28S rRNA sequences. Zool Scr 38:155–170. https://doi.org/10.1111/j.1463-6409.2008.00368.x

Dell'Ampio E, Meusemann K, Szucsich NU, Peters RS, Meyer B, Borner J, Petersen M, Aberer AJ, Stamatakis A, Walzl MG, Minh BQ, von Haeseler A, Ebersberger I, Pass G, Misof B (2014) Decisive data Sets in phylogenomics: lessons from studies on the phylogenetic relationships of primarily wingless insects. Mol Biol Evol 31:239–249. https://doi.org/10.1093/molbev/mst196

Denis JR (1928) Études sur l'anatomie de la téte de quelques Collemboles suivies de considérations sur la morphologie de la tête des insectes. Arch Zool Exp Gén 68:1–291

Dettner K, Peters W (2011) Lehrbuch der Entomologie. Springer

Dumont ER, Samadevam K, Grosse I, Warsi OM, Baird B, Davalos LM (2014) Selection for mechanical advantage underlies multiple cranial optima in new world leaf-nosed bats. Evolution 68:1436–1449. https://doi.org/10.1111/evo.12358

Dunger W (1983) Tiere im Boden. Ziemsen, Wittenberg

Eisenbeis G, Wichard W (1985) Atlas zur Biologie der Bodenarthropoden. Gustav Fischer, Stuttgart

Evans MEG, Forsythe TG (1985) Feeding mechanisms, and their variation in form, of some adult ground-beetles (Coleoptera: Caraboidea). J Zool 206:113–143. https://doi.org/10.1111/j.1469-7998.1985.tb05640.x

Fabre P-H, Herrel A, Fitriana Y, Meslin L, Hautier L (2017) Masticatory muscle architecture in a water-rat from Australasia (Murinae, Hydromys) and its implication for the evolution of carnivory in rodents. J Anat 231:380–397. https://doi.org/10.1111/joa.12639

Fleck G (2011) Phylogenetic affinities of Petaluridae and basal Anisoptera families (Insecta: Odonata). Stuttg Beitr Zur Naturkunde A 4:83–104

Folsom JW (1900) The development of the mouthparts of *Anurida maritima*. Bull Mus Comp Zool 36:87–157

François J (1968) Anatomie et morphologie céphalique des protures. Mém Muséum Natl Hist Nat Sér A 59:1–144

François J (1970) Squelette et musculature céphalique de *Campodea chardardi* CONDÉ (Diplura: Campodeidae). Zool Jahrb Anat 87:331–376

François J, Dallai R, Yin WY (1992) Cephalic anatomy of *Sinentomon erythranum* Yin (Protura: Sinentomidae). Int J Insect Morphol Embryol 21:199–213. https://doi.org/10.1016/0020-7322(92)90016-G

Fujiwara S, Kawai H (2016) Crabs grab strongly depending on mechanical advantages of pinching and disarticulation of chela. J Morphol 277:1259–1272. https://doi.org/10.1002/jmor.20573

Goyens J, Dirckx J, Dierick M, Hoorebeke LV, Aerts P (2014) Biomechanical determinants of bite force dimorphism in *Cyclommatus metallifer* stag beetles. J Exp Biol 217:1065–1071. https://doi.org/10.1242/jeb.091744

Grimaldi D, Engel MS (2005) Evolution of the insects. Cambridge University Press, Cambridge

Gröning F, Jones MEH, Curtis N, Herrel A, O'Higgins P, Evans SE, Fagan MJ (2013) The importance of accurate muscle modelling for biomechanical analyses: a case study with a lizard skull. J R Soc Interface 10:20130216. https://doi.org/10.1098/rsif.2013.0216

Habegger ML, Motta PJ, Huber DR, Deban SM (2011) Feeding biomechanics in the Great Barracuda during ontogeny. J Zool 283:63–72. https://doi.org/10.1111/j.1469-7998.2010.00745.x

Hasenfuss I (2002) A possible evolutionary pathway to insect flight starting from lepismatid organization. J Zool Syst Evol Res 40:65–81. https://doi.org/10.1046/j.1439-0469.2002.00180.x

Hofmann RW (1908) Über die Morphologie und Funktion der Kauwerkzeuge und über das Kopfnervensystem von *Tomocerus plumbeus* L. (II. Beitrag zur Kenntnis der Collembolen). Z Für Wiss Zool 89:598–689

Ikeda Y, Machida R (1998) Embryogenesis of the dipluran *Lepidocampa weberi* Oudemans (Hexapoda, Diplura, Campodeidae): external morphology. J Morphol 237:101–115. https://doi.org/10.1002/(SICI)1097-4687(199808)237:2<101::AID-JMOR2>3.0.CO;2-4

Isely FB (1944) Correlation between mandibular morphology and food specificity in grasshoppers. Ann Entomol Soc Am 37:47–67. https://doi.org/10.1093/aesa/37.1.47

Koch M (2000) The cuticular cephalic endoskeleton of primarily wingless hexapods: ancestral state and evolutionary changes. Pedobiologia 44:374–385. https://doi.org/10.1078/S0031-4056(04)70056-6

Koch M (2001) Mandibular mechanisms and the evolution of hexapods. Ann Soc Entomol Fr NS 37:129–174

Koch M (2009) Chapter 212 – Protura. In: Resh VH, Cardé RT (eds) Encyclopedia of insects, 2nd edn. Academic, San Diego, pp 855–858

Koch M (2016) Current views on the early evolution of the insect head revisited. Zitteliana 88:31

Manton SM, Harding JP (1964) Mandibular mechanisms and the evolution of arthropods. Philos Trans R Soc Lond Ser B Biol Sci 247:1–183. https://doi.org/10.1098/rstb.1964.0001

Misof B, Liu S, Meusemann K, Peters RS, Donath A, Mayer C, Frandsen PB, Ware J, Flouri T, Beutel RG, Niehuis O, Petersen M, Izquierdo-Carrasco F, Wappler T, Rust J, Aberer AJ, Aspöck U, Aspöck H, Bartel D, Blanke A, Berger S, Böhm A, Buckley TR, Calcott B, Chen J, Friedrich F, Fukui M, Fujita M, Greve C, Grobe P, Gu S, Huang Y, Jermiin LS, Kawahara AY, Krogmann L, Kubiak M, Lanfear R, Letsch H, Li Y, Li Z, Li J, Lu H, Machida R, Mashimo Y, Kapli P, McKenna DD, Meng G, Nakagaki Y, Navarrete-Heredia JL, Ott M, Ou Y, Pass G, Podsiadlowski L, Pohl H, von Reumont BM, Schütte K, Sekiya K, Shimizu S, Slipinski A, Stamatakis A, Song W, Su X, Szucsich NU, Tan M, Tan X, Tang M, Tang J, Timelthaler G, Tomizuka S, Trautwein M, Tong X, Uchifune T, Walzl MG, Wiegmann BM, Wilbrandt J, Wipfler B, Wong TKF, Wu Q, Wu G, Xie Y, Yang S, Yang Q, Yeates DK, Yoshizawa K, Zhang Q, Zhang R, Zhang W, Zhang Y, Zhao J, Zhou C, Zhou L, Ziesmann T,

Zou S, Li Y, Xu X, Zhang Y, Yang H, Wang J, Wang J, Kjer KM, Zhou X (2014) Phylogenomics resolves the timing and pattern of insect evolution. Science 346:763–767. https://doi.org/10.1126/science.1257570

Olsen AM (2017) Feeding ecology is the primary driver of beak shape diversification in waterfowl. Funct Ecol 31:1985–1995. https://doi.org/10.1111/1365-2435.12890

Paclt J (1956) Biologie der primär flügellosen Insekten. Gustav Fischer Verlag, Jena

Pavoine S, Ricotta C (2013) Testing for phylogenetic signal in biological traits: the ubiquity of cross-product statistics. Evolution 67:828–840. https://doi.org/10.1111/j.1558-5646.2012.01823.x

Revell LJ (2009) Size-correction and principal components for interspecific comparative studies. Evol Int J Org Evol 63:3258–3268. https://doi.org/10.1111/j.1558-5646.2009.00804.x

Revell LJ (2012) Phytools: an R package for phylogenetic comparative biology (and other things). Methods Ecol Evol 3:217–223. https://doi.org/10.1111/j.2041-210X.2011.00169.x

Richards WR (1979) Collembola. Mem Entomol Soc Can 111:300–303. https://doi.org/10.4039/entm111108300-1

Richter S, Edgecombe G, Wilson G (2002) The lacinia mobilis and similar structures – a valuable character in arthropod phylogenetics? Zool Anz J Comp Zool 241:339–361. https://doi.org/10.1078/0044-5231-00083

Rühr PT, Koch M, Misof B, Blanke A (in preparation) The head of *Pogonognathellus flavescens* (Hexapoda: Collembola) and its implications on the early evolution of the hexapod endoskeleton

Sakamoto M (2010) Jaw biomechanics and the evolution of biting performance in theropod dinosaurs. Proc R Soc Lond B Biol Sci rspb20100794. doi: https://doi.org/10.1098/rspb.2010.0794

Schneider C, D'Haese CA (2013) Morphological and molecular insights on Megalothorax: the largest *Neelipleona* genus revisited (Collembola). Invertebr Syst 27:317–364

Senawi J, Schmieder D, Siemers B, Kingston T (2015) Beyond size – morphological predictors of bite force in a diverse insectivorous bat assemblage from Malaysia. Funct Ecol 29:1411–1420. https://doi.org/10.1111/1365-2435.12447

Shi J, Curtis N, Fitton LC, O'Higgins P, Fagan MJ (2012) Developing a musculoskeletal model of the primate skull: predicting muscle activations, bite force, and joint reaction forces using multibody dynamics analysis and advanced optimisation methods. J Theor Biol 310:21–30. https://doi.org/10.1016/j.jtbi.2012.06.006

Simon S, Blanke A, Meusemann K (2018) Reanalyzing the Palaeoptera problem – the origin of insect flight remains obscure. Arthropod Struct Dev. https://doi.org/10.1016/j.asd.2018.05.002

Snively E, Cotton JR, Ridgely R, Witmer LM (2013) Multibody dynamics model of head and neck function in Allosaurus (Dinosauria, Theropoda). Palaeontol Electron 16:1–29. https://doi.org/10.26879/338

Staniczek AH (2000) The mandible of silverfish (Insecta: Zygentoma) and mayflies (Ephemeroptera): its morphology and phylogenetic significance. Zool Anz 239:147–178

Staniczek AH (2001) Der Larvenkopf von *Oniscigaster wakefieldi* McLachlan, 1873 (Insecta: Ephemeroptera: Oniscigastridae). Ein Beitrag zur vergleichenden Anatomie und Phylogenie der Eintagsfliegen. PhD thesis, Eberhard-Karls-Universität Tübingen

von Lieven AF (2000) The transformation from monocondylous to dicondylous mandibles in the Insecta. Zool Anz 239:139–146

Watson PJ, Gröning F, Curtis N, Fitton LC, Herrel A, McCormack SW, Fagan MJ (2014) Masticatory biomechanics in the rabbit: a multi-body dynamics analysis. J R Soc Interface 11:20140564. https://doi.org/10.1098/rsif.2014.0564

Weihmann T, Kleinteich T, Gorb SN, Wipfler B (2015a) Functional morphology of the mandibular apparatus in the cockroach *Periplaneta americana* (Blattodea: Blattidae) – a model species for omnivore insects. Arthropod Syst Phylogeny 73:477–488

Weihmann T, Reinhardt L, Weißing K, Siebert T, Wipfler B (2015b) Fast and powerful: biomechanics and bite forces of the mandibles in the American Cockroach *Periplaneta americana*. PLoS One 10:e0141226. https://doi.org/10.1371/journal.pone.0141226
Westneat MW (1994) Transmission of force and velocity in the feeding mechanisms of labrid fishes (Teleostei, Perciformes). Zoomorphology 114:103–118. https://doi.org/10.1007/BF00396643
Westneat MW (1995) Feeding, function, and phylogeny: analysis of historical biomechanics in labrid fishes using comparative methods. Syst Biol 44:361–383. https://doi.org/10.1093/sysbio/44.3.361
Westneat MW (2004) Evolution of levers and linkages in the feeding mechanisms of fishes. Integr Comp Biol 44:378–389. https://doi.org/10.1093/icb/44.5.378
Wheater CP, Evans MEG (1989) The mandibular forces and pressures of some predacious Coleoptera. J Insect Physiol 35:815–820. https://doi.org/10.1016/0022-1910(89)90096-6
Wipfler B, Machida R, Müller B, Beutel RG (2011) On the head morphology of Grylloblattodea (Insecta) and the systematic position of the order, with a new nomenclature for the head muscles of Dicondylia. Syst Entomol 36:241–266. https://doi.org/10.1111/j.1365-3113.2010.00556.x
Xiong Y, Gao Y, Yin W, Luan Y (2008) Molecular phylogeny of Collembola inferred from ribosomal RNA genes. Mol Phylogenet Evol 49:728–735. https://doi.org/10.1016/j.ympev.2008.09.007
Yanoviak SP, Kaspari M, Dudley R (2009) Gliding hexapods and the origins of insect aerial behaviour. Biol Lett 5:510–512. https://doi.org/10.1098/rsbl.2009.0029
Yuasa H (1920) The anatomy of the head and mouth-parts of Orthoptera and Euplexoptera. J Morphol 33:251–307. https://doi.org/10.1002/jmor.1050330202
Yushkevich PA, Piven J, Hazlett HC, Smith RG, Ho S, Gee JC, Gerig G (2006) User-guided 3D active contour segmentation of anatomical structures: significantly improved efficiency and reliability. NeuroImage 31:1116–1128. https://doi.org/10.1016/j.neuroimage.2006.01.015

# Chapter 7
# The Generalized Feeding Apparatus of Cockroaches: A Model for Biting and Chewing Insects

**Tom Weihmann and Benjamin Wipfler**

**Abstract** The morphology, musculature, and function of the feeding apparatus of cockroaches is described in detail and compared with other insects with biting and chewing mouthparts. The mouthparts of cockroaches represent, in most cases, the ancestral condition for winged and neopteran insects. Their head capsule is flattened in a posterior-anterior direction and very similar among the studied species. The right mandible is very constant in shape, while the number of distal incisivi in left mandibles varies among species. With the exception of *Tivia* sp. (Corydiidae), primary mandibular adductor of the mandible has eight distinct compartments in all studied roaches, for which functional cross section and volume are provided. In all these specimens, the left adductor is smaller than the right one. Bite forces and muscle properties are discussed for *Periplaneta americana*. The maxilla, labium, and hypopharynx are also highly similar among cockroaches and close to the pterygotan ground plan. The same also applies to the associated musculature for which we also provide functional and kinematic considerations. Cockroaches salivate food outside the mouth cavity before cutting it with the mandibles. The maxillae transport food into the cibarium where the hypopharynx is involved in transporting it between the grinding mandibular molae. The crushed food is sucked into the pharynx via dilation. During the feeding process, most mouthparts exhibit highly concerted activities. This process generally follows the ground pattern for insects with biting and chewing mouthparts, although some salivation processes may differ.

---

T. Weihmann (✉)
University of Cologne, Cologne, Germany
e-mail: tom.weihmann@uni-koeln.de

B. Wipfler
Zoological Research Museum Alexander Koenig, Bonn, Germany
e-mail: benjamin.wipfler@leibniz-zfmk.de



H. W. Krenn (ed.), *Insect Mouthparts*, Zoological Monographs 5,
https://doi.org/10.1007/978-3-030-29654-4_7

## 7.1 Introduction

Cockroaches[1] comprise about 4600 mostly nocturnal species. They can be found in habitats ranging from tropical rainforest to grassland and deserts, although they are concentrated in the subtropics and tropics of Africa and the Americas (Bell et al. 2007). A few species (approx. 1%) are associated with human habitations and some of them such as *Periplaneta americana* or *Blattella germanica* are important pests. Cockroaches are paraphyletic and form together with the termites (Isoptera) the clade Blattodea (Evangelista et al. 2019; Inward et al. 2007). The relationships among the extant blattodean lineages are discussed in Evangelista et al. (2019). It is a common misconception that cockroaches are a comparatively old group. Blattodea separated around 260 million years ago from the mantids (together they form the clade Dictyoptera) (Evangelista et al. 2019). Extant cockroach lineages evolved around 205 million years ago (Evangelista et al. 2019). The cockroach-like animals with long ovipositors, the so-called roachoids that were some of the most abundant animals in Carboniferous "coal" swamps, are most likely dictyopteran stem-group species (Grimaldi and Engel 2005).

Usually, cockroaches are described as omnivores or "classical generalists." This is true for the pest species associated with humans like *Periplaneta*,[2] which are known to feed on almost anything (Bell et al. 2007). However, most cockroaches prefer a more selective diet and even can be food specialists. Since most of them feed on decomposing organic materials, they are probably best described as detritivores. This biased view on cockroach diet is related to the lack of studies concerning gut content in their natural habitat and the fact that some species eat almost anything under laboratory conditions but are highly selective in their natural habitat (Bell et al. 2007).

Several studies have dealt with the morphology of cockroach mouthparts (Buder and Klass 2013; Gangwere 1965; Mangan 1908; Miall and Denny 1886; Popham 1961a; Pradl 1971; Roberts 1972; Strenger 1939; Wipfler et al. under review, 2016; Yuasa 1920; Zhuzhikov 2007), their musculature[3] (Dorsey 1943; Snodgrass 1943, 1944; Weihmann et al. 2015a; Wipfler et al. 2016, under review), cephalic sensillae (Altner 1975; Bland et al. 1998; Lambin 1973; Moulins 1967; Prakash et al. 1995; Ramaswamy and Gupta 1981; Seidl 1991), and glands (Mkhize and Kumar 1972; Suslov 1912).

In addition to these morphological descriptions, several studies have addressed the function of the feeding apparatus of *Periplaneta* (Blattidae). Popham (1961a) provided a detailed account of the mouthparts and their movement during feeding. General reviews of the feeding mechanisms include Smith (1985) and Chapman (1995). Roberts (1972) described the structure and function of the maxillary

---

[1]The terms "cockroach" and "roach" are used differently in the literature. We decided to use the term "cockroach" and define it as all representatives of Blattodea excluding termites.

[2]For the sake of readability, we will refer to already introduced species by their generic name only.

[3]In the present contribution, we follow the muscular terminology of Wipfler et al. (2011).

appendages of *Periplaneta*. Schmitt et al. (2014) used cineradiography to study the kinematics of the mouthparts. Weihmann et al. (2015b) measured the bite forces of this species, and Weihmann et al. (2015a) dealt with the functional morphology of the mandibular apparatus. Additionally Wipfler et al. (2016) gave a detailed account of the cephalic morphology including the musculature of *Periplaneta*, and Willey (1961) described its cephalic nervous system. In addition to *Periplaneta*, Strenger (1939) also studied the functional morphology of the cockroach *Blatta orientalis* (Blattidae).

We compiled available information concerning functional, mechanical, kinematic, and morphological aspects of the generalized feeding apparatus of cockroaches and added various information, especially concerning the mandibular apparatus and the muscular control of the mouthparts, and compared the biting and chewing apparatus of cockroaches to those of other insects and arthropods. Although it is generally accepted that termites form the sister group to the cockroach clade Cryptocercidae (e.g., Evangelista et al. 2019; Inward et al. 2007; Misof et al. 2014) and cockroaches thus are not a natural (monophyletic) group, we restrict the major focus of the present contribution to the old "Blattaria," i.e., the paraphyletic cockroaches excluding termites. This restriction is justified because termites strongly differ in head and mouthpart anatomy from cockroaches: while the former are prognathous (i.e., the mouthparts are oriented to the front), all cockroaches are orthognathous (i.e., the mouthparts point toward the ground). Details about the mouthparts of termites and the differences to cockroaches are provided in Fontes (1986), Hare (1937), Seid et al. (2008), Vishnoi (1956, 1962), Walker (1933), Weesner (1969), and Wipfler et al. (2016).

## 7.2 The Feeding Apparatus of Cockroaches

### *7.2.1 The Head Capsule*

#### 7.2.1.1 Morphology of the Head Capsule

The head capsule of cockroaches (Fig. 7.1) is a strongly sclerotized structure that is in most species reversely drop-shaped and flattened in anterior-posterior direction. It is covered to variable degrees by the pronotum. Cockroaches are best described as orthognathous, i.e., the mouthparts point toward the ground. Cockroaches have a highly moveable head that can be bent under the prothorax, which is also the typical position taken in death (Wipfler et al. 2016). This coincidence caused some researchers to assume that this position represents the natural condition (Popham 1961a; Wipfler et al. 2011). However, living cockroaches regularly hold their heads in an orthognathous position. Termites in contrast are always prognathous (Vishnoi 1956, 1962; Walker 1933). Orthognathy is also observed in the closely related mantids (Wipfler et al. 2012) and in the ground plan of Polyneoptera and Pterygota (Wipfler et al. 2019).

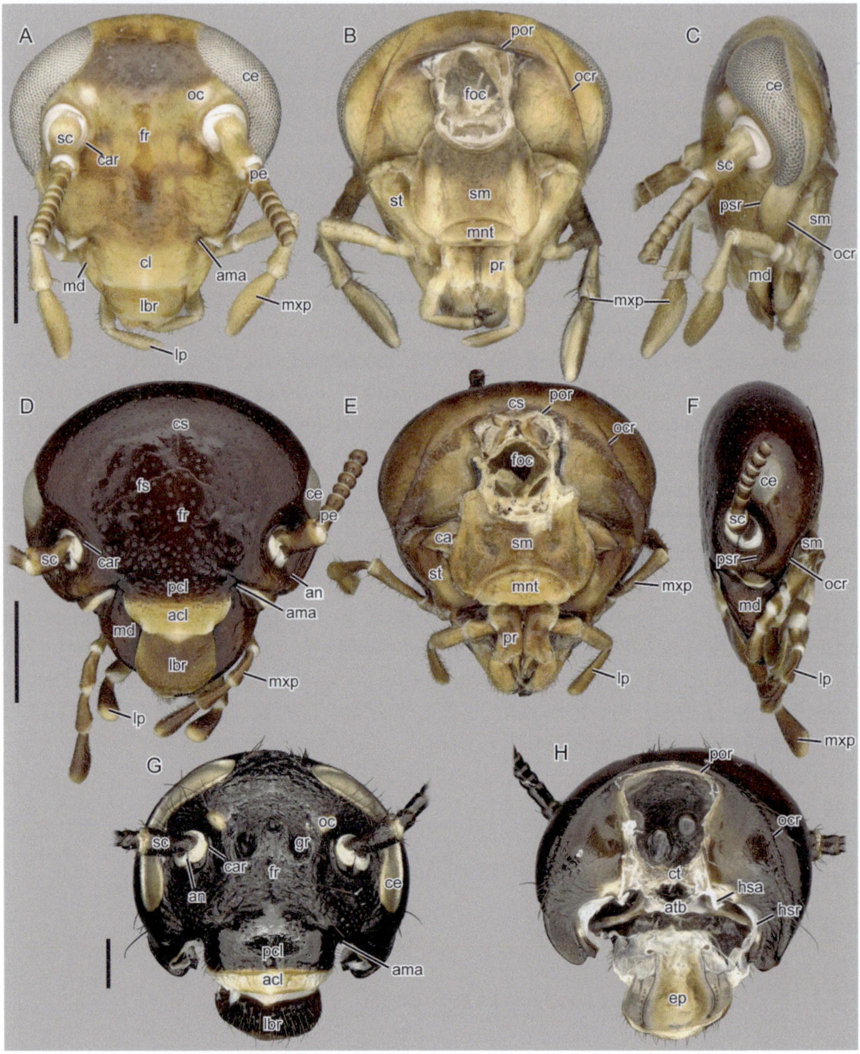


**Fig. 7.1** Head capsules of females of *Blattella germanica* (**a**–**c**), *Salganea rossi* (**d**–**f**), and *Cryptocercus* sp. (**g** and **h**) in frontal (**a**, **d**, **g**), posterior (**b**, **e**, **h**), and lateral (**c**, **f**) view. *acl* anteclypeus, *ama* anterior mandibular articulation, *an* antennifer *atb* anterior tentorial bridge, *car* circumantennal ridge, *ce* compound eye, *ct* corpotentorium, *ep* epipharynx, *foc* foramen occipitale, *fr* frons, *fs* frontal cleavage line, *gr* frontal pit, *hsa* hypostomal apodeme, *hsr* hypostomal ridge, *lbr* labrum, *lp* labial palp, *md* mandible, *mnt* mentum, *mxp* maxillary palp, *oc* ocellus, *ocr* occipital ridge, *pcl* postclypeus, *pe* pedicle, *por* postoccipital ridge, *pr* praementum, *psr* pleurostomal ridge, *sc* scape, *sm* submentum, *st* stipes. Scale bars: 1 mm

The head capsule of cockroaches has several openings that are all surrounded by strengthening ridges to prevent the cuticle from splitting. The posterior opening is the foramen occipitale (foc, Figs. 7.1 and 7.2), which provides the connection to the

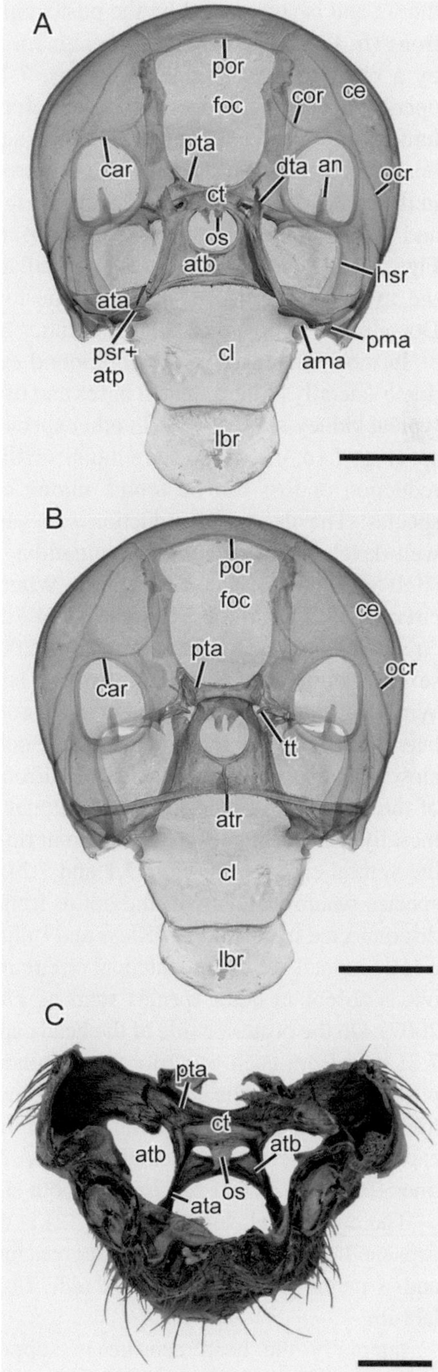


**Fig. 7.2** Tentorium of *Periplaneta americana* (**a**, **b**; taken from Wipfler et al. 2016) and *Ergaula capucina* (**c**; taken from Wipfler et al. under review). (**a**): frontal view, 3-dimensional reconstruction; (**b**): posterior view; 3-dimensional reconstruction; (**c**): dorsal view; photograph. *ama* anterior articulation of the mandible, *an* antennifer, *ata* anterior tentorial arm, *atb* anterior tentorial bridge, *atp* anterior tentorial pit, *atr* anterior tentorial ridge, *car* circumantennal ridge, *ce* compound eye, *cl* clypeus, *cor* circumocular ridge, *ct* corpotentorium, *dta* dorsal tentorial arm, *foc* foramen occipitale, *hsr* hypostomal ridge, *lbr* labrum, *ocr* occipital ridge, *os* oesotendons, *pma* posterior articulation of the mandible, *por* postoccipital ridge, *pta* posterior tentorial arm. Scale bars: 1 cm

thorax and is surrounded by the postoccipital ridge (por, Figs. 7.1 and 7.2). On the frons (fr, Fig. 7.1), there are openings for each antenna, which are both surrounded by a circumantennal ridge (car, Figs. 7.1 and 7.2). On the ventral side of these openings, a distinct process, the antennifer (an, Figs. 7.1 and 7.2), protrudes into the lumen. It serves as articulation for the scape (sc, Fig. 7.1). Ventrally there is the opening for the mouthparts, which is surrounded by the subgenal ridge. It is divided in the anterior pleurostomal ridge (psr, Fig. 7.1) which articulates with the mandible and is delimited by the two mandibular articulations and the hypostomal part (hsr, Figs. 7.1 and 7.2), which serves as articulation for the maxillae. At the dorsal edge of the hypostomal ridge, a distinct hypostomal apodeme (hsa, Fig. 7.1) is present. Dorsally the hypostomal ridge continues into the postoccipital ridge.

In most cockroaches, the compound eyes (ce, Figs. 7.1 and 7.2) are positioned dorso-laterally of the antennal bases and usually surround them which gives them the typical kidney-shaped look. In other species such as the wood feeding *Cryptocercus* sp. (Fig. 7.1d, e), the eyes are much smaller and almost round. However, even eye reduction or loss can be found among cockroaches, especially in cave-dwelling species. The degree of reduction can vary strongly within a species (e.g., from well-developed eyes to just ommatidia in males of *Alluaudellina cavernicola*) (Bell et al. 2007; Chopard 1932). When present, each eye is surrounded by a circumocular ridge (cor, Figs. 7.1 and 7.2); in many species, it is fused with the circumantennal one (Klass and Eulitz 2007). The size of eyes in many species is sexually dimorphic with males having larger eyes than females (Bell et al. 2007; Wipfler et al. under review). Next to the compound eyes, most cockroaches have two lateral ocelli on the frons. The median ocellus is always absent (but present in the closely related mantids). The ridges surrounding the openings or weakenings (eyes) of the cuticle are supported by additional strengthening lines. The epistomal ridge mesally connects the two pleurostomal ridges, thus separating the dorsal frons from the ventral clypeus (cl, Figs. 7.1 and 7.2). *Therea bernhardti* is the only described species where it runs over the entire frons, while in all other studied species it is disconnected in the middle (Klass and Eulitz 2007). In some species, the subantennal ridge connects the circumantennal or circumocular ridge with the pleurostomal ridge but is absent in other species such as *Therea* or *Cryptocercus* (Klass and Eulitz 2007). On the posterior side of the head capsule, an occipital ridge (ocr, Figs. 7.1 and 7.2) runs from each posterior mandibular articulation dorsally and fuses with the postoccipital one. The coronal cleavage line (cs, Fig. 7.1) originates at the middle of the dorsal margin of the foramen occipitale and continues over the dorsal head capsule. On the anterior dorsal head capsule, it separates into two frontal cleavage lines (fs, Fig. 7.1) that continue on both sides toward the ocelli.

The clypeus (cl, Figs. 7.1 and 7.2) is the ventral-most part of the anterior head capsule. It is divided to different degrees into a sclerotized postclypeus (pcl, Fig. 7.1) and a membranous anteclypeus (acl, Fig. 7.1). Ventrally it articulates with the labrum.

Internally the head capsule is supported by the cephalic endoskeleton or tentorium. Klass and Eulitz (2007) described it for various species from all major groups of cockroaches. It is formed by invaginations of the head capsule. The anterior tentorial pits (atp, Fig. 7.2) are located on the mesal pleurostomal ridge. Internally they give rise to the anterior tentorial arms (ata, Fig. 7.2), which are twisted in all cockroaches, the left one always counterclockwise and the right one always clockwise when seen from behind (Klass and Eulitz 2007). In their posterior half, these anterior arms are mesally fused and form the anterior tentorial bridge (atb, Fig. 7.2), whose length varies among cockroaches. On its ventral side, it bears a longitudinal ridge that is only absent in *Polyphaga aegyptiaca* and *Ectobius sylvestris* among the examined species (Klass and Eulitz 2007). Before fusing with the corpotentorium (ct, Fig. 7.2), the anterior tentorial arms separate again, thus forming the "perforation" of the tentorium. Long tendons, the oesotendons (os, Fig. 7.2), originate from the corpotentorium into the "perforation" and serve as an attachment site for muscles (see below). They are present in all Blattodea except *Cryptocercus* and Termites (Klass and Eulitz 2007). The corpotentorium is connected via the posterior tentorial arms (pta, Fig. 7.2) with the head capsule. A small ventral process, the *trabeculae tentorii* (tt, Fig. 7.2), is located on each posterior tentorial arm. The connection points between the posterior arms and the head capsules are the posterior tentorial invaginations. The dorsal tentorial arms (dta, Fig. 7.2) are slender and fairly soft projections with a widened basal part, which originate from the anterior tentorial arms (often at the level of the "perforation") (Klass and Eulitz 2007). Dorsal arms are absent in *Ergaula capucina* (Wipfler et al. under review).

#### 7.2.1.2 Functional Implications

The solid head capsule provides a stable outer hull deflecting stress and strain from the mandibular articulations occurring during biting and chewing. The various cephalic ridges, which are infoldings of the cuticle, apparently stabilize the head capsule under load (Blanke et al. 2018). Additionally all openings and weakenings such as the antennal and occipital foramen and the compound eyes are surrounded by strengthening ridges. In contrast, the cleavage lines are zones of weakening where the cuticle breaks open during ecdysis, thus allowing the animal to emerge from the old cuticle. They are still present in most adult insects, although they no longer shed their cuticle.

Presumably, the tentorium provides additional support for the head capsule by deflecting stress and counteracting lateral forces from the mandibular articulations during biting and chewing (Blanke et al. 2018). Additionally, various muscles of the antennae, mandibles, maxillae, hypopharynx, labium, pharynx, and prothorax attach at the tentorium. According to Popham (1961a), the "perforation" of the tentorium is caused by the anterio-posterior flattening of the head capsule, which resulted in a position of the corpotentorium mesad to the maxillae. To allow the tentorio-maxillary muscles to achieve a transverse and forward movement, they must originate on the anterior tentorial arms near the midline. As a result, the tentorial arms form mesal extensions that eventually fused to the anterior tentorial bridge.

## 7.2.2 *The Labrum and the Epipharnyx*

### 7.2.2.1 Morphology of the Labrum and Epipharynx

The anterior side of the labrum (lbr, Figs. 7.1 and 7.2) comprises a single sclerite in all cockroaches. It can take various forms ranging from nearly squared (e.g., in *Polyphaga*) over semicircular at the anterior margin (*Cryptocercus relictus*) to deep distal incisions or notches (e.g., in *Periplaneta*; Fig. 7.3). Zhuzhikov (2007) describes its shape for a range of species. The labrum is covered with several spines whose number ranges from four (*Ectobius*) to more than 50 (*Gromphadorhina portentosa*) (Zhuzhikov 2007). The inner (posterior) side of the labrum is formed by the membranous epipharynx (Fig. 7.3). In cockroaches, the central area of the epipharynx is covered by the epipharyngeal brush (epb, Fig. 7.3), a field of dorsally

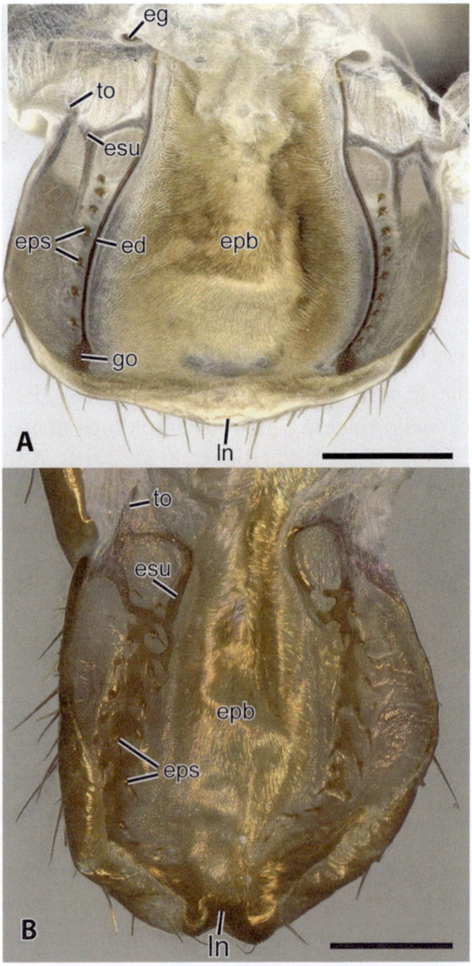


**Fig. 7.3** Epipharynx of (**a**) *Ergaula capucina* (taken from Wipfler et al. under review) and (**b**) *Periplaneta americana* (taken from Wipfler et al. 2016) in posterior view. *ed* duct of the epipharyngeal gland, *eg* epipharyngeal gland, *epb* epipharyngeal brush, *eps* epipharyngeal spines, *esu* epipharyngeal suspensorium, *go* opening of the epipharyngeal gland, *ln* labral notch, *to* tormae. Scale bars: 500 μm

orientated setae. Laterally on both sides of the epipharyngeal brush, epipharyngeal spikes (eps, Fig. 7.3) are located. They can be arranged in a single row (*Ergaula*; Fig. 7.3a) or in arrays of several rows (as in *Periplaneta*, Fig. 7.3b). Laterally at the base of the labrum, the tormae (to, Fig. 7.3) attach and run dorsally on the epipharynx. They serve as attachment sites for Musculus (M). frontoepipharyngalis. Mesally to the tormae the epipharyngeal suspensorium (esu, Fig. 7.3) continues; it is quite variable in shape. Zhuzhikov (2007) provided an overview of its form in different groups of cockroaches. The epipharynx holds a wide array of different sensory organs including contact chemoreceptors (Moulins 1971b) and stretch receptors (Moulins 1974). In corydiid cockroaches such as *Ergaula* (Fig. 7.3a), *Polyphaga* (Zhuzhikov 2007), or *Arenivaga investigata* (O'Donnell 1977), an epipharyngeal gland (e.g., Fig. 7.3a) or frontal body is present on each side of the epipharynx. They are located in a membranous fold on the dorsal epipharynx. From each gland, an epipharyngeal duct (ed, Fig. 7.3a) runs toward a glandular opening (go, Fig. 7.3a) on the distal epipharynx. These glands produce a fluid that is conveyed to the surface of the protruded hypopharyngeal bladder to absorb atmospheric water (O'Donnell 1981, 1982).

#### 7.2.2.2 The Musculature of the Labrum and Epipharynx

M. frontolabralis/0 lb1 (Fig. 7 in Wipfler et al. 2016)[4]: This muscle originates mesally on the frons, on the level of the tips of the antennifers, and inserts mesally on the baso-frontal wall of the labral sclerite. It functions as levator of the labrum and is antagonized by M. frontoepipharyngalis. It is present in all described cockroaches.

M. frontoepipharyngalis/0 lb2 (Fig. 7 in Wipfler et al. 2016): This muscle originates mesally on the frons, laterally and slightly posteriorly of M. frontolabralis. It inserts at the tormae and acts as depressor of the labrum. Its antagonist is M. frontolabralis. It is present in all described cockroaches.

M. labralis transversalis/0 lb4: This muscle originates at the ventral labral wall and inserts on the same structure on the opposite side. It acts as compressor of the labrum. It is present in all described cockroaches with the exception of *Blattella*, where it is not described (Snodgrass 1943, 1944).

M. labroepipharyngealis/0 lb5 (Fig. 7.4): This muscle originates on the inner labral wall and inserts laterally and ventrally of the origin on the inner epipharyngeal wall. It acts as dilator of the cibarium (i.e., the anterior part of the mouth cavity, between the labrum and the hypopharynx). It is present in all described cockroaches.

M. clypeopalatalis/0ci1 (Fig. 7.4): This muscle originates slightly laterally of the midline of the clypeus and inserts on the epipharynx, shortly before the functional mouth opening. It acts as dilator of the cibarium. It is present in all described cockroaches.

[4]In the present contribution, we follow the muscular terminology of Wipfler et al. (2011).

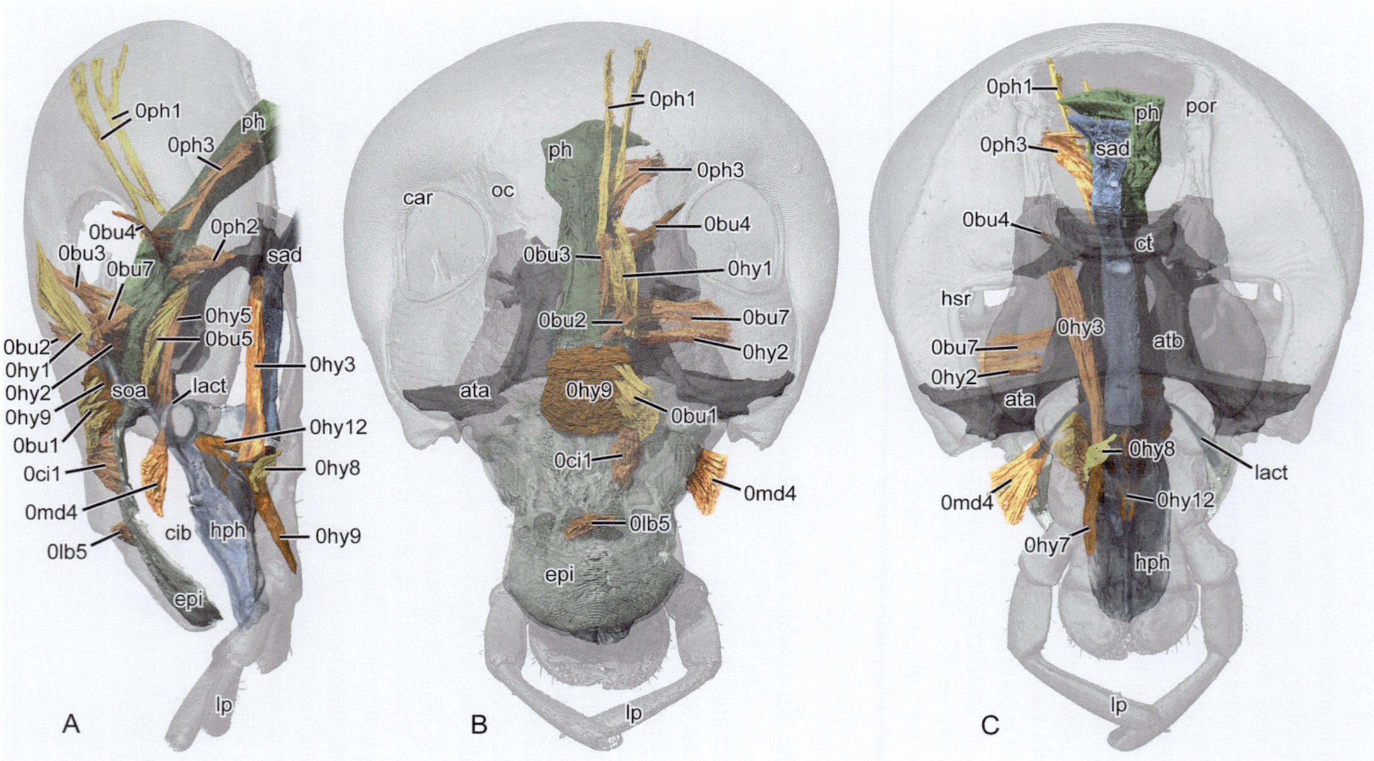


**Fig. 7.4** Three-dimensional reconstruction of musculature of hypopharynx, buccal cavity, and foregut of *Periplaneta americana* (taken from Wipfler et al. under review), cuticle rendered transparent, all mouthparts except hypopharynx and labium removed. (**a**) left-lateral view; (**b**) anterior view; (**c**) posterior view. *ata* anterior tentorial arm, *atb* anterior tentorial bridge, *car* circumantennal ridge, *cib* cibarium, *ct* corpotentorium, *epi* epipharynx, *hph* hypopharynx, *hsr* hypostomal ridge, *lact* linguacutal tendon, *lp* labial palp, *oc* ocellus, *ph* pharynx, *pocr* postoccipital ridge, *sad* salivary duct, *SOA* oral arm of hypopharyngeal suspensorium, muscle nomenclature see Sect. 7.2.2.2

#### 7.2.2.3 The Function of the Labrum

The labrum forms the anterior wall of the cibarium, ensuring that food particles do not fall out. It can be moved anteriorly (by M. frontolabralis) and posteriorly (by M. frontoepipharyngalis); additionally it can be compressed (by M. labralis transversalis), thus ensuring that the mouth cavity is always closed. There is no muscular antagonist to M. labralis transversalis, but it is likely that the compressor deforms the labrum against internal elasticities, restoring its original shape when muscle activity ceases.

The epipharynx can be moved anteriorly by M. clypeopalatalis, which dilates the cibarium. Additionally the epipharynx with its brush and spikes aids in pushing the food particles in between the mandibles and then toward the anatomical mouth opening. The number of spikes is thereby correlated with the length of the labrum (Zhuzhikov 2007). Additionally it has important sensory functions.

### *7.2.3 The Mandible*

#### 7.2.3.1 Morphology of the Mandible

The mandibles of cockroaches (Fig. 7.5) are robust grinding and cutting tools composed of a single sclerite each (Wipfler et al. 2016; Zhuzhikov 2007). In all studied species, they are asymmetrical with respect to their inner margin and sometimes also with respect to their size (the left one being larger than the right one) (Zhuzhikov 2007). The distal part of the mandibles bends posteriorly. When closed the left mandible overlaps the right one.

The mesal margins of the mandibles are divided into several functional parts. The dent-like incisivi (I–IV, Fig. 7.5) are located distally. They grab and cut food particles during biting and chewing. All studied specimens have three incisivi in the right mandible. In some species, the distal incisivus is the largest while in others the apical and the subapical ones are equally sized (Zhuzhikov 2007). The left mandibles are more variable among different species and have between four and six apical dents, which are subcategorized by some authors into incisivi and denticles (e.g., Zhuzhikov 2007). The molar grinding area (mo, Fig. 7.5) is proximal to the incisivi. Its shape can vary but usually the opposite sides fit each other perfectly, thus forming an effective grinding apparatus. In most species, one side is convex with a median ridge and the other one concave. In the wood-feeding genus *Cryptocercus* (Fig. 7.5h, k) the molar surfaces have lamellae. The basal area of the mandible is formed by the postmola (pm, Fig. 7.5), a membranous area with dense setation that is found in all cockroaches. At its basal margin, a small additional sclerite, the basatendon, can occur (Wipfler et al. under review; Zhuzhikov 2007). Each mandible has rows of setae directly lateral of the mandibular mola and the proximal incisivi on both the anterior and posterior side. Additionally some species have isolated setae on the lateral mandibular margin. Other insects have metals embedded in the

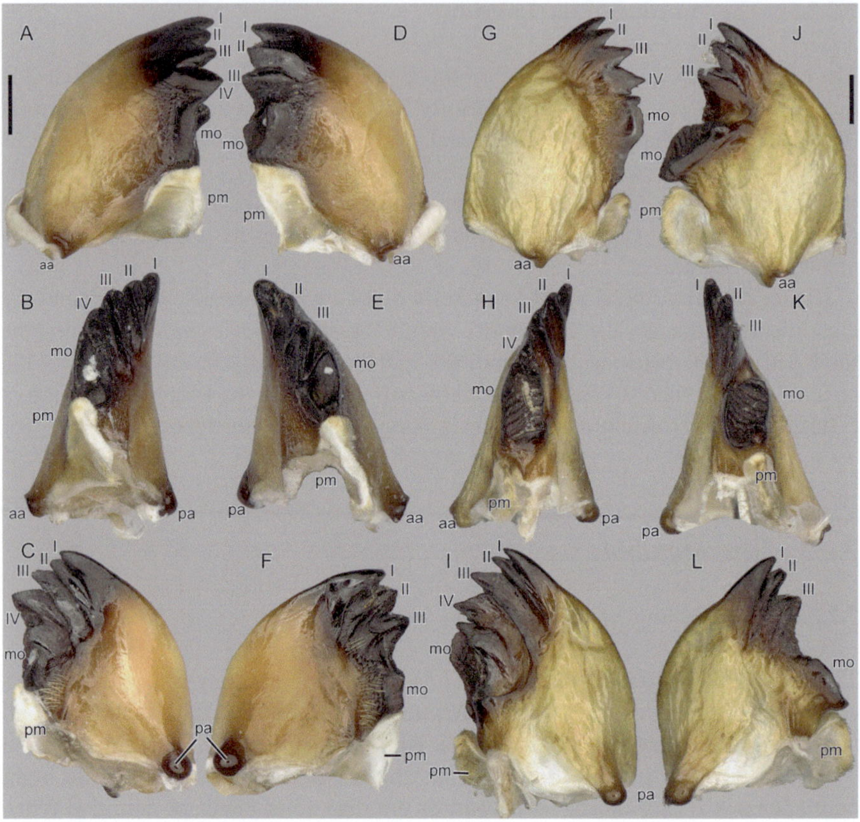


**Fig. 7.5** Photographs of mandibles of *Nauphoeta cinerea* (**a–f**) and *Salganea rossi* (**g–l**). Left mandibles (**a–c** and **g–i**) and right mandibles (**d–f** and **j–l**) in anterior (**a**, **d**, **g**, **j**), mesal (**b**, **e**, **h**, **k**), and posterior (**c**, **f**, **i**, **l**) view. *I–IV* mandibular incisivi, *aa* anterior mandibular condyle, *mo* molar region, *pa* posterior mandibular condyle, *pm* postmola. Scale bars 2 mm

mandibular cuticle that increase the hardness and elastic modulus (Cribb et al. 2008; Schofield et al. 2002), which probably also applies to cockroaches.

#### 7.2.3.2 The Musculature of the Mandible

M. craniomandibularis externus posterior/0md3 (Fig. 7.6): This muscle attaches with a tendon (abt, Fig. 7.6) on the lateral basal margin of the mandible and runs toward the lateral wall of the head capsule slightly posterior of the compound eye. It is present in all examined cockroaches (Dorsey 1943; Snodgrass 1943, 1944; Wipfler et al. 2016) and in all other insects with biting and chewing mouthparts (Matsuda 1965; Wipfler et al. 2011). It acts as the abductor of the mandible and thus opposing

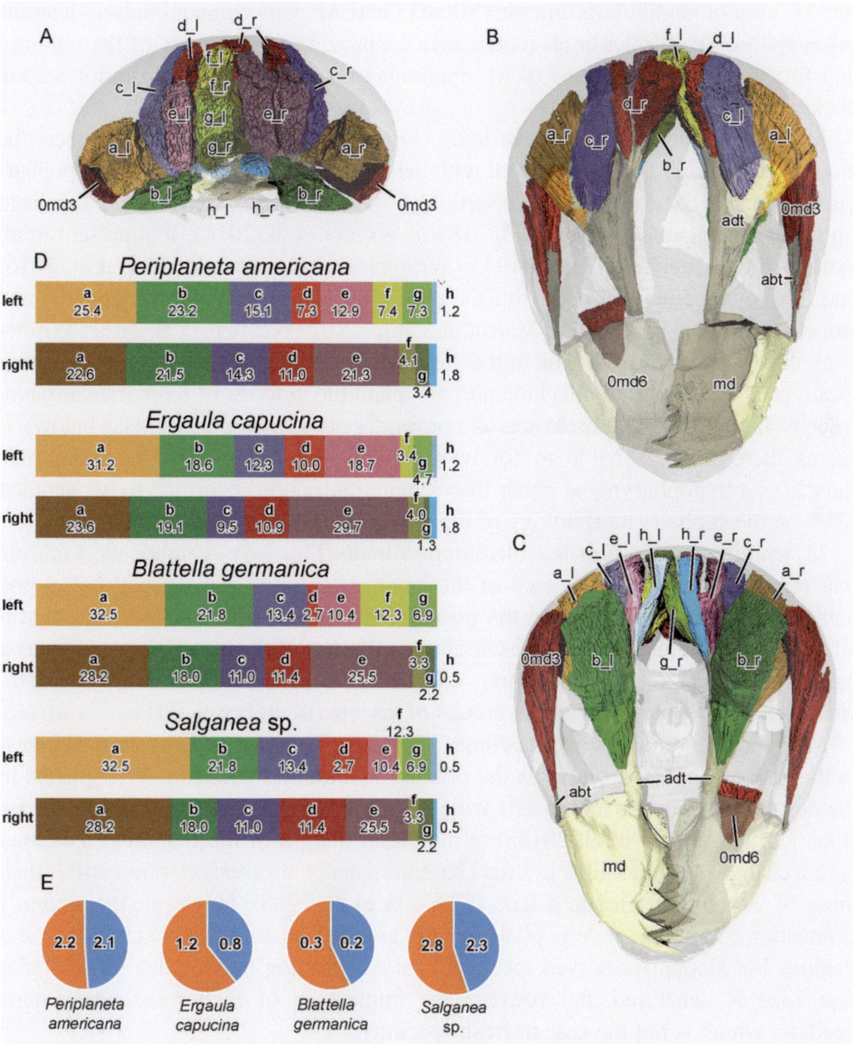


**Fig. 7.6** Mandibular musculature of *Periplaneta americana*. (**a**) dorsal view, three-dimensional reconstruction with cuticle rendered transparent; (**b**) frontal view, three-dimensional reconstruction with cuticle and right mandible rendered transparent; (**c**) posterior view, three-dimensional reconstruction with cuticle and right mandible rendered transparent; (**d**) effective cross-section areas of the bundles of Musculus craniomandibularis internus (0md1) in percentage of the total area of different species of cockroaches. Color code matches the one used in (**a–c**). Based on the values provided in Wipfler et al. (under review). (**e**) Comparison between the total effective cross-section areas of the left (orange) and right (blue) Musculus craniomandibularis internus (0md1) in different species of cockroaches. Based on the values provided in Wipfler et al. (under review). *0md3* M. craniomandibularis externus posterior, *0md6* M. tentoriomandibularis lateralis inferior, *a_l–h_l* bundles of M. craniomandibularis internus on the left side, *a_r–h_r* bundles of M. craniomandibularis internus on the right side, *abt* abductor tendon of the mandible, *adt* adductor tendon of the mandible, *md* mandible. Origin and insertion of the bundles of M. craniomandibularis internus are provided in Table 7.1

the M. craniomandibularis internus (0md1) and M. tentoriomandibularis lateralis inferior (0md6). Wipfler et al. (under review) provide an overview of the volumes and functional cross sections of M. craniomandibularis externus posterior among cockroach species.

M. hypopharyngomandibularis/0md4 (Fig. 7.4): This thin muscle connects the antero-lateral inner mandibular wall with the linguacutal apodeme of the hypopharynx. In cockroaches, the point of insertion on the mandible is far ventral to the distal tip of the linguacutal apodeme (Fig. 18a in Wipfler et al. 2016). It is present in all examined cockroaches (Dorsey 1943; Snodgrass 1943, 1944; Wipfler et al. 2016) and all polyneopteran insects with the exception of the earwigs, zorapterans, and stoneflies (Blanke et al. 2012; Matsumura et al. 2015; Neubert et al. 2017; Wipfler et al. 2011, under review). Due to the delicate structure of this muscle and its rather disadvantageous point of attachment to the mandible in terms of force transmission, it seems likely that this muscle acts as a protruder of the hypopharynx (see below). It seems therefore reasonable to follow Chaudonneret (1951) in considering this muscle as a hypopharyngeal rather than a mandibular one according to his detailed work on the cephalic morphology of the silverfish *Thermobia domestica*.

M. tentoriomandibularis lateralis inferior/0md6 (Fig. 7.6): A rather short muscle that runs from the posterior side of the anterior tentorial arm directly before the anterior tentorial bridge toward the posterior inner wall of the mandible where it attaches in the mesal half. This muscle acts as adductor of the mandible; it may play a minor functional role in the biting and chewing process (David et al. 2016). However, it is reduced in various groups of insects (Beutel et al. 2011).

M. craniomandibularis internus/0md1 (Fig. 7.6): This muscle is by far the biggest in the insect head and comprises the primary mandibular adductor. It originates in the dorsal head capsule and inserts with a thick, stiff tendon that has several wings (Fig. 7.6; Weihmann et al. 2015a) at the mesal margin of the mandible. Potential length changes are negligible as insect tendons usually are about 40 times stiffer than those of vertebrates (Bennet-Clark 1975; Ker et al. 1988). However, the tendon's connection to the mandible is pliable as the articulation zone is not sclerotized and flexible. For alcohol-preserved specimens, manual testing showed that this articulation zone is denatured and conveys the impression of high resistance against bending, which is not the case in fresh specimens.

In almost all studied cockroaches (i.e., *Periplaneta*, *Ergaula*, *Blattella*, *Salganea rossi*, *Diploptera punctata*, *Nocticolla* sp.), M. craniomandibularis internus is composed of eight distinct bundles that can be characterized according to their origin and insertion (Fig. 7.6). Table 7.1 provides an overview of the insertion and origin of these bundles. The only species that differs from this pattern is *Tivia* sp. where the M. craniomandibularis internus is distinctly reduced. A striking feature in the mandibular adductor of cockroaches is the asymmetry between the right and the left half of the head. In general, the right adductor is bigger than the left one and several bundles of the right adductor muscle enter the left hemisphere (Fig. 7.6). The most extreme observed case is bundle h in *Blattella* where the muscle inserting on the left side of the head capsule also originates in the right hemisphere. The right/left asymmetry is also reflected in the effective cross-section areas and the volumes of the muscle and its individual bundles. Wipfler et al. (under review) provide these

**Table 7.1** Origin and insertion of the bundles of Musculus craniomandibularis internus (0md1) found in cockroaches

| | Muscle bundle | Origin | Insertion | Comment |
|---|---|---|---|---|
| Right | a_r | Laterally on the basal and mesal wing | Latero-posterior vertex, directly posterior of the compound eye | – |
| | b_r | Along the posterior side of the besal wing inclusing the ventral parts of the lateral and mesal wing | Postgena, laterally of the foramen occipitale | – |
| | c_r | Mesal side of the lateral wing | Dorsal vertex, mesal to a_r and laterally of d_r and e_r | – |
| | d_r | In 3 bundles along the anterior side of the mesal wing | Anterior vertex, two lateral bundles in right hemisphere, mesal bundle in left hemisphere; two lateral bundles anterior of e_r, mesal bundle anterior of f_l , mesal to c_r | – |
| | e_r | Lateral side of the mesal wing | Posterior vertex, mesal to bundle c_r, posterior to bundle d_r, lateral of bundles f_r, g_r & h_r | – |
| | f_r | Mesal and lateral side of the mesal wing, in between d_r and g_r | Dorsal vertex in the left hemisphere, posterior to bundle g_l | – |
| | g_r | Mesal side of the the mesal wing in between f_r and h_r | Posterior vertex in the left hemisphere; posterior of g_l | – |
| | h_r | Mesal side of the mesal wing, posterior to h_r | A thin bundles on the dorsal edge of the foramen occipitale | In *Blattella germanica*, the bundle that inserts in the left body side also originates from the right side. |
| Left | a_l | Laterally on the basal and mesal wing | Latero-posterior vertex, directly posterior of the compound eye | – |
| | b_l | Along the posterior side of the besal wing inclusing the ventral parts of the lateral and mesal wing | Postgena, laterally of the foramen occipitale | – |
| | c_l | Mesal side of the lateral wing | Dorsal vertex, mesal to a_l and laterally of d_l and e_l | – |

(continued)

**Table 7.1** (continued)

| | Muscle bundle | Origin | Insertion | Comment |
|---|---|---|---|---|
| | d_l | One bundle on the anterior side of the mesal wing, similar position as the lateral bundle of d_r | Laterally on the anterior vertex, corresponding to the lateral bundle of d_r | – |
| | e_l | Lateral side of the mesal wing | Posterior vertex, mesal to bundle c_l, posterior to bundle d_l, lateral of bundles f_l & g_l | – |
| | f_l | Mesal side of the mesal wing, in between d-l and g_l | Dorsal hemisphere in the left hemisphere, in beetwnn | – |
| | g_l | Mesal side of the mesal wing, posteriorly of f_l | Posterior vertex in the left hemisphere; posterior of f-r, anterior of g_r | – |
| | h_l | Mesal side of the mesal wing, posterior to h_r | A thin bundles on the dorsal edge of the foramen occipitale | Missing in *Blattella germanica* as the left bundle also attaches on the right side |

These bundles are illustrated in Fig. 7.6 where also some information about their effective cross-section area is provided (based on the values in Wipfler et al. under review). Details about the morphology of the tendon are provided in Wipfler et al. (2016, under review). This pattern is found in all studied cockroaches with the exception of *Tivia* sp.

values for several species, and Fig. 7.6 summarizes the most important results. In the studied species, the ratio between the total effective cross-section areas of right and left muscles ranges between 1.05 (*Periplaneta*) and 1.56 (*Ergaula*) (Fig. 7.6). When considering the individual bundles, there are also differences between the right and left side. In general, bundle e is between 2.49 (*Blattella*) and 1.58 (*Ergaula*) times larger in the right hemisphere than in the left one (Fig. 7.6).

Most of the fiber bundles gain attachment area by spreading out in an anterio-posterior direction at the curved dorsal wall of the head capsule (Fig. 7.6). In this way, the attachment areas largely match the effective cross-sectional areas. Only bundle b and h attach almost exclusively at the posterior wall of the head capsule. Their attachment areas are significantly larger than the cross-sectional areas. Accordingly, the forces generated by these muscle bundles are distributed over relatively larger areas, which reduces tensile loading on the comparatively flat and probably not load-optimized posterior wall of the head capsule.

#### 7.2.3.3 Muscle Fiber Length and Angles

These parameters are only studied in *Periplaneta* (Weihmann et al. 2015a). With closed mandibles, the mean weighted fiber angle in this species is about 34°, whereas single fiber bundles deviate markedly from this value (Weihmann et al. 2015a). The

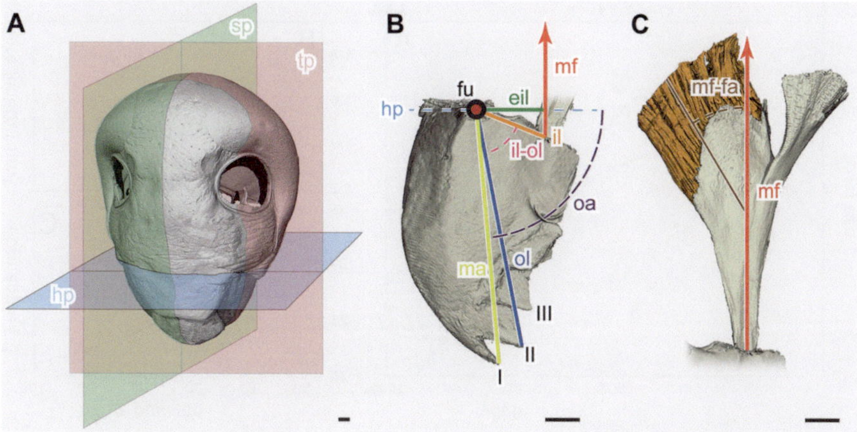


**Fig. 7.7** (**a**) Coordinate system of the head capsule of *Periplaneta americana*. (**b**) Distances, points, and angles in the right mandible of *Periplaneta americana*. (**c**) Angles and directions in the tendon of M. craniomandibularis internus of *Periplaneta americana* (modified from Weihmann et al. 2015a). *I–IV* incisivi, *eil* effective inner lever, *hp* horizontal plane, *il* inner lever, *il-ol* angle between outer and inner lever for the second incisivi, *mf* main direction of muscle force, *ma* mandibular axis or outer lever of the first incisivus, *mf-fa* angle between the main direction of the muscle force and one muscle fiber, *oa* opening angle of the mandible, *ol* outer lever of the second incisivi, *sp* sagittal plane, *tp* transverse plane. Scale bars: 0.5 mm

mean weighted fiber length is about 1.24 mm (i.e., about ¼ of the specimen's mean maximum head width of about 5.1 mm) while the mean fiber lengths of the single bundles ranged from 1 to 1.52 mm.

When the cockroach mandibles open, the length of the effective inner lever (Fig. 7.7), the mean fiber length, and the mean fiber angle change markedly (Fig. 7.8). Although these changes are caused by the rotation of the inner lever around the axis of the mandible joint, the mean fiber length increases nearly linearly from 1.24 mm with closed mandibles to 1.93 mm at 100° mandible opening. Fiber length and opening angle are almost linearly related. Thus, the ascending limb of the relationship of bite force and opening angle, which occurs between 55° and 62°, corresponds to a mean muscle fiber length ranging from 1.36 mm to 1.46 mm (Fig. 7.8). The bite force plateau, between 62° and about 75°, corresponds to fiber lengths from 1.46 mm to 1.63 mm and the descending limb to fiber length between 1.63 mm and about 1.76 mm (Weihmann et al. 2015a). Starting from closed mandibles the mean fiber angle decreases from about 34° to about 21° at maximally opened mandibles (Fig. 7.8). At mandible opening from 55° to 85°, i.e., in the range where significant bite forces were obtained (Fig. 7.8), the mean fiber angle changes from about 31° to 23° with a change in fiber length from about 1.36 mm to 1.76 mm. These changes correspond to relative increases of 35% and 23%, respectively. Muscle pennation results in muscle fiber stresses higher than whole muscle stresses (Paul and Gronenberg 1999). According to its dependency on the cosine of the pennation angle, the differences are larger the higher the muscle pennation angle.

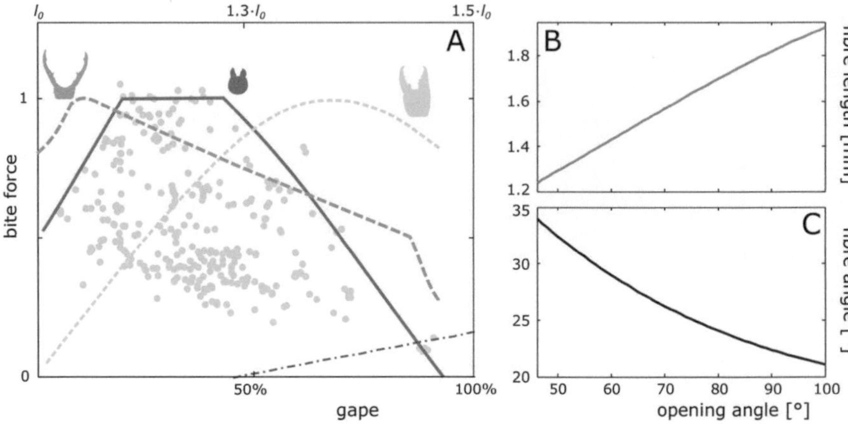


**Fig. 7.8** (**a**) Specific bite forces for differently adjusted biting and chewing structures plotted against gape (lower abscissa) and—for the cockroach data—against mean relative fiber length of the adductor muscles (upper abscissa). $l_0$ corresponds to the resting length of 1.24 mm with closed mandibles. According to the color of the icons (depictions are not to scale), scatter and solid dark gray line depicts bite forces of *Periplaneta americana* (Weihmann et al. 2015b). The red, dash-dotted line depicts the passive forces of the mandible joint resisting external mandible opening. The medium gray, dashed line shows the approximate bite force trajectory of male stag beetles after Goyens et al. (2014). Note that these animals use their much stronger mandibles primarily for clasping rivals during male–male competitions. The light gray, dashed line shows the conditions in the chelae of mud crabs (Yap et al. 2013). Unfortunately, in stag beetles and mud crabs no passive force components were examined. Moreover, it remains unclear from the aforementioned publications whether the experiments in stag beetles and mud crabs did cover the full voluntary gape range of the species. (**b**) Mean fiber length of the mandible adductors against the opening angle in *Periplaneta* (see Fig. 7.7). (**c**) Mean weighted fiber angle of the mandible adductors of *Periplaneta* against the opening angle of the mandibles

The pennation of the mandibular adductor is maximal when the mandibles are closed and decreases with increasingly opened mandibles. Thus, muscle fiber stress is up to 20% higher than whole muscle stress when mandibles are closed and the mean pennation angle gains its maximum value of 34°. At maximally opened mandibles, the mean pennation angle is only about 21° and the surplus in fiber stress, thus, is only about 7%. In the range from 55° to 85° mandible opening, the pennation decreases from 31° to 23° which results in fiber stress values exceeding that of the whole muscle by about 17% and 9%, respectively.

#### 7.2.3.4 Function of the Mandible

Cutting Mechanism

The primary task of the mandibles is to cut and grind food particles. Therefore, mandibles are by far the most powerful mouthparts. Each neopteran mandible is connected to the head capsule via two firm joints (Chapter 6 describes the evolution of the mandibular joint). These two joints restrict the mandible movements to a single plane, similar to a door hinge.

The independent axes of rotation usually do not yield a point of intersection between the two mandible blades if the animal grasps a piece of food. Nevertheless, mandible action can roughly be approximated as scissor-like. Depending on the structure, thickness, and shape of the substrate and determined by the clearance of the two blades, individual mandibles cut either more or less independently by driving apart the material and generating mode I fractures (see Clissold 2007), or the blades act jointly on thinner substrates and tear apart the material parallel to the plane of the crack. Then, the mandibles generate mode III fractures (Clissold 2007).

Among cockroaches, the mandibles are generally morphologically very similar (see above). Their distal parts are characterized by sharp edges and teeth, while the more proximal parts are blunt and not suitable to generate shear forces. Thus, the mechanism of the distal part of cockroach mandibles is rather analogous to staggered pairs of parrot beak pruners. Initially, the tips of the mandible teeth perforate the outer surface of a food item and then the proximal edges of the teeth cut apart the material (Fig. 7.5). In *Periplaneta*, the second right and the third left teeth have the particular capacity to form a structure similar to the carnassial structure of carnivorous mammals enabling the animals to cut up stringy matter such as fibrous plant and animal materials. When the distal teeth have broken the mechanical resistance of the substrate, the mandibles can further close such that the two molar regions contact one another, and can further grind up the food. Grinding is facilitated here by the proximal position of the molar regions and the consequently greater mechanical advantage. Moreover, the axes of the mandible joints are tilted with regard to the length axis of the head: in *Periplaneta* by about 17°. In contrast, the blades and cutting edges of the mandibles are almost aligned in parallel to the transverse plane (Fig. 7.6). Therefore, a fraction of about 6% of the forces generated by the mandible closer muscles is redirected in posterior directions (Weihmann et al. 2015a). These force components, however, primarily seem to facilitate shoveling of the reduced food toward the esophagus. Nevertheless, they can also contribute to tearing off pieces from a larger food item.

During feeding, the mandibles of cockroaches are strongly coupled with the other mouthparts (see below), which is not necessarily the case during other tasks such as grooming.

### Opening Range

So far, the opening angle or range (oa, Fig. 7.7) has only been studied for the cockroach *Periplaneta* (Weihmann et al. 2015a, b). In this species, voluntary mandible opening of up to 103° could be observed, with three of eight specimens opening their mandibles to more than 99° (Weihmann et al. 2015b). However, significant bite forces were obtained only up to 85° of mandible opening (Fig. 7.7). When mandibles are closed, the opening angle was about 47° in the right and 49° in the left mandible (Weihmann et al. 2015a).

Bite Duration

In experiments for measuring voluntary bite forces in *Periplaneta* with a relatively stiff force transducer, bite durations ranged from 0.24 s to 1.52 s (Weihmann et al. 2015b). When ingesting liquids, Schmitt et al. (2014) found complete biting and chewing cycle durations of only about 0.5 s, indicating considerably shorter closing motions. Particularly strong bites, i.e., those data points shaping the upper edge of the scatter diagram (Fig. 7.8) and, therefore, indicating the maximum forces for given mandible opening angles, lasted markedly longer than weaker bites (0.99 s ± 0.21 s vs. 0.49 s ± 0.13 s). This result seems to indicate higher time constants for reaching maximum bite forces. Likewise, Seath (1977) showed significantly increased bite durations for desert locusts (*Schistocerca gregaria*) with artificially loaded mandibles if compared to chewing on soft wheat leaves (0.75 s ± 0.15 s vs. 0.47 s ± 0.1 s).

Bite Forces

So far, *Periplaneta* is the only cockroach species, for which biting and chewing forces have been measured (Weihmann et al. 2015b). In this species, the upper edge of the scatter of voluntary bite forces plotted over the opening range of the mandibles (Fig. 7.8) resembles a typical force–length relationship as found in many skeletal muscles (e.g., Siebert et al. 2010). It is characterized by an ascending limb at low opening angles, a force plateau at intermediate, and a descending limb at high angles. The plateau occurs at opening angles between about 60° and 75° with bite forces of about 0.5 N. According to Schmitt et al. (2014), during ingestion, the mandibles open to about 60° to 65° which indicates that the mandibles of *Periplaneta* work mainly on the ascending limb of the moment–angle relation and the force–length relation of the mandible adductors.

Similar activity patterns were reported for leg muscles in stick insects and spiders as well as a number of vertebrates (Guschlbauer et al. 2007; Herzog et al. 1992; Siebert et al. 2010; Weihmann et al. 2009). Under static conditions activity on the ascending limb results in self-stabilization of the muscle properties (Wagner and Blickhan 1999). The self-stabilization of the muscle properties, in turn, may allow an actively closed mandible affected by external—opening—forces to recover a functional position without additional neuronal control. Furthermore, there are also indications that the activity on the ascending limb of the force–length relation contributes to the self-stabilization of the system under dynamic conditions (Wagner and Blickhan 2003).

The use of passive joint forces is another mechanism allowing for reduced control effort (Haeufle et al. 2014) and reduced complexity of musculoskeletal systems. In small animals, passive forces result from material properties of the joint structures and from the passive elastic properties of the driving muscles. So far, their role in movement initiation and control has been examined primarily in locomotor apparatuses (Ache and Matheson 2013; Hooper et al. 2009; Weihmann 2018). While in walking legs, the resting position of a joint occurs mostly at intermediate joint angles, mandibles usually rest in a closed position, i.e., when maximally flexed. In *Periplaneta*, significant passive forces resisting further opening occur when mandibles are wide apart. At opening angles larger than 85°, those forces reach values of 0.09 N, which is about 18% of the bite force maximum. In contrast to legs (Ache and Matheson 2013), the passive forces in cockroach mandibles complement the activity of the stronger of two antagonistic muscles, i.e., the adductor. However, in the opening range where passive forces become significant, the mandible adductors are far on the descending limb of the force–length relation with only a small potential for force generation. Moreover, the effective in-lever is by about 30% shorter than in closed mandibles reducing the effective mechanical advantage of the mandible (Weihmann et al. 2015a). Thus, here, the passive forces antagonize the abductors, which can open the mandibles up to 100° and are most likely the stronger of the two antagonistic muscles at large gape angles. Mandible adduction at large opening angles that occur regularly in intraspecific aggression (Stevenson and Rillich 2016) or grasping is controlled only via the activity of the abductor muscle while closing movements rely entirely on passive mechanisms. Consequently, the range actively driven by the adductor can be kept small, which also limits the length range in which muscle fibers have to generate significant forces. Accordingly, biting and chewing can be actuated by only one adductor muscle while otherwise additional structures might be necessary to cover the full range of gape angles [cp. the complexity of the closer muscles in the chelicerae of solifuges (van der Meijden et al. 2012)].

During prolonged sequences of biting and chewing, in *Periplaneta*, residual forces, i.e., the baseline of the force traces, happened to gradually increase even when opening angles and therefore passive joint forces were relatively small. Hence, passive joint forces provide only a small part of these residual forces. Instead, residual forces must be generated by the adductor muscles, which are corroborated by decreasing residual forces when the opening angles and muscle lengths increase. However, it has been long known that specialized muscle fibers with long sarcomeres can maintain residual tension after previous neuronal activation for minutes without any further neuronal signal (Hoyle 1978). Chesler and Fourtner (1981) could demonstrate the effect even for *Periplaneta* coxa muscles. Moreover, specialized muscle fibers with relatively long sarcomeres have been demonstrated for the mandible adductor muscles in a number of ant species (Gronenberg et al. 1997). Therefore, it seemed likely that specialized slow muscle fibers also occur in the mandible adductors of cockroaches whose activity results in increasing residual forces (Weihmann et al. 2015b). However, in preliminary experiments aiming at differential visualization of the ATPase activity of the mandible muscles in *Periplaneta* (Stokes et al. 1979), i.e., in the attempt to determine the position of slow and fast muscle fibers, we did not find any differentiation for the adductor (TW unpublished material); that is, all muscle fibers showed intermediate myosin ATPase activity. Only the abductor showed muscle fibers varying in their activity level. However, during our force measurements, the mandibles of the specimens were forced to adopt specific opening angles by the needle tip of a force transducer

(Weihmann et al. 2015b). Forced stretches of activated, i.e., stimulated, muscles however result in force enhancement, which can also last when muscle activation ceases (Campbell and Campbell 2011). Since the cockroaches often tried to resist the imposed mandible manipulations, the adductors were most likely activated during the adjustment of the opening angles. Accordingly, the mechanism can appropriately explain increasing residual forces without the necessity for specific muscle fibers.

Irrespective of the underlying mechanism, increased residual forces seem to have particular significance when the animals chew on tough or resilient materials without the necessity of active mandible opening. Under these circumstances, both stretch-induced force enhancement and the employment of tonic muscle fibers would provide distinctly enhanced bite forces.

#### Mechanical Advantage

The ratio of the distances between pivot and tendon attachment on the one side (inner lever, il, Fig. 7.7) and pivot and the respective mandibular incisive (outer lever, oa, Fig. 7.7) on the other side are constant and define the mechanical advantage (MA) of specific teeth. Depending on the tooth considered, in *Periplaneta* MA ranged from 0.37 to 0.47. Here, the position of the second right and third left teeth seems to be of particular significance. Both teeth are well developed and their proximal rims are cutting edges; their MA is about 0.39. Accordingly, in the range of these sectorial teeth, the effective mechanical advantage (EMA) of the two mandibles was about 0.4. It was calculated as the quotient of the length of the effective input lever and the length of the external lever. The effective input lever corresponds to the projection of the anatomical input lever, onto the horizontal line (Fig. 7.7). In closed mandibles, the angle between both inner levers and the horizontal line is about 0° and increases with increasing mandible opening. Consequently, the effective input lever shortens according to the cosine of the included angle and the EMA increases. The effective lever, i.e., the horizontal distance between the pivot and the attachment of the tendon, reduces by about 30% when considering the range from closed to maximally open mandibles. If we consider those opening angles, where considerable bite forces were measured, i.e., from about 55° to 85°, the effective lever decreased by only 18%.

The MA values for mandibular teeth in other dictyopteran species do not deviate much from the situation in *Periplaneta* (Table 7.2). However, in contrast to the ominivorous cockroach species, food specialists and the generally dietary deviating termites and mantids seem to have somewhat stouter mandibles, which probably also result in higher effective mechanical advantages.

While mean muscle fiber angles may provide a general understanding of muscle function (see Sect. 7.2.3.3), the orientation of single muscle fibers may deviate significantly from mean values and the current direction of the main muscle force. In *Periplaneta* the fiber angles of the bundles f and a deviate up to 60° (Fig. 7.6) in the transverse plane (Fig. 7.7). Therefore, depending on the opening angle of the mandibles, differential or sequential activation patterns are conceivable, i.e., lateral

**Table 7.2** Mechanical advantage of selected dictyopteran species *Blattella germanica*, *Ergaula capucina*, *Periplaneta americana* (taken from Weihmann et al. 2015a), *Tivia* sp., *Salganea rossi*, *Mastotermes darwiniensis*, *Mantoida maya*, and *Eremiaphila* sp.

| | | Blattodea | | | | | Isoptera | Mantodea | |
|---|---|---|---|---|---|---|---|---|---|
| | | *Blattella* | *Ergaula* | *Periplaneta* | *Tivia* | *Salganea* | *Mastotermes* | *Mantoida* | *Eremiaphila* |
| Right | Incisivus 1 | 0.41 | 0.37 | 0.37 | 0.44 | 0.42 | 0.51 | 0.47 | 0.42 |
| | Incisivus 2 | 0.42 | 0.39 | 0.39 | 0.47 | 0.44 | 0.52 | 0.49 | 0.43 |
| | Incisivus 3 | 0.46 | 0.45 | 0.46 | – | 0.49 | 0.55 | – | 0.45 |
| Left | Incisivus 1 | 0.38 | 0.37 | 0.37 | 0.37 | 0.34 | 0.46 | 0.41 | 0.38 |
| | Incisivus 2 | 0.38 | 0.40 | 0.39 | 0.39 | 0.39 | 0.49 | 0.43 | 0.39 |
| | Incisivus 3 | 0.39 | 0.41 | 0.40 | 0.42 | 0.43 | 0.55 | – | – |
| | Incisivus 4 | 0.44 | 0.46 | 0.47 | 0.45 | 0.45 | – | – | – |
| | Incisivus 5 | – | 0.52 | – | 0.49 | 0.49 | – | – | – |

Note that the incisivi might not be homologues

bundles are active primarily when opening angles are small and medial fiber bundles show the highest activity when opening angles are high. However, it has been previously found in ants (Paul and Gronenberg 2002) that posterior and lateral fibers are probably not recruited differentially, although fast muscle fibers, lumped together in specific muscle subunits, can be activated independently.

## 7.2.4 *The Maxilla*

### 7.2.4.1 Morphology of the Maxilla

The maxilla (Fig. 7.9) is composed of several sclerites that are located mostly on the postero-lateral side, while the antero-mesal side that faces the head capsule is mostly membranous. It is morphologically quite homogenous among cockroaches (Crampton 1923; Pradl 1971; Walker 1933; Wipfler et al. under review, 2016; Zhuzhikov 2007).

The cardo (ca, Fig. 7.9), a boat-shaped transversely orientated sclerite, is most proximal. The mesal edge of the cardo rests against an apodeme of the hypostomal ridge (hpa, Fig. 7.1), thus forming the maxillary articulation. The cardinal ridge (cr, Fig. 7.9) divides the cardo into a mesal and a lateral part. The long tendon of M. craniocardinalis attaches on the mesal part of the cardo.

The stipes (st, Fig. 7.9) attaches distal to the cardo, connected via the cardo-stipital articulation (csa, Fig. 7.9) on the posterior surface. In this articulation, the two sclerites are separated by a membranous band, the cardo-stipital syndesis (css, Fig. 7.9). The stipes is a nearly rectangular massive sclerite on the posterior side on the maxilla. Internally it is reinforced by the externally visible stipital ridge that serves as an attachment side for muscles (sr, Fig. 7.9). The two lobes of the maxilla, the galea (ga, Fig. 7.9) and lacinia (la, Fig. 7.9), articulate distally. The five-segmented maxillary palp articulates laterally to the anterior side of the stipes. Proximal to the base of the maxillary palp, a small sickle-shaped sclerite, the palpifer (pf, Fig. 7.9), is found in all examined cockroaches. It articulates with the basal palpomere, thus forming the second joint of the palp. This palp is five-segmented in all cockroaches (I–V, Fig. 7.9). The first palpomere (I, Fig. 7.9) is approximately as long as wide. The second one (II, Fig. 7.9) is similar to the first in length, but is much more slender. Palpomeres two to five widen distally. The distal three palpomeres (III–V, Fig. 7.9) are distinctly longer than the first two. In cave-dwelling cockroaches, the palp can be strongly elongated (Bell et al. 2007). The maxillary palp is densely covered with setae. Additionally, as in other insects, it contains a dense array of different sense organs and sensillae (Altner and Stetter 1982; Bland et al. 1998; Burry and Moran 1973; Prakash et al. 1995; Pringle 1938; Ramaswamy and Gupta 1981; Seidl 1991). On the mesal surface of the distal palpomere, a membranous sensory field (smp, Fig. 7.9) with different types of sensillae is located in all examined cockroaches. A similar field has not been found in other closely related insects. In *Periplaneta* its area covers about 1.26 $mm^2$ and contains ca. 2300

**Fig. 7.9** The maxilla of *Periplaneta americana* (**a**, **b**; digital micrographs; taken from Wipfler et al. 2016), *Nauphoeta cinerea* (**c**, **d**; confocal laser scanning micrographs), and *Ergaula capucina* (**e**, **f**, **g**, **h**; digital micrographs; taken from Wipfler et al. under review) in posterior (**a**, **c**, **e**) and anterior (**b**, **d**, **f**) view. (**g**) Anterior side of the lacinial setation; (**h**) posterior view of the sensory field on palpomere V. *bga* basigaleal sclerite, *ca* cardinal sclerite, *cr* cardinal ridge, *css* cardino-stipital syndesis, *dga* distigaleal sclerite, *ga* galea, *gb* galeal brush, *gs* galeal setae, *I–V* maxillary palpomeres, *la* lacinia, *lac* lacinula, *las* lacinial syndesis, *ls* lacinial setae, *li* teeth or incisivi of lacinia, *mga* mesogaleal sclerite, *pf* palpifer, *sd* stipital disc, *smp* sensory field on the maxillary palp, *sls* strong lacinial setae, *sr* stipital ridge, *st* stipital sclerite. Scale bars: 500 μm

sensillae, thus being the most sensitive surface of the cockroach body (Burry and Moran 1973). In the cave-dwelling cockroach *Paratemnopteryx*, the number of sensillae ranges between 1045 (males) and 1486 (females) (Bland et al. 1998).

The galea (ga, Fig. 7.9) is a mostly membranous structure that forms the lateral maxillary lobe. It is connected to the stipes via membranes. In all studied cockroaches, the galea holds three sclerites, the basi- (bga, Fig. 7.9), disti- (dga, Fig. 7.9), and mesogaleal sclerite (mga, Fig. 7.9). They have not been described for any closely related insects. The basigaleal sclerite is positioned on the galeal base on the posterior and lateral side of the maxilla. The mesogaleal sclerite is found on the anterior galeal side; it is L-shaped, with one part attaching baso-laterally to the basogaleal sclerite, while the other part runs proximo-distally on the mesal side of the galea. In some species (e.g., *Ergaula*, Fig. 7.9f) these two parts are separated by a membrane while in others such as *Periplaneta* (Fig. 7.9b) they are fused. Distally the second part bears long setae (ls, Fig. 7.9). The disto-galeal sclerite covers the lateral and anterior parts of the distal galea. It attaches distally to the basigaleal sclerite. It can be a uniformly sclerotized area as in *Periplaneta* (Fig. 7.9b) or have multiple sclerotized areas as in *Nauphoeta cinerea* (Fig. 7.9d). The galeal brush (gb, Fig. 7.9) is positioned meso-distally on the galea. It is separated by membranes from the galeal sclerites. In cockroaches, the galea forms a cavity that holds the lacinia.

The lacinia (la, Fig. 7.9) is a single sclerite that, in most species, is somewhat triangular or sail-shaped. It is located postero-mesad of the galea and lies in a galeal cavity when abducted. It articulates with the stipital sclerite via the stipital disc (sd, Fig. 7.9), a strongly sclerotized area on the middle area of the posterior side of the maxilla. The stipital disk is fused with the basal stipes and the mesal lacinia. The remaining parts of the lacinia are separated from the stipes via a thin membranous band the lacinial syndesis. The distal tip of the lacinia is formed by the strongly sclerotized lacinial incisivi (li, Fig. 7.9). In cockroaches, they number between one (e.g., *Gromphadorina* or *Schulthesia*) and two (e.g., *Periplaneta* or *Ergaula*) (Zhuzhikov 2007). At the mesal margin at the base of the lacinial incisivi, all cockroaches with the exception of Corydiidae (Wipfler et al. under review; Zhuzhikov 2007) have a lacinula (lac, Fig. 7.9), a small sometimes sclerotized structure that can vary distinctly in shape. Zhuzhikov (2007) distinguished four distinct shapes: (I) a rigid pigmented process resembling lacinial incisivi (as found in *Cryptocercus*); (II) a row of 2–3 low basal elements with a soft process each (as found in Epilamproidea); (III) a non-sclerotized and soft base in the form of a column that bears several blunt and short posteriorly orientated spines (as found in Blattoidea, e.g., Fig. 7.9a); and (IV) a conical base bearing thinner apical setae (as found in Blaberoidea, e.g., Fig. 7.9c). The possible homology of these structures with those of other insects is discussed in Wipfler et al. (2016) and Wipfler et al. (under review). In all cockroaches, the mesal area of the lacinia proximally to the lacinula is covered with a dense field of long lacinial setae (ls, Fig. 7.9) that are arranged in several rows. In the corydiid cockroach *Ergaula*, they are interspersed by shorter and stronger setae (sls, Fig. 7.9g) in the distal half.

#### 7.2.4.2 The Musculature of the Maxilla

M. craniocardinalis/0mx1 (Fig. 7.10): This muscle has two distinct bundles that originate on the lateral postgena and on the lateral vertex (i.e., the dorsal part of the head capsule). They both insert on a long tendon that attaches to the distal tip of the cardinal tendon. The function of this muscle is discussed below. The muscle is bipartite in the examined cockroaches as well as in most other polyneopteran insects (Matsuda 1965; Wipfler et al. 2011).

M. craniolacinialis/0mx2 (Fig. 7.10): This long and slender muscle originates on the lateral vertex, directly antero-mesally of the origin of bundle 2 of M. craniocardinalis. It inserts on the anterior lacinial margin, together with M. stipitolacinialis. It serves as adductor of the lacinia and retractor of the maxilla.

M. tentoriocardinalis/0mx3 (Fig. 7.10): This muscle originates on the lateral areas of the anterior bridge of the cephalic endoskeleton. It originates along the entire posterior inner wall of the cardo on both sides of the cardinal ridge. It functions as the adductor of the cardo and the protractor of the maxilla.

M. tentoriostipitalis anterior/0mx4 (Fig. 7.10): This muscle originates on the mesal areas of the anterior tentorial bridge including the mesal ridge. One bundle attaches on the lateral side of the corpotentorium in all studied species of cockroaches (Dorsey 1943; Snodgrass 1943, 1944; Wipfler et al. 2016). It inserts along the entire mesostipes and the mesal wall of the stipital ridge. It functions as adductor of the maxilla and the lacinia.

M. tentoriostipitalis posterior/0mx5 (Fig. 7.10): This muscle originates on the anterior area of the anterior tentorial bridge, in between M. tentoriocardinalis and M. tentoriostipitalis anterior. It originates on the dorsal inner stipital wall close to the stipito-cardinal ridge in between bundles of M. stipitolacinialis. It functions as adductor of the stipes and protractor of the maxilla.

M. stipitolacinialis/0mx6 (Fig. 7.10): This muscle has three distinct bundles. Two attach on the lateral inner wall of the stipes and the third one on the postero-mesal inner stipital wall. All bundles insert with M. craniolacinialis on the anterior lacinial margin. M. stipitolacinialis acts as adductor of the lacinia.

M. stipitogalealis/0mx7 (Fig. 7.10): This muscle originates on the dorsal inner stipital wall, directly ventrally of the insertion of M. tentoriostipitalis posterior. It inserts baso-laterally on the distogalea and functions as abductor of the galea.

M. stipitopalpalis externus/0mx8 (Fig. 7.10): This muscle originates dorsally on the lateral wall of the inner stipital ridge. It inserts dorsally on the base of maxillary palpomere I and functions as abductor of the maxillary palp.

M. stipitopalpalis medialis/0mx9 (Fig. 7.10): This muscle originates on the lateral wall of inner stipital ridge, ventrally of M. stipitopalpalis externus, and runs toward the anterior basal margin of maxillary palpomere I. Wipfler et al. (2011) describe it in one studied specimen of *Periplaneta* but report its absence in another representative of the same species. It is not reported by any other study on cockroaches or the termite *Odontotermes obesus* (Dorsey 1943; Snodgrass 1943, 1944; Wipfler et al. under review, 2016) but is present in all studied mantodeans (Wipfler et al. 2012;

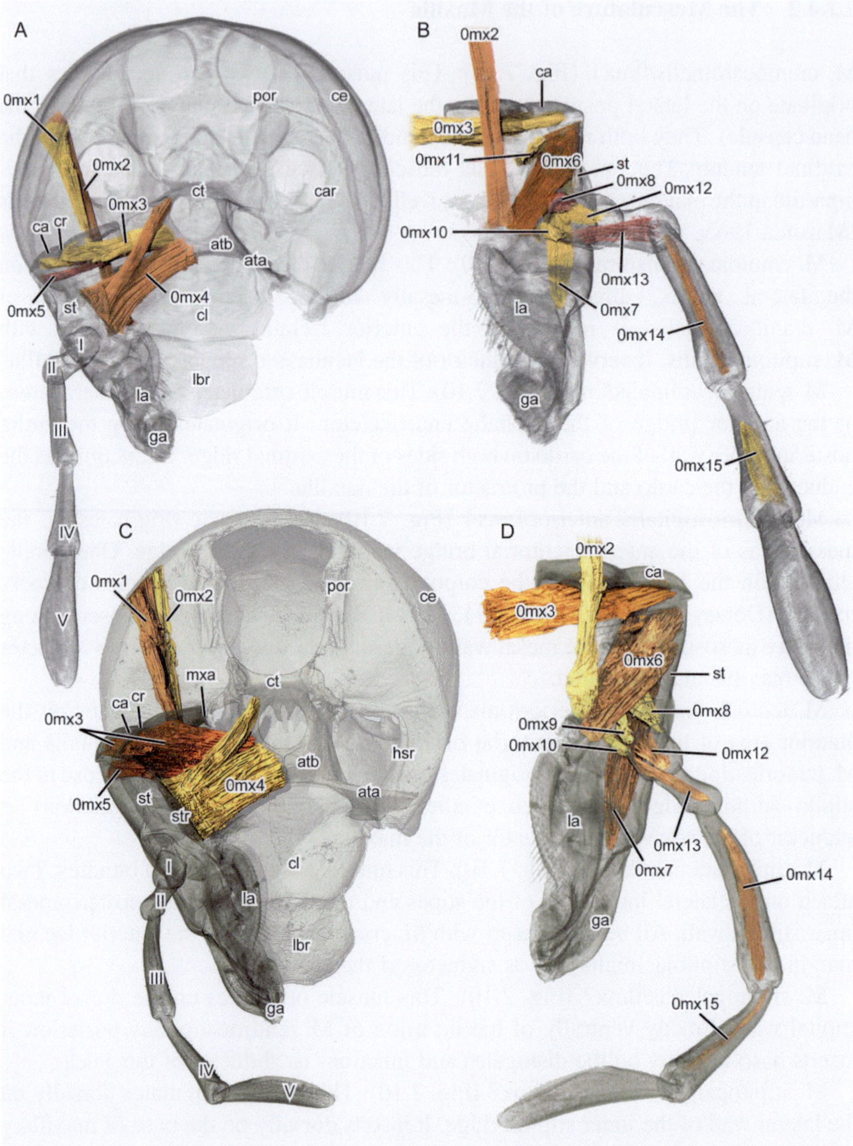


**Fig. 7.10** Three-dimensional reconstructions of the maxillary musculature of the left maxilla of *Ergaula capucina* (**a**, **b**; taken from Wipfler et al. under review) and *Periplaneta americana* (**c**, **d**; taken from Wipfler et al. 2016). (**a** and **c**) Left maxilla and head capsule in posterior view, cuticle rendered transparent and other mouthparts removed; (**b** and **d**) left maxilla in anterior view, cuticle rendered transparent. *0mx1* M. craniocardinalis, *0mx2* M. craniolacinialis, *0mx3* M. tentoriocardinalis, *0mx4* M. tentoriostipitalis anterior, *0mx5* M. tentoriostipitalis posterior, *0mx6* M. stipitolacinialis, *0mx7* M. stipitogalealis, *0mx8* M. stipitopalpalis externus, *0mx9* M. stipitopalpalis medialis, *0mx10* M. stipitopalpalis internus, *0mx11* M. stipitalis transversalis, *0mx12* M. palpopalpalis maxillae primus, *0mx13* M. palpopalpalis maxillae secundus, *0mx14*

Levereault 1938; Carbonaro 1949). It functions as additional adductor or flexor of the maxillary palp.

M. stipitopalpalis internus/0mx10 (Fig. 7.10): This muscle originates on the ventral lateral wall of the inner stipital ridge, ventrally of M. stipitopalpalis medialis. It inserts ventrally at the base of maxillary palpomere I and acts as abductor of the maxillary palp.

M. stipitalis transversalis/0mx11 (Fig. 7.10): This delicate muscle originates on the outer stipital wall, close to the stipito-cardinal border and runs toward the stipital ridge. Similar to M. stipitopalpalis medialis, the presence of this structure varies among representatives of *Periplaneta* (Wipfler et al. 2016) and is not reported for any other cockroach. However, due to its delicate nature it might have been overlooked in various reports.

M. palpopalpalis maxillae primus/0mx12 (Fig. 7.10): This muscle originates basally on the anterior wall of maxillary palpomere I and inserts meso-basally on maxillary palpomere II. It functions as adductor of maxillary palpomere II.

M. palpopalpalis maxillae secundus/0mx13 (Fig. 7.10): This muscle originates basally on the anterior wall of maxillary palpomere II and inserts latero-basally on maxillary palpomere III. It functions as abductor of maxillary palpomere III.

M. palpopalpalis maxillae tertius/0mx14 (Fig. 7.10): This muscle originates basally on the anterior wall of maxillary palpomere III and inserts meso-basally on maxillary palpomere IV. It functions as adductor of maxillary palpomere IV.

M. palpopalpalis maxillae quartus/0mx15 (Fig. 7.10): This muscle originates basally on the anterior wall of maxillary palpomere IV and inserts meso-basally on maxillary palpomere V. It functions as adductor of maxillary palpomere V.

#### 7.2.4.3 The Function of the Maxilla

The maxillae grasp and transport food toward the mouth (Betz et al. 2003; Chapman 1995; Popham 1961a; Schmitt et al. 2014). Additionally they have important sensory functions that are located on the maxillary palps and the galea.

The movement of the maxilla in *Periplaneta*, excluding the palps, is highly rhythmic during feeding and strongly synchronized with the other mouthparts (Schmitt et al. 2014). However, during other tasks such as grooming, this correlation is much weaker (Klein 1982). During normal biting and chewing, the maxillae move strongly antiphasic with respect to the mandibles (Popham 1961a; Schmitt et al. 2014), i.e., during the adduction of the mandibles, the maxillae abduct. The maxillae

**Fig. 7.10** (continued) M. palpopalpalis maxillae tertius, *Mx15* M. palpopalpalis maxillae quartus, *I–V* maxillary palpomeres, *ata* anterior tentorial arm, *atb* anterior tentorial bridge, *ca* cardo, *car* circumantennal ridge, *ce* compound eye, *cl* clypeus, *cr*: cardinal ridge, *ct*: corpotentorium, *epi* epipharynx, *ga* galea, *hsr* hypostomal ridge, *la* lacinia, *lbr* labrum, *mxa* maxillary articulation, *oc* ocellus, *por* postoccipital ridge, *st* stipes, *str* stipital ridge

have a higher degree of freedom among the different sclerotized parts of the maxillary body than the one-piece mandibles (see above). However, the rhythmicity of movement between these paired parts on both sides is highly correlated (Schmitt et al. 2014).

Each maxilla is firmly attached to the head capsule at the maxillary articulation with the hypostomal ridge. The hypostomal apodeme (hpa, Fig. 7.1) acts as a counter bearing for the cardo. The maxillary movement is controlled by a system of muscles that move the cardo against the stipes (Fig. 7.11). By changing the angle between these two structures, the maxilla can be either protracted (and their distal parts simultaneously adducted) or retracted (and abducted). This movement is rhythmically repeated in synchronization with the mandible and occurs in four phases (Fig. 7.11; described in detail in Schmitt et al. 2014). In phase 1, M. tentoriocardinalis (0mx2), M. tentoriostipitalis anterior (0mx3), and M. tentoriostipitalis posterior (0mx4) are maximally contracted, thus adducting the cardo and stipes. As a result the cardo-stipital symphysis opens, i.e., the angle between cardo and stipes widens and both parts are completely extended. The entire maxilla is protracted and the distal parts (lacinia and galea) are pushed toward the central longitudinal line (i.e., adducted). The maxillary palp is maximally retracted. Phase 2 is a transitional one. M. tentoriocardinalis (0mx2), M. tentoriostipitalis anterior (0mx3), and M. tentoriostipitalis posterior (0mx4) start to relax and thus stop the strong protraction and adduction of the maxillary parts. At the same time, M. craniocardinalis (0mx1) and M. craniolacinialis (0mx2) start to contract. The maxillary palps start to move again anteriorly. In phase 3, this development is completed and the maxilla-tentorial muscles are fully relaxed, while the muscles between the head capsule and the maxilla are fully contracted. This pushes the cardo and the stipes dorsally toward the head capsule. Due to the counter bearing at the hypostomal process, which prevents the cardo from moving mesally or dorsally, the maxilla is compressed, i.e., the cardo-stipital symphysis closes (the angle between them decreases). At this stage, the cardo and the basal stipes are maximally abducted. It seems likely that the resulting compression loads the resilin-containing membrane between the cardo and the stipes with elastic energy that restores the initial angular position in the subsequent phase (Schmitt et al. 2014). As a result, the entire maxilla retracts and the lacinia and galea are completely abducted. The last phase (4) is a transition between phase 3 and 1. M. craniocardinalis (0mx1) and M. craniolacinialis (0mx2) begin to relax. At the same time, M. tentoriocardinalis (0mx2), M. tentoriostipitalis anterior (0mx3), and M. tentoriostipitalis posterior (0mx4) begin to contract. The contraction of M. tentoriostipitalis anterior (0mx3) pushes the maxillary lobes inward (to the extent that the galeae touch each other mesally), while M. tentoriocardinalis (0mx2) and M. tentoriostipitalis posterior (0mx4) contract, the cardo-stipidal symphysis opens and the entire maxilla is pushed forward. Based on the observations of Schmitt et al. (2014), this phase is not solely muscle-effected but may also involve a nonmuscular mechanism caused by the elasticity of the membrane in the cardo-stipital joint. They found significant

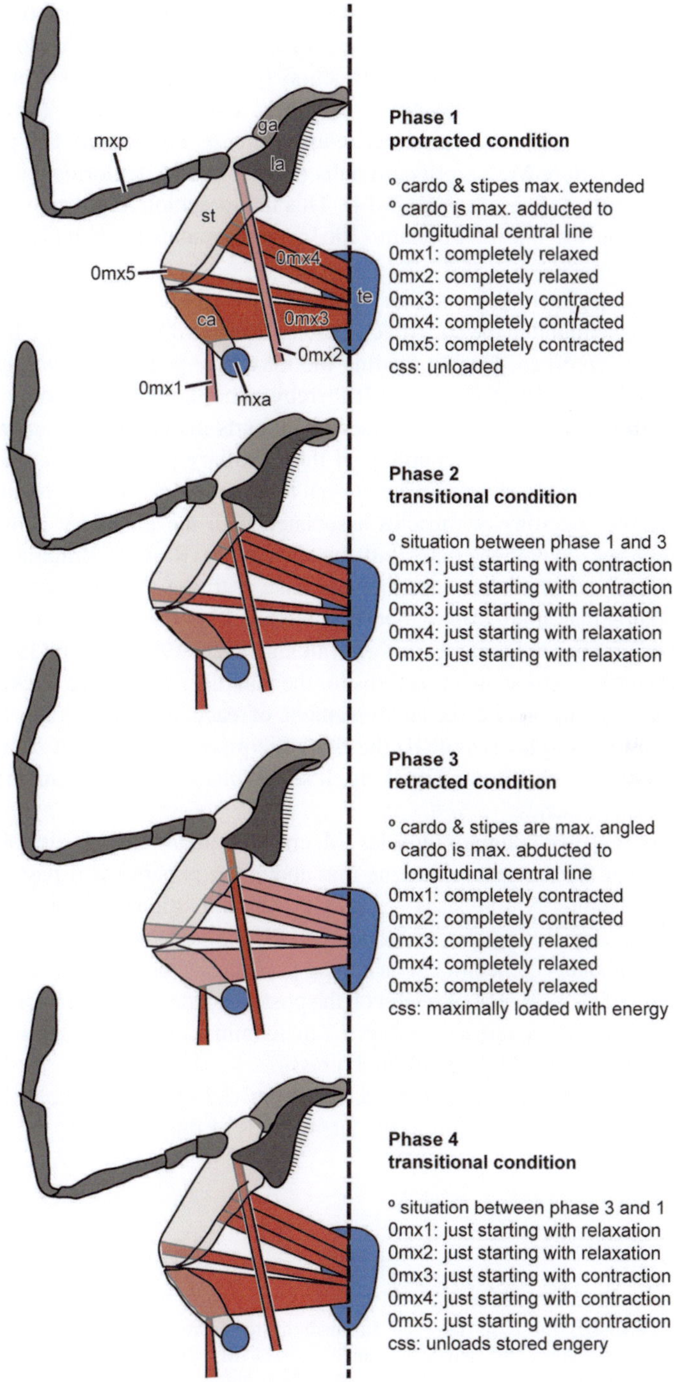


**Fig. 7.11** Movement of the maxilla of *Periplaneta americana* with the four distinct phases (redrawn from Schmitt et al. 2014). Contracted muscles in dark red, relaxed muscles in light red,

autofluorescence under UV light in this area that indicates the presence of resilin, an extremely elastic and flexible protein (e.g., Andersen and Weis-Fogh 1964; Michels and Gorb 2012). During maximum retraction (phase 3), energy is stored in this elastic protein. Upon relaxation of M. craniocardinalis (0mx1) and M. craniolacinialis (0mx2), the stored elastic energy might start the process of protraction just before M. tentoriocardinalis (0mx2) and M. tentoriostipitalis posterior (0mx4) contract (Schmitt et al. 2014). This interpretation might provide a better explanation than the one of Popham (1961a) who assumed a hemolymph-driven process.

In its distal segment, the maxillary palp holds a sensory field with the densest coverage of setae found in the cockroach cuticle (Burry and Moran 1973). It acts as an important sensory organ. During feeding the maxillary palp of *Periplaneta* moves back and forth (Schmitt et al. 2014). It thereby exhibits more autonomy from the rhythmic feeding activities than the other mouthparts including the maxillary body (Schmitt et al. 2014). This decoupling of the maxillary palp from the mandibular movement has also been shown on a neuronal level in the locust *Locusta migratoria*, where the motor neurons of muscles associated with the maxillary palp are only weakly synchronized with the mandibular motor pattern (Rast and Bräunig 2001a, b). This functional independence seems to correlate with the various functions of the maxillary palps that apart from supporting feeding also participate for example in grooming or probing movements during walking (Klein 1982). Klein (1982) distinguishes two different movement patterns for the maxillary palp in the cricket *Gryllus bimaculatus*. (I): the goal-directed movement of each individual palp (found in grooming and food uptake) and (II) the rhythmic probing movement of both palps during exploratory behavior and walking. It seems plausible that a similar pattern is observed in cockroaches.

The lacinia has two adductor muscles (M. craniolacinialis and M. stipitolacinialis) but no abductor. We therefore assume that abduction is performed passively. It is likely that the lacinial symphysis is—similar to the cardo-stipital symphysis—loaded with elastic energy during the adduction. When the adductor muscles relax, this stored energy abducts the lacinia in its resting position. Although there is no account or description of the presence of resilin on the posterior side of the lacinia, it is present in high amounts on the anterior side (Fig. 7.7 in Schmitt et al. 2014). Alternatively, the lacinia could be retracted by hemolymph pressure.

The galea has only an abductor muscle, but Fig. 7.7 of Schmitt et al. (2014) also indicates high amounts of resilin in the lateral parts of this structure.

---

**Fig. 7.11** (continued) those in between in medium red. *0mx1* M. craniocardinalis, *0mx2* M. craniolacinialis, *0mx3* M. tentoriocardinalis, *0mx4* M. tentoriostipitalis anterior, *0mx5* M. tentoriostipitalis posterior, *ca* cardo, *ga* galea, *la* lacinia, *mxa* maxillary articulation, *mxp* maxillary palp, *st* stipes, *te* tentorium

## 7.2.5 *The Labium*

### 7.2.5.1 Morphology of the Labium

The labium of cockroaches (Fig. 7.12) is generally quite uniform in shape and follows the general ground plan of winged insects (cp. Matsuda 1965; Yuasa

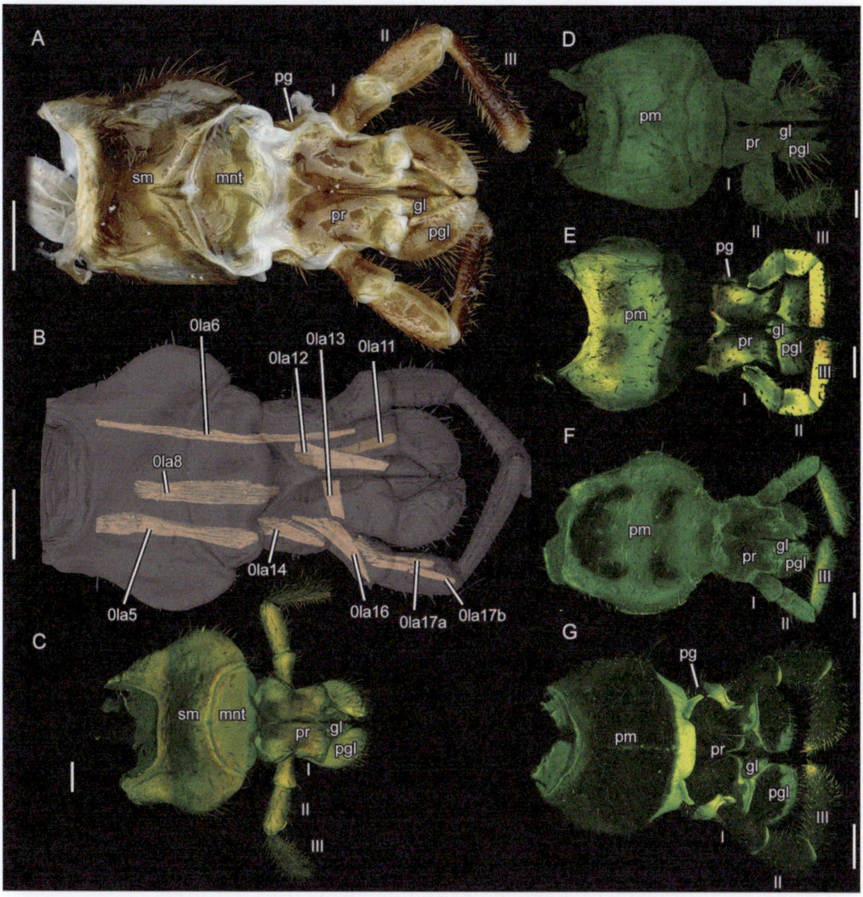


**Fig. 7.12** The labium and its musculature of different species of roaches in posterior view. (**a**) *Periplaneta americana*, digital micrograph, taken from Wipfler et al. (2016); (**b**) *Periplaneta americana*, three-dimensional reconstruction, cuticle rendered transparent to show musculature, taken from Wipfler et al. (2016); (**c**) *Eurycotis floridana*, confocal laser scanning micrograph; (**d**) *Salganea rossi*, confocal laser scanning micrograph; (**e**) *Nauphoeta cinerea*, confocal laser scanning micrograph; (**f**) *Supella* sp., confocal laser scanning micrograph; (**g**) *Therea bernhardti*, confocal laser scanning micrograph. Orange colors in the confocal laser scanning micrographs indicate sclerotization. *0la11* M. praementoparaglossalis, *0la12* M. praementoglossalis, *0la13* M. praementopalpalis internus, *0la14* M. praementopalpalis externus, *0la16* M. palpopalpalis labii primus, *0la17* M. palpopalpalis labii secundus, *0la5* M. tentoriopraementalis inferior, *0la6* M. tentoriopraementalis superior, *0la8* M. submentopraementalis, *gl* glossa, *I/II/III* labial palpomeres I/II/III, *mnt* mentum, *pg* palpiger, *pgl* paraglossa, *pm* postmentum, *pr* praementum, *sm* submentum. Scale bars: 500 μm

1920). It comprises a basal postmentum (pm, Fig. 7.12) and a distal praementum (pr, Fig. 7.12). In some species, the postmentum is divided into a mentum (mnt, Fig. 7.12) and submentum (sm, Fig. 7.12). The latter articulates with the three-segmented palp and the palpiger (pg, Fig. 7.12), and two distal lobes, the glossae (gl, Fig. 7.12) and paraglossae (pgl, Fig. 7.12).

The postmentum varies in shape and sclerotization (shown in the confocal laser scanning images in Fig. 7.12) among different cockroaches (Zhuzhikov 2007) and covers the base of the maxilla. The length to width ratio varies between 1.25 in *Ectobius* sp. and 0.3 in *Periplaneta brunnea* (Zhuzhikov 2007). The separation of the postmentum in mentum and submentum seems to be correlated with this ratio. According to the data collected by Zhuzhikov (2007), all species—independent of phylogeny—with a length to width ratio higher than 0.65 have an undivided postmentum, while it is subdivided in all species with a lower value. The post- or submentum can hold different degrees of setation among cockroach species. The praementum is separated from the postmentum (or the mentum in case of a subdivided postmentum) by a thick band of membrane.

The praemental sclerite is a sparsely setose and nearly rectangular plate. It is cut mesally in its distal part by the praemental cleft. Basally, the cleft continues as a median ridge. The palpiger is positioned laterally on the praementum, forming an articulation with the labial palp. In some species such as *Salganea* (Fig. 7.12d), it is quite indistinct and may even be absent. The praementum on each side has a concave emargination that holds the base of the labial palp and forms its second articulation.

The palp is three-segmented in all cockroaches (I, II, III in Fig. 7.12). The individual palpomeres vary in length and diameter among species. In *Therea* (Fig. 7.12g) the palpomeres are extremely broad and massive, while in *Periplaneta* (Fig. 7.12a) or in *Ergaula* (Wipfler et al. under review) they are rather slender. Labial palpomere I in most cases is the shortest one, while palpomere III is always the longest. A sensory field is distally located on palpomere III. Prakash et al. (1995) and Ramaswamy and Gupta (1981) described the sensillae on the labial palps of different species of cockroaches.

The paraglossae are located disto-laterally on the praementum and form the lateral lobes of the labium. The proximal two-thirds are sclerotized in most species, while the distal third is membranous. They are covered with setae and bear a distal field of tapering microtrichia. In most cockroaches, the paraglossae are larger than the glossae and overlap them medially (an exception is *Salganea*, where both glossae and paraglossae are equally long and of similar shape, Fig. 7.12d).

The glossae are composed of a single sclerite that narrows distally. It bears a field of tapering microtrichia distally.

#### 7.2.5.2 The Musculature of the Labium

The labial musculature of cockroaches is highly conservative and follows the ground pattern for winged insects (Wipfler et al. 2011). Among the studied species, no differences have been found.

M. tentoriopraementalis inferior/0la5 (Fig. 7.12): This muscle originates mesally on the anterior part of the trabeculae tentorii and inserts laterally on the basal praementum. Its functions as retractor of the praementum.

M. tentoriopraementalis superior/0la6 (Fig. 7.12): This muscle originates on the anterior part of the trabeculae tentorii, directly laterad M. tentoriopraementalis inferior. It inserts anteriorly on the praementum, in between the labial palp and the paraglossa. It functions as retractor of the praementum.

M. submentopraementalis/0la8 (Fig. 7.12): This muscle originates mesally on the submentum or postmentum and inserts meso-basally on the praementum. It functions as retractor of the praementum.

M. praementoparaglossalis/0la11 (Fig. 7.12): This muscle originates on the inner praemental wall, dorso-laterally of the end of the praemental cleft, directly dorsad of M. praementoglossalis. It inserts on the baso-mesal edge of the paraglossa and acts as adductor of the paraglossa.

M. praementoglossalis/0la12 (Fig. 7.12): This muscle originates on the inner praemental wall, slightly laterad the end of the praemental cleft, directly ventrally of M. praementoparaglossalis. It inserts on the baso-mesal edge of the glossa. This muscle functions as the adductor of the glossa.

M. praementopalpalis internus/0la13 (Fig. 7.12): This muscle inserts on the inner wall of the praemental cleft, near its end, and inserts at the ventral basal margin of the first labial palpomere. It is the adductor of the labial palpomere I.

M. praementopalpalis externus/0la14 (Fig. 7.12): This muscle originates on the anterio-lateral part of the base of the praementum. It inserts on the dorsal basal margin of the first labial palpomere and functions as abductor of the labial palpomere I.

M. palpopalpalis labii primus/0la16 (Fig. 7.12): This muscle originates on the anterio-lateral part of the base of the praementum, directly ventrally of M. praementopalpalis externus. It inserts at the lateral basal margin of the labial palpomere II. This muscle functions as the adductor of the second palpomere.

M. palpopalpalis labii secundus/0la17 (Fig. 7.12): This muscle originates on the meso-basal wall of the labial palpomere II. It inserts in two bundles, one mesally (0la17a) and one laterally (0la17b) on the basal edge of the labial palpomere III. It moves the third palpomere.

#### 7.2.5.3 The Function of the Labium

The labium forms the posterior closure of the cibarium. The praementum protracts and retracts during feeding (Evans 1964; Schmitt et al. 2014). The retraction of the praementum is achieved by three muscles that are present in all cockroaches: M. tentoriopraementalis inferior, M. tentoriopraementalis superior, and M. submentopraementalis. With the retraction, the membrane between the praementum and the mentum or postmentum (depending on the subdivision of the postmentum; see above) is compressed and folded. It seems likely that elastic energy is stored during this compression, which powers the protraction of the praementum when the retractors relax. In contrast to the rhythmic and strongly synchronized movements of the maxillary body, the hypopharynx, and the mandible, during feeding the labium moves much more independently. The labial kinematics can be rhythmic and consistent but can also significantly differ in terms of both their temporal coordination and amplitude (Schmitt et al. 2014). In carabid beetles, the labium only retracts when there is a sufficient amount of food in the cibarium (Evans 1964).

Similar to the maxillary lacinia, the glossa and paraglossa only have an adductor muscle. It is likely that the abduction is driven by a passive mechanism (similar to the cardo-stipital joint or the lacinial abduction; see above) or hemolymph pressure. Both lobes aid in ensuring that no food falls out of the cibarium during the transport toward the pharynx (Smith 1985).

It is possible that the saliva runs through the praemental cleft during feeding.

## 7.2.6 *The Hypopharynx*

### 7.2.6.1 Morphology of the Hypopharynx

The hypopharynx of various species of cockroaches (Fig. 7.13) was described in detail in Buder and Klass (2013), Wipfler et al. (2016), and Wipfler et al.

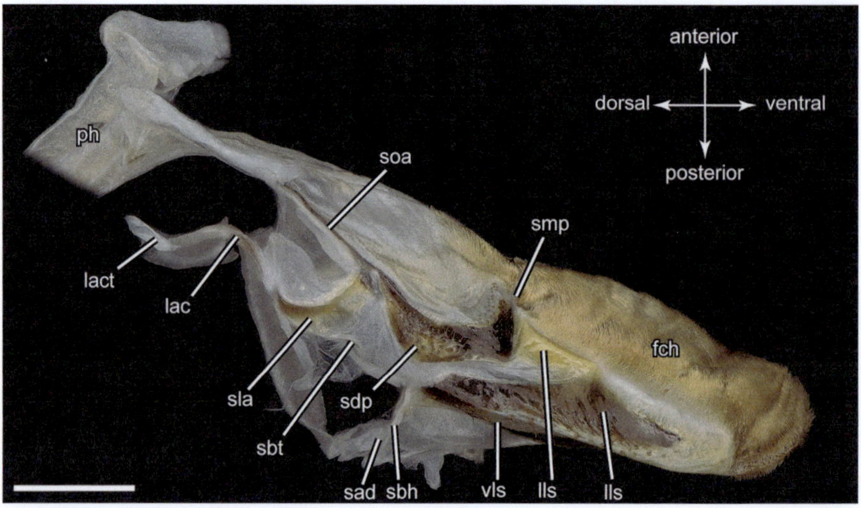


**Fig. 7.13** Digital micrograph of the hypopharynx of *Periplaneta americana* in lateral view (taken from Wipfler et al. 2016). *fch* field of microtrichia on dorsal and distal surface, *lac* linguacutal sclerite, *lact* linguacutal tendon/apodeme, *lls* lateral lingual sclerite of hypopharynx, *ph* pharynx, *sad* salivary duct, *sbh* arm at base of ventral lingual sclerite, *sbt* basitendinal sclerite of hypopharynx, *sdp* plate-like distal part of hypopharyngeal suspensorium, *sla* loral arm of hypopharyngeal suspensorium, *smp* arm-like disto-mesal part of hypopharyngeal suspensorium, *soa* oral arm of hypopharyngeal suspensorium, *vls* ventral lingual sclerite of hypopharynx. Scale bar: 500 μm

(under review). In general, the hypopharynx of cockroaches is morphologically quite constant (Buder and Klass 2013). It is a tongue-shaped and mostly membranous structure located centrally in the mouth cavity similar to the human tongue. It forms a slope toward the anatomical mouth opening. Its posterior-basal membrane continues into the labium, the lateral membranes into the maxillae and the mandibles, and the anterio-basal membrane into the pharynx. The hypopharynx thus forms—with the exception of the entrance into the pharynx—the dorsal boundary of the oral cavity, the anterior boundary of the salivarium, and the posterior boundary of the cibarium. The distal half of the anterior side and the tip of the hypopharynx are densely covered by a large field of dorsally pointing microtrichia (fch, Fig. 7.13). They push the food toward the mouth opening. Several sclerites are embedded within the ventral and lateral membranes of the hypopharynx. They are separated into the proximal suspensorial and the distal lingual ones. The somehow V-shaped lateral lingual sclerite (lls, Fig. 7.13) is found in the lateral distal area of the hypopharynx and reaches distally and anteriorly almost to the field of microtrichia. It continues proximo-posteriorly into the ventral lingual sclerite, which lies on the posterior hypopharyngeal wall. Between species, this sclerite can vary strongly in shape and the sclerites of both body halves may or may not be interconnected via the midline (Buder and Klass 2013). At the base of the ventral lingual sclerite, a postero-mesally directed arm (SBH, Fig. 7.13) attaches. In some cockroaches it is not clearly connected with the ventral lingual sclerite (Buder and Klass 2013). In between these arms, the opening the salivary duct is positioned. The proximal suspensorial sclerites are composed of the plate-like distal part (SDP, Fig. 7.13), a wide loral arm (SLA, Fig. 7.13), a linguacutal sclerite (LAC, Fig. 7.13), and a slender oral arm (SOA, Fig. 7.13). The plate-like distal part of the hypopharynx is the distal-most suspensorial sclerite and fused with the lateral lingual sclerite. From the point where they fuse, an arm-like sclerotization (SDP, Fig. 7.13) continues toward the anterior hypopharyngeal surface. The oral arm is fused with the plate-like sclerite and continues dorsally to frame the functional mouth opening laterally. Additionally it serves as attachment site for different muscles. The loral arm is a sickle-shaped sclerite connected to the oral arm. Dorsally it continues into the curved linguacutal sclerite (lact, Fig. 7.13) that enters the mandible. In the illustration of dissected specimens (Buder and Klass 2013; Fig. 7.13), this sclerite appears to run dorsally as a continuation of the loral arm. However, in three-dimensional reconstructions of non-dissected animals, this sclerite bends clearly ventrally (Fig. 7.4; Wipfler et al. 2016, under review), which is important when considering the function of the associated muscle (see below). It is likely that the orientation of this sclerite changes during the removal of the hypopharynx. In the corydiid cockroach *Arenivaga investigata*, the hypopharynx is inflatable and contains lateral bladders. Their function is discussed below.

The hypopharynx of cockroaches is—similar to other insects—covered with different sensory organs including contact chemoreceptors and stretch sensors (Moulins 1967, 1971a, b, 1974).

### 7.2.6.2 The Musculature of the Hypopharynx

M. frontooralis/0hy1 (Fig. 7.4): This muscle originates on the frons, directly laterally of *M.* frontolabralis. It inserts on the distal tip of the oral arm of the hypopharynx. It functions together with M. tentoriohypopharyngalis as retractor and depressor of the hypopharynx.

M. tentoriooralis/0hy2 (Fig. 7.4): This muscle originates on the subgena, in the middle between the antennal base and the anterior mandibular articulation. It inserts laterally on the oral arm of the hypopharynx, slightly proximal of the insertion of M. frontooralis. It functions as dilator of the anatomical mouth opening and widener of the hypopharynx, antagonizing M. oralis transversalis.

M. tentoriohypopharyngalis/0hy3 (Fig. 7.4): This muscle originates on the anterior part of the posterior processes of the corpotentorium (trabeculae tentorii), directly laterally of M. tentoriopraementalis superior. It inserts on the dorsal side of the arm at the base of the ventral lingual sclerite. It functions as retractor of the hypopharynx.

M. tentoriosuspensorialis/0hy5 (Fig. 7.4): This muscle originates on the oesotendons of the tentorium. It inserts mesally on the hypopharyngeal membrane, slightly ventrally of the anatomical mouth opening. It functions as retractor of the hypopharynx and dilator of the cibarium.

M. praementosalivarialis anterior/0hy7 (Fig. 7.4): This muscle originates on the posterior wall of the praementum, mesally of the articulation of the labial palp. It inserts proximally on the ventral side of the basal hypopharyngeal apodeme. It functions as protractor of the hypopharynx.

M. praementosalivarialis posterior/0hy8 (Fig. 7.4): This muscle originates on the anterio-lateral wall of the base of the praementum, slightly mesad of M. praementopalpalis externus. It inserts distally on the ventral side of the arm at the base of the ventral lingual sclerite. It functions as widener of the opening of the salivary ductus and/or the hypopharynx.

M. oralis transversalis/0hy9 (Fig. 7.4): This muscle originates on the oral arm of the hypopharynx and inserts on the oral arm of the opposite side. Fibers of this muscle run both above and below the pharynx. It functions as contractor of the anatomical mouth opening and narrows the hypopharynx; it is the antagonist of M. tentoriooralis and M. clypeobuccalis.

M. loroloralis/0hy10 (Fig. 7.4): This structure originates on the mesal wall of lateral hypopharyngeal sclerite. It inserts on the mesal wall of the lateral hypopharyngeal sclerite of the opposite side. It is not completely clear if this structure has muscular fibers in all species or if it in some cases comprises a tendon-like band. It narrows the hypopharynx.

M. hypopharyngosalivarialis/0hy12 (Fig. 7.4): This muscle originates anteriorly on the loral sclerite of the hypopharyngeal suspensorium and inserts with several bundles across the entire membranous floor of the hypopharynx. It functions as dilator of the salivarium.

M. hypopharyngomandibularis/0md4 (Fig. 7.4): This muscle is traditionally assigned to the mandible, but it most likely moves the hypopharynx rather than the mandible. Its origin and insertion are provided in the chapter on the mandibular muscles.

Moulins (1971a) describes two additional structures in the hypopharynx of *Blaberus craniifer*, which Buder and Klass (2013) considered as "muscles" (however, they placed them in quotes). Similar to M. loroloralis, it is likely that they are at least in some cases rather membranous or tendon-like bands than actually contractible structures.

### 7.2.6.3 The Function of the Hypopharynx

The hypopharynx works as a mostly membranous tongue-like structure that transports the food within the mouth cavity. Additionally it divides the mouth cavity into two chambers: (I) The frontal cibarium, i.e., the space between the hypopharynx and the epipharynx, in which the mandibles and the maxillae operate and that ends dorsally with the anatomical mouth opening. Here, the chewing and grinding of the food takes place, and finally, the processed diet is sucked into the pharynx. (II) The much smaller posterior salivarium, located between the hypopharynx and the labium. The salivarium is dorsally delimited by the opening of the salivary duct. The dilation of the salivarium allows saliva to flow into the cibarium (or onto food particles outside the mouth cavity) while its compression stops this process.

The kinematics and movement of the hypopharynx have received very little attention, most likely due to the fact that it is completely internal and cannot be observed properly in its functioning. According to Popham (1961a), the hypopharynx of *Periplaneta* protracts and retracts and also performs anterior and posterior movements during feeding. He also provides some indications on how these movements are controlled by muscles. However, in *Periplaneta* he only described two muscles (his "dorsal" and "ventral hypopharyngeal muscle"), both of which comprise the bundles of M. hypopharyngosalivarialis. He thus only explains the opening and closing of the salivary duct and the salivarium (i.e., the anterior and posterior movements of the hypopharynx). Popham (1959) mentioned a hypopharyngeal depressor muscle (his muscle "HDM") that connects the ventral posterior hypopharynx with the mentum in the prognathous earwig *Forficula auricularia*. However, this muscle is homologous to M. tentoriohypopharyngalis since in earwigs (Dermaptera) the origin of this muscle moved from the tentorium to the labium (Neubert et al. 2017). With the change of origin, this muscle takes a different attachment angle in earwigs than in other insects, which might also alter its function. In the description of the feeding process of dragonflies (Popham and Bevans 1979) and other earwigs (Popham 1962), no hypopharyngeal muscles were described. Consequently, Popham explains all other motions of the hypopharynx as caused by hemolymphic fluid pressure (Popham 1959, 1961a). Based on the subsequent discovery of more extensive musculature of the hypopharynx (see above), we

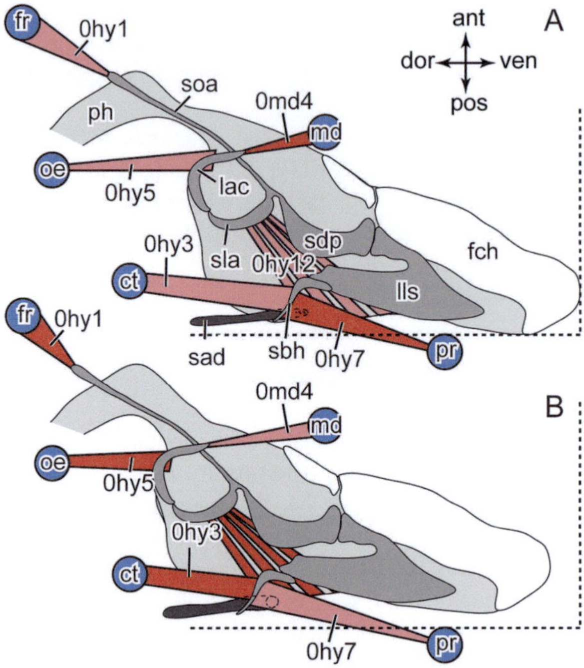


**Fig. 7.14** Proposed muscular control of the protraction and posterior movement (**a**) and retraction and anterior movement (**b**) of the hypopharynx *Periplaneta americana*. Contracted muscles in dark red and relaxed ones in light red. *0hy1* M. frontooralis, *0hy12* M. hypopharyngosalivarialis, *0hy3* M. tentoriohypopharyngalis, *0hy5* M. tentoriosuspensorialis, *0hy7* M. praementosalivarialis anterior, *0md4* M. hypopharyngomandibularis, *ant* anterior, *ct* corpotentorium, *dor* dorsal, *fch* field of microtrichia on dorsal and distal surface, *fr* frons, *lac* linguacutal sclerite, *lls* lateral lingual sclerite of hypopharynx, *md* mandible, *oe* oesotendons, *ph* pharynx, *pos* posterior, *pr* praementum, *sad* salivary duct, *sbh* arm at base of ventral lingual sclerite, *sdp* plate-like distal part of hypopharyngeal suspensorium, *sla* loral arm of hypopharyngeal suspensorium, *soa* oral arm of hypopharyngeal suspensorium, *ven* ventral

propose the following muscular mechanisms, which are illustrated and summarized in Fig. 7.14.

**Pro- and Retraction** Based on the angle of attachment, we assume that several muscles take part in the protraction and retraction of the hypopharynx. The protraction, i.e., the ventrad movement of the hypopharynx (Fig. 7.14a), is reached by the contraction of M. hypopharyngomandibularis that attaches to the anterior part of the hypopharynx and M. praementosalivarialis anterior at the posterior hypopharynx. In both cases, the origin of the muscle is clearly ventral of the insertion, which would push the hypopharynx down, i.e., protract it. During feeding, the protraction of the hypopharynx is synchronized with the opening of the mandibles and the posterior movement of the hypopharynx, i.e., the closure of the salivarium and the subsequent stop of saliva flow (Popham 1961a). The antagonistic dorsal movement, i.e., the retraction (Fig. 7.14b), is achieved by M. tentoriohypopharyngalis (which attaches directly at the same insertion as M. praementosalivarialis anterior but continues dorsally), by M. tentoriosuspensorialis (attaching at the same level as M. hypopharyngomandibularis), and by M. frontooralis who pulls at the oral arm of the suspensorium. It is possible that M. tentoriosuspensorialis also dilates the cibarium as it attaches primarily at the membrane of the hypopharynx. The retraction is synchronized with the closure of the mandibles and the anterior movement of the hypopharynx that opens the salivarium and allows the saliva to flow. The retraction (in combination with the anterior movement of the hypopharynx; see below) pushes food particles upward in the cibarium in between the closing mandibles, where it is ground up. Alternating contractions and relaxations of these two muscle groups would cause a rhythmic movement of the hypopharynx as observed by Popham (1961a).

**Anterior and Posterior Movements** Popham (1961a) identified two muscles that are mainly in charge of this movement, the "dorsal" and "ventral hypopharyngeal muscle." Both these muscles are homologous with M. hypopharyngosalivarialis in the present contribution. Upon contraction (see Fig. 7.14b), this muscle lifts the posterior wall of the hypopharynx, thus dilating the salivarium and causing saliva to flow into the mouth cavity. It is likely that it is supported in this function by M. frontooralis. Both muscles must contract at the same time to initiate the simultaneous retraction and anteriad movement of the hypopharynx as observed by Popham (1961a) (Fig. 7.14b). These muscles' contractions would also ensure that the suspensorial part of the hypopharynx is not pushed ventrally by the contraction of M. hypopharyngosalivarialis, but instead the entire hypopharynx rotates anteriad, thus moving toward the epipharynx and aiding in pushing the food between the mandibles. The contraction of M. hypopharyngosalivarialis is synchronized with the closure of the mandible and the retraction of the hypopharynx (Popham 1961a), which further supports the flow of saliva. Additionally it seems plausible that it is temporally coupled with the lateral dilation of the salivary opening by M. praementosalivarialis posterior (see below). The antagonist of M. hypopharyngosalivarialis is at least partially the M. praementosalivarialis anterior (Fig. 7.15b) that protracts (see above) the hypopharynx but also moves it slightly posteriorly. It is likely that it is supported by hemolymphic pressure as proposed by (Popham 1961a) or by a passive elastic mechanism. When the hypopharynx moves posterior, the salivarium is closed and the saliva flow ends.

**Lateral and Median Movements** These types of movements were not observed by (Popham 1961a), but the muscular equipment indicates that they can be performed to a certain degree at two points of the hypopharynx. The Musculi tentoriooralis push the oral arms toward the sides, thus widening the functional mouth opening that is placed between them. Since we have no kinematic data on the hypopharynx, we cannot address whether these movements also widen the remaining hypopharynx. However, if this were the case, it is likely that M. loroloralis would act as a compressor of the middle part of the hypopharynx. In some cases, this structure

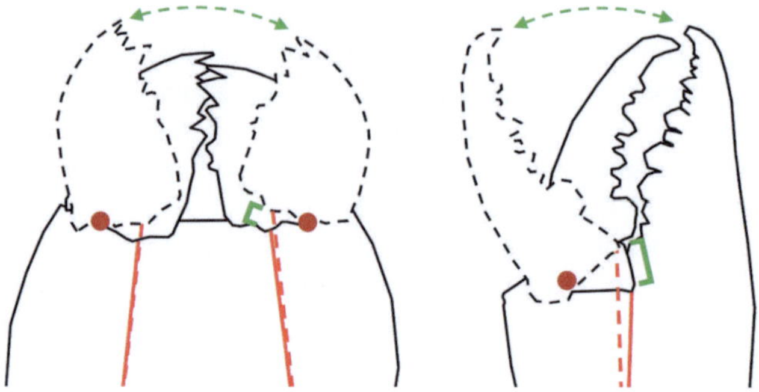

**Fig. 7.15** Length changes (green brackets) of the adductor muscles in mandibular biting and chewing systems and chelae using the examples of the American cockroach (left) and a chela of the mud crab (right). Given gape angles (dashed arrowed arc) are achieved by the rotation of two moving parts in the former and by the rotation of only one moving part in the latter around joint axes (dark red dots). Fiber length change and the muscles contraction speed, therefore, are lower in the mandibular system (see main text for further explanations). Open: dashed mandibles and mean directions of the adductor muscles; closed: solid line drawings

does not contain muscle fibers but appears to be purely membranous or tendon-like. It is therefore likely that it is rather a passive mechanism that pushes the hypopharynx back. The direct antagonistic movement, i.e., the medial compression of the oral arms is performed by M. oralis transversalis that connects the arms mesally over and under the functional mouth opening. Another lateral contraction is performed by M. praementosalivarialis posterior, which—upon contraction—pushes the two arms at the base of the ventral lingual sclerite apart. It is likely that the opening of the salivary duct between them is dilated and very slightly protracted by this movement. Additionally this movement might widen the entire posterior and ventral hypopharynx. There is no direct muscular antagonist to this movement, but similar to the lateral movement of M. tentoriooralis, M. loroloralis would act as a hypopharyngeal compressor when it is widened by M. praementosalivarialis posterior.

In addition to aiding the movement of food toward the mandibles and mouth opening, the hypopharynx has important sensory functions that are performed by various sensillae (see above). The desert cockroach *Arenivaga investigata* (Corydiidae) also uses its hypopharynx to absorb condensed water vapor from the air (O'Donnell 1977). The actual condensation of water takes place on hypopharyngeal bladders and is supported by a fluid produced by the epipharyngeal gland (see above) that is transported via epipharyngeal ducti toward the bladder (O'Donnell 1981, 1982). As similar glands are also present in other corydiid cockroaches such as *Ergaula* (Fig. 7.3a; Wipfler et al. under review) and *Polyphaga* (Zhuzhikov 2007), it is likely that these species do also exhibit a protrusible hypopharynx. This group of cockroaches constitutes the only insects that absorb atmospheric water with the mouth (O'Donnell 1981).

## 7.2.7 *The Pharynx*

### 7.2.7.1 Morphology of the Pharynx

The cephalic pharynx (ph, Fig. 7.4) is slightly wider than high in cross section in the studied cockroaches. The functional mouth opening is located slightly ventral to the frontal ganglion and forms the dorsal end of the cibarium. The oral arms of the hypopharyngeal suspensorium run on each of its sides; between them the M. oralis transversalis spans around the pharynx. The posterior wall of the pharynx continues into the hypopharyngeal membrane, while the anterior wall continues into the epipharynx. The pharyngeal wall is folded inwardly in longitudinal direction, which allows for considerable expansion when food is taken up.

### 7.2.7.2 Muscles of the Pharynx

M. clypeobuccalis/0bu1 (Fig. 7.4): This muscle originates on the clypeus, directly dorsally of M. clypeopalatalis. It inserts behind the functional mouth opening on the anterior pharynx and in between the bundles of M. oralis transversalis. It functions as dilator of the anterior pharynx and antagonist of M. oralis transversalis.

M. frontobuccalis anterior/0bu2 (Fig. 7.4): This muscle originates mesally on the frons. It inserts on the anterior surface of the pharynx, directly dorsally of the frontal connective. It functions as dilator of the pharynx and antagonist of M. anularis stomodaei.

M. frontobuccalis posterior/0bu3 (Fig. 7.4): This muscle originates mesally on the frons, directly dorsally of the origin of M. frontolabralis. It inserts in the anterior surface of the pharynx, dorsally of the insertion of M. frontobuccalis anterior and ventrally of the brain. It acts as dilator of the pharynx, antagonizing M. anularis stomodaei.

M. tentoriobuccalis lateralis posterior/0bu4 (Fig. 7.4): This muscle originates at the dorsal tentorial arm or corpotentorium and inserts on the lateral pharyngeal wall, directly dorsally of the circumesophageal connective. It functions as dilator of the pharynx and antagonist of M. anularis stomodaei.

M. tentoriobuccalis anterior/0bu5 and M. tentoriobuccalis posterior/0bu6 (Fig. 7.4): These muscles originate on the oesotendons, together with M. tentoriosuspensorialis. They insert along the posterior wall of the pharynx, slightly ventrally of the level of the insertion of M. frontobuccalis anterior. They function as dilators of the pharynx and antagonize M. anularis stomodaei.

M. tentoriobuccalis lateralis anterior/0bu7 (Fig. 7.4): This muscle originates on the subgena, in the middle between the antennal base and the anterior mandibular articulation, directly dorsally of M. tentoriooralis. It inserts on the lateral pharyngeal wall, directly dorsally of the insertion of M. tentoriooralis. It acts as dilator of the pharynx and antagonist of M. anularis stomodaei. M. tentoriobuccalis lateralis anterior is only described for the cockroaches *Periplaneta* and *Ergaula* but not for any other insect (Wipfler et al. 2011, 2016, under review).

M. verticopharyngalis/0ph1 (Fig. 7.4): This muscle originates in two bundles on the vertex, in between the bundles of M. craniomandibularis internus. It inserts with two bundles on the anterior surface of the pharynx. The first bundle is on the same level as the insertion of M. tentoriobuccalis lateralis posterior, directly dorsally of the brain. The second bundle inserts further dorsally. This muscle functions as dilator of the pharynx and antagonist of M. anularis stomodaei.

M. tentoriopharyngalis/0ph2 (Fig. 7.4): This muscle originates on the ventro-anterior margin of the posterior tentorial arms and the corpotentorium. It inserts on the posterior wall of the pharynx on the same level as the insertion of M. verticopharyngalis. It is a dilator of the pharynx and antagonizes M. anularis stomodaei.

M. postoccipitopharyngealis/0ph3 (Fig. 7.4): This muscle originates on the postoccipital ridge on the lateral side of the foramen occipital. It inserts broadly along the lateral pharyngeal wall, ranging from the insertion of the dorsal bundle of M. verticopharyngalis to the point where the pharynx leaves the head capsule. It acts as dilator of the pharynx and antagonist of M. anularis stomodaei.

M. anularis stomodaei/0st1 (Fig. 7.4): This muscle forms a ring muscle layer around the pharynx. It ends at the level where the pharynx enters the thorax. It functions as constrictor of the pharynx and is the antagonist to the pharyngeal dilators. This muscle is not described for *Blattella* (Snodgrass 1943, 1944) but is most likely also present in this species.

M. longitudinalis stomodaei/0st2 (Fig. 7.4): This muscle is composed of longitudinal fibers along the pharynx. It ends at the level where the pharynx enters the thorax and is the contractor of the pharynx. This muscle is not described for *Blattella* (Snodgrass 1943, 1944) but is most likely also present in this species.

#### 7.2.7.3 Function of the Pharynx

The main task of the pharynx is to swallow processed food and transport it into the digestive tract through peristalsis (Seifert 1970). Food is sucked out of the cibarium in the same way (shown in the radiograms in Schmitt et al. (2014). This movement is caused by a whole set of pharyngeal dilators which widen the pharynx. They attach over the entire length of the pharynx on the frontal, posterior, and lateral sides (see above and Fig. 7.4). Their antagonists are the M. oralis transversalis in the area of the functional mouth opening and the M. anularis stomodaei dorsally on the entire following pharynx.

### *7.2.8 The Feeding Process in Cockroaches*

Before feeding starts, the potential food is inspected with receptors on the tarsi and labial and maxillary palps (Bernays 1985; Bernays and Simpson 1982). If it is found to be acceptable, the head is brought in a proper position and feeding starts (Bernays and Simpson 1982). During feeding, all mouthparts act together as a functional system to transport food into the mouth opening where it is chewed and ground. The movements of the mandibles, maxillae, and the hypopharynx are highly rhythmic and strongly synchronized with each other during feeding (Popham 1961a; Rast and Bräunig 2001a; Schmitt et al. 2014). This synchronization is less marked in the labium and the maxillary palp, suggesting a higher flexibility and context-dependent control of these components (Schmitt et al. 2014). The coupling, however, is based on motor neuronal correlations of the mouthparts with the mandibular motor pattern generator on the level of the suboesophageal ganglion as well as on bilateral coupling (Rast and Bräunig 2001a, b). Thus, no sensory feedback is required for a stable coordination of the feeding process (compare with Rast and Bräunig 2001a). However, during other tasks such as grooming, the mouthparts act independently from each other and show a considerably lower degree of coupling. Popham (1961a), Roberts (1972), Smith (1985), and Schmitt et al. (2014) described the interaction of the mouthparts for the cockroach *Periplaneta*. Salivated food, which mostly comprises soft organic matter, is swept under the head with the galeae and the maxillae. Then the lacinial incisivi are driven into the salivated food mass during the adduction and protraction of the maxillae. With the help of the lacinial and galeal setae, the food is transported via retraction of the maxillae into the cibarium and lifted over the anterior surface of the hypopharynx (phases 2 and 3 of the maxillary movement); simultaneously, the mandibles are opened. The passage through the cibarium is assisted by orally directed setae on the glossae, paraglossae, hypopharynx, and epipharynx. The maxillary galeae act as scoops, while the labrum, the labium, and the lacinial and galeal setae ensure that no food falls back out of the mouth during this transport. As the mandibles close, the maxillae are abducted and protracted to make space in the cibarium. At the same time, the hypopharynx is rotated forward and upward, thus pushing the food between the mandibles. This hypopharyngeal movement opens the salivarium and saliva flows through the opened maxillae onto food material beneath the head. The mandibular setae on the posterior surface help to hold the food in place as the lacinial incisivi are withdrawn. Setae at the dorsal end of the mandible push the masticated food into the mouth as the mandibles fully close. Finally the ground food is sucked into the pharynx via dilation of the cibarium and esophagus (shown on radiographs in Schmitt et al. (2014). As the mandibles open, the maxillae grab new food. Additionally the hypopharynx moves backward and downward, thus closing the salivarium via pressure and stopping the saliva flow.

## 7.3 The Feeding Apparatus of Cockroaches as a Model for Biting and Chewing Insects

Cockroaches are considered to have an unspecialized and generalized feeding apparatus (Seifert 1970; Weihmann et al. 2015a). This is true in various respects as, for example, their mandibles are typically dicondylic and show no extreme specialization as observed in herbivores or carnivores (discussed below). The muscular equipment of the mouthparts is typical for neopteran insects (Weihmann et al. 2015b; Wipfler et al. 2011). Additionally the cockroaches that are most studied (i.e., *Periplaneta*, *Blatta orientalis*, or *Blattella*) are human-associated species that feed on virtually anything. These pest species are true omnivores, although most other cockroach species are best described as detritivores. This generalized feeding apparatus (next to the easy access and breeding) might be one of the reasons why the feeding apparatus of the American cockroach *Periplaneta* might be the best studied biting and chewing insect. Several studies addressed different functional, kinematic, and morphological aspects (see introduction for a detailed list). In the following sections, we compare the feeding apparatus of cockroaches with those of other insects and arthropods to evaluate shared anatomical and functional traits. Additionally, we will examine advantages and disadvantages of mandibles compared to competing designs.

### *7.3.1 Comparison with Other Biting and Chewing Insects*

The general design of the head capsule is quite similar in biting and chewing mouthparts of winged insects. This includes the openings, the presence of a tentorium, a comparable set of cephalic strengthening ridges, and the articulation of the mouthparts (e.g., a dicondylic mandible) (Blanke et al. 2012; Yuasa 1920). Cockroaches (and their close relatives, the mantids) have a highly moveable head capsule that is usually held in an orthognathous position (see above). This flexibility distinguishes them from most other insects where head mobility is much more restricted. Orthognathy represents the ancestral condition for both winged insects in general and neopterans, i.e., those insects that can fold their wings (Wipfler et al. 2019). Prognathy, i.e., frontally orientated mouthparts, or hypognathy (posteriorly orientated mouthparts) thus represent derived conditions that evolved independently several times (Wipfler et al. 2019). Generally, it is considered that prognathy resembles an adaptation toward a raptorial lifestyle as it allows to catch prey with the mandibles (Beutel et al. 2014). However, in many cases herbivores such as stick and leaf insects or wood feeders such as termites are also prognathous (Friedemann et al. 2012; Vishnoi 1956, 1962). Additionally raptorial groups that catch prey with their legs such as mantids or dragon- and damselflies are typically orthognathous (Blanke et al. 2012; Wipfler et al. 2012). Changes in the head position have very little effect on the cephalic musculature and general composition (Wipfler et al. 2011; Yuasa 1920) but can lead to far-reaching modifications in the mechanics of the head capsule (Blanke et al. 2018).

Some cockroaches such as *Periplaneta* have an emarginated distal labrum of unclear function. In leaf-feeding grasshoppers, this structure is associated with margin-feeding (in contrast to center-feeders that lack it) (cp. Gangwere 1965), while in the prognathous beetle *Philonthus decorus* (Staphylinidae), food is guided toward this notch, from which it falls downward between the mandibles and maxillae (Evans 1964).

A derived feature of the cockroach head is the "perforate" tentorium, which is achieved through the fusion of the anterior tentorial arms. According to Popham (1961b), it is a result of the flattening of the head capsule in anterio-posterior direction, which leads to a corpotentorium that is mesal of the maxilla. To ensure the proper function of the tentorium, the tentorial muscles move anteriorly and mesally, eventually resulting in the evolution of the anterior tentorial bridge. In other respects including its general composition, the tentorium of cockroaches reflects the pterygotan ground plan.

Some cockroach species show a strong sexual dimorphism. In the head capsule, this affects most strongly the overall size of the head and the shape of the compound eyes. This dimorphism is most pronounced in species where males are active and flying, but females have reduced or absent wings (Bell et al. 2007). The eyes of some male cockroaches are larger and protruding (Bell et al. 2007). In males of species such as *Periplaneta* or *Ergaula*, they reach ventrally toward the subgemal ridge, while the eyes of females end distinctly further dorsally (Wipfler et al. under review, 2016). The sexual dimorphism is also found in other sense organs such as the antenna (Lambin 1973) and the maxillary palps (Bland et al. 1998). Next to these morphological differences, some cockroaches also show sex-related differences concerning diet (Bell et al. 2007). It is likely that the morphological and physiological differences between the sexes also have functional and mechanical consequences for the animals. So far, no study addresses these issues specifically although the study of Blanke et al. (2018) seems to indicate that the presence of the compound eye in general has little effect on the distribution of mechanical strain in the head capsule as long as the eyes are surrounded by circumocular ridges. So far, these ridges were found in all studied cockroaches (Klass and Eulitz 2007) and most other insects (Matsuda 1965). However, sex-specific differences and dimorphisms are commonly found among almost all insect lineages (Stillwell et al. 2010).

With respect to the principal movement of the mouthparts, the feeding process of cockroaches follows the general pattern as found in other biting and chewing insects such as earwigs (Popham 1959, 1962), odonates (Popham and Bevans 1979), beetles (Evans 1964, 1965; Evans and Forsythe 1985), and grasshoppers (Gangwere 1960, 1965, 1967, 1973, 2017). However, some differences were observed. Cockroaches salivate the food outside the mouth cavity, which is achieved by dropping the saliva on the food particles before transporting them in the cibarium. Similar extra-oral digestion is also found in other insect groups such as Neuroptera, Coleoptera, and among at least 79% of all predaceous land-dwelling arthropods (Cohen 1995; Evans 1964; Evans and Forsythe 1985). In other groups such as Dermaptera, Odonata, and Orthoptera, salivation takes place between the mandibles (Cohen 1995; Gangwere 1960, 1967; Popham and Bevans 1979; Popham 1959). For a range of ground beetle species, Evans and Forsythe (1985) demonstrated a smooth transition between the two extreme positions of fluid and solid food intake. The intermediate stages are described as semi-fluid, mush, or fragment feeders, depending on the degree of extra-oral digestion. *Periplaneta*, the only cockroach for which the feeding process is known, salivates pre-orally but feeds on at least partially solid food (Popham 1961a) and thus also belongs to these intermediate stages.

During feeding, the mouthparts of *Periplaneta* act in a highly coordinated manner and the movements of the maxillae alternate with that of the mandibles in order to push the food into the cibarium. A similar pattern is also found in various other insects, but there are exceptions. In orthopterans and lepidopteran larvae for example, the maxillae play only a small role in collecting food (Popham 1959). Instead, the food is forced through the cibarium by depressing the head (Popham 1959). The tiger beetle *Cicindela hybrida*, which relies on liquid feeding, chews the soft parts of the prey with the mandibular molae. During this chewing, the maxillae move only slightly and, with the help of the labial palps, hold the food in place (Evans 1965). A similar decoupling of the mandibles and the maxillae might also be found in other liquid-feeding insects with biting and chewing mouthparts.

A major difference between cockroaches and many predators such as some beetles is that the latter use their mandibles to catch and hold prey. Other carnivores such as mantids grab and hold the food with their raptorial forelegs (Wieland 2013). The mouthparts are thus not involved in fixing the food and the animal can feed in any head position. The same is observed in adult dragon- and damselflies, which also use their legs for this purpose (Popham and Bevans 1979).

The general construction of the mandibular apparatus of neopteran insects with biting and chewing mouthparts is highly conserved. The mandibles are connected to the head via two ball and socket joints. In this respect, cockroaches reflect the neopteran ground plan (Wipfler et al. 2019). The non-neopteran winged insects—Odonata and Ephemeroptera—have less fixed mandibular joints and additional mandibular muscles (Blanke et al. 2012; Staniczek 2001). Moreover, cockroach mandibles are often asymmetrical, a feature which is also found in many other winged insects, with the left mandible constantly overlapping the right one (Chapman and de Boer 1995). Similarities are also found with regard to the setation of the mandibles of various other insects (Yuasa 1920). A derived feature of the mandible that cockroaches share with termites and mantids is the membranous postmola (Wipfler et al. 2016). The function of this structure is not yet clear.

Nevertheless, to some degree mandibles can reflect the dietary preferences of an animal. This applies particularly to the mesal mandibular edge including the incisivi and molar area [e.g., Bernays (1991) or Paranjape (1985)] and the length to width ratio [e.g., Evans and Forsythe (1985)]. Cockroaches have typical "omnivorous-type" mandibles with a variable number of sharp incisivi of uniform length and usually a flattened molar area, often delicately ridged (Gangwere 1965). Similar mandibles are also found in Tetrigidae and Tridactylidae (Gangwere 1965). Predatory species often use their mandibles for prey capture and show a wide set of adaptations toward the different diets (Evans and Forsythe 1985). However, in scavengers or carnivorous groups where prey is caught, subdued and held with specialised legs such as Mantodea, the mandibles have only very short incisivi and the molar region is strongly reduced and replaced by a long and sharp mesal ridge (Wipfler et al. 2011, 2012). The two mandibles thus act as the blades of a pair of scissors. The adaptation toward herbivory, forbivory, graminivory, seminivory, and dendrophagy is summarized in Gangwere (1965).

The mandibular musculature of cockroaches is identical to other lower neopteran insects (Wipfler et al. 2011). However, the non-neopteran pterygotan groups Odonata and Ephemeroptera have additional mandibular muscles that are missing in neopterans (Blanke et al. 2012, 2013; Staniczek 2001). Holometabolan groups usually lack M. hypopharyngomandibularis (which is functionally not a mandibular muscle; see above) and M. tentoriomandibularis lateralis inferior (Beutel et al. 2011). In cockroaches the mandibular adductor, M. craniomandibularis internus, is usually divided into eight bundles. *Tivia* is an exception from this rule; however, the entire muscle is strongly reduced here and deviates strongly from other cockroaches. Other insects with biting and chewing mouthparts show a similar compartmentalization, but the bundles are not identical and thus not comparable. A peculiarity of the cockroach M. craniomandibularis internus is the asymmetry of the two sides. The right muscle is usually larger than the left one and also enters the left half of the head capsule. This different size ratio has not been observed, described, or measured for closely related groups (Blanke et al. 2012; Friedemann et al. 2012; Wipfler et al. 2011, 2012).

One of the most significant measures of mandibular performance is the maximum voluntary bite force. This force is of vital importance specifically during chopping and grinding of tough food items, and during defensive behavior and prey capture. Here they determine the maximum toughness of exploitable food or prey and whether opponents can be fended off by biting and chewing or if it is better to elope. In rove and ground beetles with head widths between 2 and 4.25 mm, bite forces range between 0.13 N and 0.39 N, whereas bite forces do not correlate strictly to head widths (Wheater and Evans 1989). In much larger stag beetles with a head width of about 11.5 mm in males and 5.2 mm in females, maximum bite forces were also much larger (Goyens et al. 2014). Thus, males reached values of about 6.5 N. However, even the females, with their smaller mandibles and head widths generated bite forces of about 1.1 N which is about twice the value of the similarly sized American cockroach. Since the mandible adductors of female stag beetles have an effective cross-section area of about 14 mm$^2$ (Goyens et al. 2014), they can reach such strong bite forces even with relatively weak muscle fibers with a maximum muscle stress of only about 17 N cm$^{-2}$.

Compared to other hemimetabolous species, omnivorous cockroaches seem to have particularly small mandible adductors (cp. Weihmann et al. 2015a). In *Periplaneta*, the mandible adductors represented about 14.6% $\pm$ 3.8% of the entire volume of the head capsule; in the likewise omnivorous cockroach *Ergaula*, these muscles occupy only 11.9% of the head volume. With the exception of the mantid *Hymenopus coronatus* (16.5%), all other xylophagous, insectivorous, and herbivorous species studied in Weihmann et al. (2015a) had distinctly higher values. In *Hymenopus*, the low value most likely results from the large cone-like compound eyes, which strongly increase the volume of the head capsule (Wipfler et al. 2012), skewing the ratio downward. In Weihmann et al. (2015a), the highest muscle volumes were found in the wood-feeding cockroach *Salganea* (30%) and the herbivorous phasmatodean *Phyllium siccifolium* (29%). Similar values were documented by Li et al. (2011) for carnivorous rove beetles. They seem to also have relatively large mandibular adductor muscles that take up about 26% to 33% of the head capsule's volume. Even larger muscles were found in ant workers (Paul 2001), where the adductors can occupy up to 66% of the head capsule volume.

The ratios between the volumes of the adductor and the abductor muscle vary considerably among differently adapted species. Higher ratios seem to be reserved for species specialized in tough food. The highest ratio between the volumes of the adductor and abductor muscles was found in herbivorous phasmatodeans (23 in *Phyllium* and 20 in *Timema*) followed by the xylophagous cockroach *Salganea* (18) (Weihmann et al. 2015a). In the omnivorous roach *Ergaula* as well as the xylophagous and herbivorous species, the adductor is at least 11 times bigger than the abductor. Relatively low ratios of about 8 were found in *Periplaneta* and in the insectivorous mantid *Hymenopus* (about 9). Since lower ratios and accordingly relatively stronger mandible abductor muscles might facilitate higher bite frequencies when repeated biting and chewing is required, or faster reopening when a predatory strike missed, low ratios are probably particularly useful in species with a diet comprising a significant part of mobile animal-source food. Li et al. (2011) found volume ratios between 6.3 and 12.2 in three species of predatory rove beetles with distinctly differing mandible morphologies and diets.

Variations about the ratios between in- and out-levers of the mandibles, i.e., the mechanical advantage (MA), used by insects does not generally seem to clearly reflect feeding habits. Thus, in omnivorous, carnivorous, and xylophagous dictyopterans, the values for specific incisivi range between 0.55 for proximal teeth and 0.37 for distal teeth (Table 7.2). Nevertheless, wood eating of the termite *Mastotermes* might be also facilitated by high lever ratios in the mandibles (Table 7.2). For a given muscle force, small MAs result in smaller bite forces. Thus, distal teeth or generally slender mandibles have relatively low capacities for generating strong bite forces. However, the potential maximum velocity of the mandible tip increases. Higher tip speeds in turn do not affect the time needed to close the mandibles. For higher closing rates, muscles with higher contraction speeds are required (see above). Nevertheless, to some degree, MA values seem to allow insights into the preferred ways mandibles are used. The detritivore larva of the beetle *Liocola* (Scarabaeidae) exhibits a rather high MA of 0.54 (Gorb and Beutel 2000). In contrast, the mandibles of predatory aquatic beetle larvae of the species *Hydrophilus* (Hydrophilidae) and *Cybister* (Dytiscidae), relying on fast attacks on rather soft-bodied prey, are comparatively slender and have low MA values of 0.28 and 0.26, respectively (Gorb and Beutel 2000). Additionally, the mandibles of *Hydrophilus* larvae have a prominent cutting ridge (retinaculum) at their median central part, which allows the cracking of snail shells by taking advantage of the much higher mechanical advantage (Gorb and Beutel 2000). Male stag beetles (Goyens et al. 2014), which use their elongated mandibles in their notorious fights for mating opportunities, have also the need for fast actions and a wide as possible gape. Thus, depending on the bite position, MA values of male mandibles range from 0.13 to 0.28, while the MA of female mandibles is about 0.34. In carnivorous ground and rove beetles, MA values range between 0.35 and 0.59 (Wheater and Evans 1989) and 0.18 and 0.46 (Li et al. 2011), respectively. Both examinations present the MA of the distal-most teeth of the mostly single-toothed predatory mandibles.

The epipharynx of cockroaches generally follows the pterygotan ground plan in being membranous, having microtrichia, an epipharyngeal brush, and baso-lateral tormae, although very little is known about this structure in many groups. However, the presence of an epipharyngeal suspensorium and spikes is not present in other studied polyneopteran insects.

The morphology of the cockroach maxillae also follows the general design of winged insects with the exception of the lacinia that lies in a galeal cavity that is only found in dictyopterans. The same applies to the movement of the maxilla including the indirect protraction that was also described for other insects including earwigs (Popham 2000), beetles (Evans 1964), and also collembolans (Singh 1963; cited in Evans 1964). An interesting feature of the cockroach lacinia is the presence of the lacinula. Similar structures are also found in Orthoptera and Plecoptera (Crampton 1923; Eidmann 1923), and their homology with a lamella in apterygote insects and the dentisetae of Odonata and Ephemeroptera has been discussed (Crampton 1923; Wipfler et al. 2016). Thus, cockroaches would represent the pterygotan ground plan also in this respect. The lacinial incisivi in cockroaches are elongate and sharp. According to Gangwere (1965), this shape is found in omnivore (detritovore), forbivore, and carnivore species while they are blunt in graminivores, seminivores, and clendrophages. Additionally the structure of the galeae (at least in grasshoppers) seems to also correlate with food habits (Gangwere 1960, 1965). The maxillary musculature, however, follows the pterygotan ground plan (Blanke et al. 2012; Wipfler et al. 2011).

The insect labium is generally quite uniform in most insects with biting and chewing mouthparts (Yuasa 1920), and the morphology in cockroaches can be generalized to other groups. Some insects including cockroaches have a labium with a median cleft. Unlike the labral emargination, this cleft does not seem to be correlated with diet or margin feeding but rather is a result of the organ's dual origin from appendages of the two body sides (Gangwere 1965). The same seems to apply to the postmentum and its division in some cockroaches and other insects (Yuasa 1920). Rather than being based on a dietary difference, this division rather correlates with the ratio between width and length of the structure (see above; Zhuzhikov 2007). The labial musculature in cockroaches including the functional implications also reflects the ground plan for Pterygota and Neoptera, although Odonata show some modifications (Blanke et al. 2012).

The hypopharynx is arguably the least studied of all mouthparts in both morphology and function. To our knowledge, there is no detailed study concerning its movement and this study is the first one to specifically address the role of the hypopharyngeal muscles and the movement they may cause. As the musculature in cockroaches is representative of various other groups of insects with biting and chewing mouthparts (Wipfler et al. 2011), we assume that the movement pattern we propose here can be considered as the ground pattern for neopteran or even winged insects. However, detailed kinematic studies of the hypopharynx should be conducted to verify our theoretical assumptions.

As explained in the introduction, the diet of most cockroaches can be best summarized as detritivorous, although many species are strongly specialized in their feeding behavior (Bell et al. 2007). A similar detritivore diet is found in primarily apterygote insects such as silverfish and bristletails, and it is an intriguing question whether this also represents the ancestral condition for the winged insects. However, the question is difficult to address as many groups at the base of the winged insects show a highly derived condition: dragon- and damselflies are strongly specialized aerial predators and adult mayflies usually do not feed at all. An obvious difference related to diet between cockroaches and other insects including most termites is the presence of the intestinal symbiont *Blattabacterium*. It is an endocellular bacteroid that supplies cockroaches with amino acids and transforms nitrogenous waste products into usable substrates (Patino-Navarrete et al. 2013; Sabree et al. 2009). The constant access to these resources is of great benefit in a nitrogen-poor detritivores diet. Its distribution and evolutionary origin among cockroaches are discussed in Evangelista et al. (2019).

In summary, the cockroach biting and chewing apparatus represents many of the ancestral conditions for winged insects with adaptations toward a detritivore or general diet. It thus represents a valuable model system for understanding the function and the mechanics of insect feeding. This is also reflected by the fact that the American cockroach *Periplaneta* might be the best studied insect with respect to its feeding apparatus.

### 7.3.2 *The Insect Feeding Apparatus in Comparison to Other Biting and Chewing Structures Found in Arthropods*

Mandibles are the determining characteristic of Mandibulata, comprising myriapods, crustaceans, and insects. Nevertheless, in several crustacean groups mandibles are not the strongest biting and chewing structures. Many decapod species have chelipeds ending in strong pincers that can exert considerable forces. Due to relatively large body sizes, some of these species can generate the strongest absolute bite forces of all arthropods (Taylor 2000). Mechanically, these chelae have many similarities to the pincers of scorpions and the scissor-shaped chelicerae of solifuges, i.e., the mouthparts of the latter. In insects and myriapods as well as in many smaller crustaceans, however, mandibles are the only mouthparts used for grinding larger food items. Therefore, mandibles, i.e., two independently driven blades working against each other, should provide some advantages over biting and chewing systems like chelicerae and chelae that possess only one moving component. Thus, the independent but highly synchronized activity of mandible adductors reduces the required length changes of the driving muscles when the animals encompass and cut food pieces of a certain size. That is because the working ranges of both sides add up, and high opening angles can be achieved with only half the muscular length change required by structures like chelae or chelicerae (Fig. 7.15). As a consequence, mandibular adductors can act closer to their optimum length, which results in increased efficiency. Moreover, the effective closing velocity is higher for a given mean sarcomere and fiber length because the velocities of both sides also add up. If no particularly fast actions are needed, the muscles can still act closer to isometric conditions, which increase the maximum biting and chewing force or efficiency again (Crow and Kushmerick 1982; Hill 1938; Wendt and Gibbs 1974).

The volume of driving muscles cannot easily be increased in mandibular biting and chewing systems without concurrently changing head dimensions in general (see Goyens et al. (2014) for an extreme example: the evolution of oversized mandibles in stag beetles is driven by a strong selection pressure due to male–male fights for reproductive success). As mandible muscles are situated within the head capsule, they may compete for space with other essential organs, like the central nervous, the digestive, or the endo- and exocrine system. Moreover, hypertrophic mandible muscles can also lead to static imbalances, impeding the locomotor skills of an organism (Goyens et al. 2015). The driving muscles of pincers, chelae, and chelicerae are found outside the head capsule. Therefore, head size is not affected by changing demands in these biting and chewing systems. Moreover, the pincers of scorpions and decapod crustaceans form the distal-most parts of multi-segmented limbs, which significantly increase their reach. However, scorpions and decapod crustaceans also have mouthparts for grinding up food items in small, digestible pieces; in the crustaceans, these specialized mouthparts are mandibles.

In vertebrates, maximum muscle stress values are very similar for all striated skeletal muscles and always close to 30 N/cm$^2$ (Alexander 1985). In arthropods, however, maximum stress varies considerably even within a single specimen, depending on the major purpose of a certain muscle (Ahn and Full 2002; Guschlbauer et al. 2007; Siebert et al. 2010). Their values range from about 8 N/cm$^2$ in the legs of stick insects up to more than 100 N/cm$^2$ in crayfish pincers (Siebert et al. 2010; Taylor 2000; van der Meijden et al. 2012). Particularly, muscles that drive biting and chewing devices like mandibles, chelae, or chelicerae are often characterized by high maximum stress values. Thus, the chelicerae of a soil digging solifugid are driven by muscles with a maximum stress value of about 90 N/cm$^2$ (van der Meijden et al. 2012), and the closer muscles of some decapod chelae may even exceed 100 N/cm$^2$ (Taylor 2000). In these decapods' chelae, the muscles' sarcomeres can reach lengths of up to 17 μm. However, sarcomere lengths in the mandible closer muscles of insects are mostly shorter (e.g., Gronenberg et al. 1997). For example, the sarcomere length in the closer muscle of *Periplaneta* is about 5 μm and those in the closer muscle of *Schistocerca* are about 7 μm (TW unpublished material). Accordingly, maximum stress values of insect mandible adductors do not reach such high values. Nevertheless, even omnivore feeders like American cockroaches reach values of up to 58 N/cm$^2$ (Weihmann et al. 2015b). In the mandibular system of male stag beetles, apart from strength, contraction speed is also of importance. Since faster muscles have shorter sarcomeres (e.g., Crow and Kushmerick 1982; Wendt and Gibbs 1974) allowing for concatenation of a high number of contractile unites, maximum stress values of these muscles are lower. Thus, as fast actions are rewarded in fighting male stag beetles, maximum stress values of the mandible adductors are as low as 17–18 N/cm$^2$ (Goyens et al. 2014) requiring the development of very large muscles and heads to reach sufficiently high absolute force values.

**Acknowledgments** We would like to thank Harald Krenn for the invitation to write this chapter. Additional thanks go to Kevin Weißing, Tobias Theska, and Alexander Blanke. The present study was supported by the Deutsche Forschungsgemeinschaft (WE 4664/4-1 & WI 4324/1-1) and a Postdoc stipend of the Daimler und Benz Stiftung (32-10/12 to BW).

## References

Ache J, Matheson T (2013) Passive joint forces are tuned to limb use in insects and drive movements without motor activity. Curr Biol 23:1418–1426

Ahn AN, Full RJ (2002) A motor and a brake: two leg extensor muscles acting at the same joint manage energy differently in a running insect. J Exp Biol 205:379–389

Alexander RM (1985) The maximum forces exerted by animals. J Exp Biol 115:231–238

Altner H (1975) The microfiber texture in a specialized plastic cuticle area within a sensillum field on the cockroach maxillary palp as revealed by freeze fracturing. Cell Tissue Res 165:79–88

Altner H, Stetter H (1982) The sensory system of the maxillary palps in the cockroach *Periplaneta americana*. Verh Dtsch Zool Ges 75:247

Andersen SO, Weis-Fogh T (1964) Resilin. A rubberlike protein in arthropod cuticle. Adv Insect Physiol 2:1–65

Bell WJ, Roth LM, Nalepa CA (2007) Cockroaches: ecology, behavior, and natural history. Johns Hopkins University Press, Baltimore

Bennet-Clark HC (1975) The energetics of the jump of the locust *Schistocerca gregaria*. J Exp Biol 63:53–83

Bernays E (1985) Regulation of feeding behaviour. In: Kerkut G, Gilbert L (eds) Comprehensive insect physiology biochemistry and pharmacology, vol 4; Regulation: digestion, nutrition, excretion. Pergamon, Oxford, pp 1–32

Bernays EA (1991) Evolution of insect morphology in relation to plants. Philos Trans R Soc Lond B 333:257–264

Bernays E, Simpson S (1982) Control of food intake. Adv Insect physiol 16:59–118

Betz O, Thayer MK, Newton AF (2003) Comparative morphology and evolutionary pathways of the mouthparts in spore-feeding Staphylinoidea (Coleoptera). Acta Zool Stockholm 84:179–238

Beutel RG et al (2011) Morphological and molecular evidence converge upon a robust phylogeny of the megadiverse Holometabola. Cladistics 27:341–355

Beutel RG, Friedrich F, Yang X-K, Ge S-Q (2014) Insect morphology and phylogeny: a textbook for students of entomology. Walter de Gruyter, Berlin

Bland RG, Slaney DP, Weinstein P (1998) Mouthpart sensilla of cave species of Australian *Paratemnopteryx* cockroaches (Blattaria: Blattellidae). Int J Insect Morphol Embryol 27:291–300

Blanke A, Wipfler B, Letsch H, Koch M, Beckmann F, Beutel R, Misof B (2012) Revival of Palaeoptera – head characters support a monophyletic origin of Odonata and Ephemeroptera (Insecta). Cladistics 28:560–581

Blanke A, Greve C, Wipfler B, Beutel RG, Holland BR, Misof B (2013) The identification of concerted convergence in insect heads corroborates Palaeoptera. Syst Biol 62:250–263

Blanke A, Pinheiro M, Watson PJ, Fagan MJ (2018) A biomechanical analysis of prognathous and orthognathous insect head capsules: evidence for a many-to-one mapping of form to function. J Evol Biol 31:665–674

Buder G, Klass K-D (2013) A comparative study of the hypopharynx in Dictyoptera (Insecta). Zool Anz 252:383–403

Burry RW, Moran D (1973) Sense organs on the cockroach maxillary palp. Naturwissenschaften 60:521–521

Campbell SG, Campbell KS (2011) Mechanisms of residual force enhancement in skeletal muscle: insights from experiments and mathematical models. Biophys Rev 3:199–207

Carbonaro I (1949) I muscoli del capo di *Mantis religiosa.* Annuario dell'Istituto e Museo di Zoologia dell'Università 1:1–15

Chapman R (1995) Mechanics of food handling by chewing insects. In: Chapman R, de Boer G (eds) Regulatory mechanisms in insect feeding. Chapman & Hall, New York, pp 3–31

Chapman RF, de Boer G (1995) Regulatory mechanisms in insect feeding. Springer, Heidelberg

Chaudonneret J (1951) La morphologie céphalique de *Thermobia domestica* (Packard) (Insects Aptérygote Thysanoure). Ann Sci Nat 12:145–302

Chesler M, Fourtner CR (1981) Mechanical properties of a slow muscle in the cockroach. J Neurobiol 12:391–402

Chopard L (1932) Un cas de micropthalmie liée a l'atrophie des ailes chez une blatte cavernicole. Livre du Centenaire, Société Entomologique de France, pp 485–496

Clissold FJ (2007) The biomechanics of chewing and plant fracture: mechanisms and implications. Adv Insect Physiol 34:317–372

Cohen AC (1995) Extra-oral digestion in predaceous terrestrial Arthropoda. Annu Rev Entomol 40:85–103

Crampton G (1923) A phylogenetic comparison of the maxillae throughout the orders of insects. J NY Entomol Soc 31:77–107

Cribb BW, Stewart A, Huang H, Truss R, Noller B, Rasch R, Zalucki MP (2008) Insect mandibles—comparative mechanical properties and links with metal incorporation. Naturwissenschaften 95:17–23

Crow MT, Kushmerick MJ (1982) Chemical energetics of slow- and fast-twitch muscles of the mouse. J Gen Physiol 79:147–166

David S, Funken J, Potthast W, Blanke A (2016) Musculoskeletal modelling of the dragonfly mandible system as an aid to understanding the role of single muscles in an evolutionary context. J Exp Biol 219:1041–1049

Dorsey CK (1943) The musculature of the labrum, labium, and pharyngeal region of adult and immature Coleoptera. Smithson Misc Collect 103:1–40

Eidmann H (1923) Das sogenannte 'Praemaxillare'der Insekten. Zool Anz 58:43–52

Evangelista DA, Wipfler B, Bethoux O, Donath A, Fujita M, Kohli MK, Legendre F, Liu S, Machida R, Misof B, Peters RS, Podsiadlowski L, Rust J, Schuette K, Tollenaar W, Ware J, Wappler T, Zhou X, Meusemann K, Simon S (2019) An integrative phylogenomic approach illuminates the evolutionary history of cockroaches and termites (Blattodea). Proc Biol Sci 286 (1895):20182076. https://doi.org/10.1098/rspb.2018.2076

Evans M (1964) A comparative account of the feeding methods of the beetles *Nebria brevicollis* (F.) (Carabidæ) and *Philonthus decorus* (Grav.) (Staphylinidæ). Earth Environ Sci Trans R Soc Edinb 66:91–109

Evans M (1965) The feeding method of *Cicindela hybrida* L.(Coleoptera: Cicindelidae). Proc Royal Entomol Soc 40:61–66

Evans M, Forsythe T (1985) Feeding mechanisms, and their variation in form, of some adult ground-beetles (Coleoptera: Caraboidea). J Zool 206:113–143

Fontes LR (1986) Morphology of the alate and worker mandibles of the soil-feeding nasute termites (Isoptera, Termitidae, Nasutitermitinae) from the Neotropical region. Rev Bras Zool 3:503–531

Friedemann K, Wipfler B, Bradler S, Beutel RG (2012) On the head morphology of *Phyllium* and the phylogenetic relationships of Phasmatodea (Insecta). Acta Zool-Stockholm 93:184–199

Gangwere S (1960) The use of the mouthparts of Orthoptera during feeding. Entomol News 71:193–206

Gangwere S (1965) The structural adaptations of mouthparts in Orthoptera and Allis. Eos 41:65–87

Gangwere S (1967) The feeding behavior of *Atlanticus testaceus* (Orthoptera: Tettigoniidae). Ann Entomol Soc Am 60:74–81

Gangwere S (1973) Food selection and feeding behaviour in Iberian Orthopteroidea. Anales del Instituto Nacional de Investigaciones Agrarias Ser Produccion Vegetal

Gangwere S (2017) Relationships between the mandibles, feeding behavior, and damage inflicted on plants by the feeding of certain acridids (Orthoptera). Great Lakes Entomol 1:3

Gorb S, Beutel RG (2000) Head-capsule design and mandible control in beetle larvae: a three-dimensional approach. J Morphol 244:1–14

Goyens J, Dirckx J, Dierick M, Van Hoorebeke L, Aerts P (2014) Biomechanical determinants of bite force dimorphism in *Cyclommatus metallifer* stag beetles. J Exp Biol 217:1065–1071. https://doi.org/10.1242/jeb.091744

Goyens J, Dirckx J, Aerts P, Davidowitz G (2015) Costly sexual dimorphism in *Cyclommatus metallifer* stag beetles. Funct Ecol 29:35–43

Grimaldi D, Engel MS (2005) Evolution of the insects. Cambridge University Press, New York, NY

Gronenberg W, Paul J, Just S, Hölldobler B (1997) Mandible muscle fibers in ants: fast or powerful? Cell Tissue Res 289:347–361

Guschlbauer C, Scharstein H, Büschges A (2007) The extensor tibiae muscle of the stick insect: biomechanical properties of an insect walking leg muscle. J Exp Biol 210:1092–1108

Haeufle DFB, Günther M, Wunner G, Schmitt S (2014) Quantifying control effort of biological and technical movements: an information-entropy-based approach. Phys Rev E 89:012716

Hare L (1937) Termite phylogeny as evidenced by soldier mandible development. Ann Entomol Soc Am 30:459–486

Herzog W, Leonard T, Renaud J, Wallace J, Chaki G, Bornemisza S (1992) Force-length properties and functional demands of cat gastrocnemius, soleus and plantaris muscles. J Biomech 25:1329–1335

Hill AV (1938) The heat of shortening and the dynamic constants of muscle. Proc R Soc London B 126:136–195

Hooper SL, Guschlbauer C, Blümel M, Rosenbaum P, Gruhn M, Akay T, Büschges A (2009) Neural control of unloaded leg posture and of leg swing in stick insect, cockroach, and mouse differs from that in larger animals. J Neurosci 29:4109–4119

Hoyle G (1978) Distributions of nerve and muscle fibre types in locust jumping muscle. J Exp Biol 73:205–233

Inward D, Beccaloni G, Eggleton P (2007) Death of an order: a comprehensive molecular phylogenetic study confirms that termites are eusocial cockroaches. Biol Lett 3:331–335

Ker RF, Alexander RM, Bennett MB (1988) Why are mammalian tendons so thick? J Zool Lond 216:309–324

Klass K-D, Eulitz U (2007) The tentorium and anterior head sulci in Dictyoptera and Mantophasmatodea (Insecta). Zool Anz 246:205–234

Klein U (1982) The articulation of cricket palps: morphology and movement patterns in behaviour. Physiol Entomol 7:297–314

Lambin M (1973) Les sensilles de l'antenne chez quelques blattes et en particulier chez *Blaberus craniifer* (Burm.). Z Zellforch Microsk Anat 143:183–206

Levereault P (1938) The morphology of the Carolina mantis. Section II: The musculature. Kans Univ Sci Bull 25:578–633

Li D, Zhang K, Zhu P, Wu Z, Zhou H (2011) 3D configuration of mandibles and controlling muscles in rove beetles based on micro-CT technique. Anal Bioanal Chem 401:817–825

Mangan J (1908) On the mouth-parts of some Blattidae. Proc R Ir Acad B 27:1–10

Matsuda R (1965) Morphology and evolution of the insect head. Mem Am Entomol Inst 4:1–334

Matsumura Y et al (2015) Cephalic anatomy of *Zorotypus weidneri* New, 1978: new evidence for a placement of Zoraptera. Arthropod Syst Phylogeny 73:85–105

Miall LC, Denny A (1886) The structure and life-history of the cockroach *Periplaneta orientalis*, vol 3. L. Reeve & Company, London

Michels J, Gorb S (2012) Detailed three-dimensional visualization of resilin in the exoskeleton of arthropods using confocal laser scanning microscopy. J Microsc 245:1–16

Misof B, Liu S, Meusemann K, Peters RS, Donath A, Mayer C, Frandsen PB, Ware J, Flouri T, Beutel RG, Niehuis O, Petersen M, Izquierdo-Carrasco F, Wappler T, Rust J, Aberer AJ, Aspöck U, Aspöck H, Bartel D, Blanke A, Berger S, Böhm A, Buckley T, Calcott B, Chen J, Friedrich F, Fukui M, Fujita M, Greve C, Grobe P, Gu S, Huang Y, Jermiin LS, Kawahara AY, Krogmann L, Kubiak M, Lanfear R, Letsch H, Li Y, Li Z, Li J, Lu H, Machida R, Mashimo Y, Kapli P, McKenna DD, Meng G, Nakagaki Y, Navarrete-Heredia JL, Ott M, Ou Y, Pass G, Podsiadlowski L, Pohl H, von Reumont BM, Schütte K, Sekiya K, Shimizu S, Slipinski A, Stamatakis A, Song W, Su X, Szucsich NU, Tan M, Tan X, Tang M, Tang J, Timelthaler G, Tomizuka S, Trautwein M, Tong X, Uchifune T, Walzl MG, Wiegmann BM, Wilbrandt J, Wipfler B, Wong TK, Wu Q, Wu G, Xie Y, Yang S, Yang Q, Yeates DK, Yoshizawa K, Zhang Q, Zhang R, Zhang W, Zhang Y, Zhao J, Zhou C, Zhou L, Ziesmann T, Zou S, Li YXX, Zhang Y, Yang H, Wang J, Wang J, Kjer KM, Zhou X (2014) Phylogenomics resolves the timing and pattern of insect evolution. Science 346:763–767

Mkhize S, Kumar R (1972) Histology and physiology of the mandibular glands in Dictyoptera. J Entomol A 46:161–165

Moulins M (1967) Les sensilles de l'organe hypopharyngien de *Blabera craniifer* Burm. (Insecta, Dictyoptera). J Ultrastruct Res 21:474–513

Moulins M (1971a) La cavite preorale de *Blabera craniifer* Burm. (Insecte, Dictyoptere) et son innervation: etude anatomo-histologique de l'epipharynx et l'hypopharynx. Zool Jahrb Abt Allg Zool Physiol Tiere 88:527–586

Moulins M (1971b) Ultrastructure et physiologie des organes épipharyngiens et hypopharyngiens (chimiorécepteurs cibariaux) de *Blabera craniifer* Burm (Insecte, Dictyoptère). Z Vgl Physiol 73:139–166

Moulins M (1974) Récepteurs de tension de la région de la bouche chez *Blaberus craniifer* Burmeister (Dictyoptera: Blaberidae). Int J Insect Morphol Embryol 3:171–192

Neubert D, Simon S, Beutel RG, Wipfler B (2017) The head of the earwig *Forficula auricularia* (Dermaptera) and its evolutionary implications. Arthropod Syst Phylogeny 75:99–124

O'Donnell MJ (1977) Hypopharyngeal bladders and frontal glands: novel structures involved in water-absoprtion in the desert cockroach *Arenivaga investigata*. Am Zool 17:902–902

O'Donnell M (1981) Fluid movements during water-vapour absorption by the desert burrowing cockroach, *Arenivaga investigata*. J Insect Physiol 27:877–887

O'Donnell M (1982) Water vapour absorption by the desert burrowing cockroach, *Arenivaga investigata*: evidence against a solute dependent mechanism. J Exp Biol 96:251–262

Paranjape S (1985) Behavioural analysis of feeding and breeding in Orthopteran insects. Proc Anim Sci 94:265–282

Patino-Navarrete R, Moya A, Latorre A, Pereto J (2013) Comparative genomics of *Blattabacterium cuenoti*: the frozen legacy of an ancient endosymbiont genome. Genome Biol Evol 5:351–361

Paul J (2001) Mandible movements in ants. Comp Biochem Physiol A Mol Integr Physiol 131:7–20. https://doi.org/10.1016/S1095-6433(01)00458-5

Paul J, Gronenberg W (1999) Optimizing force and velocity: mandible muscle fibre attachments in ants. J Exp Biol 202:797–808

Paul J, Gronenberg W (2002) Motor control of the mandible closer muscle in ants. J Insect Physiol 48:255–267

Popham EJ (1959) The anatomy in relation to feeding habits of Forficula auricularia L. and other Dermaptera. Proc Zool Soc Lond 133:251–300

Popham E (1961a) The functional morphology of the mouthparts of the cockroach *Periplaneta americana* L. Entomologiste 94:185–192

Popham EJ (1961b) On the systematic position of *Hemimerus* Walker—a case for ordinal status. Syst Entomol 30:19–25

Popham EJ (1962) The anatomy related to the feeding habits of *Arixenia* and *Hemimerus* (Dermaptera). Proc Zool Soc Lond 139:429–450

Popham EJ (2000) The geographical distribution of the Dermaptera (Insecta) with reference to continental drift. J Nat Hist 34:2007–2027

Popham E, Bevans E (1979) Functional morphology of the feeding apparatus in larval and adult *Aeshna juncea* (L.) (Anisoptera: Aeshnidae). Odonatologica 8:301–318

Pradl W-D (1971) *Blaberus giganteus*: Schaben; eine Präparationsanleitung im Vergleich mit anderen Blattodea. Gustav Fischer Verlag, Stuttgart

Prakash S, Mendki M, Rao K, Singh K, Singh RN (1995) Sensilla on the maxillary and labial palps of the cockroach *Supella longipalpa* Fabricius (Dictyoptera: Blattellidae). Int J Insect Morphol Embryol 24:13–34

Pringle J (1938) Proprioception in insects: I. A new type of mechanical receptor from the palps of the cockroach. J Exp Biol 15:101–113

Ramaswamy S, Gupta A (1981) Sensilla of the antennae and the labial and maxillary palps of *Blattella germanica* (L.) (Dictyoptera: Blattellidae): their classification and distribution. J Morphol 168:269–279

Rast G, Bräunig P (2001a) Insect mouthpart motor patterns: central circuits modified for highly derived appendages? Neuroscience 108:167–176

Rast GF, Bräunig P (2001b) Feeding-related motor patterns of the locust suboesophageal ganglion induced by pilocarpine and IBMX. J Insect Physiol 47:43–53

Roberts M (1972) Structure and function of the maxillary appendages in the American cockroach, *Periplaneta americana* (L.). Forma et Functio Int J Funct Biol 5:253–272

Sabree ZL, Kambhampati S, Moran NA (2009) Nitrogen recycling and nutritional provisioning by *Blattabacterium*, the cockroach endosymbiont. Proc Natl Acad Sci U S A 106:19521–19526

Schmitt C, Rack A, Betz O (2014) Analyses of the mouthpart kinematics in *Periplaneta americana* (Blattodea, Blattidae) using synchrotron-based X-ray cineradiography. J Exp Biol 217:3095–3107

Schofield RM, Nesson MH, Richardson KA (2002) Tooth hardness increases with zinc-content in mandibles of young adult leaf-cutter ants. Naturwissenschaften 89:579–583

Seath I (1977) The effects of increasing mandibular load on electrical activity in the mandibular closer muscles during feeding in the desert locust, *Schistocerca gregaria*. Physiol Entomol 2:237–240

Seid MA, Scheffrahn RH, Niven JE (2008) The rapid mandible strike of a termite soldier. Curr Biol 18:R1049–R1050

Seidl S (1991) Structure and differentiation of the sensilla of the ventral sensory field on the maxillary palps of *Periplaneta americana* (Insecta, Blattodea), paying special attention to the ciliogenesis of the sensory cells. Zoomorphology 111:35–47

Seifert G (1970) Entomologisches Praktikum, vol QL494. S44 1975. Georg Thieme Verlag, Stuttgart

Siebert T, Weihmann T, Rode C, Blickhan R (2010) *Cupiennius salei*: biomechanical properties of the tibia-metatarsus joint and its flexing muscles. J Comp Physiol B 180:199–209

Singh SB (1963) Studies on the feeding mechanisms of Collembola. University of Wales

Smith J (1985) Feeding mechanisms. In: Kerkut G, Gilbert L (eds) Comprehensive insect physiology biochemistry and pharmacology, vol 4, Regulation: digestion, nutrition and excretion. Pergamon, Oxford, pp 33–68

Snodgrass RE (1943) The feeding apparatus of biting and chewing and disease carrying fleas: a war-time contribution to medical entomology. Smithson Misc Collect 104:1–51

Snodgrass RE (1944) The feeding apparatus of biting and chewing and sucking insects affecting man and animals. Smithson Misc Collect 104:1–102

Staniczek AH (2001) Der Larvenkopf von *Oniscigaster wakefieldi* McLachlan, 1873 (Insecta: Ephemeroptera: Oniscigastridae); ein Beitrag zur vergleichenden Anatomie und Phylogenie der Eintagsfliegen. Eberhard-Karls-Universität

Stevenson PA, Rillich J (2016) Controlling the decision to fight or flee: the roles of biogenic amines and nitric oxide in the cricket. Curr Zool 62:265–275

Stillwell RC, Blanckenhorn WU, Teder T, Davidowitz G, Fox CW (2010) Sex differences in phenotypic plasticity affect variation in sexual size dimorphism in insects: from physiology to evolution. Annu Rev Entomol 55:227–245

Stokes DR, Vitale AJ, Morgan CR (1979) Enzyme histochemistry of the mesocoxal muscles of *Periplaneta americana*. Cell Tissue Res 198:175–189

Strenger A (1939) Funktionelle Analyse des Orthopterenkopfes. Zool Jahrb Abt Allg Zool Physiol Tiere 75:1–72

Suslov S (1912) Über die Kopfdrüsen einiger niederer Orthopteren. Zool Jahrb Abt Allg Zool Physiol Tiere 34:96–120

Taylor GM (2000) Maximum force production: why are crabs so strong? Proc Biol Sci 267:1475–1480

van der Meijden A, Langer F, Boistel R, Vagovic P, Heethoff M (2012) Functional morphology and bite performance of raptorial chelicerae of camel spiders (Solifugae). J Exp Biol 215:3411–3418

Vishnoi H (1956) The structure, musculature and mechanism of the feeding apparatus of the various castes of the termite *Odontotermes obesus* (Rambur). Part I. Clypeo-labrum. J Zool Soc India 8:1–18

Vishnoi H (1962) Musculature of a common Indian termite *Odontotermes obesus* (Rambur) (Isoptera, Termitidae). Bull Entomol (Loyola College, Madras) 3:13–20

Wagner H, Blickhan R (1999) Stabilizing function of skeletal muscles: an analytical investigation. J Theor Biol 199:163–179

Wagner H, Blickhan R (2003) Stabilizing function of antagonistic neuromusculoskeletal systems: an analytical investigation. Biol Cybern 89:71–79

Walker E (1933) On the anatomy of *Grylloblatta compodeiformis* Walker. 2. Comparisons of head with those of other orthopteroid insects. Ann Entomol Soc Am 26:309–344

Weesner FM (1969) External anatomy. In: Krishna K, Weesner FM (eds) Biology of termites, vol 1. Academic, New York, pp 19–47

Weihmann T (2018) Leg force interference in polypedal locomotion. Sci Adv 4:eaat3721

Weihmann T, Siebert T, Blickhan R (2009) Muscle properties of the tibia–metatarsus joint flexors in the labidognath spider *Cupiennius salei*. Comp Biochem Phys A 153:124–125

Weihmann T, Kleinteich T, Gorb S, Wipfler B (2015a) Functional morphology of the chewing apparatus in the cockroach *Periplaneta americana* (Blattodea, Blattidae) – a model species for generalist omnivore insects. Arthropod Syst Phylogeny 73:477–488

Weihmann T, Reinhardt L, Weissing K, Siebert T, Wipfler B (2015b) Fast and powerful: biomechanics and bite forces of the mandibles in the American cockroach *Periplaneta americana*. PLoS One 10:e0141226. https://doi.org/10.1371/journal.pone.0141226

Wendt IR, Gibbs CL (1974) Energy production of mammalian fast- and slow-twitch muscles during development. Am J Phys 226:642–647

Wheater CP, Evans MEG (1989) The mandibular forces and pressures of some predacious coleoptera. J Insect Physiol 35:815–820

Wieland F (2013) The phylogenetic system of Mantodea (Insecta: Dictyoptera). Species Phylogeny Evol 3:3–222

Willey RB (1961) The morphology of the stomodeal nervous system in *Periplaneta americana* (L.) and other Blattaria. J Morphol 108:219–261

Wipfler B, Machida R, Muller B, Beutel RG (2011) On the head morphology of Grylloblattodea (Insecta) and the systematic position of the order, with a new nomenclature for the head muscles of Dicondylia. Syst Entomol 36:241–266

Wipfler B, Wieland F, DeCarlo F, Hörnschemeyer T (2012) Cephalic morphology of *Hymenopus coronatus* (Insecta: Mantodea) and its phylogenetic implications. Arthropod Struct Dev 41:87–100. https://doi.org/10.1016/j.asd.2011.06.005

Wipfler B, Weissing K, Klass KD, Weihmann T (2016) The cephalic morphology of the American cockroach *Periplaneta americana* (Blattodea). Arthropod Syst Phylogeny 74:267–297

Wipfler B, Letsch H, Frandsen PB, Kapli P, Mayer C, Bartel D, Buckley TR, Donath A, Edgerly-Rooks J, Fujita M, Liu S, Machida R, Mashimo Y, Misof B, Niehuis O, Peters RS, Peterson M, Podsiadlowski L, Schütte K, Shimizu S, Uchifune T, Wilbrandt J, Yan E, Zhou X, Simon S (2019) Evolutionary history of Polyneoptera and its implications for our understanding of early winged isects. Proc Natl Acad Sci U S A. https://doi.org/10.1073/pnas.1817794116

Wipfler B, Triesch F, Evangelista D, Weihmann T (under review) Morphological, functional and systematic aspects of the head capsule of the roach *Ergaula capucina* (Insecta/Blattodea)

Yap NWL, Lin Y, Todd PA (2013) Chelae force generation at variable gape sizes in the mud crab, *Scylla olivacea* (Brachyura: Portunidae). Nat Singapore 6:179–185
Yuasa H (1920) The anatomy of the head and mouth-parts of Orthoptera and Euplexoptera. J Morphol 33:251–307
Zhuzhikov D (2007) The mouthparts of cockroaches (Blattodea). Entomol Rev 87:25–37

# Chapter 8
# Physical Determinants of Fluid-Feeding in Insects

**Konstantin G. Kornev and Peter H. Adler**

**Abstract** Fluid feeders represent more than half of the world's insect species. We review current understanding of the physics of fluid-feeding, from the perspective of wetting, capillarity, and fluid mechanics. We feature butterflies and moths (Lepidoptera) as representative fluid-feeding insects. Fluid uptake by live butterflies is experimentally explained based on X-ray imaging and high-speed optical microscopy and is augmented by modeling and by mechanical and physicochemical characterization of biomaterials. Wetting properties of the lepidopteran proboscis are reviewed, and a classification of proboscis morphology and wetting characteristics is proposed. The porous and fibrous structure of the mouthparts is important in determining the dietary habits of fluid-feeding insects. The fluid mechanics of liquid uptake by insects cannot be explained by a simple Hagen–Poiseuille flow scenario of a drinking-straw model. Fluid-feeding insects expend muscular energy in moving fluid through the proboscis or through the sucking pump, depending primarily on the ratio of the proboscis length to the food canal diameter. A general four-step model of fluid-feeding is proposed, which involves wetting, dewetting, absorbing, and pumping. The physics of fluid-feeding is important for understanding the evolution of sucking mouthparts and, consequently, insect diversification through development of new fluid-feeding habits.

## 8.1 Introduction

The success of fluid-feeding insects on our "buggy" planet is unprecedented; more than half of all known insects on Earth—over 500,000 species—are fluid feeders (Grimaldi and Engel 2005; Adler and Foottit 2009; Shaw 2014). For about 300 million years, insects that feed on fluids have used unique feeding devices, termed "proboscises" to

K. G. Kornev (✉)
Department of Materials Science and Engineering, Clemson University, Clemson, SC, USA
e-mail: kkornev@clemson.edu

P. H. Adler
Department of Plant and Environmental Sciences, Clemson University, Clemson, SC, USA
e-mail: padler@clemson.edu

 
H. W. Krenn (ed.), *Insect Mouthparts*, Zoological Monographs 5,
https://doi.org/10.1007/978-3-030-29654-4_8

acquire water and nutriment (Kingsolver and Daniel 1995; Chapman 2013). Anatomical features of the proboscis are detailed in Chap. 3.

Fluid feeders have evolved to exploit a remarkable diversity of food resources including nectar, phloem, and xylem of plants, and blood, carrion, dung, sweat, tears, and urine of animals (Labandeira 1997, 2010; Misof et al. 2014). Lepidoptera provide many examples of opportunistic fluid-feeding (Fig. 8.1). Feeding devices of insects have been popular evolutionary subjects since Charles Darwin predicted that a sphinx moth with a 30-cm-long proboscis feeds from the long nectar spur of the orchid *Angraecum sesquipedale* (Darwin 1904; Arditti et al. 2012). Among the most attractive and demanding areas of research are the evolution and diversification of insect feeding organs within the context of their morphology, materials properties, and functional performance (Russell 1916; Lauder 2003; Krenn and Aspöck 2012).

The feeding devices of all fluid-feeding insects consist of a proboscis paired with a sucking pump. The geometry of sucking pumps is complex and many details are poorly understood (Bennet-Clark 1963; Davis and Hildebrand 2006; Bauder et al. 2013; Karolyi et al. 2013, 2014). The sucking pump, the proboscis, and their geometries exhibit a range of sizes, as represented by a selection of eight species with proboscises ranging from 0.3 mm to about 70 mm long (Table 8.1). Although the pump of these species is sizable relative to the head, the proboscis length varies greatly. Accordingly, the ratio of maximum size of the pump chamber to proboscis length changes from infinity to zero, suggesting that the mechanism of energy dissipation changes from one species to another.

Proboscises of fluid-feeding insects, such as butterflies, moths, house flies, and mosquitoes, have multiple functions enabling them to access floral nectaries, mop up liquid films and droplets, and pierce plant and animal tissues. Topographically sculptured surfaces, together with a fibrous and porous structure, are critical for multifunctionality of proboscises (Snodgrass 1935; Schmitt 1938; Eastham and Eassa 1955; Hepburn 1971, 1985; Harder 1986; Cheer and Koehl 1987; Hainsworth et al. 1991; Koehl 2001; Corbet 2000; Krenn 2010; Monaenkova et al. 2012; Abou-Shaara 2014; Tsai et al. 2014; Chen et al. 2015; Zhu et al. 2016b). This architecture offers a large surface area, suggesting that capillary and wetting forces, which pull water into a dry sponge, should also play an integral role in the fluid uptake strategies of sucking insects.

Insect evolution has involved increases and decreases of organ size (Rensch 1948; McMahon and Bonner 1983; Hanken and Wake 1993; Shaw 2014; Polilov 2015). Accordingly, feeding devices of insects exhibit a broad range of sizes, from extremely small in insects such as aphids (Barber 1924; Auclair 1963) to extraordinarily long in some sphinx moths (Arditti et al. 2012). Different behavioral strategies and physical and materials organization of the feeding devices are associated with this large size range.

The proboscis of fluid-feeding insects differs structurally among species. For example, the butterfly proboscis consists of two elongated components (i.e., galeae), joined together by linking mechanisms made of fence-like structures (Fig. 8.2, and Chap. 3). Fluid is transported through a food canal between the galeae (Eastham and Eassa 1955; Krenn 2010). The proboscises of bees, house flies, and mosquitoes have

**Fig. 8.1** Lepidoptera feeding opportunistically from various substrates in Peru, Madre de Dios Region, Las Piedras, ca. 235–275 m above sea level. (**a**) *Euselasia* sp. (Riodinidae) probing beneath the margin of a fingernail. (**b**) *Heliconius* sp. gathering pollen on its proboscis from a species of *Passiflora*. (**c**) *Euselasia toppini* (Riodinidae) probing on a leaf of the palm *Geonoma* sp. (**d**) *Syllectra* sp. (Erebidae; identified by P.Z. Goldstein) probing on drying laundry at night. (**e**) *Ithomia lichyi* (Nymphalidae) feeding on a dead caterpillar; a bead of saliva is present proximal to the "knee" bend. (**f**) The skipper *Tarsoctenus papias* (Hesperiidae) feeding on antbird droppings near an army ant raid (Photos courtesy of Allison Stoiser)

**Table 8.1** Characteristics of the proboscis and sucking pump of selected fluid-feeding insects

| Species | Proboscis length, $L_p$ (mm) | Food canal diameter, $d_p$ (μm) | Pump length[a], $L$ (μm) | Pump width[b], $W$ (μm) | Pump height[c], $h$ (μm) | $d_p/L$ |
|---|---|---|---|---|---|---|
| *Acherontia atropos*[d] | 10.1 | 494.6 | 2417.6 | 2030.3 | 765.9 | 0.2 |
| *Danaus plexippus*[e] | 14.4 | 35.0 | 886.2 | 606.3 | 145.6 | 0.03 |
| *Manduca sexta*[e] | 50–70 | 82.5 | 1834.5 | 2272.2 | 1227.4 | 0.04 |
| *Nadata gibosa* | 3.58 | 25.4 | 598.3 | 362.8 | 240.4 | 0.04 |
| *Symmerista albifrons*[f] | 0.35 | 17.3 | 298.8 | 358.8 | 91.0 | 0.12 |
| *Rhodnius prolixus*[g] | 5.2 | 8–10 | 3000–5000 | 280 | 160 | 0.02–0.03 |
| *Pediculus humanus*[h] | 0.4(f) 0.4(m) | 7.2 (f) 6.8(m) | 50(f) 43(m) | 50(f) 43(m) | 30(f) 30(m) | 0.14(f) 0.17(m) |
| *Cimex lectularius*[h] | 1.23(f) 1.00(m) | 16(f) 14(m) | 134(f) 114(m) | 134(f) 114(m) | 75(f) 75(m) | 0.12(f) 0.12(m) |

[a]Longest distance within the chamber, parallel to the floor of the pump
[b]Greatest width within the chamber, perpendicular to the length of the pump chamber
[c]Greatest height, dorsal to ventral, along the mid-sagittal plane and perpendicular to the length of the pump
[d]Data from Figs. 2D, E in Brehm et al. (2015)
[e]Proboscis lengths for *D. plexippus* and *M. sexta* and food canal diameter for *D. plexippus* are from Campos et al. (2015) and Lehnert et al. (2016)
[f]Diameter of the food canal of a single galea. The two galeae do not meet; therefore, a single food canal is not formed; the food canal is assumed to be circular
[g]Data from Bennet-Clark (1963)
[h]Data from Tawfik (1968); f-female, m-male. Pump height is interpreted as the "travel of diaphragm" values in this reference; the pump is considered a cylinder

different structure, often independently evolved, implying that different physical mechanisms for fluid uptake and transport might be involved (Smith 1985; Kingsolver and Daniel 1995; Kim and Bush 2012). Yet, the feeding efficiency for all proboscises is remarkably high. We suggest that common principles of fluid uptake can explain the structure and function of mouthparts across fluid-feeding insect groups. The proboscises of fluid feeders, for example, share a ground plan of their materials organization: the drinking regions are fibrous and porous, pointing to the importance of capillary and wetting forces in the feeding strategies of all fluid-feeding insects (Monaenkova et al. 2012). We focus on the lepidopteran proboscis as a model to explore the physical aspects and constraints of fluid-feeding in insects.

Proboscises present a functional paradox: insects must acquire fluids, such as sticky sugars or blood, while also keeping their proboscises clean. Another paradox is associated with mechanics of fluid uptake: many insects are able to acquire thick, viscous liquid, such as honey, and thin almost inviscid mineral water at high rates

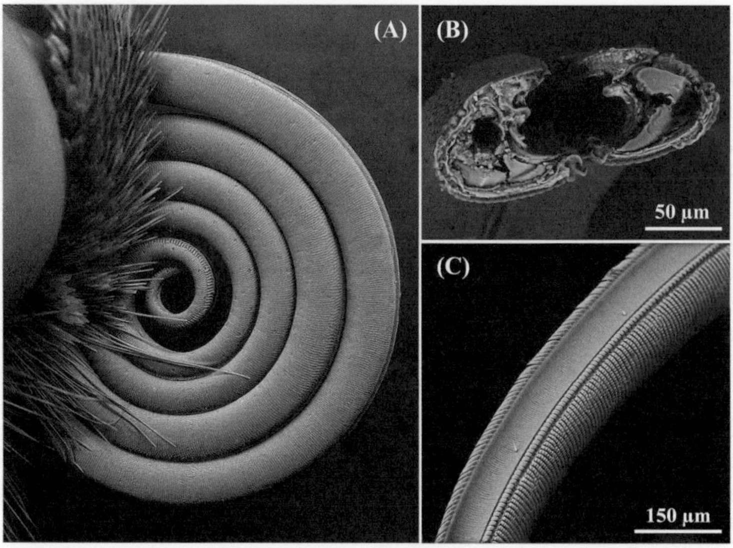


**Fig. 8.2** Photomicrographs of butterfly proboscises (SEM). (**a**) Coiled proboscis of the tiger swallowtail (*Papilio glaucus*). (**b**) Cross section of the proboscis of the monarch butterfly (*Danaus plexippus*) with a circular hole, the food canal, formed by the two halves (galeae). The top and bottom fence-like structures (legulae) link the two galeae together. (**c**) A single galea of the monarch butterfly resembles a C-shaped fiber with a semicircular C-channel, which, when united with the other half, forms a food canal (Reproduced by SPIE permission from Kornev et al. 2016)

(Wigglesworth 1972; Smith 1979; Adler 1982; Smith 1985; Kingsolver and Daniel 1995; Eisner 2005; Borrell and Krenn 2006; Krenn 2010). Explanations of these paradoxes need an integrated approach that brings together the science of wetting and adhesion, capillarity, biofluid mechanics, and quantitative biology.

## 8.2 Wetting and Capillary Forces

### *8.2.1 Introduction to Wettability and Adhesion*

When an insect immerses its proboscis in a liquid, an interface separating the cuticular solid from the air is replaced by an interface separating the same solid from a liquid. Thus, a solid/liquid interface is created in place of the solid/air interface. As stated by Thomas Young and Pierre-Simon Laplace (Young 1805; Laplace 1966), the interfaces separating distinct phases (solids, gases, and liquids) are under tension. That is, each interface has its own liquid/solid ($\sigma_{SL}$), solid/air ($\sigma_{SA}$), and liquid/air ($\sigma_{LA}$) surface tension. These surface tensions represent molecular forces acting at the interfaces and resulting from breaking the symmetry of the force acting on a molecule from its neighbors. If we consider a given phase as a continuum, the surface tension represents the force acting parallel to the surface of a drop per unit length of the contact line (Fig. 8.3a–c). In the International System of Units (SI), surface tension is measured in N/m; for example, the surface tension of the air/water interface is high, $\sigma_{LA} = 73$ mN/m. Lipids adsorbing at the interface and acting as surfactants reduce the surface tension to $\sigma_{lipids} = 20-40$ mN/m. Salts increase the air/water surface tension up to 10–15%. In contrast, addition of sucrose to water has little effect on surface tension: an increase of sucrose concentration up to 50% increases the surface tension of the solution to only about 5% (Van Oss 1994; Reiser et al. 1995; Docoslis et al. 2000; Aroulmoji et al. 2004). For example, at 21 °C, 10 g of sucrose in 100 g of its aqueous solution gives $\sigma_{LA} = 73.51$ mN/m, whereas 50 g of sucrose in 100 g of its aqueous solution gives $\sigma_{LA} = 78.27$ mN/m (Reiser et al. 1995).

The Young–Laplace equation represents the balance of forces acting at the contact line in the plane parallel to the solid surface (Fig. 8.3a–c); it introduces important physicochemical characteristics of the three dissimilar solid, liquid, and air phases meeting at a contact line: the surface tensions of the solid/air $\sigma_{SA}$ and solid/liquid $\sigma_{SL}$ interfaces and contact angle $\theta$:

$$\sigma_{SA} - \sigma_{SL} = \sigma_{LA} \cos\theta. \tag{8.1}$$

Good wetting assumes that the contact angle formed by a droplet is acute; poor wetting assumes that the drop beads up on the surface, with a contact angle $\theta$ greater than 90° (Fig. 8.3c). The differential $\sigma_{SA} - \sigma_{SL}$ reaches its limiting value, $\sigma_{SA} - \sigma_{SL} = \sigma_{LA}$, when the material becomes completely wettable by the liquid; the contact angle goes to zero and the drop spreads over a flat substrate to form a thin

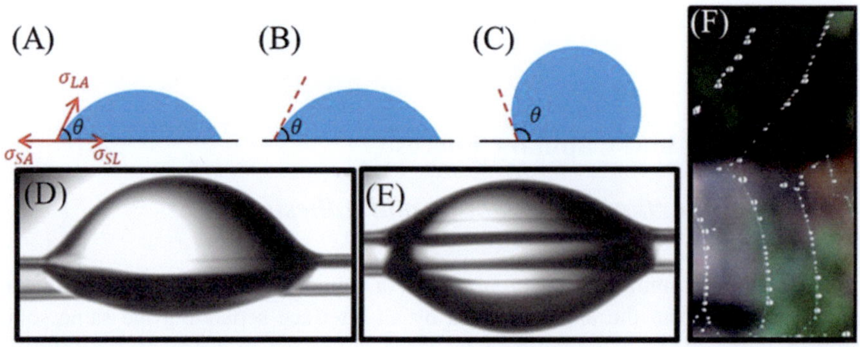


**Fig. 8.3** Drop shapes for different wetting scenarios: (**a**) Three surface tensions acting at the contact line are counterbalanced. (**b**) Good wetting. (**c**) Poor wetting. (**d**) Wetting of cylindrical fiber: when a drop is small, it is shaped like a clamshell. (**e**) When drop size increases, the drop is shaped like a barrel embracing the fiber. (**f**) Rain droplets on a spider web. (**a**–**c** and **f** by the authors, **d** and **e** courtesy of T. Andrukh)

film with its air/liquid interface parallel to the substrate surface (Adamson and Gast 1997). For example, the contact angle that a water droplet makes with a dry layer of glucose immediately after deposition is about 21°, whereas for fructose it is about 22°, and for sucrose about 19° (Van Oss 1994). These values are similar and small, as for many hydrophilic materials on which water is prone to spread. In contrast, the contact angle that a water droplet makes with hexatriacontane ($C_{36}H_{74}$), a waxy hydrocarbon of plants, is about 105°–111°. Thus, water droplets bead up on hexatriacontane substrates and tend to roll down when the substrate is tilted (Fadeev 2006). A similar tendency is observed for many hydrophobic cuticles of insects (Holdgate 1955; Watson et al. 2010, 2017; Cao and Jiang 2016).

Surface tension can be viewed conceptually from an energetic perspective, helping to understand the stickiness of some fluids and repellence of others from a chemical point of view. Particles in fluids and solids are constantly moving. Particles in fluids actively collide with one another, exchanging their positions; particles in solids actively oscillate around their equilibrium positions, engaging the surrounding particles in oscillations. Thus, the particles exchange their kinetic energies. As one moves from the bulk of the material to the interfacial layer separating two different phases, the packing density of particles decreases and the interparticle distance increases; hence, the interparticle potential energy associated with the chemical structure of the constituents decreases. Therefore, it is natural to introduce a surface energy density as the energy of an infinitesimally thin interfacial layer per unit area of this layer with units J/m$^2$, where $J = N \cdot m$. As shown in surface thermodynamics, this energy density is equal to the magnitude of the surface tension. Athanase Dupré (1808–1869), a French mathematician and physicist, derived an equation, named for him, which shows that the work required to separate a drop from the substrate, say a proboscis surface, is equal to

$$W = \sigma_{\mathrm{LA}}(1 + \cos\theta). \tag{8.2}$$

This equation, which defines the work of adhesion, $W$, suggests that the smaller the contact angle the droplet makes with the substrate, the more difficult it is to detach. Therefore, surfaces offering a high contact angle, $\theta$, are considered to have low surface energy because of their low energy of adhesion. Roughness of the proboscis surface significantly changes the surface energy, decreasing it in the case of waxed hydrophobic patches and increasing it in the case of protein-rich hydrophilic patches (Cassie and Baxter 1944; Cassie 1948; Quere 2008; Liu et al. 2010; Bhushan and Jung 2011).

Curvature of substrates significantly affects the wetting behavior of liquids. Liquid that wets a flat substrate, completely spreading over it and forming a film, as water does on a metal surface, does not spread on a cylindrical wire (Fig. 8.3e–f). Small droplets sit like clamshells on the fiber, whereas large droplets sit like little barrels. When the volume becomes larger, the barrel-like droplets sag under their weight and fall (Adam 1937; Carroll 1976; Zhang et al. 2018b). This paradoxical wetting/dewetting dichotomy of fibers has rich implications in biology and will be discussed in subsequent sections.

### 8.2.2 Introduction to Capillarity

The mechanism of fluid acquisition by fluid-feeding insects was historically based on a drinking-straw model (Daniel and Kingsolver 1983; Daniel et al. 1989; Kingsolver and Daniel 1995; Kim et al. 2011). According to this model, when an insect first immerses its proboscis in a liquid, for example, when a butterfly dips its proboscis in a flower, nectar rises up the food canal, owing to capillary action. The capillary action is associated with two types of forces: the surface tension of the dry part of the food canal and the force due to pressure acting at the leading edge of the liquid surface, which deforms into a meniscus.

The shape of a meniscus is determined by the Laplace law of capillarity (Laplace 1806), which states that the pressure under a meniscus, $P_l$, is different from atmospheric pressure, $P_g$, and the pressure differential is determined by the surface tension $\sigma$ and the mean curvature of the meniscus surface. We will illustrate capillary forces using two representative meniscus shapes: a spherically capped meniscus, which can be viewed as a part of a drop suspended in air, and a meniscus in the form of a circular cylindrical column, which can be viewed as a part of a jet of water in air. The Laplace equation of capillarity is derived by constructing a free-body diagram by cutting the drop or cylindrical column along the diameter or the axis, respectively. Setting all forces acting on the cut to zero (Fig. 8.4), we obtain the Laplace equations for a spherical drop of radius $R$ as

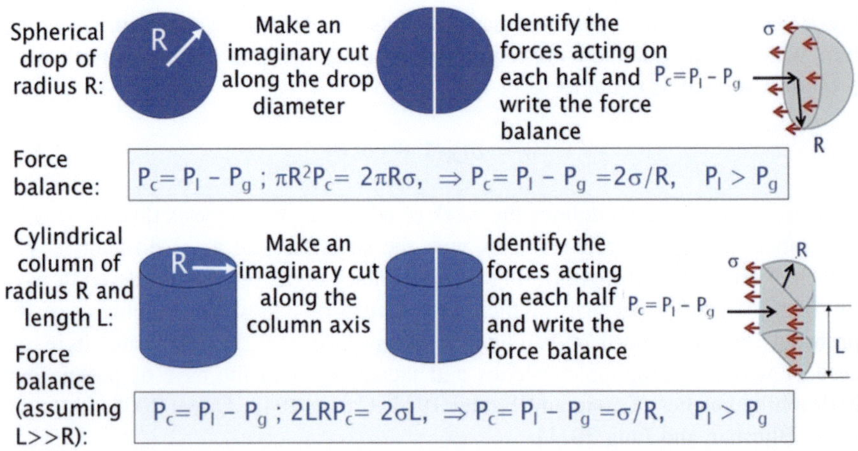


**Fig. 8.4** Derivation of the Laplace equation by cutting a drop or a liquid column and setting up an equivalent system of forces acting on the cut to equilibrate the right sides of the liquid bodies. The surface tension acts along the boundary of the spherical segment and along the sides of the rectangle. Assuming that the liquid column is long, $L > > R$, the contributions of surface tension acting along the two edges at the ends of the column are much smaller than those of the two column edges that are parallel to the column axis, $4\sigma R \ll 2\sigma L$; hence, we neglect these two contributions. The force due to pressure acting on the circular disk and rectangle is shown by arrows; the pressure in the liquid $P_l$ is measured with respect to the gas pressure $P_g$; hence, the force due to pressure is defined as $(P_l{-}P_g) \times$(*cross-sectional area of the cut*) (Figures by authors)

$$P_l - P_g = 2\sigma/R, \tag{8.3}$$

and for a cylindrical column as

$$P_l - P_g = \sigma/R. \tag{8.4}$$

Equation (8.3) can be used to evaluate the height $Z_c$ of a nectar column created in the food canal upon immersion of a proboscis in a flower. Following the drinking-straw model, we assume that the food canal is a cylinder (Fig. 8.5c). For a spherically capped meniscus meeting the food canal at contact angle $\theta$, Eq. (8.3) gives the pressure under the meniscus: $P_l = P_g - 2\sigma/R$. On the other hand, the pressure in the nectar column is hydrostatic; hence, $P_l = P_g - \rho g Z_c$, where $\rho$ is the density of liquid,

**Fig. 8.5** Nectar spontaneously rises to different heights in lepidopteran food canals of different diameters. (**a**) Skipper (Hesperiidae) feeding on *Lantana* floret. (**b**) Painted lady (*Vanessa cardui*) feeding on *Lantana* floret. (**c**) The height of the nectar column $z_c$ in the slender skipper proboscis is greater than that in the painted lady proboscis. (**d**) Schematic specifying the relation between the radius of a circular tube, *r*, the radius of a spherically capped meniscus, *R*, and the contact angle $\theta$ as $R = r/cos\theta$. (**e**) A model cylindrical proboscis threaded through a liquid bridge of length L. The liquid bridge is trapped inside a cylindrical tube of radius R. The proboscis is separated from the walls by a liquid film of thickness *h*. When the proboscis moves, the velocity profile through the film is assumed to be linear, as indicated by arrows. (**f**) When the proboscis is not completely submersed in the liquid and is on its way to the tube bottom, backup pressure is created, which significantly changes the flow and friction force. (**g**) When the proboscis is pulled upward, fluid flow in the film and the nectar remaining between the proboscis tip and the bottom of the flower corolla depends on the tip position; hence, the force required to withdraw the proboscis depends on the tip position. (Figs. **a** and **b** courtesy of Carmony Corley, **c–g** by the authors)

and $g$ is acceleration due to gravity. These two equations then give $\rho g Z_c = 2\sigma/R$. Using the schematic in Fig. 8.5d, we can relate the meniscus radius $R$ with the food canal radius $r$ as $R = r/cos\theta$. Finally, the height of the nectar column is estimated as

$$Z_c = 2\sigma cos\theta/(\rho g r). \tag{8.5}$$

If we assume that the water column makes contact angle $\theta = 0$ with the food canal, this formula allows an upper estimate of the height of a water column in the food canal of a proboscis. In a 2-μm diameter food canal, this height is $Z_c = 14.7$ m; in a 200-μm diameter food canal, it rises to 14.7 cm above the free water surface of the reservoir. This estimate also can be applied to nectar, suggesting that the insect does not need to apply suction pressure to make a first sip: the estimated height is much greater than a typical proboscis length. The estimate brings about another question: Is it realistic to assume that the insect relies on acquiring its food by dipping the proboscis into a drop and swallowing the acquired liquid column?

## 8.3 Proboscis Wetting as the First Step of Fluid Acquisition

### *8.3.1 Cuticular Patterns of Proboscis Surface in Relation to Wettability*

Fluid feeders use their proboscises to acquire liquids; hence, one would expect the proboscis to be hydrophilic. Yet, it would be advantageous to have cuticle that minimizes adhesion of pollen grains and other debris. Chemical analysis shows that the proboscis surface consists of hydrophobic chitin and different types of proteins, lipids, and waxes (Hepburn 1985; Vincent and Wegst 2004; Vincent 2012). Chitinous and waxed surfaces have high contact angles with water droplets, which facilitate cleaning (Beament 1961; Bush et al. 2007; Forbes 2008; Genzer and Marmur 2008) but prevent water-like fluids from wetting them. The structure of the proboscis surface is complex (Davis 1986; Krenn and Mühlberger 2002; Krenn and Aspöck 2012; Lehnert et al. 2016); however, with respect to wetting and fluid uptake properties, it can be organized into different structural categories, with variations on each theme: bumps–valleys (Fig. 8.6a), spikes (Fig. 8.6b), and shingles (Fig. 8.6c). Figure 8.7 illustrates this categorization, with examples representing extreme cases of short and long proboscises.

The bump–valley topography of short proboscises of notodontid moths is characterized by a dense system of bumps separated by narrow, groove-like valleys (Fig. 8.7b). The long proboscis of sphinx moths has a smooth surface with leveled bumps separated by narrow valleys (Fig. 8.7e). And hairy proboscises of the ancient moth family Eriocraniidae (Kristensen 1968; Monaenkova et al. 2012) exemplify spiky surface topography.

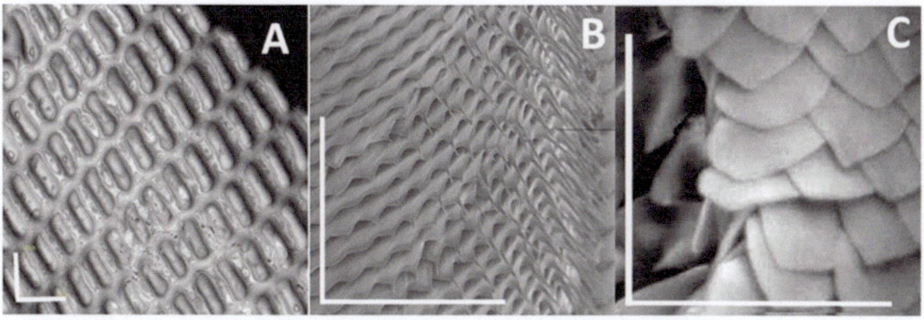


**Fig. 8.6** Structural categories of external surface sculpture of lepidopteran proboscises. (**a**) Smooth bumps interspersed with valleys, monarch butterfly (*Danaus plexippus*); scale bar = 10 μm. (**b**) Ridges with spiky edges separated by deep grooves, cabbage butterfly (*Pieris rapae*); scale bar = 40 μm. (**c**) Shingles, gemmed satyr butterfly (*Cyllopsis gemma*); scale bar = 25 μm (Photos by the authors)

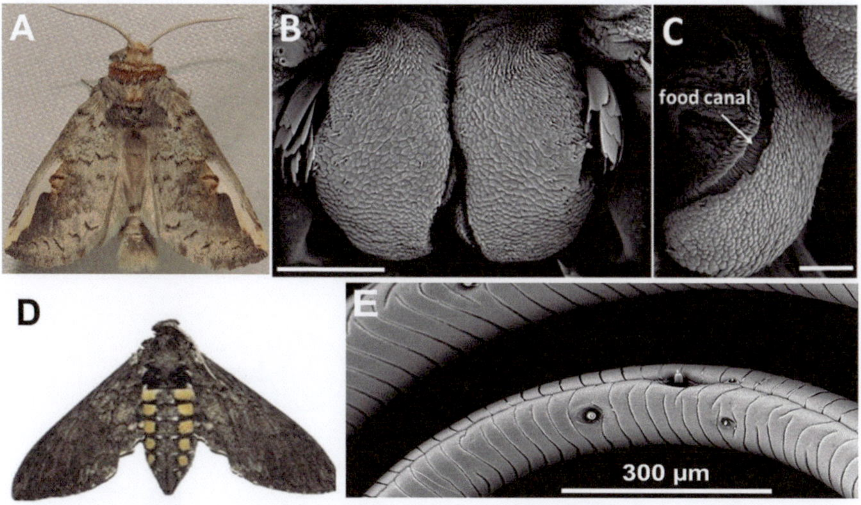


**Fig. 8.7** (**a**) Moth *Symmerista albifrons* (Notodontidae), with (**b**) the proboscis; scale bar = 150 μm. (**c**) half of the proboscis; scale bar = 50 μm. This surface microsculpture typifies a limiting case of smooth bumps interspersed with narrow valleys. (**d**) Carolina sphinx moth (*Manduca sexta*). (**e**) Part of its ~ 7-cm-long proboscis showing smoothed bumps interspersed with wide valleys. (**a** by Andy Reago and Chrissy McClarren [CC BY 2.0 (https://creativecommons.org/licenses/by/2.0)], via Wikimedia Commons; **b**, **c**, from Kornev et al. 2017 [CC BY 4.0 (https://creativecommons.org/licenses/by/4.0)]; **d** courtesy of C. Zhang; **e** by authors)

These topographies produce a rich behavior of droplets on the proboscis surfaces. Capillary rise experiments provide an informative picture of the large-scale wetting behavior of proboscises (Lehnert et al. 2013). When a proboscis breaches the air–water interface, a meniscus with a given contact angle forms around it. The shape and height of the meniscus are important physicochemical parameters that permit evaluation of large-scale proboscis wettability. A capillary-rise technique developed by Lehnert et al. (2013) is presented in Fig. 8.8. A camera records the meniscus profile during the experiment; the maximum height of meniscus *H* is defined by the contact line meeting the proboscis at the proboscis sides when the camera faces either the dorsal or ventral side.

By raising the height of the water in the dish, the contact line of the meniscus moves up, and the initial acute contact angle of the meniscus shifts to 90°, indicated by a flat water–air interface (Lehnert et al. 2013). As the water level rises, the contact line moves up, but the meniscus forms a circular dimple. Hence, the overall surface wettability of the proboscis shifts from hydrophilic to hydrophobic, and the meniscus forms a contact angle greater than 90°. The sharp demarcation between wetting and non-wetting regions allows quantitative classification of the hydrophilic section as a drinking region (5–17% of the proboscis length, depending on species) and the hydrophobic section as a non-drinking region (Lehnert et al. 2013, 2016). The shape of the meniscus is informative and allows one to quantitatively evaluate contact angles that the water menisci make with the proboscis surface (Lehnert et al. 2013).

The wetting/dewetting dichotomy of the proboscis has been confirmed for five different butterflies: monarchs (*Danaus plexippus*), painted ladies (*Vanessa cardui*), eastern tiger swallowtails (*Papilio glaucus*), red-spotted purples (*Limenitis arthemis astyanax*), and question marks (*Polygonia interrogationis*) (Monaenkova et al. 2012). In contrast, sphinx moths with long, smooth proboscises do not show a wetting dichotomy; the entire surface of the proboscises is wettable by water, as indicated by the acute contact angles that the water meniscus makes with the proboscis along its length (Fig. 8.9).

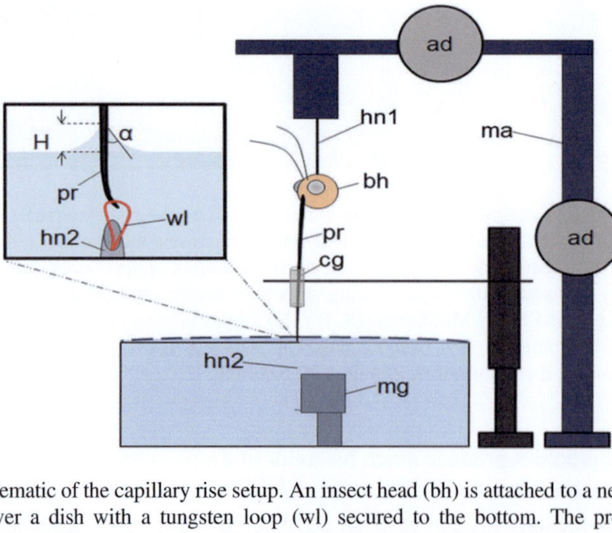


**Fig. 8.8** Schematic of the capillary rise setup. An insect head (bh) is attached to a needle (hn1) and positioned over a dish with a tungsten loop (wl) secured to the bottom. The proboscis (pr) is straightened in a capillary tube (cg) and held against the tungsten loop, which is in a second needle (hn2) held by magnets (mg); the dish is gradually filled with distilled water. (Reproduced by permission from Lehnert et al. 2013 [CC BY 4.0 (https://creativecommons.org/licenses/by/4.0)])

To absorb liquid and restrict entry of debris, lepidopteran proboscises have different structural and chemical functionalities. Detailed investigation of the shapes of the meniscus and contact line reveals that the contact line forms cusps at the dorsal and ventral legular bands of many butterflies (Fig. 8.9). In contrast, the contact line on the surface of sphinx moth (*Manduca sexta*) proboscises is smooth. The cusped menisci suggest that the bands of the linking structures (legulae) of the galeae are hydrophilic, paving the way for liquid to enter the food canal through pores between the legulae (Monaenkova et al. 2012).

The non-drinking region of the proboscis has long been considered sealed, with the galeae "meeting in such a manner as to form a tube quite air-tight between the two lateral ones" (Gosse 1993). The idea of a sealed tube persisted (Eastham and Eassa 1955; Krenn 2010) until experiments showed that water can enter the food canal through interlegular spaces proximal to the drinking region in some butterflies (Monaenkova et al. 2012); the mechanism of fluid entry was further studied by Lee et al. 2014a, b; Lee and Lee 2014. Additional experiments with fluorescent dye (Nile red) selectively adsorbing to the hydrophilic surfaces help to bridge large-scale wettability with small-scale features of proboscis surface sculpture (Fig. 8.10a). Staining experiments suggest that the legulae and chemosensilla are hydrophilic (Lehnert et al. 2013). A mosaic pattern of hydrophilic microbumps and less hydrophilic valleys on the butterfly proboscis has been discovered (Fig. 8.10b). This

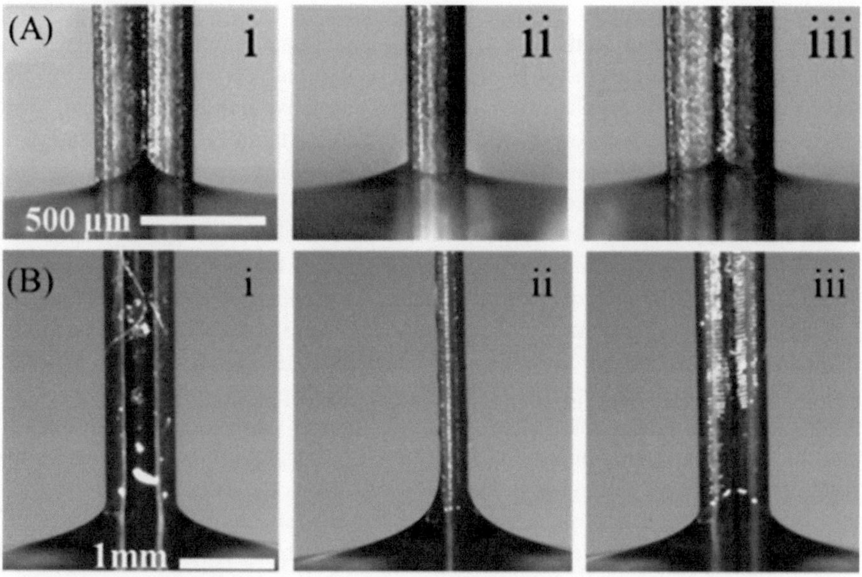


**Fig. 8.9** Menisci on the proboscises of (**a**) monarch butterfly (*Danaus plexippus*) at 3–5 mm above the immersed tip (the entire proboscis length is ~2 cm) and (**b**) sphinx moth (*Manduca sexta*) at ~4 cm above the immersed tip (the entire proboscis length is ~8 cm). The contact line profile is shown on the (i) dorsal legular band, (ii) side of the proboscis, and (iii) ventral legular band. (Courtesy of C. Zhang)

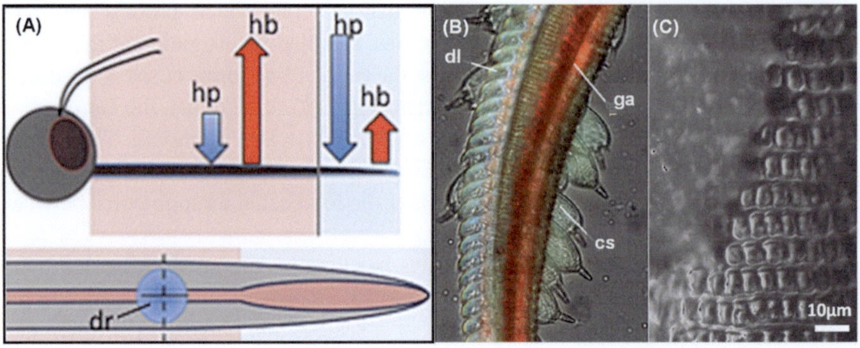


**Fig. 8.10** Overall wettability of the proboscis. (**a**) Summary hydrophobic (hb) and hydrophilic (hp) forces in the non-drinking (pink) and drinking regions (blue). Arrows representing forces are not to scale. Lower schematic shows a droplet (dr) on the non-drinking region. The drop will sink into the food canal through the pink legular band (Lehnert et al., 2013). (**b**) Proboscis of the red-spotted purple (*Limenitis arthemis astyanax*) stained with Nile red, showing hydrophilic chemosensilla (cs) and dorsal legulae (dl; blue) and overall hydrophobic galeae (ga; red). (**c**) Dark-field image of a wetting front moving from top left to bottom right over the galea of a monarch (*Danaus plexippus*) proboscis; fingers of water move along the microvalleys, leaving menisci that arc over the microbumps (**a–b** reproduced by permission from Lehnert et al. 2013, [CC BY 4.0 (https://creativecommons.org/licenses/by/4.0)], (**c**) courtesy of M. Nave)

surface arrangement of wettability is unusual and counterintuitive, as it would be expected to hinder liquid channeling through a continuous network of valleys toward the dorsal and ventral legulae where the liquid can move into the food canal. The case of the sphinx moth proboscis with its smoothed-out hydrophilic bumps suggests that the degree of fluid attraction varies significantly from species to species and depends on surface sculpture.

This mosaic pattern of wettability resembles the self-cleaning (superhydrophobic) elytral surface of *Stenocara* desert beetles (Parker and Lawrence 2001; Zhai et al. 2006) and some other arthropods (Sun et al. 2017). *Stenocara* beetles collecting fog droplets on their backs (Parker and Lawrence 2001) are probably the most popular (Cao and Jiang 2016), as they inspired materials scientists to develop versatile technology for making superhydrophobic–superhydrophilic surfaces (Zhai et al. 2006). *Stenocara* beetles take advantage of this mosaic pattern to fix the droplet in place using hydrophilic bumps and to limit the droplet from spreading using hydrophobic valleys. Tilting its body, the insect causes the droplets to roll toward the mouth. This mechanism suggests that the mosaic bump–valley surface of proboscises supports self-cleaning by facilitating shake-off of droplets (Quere 2008).

However, the idea of hydrophilic bumps and hydrophobic valleys makes the mechanism of droplet movement toward the permeable legular bands of the lepidopteran proboscis difficult to explain. Fluid should adhere to the hydrophilic islands, but the hydrophobic valleys would limit the adhesive area and create air pockets. This scenario, however, does not apply to lepidopteran proboscises:

dark-field imaging of movement of the contact line over the bumpy surface of the monarch butterfly's proboscis reveals that the water fingers propagate through the valleys between the bumps; hence, the valleys are not completely hydrophobic (Fig. 8.10c).

Further investigations are needed to identify the physical determinants of self-cleaning at the microscale level of bumps. To our knowledge, the other two types of surface sculpturing, spikes and shingles, have never been studied from a materials science perspective; hence, the mechanism of wetting of these surfaces remains unknown.

## 8.4 Role of Cross-Sectional Shapes of Proboscises and Food Canals in Fluid Uptake

### *8.4.1 Lack of a Food Canal Opening at the Proboscis Tip and Effect of Curvature of the External Surface of the Proboscis*

The level of insertion of the proboscis into a food source determines the extent of the proboscis exposed to fluid. Even if the proboscis surface is wettable, the insect has to deliver the liquid to the food canal. When the food canal opening is accessible to liquid, as in the case of Diptera, fluid delivery is not a problem: the fluid is drawn up by the sucking pump, as one would expect with the drinking-straw model. However, in many cases, the drinking-straw model of insect proboscises fails to describe the process of fluid delivery. For example, lepidopteran proboscises do not have a sizable opening to the food canal at their apices (Fig. 8.11).

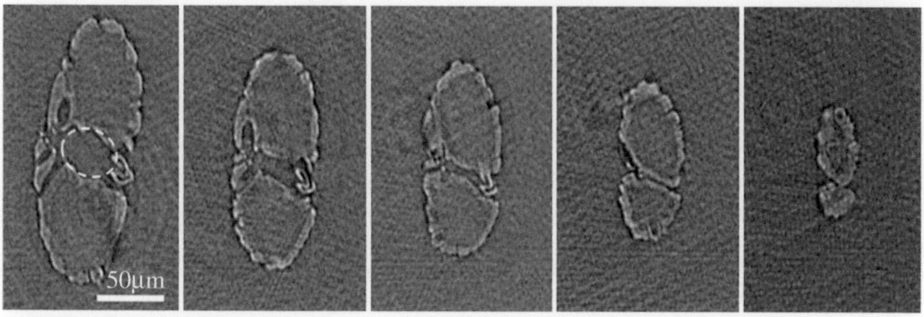


**Fig. 8.11** Series of X-ray tomography cross sections of a monarch butterfly (*Danaus plexippus*) proboscis at distances (left to right) of 141 μm, 69.75 μm, 45 μm, 24 μm, and 9.75 μm from the proboscis tip, and reduction of an apical opening (Monaenkova et al. 2012). The food canal, shown as a dashed line, starts to appear at a distance of ~35.25 μm from the proboscis tip. (Courtesy of D. Monaenkova and W.-K. Lee)

Lepidoptera use their legulae to hold the two galeae together and to facilitate fluid entry while restricting the entry of debris (Fig. 8.12d, e). The legulae are next to one another or overlap, but the legular bands are not sealed, allowing liquid to move through the interlegular slits to the food canal (Tsai et al. 2011, 2014; Monaenkova

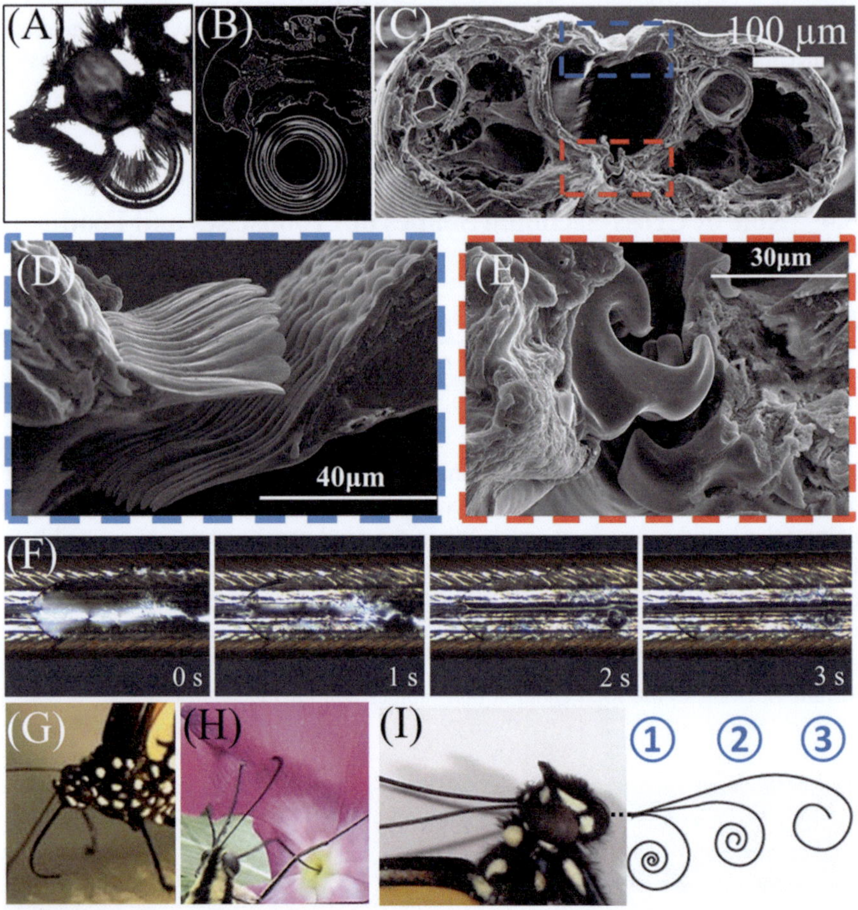


**Fig. 8.12** (**a**) Coiled proboscis of a monarch butterfly (*Danaus plexippus*). (**b**) Longitudinal section of the proboscis revealed by micro-CT. (**c**) Cross section of the proboscis of a sphinx moth (*Manduca sexta*); areas in the dashed boxes are magnified in (**d**) and (**e**). (**d**) Dorsal legulae, showing their overlapping configuration. (**e**) Ventral legulae, showing their interdigitation. (**f**) Movement of methylene blue-dyed water through interlegular dorsal pores of the sphinx moth (*Manduca sexta*) proboscis. (**g**) J-configuration of the proboscis of a monarch butterfly (*Danaus plexippus*). (**h**) Eastern tiger swallowtail (*Papilio glaucus*) dipping its proboscis into a flower corolla. When the butterfly pulls the proboscis out, nectar remains on the external surface of the proboscis, which then moves to the permeable dorsal and ventral legular bands. (**i**) Monarch butterfly (*Danaus plexippus*), showing schematic of different stages of proboscis coiling and uncoiling during feeding. (Reproduced by permission from Zhang et al. 2018b, CC BY 4.0 (https://creativecommons.org/licenses/by/4.0))

et al. 2012; Lehnert et al. 2013, 2016; Lee et al. 2014a, b; Lee and Lee 2014). Accordingly, Lepidoptera capitalize on capillarity via the interlegular slits that pull fluid from pores in wet surfaces into the food canal.

Butterflies feeding from flowers often flex their proboscises at the bend region and insert only the distal portion into the corolla, which leaves the non-drinking region largely unexposed to fluid (Krenn 2010). However, an analysis (Tiple et al. 2009) of 108 species of butterflies feeding from the flowers of 20 species of plants showed that even when proboscises were longer than corollas, greater lengths of proboscises were inserted into some flowers, constraining the degree of flexing at the bend, especially when feeding from narrow corollas (Barth 1991; Tiple et al. 2009; Arditti et al. 2012; Campos et al. 2015; Wozniak and Sicard 2018). The implication is that the legular fence would be sealed in the straightened portion of the non-drinking region while feeding from flowers with narrow corollas (Monaenkova et al. 2012). Floral visitations for butterflies are usually brief (e.g., typically less than 10 s per flower) (Kunte 2007), which is often not enough time for fluid to enter the food canal in the non-drinking region of a straightened proboscis.

Therefore, nectar or other liquid deposited over the proboscis as a film has to be collected at the hydrophilic bands of the linking mechanism to enter the food canal through the interlegular gaps. An ellipsoidal proboscis ensures that the film will gather at the legular bands. The radius of curvature of galeae far from the legular bands is much greater than that at the legular bands where the proboscis is flattened. When the film is thin and its free surface is equidistant to the proboscis surface, the pressure through the film is almost constant but changes along the proboscis. As seen from the analysis of the Laplace equation of capillarity, Eq. (8.4), capillary pressure at the side poles of the ellipse is greater and forces the liquid toward the linking mechanism of the dorsal and ventral flattened sides of the proboscis (Fig. 8.13k–m).

These physical arguments of fluid movement have been confirmed in experiments on the Carolina sphinx moth (*Manduca sexta*) proboscis (Fig. 8.13) (Zhang et al. 2018b). The proboscis of the sphinx moth was inserted in a capillary tube with black ink. The proboscis leaving the tube at 1 mm/sec was covered with an ink film that flowed from the sides with a smaller radius of curvature to the dorsal legular band, leaving the proboscis sides dewetted. At the same time, liquid was drawn into the food canal through the porous legular band. The film was broken, with formation of four nearly straight contact lines receding toward the two legular bands. Liquid from the most curved parts of the proboscis moved to the less curved sides, and eventually moved completely into the food canal through the legular band.

## 8.4.2 *Fluid Delivery to the Food Canal: Physical Determinants of Puddling Habits*

Droplets or films collected at the legular bands of the proboscis move into the food canal spontaneously due to capillary action of spaces (i.e., pores) between the dorsal legulae

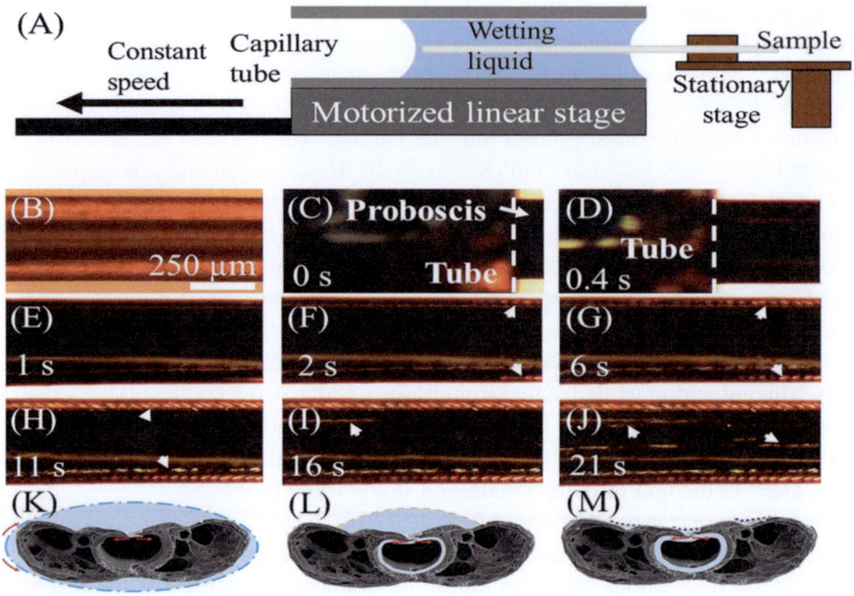


**Fig. 8.13** (**a**) Experimental setup for coating a proboscis with ink. The tube with ink is moved at a constant speed, and the meniscus leaves a film coating the proboscis surface. (**b**) A straightened proboscis before it is coated with ink. The dorsal legular band running along the proboscis appears darker compared to the rest of the proboscis. (**c**, **d**) Film deposition process. The tube of ink (left of the dashed line) is moved to the left and a black film is deposited on the proboscis. (**e–h**) The film flows from the sides to the legular band. (**e**) The proboscis is completely covered with the film. (**f**) Two contact lines (arrows) form after dewetting of the proboscis sides. (**g**, **h**) The contact lines recede from the proboscis sides toward the legular band. (**i**, **j**) The legular bands dewet as the film moves into the food canal. The bright bands (shown by arrows) are the grooved features of this band. (**k–m**) Schematic illustration of movement of the external film to the food canal caused by capillary pressure set by the wetting film formed on the wall of the food canal. Schematic (**k**) describes images (**e**) and (**f**), schematic (**l**) describes images (**g**) and (**h**), and schematic (**m**) describes images (**i**) and (**j**). (Reproduced by permission from Zhang et al. 2018b, CC BY 4.0 (https://creativecommons.org/licenses/by/4.0)

(Monaenkova et al. 2012; Kwauk et al. 2014; Kornev et al. 2016) (Fig. 8.14a–c). The pores are located between adjacent legulae (interlegular slit-like pores) and between the two rows of dorsal legulae of the two galeae (Schmitt 1938; Eastham and Eassa 1955; Snodgrass 1961; Hepburn 1971), forming two fences with pores spaced a few to several hundred nanometers apart (Monaenkova et al. 2012).

Slit-like pores in the legular bands enable lepidopterans to feed opportunistically from many food sources including porous substrates such as rotting fruit and wet soil. Males of many species of butterflies and moths, including those that routinely imbibe floral nectar, often drink from mud puddles and soil that is either damp or dry; if dry, the nutrients are solubilized with saliva or a drop of liquid expelled from the gut (Norris 1936; Arms et al. 1974; Adler 1982; Adler and Pearson 1982; Smedley and Eisner 1995, 1996; Molleman 2010; Snell-Rood et al. 2014). Trace

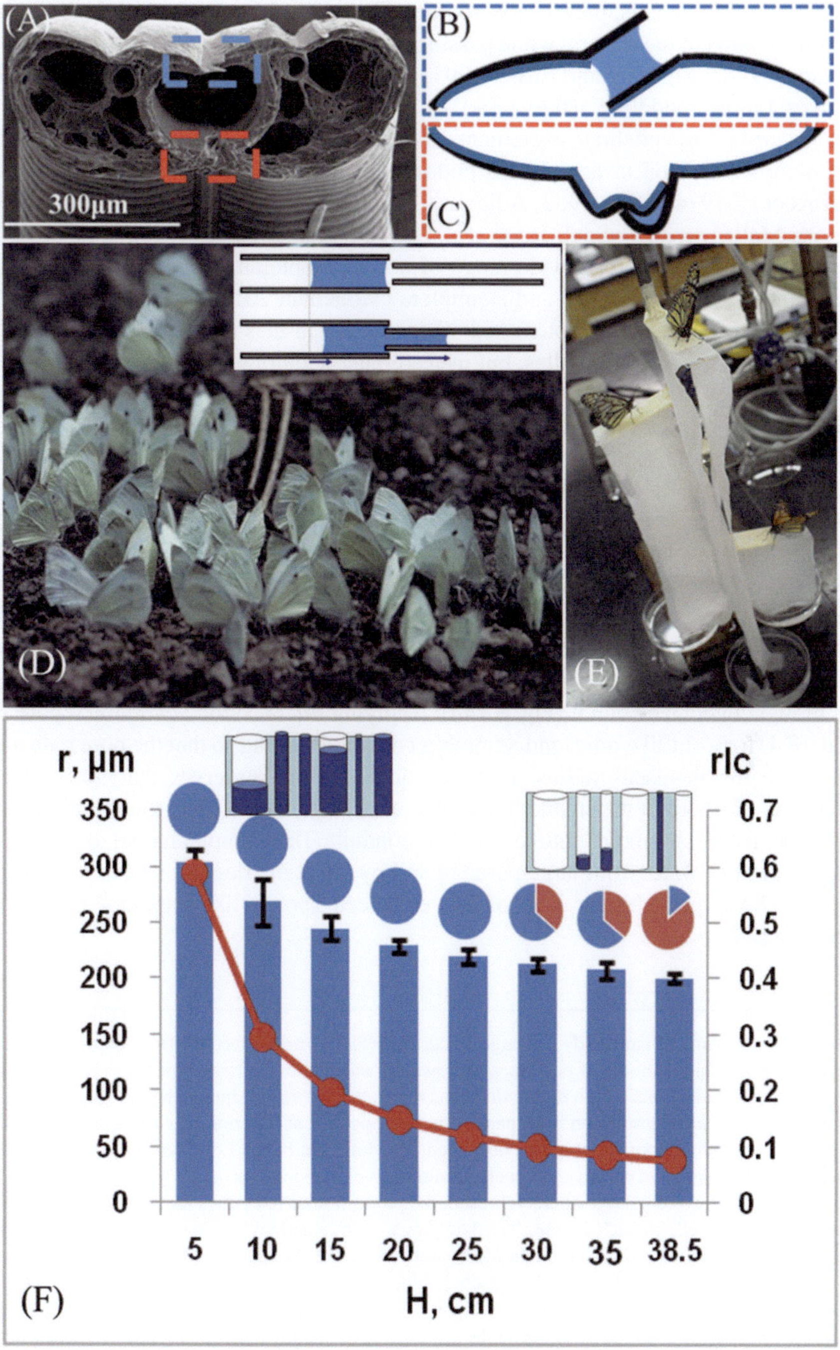


**Fig. 8.14** (**a**) Scanning electron micrograph of the proboscis cross section of *Manduca sexta*; the dorsal legular band is blue boxed and the ventral legular band is red boxed. (**b**) and (**c**) illustrate capillary adhesion of liquid bridges trapped by legulae in the dorsal and ventral bands, respectively. These liquid bridges coexist in equilibrium with a thin liquid film covering the wall of the food

nutrients and minerals, particularly sodium, are sought by males, which lose significant amounts of the mineral while passing spermatophores to the females (Adler and Pearson 1982; Smedley and Eisner 1995, 1996). The behavior of gathering at soil is referred to as "puddling" (Fig. 8.14d), a term loosely extended to include feeding from carrion and vertebrate excrements and secretions (Molleman 2010). The choice of damp or dry soil or actual mud puddles is often species-specific (Norris 1936; Arms et al. 1974; Adler 1982; Adler and Pearson 1982; Smedley and Eisner 1995, 1996; Molleman 2010; Snell-Rood et al. 2014). Puddling has also been documented, often for both sexes, for a wide range of non-lepidopteran insects with haustellate mouthparts, such as Diptera and Hemiptera (Molleman 2010).

Many species of Lepidoptera and other sucking insects do not necessarily drink from the open water of puddles, but instead must withdraw water from pores of the soil. However, soil is not necessarily fully saturated with water. Thus, fluid uptake is hindered by capillary pressure created by liquid menisci sitting in the pores.

Butterflies and moths feeding from porous substrates, such as rotten fruits and soils (Norris 1936; Adler 1982; Büttiker et al. 1996; Krenn et al. 2001; Knopp and Krenn 2003), press the dorsal, distal portion of their proboscises to the substrate. The spaces between legulae in the drinking region are $2.6 \pm 0.12$ μm for monarch butterflies—about 15 times smaller than the average radius (35 μm) of the food canal.

A liquid bridge sitting in a large pore (e.g., in soil), modeled as either a tube or a slit, cannot remain in the pore when a smaller tube or a slit (e.g., an interlegular pore) is brought in contact with it (Fig. 8.14d). Applying Eq. (8.3) for a cylindrical pore or Eq. (8.4) for a slit-like pore, and assuming complete wetting so that the pore radius is equal to the meniscus radius in these equations, one observes that the suction capillary pressure in the smaller conduit is always greater. Hence, the liquid bridge will be forced to move into a smaller conduit. This simple model of passive withdrawal of water from large pores of soil by small interlegular slits in the dorsal legular band provides an idea of the enhanced sponge-like capillary action of proboscises.

---

**Fig. 8.14** (continued) canal. (**d**) Cabbage butterflies (*Pieris rapae*) feeding from damp soil. The insert shows a simple model of passive withdrawal of water from the large pores of soil by a small interlegular slit-like pore. (**e**) To mimic this behavior in the laboratory, monarch butterflies were fed a 25% aqueous sucrose solution from paper towels suspended at different heights to determine the limiting pore radius from which they could feed. (**f**) Minimum radius r of open pores in the paper towel (filled circles solid curve) and towel-saturation level (bars) versus height (H) of the stage from which butterflies were fed. Pie charts represent the proportion of butterflies that drank (blue) or did not drink (dark red) at each height. The bar graph (with standard deviations) shows the liquid content at each height relative to the liquid content in the completely saturated towel. The relative liquid saturation (rlc) of the towel at each height (H) is calculated as $\mathrm{rlc} = (M(H) - M_{dry})/(M_{saturated}) - M_{dry})$, where M is sample weight. The inserts show schematically a bundle-of-tubes model of the paper towel; liquid in each tube is shown in dark blue. If the tube is partially filled with liquid at a particular height H, the tube is partially colored. White tubes correspond to empty pores. (**a**, **e** by permission from Zhang et al. 2018b, CC BY 4.0 (https://creativecommons.org/licenses/by/4.0, **b**, **c**, **d**, and **f** by the authors)

However, the sponge (i.e., proboscis) has to give up the acquired fluid to the food canal. The hypothesis that lepidopterans withdraw liquid from only a specific size range of pores was tested by placing monarch butterflies on paper towels saturated in a 25% sucrose solution (Monaenkova et al. 2012) (Fig. 8.14e). Before engaging butterflies in feeding, the saturated towels were suspended at different heights, $H$, with their ends submersed in the sucrose solution until a hydrostatic equilibrium was reached. This equilibrium implies that at each stage, the pores of different sizes are filled with the liquid. The pores in the paper towels are modeled as a bundle of tubes of different diameters (Fig. 8.14f). Applying Eq. (8.5), we observe that at stage height $H$, only pores offering $Z_c > H$ will be filled with sucrose solution. The stage height and the size of the pores retaining liquid are inversely related: the greater the stage height, the smaller the size of pores that still hold liquid. For example, at a stage height of $H = 5$ cm, only pores with radii less than $r = 300$ μm should be filled with liquid and at $H = 15$ cm only pores smaller than $r = 250$ μm should contain liquid (Fig. 8.14f). At heights of 5–25 cm, all monarch butterflies fed. At 38.5 cm, only two of 14 butterflies fed. The pore radius in the paper towel, $H_c = 38.5$ cm, is considered the critical minimum pore radius, $r_m$, from which a butterfly is able to withdraw liquid. Above this height, butterflies never drank. The pore radius $r_m$ is estimated from Eq. (8.5), assuming complete wetting, $\theta = 0$, of the paper towel by the sucrose solution, $r_m = 2\sigma/(\rho g H_c) \approx 38$ μm. This radius is about the average radius, $R_p$, of the food canal in the distal region of the proboscis (Monaenkova et al. 2012).

These findings created a new paradox: the approximate equality between $r_m$ and $R_p$ suggests that the butterflies would rely on capillary rise in the tubular food canal, as explained earlier with the drinking-straw model. However, in the drinking region, the interlegular pores are about 15 times smaller than the food canal radius. Accordingly, we would expect that once menisci have been spontaneously formed in the interlegular slits, they should remain in the slits.

Schematics in Fig. 8.14b, c help to resolve this paradox. Lepidoptera typically keep the food canal wet by moving a film of saliva back and forth (Zhang et al. 2018a). The film is connected to liquid bridges (Fig. 8.14b, c). The wall of the food canal is almost cylindrical; hence, the pressure in the wetting film is below atmospheric pressure, Eq. (8.4), and the pressure in the liquid bridges is also below atmospheric pressure. By pressing the proboscis to the liquid-saturated substrate or submersing it in a drop, the equilibrium breaks up, eliminating the air/liquid menisci in the liquid bridges sitting between legulae (Fig. 8.14b, c). Capillary suction pressure of the internal film on the surface of the food canal pulls the liquid from the substrate pores or drop while the liquid bridges sitting between the legulae work as conduits transferring the liquid from the pores or from the drop to the food canal (Monaenkova et al. 2012; Kornev et al. 2016). This scenario sets a physical metric for the ability of the insect to passively pull the liquid from the pores of a food source: the pore radius has to be smaller than the radius of the food canal. Thus, the criterion for the minimum pore size is $r_m = R_p$.

This capillary pull is quite strong and enables insects to acquire thin, watery and thick, honey-like fluids (Norris 1936). The details of this phenomenon were discovered in 2011 using X-ray phase-contrast imaging at Argonne National Laboratory, USA (Fig. 8.15) (Monaenkova et al. 2012). For butterflies drinking from paper

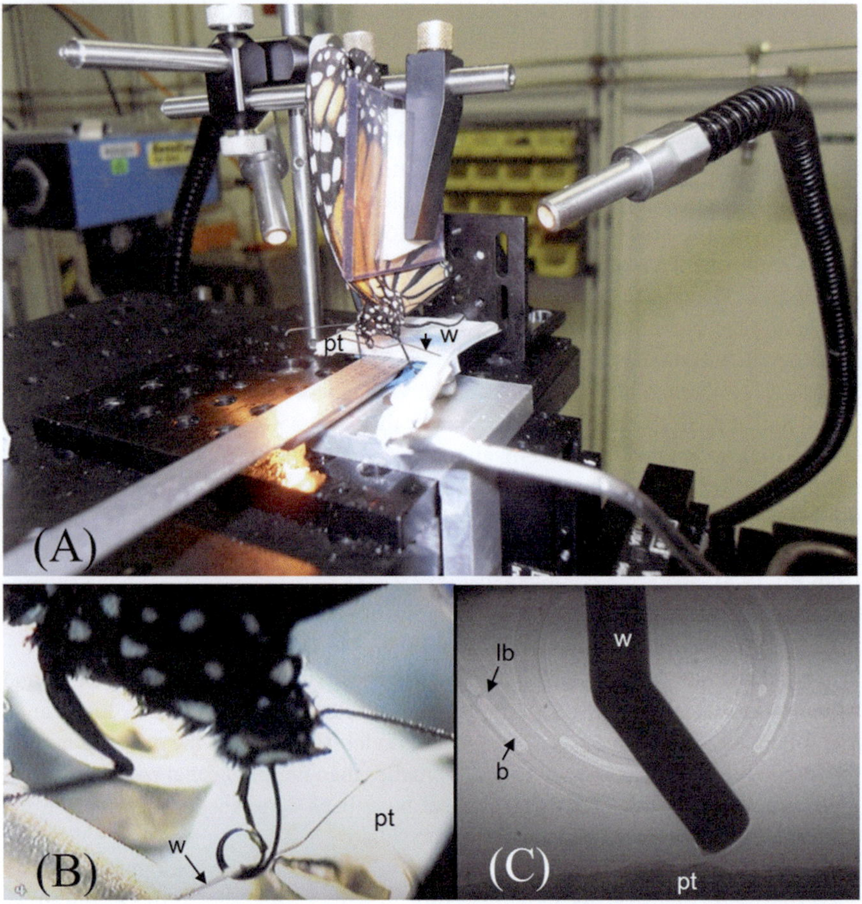


**Fig. 8.15** (**a**) A monarch butterfly (*Danaus plexippus*) held by the wings at the X-ray stage of the Advanced Photon Source in Argonne National Laboratory, Argonne, IL. (**b**) Coiled proboscis slightly widened by a wire (w) threaded through the coil and connected to a manipulator (not shown). (**c**) X-ray image of a coiled proboscis partially filled with liquid after the butterfly begins drinking from a paper towel (pt) saturated with a sucrose solution. A liquid bridge (lb) and trapped bubble (b) are indicated by arrows. The brighter domains are bubbles and the darker domains are liquid bridges of different lengths. The wire is darkest. (Courtesy of D. Monaenkova and W.-K. Lee)

towels under the X-ray beam, the liquid penetrates the dorsal legular band and collects on the back of this band from inside, forming a bulge that transforms into a liquid bridge (Fig. 8.16). Calculations of the pressure in the internal film, using a simple model of a hollow torus, show that by bending and unbending the proboscis, Lepidoptera can control fluid collection and formation of liquid bridges during feeding from droplets or porous substrates (Zhang et al. 2018b). The higher the pressure difference, the faster liquid collection will occur.

When a butterfly drinks from a film, the food canal is not entirely filled with liquid; rather, in all studied cases, the internal film coating the food canal forms multiple bulges that transform into liquid bridges separating trapped bubbles (Fig. 8.16). The liquid bridges are formed in straight and bent proboscises with the same effectiveness; that is, the phenomenon is mostly associated with hydrodynamic instability of the coating film (Monaenkova et al. 2012; Kornev et al. 2016).

Partitioning liquid droplets and films as liquid bridges and then drinking the bubble trains represents a newly discovered mechanism of liquid transport in insects (Monaenkova et al. 2012; Kornev et al. 2016). We observed liquid bridges in the proboscises of butterflies drinking water and sugar solutions and producing saliva. We hypothesize that Lepidoptera take advantage of this spontaneous formation of

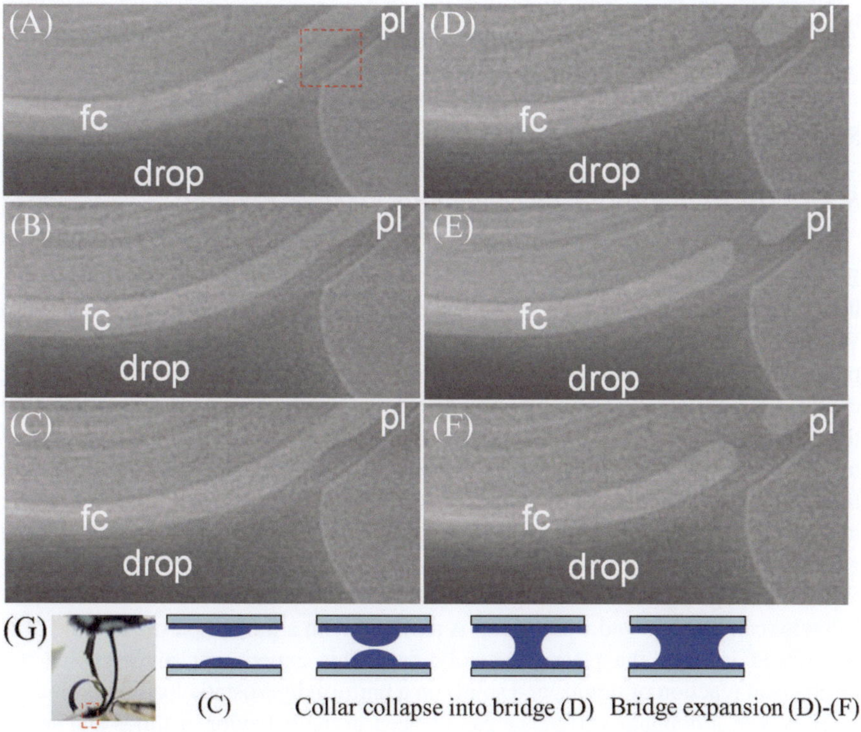


**Fig. 8.16** Monarch butterfly (*Danaus plexippus*) drinking from a drop. The proboscis was pushed toward the drop by a wire as shown in Fig. 8.15b and in Fig. 8.16g. (**a–f**) X-ray phase-contrast images showing an outmost proboscis loop (pl) touching a liquid drop. The liquid enters the food canal (fc) and forms a bulge on the inner surface of the dorsal legular band. The area of nucleation of the liquid bump is boxed in (**a**). (**b–c**) The bulge pulls liquid and enlarges. At the moment (**c**), it forms a collar or liquid ring that collapses into a liquid bridge almost instantaneously (ca. 0.15 s). (**d–f**) The liquid bridge increases in size by acquiring liquid from the droplet through the legular band. (**g**) Illustration of the field of view, with the boxed area imaged in (**a–f**) and the stage-by-stage schematic of collar formation and collapse happening between frames (**c–f**). (Courtesy of D. Monaenkova and W.-K. Lee)

liquid bridges to significantly broaden the range of fluids in their diets, from watery fluids to highly viscous liquids (Wigglesworth 1972; Kingsolver and Daniel 1995; Eisner 2005; Krenn 2010; Shaw 2014). Fluid entry into the food canal, at least in the non-drinking region, is facilitated by expansion of the legular spaces in the dorsal legular band as the proboscis transitions from straight to flexed. Fluid penetration through the non-drinking region is species-specific, possibly reflecting differences in legular spacing (Monaenkova et al. 2012; Kwauk et al. 2014; Lehnert et al. 2016). In *D. plexippus*, legular spacing in the non-drinking region of the dorsal legular band ranges from 96 nm near the head to 162 nm more distally (Monaenkova et al. 2012). Thus, closer to the head, the legular band is less permeable relative to the distal region. A drop of liquid sinks into the food canal more readily in the buckeye butterfly (*Junonia coenia*) than in the eastern tiger swallowtail (*Papilio glaucus*). Differences in legular spacing among species might reflect differences in feeding behavior.

## 8.5 Viscous Forces

### *8.5.1 Introduction to Viscosity and Viscous Friction*

Before introducing quantitative metrics for transport properties of insect proboscises, it is necessary to introduce an important materials characteristic of fluid viscosity. When you stop stirring coffee, the beverage eventually stops moving. A spectacular behavior of fluid particles attempting to stop can be observed by placing a drop of cream in the coffee; the creamy fingers eventually cease perceptible flow, indicating the slowing of particle movement. Thus, fluids only move if they are forced to flow by some external force. Surprisingly, we do not see a difference between solids and fluids when these materials are subjected to a uniform pressure. For example, water in a bottle can withstand high pressure exerted by a cork when it is pushed into the neck of the bottle; water does not flow if the cork is moved slowly. Pushing the cork farther could generate sufficient pressure in the water to break the bottle. A similar behavior can be observed when water is replaced with a molten plastic and the cork is screwed in once the plastic has solidified. This experiment suggests that the mechanical reaction of liquids and solids on a uniform hydrostatic loading is similar.

However, a dramatic difference can be seen in the behavior of fluids and solids when the force is applied tangentially to the object's surface. We can walk over a solid floor, but slip on a floor with spilled oil. This unpleasant experience suggests that solids can withstand tangential forces while fluids cannot. Therefore, we can place liquids in a special class of materials sensitive to tangential or shear forces.

A materials parameter of liquids characterizing the ease of flow upon shearing is viscosity, $\eta$. An example of the spill is relevant to the analysis of mechanisms of proboscis dipping or withdrawal from a floral corolla with nectar, or proboscis withdrawal from animal tissue by a blood-feeding insect. In these examples, the moving proboscis plays the role of a shoe in the oil-spill experiment.

By modeling the insect proboscis as a circular cylinder, one can estimate the relationship between proboscis velocity and the force required to submerse and withdraw the proboscis from a food source. The proboscis is modeled as a circular cylinder of radius $R$ filled with a liquid (Fig. 8.5e–g). In the simplest model, the proboscis as it is moving is assumed to tightly fit a cylindrical tube, leaving a thin lubricating film of thickness $h$ between the proboscis and the tube wall. In the thin film approximation, $h < < R$, when the proboscis is pulled in and out of the tube with a force $F$, the velocity of simple fluids, such as aqueous sugar solutions, takes on a liner profile (Fig. 8.5e–g) (Vogel 1996, 2003). At the proboscis surface, the liquid particles move with the velocity of the proboscis $v$, and at the tube wall the liquid particles are not moving at all (i.e., their velocity is zero). Thus, the established velocity profile is $(R - r)/h$, where $r$ is the radial coordinate measured from the common axis of these two cylinders.

The ratio $\gamma = v/h$ is measured in reciprocal seconds and is called the "shear rate"; it is a metric for how fast the velocity changes through the thickness of the layer. In this scenario, proboscis velocity, $v$, is related to force, $F$, through Newton's law of friction as (Vogel 1996, 2003)

$$v = \left(\frac{F}{A}\right)\frac{h}{\eta}, \tag{8.6}$$

where $A$ is the area of the wet part of the proboscis at the given moment of time, $t$. The fluid is uniquely characterized by its viscosity, η, measured in SI units as $Pa \cdot s$.

Assume that a foraging butterfly finds a corolla with a nectar bridge of length $L > > R$ trapped midway in the corolla and not touching the bottom (Fig. 8.5f). The butterfly dips its proboscis into the nectar bridge and pulls it back through the bridge. Further assume that the butterfly applies a constant force as it extracts its proboscis from the corolla. From Eq. (8.6), we can infer that the velocity should be constant. Specifying the surface area for this particular case, $A = 2\pi RL$, we have $v = Fh/(2\pi R\eta L)$. Thus, the thinner the lubricating film, the smaller the velocity. This equation prompts the design of a simple experiment to evaluate fluid viscosity by placing a tube with a liquid bridge vertically and applying a needle of known weight. Engineers designed special instruments, falling cylinder viscometers, which take advantage of this simple idea to study viscosity (Irving 1972; Gui and Irvine 1994).

In many cases, viscosity measurements are not straightforward, especially when the amount of available fluid is minute. This limitation, in particular, explains the dearth of viscosity data on insect saliva. Table 8.2 lists viscosities of some fluids relevant to analysis of insect feeding.

To appreciate viscous friction, consider an insect pulling its proboscis from a hole of radius $R = 100$ μm with a nectar bridge of length $L = 1$ mm. Assume that thickness of the lubricating film separating the proboscis from the hole surface is $h = 10$ μm. Typically, a nectar-feeding insect can move its proboscis in and out of the corolla with a frequency up to 10 times per second. For a conservative estimate of the dipping/withdrawing velocity, $v$, we use a frequency of 1 dip per second, and obtain the velocity $v \sim L \cdot 1 = 10^{-3}$m/s. Rewriting Newton's law of friction as $F = 2\pi v R\eta L/h$ and plugging these parameters into the force equation, we obtain $F \approx 6 \cdot 10^{-5}\eta$(N).

**Table 8.2** Viscosity of different fluids

| Liquid | Viscosity, mPa·s |
|---|---|
| Monarch (*Danaus plexippus*) saliva | 3.9 ± 0.7 (24 °C) |
| Tiger swallowtail (*Papilio glaucus*) saliva | 5.0 ± 0.5 (24 °C) |
| Painted lady (*Vanessa cardui*) saliva | 5.2 ± 1.0 (24 °C) |
| Water | 1.00 (24 °C) |
| 10% sucrose in water | 1.3 ± 0.2 (24 °C) |
| 20% sucrose in water | 1.7 ± 0.2 (24 °C) |
| 30% sucrose in water | 3.8 ± 0.4 (24 °C) |
| 40% sucrose in water | 6.9 ± 0.7 (24 °C) |
| Peanut oil | 56.5 (21 °C)<br>38.7 (38 °C) |
| Safflower oil | 55.2 (25 °C)<br>35.3 (35 °C) |
| Soybean oil | 40.6 (30 °C)<br>38.6 (35 °C) |
| Human blood plasma/serum | 1.2 (37 °C) |
| Whole human blood (shear rate > 100 $s^{-1}$) | 3-4 (37 °C) |

(Noureddini et al. 1992; Steffe 1996; Fasina and Colley 2008; Fournier 2011; Tokarev et al. 2013); concentration of sucrose solutions is given as weight/weight percent

As an estimate of nectar viscosity, we use the value from Table 8.2 equivalent to a 40% sucrose solution, $\eta \approx 7 \cdot 10^{-3}$ Pa·s; the result is $F \approx 42 \cdot 10^{-8} N$. For an insect of mass $m = 0.1\ mg$ weighing $W \approx 10^{-6} N$, the force required to dip and withdraw the proboscis from the nectar is $F \approx 0.4$W. This value would be equivalent to an 80-kg man lifting a 32-kg weight every second for more than an hour! If we take the dipping frequency of 10 times per second, the force increases tenfold. This estimate allows us to appreciate the challenges associated with feeding by small insects on viscous fluids.

Some blood-feeding insects (e.g., mosquitoes, bed bugs, and sucking lice) and plant-sucking bugs (e.g., aphids and leafhoppers) that pierce soft tissues have little room for blood and sap flow, and they form much thinner lubricating films, compared with Lepidoptera feeding from floral tubes. Consequently, they are strongly challenged when they are about to withdraw their proboscises. The female mosquito *Aedes aegypti*, for example, has a sophisticated procedure for withdrawing its proboscis from the skin after blood feeding (Jones and Pilitt 1973). It straightens its front legs, lifts its head and thorax, and inclines its body backward. The extraction takes 2–10 s, which is 1/4 – 1/10 the time required for insertion (Jones and Pilitt 1973), suggesting that the insect is able to overcome significant friction.

Many blood-feeding insects (e.g., ceratopogonid midges, black flies, and horse flies) do not use the mosquito's hypodermic needle style of blood feeding. Instead, they cut and lacerate host tissue and feed from the resulting pool of blood. These pool feeders, called "telmophages," therefore, do not encounter the challenges of piercing tissue and withdrawing the proboscis through the small diameter of the feeding hole, as do the vessel feeders, the "solenophages." This alternative blood-feeding strategy minimizes viscous friction and the physical constraints imposed by feeding from a hole of small diameter. The blood-feeding vampire moths (*Calyptra* sp.) perhaps have elements of both the solenophages and the telmophages. The tearing hooks, rasping spines, and erectile barbs of the vampire moth's proboscis, which can penetrate host tissue to a depth of 6 mm (Bänziger 1971), might widen the feeding hole and ease the challenge of removing the proboscis from a narrow hole.

Viscosity depends on temperature: most liquids become stickier and flow less easily when they become colder. Saliva viscosity of butterflies is 4–5 times greater than that of water (Table 8.2). Saliva of insects lubricates the mouthparts, aids digestion, and dissolves viscous and dried substances. Saliva should not be needed for liquefying nectars with sugar concentrations up to 30–40%; viscosity stratification would not be expected when butterflies feed on nectar with 30–40% sugar concentrations (Fig. 8.17).

## 8.5.2 *Metrics for Pulsatile Flow and Inertial Effects*

In biofluid mechanics dealing with flow through conduits, such as arteries and veins or proboscises, an important indicator of the frequency of the pulsatile flow (unsteadiness) is the Womersley number (Womersley 1955; Daniel et al. 1989; Loudon and Tordesillas 1998; Vogel 2007). The number was introduced by John R. Womersley as $\mathrm{Wo} = h\sqrt{2\pi f\rho/\eta}$, where $\rho$ is the density of the liquid, $\eta$ is the dynamic viscosity of the liquid, and $f$ is the beating frequency, for example, of the heart or sucking pump.

For small Womersley numbers, $\mathrm{Wo} < 1$, unsteady effects of flow can be neglected, but for large Womersley numbers, $\mathrm{Wo} > 1$,these effects become important. In applications to fluid-feeding mechanics, the Womersley number is typically less than one. For example, Bennet-Clark (1963) observed a beating frequency of $f = 3$ beats/second in the pump of a blood feeder, *Rhodnius prolixus*. Taking water as the reference for viscosity and density of blood, we see that $\sim 3 \cdot 10^3 \cdot h$, where $h$ is pump height, which cannot exceed $h_{\max} \sim 0.16 \cdot 10^{-3}$ m. Thus, during the expansion–contraction cycle when

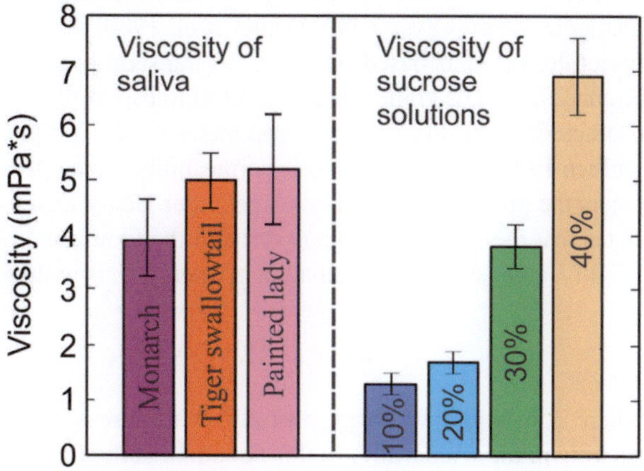


**Fig. 8.17** Comparison of viscosities of butterfly saliva and sucrose solutions of different concentrations (Tokarev et al. 2013) (Reproduced by AIP permission from Tokarev et al. 2013)

pump height is much smaller than $h_{max}$, the Womersley number is much smaller than 1, implying that unsteady effects can be neglected for this insect. The same conclusion has been drawn for Lepidoptera (Daniel et al. 1989; Lee et al. 2014a; Kornev et al. 2017) and mosquitoes (Lee et al. 2009; Kikuchi et al. 2018).

The Womersley number is related to another dimensionless parameter, the Reynolds number, named after Osborn Reynolds, who discovered many effects caused by fluid viscosity. In particular, Reynolds (1886) explained the mechanism of fluid lubrication. In pulsatile flows or when an insect moves its proboscis back and forth through a liquid-filled hole, if the fluid is inviscid, each stroke would push a liquid layer to move at the same velocity, $v$. In the case of a moving proboscis, the inertial force acting parallel to the proboscis surface through the film of thickness $h$ is $F_i = 2\pi\rho v^2 Rh$ (Vogel 1996). This force acts to resist the flow. Due to its viscosity, the fluid velocity changes through the film thickness, generating a shear force, Eq. (8.6), which also resists the flow and acts parallel to the proboscis surface, $F = 2\pi v R \eta L/h$. The ratio of these two forces is defined as the Reynolds number, $\mathrm{Re} = F_i/F = \rho v h^2/(\eta L)$. When the Reynolds number is greater than one, $\mathrm{Re} > 1$, the inertial forces are appreciable and fluid moves as though it is almost inviscid. In the opposite case, with small Reynolds numbers, shear forces prevail and inertia effects can be ignored. In other words, liquid metals having high densities would flow like water with low density but with almost the same viscosity.

The Reynolds number for flow through proboscises is introduced similarly as $\mathrm{Re} = \rho v d_p^2/(\eta L_p)$. It is typically small; hence, the fluid density plays an insignificant role unless the effects are related to capillary rise. Thus, flows with small Womersley and Reynolds numbers can be considered quasi-static and the effects of flow unsteadiness can be neglected.

### 8.5.3 Introduction to Flows Through Tubes and Pores

Seminal works (Bennet-Clark 1963; Tawfik 1968; Kingsolver and Daniel 1979, 1995; Daniel and Kingsolver 1983; Daniel et al. 1989) set up an index for performance of insect fluid feeders based on proboscis transport properties (Borrell and Krenn 2006; Krenn 2010). The ability of a conduit to transport fluid at a fixed pressure gradient is characterized by the materials parameter called "permeability" (Scheidegger 1974; Dullien 1991), denoted here as $k$; permeability is measured in units of $m^2$. Permeability enters the problem via the law named after Henry Darcy (1865). Darcy's law, which states that the flow discharge $Q$ through the food canal of cross-sectional area $A$ and length $L$ and associated dissipation of energy $\dot{E}$, is written as

$$Q = (kA/\eta)(\Delta P/L), \quad \dot{E} = Q\Delta P, \tag{8.7}$$

where $\eta$ is liquid viscosity and $\Delta P$ is the pressure differential generated by the sucking pump. According to this equation, at the fixed pressure gradient ($\Delta P/L$) and the cross-sectional area of the food canal, the insect with the greatest permeability of the food canal will get the greatest amount of food per second.

In the drinking-straw model of the proboscis, considered as a cylindrical tube of radius $r$, the permeability $k$ is derived from the Hagen–Poiseuille law as $k = r^2/8$; for a slit-like pore with an opening gap, H, permeability is given by the Boussinesq formula as $k = H^2/12$ (Vogel 1996, 2003). For a more complex geometry, the material parameter $k$ can be obtained by directly measuring the total flow rate $Q$ and pressure gradient ($\Delta P/L$) and knowing the cross-sectional area of the food canal and viscosity of the liquid.

Permeability is a convenient metric for characterizing transport properties not only of proboscises but also of porous and fibrous materials. For example, the transport or permeation ability of an interlegular pore structure of lepidopteran proboscises can be characterized by introducing the permeability metric (Monaenkova et al. 2012).

### *8.5.4 Viscous Adhesion (Stefan Adhesion)*

The specific features of thin lubricant films are manifested through the enhanced friction when a proboscis slides over a film surface. Additionally, when two surfaces move toward one another, squeezing out the liquid from a thin lubricating layer, the velocity through the film thickness is typically much slower than that along the surfaces. For example, by pressing its proboscis to the wall of a flower tube, a butterfly forces the nectar to flow along the legular band much faster, exposing more legular gaps through which nectar can flow into the food canal. An estimate of the relation between the velocity components can be made by considering the following imaginary experiment. Assume that a nectar bridge of length $L$ and thickness $h$ is spread along a legular band of width $W$, separating the proboscis from the flower. The nectar volume $V$ in a long liquid bridge is estimated, ignoring the contribution of the side menisci, as $V = LWh$. When the insect presses its proboscis against the flower, forcing the nectar bridge to decrease its thickness, with velocity $v_h = dh/dt$, the conservation of nectar volume gives $dV/dt = d(LWh)/dt = 0$, or, using the chain rule of differentiation, $hdL/dt + Ldh/dt = 0$. Thus, the nectar bridge elongates with velocity, $v_L = dL/dt = -(L/h)v_h$; the smaller the thickness $h$, the greater the ratio of the velocity components $v_L/v_h = -L/h$. Thus, the liquid particles move along the legular band ($L/h$) times faster than they move through the film thickness. The same argument applies when the insect lifts the proboscis from the flower; the sign of velocity $v_h$ changes from negative to positive, leading to contraction of the liquid bridge, $v_L < 0$; yet the same factor ($L/h$) emphasizes the velocity difference.

Thus, even slow relative motion of surfaces, one toward the other, will give a high shear rate, causing a high friction force and, hence, a strong pressure gradient, Eq. (8.7). The faster the rate of separation/attachment, the higher the rate of shear, and the stronger the suction/injection pressure gradient.

A quasi-steady flow through thin conduits formed by two closely positioned surfaces, where in-plane velocity is much greater than trans-plane velocity, is called

Hele-Shaw flow, named after its discoverer Henry Selby Hele-Shaw (1898). In Hele-Shaw flow, as soon as the thickness of the liquid layer becomes much smaller than the longitudinal scale of flow, the pressure through the conduit thickness is nearly independent of the z-coordinate perpendicular to the surfaces. The in-plane flow is described by Darcy's law, Eq. (8.7) (Reynolds 1886; Hele-Shaw 1898).

The same Hele-Shaw analysis can be used for understanding flow of liquids in the sucking pump of many insects. Bennet-Clark (1963) reported blood suction by the bug *Rhodnius prolixus* at the rate of $Q{\sim}0.33\ \text{mm}^3/\text{s}$. The cavity of the sucking pump of *R. prolixus* is almost rectangular, with width $W$ and length $L$ (Table 8.1). When the cavity of the sucking pump is contracted so that the thickness $h$ of the blood layer in the chamber decreases and the roof approaches the floor to close the chamber, the in-plane components of the velocity vectors $V_x$ and $V_y$ are much greater than the trans-plane velocity component $V_z$. The longitudinal velocity in the pump of *R. prolixus* is estimated as $V_x{\sim}Q/(Wh)$. Substituting the flow rate reported by Bennet-Clark (1963) and geometrical parameters of the sucking pump from Table 8.1, we obtain $V_x \sim 7$ mm/s. Assuming that blood is incompressible, like any liquid at the pressure of interest (Vogel 1996), and using the balance of blood volume, which shows that the amount moving in the longitudinal direction, $V_x Wh$, and leaving the chamber is approximately equal to the amount displaced in the $z$-direction, $V_z WL$, we confirm that in-plane velocity is much greater than trans-plane velocity, $V_x{\sim}\left(\frac{L}{h}\right) V_z{\sim}30V_z$ (Kornev et al. 2017).

Because the in-plane velocity is typically at least one order of magnitude greater than the trans-plane velocity, the flow in the sucking pump and in the liquid bridges sitting between a proboscis and a substrate, such as a corolla, soil, or rotten fruit, can be considered almost in-plane two-dimensional flow. Joseph Stefan (1874) carried out a fluid mechanics analysis of a general problem of viscous adhesion to relate the rate of separation/attachment of the object with the applied force. He used Eq. (8.7) to evaluate a nonuniform in-plane distribution of pressure in the moving fluid caused by suction/ejection of the fluid from the separation layer. The Stefan law of viscous adhesion, named after him, states that separation/attachment of closely positioned objects separated by a layer of viscous fluid requires a force, $F$, that is directly proportional to the velocity of the moving object and to fluid viscosity and inversely proportional to the layer thickness cubed, $F \propto \eta(dh/dt)h^{-3}$. We will use Stefan's law of viscous adhesion to describe the interaction of moving liquid in the sucking pump with the dilator muscles.

## 8.6 Mechanics of Fluid Uptake

### *8.6.1 Paradox of the Drinking-Straw Model*

Fluid uptake rates in Lepidoptera typically range from $0.2 \times 10^{-9}$ to $0.5 \times 10^{-9}\ \text{m}^3/\text{s}$ (May 1985; Tsai et al. 2014). Uptake rates for blood by *Rhodnius prolixus* are in the same range (Bennet-Clark 1963; Tawfik 1968). The rate of blood uptake by mosquitoes and sucking lice is much smaller. For example, the rate of blood uptake by the louse *Pediculus humanus* is $0.1 \times 10^{-11}$ $m^3$/s. It is slightly greater in the bed bug *Cimex lectularius*, $0.2 \times 10^{-11}$ $m^3$/s, and even greater in the mosquito *Aedes aegypti*, $0.6 \times 10^{-11}$ $m^3$/s (Tawfik 1968). Mosquitoes have a distinctive method of fluid uptake that exploits their two-pump system (Kikuchi et al. 2018). In a more economical, continuous mode of drinking, the mosquito uses continuous reciprocation between pumps with multiple small strokes, and it uses a newly discovered "burst mode" involving a large-volume stroke (Kikuchi et al. 2018). The Asian tiger mosquito *Aedes albopictus* drinks in the continuous mode at a slow rate of $0.56 \times 10^{-12}$ $m^3$/s, but in the burst mode, the rate increases to $0.16 \times 10^{-11}$ $m^3$/s (Kikuchi et al. 2018). The sap-feeding spittlebug *Philaenus spumarius* extracts sap at rates from $2 \times 10^{-11}$ to $4 \times 10^{-11}$ $m^3$/s (Malone et al. 1999). The broad range of flow rates in fluid-feeding insects suggests that these insects have evolved special sucking pumps to accommodate their feeding habits.

The drinking-straw model of the proboscis as a straight cylindrical tube has populated the biological literature and has been used as a baseline for identification of physical determinants of uptake efficiency during feeding from pools of liquid (Bennet-Clark 1963; Tawfik 1968; Kingsolver and Daniel 1979, 1995; Daniel and Kingsolver 1983; Daniel et al. 1989; Kim and Bush 2012; Kornev et al. 2017; Kikuchi et al. 2018). Eq. (8.7) constitutes the basis for the drinking-straw model of insect proboscises. This model takes into consideration (a) the food canal radius $R_p = d_p/2$, (b) the proboscis length $L_p$, and (c) the reported uptake rates for the given insect. The model allows an estimate of the pressure $P_t = \Delta P$ required by a sucking pump to support this flow rate. The results for monarch butterflies are given in Fig. 8.18. Further analysis of energy consumption during drinking can be used to relate the energy required to support the compressor and dilator muscles of the pump (Pivnick and McNeil 1985).

A tapered tube more accurately approximates actual proboscis structure, but it requires measurements of the taper angle φ, height $H$, and the radius of the food canal opening during feeding, $R_0$. The taper angle can be obtained from scanning electron micrographs of actual proboscises, but the food canal opening $R_0$ and height H require analysis of the permeability of the drinking region of the proboscis and interpretation of the measurements, using a model of a composite conduit consisting of two tubes connected in a sequence. This analysis has been done by Tsai et al. (2014) and is summarized in Fig. 8.18. The tapered proboscis significantly changes the results for suction pressure based on the straight tube model: the pressure drops significantly in the drinking region where the food canal tapers. The proximal portion of the proboscis makes a minor contribution to the pressure differential; pressures at the drinking region ($P_H$) and in the sucking pump ($P_p$) are similar to one another (Fig. 8.18b). On the other hand, the required pressure can be significantly reduced when the taper is ignored; in most cases, the required pressure drops below atmospheric pressure for the same flow rate.

The model of a tapered proboscis with measured φ, $H$, and $R_0$ values provides an important estimate of the pressure differential $P_p$ in the sucking pump: at the observed flow rates it should be greater than 1 atm. Even for dilute sucrose solutions

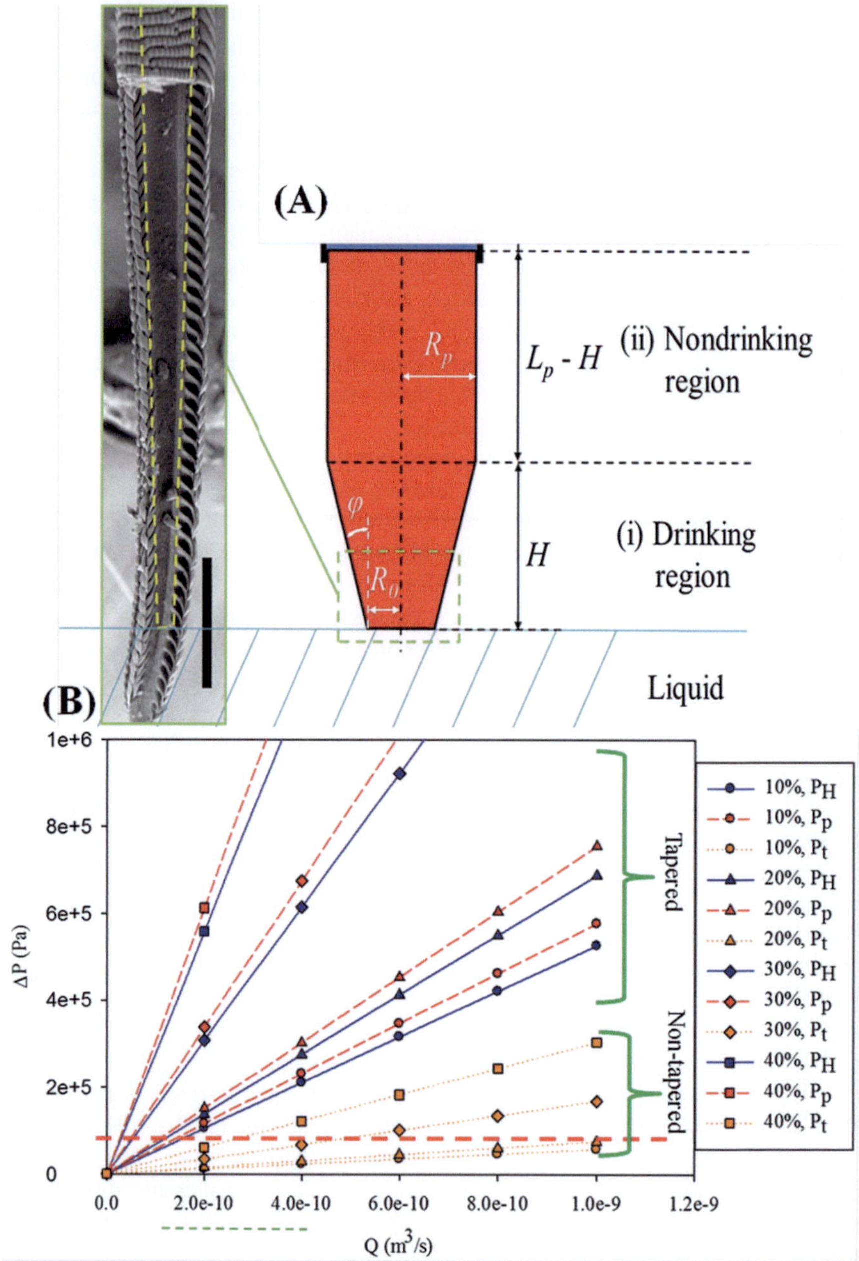


**Fig. 8.18** (**a**) Scanning electron micrograph of a galeael tip showing the tapered food canal of a monarch butterfly (*Danaus plexippus*) proboscis; scale bar = 1 mm. Right image: Schematic of food canal model. $L_p$ is the proboscis length, $H$ is the length of the drinking region, $R_p$ is the radius of the straight section of the food canal, $R_0$ is the radius of the opening, and $\varphi$ is the taper angle. (**b**) Pressure required to provide flow in tapered and non-tapered theoretical models of butterfly proboscises. $\Delta P = P_H$ is the pressure at the junction H where the drinking (tapered) and non-drinking (straight) regions merge in figure (**a**); $\Delta P = P_p$ is the pressure in the sucking pump

(e.g., 10%), the pressure differential is expected to be greater than 1 atm. Estimates for blood- and sap-feeding bugs support the need for the same above-one-atmosphere pressure differentials (Bennet-Clark 1963; Tawfik 1968; Malone et al. 1999). However, no vacuum pump can produce a pressure differential greater than 1 atm in surface atmosphere (i.e., in an open system).

A contradiction arises when uptake is optimized from films versus pools. Capillarity is enhanced when the lumen is small, but flow is enhanced when it is large. Thus, a paradox arises. The enormous pressure differential is avoided by a simple drinking-straw model that ignores structural features of the proboscis; however, the pressure rises steeply when the actual taper of the proboscis is incorporated into the model.

A tapered food canal might be characteristic of most fluid-feeding insects with an elongated proboscis. Thus, insects such as blood-sucking true bugs and mosquitoes, sap-feeding planthoppers, and nectar-feeding flies all have a tapered food canal (Snodgrass 1935; Robinson 1939; Surtees 1959; Bennet-Clark 1963; Christophers 1960; Malone et al. 1999; Karolyi et al. 2012, 2013; Chapman 2013). Therefore, the paradox of the drinking-straw model applies not only to Lepidoptera but also to most fluid-feeding insects with high flow rates, if feeding in an open system.

### 8.6.2 *Resolving the Drinking-Straw Paradox Through Behavioral Strategies*

The drinking-straw paradox can be resolved by a suite of behavioral strategies used by lepidopterans (Kwauk 2012; Tsai et al. 2014). One of the most common is galeal sliding (Fig. 8.19), also referred to as "anti-parallel movements," which may adjust the fluid-pressure differential by changing the size of the interlegular slits and terminal opening and by reducing the active, tapered length of the food canal (Kwauk 2012; Tsai et al. 2014). Galeal sliding is also used in self-assembly of the proboscis after a lepidopteran emerges from the pupa (Krenn 1997; Zhang et al. 2018a). Moths that pierce animal and plant tissues also use this strategy (Büttiker et al. 1996). Galeal sliding might facilitate meniscus formation and transport if the amount of fluid is small and a droplet is trapped in the food canal. And galeal sliding might help remove debris from the legular bands.

**Fig. 8.18** (continued) in a tapered food canal (position $L_p$ at the top of the straight tube in (**a**)); $\Delta P = P_t$ is the pressure in the sucking pump in the drinking-straw (non-tapered) model when the radius ($R_p$) of the food canal does not change along the proboscis. Percentages indicate aqueous sucrose concentrations. The red horizontal dashed line at 1 atm indicates the maximum limit of pressure differential that could be produced by the sucking pump. The dashed line below the X-axis indicates the measured flow rates for monarch butterflies. (Reproduced from Tsai et al. 2014, by permission from The Company of Biologists)

**Fig. 8.19** Illustration of galeal sliding with the image sequences from a video of the apex of the proboscis of a monarch butterfly (*Danaus plexippus*) during feeding (Reproduced from Tsai et al. 2014, by permission from The Company of Biologists)

Galeal splaying opens the distal end of the proboscis (Kwauk 2012; Tsai et al. 2014). It could reduce the pressure differential by increasing the diameter of the tapered region. Galeal pulsing might be controlled by hemolymph pressure, similar to processes involved in coiling and uncoiling the proboscis. In Fig. 8.20, we reproduce the flow maps around proboscises of butterflies and mosquitoes drinking from a pool of water (Lee et al. 2014b). Water enters the food canal from the interlegular slits at the proximal part of the drinking region. The authors did not study the cause of the flow in this part of the drinking region; however, the opening of interlegular slits in this area might be explained by galeal pulsing. In contrast, mosquitoes have an opening at the tip of the proboscis and liquid enters the food canal, following predictions of the drinking-straw model. The rate of fluid uptake is slower, however; hence, the model explains the main features of flow (Kikuchi et al. 2018). Pressing the proboscis to the substrate could change the diameter of the food canal, the legular spacing, and maybe the contact area with the substrate for enhanced capillarity.

We previously showed that a full cycle of the sucking pump in Lepidoptera takes 0.45–0.70 s at 25 °C (Monaenkova et al. 2012). Galeal sliding is typically slower (0.1–90.0 s) and, therefore, can be independent of the pump; thus, the proboscis could remain open for multiple pump cycles. Galeal pulsing occurs faster than the contraction–expansion rate of the sucking pump, possibly facilitating flow during the first half of the cycle when the pump is open. Figure 8.20 supports this hypothesis, but more experiments are required to test it.

The drinking-straw model implicitly assumes that unlimited liquid is available. However, it is not clear when the liquid source can be considered a large pool versus

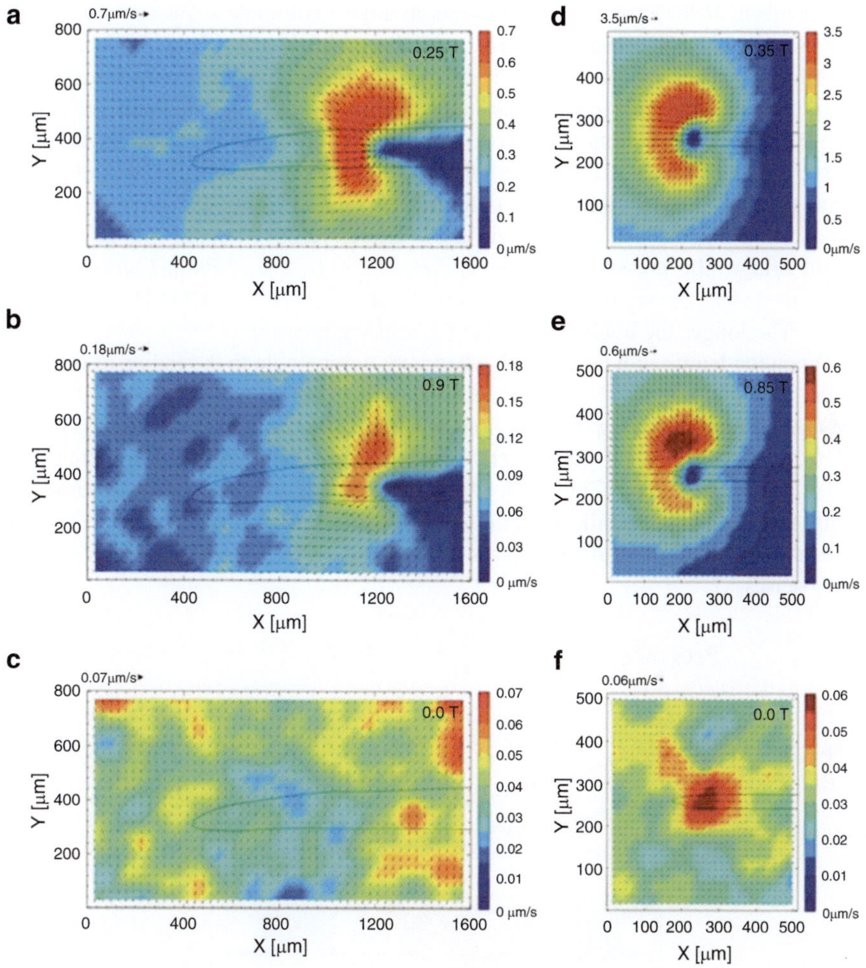


**Fig. 8.20** Flow maps around the tip of a proboscis with fluorescent particles for the butterfly *Pieris rapae* and the mosquito *Aedes togoi* (female) (Lee et al. 2014b). Color bars on the right indicate velocity; maximum velocity is specified at the upper left of each frame. $T = 280$ s. (**a–c**) The phase-averaged velocity field around the butterfly proboscis (**a**) when the intake rate is maximum, (**b**) when the butterfly pushes water out at a maximum rate, and (**c**) when the butterfly is at rest and neither takes water in nor out. (**d–f**) Phase-averaged velocity fields around the proboscis of a mosquito, corresponding to intake, discharge, and resting, respectively (Reproduced from Lee et al. 2014b by permission of Elsevier)

a limited volume. X-ray imaging of liquid uptake by butterflies suggests that the insects partition continuous liquid columns in the food canal into a system of air bubbles (Fig. 8.15c, 8.16a–f). The idea behind this partitioning can be understood based on Eq. (8.7), assuming that the length of the liquid bridge $L_{lb}$ is much greater than the radius of the food canal, $L_{lb} >> R_p$, to neglect the effect of fluid flow at the end menisci. If the insect drinks continuous liquid columns, a pressure drop, $\Delta P$, results, which is distributed linearly over the entire length of the proboscis, $L_p$, creating a constant pressure gradient, $\Delta P/L_p$ (Fig. 8.21). If the insect drinks a bubble train of $N$ liquid bridges, each of $L_{lb}$ length, separated by $N + 1$ bubbles, each of $L_b$ length, the pressure drops only over the liquid bridge with the gradient $\Delta P_{bt}/L_b$, where $\Delta P_{bt} = P_n - P_{n+1}$. Assuming that the sucking pump takes up the liquid at the same rate for both a continuous column and bubble trains, we have $(kA/\eta)\Delta P/L_p = (kA/\eta)\Delta P_{bt}/(L_{lb})$, or $\Delta P/L_p = \Delta P_{bt}/L_{lb}$. Therefore, for a bubble with an adjacent liquid bridge, the pressure gradient is $\Delta P_{bt}/(L_b + L_{lb}) = (\Delta P/L_p)[L_{lb}/(L_b + L_{lb})]$. Thus, the insect can significantly decrease the sucking pressure up to $[L_{lb}/(L_b + L_{lb})]$ times. The longer the bubbles, the less the sucking pressure.

When the length of a liquid bridge becomes comparable to the radius of the food canal, the flow deforming the end meniscus might change the given estimate, as in the case of foams moving through porous media (Kornev et al. 1999; Fujioka et al. 2016). The possibility of surface-active chemosensory proteins in lepidopteran food canals (Liu et al. 2014; Zhu et al. 2016a) favors this hypothesis, suggesting that liquid bridges can be stabilized by these surfactants. These effects need to be studied to evaluate the physical determinants of the feeding features of insects with long proboscises.

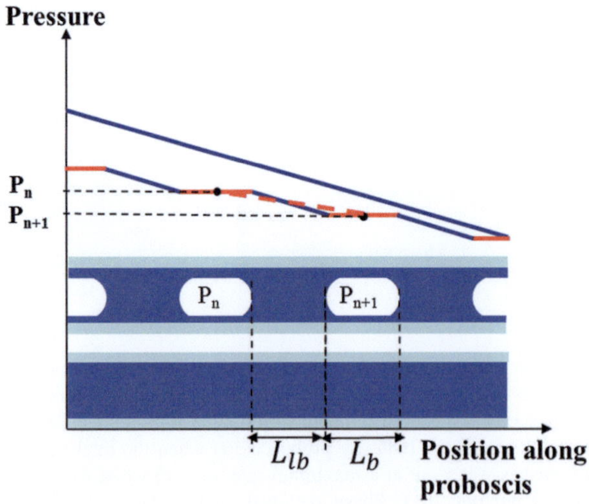


**Fig. 8.21** Schematic of the pressure distribution in a proboscis with a bubble train and with a continuous liquid column, assuming that the mean velocity of the liquid bridges is the same as that of the column. In each bubble, the pressure is constant, and the n-th bubble has pressure $P_n$. The liquid bridge has length $L_{lb}$ and the bubble has length $L_b$. The blue line is the pressure along the liquid column; it linearly decreases toward the insect head, with slope $\Delta P/L_p$, where $\Delta P$ is the pressure differential generated by the sucking pump. The polygonal line represents the pressure distribution along the bubble train. To have the same mean velocity, the slopes $(P_n - P_{n+1})/L_{lb}$ and $\Delta P/L_p$ must be equal. However, the slope of the dashed line situated between the centers of adjacent bubbles is smaller suggesting that the pressure required to move the bubble train is lower than that needed for transport of continuous liquid columns (Schematic by authors)

### 8.6.3 Pairing of the Proboscis and Sucking Pump

Fluid-feeding insects evolved a unique strategy to apportion the work of a proboscis and a sucking pump. However, until recently, the sucking pump was largely ignored in analyses of insect feeding mechanisms. Kingsolver and Daniel hypothesized that an insect with a long proboscis expends its muscular energy mostly on combating viscous friction of fluid moving through the proboscis (Daniel and Kingsolver 1983; Daniel et al. 1989). In fluid-feeding insects, however, proboscis length varies over a wide range of scales and the viscous dissipation of moving fluid in the pump cannot always be negligible.

The pump mechanics were first explained by Bennet-Clark (1963), who proposed an informative model of the buccal chamber of *Rhodnius prolixus* as a U-shaped dish covered by a piston (plunger) moving up and down through the central opening of the dish (Fig. 8.22). This model was generalized to other fluid-feeding insects, with the U-shaped cross section of the pump as the main geometrical motif (Daniel et al. 1989; Lehane 2005; Vogel 2007; Bach et al. 2015), but with cylindrical and rectangular buccal chambers treated separately (Fig. 8.22). The plunger is assumed to fit the U-shaped floor tightly so that the pump height, $h$, in the z-direction perpendicular to the floor remains smaller than other scales (Table 8.1). Bennet-Clark (1963) and others did not discuss the mechanism of suction pressure generation, assuming that the pressure in the pump is uniform and its magnitude is somehow related to the plunger position. The model of a time-dependent chamber pressure has been applied to mosquitoes, with the chamber modeled as an ellipsoid of revolution (Kikuchi et al. 2018).

To identify the chamber pressure–plunger position relationship, one has to consider specifics of flow in the buccal chamber. When the buccal chamber is almost closed and only a thin layer of liquid separates the plunger from the chamber bottom, and the plunger is set to move, due to the cohesion of the liquid particles, the whole liquid layer is engaged in the flow. A theoretical analysis of flow in the buccal chamber reveals that movement of the plunger establishes a nonhomogeneous pressure distribution in the pump (Kornev et al. 2017) (Fig. 8.23a, b). When the plunger is moving up and opening the chamber, and the distance between the plunger and chamber floor remains small, $h/R \to 0$, it generates a suction (negative with respect to the atmospheric) pressure written as

$$p = \eta A(dh/dt)\left(12/h^3\right) U(X,Y) - \eta A(dh/dt)\left(128 B L_p / d_p^4\right), \qquad (8.8)$$

where, for a cylindrical chamber $A = B = R^2$ , $X = x/R$, $Y = y/R$, $R$ is the chamber radius, and for a rectangular chamber $A = L^2$ and $B = W/(\pi L)$, $X = x/L$, $Y = y/L$, and $L$, $W$ are the chamber length and width, respectively. Other parameters are defined in Fig. 8.22. The dimensionless function $U$ is plotted in Fig. 8.23a for a cylindrical chamber; more details on the behavior of pressure in cylindrical and rectangular chambers are given by Kornev et al. (2017). The function $U$ does not depend on any

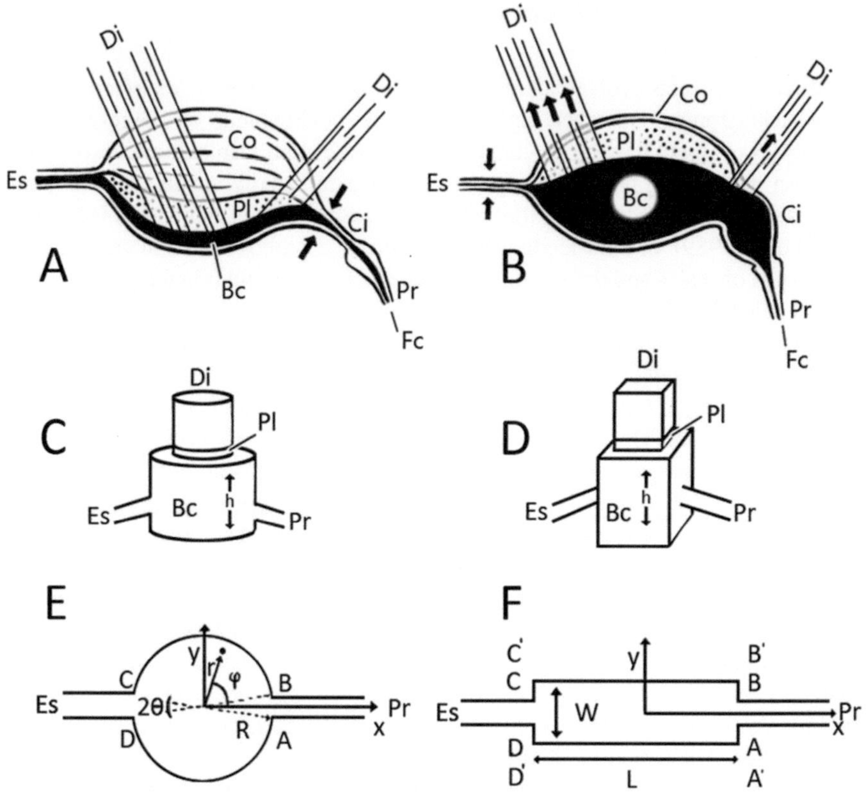


**Fig. 8.22** Schematic of lepidopteran sucking pump. (**a**, **b**) Lateral view of the sucking pump, consisting of the buccal chamber (Bc) and cibarium (Ci). Flow of liquid is from right to left, through the food canal (Fc) of the proboscis (Pr), into the sucking pump (Ci and Bc), and exiting via the esophagus (Es). Arrows indicate direction of muscle contraction. In A, the dilator muscles (Di) have relaxed and the compressor muscle (Co) of the pump has contracted, forcing the plunger (Pl) toward the chamber floor. In B, the compressor muscle of the pump has relaxed and the dilator muscles (Di) have contracted, drawing the plunger toward the dorsum of the buccal chamber. (**c**–**f**) Models of the sucking pump (not to scale). (**c**) The Daniel and Kingsolver (1983) model of a sucking pump. The buccal chamber is modeled as a cylindrical chamber and the circular plunger fits the chamber firmly. When the plunger moves in the vertical direction, it changes the expansion h. The proboscis and esophagus are attached to the buccal chamber. (**d**) The Bennet-Clark (1963) model of a sucking pump. The buccal chamber is modeled as a rectangular box and the plunger fits the box firmly. When the plunger moves in the vertical direction, it changes the expansion h. The proboscis and esophagus are attached to the buccal chamber. (**e**) Circular (radius R) lengthwise cross section of a model pump. Ratio AB/R is equal to 2θ. The model pump has opening AB connecting the chamber with the proboscis and opening CD connecting the chamber with the esophagus. (**f**) Rectangular lengthwise cross section of width W and length L of a model pump. Any point on the chamber floor can be specified by either Cartesian coordinates (x, y) for rectangular U-chambers or cylindrical coordinates (r, φ) for circular U-chambers. (Reproduced by permission from Kornev et al. 2017, CC BY 4.0 (https://creativecommons.org/licenses/by/4.0)

physical parameters of the fluid, but depends only on the ratio of the food canal diameter $d_p = |AB|$ to the cylinder radius *R*; for a rectangular chamber, this function depends on the chamber width to length ratio *W*/*L* and food canal diameter $d_p = |AB|$ to the chamber length ratio |*AB*|/*L* (Fig. 8.22).

Equation (8.8) demonstrates factorization of the time-dependent terms involving the time-dependent height of the plunger, *h*(*t*), and its velocity, *dh*/*dt*, from function *U*, which depends only on the in-plane coordinates *x* and *y*. This factorization suggests that the in-plane pressure pattern remains universal; the rate of plunger movement and the instantaneous height of the plunger affect the magnitude of the generated pressure, but they do not change the shape and positions of the lines of equal pressure in Fig. 8.23. As follows from Darcy's law, the spatial pattern of the in-plane fluid velocity, which depends on the pressure gradient, remains the same during the suction stroke; only the magnitude of velocity changes.

For long proboscises, the second term, $\eta A(dh/dt)\left(128BL_p/d_p^4\right)$, in Eq. (8.8) is much greater than the first, $\eta A(dh/dt)(12/h^3)\,U(X, Y)$, implying that the pressure in the chamber has time to relax to a constant quasi-static value during the suction stroke. This is the Daniel–Kingsolver scenario for fluid uptake by insects with long proboscises (Daniel and Kingsolver 1983; Daniel et al. 1989). However, at any given time moment *t*, the pressure in the chamber depends on the rate of the plunger pull, $p \propto dh/dt$.

Even for a long-proboscis insect, for which the flow in the chamber is expected to be much faster than that through the proboscis, at the first time moments of the stroke, the pressure cannot be considered quasi-static: the first term, $\eta A(dh/dt)(12/h^3)\,U(X, Y)$, in Eq. (8.8) dominates the second term, $\eta A(dh/dt)\left(128BL_p/d_p^4\right)$, and the pressure follows the Stefan–Reynolds law, $p \propto h^{-3}dh/dt \to \infty$ (Stefan 1874; Reynolds 1886). The suction pressure is large at the proboscis opening and goes to zero at the opposite side of the chamber near the closed esophagus. Accordingly, the fluid velocity at the proboscis opening is large and decreases to zero toward the esophagus. As the plunger moves farther up, the suction pressure steeply decreases $\propto h^{-3}$. For example, doubling the plunger height reduces the pressure gradient and fluid velocity in the chamber eightfold. This steep dependence on plunger height supports the hypothesis of a fast establishment of a quasi-static uniform pressure distribution in the chamber. Thus, for a long-proboscis insect with a large chamber, the pressure inside the chamber is almost constant during the stroke and one can safely neglect any flow from the chamber opening to the esophagus. However, for insects with a small plunger height to chamber size ratio, this hypothesis is probably incorrect. In any situation, the pressure in the chamber cannot be assumed to be time-independent.

The same rationale is applicable for the swallowing stroke when the fluid moves to the open esophagus. In Eq. (8.8), the food canal diameter $d_p = |AB|$ and its length $L_p$ would be replaced with the esophagus diameter and length (Kornev et al., 2017).

An organism's performance is based on its physiological and anatomical features (Wainwright et al. 1982; Daniel et al. 1989; Barth 1991; Pass 2018). Accordingly,

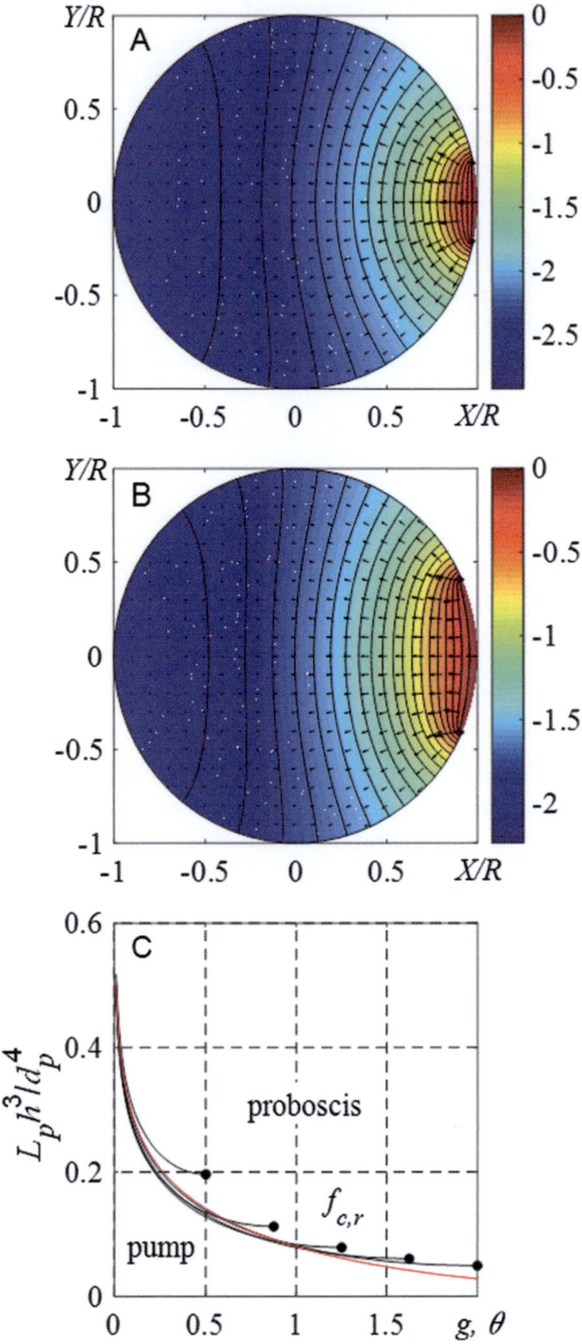


**Fig. 8.23** Dimensionless in-plane pressure distribution $U$ ($x/R$, $y/R$) for a circular chamber with a food canal opening angle (**a**) $\theta = 2\pi/15$ and (**b**) $\theta = 4\pi/15$. The color bar sets the pressure level. Black lines indicate constant pressure; the two nearest lines have a dimensionless pressure difference of 0.2. (**c**) Classification of the pump–proboscis pair with respect to energy dissipation mechanisms. The red curve indicates a circular pump, and the black curves indicate rectangular pumps with an increasing W/L ratio, with uniform steps of 0.5, starting from 0.5 and ending at 2. Circles at the end of each curve correspond to limiting cases of pumps with $W = AB$. (Courtesy of A. Salamatin)

fluid mechanics of the sucking pump–proboscis pair are related to the specific arrangement of the dilator and compressor muscles and the properties of the plunger (Tawfik 1968; Daniel et al. 1989; Kingsolver and Daniel 1995). A useful description of fluid interactions, with dilator muscles originating on the head capsule and inserting on the flexible plunger of the sucking pump, is given in terms of impulse of force, $\Pi$, which contains more physiological information on functioning of the dilator muscles than on force itself (Kornev et al. 2017). Only in a special case when dilator muscles exert a constant force, $F$, on the plunger at each time moment, $t$, can the impulse of force be factorized as $\Pi = \int_0^t F dt = F \cdot t$. In a general case, the dilator and compressor muscles should not be assumed to exert a time-independent force; the impulse of force has to be considered as a physiologically relevant mechanical determinant of the plunger. The reaction force, that is, the force of viscous drag acting on the rectangular ($F_r$) and cylindrical ($F_c$) plungers, is calculated, using Eq. (8.8), as

$$F_r(t) = \left(12\eta/h^3\right)(dh/dt)A^2D_{\mathrm{r}} - (dh/dt)\frac{128\pi\eta L_p}{d_p^4}(AB)^2, \tag{8.9}$$

$$F_c(t) = \left(12\eta/h^3\right)(dh/dt)A^2D_{\mathrm{c}} - (dh/dt)\frac{128\eta\pi L_{\mathrm{p}}}{d_p^4}(AB)^2, \tag{8.10}$$

where constants $D_{\mathrm{r}}$ and $D_{\mathrm{c}}$ depend only on the food canal diameter through the ratios $|AB|/L$ for a rectangular and $|AB|/(2\mathrm{R})$ for a circular pump and on the elongation of a rectangular pump, $W/L$ (Kornev et al. 2017), and $A = B = R^2$ for a cylindrical chamber, and $A = L^2$ and $B = W/(\pi L)$ for a rectangular chamber. The first terms, $(12\eta/h^3)(dh/dt)A^2D_{\mathrm{i}}$, $i = r, c$, of these equations resemble the familiar Stefan law, suggesting that the force generated by the dilator muscles at the first moments of the suction stroke has to be extremely high, $F \propto h^{-3}dh/dt$, to detach the plunger from the chamber floor. The Stefan force decreases steeply as the plunger moves up, and the second term, $(dh/dt)\left(128\pi\eta L_{\mathrm{p}}/d_p^4\right)(AB)^2$, is expected to prevail over the rest of the stroke. Still, the drag force depends on the rate of the plunger movement; hence, the impulse of drag force can be factorized only in the case $dh/dt = const.$

Energy dissipation in the proboscis–sucking pump complex is associated with the viscous drag of liquid moving through the proboscis ("proboscis dissipation") and energy dissipation from the viscous drag of liquid on the moving pump plunger ("pump dissipation"). Pump dissipation is caused by the Stefan viscous adhesion of the plunger to the floor of the pump chamber, whereas proboscis dissipation is related to viscous friction of liquid moving through the proboscis. These two mechanisms control the division of labor between the proboscis and sucking pump.

Insects can be classified according to their mechanism of energy dissipation by introducing two dimensionless parameters, $f = L_{\mathrm{p}}h^3/d_p^4$ and $g = d_{\mathrm{p}}/L$. Figure 8.23c specifies the insects that dissipate their muscular energy mostly in transporting fluids through their sucking pump versus their proboscis. The derived diagram allows the constraints of fluid mechanics on evolution of the feeding organs to be examined. Insects with a large *f*-factor expend most of their musculature energy fighting viscous drag of fluids moving through the proboscis. These insects include bed bugs, mosquitoes, planthoppers, and sucking lice (Snodgrass 1935; Robinson 1939; Surtees 1959; Christophers 1960; Bennet-Clark 1963; Tawfik 1968; Malone et al. 1999; Lee et al. 2009; Chapman 2013; Karolyi et al. 2013; Kikuchi et al. 2018). Insects with a small *f*-factor expend most of their muscular energy fighting viscous drag of fluids moving through the pump. In butterflies and moths, the structure of the sucking pump and dilator musculature is similar across species, whether they feed on watery or thick fluids. The morphology of the proboscis (e.g., its length to width ratio), however, varies across species (Adler 1982; Miller 1991). Moths with unlinked galeae (e.g., some short-proboscis Notodontidae) could alternately bring the galeae together and separate them, like a valve, permitting fluid uptake to accelerate or decelerate by switching the dissipation mechanism from proboscis-driven to pump-driven (Tsai et al. 2014; Zhang et al. 2018a). A butterfly whose galeae came apart could achieve the flip from pump to proboscis dissipation by bringing the galeae somewhat together but never linking them. In this way, it would still be able to feed even though it could never repair the proboscis completely.

## 8.7 Synthesis

### *8.7.1 Generalized Model for Fluid-Feeding*

The insect proboscis is a remarkable example of natural engineering. Its tiny, fiber-like components can take up fluids ranging from thin mineral water to thick nectar and honey. Blood-feeding insects have a remarkable ability to transport blood through food canals as narrow as a cell diameter. Equally remarkable is proboscis diversity of form and function across more than half a million fluid-feeding species. The deceptive simplicity of the proboscis belies the complexity in nature's evolutionary solutions to fluid-feeding. Yet, a small set of physical principles underlies the diversity of proboscises. We have used the lepidopteran proboscis to illustrate these fundamental principles and to describe a general model that can accommodate proboscis configurations of other fluid-feeding insects.

Fluid-feeding by insects has conventionally been generalized into three models, covering an enormous number of species, primarily in four megadivers-orders, the Diptera, Hemiptera, Hymenoptera, and Lepidoptera, which have feeding organs derived from different mouthpart structures. These models have been recognized as the drinking straw, dipstick, and sponge (Smith 1985; Kingsolver and Daniel 1995; Kim and Bush 2012). This trichotomy is based on gross structure of the mouthparts, suggesting that tubular mouthparts function as a drinking straw (e.g., bugs, butterflies, and mosquitoes) (Bennet-Clark 1963; Kingsolver and Daniel 1995) or a dipstick (e.g., bees) (Barth 1991; Abou-Shaara 2014; Zhu et al. 2016b), and that non-tubular mouthparts function as a sponge (e.g., flies) (Graham-Smith 1930; Driscoll and Condon 1994). Thus, insect mouthparts were classified largely according to hypothetical flow mechanisms suggested by fluid mechanics that considered only the shape of the fluid conduit (Smith 1985; Kingsolver and Daniel 1995; Kim and Bush 2012). Microscale visualization of fluid uptake by insects had been limited (Socha et al. 2007; Westneat et al. 2008), and the classification lacked an experimental basis. Recent experimental data have called for integration of physiology, mechanics, behavior, and neural control at nano-, micro-, and organismal scales to decipher complex mechanisms of fluid uptake.

We suggest that all fluid-feeding insects can be accommodated under a single, generalized sponge–straw model with four component steps in fluid acquisition: (1) wetting, (2) dewetting, (3) absorbing, and (4) pumping. This general model is based on a fundamental structural organization of feeding devices that consists of porous and fibrillar mouthpart structures with high surface area (Snodgrass 1935; Krenn 2010; Krenn and Aspöck 2012; Chapman 2013). Identifying the four elements of the general model, largely centered around physical principles of wetting, capillarity, and fluid mechanics, can help explain variations in structure across more than 500,000 fluid-feeding species. The physical determinants involved in the four-step process of fluid acquisition offer quantitative correlations between the structure of insect feeding devices and their performance characteristics. Thus, even small changes in proboscis structure and size and surface properties, as they affect one or more of these four steps, can lead to novel functions. By focusing on the physicochemical details of these steps, paradoxes of the proboscis can be resolved, such as simultaneous hydrophilic–hydrophobic functionality and tapering, which enhances capillarity but increases viscous friction. We present the four-step model, using Lepidoptera as the example. With slight adjustments as appropriate, the model should be applicable to most fluid-feeding insects.

**Wetting** In the initial step of fluid-feeding, an insect probes the liquid. If the available fluid is in the form of droplets, films, or pools, a layer of fluid readily covers the proboscis. On the other hand, if the fluid is held in pores of a substrate such as rotting fruit or soil, the proboscis must contact the meniscus in the substrate pores. This contact occurs when the insect presses its proboscis against the substrate, resulting in wetting of the proboscis by a film. The structure of the proboscis with its interlegular spaces promotes capillary pull of fluids from porous surfaces of many food resources, much like a sponge (Monaenkova et al. 2012). Cross-sectional shape of the proboscis plays a role in liquid acquisition. Lepidopteran proboscises are often elliptical, especially in species that feed from surfaces rather than from floral tubes. Ellipticity is enhanced by lateral brush-like arrays of chemosensilla in the distal drinking region of species that routinely feed from surface films such as sap flows (Campos et al. 2015; Lehnert et al. 2016). An elliptical proboscis, compared with a cylindrical proboscis, brings the contact line higher on the drinking region; consequently, more interlegular spaces are covered with fluid (Lehnert et al. 2013).

**Dewetting** In the second step, fluid moves toward the permeable interlegular gaps of the dorsal legular band. This directional flow of fluid—dewetting—is increased by motions of the proboscis, such as bending while feeding and coiling after feeding (Zhang et al. 2018b). Bending enhances the role of gravity in moving the fluid, and mopping and sweeping the proboscis over a food source, such as open fruit, accelerates the flow. Materials features of the proboscis surface, such as its chemistry and topography, aid directional movement. The variation in lepidopteran proboscis sculpture, which we have reduced to three categories—bumps–valleys, spikes, and shingles—suggests a variety of fluid behaviors on the proboscis surface. In other words, the different shapes and heights of the meniscus formed by these types of sculpturing will influence large-scale wettability. Proboscis surface sculpture, which varies not only among species, but also along the length and circumference of an individual proboscis, offers a rich source for future experiments on wettability and directed flow.

**Absorbing** In the third step, the proboscis again performs like a sponge, absorbing liquid into the food canal. Most fluid enters the food canal in the distinctive drinking region, typically the distal 5–17% of the proboscis (Monaenkova et al. 2012). As fluid along the dorsal legular seam is absorbed into the food canal via capillary action, a film forms on the canal wall. The film becomes a bulge that enlarges until plateau instability causes it to collapse into a liquid bridge in the food canal (Monaenkova et al. 2012). Thus, a series of liquid bridges, separated by tiny air pockets, forms in the food canal, ready for transport into the gut.

**Pumping** During the fourth and final step, fluid in the food canal moves into the gut by action of a sucking pump in the insect's head. The pump allows the acquired fluid to be swallowed. Liquid-bridge formation in the food canal is independent of the sucking pump (Monaenkova et al. 2012), as are pulsing and antiparallel sliding of the galeae, which can facilitate fluid uptake (Tsai et al. 2014). Independence of the proboscis and pump increases the breadth of functionality. Yet, the proboscis and pump also operate as a functional complex, with two mechanisms of energy dissipation—proboscis dissipation and pump dissipation—depending on the viscous drag of fluid moving through them (Kornev et al. 2017).

### *8.7.2 Evolutionary Perspectives*

The earliest Lepidoptera with a proboscis, the Glossata, are represented today by groups such as the Eriocraniidae (Regier et al. 2013) that use their short proboscises to obtain water and sap (Monaenkova et al. 2012). These contemporary representatives of ancient moth lineages provide insights into the evolution of the lepidopteran proboscis and illustrate the probable role that physical principles have played since the origin of the glossatan Lepidoptera, perhaps 150 mya. The eriocraniids have most of the fundamental features that characterize all coilable lepidopteran proboscises, suggesting that capillarity and wetting were operating early in the phylogeny of the species-rich suborder Glossata: interlegular spaces between adjacent legulae, a bristly (microtrichiated) proboscis consisting of two galeae, a ridged food canal, prominent surface sculpturing, and sensilla (Monaenkova et al. 2012). The implication is that structural features enhanced capillarity and wettability, conferring an early selective advantage in acquiring liquids. Thus, the challenges of withdrawing fluid from porous surfaces and trapped in crevices would have been overcome early in the evolution of the oldest Glossata, which, like the Eriocraniidae, presumably fed on these fluids. Lepidoptera, therefore, would have been capable not only of feeding opportunistically from gymnosperm sap, wet soil, and animal excreta when no flowering plants existed (Labandeira 2010; van Eldijk et al. 2018), but also of exploiting nectar in floral tubes and juices from rotting fruit during diversification of the angiosperms. Similar arguments could be applied to other groups of insects with spongy and fibrous proboscises (Driscoll and Condon 1994; Zhu et al. 2016b).

Early Lepidoptera that acquired fluid with their proboscises also would have had to deliver the fluid to their guts. Thus, a well-developed sucking pump, essential to swallowing acquired fluid, would have been in place early in the evolution of the Glossata. Insects with short proboscises and uncoupled galeae would have spent most energy moving fluid through the pump (Kornev et al. 2017). Capillary forces alone would have ensured liquid delivery to the pump entrance, without involving the pump. Thus, pump dissipation might have been the dominant energy dissipation mechanism for the earliest Lepidoptera with short non-coilable proboscises. For insects with a long proboscis, such as Lepidoptera that exploit nectar in floral tubes, most energy would be spent fighting frictional forces of fluid moving through the proboscis (proboscis dissipation). Lengthening the proboscis, through natural selection, to reach fluids in deeper crevices and eventually to access nectar in elongated floral tubes during angiosperm diversification might have flipped the dissipation mechanism to the proboscis mode.

Overlaid on the structure-enabled functionality of the proboscis is the insect's behavioral repertoire, which can enhance mechanical performance and ease physical constraints. For example, a butterfly can exert control over the flow on its proboscis surface by any of numerous proboscis movements (Zhang et al. 2018b). Additionally, a butterfly can reduce the viscous friction in the tapered food canal by slightly displacing the galeae, for instance by sliding them in antiparallel fashion or pulsing them (Tsai et al. 2014). To increase control over fluid uptake and saliva expression, a butterfly can alter the size of the interlegular gaps by bending and straightening the proboscis (Kwauk et al. 2014; Tsai et al. 2014). Thus, an insect's behavior, responsive to the environmental context, can significantly broaden the functionality of the feeding device. Most artificially engineered devices cannot capture the range of behaviors able to respond to a particular environmental situation. Perhaps for this reason, true biomimicry is not possible, and "bioinspiration" is the more appropriate term.

A compelling case can be made for the role of the proboscis in driving diversification and adaptive radiation of fluid-feeding insects (Monaenkova et al. 2012). Although Ehrlich and Raven (1964) claimed that "the plant-herbivore interface may be the major zone of interaction responsible for generating terrestrial organic diversity," the multifunctionality of the proboscis might have allowed insects to move into additional adaptive zones, generating even greater biodiversity.

A foundation has been laid for the physical determinants of fluid-feeding, but much remains to be done while bridging nano-, micro-, and organismal scales. The lessons from natural proboscises have set the scene for producing microfluidic probes that can be used, for instance, in medical and forensic sciences. These artificial probes would have physical properties similar to a lepidopteran proboscis: porous, strong, flexible, self-cleaning, and with multiple functionality (Tsai et al. 2011; Wong et al. 2017; Shou and Fan 2018).The next frontier is the merger of surface chemistry and physical structure. Lipids of the insect cuticle—little studied for the proboscis—are rich in diversity, including more than 100 hydrocarbons (Wang et al. 2016). Among their various functions, they confer differential wettability to the cuticle.

**Acknowledgments** We are grateful to our students and collaborators with whom we have worked over the past 10 years on the butterfly mechanics project. Taras Andrukh, Pavel Aprelev, Charles Beard, Yu Gu, Daniel Hasegawa, Wah-Keat Lee, Matthew Lehnert, Daria Monaenkova, Maryana Nave, Suellen Pometto, Benyamin Rubin, Vijoya Sa, Arthur Salamatin, Luke Sande, Leslie Sanders, Alexander Tokarev, Chen-Chih Tsai, Chengqi Zhang, and many undergrad and high school students helped us with experiments at different stages of the research. Over these 10 years, this work was generously supported by the National Science Foundation. We thank Carmony Corley and Allison Stoiser for generously providing photos of Lepidoptera and the associated information.

## References

Abou-Shaara HF (2014) The foraging behaviour of honey bees, *Apis mellifera*: a review. Veti Med 59:1–10

Adam NK (1937) Detergent action and its relation to wetting and emulsification. J Soc Dye Colour 53:121–129

Adamson AW, Gast AP (1997) Physical chemistry of surfaces. Wiley, New York

Adler PH (1982) Soil- and puddle-visiting habits of moths. J Lepid Soc 36:161–173

Adler PH, Foottit RG (2009) Introduction. In: Foottit RG, Adler PH (eds) Insect biodiversity: science and society. Wiley, Chichester, pp 1–6

Adler PH, Pearson DL (1982) Why do male butterflies visit mud puddles? Can J Zool 60:322–325

Arditti J, Elliott J, Kitching IJ, Wasserthal LT (2012) 'Good heavens what insect can suck it' – Charles Darwin, *Angraecum sesquipedale* and *Xanthopan morganii praedicta*. Bot J Linn Soc 169:403–432

Arms K, Feeny P, Lederhouse RC (1974) Sodium: stimulus for puddling behavior by tiger swallowtail butterflies, *Papilio glaucus*. Science 185:372–374

Aroulmoji V, Aguie-Beghin V, Mathlouthi M, Douillard R (2004) Effect of sucrose on the properties of caffeine adsorption layers at the air/solution interface. J Colloid Interface Sci 276:269–276

Auclair JL (1963) Aphid feeding and nutrition. Annu Rev Entomol 8:439–490

Bach D, Schmich F, Masselter T, Speck T (2015) A review of selected pumping systems in nature and engineering – potential biomimetic concepts for improving displacement pumps and pulsation damping. Bioinspir Biomim 10:051001

Bänziger H (1971) Blood sucking moths of Malaya. Fauna Zool Mag 1:5–16

Barber HS (1924) New Ptiliidae related to the smallest known beetle. Proc Entomol Soc Wash 26:167–178

Barth FG (1991) Insects and flowers: the biology of a partnership. Princeton University Press, Princeton, NJ

Bauder JAS, Handschuh S, Metscher BD, Krenn HW (2013) Functional morphology of the feeding apparatus and evolution of proboscis length in metalmark butterflies (Lepidoptera: Riodinidae). Biol J Linn Soc 110:291–304

Beament JW (1961) The water relations of insect cuticle. Biol Rev Camb Philos Soc 36:281–320

Bennet-Clark HC (1963) Negative pressures produced in the pharyngeal pump of the blood-sucking bug, *Rhodnius prolixus*. J Exp Biol 40:223–229

Bhushan B, Jung YC (2011) Natural and biomimetic artificial surfaces for superhydrophobicity, self-cleaning, low adhesion, and drag reduction. Prog Mater Sci 56:1–108

Borrell BJ, Krenn HW (2006) Nectar feeding in long-proboscid insects. In: Herrel A, Speck T, Rowe NP (eds) Ecology and biomechanics: a mechanical approach to the ecology of animals and plants. CRC, Boca Raton, FL, pp 185–205

Brehm G, Fischer M, Gorb S, Kleinteich T, Kuhn B, Neubert D, Pohl H, Wipfler B, Wurdinger S (2015) The unique sound production of the Death's-head hawkmoth (*Acherontia atropos* (Linnaeus, 1758)) revisited. Sci Nat 102:43

Bush JWM, Hu DL, Prakash M (2007) The integument of water-walking arthropods: form and function. In: Casas J, Simpson SJ (eds) Advances in insect physiology: insect mechanics and control. Academic, Camebridge, MA, pp 117–192

Büttiker W, Krenn HW, Putterill JF (1996) The proboscis of eye-frequenting and piercing Lepidoptera (Insecta). Zoomorphology 116:77–83

Campos EO, Bradshaw HD, Daniel TL (2015) Shape matters: corolla curvature improves nectar discovery in the hawkmoth *Manduca sexta*. Funct Ecol 29:462–468

Cao MY, Jiang L (2016) Superwettability integration: concepts, design and applications. Surf Innov 4:180–194

Carroll BJ (1976) The accurate measurement of contact angle, phase contact areas, drop volume, and Laplace excess pressure in drop-fiber systems. J Colloid Interface Sci 57:488–495

Cassie ABD (1948) Contact angles. Discuss Faraday Soc 3:11–16

Cassie ABD, Baxter S (1944) Wettability of porous surfaces. Trans Faraday Soc 40:0546–0550

Chapman RF (2013) The insects: structure and function, 5th edn. Cambridge University Press, Cambridge

Cheer AYL, Koehl MAR (1987) Paddles and rakes – fluid flow through bristled appendages of small organisms. J Theor Biol 129:17–39

Chen J, Wu JN, Yan SZ (2015) Switchable wettability of the honeybee's tongue surface regulated by erectable glossal hairs. J Insect Sci 15:5

Christophers SR (1960) *Aëdes aegypti* (L.) the yellow fever mosquito: its life history, bionomics and structure. Cambridge University Press, London

Corbet SA (2000) Butterfly nectaring flowers: butterfly morphology and flower form. Entomol Exp Appl 96:289–298

Daniel TL, Kingsolver JG (1983) Feeding strategy and the mechanics of blood sucking insects. J Theor Biol 105:661–678

Daniel TL, Kingsolver JG, Meyhofer E (1989) Mechanical determinants of nectar feeding energetics in butterflies – muscle mechanics, feeding geometry, and functional equivalence. Oecologia 79:66–75

Darwin CR (1904) The various contrivances by which orchids are fertilised by insects. 7th impression of the 2nd edition. John Murray, London

Davis DR (1986) A new family of monotrysian moths from austral South America (Lepidoptera: Palaephatidae), with a phylogenetic review of the Monotrysia. Smithson Contr Zool 434:1–202

Davis NT, Hildebrand JG (2006) Neuroanatomy of the sucking pump of the moth, *Manduca sexta* (Sphingidae, Lepidoptera). Arthropod Struct Dev 35:15–33

Docoslis A, Giese RF, van Oss CJ (2000) Influence of the water-air interface on the apparent surface tension of aqueous solutions of hydrophilic solutes. Colloid Surf B-Biointerfaces 19:147–162

Driscoll CA, Condon MA (1994) Labellar modifications of *Blepharoneura* (Diptera: Tephritidae): Neotropical fruit flies that damage and feed on plant surfaces. Ann Entomol Soc Am 87:448–453

Dullien FAL (1991) Porous media. Fluid transport and pore structure. Academic, New York

Eastham LES, Eassa YEE (1955) The feeding mechanism of the butterfly *Pieris brassicae* L. Philos Trans R Soc Lond Ser B Biol Sci 239:1–43

Ehrlich PR, Raven PH (1964) Butterflies and plants: a study in coevolution. Evolution 18:586–608

Eisner T (2005) For love of insects. Belknap Press of Harvard University Press, New York, NY

Fadeev AY (2006) Hydrophobic monolayer surfaces: synthesis and wettability. In: Somasundaran P (ed) Encyclopedia of surface and colloid science. CRC, Boca Raton, FL, pp 2854–2875

Fasina OO, Colley Z (2008) Viscosity and specific heat of vegetable oils as a function of temperature: 35°C to 180°C. Int J Food Prop 11:738–746

Forbes P (2008) Self-cleaning materials. Sci Am 299:88–95

Fournier RL (2011) Basic transport phenomena in biomedical engineering. CRC, New York

Fujioka H, Halpern D, Ryans J, Gaver DP (2016) Reduced-dimension model of liquid plug propagation in tubes. Phys Rev Fluids 1:22

Genzer J, Marmur A (2008) Biological and synthetic self-cleaning surfaces. MRS Bull 33:742–746

Gosse PH (1993) Letters from Alabama (US), chiefly relating to natural history. University of Alabama Press, Originally published 1859, Morgan and Chase, London, Tuscaloosa, AL

Graham-Smith GS (1930) Further observations on the anatomy and function of the proboscis of the blow-fly, *Calliphora erythrocephala* L. Parasitology 22:47–115

Grimaldi D, Engel MS (2005) Evolution of the insects. Cambridge University Press, New York

Gui FL, Irvine TF (1994) Theoretical and experimental study of the falling cylinder viscometer. Int J Heat Mass Transf 37:41–50

Hainsworth FR, Precup E, Hamill T (1991) Feeding, energy processing rates and egg-production in painted lady butterflies. J Exp Biol 156:249–265

Hanken J, Wake DB (1993) Miniaturization of body size – organismal consequences and evolutionary significance. Annu Rev Ecol Syst 24:501–519

Harder LD (1986) Effects of nectar concentration and flower depth on flower handling efficiency of bumble bees. Oecologia 69:309–315

Hele-Shaw HS (1898) The flow of water. Nature 58:34–36

Hepburn HR (1971) Proboscis extension and recoil in Lepidoptera. J Insect Physiol 17:637–656

Hepburn HR (1985) Structure of the integument. In: Kerkut GA, Gilbert LI (eds) Comprehensive insect physiology, biochemistry, and pharmacology. Pergamon, New York, pp 1–58

Holdgate MW (1955) The wetting of insect cuticles by water. J Exp Biol 32:591–617

Irving JB (1972) Effect of nonvertical alignment on performance of a falling cylinder viscometer. J Phys D-Appl Phys 5:214

Jones JC, Pilitt DR (1973) Blood-feeding behavior of adult *Aedes aegypti* mosquitoes. Biol Bull 145:127–139

Karolyi F, Szucsich NU, Colville JF, Krenn HW (2012) Adaptations for nectar-feeding in the mouthparts of long-proboscid flies (Nemestrinidae: *Prosoeca*). Biol J Linn Soc 107:414–424

Karolyi F, Morawetz L, Colville JF, Handschuh S, Metscher BD, Krenn HW (2013) Time management and nectar flow: flower handling and suction feeding in long-proboscid flies (Nemestrinidae: *Prosoeca*). Naturwissenschaften 100:1083–1093

Karolyi F, Colville JF, Handschuh S, Metscher BD, Krenn HW (2014) One proboscis, two tasks: adaptations to blood-feeding and nectar-extracting in long-proboscid horse flies (Tabanidae, *Philoliche*). Arthropod Struct Dev 43:403–413

Kikuchi K, Stremler MA, Chatterjee S, Lee WK, Mochizuki O, Socha JJ (2018) Burst mode pumping: a new mechanism of drinking in mosquitoes. Sci Rep 8:15

Kim WJ, Bush JWM (2012) Natural drinking strategies. J Fluid Mech 705:7–25

Kim W, Gilet T, Bush JWM (2011) Optimal concentrations in nectar feeding. Proc Natl Acad Sci U S A 108:16618–16621

Kingsolver JG, Daniel TL (1979) On the mechanics and energetics of nectar feeding in butterflies. J Theor Biol 76:167–168

Kingsolver JG, Daniel TL (1995) Mechanics of food handling by fluid-feeding insects. In: Chapman RF, de Boer G (eds) Regulatory mechanisms in insect feeding. Springer, New York, pp 32–74

Knopp MCN, Krenn HW (2003) Efficiency of fruit juice feeding in *Morpho peleides* (Nymphalidae, Lepidoptera). J Insect Behav 16:67–77

Koehl MAR (2001) Transitions in function at low Reynolds number: hair-bearing animal appendages. Math Methode Appl Sci 24:1523–1532

Kornev KG, Neimark AV, Rozhkov AN (1999) Foam in porous media: thermodynamic and hydrodynamic peculiarities. Adv Colloid Interf Sci 82:127–187

Kornev KG, Monaenkova D, Adler PH, Beard CE, Lee WK (2016) The butterfly proboscis as a fiber-based, self-cleaning, micro-fluidic system. In: Martin-Palma RJ, Lakhtakia A, Knez M (eds) Bioinspiration, biomimetics, and bioreplication 2016. SPIE, Bellingham, Washington

Kornev KG, Salamatin AA, Adler PH, Beard CE (2017) Structural and physical determinants of the proboscis-sucking pump complex in the evolution of fluid-feeding insects. Sci Rep 7:6582

Krenn HW (1997) Proboscis assembly in butterflies (Lepidoptera) - a once in a lifetime sequence of events. Eur J Entomol 94:495–501

Krenn HW (2010) Feeding mechanisms of adult Lepidoptera: structure, function, and evolution of the mouthparts. Annu Rev Entomol 55:307–327

Krenn HW, Aspöck H (2012) Form, function and evolution of the mouthparts of blood-feeding Arthropoda. Arthropod Struct Dev 41:101–118

Krenn HW, Mühlberger N (2002) Groundplan anatomy of the proboscis of butterflies (Papilionoidea, Lepidoptera). Zool Anz 241:369–380

Krenn HW, Zulka KP, Gatschnegg T (2001) Proboscis morphology and food preferences in nymphalid butterflies (Lepidoptera: Nymphalidae). J Zool 254:17–26

Kristensen NP (1968) The anatomy of the head and the alimentary canal of adult Eriocraniidae (Lep. Dacnonypha). Entomologiske Meddelelser 36:239–313

Kunte K (2007) Allometry and functional constraints on proboscis lengths in butterflies. Funct Ecol 21:982–987

Kwauk KJ (2012) Fluid uptake by the lepidopteran proboscis in relation to structure. MS Thesis. Clemson University, Clemson, SC

Kwauk KJ, Hasegawa DK, Lehnert MS, Beard CE, Gerard PD, Kornev KG, Adler PH (2014) Drinking with an unsealed tube: fluid uptake along the butterfly proboscis. Ann Entomol Soc Am 107:886–892

Labandeira CC (1997) Insect mouthparts: ascertaining the paleobiology of insect feeding strategies. Annu Rev Ecol Syst 28:153–193

Labandeira CC (2010) The pollination of mid Mesozoic plants and the early history of long proboscis insects. Ann Mo Bot Gard 97:469–513

Laplace PS (1806) Mecanique Celeste. Encyclopaedia Britannica, Paris

Laplace PS (1966) Mecanique Celeste. Chelsea Publishing Company, New York

Lauder GV (2003) The intellectual challenge of biomechanics and evolution. In: Bels VL, Gasc JP, Casinos A (eds) Vertebrate biomechanics and evolution. BIOS Scientific, Oxford, pp 319–324

Lee SC, Lee SJ (2014) Uptake of liquid from wet surfaces by the brush-tipped proboscis of a butterfly. Sci Rep 4:6934

Lee SJ, Kim BH, Lee JY (2009) Experimental study on the fluid mechanics of blood sucking in the proboscis of a female mosquito. J Biomech 42:857–864

Lee SC, Kim BH, Lee SJ (2014a) Experimental analysis of the liquid-feeding mechanism of the butterfly *Pieris rapae*. J Exp Biol 217:2013–2019

Lee SJ, Lee SC, Kim BH (2014b) Liquid-intake flow around the tip of butterfly proboscis. J Theor Biol 348:113–121

Lehane MJ (2005) The biology of blood-sucking in insects. Cambridge University Press, New York

Lehnert MS, Monaenkova D, Andrukh T, Beard CE, Adler PH, Kornev KG (2013) Hydrophobic-hydrophilic dichotomy of the butterfly proboscis. J R Soc Interface 10:20130336

Lehnert MS, Beard CE, Gerard PD, Kornev KG, Adler PH (2016) Structure of the lepidopteran proboscis in relation to feeding guild. J Morphol 277:167–182

Liu K, Yao X, Jiang L (2010) Recent developments in bio-inspired special wettability. Chem Soc Rev 39:3240–3255

Liu YL, Guo H, Huang LQ, Pelosi P, Wang CZ (2014) Unique function of a chemosensory protein in the proboscis of two *Helicoverpa* species. J Exp Biol 217:1821–1826

Loudon C, Tordesillas A (1998) The use of the dimensionless Womersley number to characterize the unsteady nature of internal flow. J Theor Biol 191:63–78

Malone M, Watson R, Pritchard J (1999) The spittlebug *Philaenus spumarius* feeds from mature xylem at the full hydraulic tension of the transpiration stream. New Phytol 143:261–271

May PG (1985) Nectar uptake rates and optimal nectar concentrations of two butterfly species. Oecologia 66:381–386

McMahon T, Bonner JT (1983) On size and life. Scientific American Books - W. H. Freeman and Co, New York

Miller JS (1991) Cladistics and classification of the Notodontidae (Lepidoptera: Noctuoidea) based on larval and adult morphology. Bull Am Mus Nat Hist 204:1–230

Misof B, Liu S, Meusemann K, Peters RS, Donath A, Mayer C, Frandsen PB, Ware J, Flouri T, Beutel RG, Niehuis O, Petersen M, Izquierdo-Carrasco F, Wappler T, Rust J, Aberer AJ, Aspöck U, Aspöck H, Bartel D, Blanke A, Berger S, Böhm A, Buckley T, Calcott B, Chen J, Friedrich F, Fukui M, Fujita M, Greve C, Grobe P, Gu S, Huang Y, Jermiin LS, Kawahara AY, Krogmann L, Kubiak M, Lanfear R, Letsch H, Li Y, Li Z, Li J, Lu H, Machida R, Mashimo Y, Kapli P, McKenna DD, Meng G, Nakagaki Y, Navarrete-Heredia JL, Ott M, Ou Y, Pass G, Podsiadlowski L, Pohl H, von Reumont BM, Schütte K, Sekiya K, Shimizu S, Slipinski A, Stamatakis A, Song W, Su X, Szucsich NU, Tan M, Tan X, Tang M, Tang J, Timelthaler G, Tomizuka S, Trautwein M, Tong X, Uchifune T, Walzl MG, Wiegmann BM, Wilbrandt J, Wipfler B, Wong TK, Wu Q, Wu G, Xie Y, Yang S, Yang Q, Yeates DK, Yoshizawa K, Zhang Q, Zhang R, Zhang W, Zhang Y, Zhao J, Zhou C, Zhou L, Ziesmann T, Zou S, Li Y, Xu X, Zhang Y, Yang H, Wang J, Wang J, Kjer KM, Zhou X (2014) Phylogenomics resolves the timing and pattern of insect evolution. Science 346:763–767

Molleman F (2010) Puddling: from natural history to understanding how it affects fitness. Entomol Exp Appl 134:107–113

Monaenkova D, Lehnert MS, Andrukh T, Beard CE, Rubin B, Tokarev A, Lee WK, Adler PH, Kornev KG (2012) Butterfly proboscis: combining a drinking straw with a nanosponge facilitated diversification of feeding habits. J R Soc Interface 9:720–726

Norris MJ (1936) The feeding-habits of the adult Lepidoptera Heteroneura. Trans R Entomol Soc 85:61–90

Noureddini H, Teoh BC, Clements LD (1992) Viscosities of vegetable oils and fatty acids. J Am Oil Chem Soc 69:1189–1191

Parker AR, Lawrence CR (2001) Water capture by a desert beetle. Nature 414:33–34

Pass G (2018) Beyond aerodynamics: the critical roles of the circulatory and tracheal systems in maintaining insect wing functionality. Arthropod Struct Dev 47:391–407

Pivnick KA, McNeil JN (1985) Effects of nectar concentration on butterfly feeding – measured feeding rates for *Thymelicus lineola* (Lepidoptera, Hesperiidae) and general feeding model for adult Lepidoptera. Oecologia 66:226–237

Polilov AA (2015) Small is beautiful: features of the smallest insects and limits to miniaturization. Annu Rev Entomol 60:103–121

Quere D (2008) Wetting and roughness. Annu Rev Mater Res 38:71–99

Regier JC, Mitter C, Zwick A, Bazinet AL, Cummings MP, Kawahara AY, Sohn JC, Zwickl DJ, Cho S, Davis DR, Baixeras J, Brown J, Parr C, Weller S, Lees DC, Mitter KT (2013) A large-scale, higher-level, molecular phylogenetic study of the insect order Lepidoptera (moths and butterflies). PLoS One 8:23

Reiser P, Birch GG, Mathlouthi M (1995) Physical properties. In: Mathlouthi M, Reiser P (eds) Sucrose properties and applications. Blackie Academic and Professional, London, pp 186–222
Rensch B (1948) Histological changes correlated with evolutionary changes of body size. Evolution 2:218–230
Reynolds O (1886) On the theory of the lubrication and its application to Mr. Beauchamps Tower's experiments including an experimental determination of the viscosity of olive oil. Philos Trans R Soc Lond 177:157–234
Robinson GG (1939) The mouthparts and their function in the female mosquito, *Anopheles maculipennis*. Parasitology 31:212–242
Russell ES (1916) Form and function. John Murray, London
Scheidegger AE (1974) The physics of flow through porous media. University of Toronto, Toronto
Schmitt JB (1938) The feeding mechanism of adult Lepidoptera. Smithson Misc Collect 97:1–28
Shaw SR (2014) Planet of the bugs. The University of Chicago Press, Chicago
Shou DH, Fan JT (2018) Design of nanofibrous and microfibrous channels for fast capillary flow. Langmuir 34:1235–1241
Smedley SR, Eisner T (1995) Sodium uptake by puddling in a moth. Science 270:1816–1818
Smedley SR, Eisner T (1996) Sodium: a male moth's gift to its offspring. Proc Natl Acad Sci U S A 93:809–813
Smith JJB (1979) Effect of diet viscosity on the operation of the pharyngeal pump in the blood-feeding bug *Rhodnius prolixus*. J Exp Biol 82:93–104
Smith JJB (1985) Feeding mechanisms. In: Kerkut GA, Gilbert LI (eds) Comprehensive insect physiology, biochemistry and pharmacology. Pergamon, New York, pp 33–85
Snell-Rood EC, Espeset A, Boser CJ, White WA, Smykalski R (2014) Anthropogenic changes in sodium affect neural and muscle development in butterflies. Proc Natl Acad Sci U S A 111:10221–10226
Snodgrass RE (1935) Principles of insect morphology. McGraw-Hill Book Co, New York, NY
Snodgrass RE (1961) The caterpillar and the butterfly. Smithson Misc Collect 143:1–51
Socha J, Westneat M, Harrison J, Waters J, Lee W-K (2007) Real-time phase-contrast X-ray imaging: a new technique for the study of animal form and function. BMC Biol 5: 6
Stefan J (1874) Versuche über die scheinbare Adhäsion. Sitzber Akad Wiss Wien (Abt II Math-Phy) 69:713–735
Steffe JF (1996) Rheological methods in food process engineering. Freeman Press, East Lansing, MI
Sun MX, Chen Y, Zheng YM, Zhen MM, Shu CY, Dai ZD, Liang AP, Gorb SN (2017) Wettability gradient on the elytra in the aquatic beetle *Cybister chinensis* and its role in angular position of the beetle at water-air interface. Acta Biomater 51:408–417
Surtees G (1959) Galeal structure in adult mosquitoes. Nature 183:971–972
Tawfik MS (1968) Feeding mechanisms and the forces involved in some blood-sucking insects. Quaest Entomol 4:92–111
Tiple AD, Khurad AM, Dennis RLH (2009) Adult butterfly feeding – nectar flower associations: constraints of taxonomic affiliation, butterfly, and nectar flower morphology. J Nat Hist 43:855–884
Tokarev A, Kaufman B, Gu Y, Andrukh T, Adler PH, Kornev KG (2013) Probing viscosity of nanoliter droplets of butterfly saliva by magnetic rotational spectroscopy. Appl Phys Lett 102:33701
Tsai C-C, Mikes P, Andrukh T, White E, Monaenkova D, Burtovyy O, Burtovyy R, Rubin B, Lukas D, Luzinov I, Owens JR, Kornev KG (2011) Nanoporous artificial proboscis for probing minute amount of liquids. Nanoscale 3:4685–4695
Tsai C-C, Monaenkova D, Beard CE, Adler PH, Kornev KG (2014) Paradox of the drinking-straw model of the butterfly proboscis. J Exp Biol 217:2130–2138
van Eldijk TJB, Wappler T, Strother PK, van der Weijst CMH, Rajaei H, Visscher H, van de Schootbrugge B (2018) A Triassic-Jurassic window into the evolution of Lepidoptera. Sci Adv 4:7

Van Oss CJ (1994) Interfacial forces in aqueous media. Marcel Dekker, New York
Vincent JFV (2012) Structural biomaterials, 3rd edn. Princeton University Press, Princeton, NJ
Vincent JFV, Wegst UGK (2004) Design and mechanical properties of insect cuticle. Arthropod Struct Dev 33:187–199
Vogel S (1996) Life in moving fluids: the physical biology of flow. Princeton University Press, Princeton, NJ
Vogel S (2003) Comparative biomechanics: life's physical world. Princeton University Press, Princeton, NJ
Vogel S (2007) Living in a physical world - X. pumping fluids through conduits. J Biosci 32:207–222
Wainwright SA, Biggs WD, Currey JD, Gosline JM (1982) Mechanical design in organisms. Princeton University Press, Princeton, NJ
Wang YW, Yu ZT, Zhang JZ, Moussian B (2016) Regionalization of surface lipids in insects. Proc R Soc B-Biol Sci 283:8
Watson GS, Cribb BW, Watson JA (2010) The role of micro/nano channel structuring in repelling water on cuticle arrays of the lacewing. J Struct Biol 171:44–51
Watson GS, Watson JA, Cribb BW (2017) Diversity of cuticular micro- and nanostructures on insects: properties, functions, and potential applications. Annu Rev Entomol 62:185–205
Westneat MW, Socha JJ, Lee WK (2008) Advances in biological structure, function, and physiology using synchrotron X-ray imaging. Annu Rev Physiol 70:119–142
Wigglesworth VB (1972) The life of insects. Weidenfeld and Nicolson, New York
Womersley JR (1955) Method for the calculation of velocity, rate of flow and viscous drag in arteries when the pressure gradient is known. J Physiol 127:553–563
Wong WSY, Liu G, Tricoli A (2017) Superamphiphobic bionic proboscis for contamination-free manipulation of nano and core-shell droplets. Small 13:201603688
Wozniak NJ, Sicard A (2018) Evolvability of flower geometry: convergence in pollinator-driven morphological evolution of flowers. Semin Cell Dev Biol 79:3–15
Young T (1805) An essay on the cohesion of fluids. Philos Trans R Soc Lond 95:65–87
Zhai L, Berg MC, Cebeci FC, Kim Y, Milwid JM, Rubner MF, Cohen RE (2006) Patterned superhydrophobic surfaces: toward a synthetic mimic of the Namib Desert beetle. Nano Lett 6:1213–1217
Zhang CQ, Adler PH, Monaenkova D, Andrukh T, Pometto S, Beard CE, Kornev KG (2018a) Self-assembly of the butterfly proboscis: the role of capillary forces. J R Soc Interface 15:20180229
Zhang CQ, Beard CE, Adler PH, Kornev KG (2018b) Effect of curvature on wetting and dewetting of proboscises of butterflies and moths. R Soc Open Sci 5:171241
Zhu J, Iovinella I, Dani FR, Liu YL, Huang LQ, Liu Y, Wang CZ, Pelosi P, Wang GR (2016a) Conserved chemosensory proteins in the proboscis and eyes of Lepidoptera. Int J Biol Sci 12:1394–1404
Zhu RG, Lv HS, Liu TY, Yang YQ, Wu JN, Yan SZ (2016b) Feeding kinematics and nectar intake of the honey bee tongue. J Insect Behav 29:325–339

# Chapter 9
# Hierarchical Microstructures and Functions of the Lepidopteran Proboscis Cuticle

**Matthew S. Lehnert and Qi-Huo Wei**

**Abstract** The feeding mechanism of fluid-feeding insects has become a highly multidisciplinary topic of research. Advances in experimental techniques, such as X-ray phase-contrast imaging and micro-computed tomography, have empowered researchers with capabilities to investigate the fundamental physical mechanisms of fluid uptake in detail. Insights into the physical mechanism of fluid uptake by butterflies and moths (Lepidoptera), the most diverse group of fluid-feeding organisms, have been of particular interest to researchers, in part due to their variety of feeding habits and ability to transport nanoliter amounts of fluids over great distances. This chapter aims to link the diverse size and hierarchical structures of the lepidopteran proboscis with their wetting properties, capillary actions, and fluidic instability, which amalgamates to enable efficient fluid acquisition and transport in diverse environments.

## 9.1 Introduction

Adult butterflies and moths (Lepidoptera) are arguably the most successful fluid feeders. The vast diversity of Lepidoptera [>160,000 described species, second only to beetles (Coleoptera), Stork et al. 2015; Adler and Foottit 2009] and other insect groups is often related to insect–plant coevolution, particularly larval specialization on certain host plants (Siemann et al. 1998; Bennett and O'Grady 2012; Dinnage et al. 2012). However, the unique ability of the adult glossatan Lepidoptera to feed on diverse types of fluids, such as nectar, sap, blood, sweat, and tears, and from environments as challenging as rotting fruit, decaying animals, and wetted soil likely

M. S. Lehnert (✉)
Department of Biological Sciences, Kent State University at Stark, North Canton, OH, USA
e-mail: mlehner1@kent.edu

Q.-H. Wei
Advanced Materials and Liquid Crystal Institute, Kent State University, Kent, OH, USA

Department of Physics, Kent State University, Kent, OH, USA
e-mail: qwei@kent.edu



H. W. Krenn (ed.), *Insect Mouthparts*, Zoological Monographs 5,
https://doi.org/10.1007/978-3-030-29654-4_9

facilitated their radiation (Adler 1982; Krenn 2010; Monaenkova et al. 2012) (Fig. 9.1).

Structure–function relationships are an important aspect of evolutionary biology, because natural selection favors particular variations of structures that are chemically and morphologically modified to increase fitness (Bernays et al. 1991; Weibel et al. 1991; Kramer et al. 2018). Given the range of food sources of adult glossatan Lepidoptera, how the components of their mouthparts (modified to a proboscis) are optimized for efficient liquid acquisition and transport in diverse environments is a great illustration of natural selection. The proboscis evolved only once within the Lepidoptera; therefore, butterflies and moths share a similar proboscis composition (Krenn 2010). Proboscises among Lepidoptera with fundamentally different feeding habits, however, exhibit a diversity of specialized cuticular structures and chemical properties that are optimized for their respective food sources by reducing the time spent searching and handling food (Hirota and Obara 2000; Bauder et al. 2011) and by enabling rapid fluid uptake (Monaenkova et al. 2012; Lehnert et al. 2013).

Although the opaqueness of lepidopteran proboscises hinders applications of many advanced optical microscopic techniques, studies of proboscis structure–function relationships have greatly advanced, thanks to various new visualization techniques. For instance, X-ray phase-contrast imaging has enabled direct observation of microstructures and fluid dynamics inside living Lepidoptera (Monaenkova et al. 2012). In addition, micro-CT scans (Bauder et al. 2013; Krenn and Bauder 2017), X-ray tomography (Monaenkova et al. 2012; Lehnert et al. 2017a), and laser ablation tomography (Lehnert et al. 2017b) have been used to visualize internal structures of the feeding apparatus of Lepidoptera. These techniques have revealed that the proboscis represents an exemplary hierarchical system where what happens at the small-scale impacts that at the larger scale. The physical mechanisms underlying the feeding processes are directly related to the shape, chemistry, and structures at the nano- and microscale. This chapter focuses on how the wetting properties of the proboscis cuticle and the hierarchical organization of structures integrate to support fluid uptake from a range of environments, including pools and liquid films on porous surfaces.

## 9.2 Proboscis Hierarchical Morphology

The lepidopteran proboscis is composed of two elongated, semicircular maxillary galeae. Each galea is a functional unit that contains nerves, tracheae, muscles, and a lumen for the movement of hemolymph (Eastham and Eassa 1955; Krenn 1990, 2010; Krenn and Mühlberger 2002) (Fig. 9.2). Pressure from the hemolymph assists with the uncoiling and probing movements of the proboscis (Krenn 2000, 2010; Wannenmacher and Wasserthal 2003; Lehnert et al. 2015). Upon eclosion from the pupa, the two galeae are initially brought together by adhesive forces from droplets of saliva (Krenn 1997; Zhang et al. 2018b) and then are joined by cuticular projections, the legulae, on the dorsal and ventral sides of each galea. The linked galeae

**Fig. 9.1** Diverse feeding habits of glossatan Lepidoptera. (**a**) Eastern tiger swallowtail (*Papilio glaucus* L.) visiting a flower and likely feeding on nectar. (**b**) Question mark butterfly (*Polygonia interrogationis* Fabricius) feeding on a tree sap flow. (**c**) Two individuals of pearl crescent butterflies (*Phyciodes tharos* Drury) feeding from wetted soil (i.e., puddling). (**d**) Vampire moth

create an approximately circular conduit, the food canal, which transports fluids from liquid pools or films into the preoral cavity of the head (Krenn 1997; Borrell and Krenn 2006; Lehnert et al. 2014) (Fig. 9.2). Images of proboscis cross sections acquired with scanning electron microscopy and light microscopy via serial semithin sectioning reveal that the shape of the proboscis circumference ranges from being nearly circular to elliptical, depending on the taxa and feeding habits (Krenn and Mühlberger 2002; Lehnert et al. 2013).

The length of the proboscis, which ranges from less than 1 mm in length to approximately 25 cm in length (Monaenkova et al. 2012; Ardetti et al. 2012), can be structurally delineated into three zones, characterized by the shape, the size, and the presence/absence of the dorsal legulae (Lehnert et al. 2016) (Fig. 9.3). Zone 1 represents 80–90% of the proximal region of the proboscis and has dorsal legulae that overlap with those of the opposite galea (Krenn et al. 2001; Lehnert et al. 2016). The interlegular spacing in this region is small, approximately 100 nm in *Danaus plexippus* L. (Nymphalidae) (Monaenkova et al. 2012) (Fig. 9.2). Zone 2 represents the distal 10–20% of the proboscis length and has dorsal legulae with structural modifications and larger spaces between them (interlegular spaces, approximately 2 μm in *Danaus plexippus* L.) that facilitate the entry of fluids from pools or liquid films into the food canal (Krenn et al. 2001; Krenn 2010; Monaenkova et al. 2012; Lehnert et al. 2013, 2016). Unlike the single-layered dorsal legulae in Zone 1, the dorsal legulae of the distal region are often bifurcated with an upper and lower branch, with the upper branch extending from elongated cuticular projections near the legular seam (Lehnert et al. 2013, 2016). The dorsal legulae also might have small, groove-like conduits [e.g., *Pieris rapae* (Pieridae) (Lehnert et al. 2016)], teeth-like projections on the distal parts [e.g., *Limenitis* spp., (Nymphalidae) (Kramer et al. 2018)], and other modifications (Fig. 9.2). Zone 3 represents the distal tip of the proboscis where no dorsal legulae are found and has only been reported for lepidopteran species that visit flowers (i.e., non-flower visitors lack Zone 3 because their dorsal legulae extend to the tip) (Lehnert et al. 2016) (Fig. 9.3).

The proboscis is equipped with an array of mechano- and chemosensilla (Krenn 1998; Faucheux 2013). Arguably, the most noteworthy are the sensilla styloconica, which are found at the distal region of the proboscis of some Lepidoptera (Petr and Stewart 2004; Krenn et al. 2001; Knopp and Krenn 2003; Kramer et al. 2015; Lehnert et al. 2016) (Figs. 9.2 and 9.3). The presence or absence of sensilla styloconica, and their structural modifications, results in a diversity of proboscis forms and is a good predictor of feeding habits. Blood-feeding *Calyptra* spp. (Noctuidae), for example, have piercing proboscises

**Fig. 9.1** (continued) (*Calyptra thalictri* Borkhausen) piercing human tissue to take a blood meal [image taken by Jennifer Zaspel and used with permission (Zaspel et al. 2012)]. (**e**) An erebid moth (*Gorgone macarea* Cramer) using its proboscis to feed on the tears of a black-chinned antbird (*Hypocnemoides melanopogon* Sclater); image taken by Leandro Joao Carneiro de Lima Moraes and used with permission (de Lima Moraes 2018)

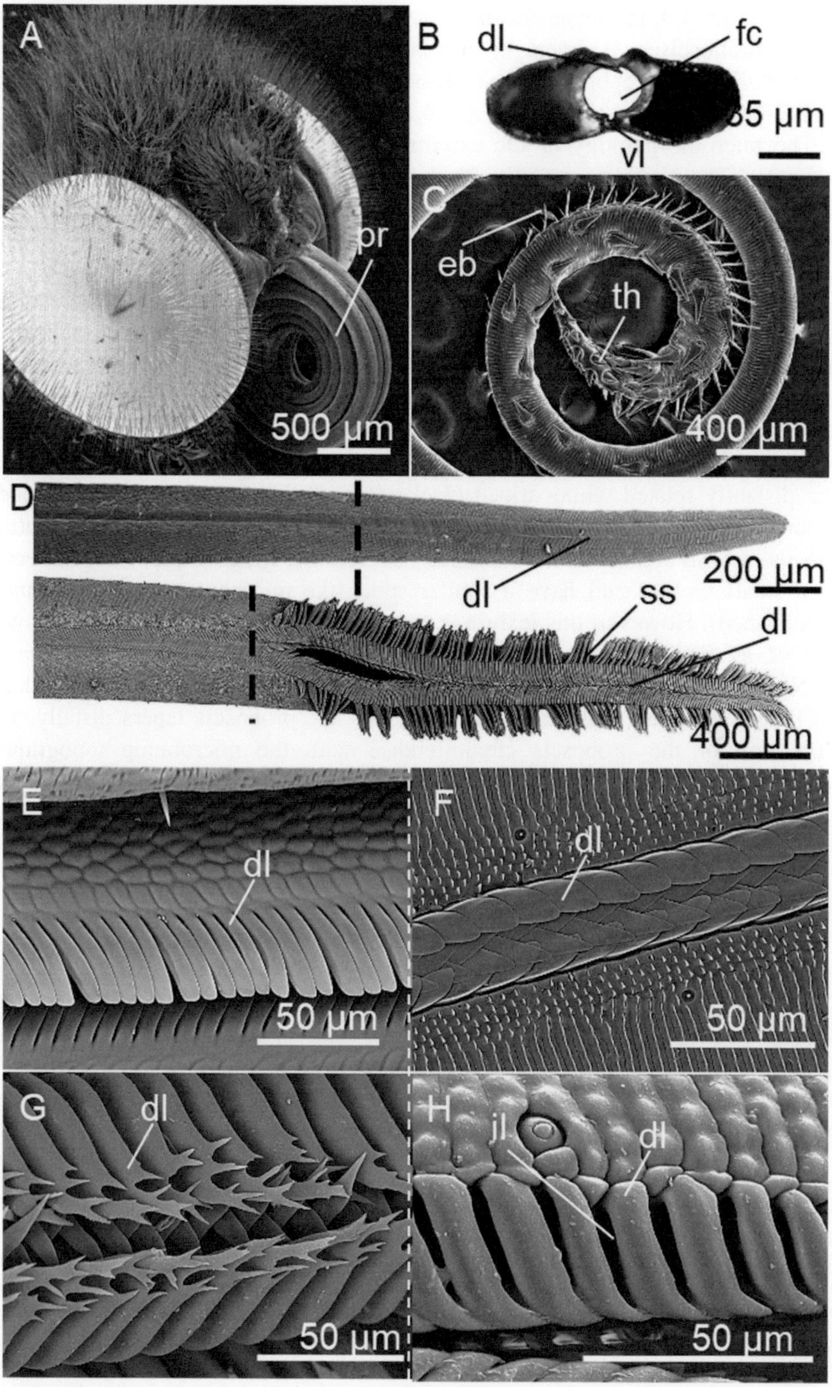


**Fig. 9.2** Morphology of lepidopteran proboscises. (**a**) Head of a painted lady butterfly (*Vanessa cardui* L.) with the coiled proboscis (pr); labial palpi are removed. (**b**) Cross section image of a proboscis of *Vanessa cardui* L. showing the two, c-shaped galeae that connect by the dorsal legulae

(Zaspel et al. 2011), nectar-feeding *Danaus* spp. (Nymphalidae) have smooth-tipped proboscises (Lehnert et al. 2016), and sap-feeding *Limenitis* spp. (Nymphalidae) have brush-tipped proboscises (Krenn et al. 2001; Molleman et al. 2005; Kramer et al. 2018) (Figs. 9.2 and 9.3).

The micromorphology of the external surface of a galea changes along its perimeter. The lateral parts, between the dorsal and ventral legulae, consist of a series of microbumps interspersed by micro- and macro-valleys (Lehnert et al. 2016) (Fig. 9.4). The microbump patterns change along the proboscis length and differ among species (Lehnert et al. 2016); for instance, the galeal surface can be nearly smooth [e.g., tobacco hornworm, *Manduca sexta* L. (Sphingidae)], wrinkled [e.g., red-spotted purple, *Limenitis arthemis astyanax* Fabricius (Nymphalidae)], and reticulated [e.g., monarch butterfly, *Danaus plexippus* L. (Nymphalidae)], consist of rings of spikes [e.g., cabbage butterfly, *Pieris rapae* L. (Pieridae)], and have other variations (Lehnert et al. 2016). Although the microbump texture might be related to common ancestry, the texture also could be related to feeding habits. The distantly related white-striped black moth, *Trichodezia albovittata* Guenée (Geometridae); cabbage butterfly, *Pieris rapae* L. (Pieridae); and eight-spotted forester, *Alypia octomaculata* Fabricius (Noctuidae), for example, feed on nectar from small flowers and have a similar spike-like microbump pattern (Lehnert unpublished). However, this texture is not necessarily representative of all flower visitors, as the microbump textures of the monarch butterfly, *Danaus plexippus* L. (Nymphalidae), and eastern tiger swallowtail, *Papilio glaucus* L. (Papilionidae), are both reticulated (Lehnert et al. 2016). As the proboscis tapers distally, the proportion of the proboscis circumference with the microbump topography decreases, and the proportion occupied by the dorsal and ventral legulae increases (Lehnert et al. 2013).

**Fig. 9.2** (continued) (dl) and ventral legulae (vl) to create a food canal (fc) (image acquired with a Dino-Lite digital microscope). (**c**) Proboscis of a vampire moth (*Calyptra thalictri* Borkhausen) that uses tearing hooks (th) and erectile barbs (eb) to pierce fruit and vertebrate tissue. (**d**) Proboscises of a monarch butterfly (*Danaus plexippus* L., top) and a red-spotted purple (*Limenitis arthemis astyanax* Fabricius, bottom). The smooth proboscis of *Danaus plexippus* L. is putatively adapted for reaching nectar in floral corollas, whereas the brushy proboscis (due to the enlarged sensilla styloconica, ss) of *Limenitis arthemis astyanax* Fabricius is adapted for feeding on liquid films, such as sap. The dashed line indicates the structural boundary between a drinking (distal) and nondrinking (proximal) region. (**e, f**) Differences in the overlapping dorsal legulae in the nondrinking region of the clear-winged sphinx moth (*Hemaris thysbe* Fabricius) shown in (**e**) and a cloudless sulfur butterfly (*Phoebis sennae* L.) (**f**), which has its dorsal legulae arranged similar to a zipper. (**g, h**) Dorsal legulae in the drinking region that are modified to support fluid uptake (viceroy butterfly, *Limenitis archippus* Cramer, shown in (**g**), and *Danaus plexippus* L. shown in (**h**). All images, except for (**b**), were acquired with a scanning electron microscope (SEM)

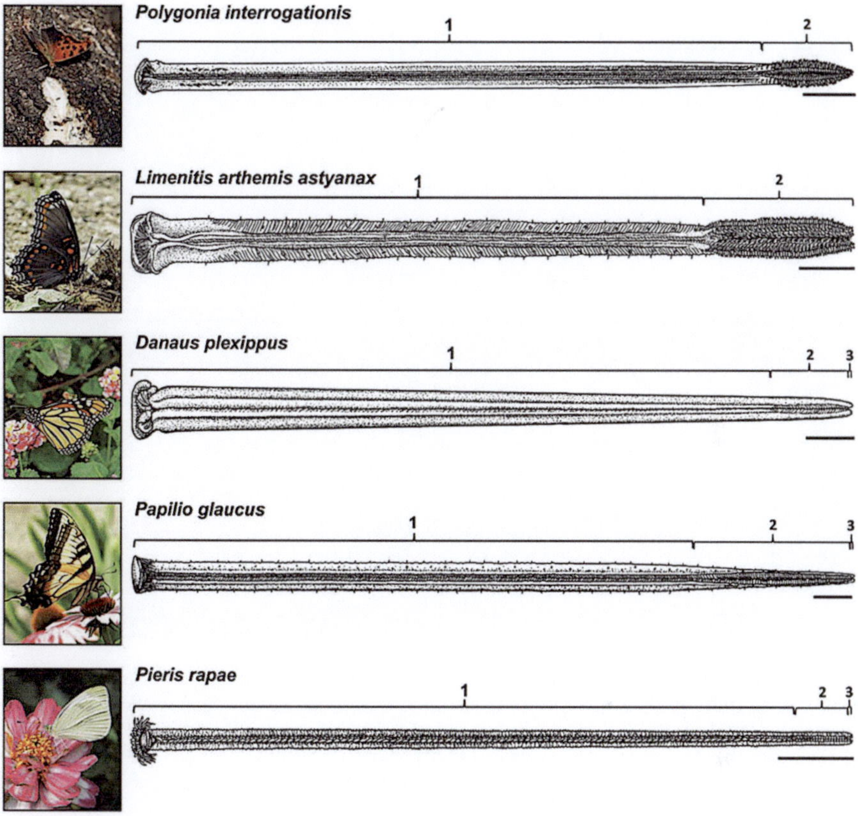


**Fig. 9.3** Flower-visiting and non-flower-visiting butterflies and proboscis zones. All studied Lepidoptera have a proboscis that has Zones 1 and 2, which are delineated by the structural modifications of the dorsal legulae. The bottom three butterflies (*Danaus plexippus* L., *Papilio glaucus* L., and *Pieris rapae* L.) are flower visitors and have a Zone 3 at the proboscis tip that is characterized by a lack of dorsal legulae. The proboscises of the top two butterflies (*Polygonia interrogationis* Fabricius and *Limenitis arthemis astyanax* Fabricius) are non-flower-visiting butterflies, lack Zone 3, and have a brushy Zone 2 due to the enlarged sensilla styloconica. The bars indicate 1 mm. (Illustrations by Angela H. Newman, figure used with permission from Lehnert et al. 2016)

## 9.3 Cuticular Properties of the Proboscis

### *9.3.1 Wettability of Proboscis Cuticle*

Wetting properties of surfaces are important to liquid collection and transport. As a measure of wettability, the contact angle ($\theta$) between the liquid–vapor and the solid–liquid interfaces at the contact line (Fig. 9.5a, b) is determined by the balance between surface tensions and can be mathematically expressed with Young's equation:

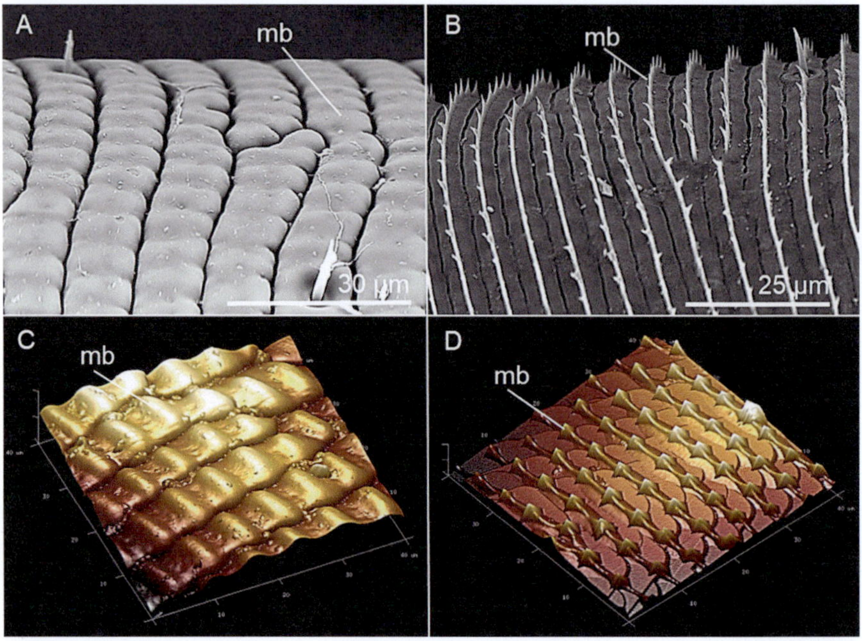


**Fig. 9.4** Variations in galeal surface roughness among butterfly species. SEM images showing the microbumps (mb) on the galea surface of the proboscis of an eastern tiger swallowtail, *Papilio glaucus* L. (Papilionidae) (**a**), and the spikey, galeal surface of the cloudless sulfur, *Phoebis sennae* L. (Pieridae) (**b**). Atomic force microscopy (AFM) images of the galeal surface of *Papilio glaucus* L. (**c**) and cabbage butterfly, *Pieris rapae* L. (Pieridae) (**d**). The AFM images represent 40 µm × 40 µm sections

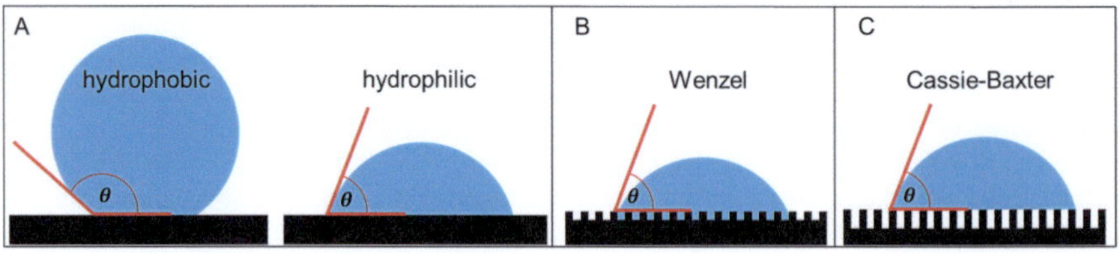


**Fig. 9.5** (**a**) Contact angles ($\theta$) on hydrophobic and hydrophilic surfaces. (**b**) Schematic Wenzel model. (**c**) Schematic of Cassie-Baxter model

$$\cos\theta = \frac{\gamma_{SG} - \gamma_{SL}}{\gamma_{LG}} \tag{9.1}$$

where $\gamma_{SG}$, $\gamma_{SL}$, and $\gamma_{LG}$ are the surface energies of smooth solid–gas, solid–liquid, and liquid–gas interfaces, respectively. For water droplets, the contact angle is smaller than 90° on hydrophilic surfaces and larger than 90° for hydrophobic surfaces (Fig. 9.5a).

It has been recognized since 1936 that surface roughness and textures affect the surface energy and thus significantly affect the wetting properties (Liu et al. 2010; Marmur 2003). Robert N. Wenzel showed that for a rough surface (shown schematically in Fig. 9.5b) the contact angle $\theta*$ needs to be modified from the Young's equation (Wenzel 1936):

$$\cos\theta^* = r\frac{\gamma_{SG} - \gamma_{SL}}{\gamma_{LG}} = r\cos\theta \qquad (9.2)$$

where $r$ is the roughness ratio, defined as the ratio of real contact area to the apparent area of the solid–liquid interface; this equation is known as the Wenzel model. Surface textures with high aspect ratios often have air trapped beneath the water droplets (Fig. 9.5c), and the contact angle is described by the Cassie-Baxter model (Cassie and Baxter 1944):

$$\cos\theta^* = -1 + \varphi_s(\cos\theta + 1) \qquad (9.3)$$

where $\varphi_s$ is the fraction of the solid area in contact with the water droplet.

As indicated in Equations 9.2 and 9.3, when $r$ or $\varphi_s$ is larger than one, the wetting properties are enhanced by surface roughness, i.e., hydrophilic surfaces become more hydrophilic with smaller contact angles, and hydrophobic surfaces become more hydrophobic with larger contact angles and self-cleaning proper, a phenomenon also named the Lotus effect (Zhang et al. 2016). A contact angle larger than 150° is superhydrophobic, and a contact angle less than 5° is superhydrophilic. As a comparison, the most hydrophobic material (Teflon) with smooth surfaces has a contact angle ~120°. Materials with superhydrophobic properties are of great interest to biologists and material scientists because of their self-cleaning abilities, where droplets of water tend to roll off surfaces while removing dirt and debris (Fürstner et al. 2005; Zhang et al. 2016, 2018c). The Cassie-Baxter and Wenzel models can be used to explain the wetting properties of the lepidopteran proboscis, where surface chemistry and roughness determine wettability.

The hydrophobicity of biological surfaces often originates from the presence of water-resistant waxes (esters of long-chain alcohols and fatty acids), which also are found on the surface of insect cuticle (Holdgate 1955; Vincent and Wegst 2004). Lipophilic fluorescent dyes (those that bind to waxes), such as Nile red, have verified the occurrence of waxes on the cuticle of tested lepidopteran proboscises and revealed inhomogeneous wetting properties of structures (Lehnert et al. 2013, 2014) (Fig. 9.6). Nile red binds to the regions of the galeae where the microbumps are located but does not bind to the legulae, food canal, or sensilla. Eastham and Eassa (1955) described small glands inside the galeae that lead to the dorsal legulae, which might either secrete waxes to provide hydrophobicity to the galeal surface or some other secretion that contributes to the hydrophilicity of dorsal legulae, which could create the inhomogeneous wetting properties, but this requires further research.

According to the Cassie-Baxter and Wenzel models, the wetting properties are defined by both structure and the presence of waxes on the galea. The hydrophobicity

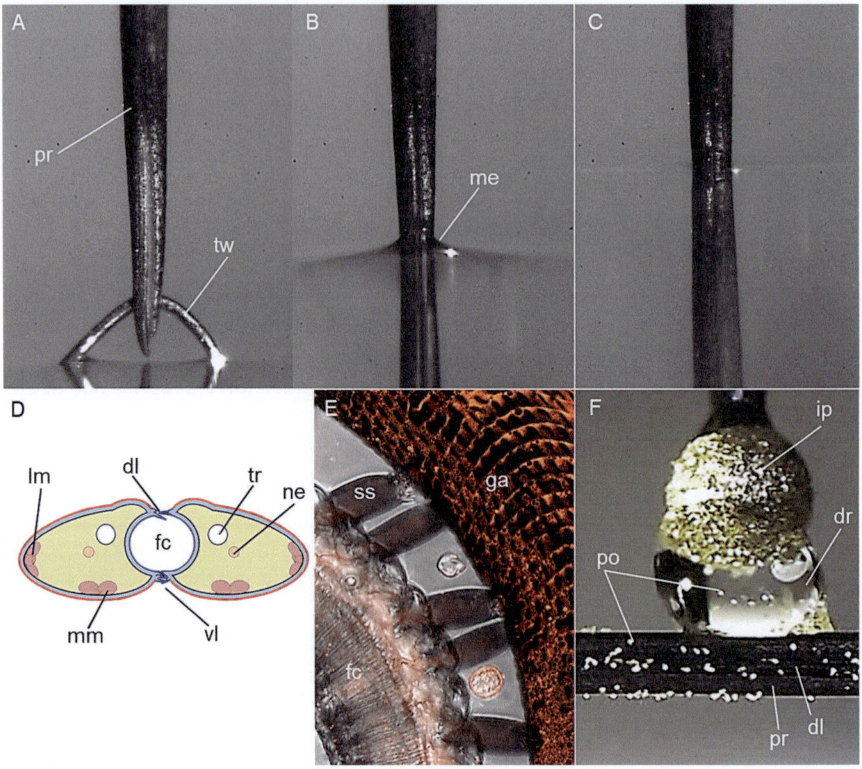


**Fig. 9.6** Wetting properties on the galeal surface of lepidopteran proboscises. (**a–c**) Images showing the wetting dichotomy of the proboscis (pr) of the monarch butterfly, *Danaus plexippus* L., using the capillary-rise technique with a tungsten wire loop (tw) used to stabilize the proboscis. The distal region shows a hydrophilic meniscus (me) (**b**), whereas the meniscus cannot be visualized on the hydrophobic proximal region (**c**). (**d**) Illustration of the proboscis cross section shows the lateral (lm) and medial (mm) intrinsic musculature, tracheae, and nerves. The dorsal legulae (dl), ventral legulae (vl), and food canal (fc) are outlined in blue to show their hydrophilic wetting properties, and the remainder of the galeal surface is outlined in red to indicate hydrophobicity. (**e**) Confocal microscopy image showing the proboscis of a painted lady butterfly, *Vanessa cardui* L., that is stained with Nile red, which binds to lipophilic surfaces. The galeal surface (ga) shows the red stain, indicating the presence of lipids on the surfaces; however, the sensilla styloconica (ss) and the food canal did not stain. (**f**) A droplet of water (dr) adhering to an insect pin (ip) was dragged across the surface of the proboscis that was dusted with pollen grains (po). The droplet does not adhere to the galeal surface, but does remove the pollen grains, demonstrating the self-cleaning properties

is then further augmented by the microbumps (Fig. 9.4) that provide the surface roughness (shown in Eq. 9.2 and 9.3 as $r$ or $\varphi_s$), therefore creating a hydrophobic surface with self-cleaning abilities. Conversely, areas where the Nile red does not bind (the food canal, sensilla, legulae) have a hydrophilic surface that is enhanced by their surface roughness, i.e., surface roughness of legulae with corresponding interlegular spaces and the food canal with its ridges. X-ray imaging of droplets on the dorsal legulae displays hydrophilic contact angles, and fluorescent microscopy imaging coupled with X-ray imaging has indicated that the food canal might be superhydrophilic (meaning that the contact angle is close to zero) (Monaenkova et al. 2012; Lehnert et al. 2014).

Given that a substantial portion of the proboscis is hydrophobic, there is a paradoxical question for the feeding mechanism of Lepidoptera–how can insects that have hydrophobic cuticle acquire fluids? This fundamental question was resolved in a study of the feeding mechanism of butterflies by Lehnert et al. (2013), which found that the inhomogeneous surface energy properties and shapes of proboscis structures influence feeding abilities and partition the proboscis into distinct drinking and nondrinking regions.

Lehnert and coworkers used a capillary-rise technique to directly measure contact angles on the proboscis cuticle and examine how the inhomogeneity of the proboscis wetting properties enables fluid feeding (Lehnert et al. 2013). The capillary-rise technique involves straightening a proboscis so that it is positioned vertically over a dish with water. The distal region of the proboscis is stabilized against a tungsten wire loop that is situated in the dish. The water level in the dish is slowly raised while a video camera records the interactions between the water and the proboscis (Fig. 9.6). Although this method does not necessarily assess wettability of individual structures at a microscale, it shows the wetting profile of the proboscis circumference and that the wetting properties of structures synergize for efficient fluid uptake. The capillary-rise technique revealed that the distal region of the proboscis has an overall hydrophilic profile (i.e., drinking region), whereas the remaining proximal region of the proboscis has a hydrophobic profile (i.e., nondrinking region (Fig. 9.6a–c) (Lehnert et al. 2013).

It was determined that the functionally defined boundary between the drinking and the nondrinking region is structurally induced. The nondrinking region is characterized by the overlapping dorsal legulae and small interlegular spaces with a large proportion of the circumference with hydrophobic microbumps. The overall wetting changes at the distal region where the hydrophilic dorsal legulae are adapted for fluid uptake (larger interlegular spaces). At this region, the proboscis rapidly tapers; however, the dorsal legulae remain approximately the same size, and therefore, the tapering results in less surface area with the microbump (hydrophobic) micromorphology, resulting in an overall hydrophilic profile (Lehnert et al. 2013). The structural boundaries between Zones 1 and 2 (Lehnert et al. 2016) of the proboscis mirror the functional boundary, thus presenting an excellent example of the hierarchy organization of structure–function relationships.

The dichotomy of the wetting profile provides the ability to acquire fluids while keeping the proboscis clean. Unlike other insects with shorter mouthparts, Lepidoptera with longer proboscises lack the ability to use their forelimbs or other appendages to manually remove debris from their mouthparts. Having a primarily hydrophobic region provides self-cleaning abilities, which can remove debris or sticky fluids from the proboscis surface allowing efficient feeding and proboscis coiling/uncoiling. Recently, Lehnert and his coworkers performed tests to observe the self-cleaning properties of proboscises dusted with pollen (Fig. 9.6). By dragging a droplet of water across the proboscis surface, they observed that the water droplet collects and engulfs pollen grains. The wetting properties also likely aid the feeding of pollen in *Heliconius* spp. by relocating and adhering pollen to regions where saliva can perform extraoral digestion (Gilbert 1972; Krenn and Penz 1998; Krenn et al. 2009; Hikl and Krenn 2011).

The capillary-rise technique also demonstrated differences in wetting properties of proboscises among species with different feeding habits. The menisci of the water in contact with brushy proboscises (non-flower-visiting butterflies with enlarged sensilla styloconica) are higher than the menisci in contact with smooth proboscises (flower-visiting butterflies) (Lehnert et al. 2013). The enlarged meniscus on proboscises from non-flower-visiting butterflies is due to the presence of the hydrophilic sensilla styloconica and the shape of the proboscis perimeter. Menisci are higher on fibers (and proboscises) that have an elliptical cross section than those that have a circular cross section. The higher meniscus of non-flower-visiting Lepidoptera is an adaptation to increase capillary pull on the viscous fluids (e.g., sap, rotting fruit) from which they feed. A higher meniscus would also cover a greater number of interlegular spaces, facilitating the entry of liquids into the food canal for subsequent fluid uptake (Lehnert et al. 2013).

### 9.3.2 *Capillarity of Fluid Acquisition*

The proboscis was modeled to function similarly to a drinking straw in the first detailed studies of fluid dynamics of nectar feeding (Kingsolver and Daniel 1979), which can be described by the Hagen–Poiseuille equation; however, the proboscis is unlike a drinking straw in that it is porous. The porosity of the proboscis, due to the interlegular spaces, hinders the efficiency of the sucking pump and indicates that the fluid uptake mechanism is more complex and includes capillarity (Monaenkova et al. 2012; Lehnert et al. 2013; Kwauk et al. 2014) and feeding behaviors that alter the shape of the proboscis (Tsai et al. 2014; Zhang et al. 2018a).

Droplets of water placed on the dorsal legulae spread along the legular seam, and the interlegular spaces act similarly to other porous, spongelike materials, such as paper towels (Monaenkova et al. 2012). The spreading of the liquids has adaptive value in partitioning the liquids along the interlegular spaces, which aids in the movement of fluids into the food canal. In addition, the capillary pressure induced by the spreading of fluids between the dorsal legulae in the nondrinking region might assist in providing a temporary seal for the food canal (Zhang et al. 2018a).

The ability for the dorsal legulae (and their interlegular spaces) to act similar to a sponge resolves one of the fundamental challenges of how Lepidoptera acquire fluids from porous surfaces, such as liquid films on rotting fruit and other wetted surfaces, such as soil. The issue with feeding from porous surfaces involves the ability for Lepidoptera to overcome the capillary pressure that keeps the fluids inside the pores, a task that is impossible by action of the sucking pump alone because it cannot create a pressure differential of this magnitude (Tsai et al. 2014). The ability to feed on fluids confined to small pores is one of the main interests that has launched many of the recent studies of lepidopteran proboscis functionality.

Studies of proboscis morphology with scanning electron microscopy have shown that the small interlegular spaces would provide the ability for capillary action, a physical mechanism of withdrawing fluids for porous surfaces. When the dorsal side of the proboscis is placed on wetted, porous surfaces, the liquids in the pores move by capillary action, filling the interlegular spaces. The employment of capillary action in order to remove fluids from pores would be favored by natural selection, because this is an energetically effective method that does not involve action of the sucking pump. In addition, the ability to employ capillary action would create opportunities to feed from new food sources, such as squashed fruit and the surface of decaying organic substances, which might have facilitated the adaptive radiation of Lepidoptera.

Capillarity is not only induced by the interlegular spaces but also by the spaces between the hydrophilic sensilla styloconica (Lehnert et al. 2013; Lee and Lee 2014). Non-flower-visiting butterflies, those characterized by a brushy proboscis tip (Krenn et al. 2001; Lehnert et al. 2013, 2016), have densely arranged sensilla styloconica. Microparticle image velocimetry provided experimental evidence that the spacing between the sensilla channels fluids from liquid films to the interlegular spaces where fluids enter the food canal (Lee and Lee 2014).

The ability to use capillarity to remove liquids from small pores raised an important question about feeding abilities—what is the smallest pore from which butterflies can feed? Initial experiments to determine a limiting pore size consisted of hanging paper towels saturated with sucrose solutions from platforms of particular heights. This experimental setup used the Jurin law of capillarity, which holds that there is a relationship between the platform height and the size of pores in the paper towels that can retain liquids before gravity overcomes the capillary pressure of the pore, i.e., smaller pores can withhold fluids and overcome gravity at greater heights than larger pores. Monarch butterflies, *Danaus plexippus* L., when placed on different stage heights were able to feed from pores that had a radius of approximately 35 μm and larger, indicating that 35 μm is the limiting pore size from which this butterfly species can feed (Monaenkova et al. 2012). Scanning electron microscopy revealed that the radius of the food canal also has a radius of 35 μm, suggesting an important relationship between food canal radius and pore size radius (Monaenkova et al. 2012).

A relationship between the smallest pore size radius from which a butterfly can feed and the food canal radius, known as the limiting pore size hypothesis, was further explored with other butterfly species (Lehnert et al. 2017a, b). In this study, fluorescent, magnetic nanoparticles were mixed with a sucrose solution and used to fill pores in different filters, each with a specified pore size. Butterflies were provided the opportunity to feed from the different filter papers and then dissected to remove the crop, which was imaged with a confocal microscope. A magnetic stir bar was moved back and forth near the crop. If nanoparticles were ingested (indicating the ability to feed from a filter with a specific pore size), they would move in unison with the movement of the magnetic stir bar. The results from this experiment supported the limiting pore size hypothesis—species of butterflies were unable to feed from pore sizes with a radius smaller than the radius of their food canal (Lehnert et al. 2017a, b).

## 9.4 Rayleigh-Plateau Instability Inside the Food Canal

The relationship between the radius of the food canal and the radius of the limiting pore size was further studied using X-ray phase-contrast imaging (Monaenkova et al. 2012; Lehnert et al. 2017a), which provides the ability to observe structures and fluids inside living insects (Kaiser et al. 2007; Westneat et al. 2008; Socha et al. 2008, 2010). Synchrotron X-ray imaging involves placing an insect in the path of an X-ray beam, which passes through a specified region of the insect and then through a scintillator that converts the X-ray to visible light before entering a video camera. Studies with Lepidoptera have involved securing butterflies in front of the X-ray beam so that the proboscis is in line with the beam, then placing droplets of various solutions (water and sucrose solutions, sometimes mixed with a contrast agent, e.g., iodine) on the legular seam, and watching the droplets enter the proboscis (Monaenkova et al. 2012). The results indicated that the droplets spread along the legular seam, move through the interlegular spaces into the food canal, and then grow into a liquid bulge that collapses into the liquid bridge (Fig. 9.6) (Monaenkova et al. 2012). Measurements of liquid bridges indicate that they form via Rayleigh-Plateau instability, which is where fluids try to minimize their surface area due to the surface tension of the fluid and form a stable configuration. As an example, Rayleigh-Plateau instability explains why, when liquids fall from a faucet, they produce the shape of a spherical droplet.

The liquid bridge formation inside a cylindrical pore originates from a classical fluid mechanical instability induced by surface tension, known as Rayleigh-Plateau instability (Goren 1962; Heil et al. 2008; Plateau 1863; Rayleigh 1879). To understand this phenomenon, we recall the Young–Laplace equation relating the pressures at two sides of curved surfaces:

$$P_{\text{in}} - P_{\text{out}} = \sigma\kappa$$

where $P_{\text{in}}$ and $P_{\text{out}}$ are the pressures inside and outside the curved surfaces, $\sigma$ is the surface tension of the liquid, and $\kappa$ is the mean curvature of the liquid–air interface averaged along two principle axes. When there is a modulation on the thickness $h$ of an annular film inside a cylindrical pore (e.g., a bulge of the liquid film on the surface of the food canal of a proboscis) (with a radius $a$), then the mean curvature is the summation of circumferential curvature $1/(a-h)$ and the axial curvature $dh^2/dz^2$, i.e., $\kappa = 1/(a - h) + d^2h/dz^2$ (Fig. 9.7). The circumferential curvature tends to close the hole of the liquid (i.e., the space in the food canal between the liquid films gets smaller), while the axial curvature terms tend to flatten out the surface in axial direction. The interplay between these two competing terms determines the onset of

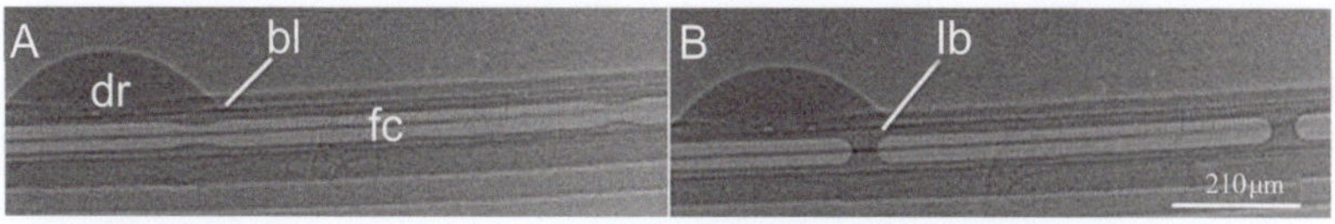


**Fig. 9.7** Liquid films inside a proboscis with X-ray phase-contrast imaging. (**a**) Modulation on the film thickness (liquid bulge, bl), and (**b**) liquid bridge (lb) formed due to the Rayleigh-Plateau instability (image acquired with permission from Monaenkova et al. 2012)

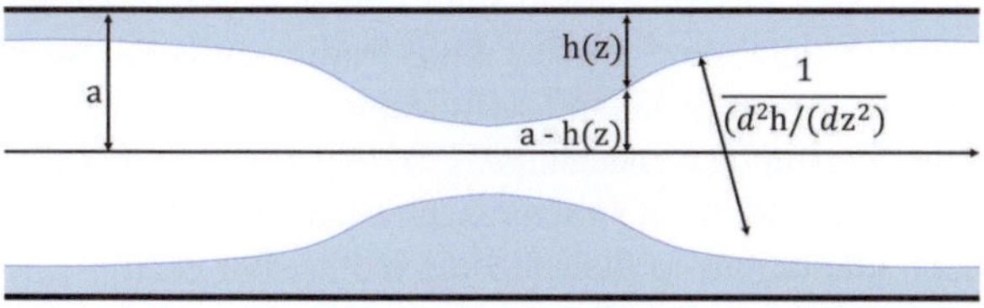


**Fig. 9.8** Schematic of Rayleigh-Plateau instability inside a proboscis food canal. The image shows a liquid film on the surface of the food canal. There is a liquid bulge in the center, which under particular conditions will collapse into a stable liquid bridge. The oblique arrow represents the axial curvature

liquid bridges. It can be shown that when the wavelength of surface modulation $L$ is larger than the circumference length, i.e., $L > 2\pi(a - h_0)$, the liquid film becomes unstable, and liquid bridges will emerge (Everett and Haynes 1972; Goren 1962; Heil et al. 2008). While when $L < 2\pi(a - h_0)$, the axial curvature term dominates, and the liquid film is stable (Everett and Haynes 1972; Goren 1962; Heil et al. 2008; Fig. 9.8).

The formation of liquid bridges explains the limiting pore size hypothesis. The pore size radius must be equal to or larger than the radius of the food canal to support the formation of a stable liquid bridge. Liquid bridge formation would be essential for Lepidoptera to feed from porous surfaces and liquid films because it would provide a liquid surface in the food canal for the sucking pump in the head to act on.

## 9.5 Conclusion and Perspective

Investigations have shown that the hierarchical microstructures, surface wetting properties, and fluid mechanical instability are optimized through natural selection for efficient liquid acquisition and transport. The wetting properties of the proboscis cuticle and shapes of structures to promote feeding is probably widespread among other fluid-feeding insects, such as flies (Diptera) that face the same selective pressures, e.g., feeding on various types of fluids from pools and porous surfaces. Studies of the dipteran feeding mechanism also have supported the limiting pore size hypothesis (Lehnert et al. 2017a, b), which suggests that they rely on the same physical mechanisms of Rayleigh-Plateau instability and liquid bridge formation.

The similar feeding mechanism between Diptera and Lepidoptera is further supported by structural parallels, for instance, Diptera have cuticular projections of their pseudotracheae, which are similar to the dorsal legulae of Lepidoptera. In addition, the conduits of the pseudotracheae are lined with ridges, similar to the ridged food canal of lepidopteran proboscises (Lehnert et al. 2017a). The convergence of fluid-feeding mechanisms might further extend to other groups, such as the tropical cricket, *Glomeremus orchidophilus* (Gryllacrididae: Orthoptera), which has mouthparts classified as biting-and-chewing; however, they also have channels and other structural adaptations that support fluid transport (Krenn et al. 2016). Future studies are needed to determine if other fluid-feeding insects also have a wetting dichotomy to support fluid uptake and self-cleaning abilities.

The studies of fluid uptake mechanisms among insects are still in its infancy; questions remain open in terms of structures, physical properties, functionalities, and their evolutionary adaptations to their feeding habits. For example, it is not clear if size gradients of structures and gradients of wettability play a role in fluid acquisition and transport as in other insects (e.g., desert beetles) and plants, such as *Nepenthes alata* (Chen et al. 2016; Nørgaard and Dacke 2010; Parker and Lawrence 2001). In addition, many other aspects of insect mouthparts are still unexplored, such as identifying the chemical components of the proboscis cuticle that provides the hydrophobic and hydrophilic foundations. The genetics and development of proboscises and the local cuticle properties also remain unexplored. Further, the proboscises also exhibit intriguing mechanical properties, such as the mechanical abilities and cuticle flexibility associated with coiling and uncoiling an elongated proboscis, such as those of *Xanthopan morganii praedicta* Rothschild and Jordan (Sphingidae) that are approximately 25 cm long (see Chap. 15). How this can be accomplished is an interesting question for future studies and might serve as bioinspiration for future microfluidic devices.

**Acknowledgments** We are thankful to the numerous students and collaborators that have contributed to this research program. In particular, we would like to thank Andrew Bennett, Aubrey Brothers, Eric Brown, Miranda Byler, Daytona Hedrick, Valerie R. Kramer, Miao Jiang, Ashley L. Lash, Wah-Keat Lee, Katherine P. Mulvane, Kristen E. Reiter, Huan Yan, and other undergraduate students and collaborators. We also would like to thank Connie Kramer for her assistance in the lab and working with students. This work was supported by National Science Foundation grants and a Kent State University SEED grant.

## References

Adler PH (1982) Soil- and puddle-visiting habits of moths. J Lepid Soc 36:161–173

Adler PH, Foottit RG (2009) Introduction to insect biodiversity. In: Foottit RG, Adler PH (eds) Insect biodiversity: science and society. Wiley, Chichester

Ardetti J, Elliott J, Kitching IJ, Wasserthal LT (2012) 'Good heavens what insect can suck it' – Charles Darwin, Angraecum sesquipedale and *Xanthopan morganii praedicta*. Bot J Linn Soc 169:403–432

Bauder JAS, Lieskonig NR, Krenn HW (2011) The extremely long-tongued Neotropical butterfly *Eurybia lycisca* (Riodinidae): proboscis morphology and flower handling. Arthropod Struct Dev 40:122–127. https://doi.org/10.1016/j.asd.2010.11.002

Bauder JAS, Handschuh S, Metscher BD, Krenn HW (2013) Functional morphology of the feeding apparatus and evolution of proboscis length in metalmark butterflies (Lepidoptera: Riodinidae). Biol J Linn Soc 110:291–304. https://doi.org/10.1111/bij.12134

Bennett GM, O'Grady PM (2012) Host-plants shape insect diversity: phylogeny, origin, and species diversity of native Hawaiian leafhoppers (Cicadellidae: *Nesophrosyne*). Mol Phylogenet Evol 65:705–717. https://doi.org/10.1016/j.ympev.2012.07.024

Bernays EA, Jarzembowski EA, Malcolm SB (1991) Evolution of insect morphology in relation to plants [and discussion]. Philos Trans R Soc B Biol Sci 333:257–264. https://doi.org/10.1098/rstb.1991.0075

Borrell BJ, Krenn HW (2006) Nectar feeding in long-proboscis insects. In: Herrel A, Speck T, Rowe N (eds) Ecology and biomechanics – a mechanical approach to the ecology of animals and plants. CRC, Boca Raton, pp 185–212. https://doi.org/10.1201/9781420001594.ch9

Cassie ABD, Baxter S (1944) Wettability of porous surfaces. Trans Faraday Soc 40:546–550. https://doi.org/10.1039/tf9444000546

Chen H, Zhang P, Zhang L, Liu H, Jiang Y, Zhang D, Han Z, Jiang L (2016) Continuous directional water transport on the peristome surface. Nature 532:85–89. https://doi.org/10.1038/nature17189

de Lima Moraes LJC (2018) Please, more tears: a case of a moth feeding on antbird tears in Central Amazonia. Ecology 100(2):e02518. https://doi.org/10.1002/ecy.2518

Dinnage R, Cadotte MW, Haddad NM, Crutsinger GM, Tilman D (2012) Diversity of plant evolutionary lineages promotes arthropod diversity. Ecol Lett 15:1308–1317. https://doi.org/10.1111/j.1461-0248.2012.01854.x

Eastham LE, Eassa YEE (1955) The feeding mechanism of the butterfly *Pieris brassicae* L. Philos Trans R Soc Lond B Biol Sci 239:1–43

Everett DH, Haynes JM (1972) Model studies of capillary condensation. I. Cylindrical pore model with zero contact angle. J Colloid Interface Sci 38:125–137. https://doi.org/10.1016/0021-9797(72)90228-7

Faucheux MJ (2013) Sensillum types on the proboscis of the Lepidoptera: a review. Ann Soc Entomol F 49:73–90. https://doi.org/10.1080/00379271.2013.767514

Fürstner R, Barthlott W, Neinhuis C, Walzel P (2005) Wetting and self-cleaning properties of artificial superhydrophobic surfaces. Langmuir 21:956–961

Gilbert LE (1972) Pollen feeding and reproductive biology of *Heliconius* butterflies. Proc Nat Acad Sci 69:1403–1407

Goren SL (1962) The instability of an annular thread of fluid. J Fluid Mech 12:309–319. https://doi.org/10.1017/S002211206200021X

Heil M, Hazel AL, Smith JA (2008) The mechanics of airway closure. Respir Physiol Neurobiol 163:214–221. https://doi.org/10.1016/j.resp.2008.05.013

Hikl A–L, Krenn HW (2011) Pollen processing behavior of *Heliconius* butterflies: a derived grooming behavior. J Insect Sci 11:1–13

Hirota T, Obara Y (2000) Time allocation to the reproductive and feeding behaviors in the male cabbage butterfly. Zool Sci 17:323–327. https://doi.org/10.2108/jsz.17.323

Holdgate MW (1955) The wetting of insect cuticles by water. J Exp Biol 32:591–617

Kaiser A, Klok CJ, Socha JJ, Lee W-K, Quinlan MC, Harrison JF (2007) Increase in tracheal investment with beetle size supports hypothesis of oxygen limitation on insect gigantism. Proc Natl Acad Sci 104:13198–13203. https://doi.org/10.1073/pnas.0611544104

Kingsolver JG, Daniel TL (1979) On the mechanics and energetics of nectar feeding in butterflies. J Theor Biol 76:167–179

Knopp MCN, Krenn HW (2003) Efficiency of fruit juice feeding in *Morpho peleides* (Nymphalidae, Lepidoptera). J Insect Behav 16:67–77

Kramer VR, Mulvane CP, Brothers A, Gerard PD, Lehnert MS (2015) Allometry among structures of proboscises of *Vanessa cardui* L. (Nymphalidae) and its relationship to fluid uptake. J Lepid Soc 69:183–191. https://doi.org/10.18473/lepi.69i3.a5

Kramer VR, Reiter KE, Lehnert MS (2018) Proboscis morphology suggests reduced feeding abilities of hybrid *Limenitis* butterflies ( Lepidoptera : Nymphalidae ). Biol J Linn Soc 125:535–546

Krenn HW (1990) Functional morphology and movements of the proboscis of Lepidoptera (Insecta). Zoomorphology 110:105–114

Krenn HW (1997) Proboscis assembly in butterflies (Lepidoptera) - a once in a lifetime sequence of events. Eur J Entomol 94:495–501

Krenn HW (1998) Proboscis sensilla in *Vanessa cardui* (Nymphalidae, Lepidoptera): functional morphology and significance in flower-probing. Zoomorphology 118:23–30. https://doi.org/10.1007/s004350050053

Krenn HW (2000) Proboscis musculature in the butterfly *Vanessa cardui* (Nymphalidae, Lepidoptera): settling the proboscis recoiling controversy. Acta Zool 81:259–266

Krenn HW (2010) Feeding mechanisms of adult Lepidoptera: structure, function, and evolution of the mouthparts. Annu Rev Entomol 55:307–327. https://doi.org/10.1146/annurev-ento-112408-085338

Krenn HW, Penz CM (1998) Mouthparts of *Heliconius* butterflies (Lepidoptera: Nymphalidae): a search for anatomical adaptations to pollen feeding behavior. Int J Insect Morphol Embryol 27:301–309

Krenn HW, Mühlberger N (2002) Groundplan anatomy of the proboscis of butterflies. Zool Anz 241:369–380. https://doi.org/10.1078/0044-5231-00078

Krenn HW, Bauder JAS (2017) Morphological fine tuning of the feeding apparatus to proboscis length in Hesperiidae (Lepidoptera). J Morphol 279:396–408. https://doi.org/10.1002/jmor.20780

Krenn HW, Zulka KP, Gatschnegg T (2001) Proboscis morphology and food preferences in nymphalid butterflies (Lepidoptera: Nymphalidae). J Zool 254:17–26. https://doi.org/10.1017/S0952836901000528

Krenn HW, Eberhard M, Eberhard S, Hikl A-L, Huber W, Gilbert LE (2009) Mechanical damage to pollen aids nutrient acquisition in *Heliconius* butterflies (Nymphalidae). Arthropod Plant Interact 3:203–208

Krenn HW, Fournal J, Bauder JAS, Hugel S (2016) Mouthparts and nectar feeding of the flower visiting cricket *Glomeremus orchidophilus* (Gryllacrididae). Arthropod Struct Dev 45:221–229

Kwauk KJ, Hasegawa DK, Lehnert MS, Beard CE, Gerard PD, Kornev KG, Adler PH (2014) Drinking with an unsealed tube: fluid uptake along the butterfly proboscis. Ann Entomol Soc Am 107:886–892. https://doi.org/10.1603/AN14027

Lee SC, Lee SJ (2014) Uptake of liquid from wet surfaces by the brush-tipped proboscis of a butterfly. Sci Rep 4:6934. https://doi.org/10.1038/srep06934

Lehnert MS, Monaenkova D, Andrukh T, Beard CE, Adler PH, Kornev KG (2013) Hydrophobic-hydrophilic dichotomy of the butterfly proboscis. J R Soc Interface 10:20130336. https://doi.org/10.1098/rsif.2013.0336

Lehnert MS, Mulvane CP, Brothers A (2014) Mouthpart separation does not impede butterfly feeding. Arthropod Struct Dev 43:97–102. https://doi.org/10.1016/j.asd.2013.12.005

Lehnert MS, Brown E, Lehnert MP, Gerard PD, Yan H, Kim C (2015) The Golden ratio reveals geometric differences in proboscis coiling among butterflies of different feeding habits. Am Entomol 61:18–26

Lehnert MS, Beard CE, Gerard PD, Kornev KG, Adler PH (2016) Structure of the lepidopteran proboscis in relation to feeding guild. J Morphol 277:167–182. https://doi.org/10.1002/jmor.20487

Lehnert MS, Bennett A, Reiter KE, Gerard PD, Wie Q-H, Byler M, Lee W-K (2017a) Mouthpart conduit sizes of fluid-feeding insects determine the ability to feed from pores. Proc R Soc B Biol Sci 284:20162026. https://doi.org/10.1098/rspb.2016.2026

Lehnert MS, Reiter KE, Bennett A, Gerard PD, Wie Q-H, Byler M, Lee W-K (2017b) The ingestion of fluorescent, magnetic nanoparticles for determining fluid-uptake abilities in insects. J Vis Exp 2017(130):e56619. https://doi.org/10.3791/56619

Liu K, Yao X, Jiang L (2010) Recent developments in bio-inspired special wettability. Chem Soc Rev 39:3240–3255. https://doi.org/10.1039/b917112f

Marmur A (2003) Wetting on hydrophobic rough surfaces: to be heterogeneous or not to be? Langmuir 19:8343–8348. https://doi.org/10.1021/LA0344682

Monaenkova D, Lehnert MS, Andrukh T, Beard CE, Rubin B, Tokarev A, Lee W-K, Kornev KG (2012) Butterfly proboscis: combining a drinking straw with a nanosponge facilitated diversification of feeding habits. J R Soc Interface 9:720–726. Retrieved from http://rsif.royalsocietypublishing.org/content/early/2011/08/09/rsif.2011.0392.abstract

Molleman F, Krenn HW, van Alphen ME, Brakefield PM, Devries PJ, Zwaan BJ (2005) Food intake of fruit-feeding butterflies: evidence for adaptive variation in proboscis morphology. Biol J Linn Soc 86:333–343. https://doi.org/10.1111/j.1095-8312.2005.00533.x

Nørgaard T, Dacke M (2010) Fog-basking behaviour and water collection efficiency in Namib Desert darkling beetles. Front Zool 7:23. https://doi.org/10.1186/1742-9994-7-23

Parker AR, Lawrence CR (2001) Water capture by a desert beetle. Nature 414:33–34. https://doi.org/10.1038/35102108

Petr D, Stewart KW (2004) Comparative morphology of sensilla styloconica on the proboscis of north American Nymphalidae and other selected taxa (Lepidoptera): systematic and ecological considerations. Trans Am Entomol Soc 130:293–409

Plateau JAF (1863) Experimental and theoretical researches on the figures of equilibrium of liquid mass withdrawn from the action of gravity. Annual Report of the Board of Regents of the Smithsonian Institutions, Washington, DC, pp 207–285

Rayleigh L (1879) On the capillary phenomena of jets. Proc R Soc Lond 29:71–97

Siemann E, Tilman D, Haarstad J, Ritchie M (1998) Experimental tests of the dependence of arthropod diversity on plant diversity. Am Nat 152:738–750. https://doi.org/10.1086/286204

Socha JJ, Lee W-K, Harrison JS, Waters JS, Fezzaa K, Westneat MW (2008) Correlated patterns of tracheal compression and convective gas exchange in a carabid beetle. J Exp Biol 211:3409–3420. https://doi.org/10.1242/jeb.019877

Socha JJ, Förster TD, Greenlee KJ (2010) Issues of convection in insect respiration: insights from synchrotron X-ray imaging and beyond. Respir Physiol Neurobiol 173:S65–S73. https://doi.org/10.1016/j.resp.2010.03.013

Stork NE, McBroom J, Gely C, Hamilton AJ (2015) New approaches narrow global species estimates for beetles, insects, and terrestrial arthropods. Proc Natl Acad Sci 112:7519–7523. https://doi.org/10.1073/pnas.1502408112

Tsai C-C, Monaenkova D, Beard CE, Adler PH, Kornev KG (2014) Paradox of the drinking-straw model of the butterfly proboscis. J Exp Biol 217:2130–2138. https://doi.org/10.1242/jeb.109447

Vincent JFV, Wegst UGK (2004) Design and mechanical properties of insect cuticle. Arthropod Struct Dev 33:187–199. https://doi.org/10.1016/j.asd.2004.05.006

Wannenmacher G, Wasserthal LT (2003) Contribution of the maxillary muscles to proboscis movement in hawkmoths (Lepidoptera: Sphingidae) - an electrophysiological study. J Insect Physiol 49:765–776. https://doi.org/10.1016/S0022-1910(03)00113-6

Weibel ER, Taylor CR, Hoppeler H (1991) The concept of symmorphosis: a testable hypothesis of structure-function relationship. Proc Natl Acad Sci 88:10357–10361. https://doi.org/10.1073/pnas.88.22.10357

Wenzel RN (1936) Resistance of solid surfaces to wetting by water. Ind Eng Chem 28:988–994. https://doi.org/10.1021/ie50320a024

Westneat MW, Socha JJ, Lee W-K (2008) Advances in biological structure, function, and physiology using synchrotron X-ray imaging. Annu Rev Physiol 70:119–142. https://doi.org/10.1146/annurev.physiol.70.113006.100434

Zaspel JM, Weller SJ, Branham MA (2011) A comparative survey of proboscis morphology and associated structures in fruit-piercing, tear-feeding, and blood-feeding moths in Calpinae (Lepidoptera: Erebidae). Zoomorphology 130:203–225. https://doi.org/10.1007/s00435-011-0132-1

Zaspel JM, Zahiri R, Hoy MA, Janzen D, Weller SJ, Wahlberg N (2012) A molecular phylogenetic analysis of the vampire moths and their fruit-piercing relatives (Lepidoptera: Erebidae: Calpinae). Mol Phylogenet Evol 65:786–791. https://doi.org/10.1016/j.ympev.2012.06.029

Zhang M, Feng S, Wang L, Zheng Y (2016) Lotus effect in wetting and self-cleaning. Biotribology 5:31–43

Zhang Z-H, Wang H-J, Liang Y-H, Li X-J, Ren J-Q, Cui Z-Q, Luo C (2018a) One-step fabrication of robust superhydrophobic and superoleophilic surfaces with self-cleaning and oil/water separation function. Sci Rep 8:3869

Zhang C, Adler PH, Monaenkova D, Andrukh T, Pometto S, Beard CE, Kornev KG (2018b) Self-assembly of the butterfly proboscis: the role of capillary forces. J R Soc Interface 15:20180229. https://doi.org/10.1098/rsif.2018.0229

Zhang C, Beard CE, Adler PH, Kornev KG (2018c) Effect of curvature on wetting and dewetting of proboscises of butterflies and moths. R Soc Open Sci 5(1):171241. https://doi.org/10.1098/rsos.171241

# Chapter 10
# Comparative Functional Morphology of Ant Mouthparts and Significance for Liquid Food Intake

**Jürgen Paul and Flavio Roces**

**Abstract** Mouthparts are to ants very versatile tools for almost any task. This is particularly true for their mandibles. Ants use them for prey-catching, fighting, leaf-cutting, brood care, and communication. The key to the versatility of mandible functions is the mandible closer, the largest muscle in ants. This muscle is generally composed of distinct muscle fiber types that differ in morphology and contractile properties. Volume proportions of the fiber types are species-specific and correlate with feeding habits. Two biomechanical models explain muscle design to fulfill force and speed requirements. Adult ants mainly feed on fluids and exchange them via trophallaxis. The glossa of ants, the distal end of the labium, is the first body part involved in liquid food intake. It works either as a passive open duct during sucking or as an up- and downward moving shovel during licking. The glossa surface is covered with small spadelike microtrichia that are regularly arranged in parallel rows. A comparison of glossa surface characteristics and volumes of muscles that control licking and sucking among four ant species differing in their feeding habits shows that morphology depends on the used feeding technique and the extent of specialization on liquid food, suggesting a general design driven by evolution.

## 10.1 Introduction: The Mandibles of Ants

Insects are equipped with very diverse feeding organs capable of ingesting solid, particulate, or liquid food in many different ways (Chapman 1998). But insects use their mouthparts not only for food intake. Mandibles are to ants what hands are to humans: body parts highly adapted to master a diversity of tasks, such as prey-catching, fighting, leaf-cutting, brood care, and communication (Hölldobler and

J. Paul (✉)
Didaktik der Naturwissenschaften, Universität Bamberg, Bamberg, Germany
e-mail: juergen.paul@uni-bamberg.de

F. Roces
Lehrstuhl für Verhaltensphysiologie und Soziobiologie, Biozentrum, Universität Würzburg, Würzburg, Germany
e-mail: roces@biozentrum.uni-wuerzburg.de



H. W. Krenn (ed.), *Insect Mouthparts*, Zoological Monographs 5,
https://doi.org/10.1007/978-3-030-29654-4_10

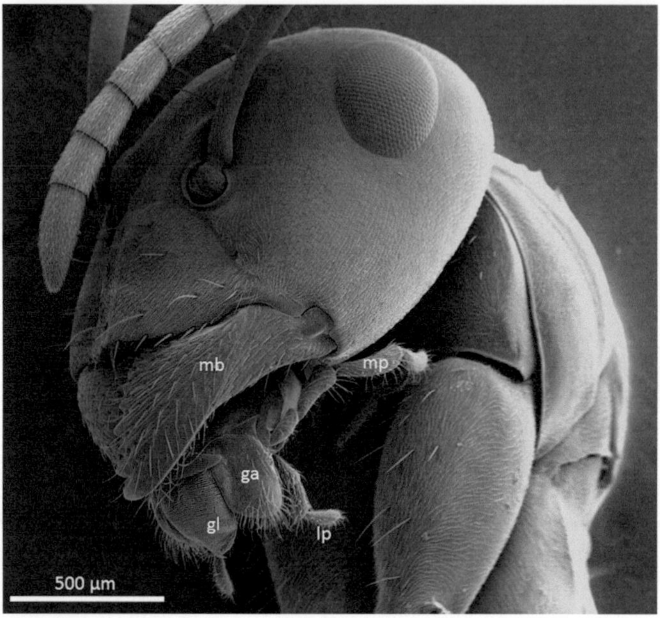


**Fig. 10.1** General view of an ant head (*Formica* sp.) to illustrate the mouthparts; abbreviations: *ga* galea; *gl* glossa; *lp* labial palpus; *mb* mandible; *mp* maxillary palpus (Scanning electron microscopical photo: Harald W. Krenn)

Wilson 1990). Adult ants possess mandibles adapted for biting and chewing and a labiomaxillary complex, which is a functional connection between maxillae and labium (Fig. 10.1). While morphology and performance of ant mandibles are known to correlate with feeding habits and ecological niche of different species (Paul 2001a), the functional morphology of other ant mouthparts was rarely investigated. Therefore, we firstly summarize some pertinent findings about underlying morphology and the control of ant mandible movements. And secondly, we present new comparative morphological data on the labiomaxillary complex of ants and derive its significance for the intake of liquid food, which is the most widespread resource for adult ants.

The key to the versatility of mandible functions is the huge mandible closer muscle. In a few specialized ants, e.g., when mandible opening is used for moving forward during tunneling (Khalife et al. 2018), the mandible opener is also exceptionally large but still smaller than the closer. The mandible closer muscle is generally composed of three distinct muscle fiber types that differ in morphology and contractile properties (Gronenberg et al. 1997). Fast-contracting fibers (first type) have short sarcomeres (2–3 µm) and attach directly to the closer apodeme, which conveys the muscle power to the mandible joint. Slow but forceful contracting fibers have long sarcomeres (5–6 µm) and attach to the apodeme either directly (second type) or via thin threadlike filaments (third type) that vary in length. The muscle fibers form different units of the same fiber type within the closer muscle.

Volume proportions of the three fiber types are species-specific and correlate with feeding habits. Predators, like *Harpegnathos saltator* or *Myrmecia* sp., depend on fast mandible movements to catch their prey. They often have long jaws equipped with piercing teeth and sharp edges. These jaws are controlled by a mandible closer muscle that mainly contains directly attached, fast-contracting fibers. Herbivorous ants, like the leaf-cutting ant *Atta sexdens* or the harvesting ant *Pogonomyrmex badius*, have mandibles suited for the particular task of processing plant material or cracking seeds, where powerful movements are needed. Their mandible closer muscle is composed of slow but forceful contracting fibers. As well as in other animals, short sarcomeres are generally associated with fast contraction, whereas long sarcomeres usually correspond to high force production (Günzel et al. 1993; Biewener and Patek 2018).

Morphological studies additionally revealed that the composition of the mandible closer muscle follows specific biomechanical and geometrical rules. Comparisons of different ant species suggest two basic principles underlying the design of mandible muscles (Paul and Gronenberg 1999): (1) Ants specialized for fast mandible movements generally feature long heads which contain long fast muscle fibers that attach to the apodeme at small angles. Their muscles comprise only a few filament-attached fibers, and they maximize speed of action at the expense of force output. (2) Ants performing particularly forceful mandible movements, such as seed cracking, rely on many short parallel muscle fibers contained within a broad head capsule. Their slower muscles incorporate a large proportion of filament-attached fibers. Two biomechanical models explain how the attachment angles are optimized with respect to force and velocity output and how filament-attached fibers help to generate the largest power output from the available head capsule volume (Paul and Gronenberg 1999).

Trap-jaw ants, like *Odontomachus chelifer*, feature highly specialized catapult mechanisms (Gronenberg 1995; Gronenberg et al. 1998a, b; Larabee and Suarez 2014; Larabee et al. 2017; Perrichot et al. 2016). Their mandible closing is known as one of the fastest movements in the animal kingdom. The mandible strike is released within less than 10 ms upon stimulation of particular mechanosensory trigger hairs. The trap-jaws themselves can be closed in less than 0.5 ms. This rapid movement is based on the storage of mechanical energy produced by the large but slow mandible closer muscle that cocks the mandible several seconds in advance of the strike. The strike is released from the catch by a small trigger muscle composed of fast-contracting tubular fibers. At first glance, trap-jaw ants seem to be an exception in terms of muscle design, since they perform fast mandible movements with a closer muscle that mainly or solely contains slow muscle fibers. Understanding the specialized catapult mechanisms though explains how the predatory trap-jaw ants rather confirm the rules mentioned above (Paul 2001a).

The mandible closer is the largest muscle in ants. Likewise, the relatively large number of 8–12 motor neurons that control this muscle reflects its importance for the behavior of ants (Just and Gronenberg 1999; Paul and Gronenberg 2002). This set of motor neurons allows the animal to independently activate different types of muscle fibers or locally distinct fiber bundles, resulting in a great variety of possible movements. The sturdy ant mandibles connected to a simple but versatile motion system are thus shaped to cope with a wide range of tasks.

## 10.2 Liquid Food Intake in Ants

Irrespective of the diversity in their feeding apparatus, insects feeding on liquid food use two basic techniques for fluid food intake, either licking or sucking, which are reflected in specific adaptations of the mouthparts. The mouthparts of sucking insects usually form a more or less elongated food canal that functions as a sucking tube (e.g., Lepidoptera: Krenn 1990; Diptera: Szucsich and Krenn 2000; Hemiptera: Labandeira and Phillips 1996; Jervis and Vilhelmsen 2000). On the other hand, licking behavior is usually associated with hairy tongues or analogous structures that enhance capillarity, a phenomenon driven by surface tension, so as to load the tongue with fluid food (e.g., bees: Snodgrass 1956; Harder 1983; Michener and Brooks 1984; Kingsolver and Daniel 1995; Kim et al. 2011; Wu et al. 2015). The two fundamental techniques of liquid food intake correspond to two basic types of anatomical design (Chapman 1998).

Although workers of some ant species frequently feed on liquid food, the ant feeding apparatus is not exclusively adapted to ingest liquids. As described above, ant mandibles can perform many other tasks. Their labiomaxillary complex is suited to ingest solid or particulate, as well as fluid food (Gotwald 1969; Paul et al. 2002). Ants are able to employ both techniques, either licking or sucking, for liquid food intake. But workers of different species use exclusively one of these two different feeding techniques when collecting liquid food during foraging. While workers of ponerine ants rely on licking, workers of formicine ants exclusively suck the fluid when collecting at an ad libitum nectar droplet (Josens and Roces 2000; Paul and Roces 2003; Paul et al. 2002).

The glossa, the distal end of the labium, works in individuals of different ant species either as a passive open duct during sucking or as an up- and downward moving "shovel" during licking (Paul and Roces 2003). For sucking, workers of the nectar-feeding ant genus *Camponotus* (Formicinae) deeply put their protracted glossa and some adjacent parts of the labiomaxillary complex into the fluid droplet, and when it is almost completely ingested, workers switch to licking behavior, since the remaining fluid built a flat film in which the glossa could not be introduced. The collected fluid is stored within the crop (Josens and Roces 2000; Paul and Roces 2003). In contrast, nectar-feeding ants of the genus *Pachycondyla* (Ponerinae) lick the fluid from the very beginning. For licking, the protracted glossa only touches the surface of the droplet, whereby the glossa is loaded with the liquid. During retraction of the glossa, by which the contact with the droplet is interrupted, the two galeae move ventrally without touching the glossa. During the following protraction of the glossa, the galeae wipe the liquid upward along the downward-moving glossa, before the glossa again get into contact with the droplet's surface. The opposite movement of galeae and glossa ensures the upward movement of the liquid, so that it accumulates as an external droplet that is transported, held by capillary forces, between the mandibles (Hölldobler 1985; Paul and Roces 2003).

In a comparative study on liquid food intake in ants, we have shown that the rates of fluid intake largely depend on both the feeding technique employed and the foraging ecology of the species under scrutiny. For similarly sized ants, workers of two nectar-feeding ant species (*Camponotus rufipes* and *Pachycondyla villosa*, which employ sucking and licking, respectively) were observed to collect fluids with the highest intake rates, while workers of leaf-cutting and predatory ants (*Atta sexdens* and *Rhytidoponera impressa*, which employ sucking and licking, respectively) did so with the lowest rate. Calculating the energy intake rates in mg sucrose per unit time, licking was shown to be a more efficient technique at higher sugar concentrations than sucking, whereas sucking provided a higher energy intake rate at lower sugar concentrations (Paul and Roces 2003). According to colony requirements, ants can vary their nectar intake rate by increasing the frequency of the sucking pump cycles (Josens et al. 2006; Falibene and Josens 2008; Falibene et al. 2009). In addition, environmental temperature affects the dynamics of ingestion by leading to an increase in the frequency of pump contractions, while the volume taken in per pumping cycle remains unaffected (Falibene and Josens 2014).

The data shown in Sect. 10.3 focuses on the comparative functional morphology of the feeding apparatus in the four ant species mentioned above and investigates the mechanisms underlying liquid food intake. Specifically, it is asked whether the extent of specialization on foraging for fluid food is reflected in specific adaptations of the ant labiomaxillary complex and associated muscles. The investigated species differ in their feeding habits. *Camponotus rufipes* represents a typical nectar-feeding ant species that collects ad libitum fluids by sucking and possesses a well-developed crop in which the collected fluid is temporarily stored (Eisner 1957; Jaffé and Sánchez 1984). Workers of *Pachycondyla villosa* exclusively lick the liquid food, which represent up to 57% of their forage (Valenzuela-González et al. 1994). In contrast, the leaf-cutting ant *Atta sexdens* and the predatory ant *Rhytidoponera impressa* are seldom observed visiting nectar sources (Littledyke and Cherrett 1976; Ward 1981). We compared the four ant species in terms of their glossa surface morphology and the relative volumes of the muscles controlling the processes of licking and sucking. To control for differences in worker size among species and to allow interspecific comparisons, we measured muscle volumes and glossa surface morphology in individuals of different sizes in the two polymorphic species, *C. rufipes* and *A. sexdens*, and used the obtained relationship between the measured variables and the head width to scale measurements to the worker size of the two monomorphic species (*P. villosa* and *R. impressa*). Hence, functional morphology of the feeding apparatus could be quantitatively compared across species for similarly sized ants in relation to the species-specific feeding habits.

## 10.3 Morphological Characteristics of the Liquid Feeding Apparatus in Ants

### *10.3.1 Studying Functional Morphology of Mouthparts*

Measurements were performed on workers of the following ant species: *Pachycondyla villosa*, *Rhytidoponera impressa*-complex (Ponerinae), *Atta sexdens rubropilosa* (Myrmicinae), and *Camponotus rufipes* (Formicinae). The ant colonies were kept in plaster-of-Paris nests under a 12 h:12 h L:D cycle, at 25 °C and 50% relative humidity. They were fed chopped cockroaches, crickets, or wingless *Drosophila* and honey-water (30%) or fresh leaves (*A. sexdens rubropilosa*).

For analysis of the glossa surface, ants were anesthetized and, after protraction of the labiomaxillary complex, decapitated. The heads were then prepared for scanning electron microscopy using standard methods (according to Paul 2001b). In order to obtain histological preparations for volume measurements and for the analysis of the head muscles, ants were also anesthetized and decapitated. After fixation, the heads were stained either with methylene blue or with osmium/ethyl gallate or silver-impregnated (according to Gronenberg et al. 1997). The heads were then dehydrated, embedded, and horizontally, sagittally, or vertically sectioned at 10–15 μm. The histological preparations were analyzed using light microscopy.

For the four investigated species, glossa surface parameters were measured from calibrated scanning electron microscopic images. We defined the surface area of the glossa that could be seen in frontal view as the "frontal surface area of the glossa." For measurements of this area, the respective images were video recorded and digitally evaluated. To determine the density of the spadelike microtrichia that cover the glossa surface, we counted the number of microtrichia in a given area at least in three different regions of the glossa, located either central, ventral, or dorsal. We defined the mean number of microtrichia per unit glossa surface area as "microtrichia density of the glossa." To determine the area of a single microtrichium, we measured its maximum and minimum width and its length. The product of the mean width by the length was defined as "microtrichium area." The single area values were averaged over a large number of measured microtrichia.

For volumetric measurements of the muscles associated with the labiomaxillary complex, light-microscopic images of the histological preparations were video recorded and digitally evaluated with a computer. From each microscopic slide, the outlines of the corresponding structures were traced on the computer screen and the areas computed. The volume of a given muscle was calculated considering the section thickness. For the four investigated species, we determined the volumes of the head capsule and of six muscles, as follows: the two largest pharyngeal muscles, i.e., the pharynx dilator and the cibarial dilator, and four labial muscles, the labium opener, the labium closer, the glossa muscle, and the paraglossa muscle. A schematic drawing of the head of an ant worker (*P. villosa*) was made from light-microscopic images using a calibrated camera lucida attached to the microscope.

To control for the differences in ant size and to be able to perform interspecific comparisons, we first plotted the relative muscle volumes and glossa surface characteristics as a function of head width for individual workers of the two polymorphic species, *C. rufipes* and *A. sexdens*. Since the measured head widths of workers of the two monomorphic species, *P. villosa* and *R. impressa*, lay within the measured range of the two polymorphic species, the scaling functions could be used to obtain a value for any head width within the range, without extrapolation. As a consequence, comparisons across species could be made independently of worker size. In contrast to the body mass that depends on variable parameters such as crop load or content of the digestive tract, the head width is a constant morphological size parameter. We therefore used head width measurements for scaling ant size. However, it is important to indicate that a significant linear relationship between head width and unloaded body mass of individual ants was found. For statistical comparisons of scaled values, the following procedure was employed. Since for the polymorphic *C. rufipes* and *A. sexdens* only one scaled value with no variability resulted from the scaling function, its error bar was calculated based on the average deviation of all measured values from the regression line. As a consequence of this estimation, no analysis of variance was possible because of the lack of a data sample for the single scaled values, so that t-tests were used instead. For a more detailed methodological description for the collection of the morphometric data presented within this chapter, see Paul (2001b).

### *10.3.2 Control of Licking and Sucking*

Although ants are able to employ licking as well as sucking, workers of a particular ant species exclusively use one technique for collecting fluids at ad libitum food sources. Workers of *P. villosa* and *R. impressa*-complex licked the sugar solution when feeding on a large droplet, whereas workers of *C. rufipes* and *A. sexdens* sucked it (Paul and Roces 2003). In the four investigated species, the glossa is covered with small spadelike microtrichia that are regularly arranged in parallel rows (Fig. 10.2). These microtrichia are bent at the tip by 90 degrees toward the interior of the mouth cavity, as observed in wasps (Kirshboim and Ishay 1998), and are expected to considerably enlarge the glossa surface, thus facilitating the adhesion of fluids as well as of solid food particles.

The whole glossa retracts upon contraction of the glossa muscle and the paraglossa muscle (muscles numbered 8 and 9, Fig. 10.3). Both muscles work as glossa retractors, whereas the protraction is based on an elastic mechanism. When both the glossa and the paraglossa muscle relax, the glossa protracts elastically (Paul et al. 2002). During licking, the glossa works as an up- and downward moving shovel. This glossa movement is supported by a synchronous movement of the entire labium (of which the glossa is the most distal part), which probably results from alternate contractions of the labium opener and closer (muscles 6 and 7, Fig. 10.3).

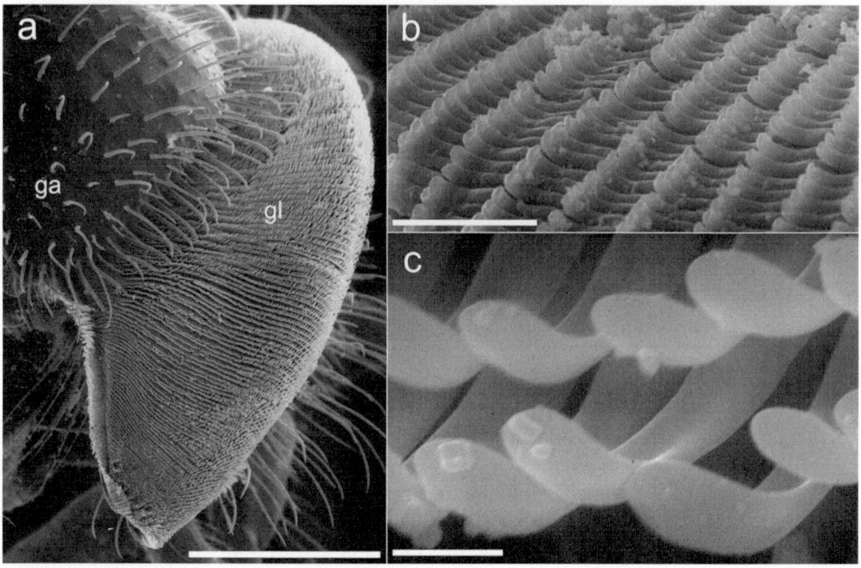


**Fig. 10.2** Glossa of various ant species (Scanning electron microscopical photo). (**a**) Lateral view of the glossa and the distal part of one galea; worker of *Camponotus rufipes*; (ga) galea, (gl) glossa; *scale bar* = 200 μm; (**b**) oblique view of the glossa surface with parallel rows of microtrichia (worker of *Atta sexdens*; *scale bar* = 10 μm). (**c**) Microtrichia on the glossa surface showing their bent tips (worker of *C. rufipes*; *scale bar* = 2 μm)

Accordingly, four different labial muscles contribute to the licking process: the labium opener and closer and the glossa and the paraglossa muscles.

For sucking, the labial muscles are not activated, because the glossa functions as a passive structure comparable to an open duct, and it is hold motionless in a protracted position. During sucking, the pharynx and cibarial dilator muscles (muscles 1 and 5, Fig. 10.3) are expected to contract periodically in a coordinated manner, thus producing and maintaining a negative pressure within the pharynx (note that even though Fig. 10.3 shows the head of a *P. villosa* worker that licks fluids at ad libitum droplets, the basic anatomical design of the involved structures is similar across species independently of the employed feeding technique). Driven by this negative pressure, the fluid is transported over glossa and hypopharynx to the actual mouth of the ant. The inner ventral sides of the closed mandibles, the upper side of the labium, and the inner sides of the maxillae form the functional sucking tube. Therefore, the sucking of fluids is mainly driven by the coordinated activity of both the pharynx dilator muscles (muscle 1, Fig. 10.3) and the cibarial dilator muscle (muscle 5, Fig. 10.3). It is important to indicate that in an early, very detailed anatomical description of *Lasius* ant queens, the muscle 5 was named "retractor of the buccal tube" (Janet 1905) and therefore not referred to as being part of a cibarial pump. However, a study using synchrotron X-ray imaging clearly showed that *Camponotus* ants have two sucking pumps, i.e., a cibarial and a pharyngeal one, which, together with complex mouthpart motions and peristalsis of esophagus and

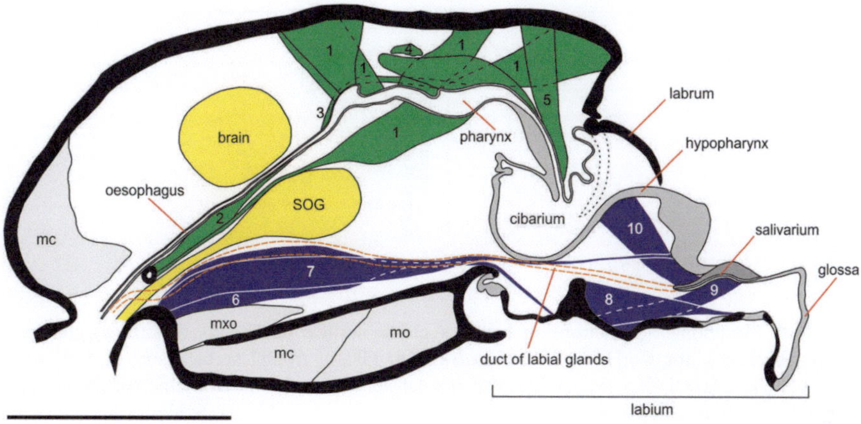


**Fig. 10.3** Sagittal section of the head of a *Pachycondyla villosa* workers (schematic drawing, several sagittal sections at the midline of two different preparations are layered). Yellow: brain and suboesophageal ganglion (SOG). Green: pharyngeal muscles; (1) pharynx dilator, (2) pharynx retractor, (3) pharynx longitudinal muscle, (4) pharynx transversal adductor, (5) cibarial dilator. Blue: labial muscles; (6) labium opener, (7) labium closer, (8) glossa muscle, (9) paraglossa muscle, (10) hypopharynx muscle. Black: sclerotized head capsule or tentorium. Gray: mandible and maxillary muscles, (mc) mandible closer, (mo) mandible opener, (mxo) maxillary opener; unsclerotized or only slightly sclerotized cuticular structures around the cibarium are also indicated in gray; (*scale bar* = 1 mm)

crop, are responsible for bolus formation and fluid ingestion rates (Cook et al. 2006). Therefore, we prefer to call the muscle 5 "cibarial dilator," as described for honeybees (Snodgrass 1956).

### *10.3.3 Glossa Surface*

We measured three glossa surface parameters, the frontal surface area, the microtrichia density, and the microtrichium area (see Fig. 10.2). For workers of the two polymorphic species (*A. sexdens* and *C. rufipes*), the values largely depended on the size of the ant. Both the frontal surface area of the glossa and the area of the microtrichia increased, whereas the microtrichia density decreased with increasing head width (examples are shown in Fig. 10.4a, b). These relationships, which held for both polymorphic species, are summarized in Table 10.1. For interspecific comparisons, we used the linear equations presented in Table 10.1 for scaling the measured glossa parameters to the respective worker size of the two monomorphic species, as their mean head widths lay within the range of the two polymorphic species (mean head width of workers of *R. impressa*-complex, 1.29 ± 0.06 mm, mean ± SD, $n = 112$; *P. villosa*, 2.37 ± 0.07 mm, $n = 115$; Table 10.1, right two columns). It is important to indicate that because of the distinct head widths of the two monomorphic species, it is not possible to illustrate the measured variables from

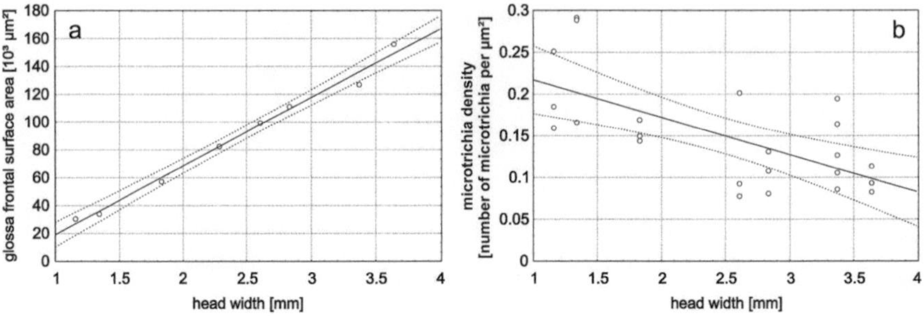


**Fig. 10.4** Morphometric relationship between glossa surface parameters (ordinates) and head width (abszissae), for workers of *Atta sexdens*. (**a**) Glossa frontal surface area × [$10^3$ µm$^2$] ($r^2 = 0.99$; $p < 0.001$; $N = 8$). (**b**) Microtrichia density [number of microtrichia per µm$^2$ surface area] ($r^2 = 0.44$; $p < 0.001$; $n = 23$; $N = 7$)

the four species in a single graph without extrapolations. Therefore, data on each of the monomorphic species are compared with the scaled values from the two polymorphic ones, i.e., data from *Pachycondyla* workers are compared with those from similarly sized, scaled *Camponotus* and *Atta* workers (Fig. 10.5) and data from *Rhytidoponera* workers with those from similarly sized, scaled *Camponotus* and *Atta* workers (Fig. 10.6).

When scaled to the mean head width of a *Pachycondyla* worker, the values of the glossa parameters were observed to depend on the species considered. In *P. villosa*, the microtrichia on the glossa surface were more than three times larger than in *C. rufipes* and *A. sexdens* (Fig. 10.5a). The opposite was true for the microtrichia density, which was much higher in the two polymorphic species (Fig. 10.5b). Putting these two parameters together, we calculated a "factor of surface enlargement" as the product of microtrichium area by microtrichia density. This factor was largest in *P. villosa*, averaging 3.2 times, and smallest in *A. sexdens* (Fig. 10.5d). Regarding the glossa frontal surface area, it was similar in *Camponotus* and *Atta*, whereas it was 50% larger in *Pachycondyla* (Fig. 10.5c). To quantify the total surface that is expected to get in contact with the fluid, i.e., a measure of the glossa loading capacity, we defined the "active glossa surface area" as the product of microtrichium area, microtrichia density, and frontal surface area. As for the factor of surface enlargement, the active glossa surface area was largest in *P. villosa* (mean: 0.44 mm$^2$), intermediate in *C. rufipes* (0.25 mm$^2$), and smallest in *A. sexdens* (0.17 mm$^2$; Fig. 10.5e). These values were statistically different ($p < 0.02$).

When scaled to the mean head width of a *Rhytidoponera* worker, the microtrichia area was similar in *Camponotus* and *Rhytidoponera* but larger than in *Atta* (Fig. 10.6a). In contrast, the microtrichia density was statistically similar in all three species, although it tended to be smaller in *Rhytidoponera* (Fig. 10.6b). The resulting factor of surface enlargement was thus largest in *Camponotus* and smallest in *Atta*. The value for *Rhytidoponera* lay in between and did not differ significantly (Fig. 10.6d). The glossa of *Atta* workers featured the smallest frontal surface area, whereas, in *Camponotus*, it was as large as in *Rhytidoponera* (Fig. 10.6c). The

**Table 10.1** Relationships between glossa surface parameters (*Y*) and head width (*X*, in mm) of workers belonging to the two polymorphic species, *Camponotus rufipes* and *Atta sexdens*

| Variable (*Y*) | Species | *a* | *b* | $r^2$ | *p* | *n* | *N* | *Y* for *X* = 1.29 mm | *Y* for *X* = 2.37 mm |
|---|---|---|---|---|---|---|---|---|---|
| Microtrichium area [$\mu m^2$] | *A. sexdens* | 9.306 | 1.505 | 0.43 | < 0.001 | 69 | 7 | 11.25 | 12.87 |
| | *C. rufipes* | 17.104 | 1.583 | 0.14 | 0.012 | 45 | 5 | 19.15 | 20.86 |
| Microtrichia density [per $\mu m^2$] | *A. sexdens* | 0.262 | −0.045 | 0.44 | < 0.001 | 23 | 7 | 0.20 | 0.16 |
| | *C. rufipes* | 0.285 | −0.065 | 0.38 | 0.015 | 15 | 5 | 0.20 | 0.13 |
| Frontal surface area [$10^3$ $\mu m^2$] | *A. sexdens* | −30.218 | 49.338 | 0.99 | < 0.001 | 8 | 8 | 33.43 | 86.71 |
| | *C. rufipes* | −15.822 | 45.935 | 0.89 | 0.017 | 5 | 5 | 43.43 | 93.04 |

Linear regressions ($Y = a + b \times X$), all statistically significant, provided the best fit and were used for scaling glossa surface parameters. $r^2$ = coefficient of determination; $p$ = probability of error; "*n*" corresponds to either the number of measured microtrichia (microtrichium area), units of glossa surface area (microtrichia density), or glossae (frontal surface area); $N$ = number of measured individuals; range of head widths in *C. rufipes*, 1.1 to 3.6 mm; in *A. sexdens*, 1.1 to 4.2 mm

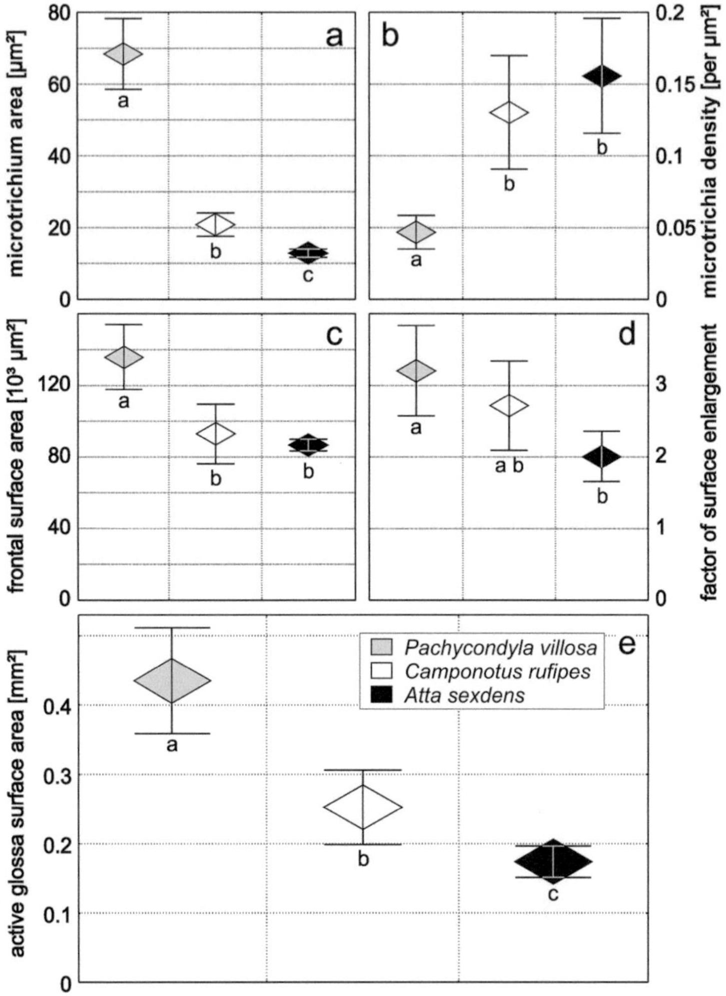


**Fig. 10.5** Glossa parameters (ordinates) among workers of *Pachycondyla villosa* (light gray), *Camponotus rufipes* (white), and *Atta sexdens* (black). For the two polymorphic species, all values are scaled to the head size of workers of *P. villosa* (mean head width ± SD = 2.37 ± 0.07 mm), and error bars represent mean SDs of the respective values calculated from the linear equations (see Table 10.1 for n- and N-numbers). Columns sharing the same letter are not statistically different. (**a**) Microtrichium area [µm$^2$] (mean ± SD, $n = 25$ microtrichia, $N = 4$ individuals for *P. villosa*). (**b**) Microtrichia density [number of microtrichia per µm$^2$ glossa surface area] (mean ± SD, $n = 10$ area units, $N = 4$). (**c**) Frontal surface area of the glossa [$10^3$ µm$^2$] (mean ± SD, $N = 4$). (**d**) Factor of surface enlargement = microtrichium area × microtrichia density. (**e**) Active glossa surface area [mm$^2$] = microtrichium area × microtrichia density × frontal surface area

calculated active glossa surface area was similar in *C. rufipes* (mean, 0.16 mm$^2$) and *R. impressa* (0.13 mm$^2$) yet larger than in *A. sexdens* (0.08 mm$^2$; Fig. 10.6e, $p < 0.001$). It is important to indicate that the active glossa surface area of a "virtual"

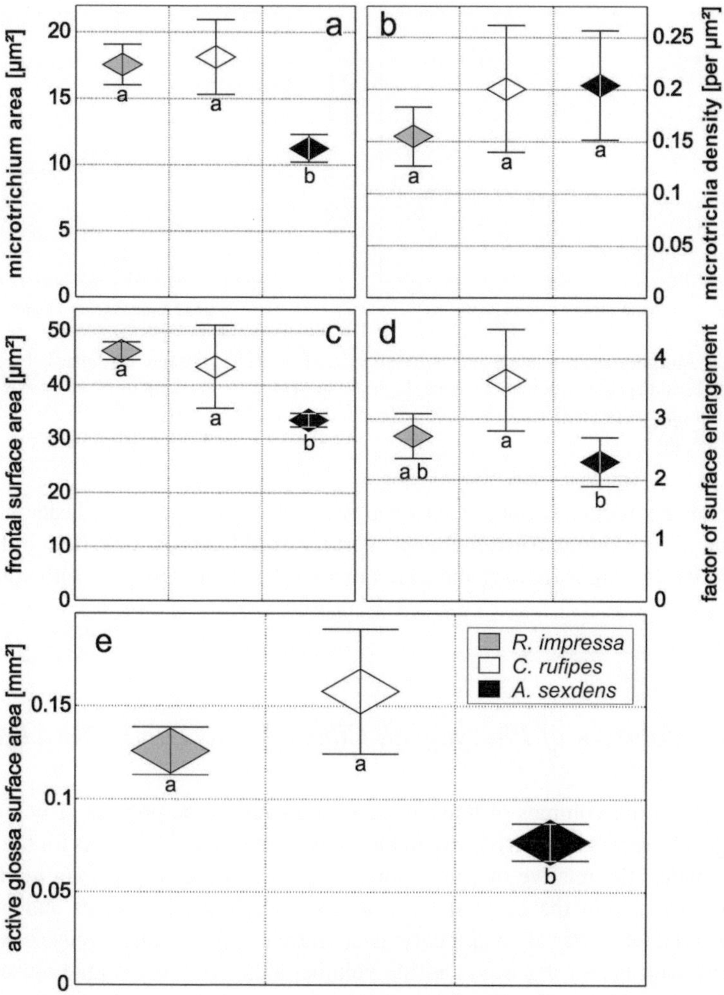


**Fig. 10.6** Glossa parameters (ordinates) among workers of *Rhytidoponera impressa*-complex (dark gray), *Camponotus rufipes* (white), and *Atta sexdens* (black). For the two polymorphic species, all values (mean ± SD, as in Fig. 10.5) are scaled to the head size of workers of *Rhytidoponera* (mean head width = 1.29 ± 0.06 mm), and error bars represent mean SDs of the respective values calculated from the linear equations (see Table 10.1 for n- and N-numbers). Columns sharing the same letter are not statistically different. (**a**) Microtrichium area [$\mu m^2$] (mean ± SD, $n = 29$ microtrichia, $N = 5$ individuals for *R. impressa*). (**b**) Microtrichia density [number of microtrichia per $\mu m^2$] (mean ± SD, $n = 10$ area units, $N = 5$). (**c**) Frontal surface area [$10^3$ $\mu m^2$] (mean ± SD, $N = 5$). (**d**) Factor of surface enlargement. (**e**) Active glossa surface area [$mm^2$]

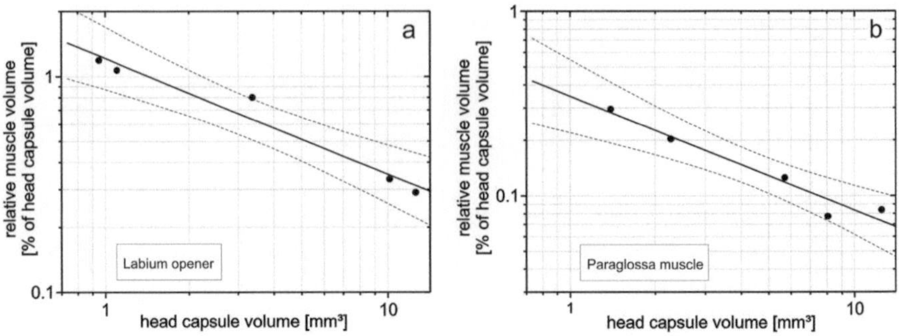


**Fig. 10.7** Morphometric relationship between relative muscle volumes (ordinates, logarithmic scale) and head capsule volume (abszissae, logarithmic scale), for workers of *Camponotus rufipes*. (**a**) Labium opener ($r^2 = 0.96$; $p < 0.005$; $N = 5$). (**b**) Paraglossa muscle ($r^2 = 0.95$; $p < 0.006$; $N = 5$)

worker of the monomorphic *Pachycondyla*, with an extrapolated head width of 1.29 mm as an average *Rhytidoponera* worker, would approximate 0.27 $mm^2$, i.e., it would be the largest among the four species examined (compare with values on Fig. 10.6e).

### *10.3.4 Volumes of Pharyngeal, Cibarial, and Labial Muscles*

We measured the volumes of those muscles involved in the process of licking and sucking and expressed them relative to the ant head capsule volume. As for the glossa characteristics, the relative muscle volumes varied considerably among workers of different size within the same species. In both polymorphic species and over all examined labial, cibarial, and pharyngeal muscles, the relative muscle volume decreased with increasing head capsule volume, although the absolute muscle volumes increased (examples for the labium opener and the paraglossa muscle are presented in Fig. 10.7a, b, respectively). This trend is brought about by the allometric growth of the mandible closer muscle, which becomes relatively larger in larger ants (Paul 2001a). The relationship between relative muscle volume and head capsule volume is shown in Table 10.2 for six muscles of the two polymorphic species, *A. sexdens* and *C. rufipes*. The slope of the linear regressions relating the relative muscle volume and the head capsule volume (log-transformed data) was negative for all measured labial, cibarial, and pharyngeal muscles (Table 10.2), i.e., the relative volume of any of the measured muscles was larger in smaller ants.

For interspecific comparisons, we used the linear equations presented in Table 10.2 for scaling the relative muscle volumes to the respective worker size of the two monomorphic species (mean head capsule volume of *Rhytidoponera* workers, 0.86 $mm^3$; of *Pachycondyla* workers, 4.01 $mm^3$; Table 10.2, right columns). As discussed above, data on each of the monomorphic species are compared

**Table 10.2** Relationships between relative muscle volumes ($Y$, as % of head capsule volume) and head capsule volume ($X$, in $mm^3$) of workers belonging to the two polymorphic species, *Camponotus rufipes* and *Atta sexdens*

| Muscle (Y) | Species | $a$ | $b$ | $r^2$ | $p$ | $N$ | $Y$ for $X = 0.86\ mm^3$ | $Y$ for $X = 4.01\ mm^3$ |
|---|---|---|---|---|---|---|---|---|
| Pharynx dilators (1) | *A. sexdens* | 0.308 | −0.452 | 0.97 | 0.003 | 5 | 2.17 | 1.09 |
| | *C. rufipes* | 0.482 | −0.467 | 0.90 | 0.014 | 5 | 3.27 | 1.58 |
| Cibarial dilator (5) | *A. sexdens* | −1.295 | −0.254 | 0.93 | 0.009 | 5 | 0.05 | 0.04 |
| | *C. rufipes* | −0.994 | −0.556 | 0.82 | 0.034 | 5 | 0.11 | 0.05 |
| Labium opener (6) | *A. sexdens* | −0.279 | −0.499 | 0.99 | 0.004 | 4 | 0.56 | 0.26 |
| | *C. rufipes* | 0.085 | −0.538 | 0.96 | 0.004 | 5 | 1.32 | 0.58 |
| Labium closer (7) | *A. sexdens* | −0.391 | −0.372 | 0.99 | 0.002 | 4 | 0.43 | 0.24 |
| | *C. rufipes* | −0.064 | −0.602 | 0.91 | 0.011 | 5 | 0.94 | 0.37 |
| Glossa muscle (8) | *A. sexdens* | −1.354 | −0.255 | 0.97 | 0.002 | 5 | 0.05 | 0.03 |
| | *C. rufipes* | −0.997 | −0.554 | 0.80 | 0.042 | 5 | 0.11 | 0.05 |
| Paraglossa muscle (9) | *A. sexdens* | −0.951 | −0.288 | 0.99 | 0.001 | 5 | 0.12 | 0.08 |
| | *C. rufipes* | −0.459 | −0.618 | 0.95 | 0.005 | 5 | 0.38 | 0.15 |

Linear regressions using log-transformed data [$\log(Y) = a + b \times \log(X)$], all statistically significant, were used for scaling relative muscle volumes. $r^2$ = coefficient of determination; $p$ = probability of error; $N$ = number of measured individuals

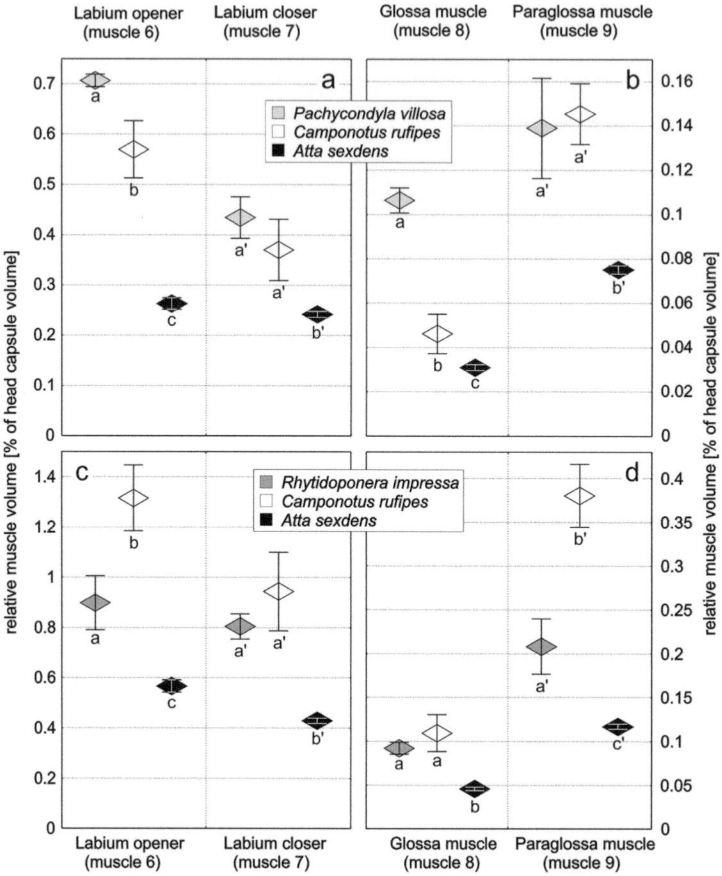


**Fig. 10.8** Relative volumes (ordinates) of labial muscles among workers of the monomorphic species *Pachycondyla villosa* (light gray) and *Rhytidoponera impressa*-complex (dark gray), and the polymorphic species *Camponotus rufipes* (white) and *Atta sexdens* (black). For the two polymorphic species, values are scaled either to the head size of a *Pachycondyla* worker (mean head capsule volume = $4.01 \pm 0.08$ mm$^3$; a, b) or that of a *Rhytidoponera* worker (mean head capsule volume = $0.86 \pm 0.02$ mm$^3$; c, d). Error bars represent mean SDs of the respective values calculated from the linear equations (see Table 10.2 for sample size). For the two monomorphic species, values are means ± SD ($N = 4$ individuals each). Data on both the labium opener (muscle 6 in Fig. 10.3) and the labium closer (muscle 7) are presented in (**a**) and (**c**). Data on glossa (muscle 8) and paraglossa muscles (muscle 9) are presented in (**b**) and (**d**). Columns sharing the same letter are not statistically different

with the scaled values from the two polymorphic ones and presented in separate graphs. Comparisons of four labial muscles are presented in Fig. 10.8, on the one side for *Pachycondyla* vs. scaled *Camponotus* and *Atta* workers (Fig. 10.8, top graphs) and on the other side for *Rhytidoponera* vs. scaled *Camponotus* and *Atta* workers (Fig. 10.8, bottom graphs).

When scaled to the head size of a *Pachycondyla* worker, the relative volumes of the four labial muscles of *Camponotus* workers were smaller or equal compared to *Pachycondyla* but larger than in *Atta* (Fig. 10.8a, b). With the exception of the labium closer and the paraglossa muscle between *Camponotus* and *Pachycondyla*, all values were statistically different ($p < 0.01$). We found the largest difference between *Camponotus* and *Pachycondyla*, the two nectar-feeding ant species, for the glossa muscle (Fig. 10.8b), which is probably most important for the process of licking (Paul et al. 2002). When scaled to the head capsule volume of a *Rhytidoponera* worker, labial muscle volumes were largest in *Camponotus* and smallest in *Atta* (Fig. 10.8c, d). Labial muscle volumes of *Rhytidoponera* workers lay in between. All values of labial muscle volumes were statistically different ($p < 0.002$), except for the labium closer and the glossa muscle between *Camponotus* and *Rhytidoponera*.

As for the labial muscles involved in licking, comparisons of those pharyngeal and cibarial muscles that control sucking are presented separately in Fig. 10.9, first for *Pachycondyla* vs. scaled *Camponotus* and *Atta* workers (Fig. 10.9a, b) and then for *Rhytidoponera* vs. scaled *Camponotus* and *Atta* workers (Fig. 10.9c, d). When scaled to the head size of a *Pachycondyla* worker, the volume of both the pharynx dilators and the cibarial dilator was largest in *C. rufipes* and smallest in *A. sexdens* (Fig. 10.9a, b). The values were statistically different ($p < 0.05$), except for the cibarial dilator between *Pachycondyla* and *Camponotus* and between *Pachycondyla* and *Atta*. When scaled to the head capsule volume of a *Rhytidoponera* worker, *Camponotus* workers had considerably larger pharyngeal and cibarial muscles than individuals of the other two species (Fig. 10.9c, d). In *Rhytidoponera*, pharyngeal and cibarial muscles were smallest. All values were statistically different ($p < 0.01$). Summing up, *Pachycondyla* workers had the largest labial muscles, whereas *Camponotus* workers had the largest pharyngeal and cibarial muscles. *Atta* workers had the smallest labial muscles, whereas *Rhytidoponera* workers had the smallest pharyngeal and cibarial muscles.

### 10.3.5 How Much Muscle per Unit Glossa Surface Area?

The active glossa surface area, as presented in Figs. 10.5 and 10.6, is a measure of the glossa loading capacity, i.e., of the total surface on which the fluid is expected to adhere. A large active glossa surface area is expected to be associated with large labial muscles for the control of licking, particularly with a large glossa muscle (muscle 8, Fig. 10.3). In fact, workers of the nectar-feeding ants *Pachycondyla* have both the largest active glossa surface area (Fig. 10.5d) and the largest glossa muscles (Fig. 10.8b).

In order to elucidate the functional design of the glossa surface and associated muscles, we investigated the extent to which the total area of the glossa is related to the size of the muscles controlling the glossa movement. For that, we compared the ratio between the glossa surface area and the glossa muscle volume across species,

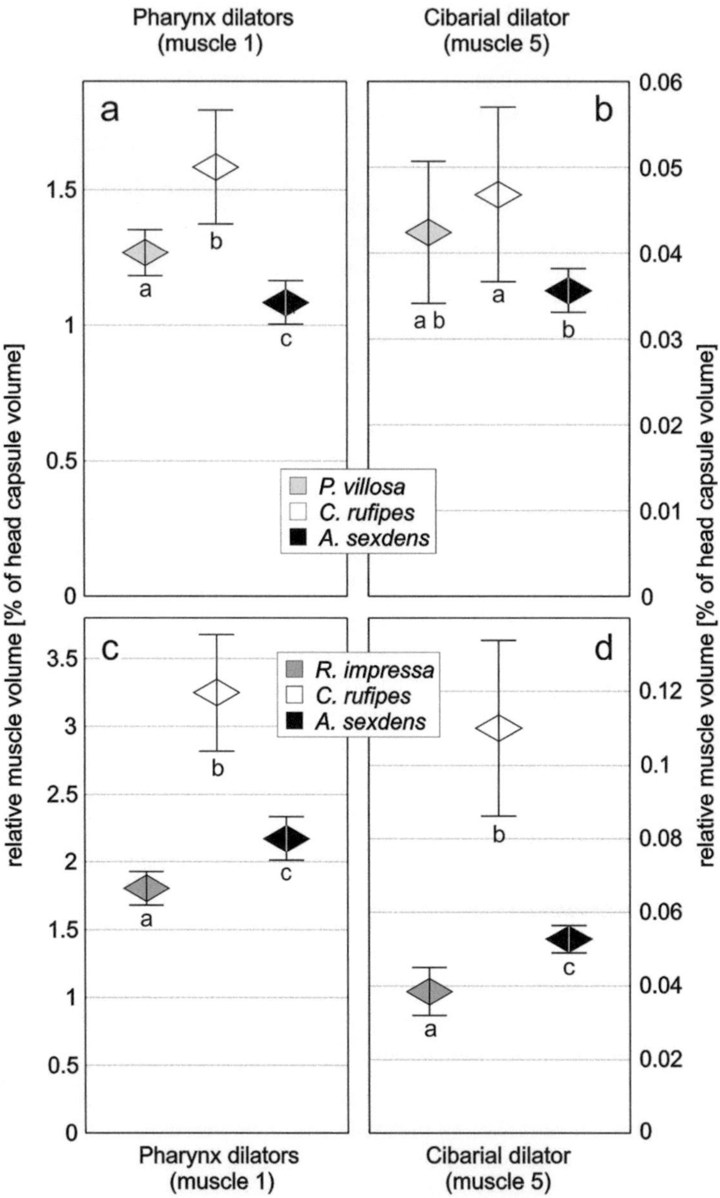


**Fig. 10.9** Relative volumes (ordinates) of pharyngeal muscles among workers of the monomorphic species *Pachycondyla villosa* (light gray) and *Rhytidoponera impressa*-complex (*dark gray*), and the polymorphic species *Camponotus rufipes* (white) and *Atta sexdens* (*black*). For the two polymorphic species, values (see Table 10.2 for N-numbers) are scaled either to the head size of a *Pachycondyla* worker (**a, b**) or that of a *Rhytidoponera* worker (**c, d**) (for head capsule volumes, see legend Fig. 10.8). Error bars represent mean SDs of the respective values calculated from the linear equations (see Table 10.2 for N-numbers). For the two monomorphic species, values are means ± SD ($N = 4$ individuals each). Data on the pharynx dilators (muscle 1 in Fig. 10.3) is presented in (**a**) and (**c**). Data on the cibarial dilator (muscle 5) is presented in (**b**) and (**d**). Columns sharing the same letter are not statistically different

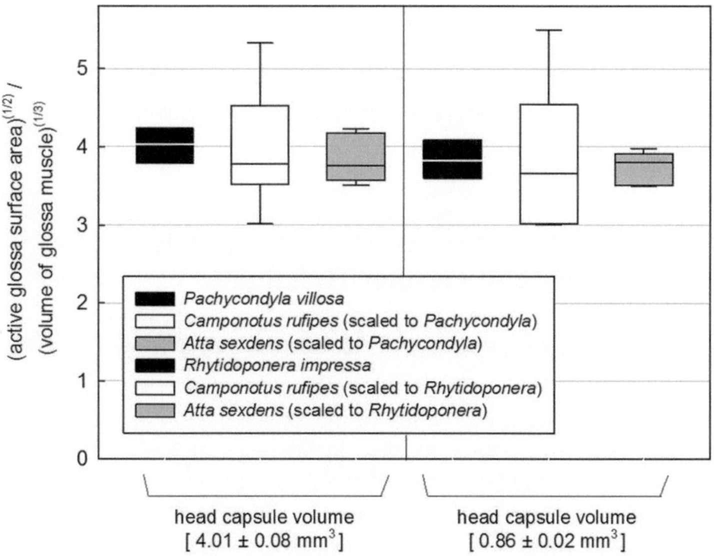


**Fig. 10.10** Ratio between the square root of active glossa surface area and the cubic root of the glossa muscle volume (muscle 8, Fig. 10.3). Black symbols correspond to the two monomorphic species, i.e., *Pachycondyla villosa* and *Rhytidoponera impressa*. For the two polymorphic species, *Camponotus* (white bars) and *Atta* (gray bars), values are scaled either to the head size of a *Pachycondyla* (head capsule volume $= 4.01 \pm 0.08$ mm$^3$, on the left) or that of a *Rhytidoponera* worker (head capsule volume $= 0.86 \pm 0.02$ mm$^3$, on the right) to allow comparisons. Data are median, quartiles, and ranges. Groups are not statistically different (Kruskal-Wallis one-way analysis of variance, $H = 1.637$, df $= 5$, $p > 0.8$; $N = 5$ for each group, except for the monomorphic species, with $N = 4$ each)

for workers of similar size. Surface areas and muscle volumes have quadratic and cubic dimensions, respectively. To calculate the ratio in linear dimensions, we first obtained, for each species, the square root of the active glossa surface area and the cube root of the glossa muscle volume. Then, the ratio between both terms was calculated and plotted for the two different average head capsule volumes of the monomorphic species. The corresponding values of the two polymorphic species were scaled. The ratio between the active glossa surface area and the glossa muscle volume was close to four and did not differ among the four species (Fig. 10.10), irrespective of the feeding technique, foraging habits, and size of the ant, even for workers largely differing in their head capsule volumes (4.01 vs. 0.86 mm$^3$). This demonstrates that a specific active surface area of the glossa corresponds to a specific volume of glossa muscle. Hence, single components of the feeding system seem to be closely adjusted to each other.

It is important to indicate that the active glossa surface area represents the product of three variables, i.e., microtrichium area, microtrichia density, and glossa frontal surface area. Species-specific differences in the active glossa surface area are mostly brought about by differences in both microtrichium area and microtrichia density, with a minor contribution of the frontal surface area (see Figs. 10.5 and 10.6). Therefore, the observed invariant ratio across species does not depend on what glossa variable (microtrichium area, microtrichia density, or frontal surface area) is responsible for the species-specific differences in active glossa surface area.

## 10.4 General Trends for the Relationship Between Morphology and Feeding Technique

### *10.4.1 Glossa Surface and the Mechanics of Fluid-Feeding*

Both the size and the surface structure of the tongue determine the performance of liquid feeding not only in ants but also in other nectar-feeding animals such as hummingbirds (Ewald and Williams 1982; Kingsolver and Daniel 1983; Tamm and Gass 1986), nectar-feeding bats (Greenbaum and Phillips 1974; Phillips et al. 1977; Griffiths 1982; Roces et al. 1993; Harper et al. 2013), and bees (Inouye 1980; Harder 1982, 1983; Michener and Brooks 1984; Zhu et al. 2016).

In foraging ants, workers of the two investigated ponerine species were observed to exclusively lick sugared solutions while loading, so that the fluid accumulates as a droplet between their mandibles. Under similar conditions, workers of the two polymorphic species exclusively suck the fluid and store it within their crop (Paul and Roces 2003). During licking, the glossa should load as much fluid as possible per single lick, so as to increase fluid uptake rates. An enlarged active surface area of the glossa is therefore expected to increase liquid adsorption, thus maximizing uptake rates during licking. Licking frequency, another variable that may contribute to increase liquid uptake rates, remains almost constant over a wide range of sucrose concentrations (ponerine ants: Dasch 1998; bumblebees: Harder 1986) and cannot be greatly increased. This is probably because the protraction of the glossa in ants is based on an elastic mechanism, and it is consequently a passive movement that cannot be precisely controlled in terms of force and velocity (Paul et al. 2002).

In contrast to licking, a large active glossa surface area is less important for sucking, since the glossa works only as a passive open duct that is deeply introduced into the fluid droplet (Paul and Roces 2003). Fluid intake rates largely depend on the activity of the sucking pump and underlying muscles (Josens 2002; Josens and Roces 2000; Falibene et al. 2009). Accordingly, workers of ant species that lick fluids are expected to rely more on large active glossa surface areas (*Pachycondyla*, *Rhytidoponera*) than workers of those species that suck liquids (*Camponotus*, *Atta*). This is the pattern we found: the active glossa surface area was shown to be largest in *Pachycondyla*, intermediate and similar in *Rhytidoponera* and *Camponotus*, and smallest in *Atta*.

If the habit of licking fluids is considered to have selected for increasing active glossa surface areas during evolution, the question arises why the glossa surface area of the sucking *Camponotus* ants was similar to that of the licking *Rhytidoponera* workers. *Camponotus* ants are typical nectar feeders that collect nectar with high intake rates at ad libitum droplets (Josens et al. 1998) and switch to licking only when the droplet is almost completely ingested or when collecting at nectar sources with low production rates (Josens and Roces 2000; Paul and Roces 2003; Schilman and Roces 2003). It therefore seems likely that a second driving force having influenced the morphology of the glossa is the extent of specialization on liquid food collection.

Considering fluid intake rates of sugared solutions ad libitum, we have found that licking is a more efficient technique than sucking when fluids of high sugar concentration, i.e., high viscosity, were collected. Conversely, sucking provided higher intake rates at lower sugar concentrations (Paul and Roces 2003; Kim et al. 2011). Fluid-licking *Pachycondyla* ants have glossae with the largest microtrichia arranged at large distances between each other. The interstices created by these microtrichia present a surface over which capillary forces may draw fluid up the structure. On the other hand, fluid-sucking *Camponotus* ants have smaller microtrichia organized at high densities, which may reduce the cohesive forces that tend to adhere the fluid at the glossa surface. A few mathematical models investigated the mechanics of fluid-feeding in insects (Kingsolver and Daniel 1995; Kim et al. 2011; Zhu et al. 2016), yet only recent studies on honeybees addressed the question whether hairs on the glossa surface influence fluid intake, depending on the drinking technique. In honeybees, the tongue surface is equipped with erectable glossal hairs, which are flattened in the protraction phase and fully erected in the retraction phase of the tongue during fluid intake (Yang et al. 2014). Physical models considering a trade-off between nectar intake and energy consumption indicated that the coordinated movements of the tongue and the unfolding of the glossal hairs are crucial for the maximization of nectar ingestion (Zhao et al. 2015; Zhu et al. 2016). To account for the adaptive value of microtrichia arrangement on the glossa of ants, an additional model that considers the differences between sucking and licking would be necessary.

Interestingly, the extremely long glossa (3–4 cm) of some tropical euglossine bee species is considerably less hairy, both in hair density and hair length, than in bumblebees and honeybees (Michener and Brooks 1984; Düster et al. 2018). Reduced hairiness in these euglossines may be important in reducing the resistance to nectar flow during unloading the long proboscis (Kingsolver and Daniel 1995). Therefore, studies on tongue morphology seem to be fundamental to understand the physiological performance of liquid feeding (Kornev et al. 2017; Zhu et al. 2016).

### 10.4.2 Muscle Volumes and Functional Design

Investigations on the morphology of muscle fibers suggest that the muscles of the labiomaxillary complex of ants feature rather slow than fast muscle characteristics (Paul et al. 2002), thus differing from the ant mandible closer muscle, which integrates both, depending on the species (see above). Since the length of fibers within one muscle of the labiomaxillary complex is roughly similar across ant species (Paul et al. 2002), differences in muscle volume mainly result from differences of the cross-sectional area, which is expected to underline both muscular force and power output. Considering the relative volumes of labial, pharyngeal, and cibarial muscles, we found significant differences among the four ant species examined. Large differences were observed between the group of nectar-feeding ants (*Camponotus* and *Pachycondyla*) and the group of ants not specialized on fluid-feeding (*Atta* and *Rhytidoponera*).

However, the largest labial muscles (muscles 6, 7, 8, 9; Fig. 10.3), which are responsible for the control of the licking process, were found in the nectar-feeding ant *Pachycondyla* that collects liquids by licking. Nectar-feeding *Camponotus* ants, which collect ad libitum fluids by sucking, have smaller labial muscles (Fig. 10.8a, b). On the other hand, *Camponotus* ants featured larger or similarly sized labial muscles than workers of *Rhytidoponera*, which indeed lick fluids but are not specialized on collecting fluid food. Finally, *Atta* leaf-cutting ants, which do not use licking at ad libitum nectar sources and are not specialized on collecting fluid food, have the relatively smallest labial muscles found (Fig. 10.8c, d).

Conversely, the pharyngeal and cibarial muscles (muscles 1 and 5; Fig. 10.3), fundamental for the sucking process, are particularly developed in ants that collect fluids by sucking (*Camponotus* and *Atta*; Fig. 10.9). These muscles occupy relatively more head capsule volume in ants of species that depend on a powerful sucking apparatus. In species that do not rely on sucking for collecting fluids, such as *Pachycondyla* and *Rhytidoponera*, the muscles were relatively small. It is important to indicate that workers of these species also ingest liquid food by sucking, but only for their individual needs and not during foraging. For collecting fluids for the whole colony, foraging workers lick liquid food and accumulate it externally as a drop between their mandibles (Paul and Roces 2003). Therefore, they only activate the sucking pump during ingestion, but not during foraging.

Taken the results of the labial, pharyngeal, and cibarial muscles together, it can be concluded that the extent of specialization on liquid food collection, more than the feeding technique, seems to have influenced the design of those muscles responsible for the mechanics of liquid feeding in ants. This view is supported by the observed differences in fluid intake rates in the species under scrutiny: for similarly sized ants, workers of the nectar-feeding ant *Camponotus* and *Pachycondyla* collected fluids with the highest intake rates, despite their different feeding techniques, while workers of the leaf-cutting ant *Atta* and the predatory ant *Rhytidoponera* did so with the lowest rate (Paul and Roces 2003). In agreement with this view, measurements of uptake rates of sugar solutions in workers of 77 ant species revealed that performance during liquid feeding was strongly associated with the presence of a highly developed proventriculus, the valve associated with an increased reliance on liquid foods in ants, thus reflecting evolutionary trade-offs related to dietary specialization or antipredator defenses (Davidson et al. 2004).

Workers of all ant species investigated so far use licking during fluid-feeding, either as the sole collecting technique during foraging or when the fluid builds a flat film in which the glossa cannot be introduced, so that sucking is not possible (Josens and Roces 2000; Paul and Roces 2003). The volume of fluid that can be loaded per licking cycle depends on the active surface area of the glossa, which is a measure of the total surface available for liquid adhesion. Interestingly, the measured ratio between the active glossa surface area and the glossa muscle volume was similar in the four investigated species, irrespective of the feeding technique used by the different ant species, their feeding habits, and the head size. It is therefore tempting to speculate that, in the course of evolution, any increase in the active glossa surface area was accompanied by a concomitant increase in the volume of the muscle responsible for glossa retraction. Hence, glossa morphology and the associated performance in liquid intake may have represented an important factor in the specialization on nectar-feeding in ants.

**Acknowledgments** We thank Bert Hölldobler and Josué A. Núñez for helpful comments and Georg Krohne and Claudia Gehrig for help with scanning electron microscopy. This work was supported by the German Research Foundation (DFG) (Graduiertenkolleg Arthropodenverhalten; SFB 554 and Gr 933/6-2). The present study complies with the NIH Publication No. 86-23 "Principles of animal care" and with the current laws of Germany, where the experiments were performed.

## References

Biewener A, Patek S (2018) Animal locomotion. Oxford University Press, Oxford

Chapman RF (1998) The insects - structure and function, 4th edn. Cambridge University Press, Cambridge

Cook SC, Socha JJ, Lee WK, Fezzaa K, Westneat MW (2006) Liquid-feeding biomechanics of an ant (*Camponotus herculeanus*) revealed using synchrotron X-ray imaging. In: 15th congress of the International Union for the Study of social insects IUSSI, Washington, USA, p 618

Dasch B (1998) Beobachtungen zur Nektaraufnahme bei *Pachycondyla villosa* (Formicidae; Ponerinae). Diploma thesis, Department of Behavioral Physiology and Sociobiology, University of Würzburg, Germany, pp. 83

Davidson DW, Cook SC, Snelling RR (2004) Liquid-feeding performances of ants (Formicidae): ecological and evolutionary implications. Oecol 139:255–266

Düster JV, Gruber MH, Karolyi F, Plant JD, Krenn HW (2018) Drinking with a very long proboscis: functional morphology of orchid bee mouthparts (Euglossini, Apidae, Hymenoptera). Arthr Struct Develop 47(1):25–35

Eisner TA (1957) Comparative morphological study of the proventriculus of ants (Hymenoptera: Formicidae). Bull Museum Comp Zool 116:439–490

Ewald PW, Williams WA (1982) Function of the bill and tongue in nectar uptake by hummingbirds. Auk 99:573–576

Falibene A, Josens R (2008) Nectar intake rate is modulated by changes in sucking pump activity according to colony starvation in carpenter ants. J Comp Physiol A 194(5):491–500

Falibene A, Josens R (2014) Environmental temperature affects the dynamics of ingestion in the nectivorous ant *Camponotus mus*. J Insect Physiol 71:14–20

Falibene A, de Figueiredo Gontijo A, Josens R (2009) Sucking pump activity in feeding behaviour regulation in carpenter ants. J Insect Physiol 55(6):518–524

Gotwald WH (1969) Comparative morphological studies of the ants, with particular reference to the mouthparts (Hymenoptera: Formicidae). Mem Cornell Univ Agric Exper Stat 408:1–150

Greenbaum IF, Phillips CJ (1974) Comparative anatomy and general histology of tongues of long-nosed bats (*Leptonycteris sanborni* and *L. nivalis*) with reference to infestation of oral mites. J Mammal 55:489–504

Griffiths TA (1982) Systematics of the New World nectar-feeding bats (Mammalia, Phyllostomidae), based on the morphology of the hyoid and lingual regions. Am Mus Novit 2742:1–45

Gronenberg W (1995) The fast mandible strike in the trap-jaw ant *Odontomachus*. I. Temporal properties and morphological characteristics. J Comp Physiol A 176:391–398

Gronenberg W, Paul J, Just S, Hölldobler B (1997) Mandible muscle fibers in ants: fast or powerful. Cell Tiss Res 289:347–361

Gronenberg W, Brandao CRF, Dietz BH, Just S (1998a) Trap-jaws revisited: the mandible mechanism of the ant *Acanthognathus*. Physiol Ent 23:227–240

Gronenberg W, Hölldobler B, Alpert GD (1998b) Jaws that snap: the mandible mechanism of the ant *Mystrium*. J Insect Physiol 44:241–253

Günzel D, Galler S, Rathmayer W (1993) Fibre heterogeneity in the closer and opener muscles of crayfish walking legs. J Exp Biol 175:267–281

Harder LD (1982) Measurement and estimation of functional proboscis length in bumblebees (Hymenoptera, Apidae). Can J Zool 60:1073–1079

Harder LD (1983) Flower handling efficiency of bumble bees: morphological aspects of probing time. Oecol 57:274–280

Harder LD (1986) Effects of nectar concentration and flower depth on flower handling efficiency of bumble bees. Oecol 69:309–315

Harper CJ, Swartz SM, Brainerd EL (2013) Specialized bat tongue is a hemodynamic nectar mop. Proc Nat Acad Sci 110:8852–8857

Hölldobler B (1985) Liquid food transmission and antennation signals in Ponerine ants. Israel J Entom 19:89–99

Hölldobler B, Wilson EO (1990) The ants. Belknap Press of Harvard University Press, Cambridge

Inouye DW (1980) The effect of proboscis and corolla tube lengths on patterns and rates of flower visitation by bumblebees. Oecol 45:197–201

Jaffé K, Sánchez C (1984) Comportamiento alimentario y sistema de reclutamiento en la hormiga *Camponotus rufipes* (Hymenoptera: Formicidae). Acta Cient Venez 35:270–277

Janet C (1905) Anatomie de la tête du *Lasius niger* reine. Ducourtieux et Gout, Limoges, p 40

Jervis M, Vilhelmsen L (2000) Mouthpart evolution in adults of the basal, symphytan, hymenopteran lineages. Biol J Linn Soc 70:121–146

Josens RB (2002) Nectar feeding and body size in the ant *Camponotus mus*. Ins Soc 49:326–330

Josens RB, Roces F (2000) Foraging in the ant *Camponotus mus*: nectar-intake rate and crop filling depend on colony starvation. J Insect Physiol 46:1103–1110

Josens RB, Farina W, Roces F (1998) Nectar feeding by the ant *Camponotus mus*: intake rate and crop filling as a function of sucrose concentration. J Insect Physiol 44:579–585

Josens RB, Falibene A, de Figueiredo Gontijo A (2006) Electrical signals during nectar sucking in the carpenter ant *Camponotus mus*. J Insect Physiol 52:1234–1242

Just S, Gronenberg W (1999) The control of mandible movements in the ant *Odontomachus*. J Insect Physiol 45:231–240

Khalife A, Keller RA, Billen J, Garcia FH, Economo EP, Peeters C (2018) Skeletomuscular adaptations of head and legs of *Melissotarsus* ants for tunnelling through living wood. Front Zool 15(1):30

Kim W, Gilet T, Bush JW (2011) Optimal concentrations in nectar feeding. Proc Nat Acad Sci 108:16618–16621

Kingsolver JG, Daniel TL (1983) Mechanical determinants of nectar feeding strategy in hummingbirds: energetics, tongue morphology, and licking behavior. Oecol 60:214–226

Kingsolver JG, Daniel TL (1995) Mechanics of food handling by fluid-feeding insects. In: Chapman RF, de Boer G (eds) Regulatory mechanisms in insect feeding. Chapman & Hall, New York, pp 32–73

Kirshboim S, Ishay JS (1998) The labium of the oriental hornet: micromorphology and functions. J Elect Microsc 47:505–515

Kornev KG, Salamatin AA, Adler PH, Beard CE (2017) Structural and physical determinants of the proboscis – sucking pump complex in the evolution of fluid-feeding insects. Sci Rep 7(1):6582

Krenn HW (1990) Functional morphology and movements of the proboscis of Lepidoptera (Insecta). Zoomorphology 110:105–114

Labandeira CC, Phillips TL (1996) Insect fluid-feeding on upper Pennsylvanian tree ferns (Palaeodictyoptera, Marattiales) and the early history of the piercing-and-sucking functional feeding group. Ann Entomol Soc Amer 89:157–183

Larabee FJ, Suarez AV (2014) The evolution and functional morphology of trap-jaw ants (Hymenoptera: Formicidae). Myrmecol News 20:25–36

Larabee FJ, Gronenberg W, Suarez AV (2017) Performance, morphology and control of power-amplified mandibles in the trap-jaw ant *Myrmoteras* (Hymenoptera: Formicidae). J Exp Biol 220:3062–3071

Littledyke M, Cherrett JM (1976) Direct ingestion of plant sap from cut leaves by the leaf-cutting ants *Atta cephalotes* (L.) and *Acromyrmex octospinosus* (Reich) (Formicidae, Attini). Bull Entom Res 66:205–217

Michener CD, Brooks RW (1984) Comparative study of the glossae of bees. Contrib Am Entomol Inst Ann Arbor 22:1–73

Paul J (2001a) Review: mandible movements in ants. Comp Bioch Physiol A 131:7–20

Paul J (2001b) The mouthparts of ants. Doctoral thesis, Department of Behavioral Physiology and Sociobiology, University of Würzburg, Germany

Paul J, Gronenberg W (1999) Optimizing force and velocity: mandible muscle fiber attachments in ants. J Exp Biol 202:797–808

Paul J, Gronenberg W (2002) Motor control of the mandible closer muscle in ants. J Insect Physiol 48:255–267

Paul J, Roces F (2003) Fluid intake rate in ants correlate with their feeding habits. J Insect Physiol 49:347–357

Paul J, Roces F, Hölldobler B (2002) How do ants stick out their tongues? J Morphol 254:39–52

Perrichot V, Wang B, Engel MS (2016) Extreme morphogenesis and ecological specialization among cretaceous basal ants. Curr Biol 26(11):1468–1472

Phillips CJ, Grimes GW, Forman GL (1977) Oral biology. In: Baker RJ, Jones JK, Carter DC (eds) Biology of bats of the New World family Phyllostomatidae, part II. Spec Publ Mus Texas Tech Univ, vol 13, pp 121–246

Roces F, Winter Y, von Helversen O (1993) Concentration preference and water balance in a flower visiting bat, *Glossophaga soricina antillarum*. In: Barthlott W, Naumann CM, Schmidt-Loske K, Schuchmann K-L (eds) Animal-plant interactions in tropical environments. Zoologisches Forschungsinstitut und Museum Alexander Koenig, Bonn, pp 159–165

Schilman PE, Roces F (2003) Assessment of nectar flow rate and memory for patch quality in the ant *Camponotus rufipes*. Anim Behav 66:687–693

Snodgrass RE (1956) Anatomy of the honey bee. Cornell University Press, New York

Szucsich NU, Krenn HW (2000) Morphology and function of the proboscis in Bombyliidae (Diptera, Brachycera) and implications for proboscis evolution in Brachycera. Zoomorphology 120:79–90

Tamm S, Gass CL (1986) Energy intake rates and nectar concentration preferences by hummingbirds. Oecol 70:20–23

Valenzuela-González J, López-Méndez A, García-Ballinas A (1994) Ciclo de actividad y aprovisionamiento de *Pachycondyla villosa* (Hymenoptera, Formicidae) en agroecosistemas cacaoteros del Soconusco, Chiapas, México. Folia Entomol Mex 91:9–21

Ward PS (1981) Ecology and life history of the *Rhytidoponera impressa* group. I. Habitats, nest sites, and foraging behavior. Psyche 88:89–108

Wu J, Yang H, Yan S (2015) Energy saving strategies of honeybees in dipping nectar. Sci Rep 5:15002

Yang H, Wu J, Yan S (2014) Effects of erectable glossal hairs on a honeybee's drinking strategy. Appl Phys Lett 104:263701

Zhao C, Wu J, Yan S (2015) Observations and temporal model of a honeybee's hairy tongue in microfluid transport. J Appl Phys 118:194701

Zhu R, Lv H, Liu T, Yang Y, Wu JYS (2016) Feeding kinematics and nectar intake of the honey bee tongue. J Ins Behav 29(3):325–339

# Chapter 11
# From Chewing to Sucking via Phylogeny—From Sucking to Chewing via Ontogeny: Mouthparts of Neuroptera

**Dominique Zimmermann, Susanne Randolf, and Ulrike Aspöck**

**Abstract** The Neuroptera are highly heterogeneous endopterygote insects. While their relatives Megaloptera and Raphidioptera have biting mouthparts also in their larval stage, the larvae of Neuroptera are characterized by conspicuous sucking jaws that are used to imbibe fluids, mostly the haemolymph of prey. They comprise a mandibular and a maxillary part and can be curved or straight, long or short. In the pupal stages, a transformation from the larval sucking to adult biting and chewing mouthparts takes place. The development during metamorphosis indicates that the larval maxillary stylet contains the Anlagen of different parts of the adult maxilla and that the larval mandibular stylet is a lateral outgrowth of the mandible. The mouthparts of extant adult Neuroptera are of the biting and chewing functional type, whereas from the Mesozoic era forms with siphonate mouthparts are also known. Various food sources are used in larvae and in particular in adult Neuroptera. Morphological adaptations of the mouthparts of adult Neuroptera to the feeding on honeydew, pollen and arthropods are described in several examples. New hypotheses on the diet of adult Nevrorthidae and Dilaridae are presented.

## 11.1 Introduction

The order Neuroptera, comprising about 5820 species (Oswald and Machado 2018), constitutes together with its sister group, the order Megaloptera (about 370 species), and their joint sister group Raphidioptera (about 250 species) the superorder Neuropterida. Neuroptera, formerly called Planipennia, are distributed worldwide and comprise 16 families of extremely heterogeneous insects. Their adults are

D. Zimmermann (✉) · S. Randolf
2nd Zoological Department, Natural History Museum Vienna, Vienna, Austria
e-mail: dominique.zimmermann@nhm-wien.ac.at; susanne.randolf@nhm-wien.ac.at

U. Aspöck
2nd Zoological Department, Natural History Museum Vienna, Vienna, Austria

Department of Integrative Zoology, University of Vienna, Vienna, Austria
e-mail: ulrike.aspoeck@nhm-wien.ac.at



H. W. Krenn (ed.), *Insect Mouthparts*, Zoological Monographs 5,
https://doi.org/10.1007/978-3-030-29654-4_11

terrestrial, as are the larvae of most families; only two families have aquatic larvae (Aspöck et al. 1980, 2001; New 1989; Aspöck and Aspöck 2007).

Biting and chewing mouthparts of adults and larvae are hypothesized for ancestors of the Neuropterida, at present comprising Raphidioptera, Megaloptera and Neuroptera. The ancestors of Neuroptera evolved larval sucking tubes. As a consequence, a change from larval sucking to adult biting and chewing mouthparts takes place in the pupal stage of each individual neuropteran life cycle. This chapter compares the morphology of larval and adult mouthparts, presents data to the transformation in the pupal stage and discusses feeding preferences and morphological adaptations of adult Neuroptera.

## 11.2 The Mouthparts of the Larvae

Already in the 1850s, the Austrian biologist Friedrich Moritz Brauer (1832–1904)—he was about 20 years old—recognized the evolutionary relevance of the larvae of the Neuroptera. He detected, studied and illustrated the larval sucking mouthparts, e.g., of Osmylidae, Chrysopidae, Mantispidae, and he established the Neuroptera as a monophylum based on the sucking feeding apparatus of the larvae (Brauer 1851a, b, 1852a, b; Brauer and Löw 1857).

The larval sucking tubes of Neuroptera are the most spectacular and most powerful autapomorphy of the order and they are highly relevant phylogenetically. Moreover, they have a high diagnostic value. Primarily and predominately, they are uniquely effective and dangerous weapons (Figs. 11.1 and 11.2).

### *11.2.1 Composition of the Sucking Tubes*

The sucking tubes—either jaws, stylets, needles, or short suckers—comprise a mandibular part which may bear teeth and a maxillary part lacking teeth.

The mandible and the distal part of the maxilla are functionally connected in different ways of coaptations, thus forming a food canal (Fig. 11.3). A venom canal runs within the maxilla of most taxa with the exception of Sisyridae (their sucking needles are used to feed on bryozoans and sponges) and with a possible exception of phytophagous Ithonidae, but detailed studies are still lacking in the latter (Wundt 1961; MacLeod 1964; Zwick 1967; Gaumont 1976; Minter 1990; Möller 2003; Grebennikov 2004; Beutel et al. 2010a; Jandausch et al. 2018).

**Fig. 11.1** Larva of *Chrysoperla plorabunda* (Chrysopidae) sucking on the aphid *Acyrtosiphon pisum*; photo: Peter Duelli

**Fig. 11.2** Larva of *Euroleon nostras* (Myrmeleontidae) sitting in its funnel, ambushing prey; photo: Heiko Bellmann

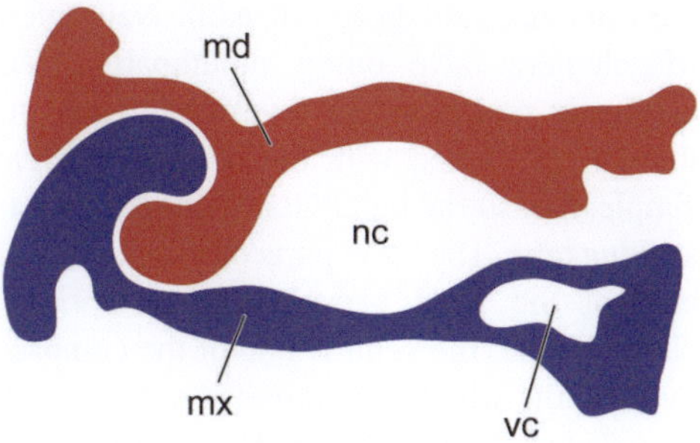


**Fig. 11.3** Cross section of larval sucking stylet of *Osmylus fulvicephalus* (Osmylidae). The mandible (md) is seamed with the maxilla (mx), thus forming a nutrition canal (nc). A venom canal (vc) runs within the maxilla; redrawn from Gaumont (1976)

### *11.2.2 The Various Types—Shapes and Functions—Of the Sucking Tubes: Pragmatic and Phylogenetic Aspects*

Shape and function of the mouthparts can be categorized into various types. "Piercing" jaws are stylets of different length, e.g. needle-like in Sisyridae and Osmylidae, or short in Coniopterygidae, Dilaridae and Berothidae in relation to the head (Fig. 11.4a–c). The representatives of all these families are special feeders preferring more or less sessile or "non-running away" prey. Apically curved jaws are weapons appropriate for catching and holding movable prey, preventing it from escaping. They are very impressive in Hemerobiidae and Chrysopidae (Fig. 11.1), and they may even be spectacular in certain Myrmeleontiformia (Figs. 11.2 and 11.4e). The larvae of Ithonidae are grub-like with corresponding stout sucking tubes hypothesized for sucking roots (Fig. 11.4f).

Curved sucking tubes are interpreted as a plesiomorphic character state and straight stylets as apomorphic adaptations to diverse feeding conditions (Aspöck 1993). This conforms with the phylogenetic results in Aspöck and Aspöck (2008). Nonetheless, this character polarity continues to be discussed controversially (Winterton et al. 2018; Engel et al. 2018). MacLeod (1964) presents a very thorough study and discussion of larval heads and mouthparts; Aspöck and Aspöck (2007) give an overview.

### *11.2.3 The Maxillary Head*

The larval heads of all three orders of Neuropterida are prognathous, in contrast to the orthognathous heads of adult Neuroptera. While larvae of Raphidioptera and Megaloptera have biting mouthparts, Neuroptera feature the above-described extremely complex and diverse sucking tubes or sucking stylets. The mandible, though co-adapted with the maxilla to yield a curved or straight shape, remains simple. Not so the maxilla: only the distal part joins the mandible for the functional sucking tube. The basal parts of the maxillae, however, are immersed into the ventral side of the head capsule and comprise two closely connected sclerites. In Nevrorthidae, the ventral side of the compact head capsule is dominated by a large oval sclerite which is interpreted as the gula (Fig. 11.4d; Aspöck and Aspöck 2007; Beutel et al. 2010a). In the Myrmeleontiformia clade, the basal parts of the maxillae are reduced and shifted terminally; the same process occurred in the gula which, in addition, became extremely reduced (Fig. 11.4e) or completely lost (Fig. 11.4a, b, c, and f; MacLeod 1964; Aspöck and Aspöck 2007).

The basal maxillary sclerites are integrated into the ventral side of the head, constituting the "maxillary head", and can be present in two different specifications: either the "bow maxillary" type ("Bogenmaxillen"-Typ, Aspöck 1993), e.g. in Osmylidae (Fig. 11.4a) and in Chrysopidae, or the "parallel maxillary" type

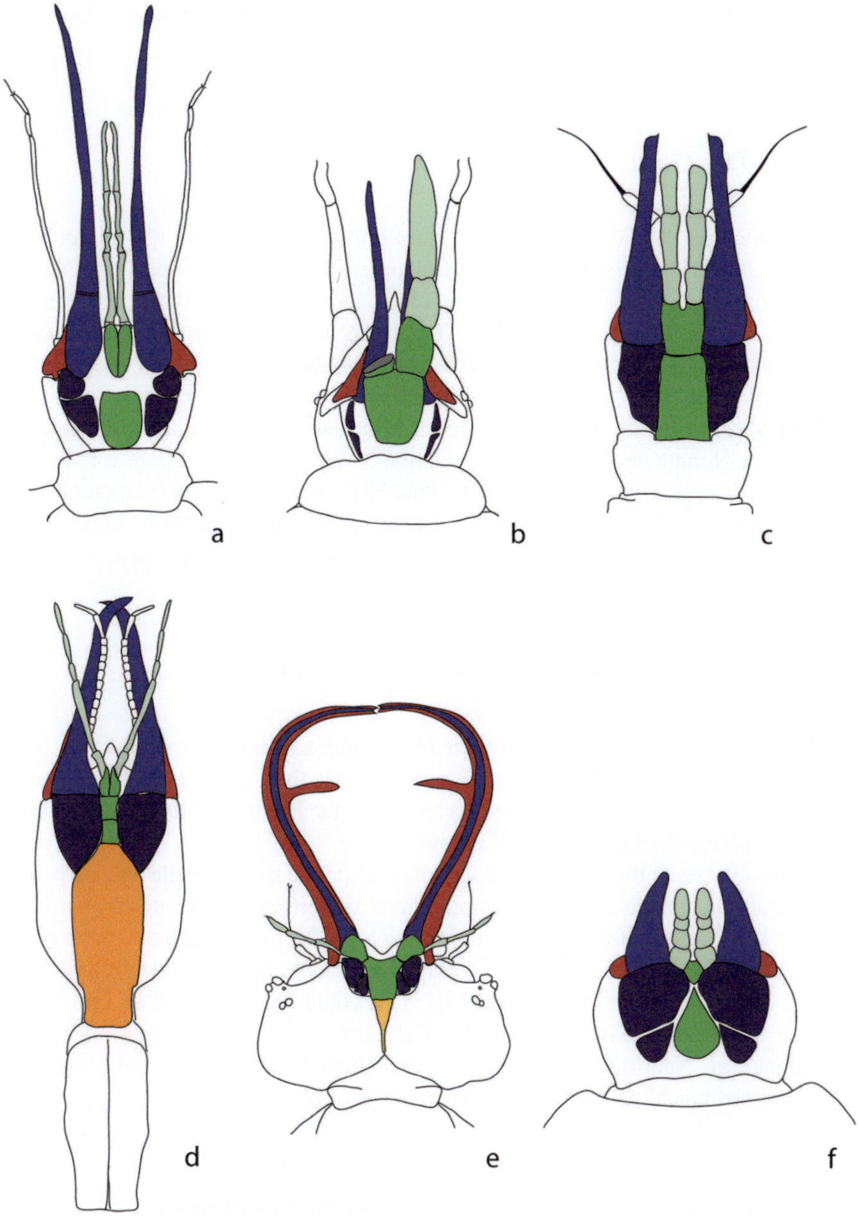


**Fig. 11.4** Larval heads of Neuroptera, ventral view; (**a**) *Osmylus fulvicephalus* (Osmylidae), (**b**) *Helicoconis* sp. (Coniopterygidae), (**c**) *Nallachius krooni* (Dilaridae), (**d**) *Nevrorthus fallax* (Nevrorthidae), (**e**) *Osmylops* sp. (Nymphidae), (**f**) *Polystoechotes punctata* (Ithonidae). Redrawn from (**a**) Wundt (1961), (**b**, **e**, **f**) MacLeod (1964), (**c**) Minter (1992), (**d**) Zwick (1967). Schematic drawings: red—mandible, blue—maxilla, dark green—labium, light green—labial palpus, orange—gula

("Parallelmaxillen"-Typ, Aspöck 1993), e.g. in Hemerobiidae, Berothidae, Dilaridae (Fig. 11.4c).

### *11.2.4 The Gula*

The presence of a gula (Fig. 11.4d, e) is traditionally considered a plesiomorphic character state that is homologous in Coleoptera and Neuropterida (e.g. Aspöck 1992; Beutel et al. 2010a). The gula "entered" neuropterology with the discovery of the larva of Nevrorthidae (Zwick 1967). The first treatment of the gula of Nevrorthidae as being homologous to the gula of Raphidioptera and Megaloptera started in the 1990s (Aspöck 1992). The compact head capsule of larval Nevrorthidae, with its large gula, represents the most archaic head type in Neuroptera. Nonetheless, recent interpretations (Winterton et al. 2018; Engel et al. 2018) treat the gula of Nevrorthidae and other Neuroptera as a "new adaptation" that evolved convergently in Raphidioptera, Megaloptera and Coleoptera. Importantly, this would imply major evolutionary transformations within Neuroptera.

### *11.2.5 Evolution of the Larval Sucking Tubes*

So far, no intermediate stage between biting and sucking mouthparts has been found, neither recent or fossil. The hypothesis that the common stem species of Megaloptera + Neuroptera evolved preconditions facilitating a future sucking construction in Neuroptera is discussed below.

The basal parts of the maxillae are withdrawn into the head capsule and are part of the ventral wall of the head. In Raphidioptera, the biting mouthparts are terminal (Aspöck and Aspöck 2007; Beutel and Friedrich 2008). In Megaloptera, parts of the basal maxillary sclerites are slightly withdrawn into the head capsule (Aspöck and Aspöck 2007; Beutel and Friedrich 2008). This could be interpreted as a precondition for evolving sucking mouthparts. The idea of a common stem species of Megaloptera + Neuroptera with slightly invaginated basal parts of the maxillae and elongated apical parts might be a hypothesis of heuristic value.

## 11.3 The Metamorphosis: A Big Leap in Small Steps

"During metamorphosis [. . .] as in a theatre break the same elements are arranged to new sceneries and the material of the costumes is draped on the same actors in a different way." (translated after Wundt 1967: 736).

In Neuroptera, metamorphosis is "slight" (Gillott 1980): In contrast to other endopterygote insects, where adult organs are formed anew from imaginal discs, which are regions with cells that remain embryonic during larval stages (Gillott 1980), in Neuroptera the strict sequential homology (sensu Svácha 1992) of larval and adult organs is always maintained. After the histolysis of the larval musculature, the musculature of the adults is exclusively built up from the larval tissue, so that the larval muscle represents the Anlagen of its imaginal homologon (Korn 1943).

The metamorphosis takes place in a cocoon spun by the last instar larva. During the last phase of the larval stage, the larva stops feeding and the transformation of the larval head and the mouthparts starts. As a first step, the epidermis detaches from the larval cuticle with the exception of the insertion sites of the musculature. Subsequently, the larva spins a cocoon with silk secreted by the Malpighian tubes wherein the last instar larva rests as a prepupa (Sundermeier 1940). The prepupal stage can vary in duration: whereas Sisyridae spend only 2 days as prepupa (Pupedis 1980), 14 days are reported for Myrmeleontidae (Sundermeier 1940). The prepupa leaves the larval cuticle and rests in the cocoon as a decticous pupa having articulated mandibles, which they use to bite a hole in the cocoon and emerge as an adult (Sundermeier 1940).

The morphological interpretation of the mouthparts was long highly speculative. The maxillary stylet was interpreted as corresponding to the lacinia by Tillyard (in Crampton 1921; Withycombe 1924; Berland and Grassé 1951 as "interne"), as corresponding to the galea by Tillyard (1922) and Weber (1933), or as fusion of a palpifer with the galea or the palpus (Crampton 1921). Alternatively, Aspöck and Aspöck (2007) interpreted, based on the situation in the neuropteran sister group Megaloptera, the maxillary stylet as being composed solely of the stipes. Therefore, they consider the sclerite described as stipes by other authors (Crampton 1921; MacLeod 1964; Matsuda 1965; Rousset 1966; Gaumont 1976) as the distal part of a subdivided cardo. The other parts of the maxilla—the palpus, the galea and/or the lacinia—were regarded as reduced by all authors. However, investigations of the musculature of the larval maxillae by Wundt (1961), MacLeod (1964), Rousset (1966) and Beutel et al. (2010a) indicated a mixed origin of the maxillary stylet.

Studies on the metamorphosis of Neuroptera focused on the whole body and on wing development (Anthony 1902; Sundermeier 1940) or, at the histological level, on the process of histolysis and histogenesis of the musculature (Korn 1943) and the transformation of the epidermis of the head appendages (Sundermeier 1940). Only two publications (Wundt 1967; Kluge 2005) aimed to identify homologous structures in larval sucking stylets and adult biting and chewing mouthparts by observing the changes occurring during metamorphosis.

### 11.3.1 The Transformation of the Maxilla

The first in-depth investigation of the maxillary stylet was done by Wundt (1967) on the larvae of the lance lacewing *Osmylus fulvicephalus* (as: *O. chrysops*). The larvae of Osmylidae are unique in Neuroptera in having a fracture line running around the mandible and the maxilla at which the stylets are broken after the cocoon is spun

(MacLeod 1964). Accordingly, Wundt (1967) treated only the basal part of the maxillary stylet. Based on the changes in the attachment sites of the musculature, he identified the homologous parts in the larval and adult maxilla: The larval maxilla is composed of two basal maxillary sclerites and an elongate distal maxillary stylet (Fig. 11.5a). Three muscles originating at the tentorium insert on the basal maxillary sclerites: M. tentoriocardinalis on the proximal and M. tentoriostipitalis on the distal one. One muscle, M. craniolacinialis, originates posteriorly on the head capsule and inserts laterally at the base of the maxillary stylet. Additionally, two intrinsic muscles connect the ventral wall of the maxillary stylet with its dorsal and dorso-lateral wall.

There are two glands basally in the maxillary stylet: a poison gland with a channel opening subapically at the tip of the stylet and a voluminous lateral gland, surrounded by a lateral fold of an extremely thin endothelium (Fig. 11.5b; Wundt 1961; Gaumont 1976; Beutel et al. 2010a).

#### 11.3.1.1 The Larval Stage

The beginning of metamorphosis is characterized by two opposite rotations in the basal maxillary stylet: the larval epithelium proximal of the lateral gland is shifted clockwise, so that the formerly laterally attached M. craniolacinialis is moved medially. More distally, the larval epithelium shifts counterclockwise, whereby the lateral fold is unfolded and forms the future maxillary palpi. The galea is formed between the lacinial part and the palpus. At this stage of metamorphosis, the appendages do not yet have a closed lumen (Fig. 11.5d) but are only bulges along longitudinal folds and in close contact with the interior of the former stylet.

The bundles of the intrinsic muscles are dedifferentiated and reduced; their course becomes oblique due to the shift of their attachment sites, but their larval arrangement is still recognizable (Fig. 11.5c). The maxillary muscles of the future cardo and stipes are still located near the basal maxillary sclerites, whereas the intrinsic muscles and their epithelium are retracted from the larval stylet.

In Chrysopidae, which retain the complete larval stylets during metamorphosis, Kluge (2005) illustrated the position of the larval epidermis in different prepupal stages (Fig. 11.6a). When the musculature is already dedifferentiated, the larval epidermis is strongly reduced and starts to take on pupal shape, still projecting into the larval cuticular stylets. The epidermis of the future pupal maxilla already has a distinguishable galea, lacinia and palpus, the latter extending into the stylet-like part of the maxilla. The epidermis gradually retracts further and further out of the larval cuticle, finally forming the pupal maxilla.

#### 11.3.1.2 The Pupal Stage

In the early pupal stage, the musculature starts differentiating again. The formerly intrinsic muscles of the maxillary stylet become recognizable as the typical muscles

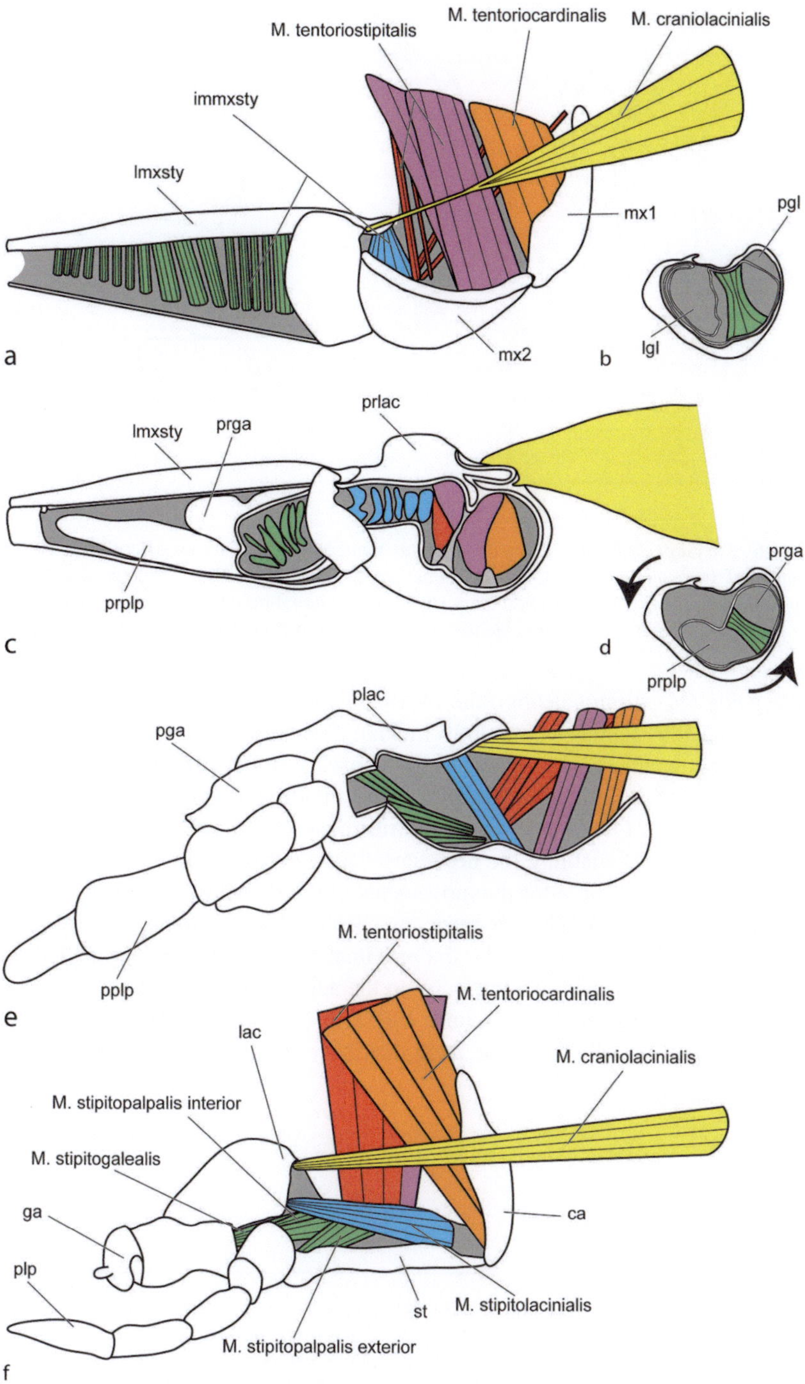


**Fig. 11.5** *Osmylus fulvicephalus* (Osmylidae), left maxilla, **a**, **c**, **e**, **f** lateral view, **b**, **d**, cross section, posterior view. (**a**, **b**) larva immediately after the breakup of the stylets, (**c**, **d**) prepupa, (**e**)

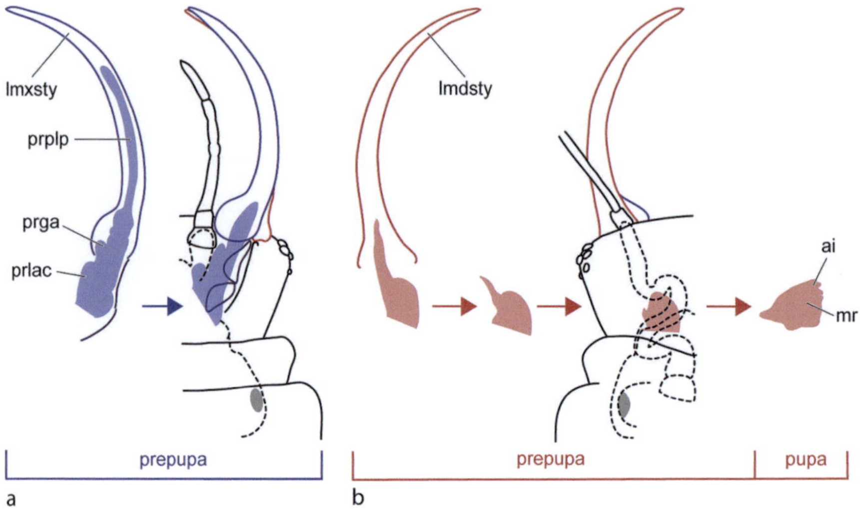


**Fig. 11.6** *Chrysoperla carnea* (Chrysopidae), (**a**) transformation of the maxilla (blue), (**b**) transformation of the mandible (red). Continuous line: larval cuticle, dotted line: prepupa; abbreviations: *ai* apical incisor, *lmdsty* larval mandibulary stylet, *lmxsty* larval maxillary stylet, *mr* molar region, *prga* prepupal galea, *prlac* prepupal lacinia, *prplp* prepupal palpus. Redrawn from Kluge (2005)

of the galea (M. stipitogalealis), the maxillary palpus (M. stipitopalpalis interior/ exterior) and the lacinia (M. stipitolacinialis). The shift of the muscle attachment sites during metamorphosis changes the muscles' mode of operation, and the transformation of the epidermis from a stylet to an orthopteroid maxilla also changes their function (Fig. 11.5f): in larvae, the intrinsic muscles function as compressors of the maxillary stylet, enabling the necessary depression for sucking. In contrast, in adults they are responsible for the movements of the maxillary appendages.

In conclusion, the larval one-piece maxillary stylet contains the Anlagen of different parts of the adult maxilla: the proximal part of the maxillary stylet forms the adult lacinia and galea. It contributes to the adult stipes and forms the complete maxillary palpus in Osmylidae (Wundt 1967). In Chrysopidae, the long stylet-like part of the larval maxilla forms the adult maxillary palpus (Kluge 2005).

---

**Fig. 11.5** (continued) pupa, (**f**) adult; abbreviations: *ca* cardo, *ga* galea, *immxsty* intrinsic muscles of the maxillary stylet, *lac* lacinia, *lgl* lateral gland, *lmxsty* larval maxillary stylet, *mx1* proximal basal maxillary sclerite, *mx2* distal basal maxillary sclerite, *pga* pupal galea, *pgl* poison gland, *plac* pupal lacinia, *plp* palpus, *pplp* pupal palpus, *prga* prepupal galea, *prlac* prepupal lacinia, *prplp* prepupal palpus, *st* stipes, redrawn from Wundt (1967) and Gaumont (1976)

### *11.3.2 The Transformation of the Mandible*

The metamorphosis starts in the head and its appendages and continuously progresses backwards through the body (Korn 1943). At the same time when the prepupal maxilla retracts from the larval cuticle, the epidermis of the mandibular stylet also starts its transformation (Fig. 11.6b). During this development, the epidermis in the mandibular stylet becomes successively shorter until it finally disappears. The median edge differentiates into the pupal mandible with an apical incisor and a molar region, which is used to bite a hole in the cocoon for the eclosion of the adult (Sundermeier 1940).

In conclusion, the larval mandibular stylet is a lateral outgrowth of the mandible and not homologous to the incisor of the later pupal mandibles (Kluge 2005).

## 11.4 The Mouthparts of Adult Lacewings

In contrast to the mouthparts of the larvae, which have been intensively studied and analysed, the adult mouthparts were neglected in Neuroptera. Stelzl (1992) published the only detailed comparative study focusing on neuropteran mouthparts: He discusses correlations of the mouthpart structures with the feeding habits in six European species. Otherwise, marginal treatments include a few comparative studies by Crampton (1917, 1921, 1923), a monograph on Neuroptera by Stitz (1931) and a PhD thesis on cephalic structures of Neuroptera by Shepard (1967). The search for illustrations or descriptions in taxonomic treatments is frustrating, the one exception being the beautifully illustrated contributions of Bo Tjeder (1957, 1959, 1960, 1961, 1966, 1967, 1972, 1992). The first notable mention of adult mouthparts in a phylogenetic analysis is in the head anatomical study by Beutel et al. (2010b). Apparently, these structures are believed to be too homogeneous throughout the order to provide useful taxonomic or phylogenetic information. This needs to be questioned.

### *11.4.1 Structure of the Mouthparts of Adult Lacewings*

#### 11.4.1.1 General Description

The mouthparts of Neuroptera are adapted to various food sources (Figs. 11.7 and 11.8). They are of the biting and chewing functional type with a movable labrum and a well-developed labium and maxilla which are connected by membranous areas. The mandibles exhibit a remarkable variety of forms (Fig. 11.9) correlated with the diverse feeding habits of Neuroptera (see Sect. 11.4.2). Their shape ranges from stout and massive with a broad rim between the ventral and the mesal cutting edge,

**Fig. 11.7** *Nemoptera sinuata* (Nemopteridae) feeding on pollen; photo: Harald W. Krenn

**Fig. 11.8** *Mantispa styriaca* (Mantispidae) consuming a fly; photo: Heiko Bellmann

as in the families Ascalaphidae, Nymphidae (Fig. 11.9g, h) and Coniopterygidae, to delicate with a sharply pointed incisor and lacking a rim as in Berothidae (Fig. 11.9e–f), Dilaridae (Fig. 11.9c–d) or Mantispidae. In Myrmeleontiformia, Hemerobiidae (Tjeder 1961) and Mantispidae (Ferris 1940; Lucchese 1956; Poivre 1978, 1981), the area between the molar process and the incisivus is prolonged and forms a distinct cutting edge (Fig. 11.9g–j). The mandibles are usually asymmetric in Neuroptera, with exceptions in phytophagous forms such as the ithonid *Ithone fusca* (Shepard 1967), the chrysopid *Pseudomallada prasinus* (Stelzl 1992) and Nemopteridae (Acker 1958; Tjeder 1967; Krenn et al. 2008).

The maxilla is equipped with a galea, a lacinia and five-segmented maxillary palpi. A sensory area is located distally on the last segment of the maxillary palpi in most Neuroptera. An apical finger-like process of the galea is documented for all neuropteran families (Tjeder 1957, 1959, 1961, 1966, 1967; Shepard 1967); certain taxa in the families Sisyridae, Coniopterygidae and Hemerobiidae lack this process (Shepard 1967; Meinander 1972; Randolf et al. 2013, 2014). Often a basigalea is

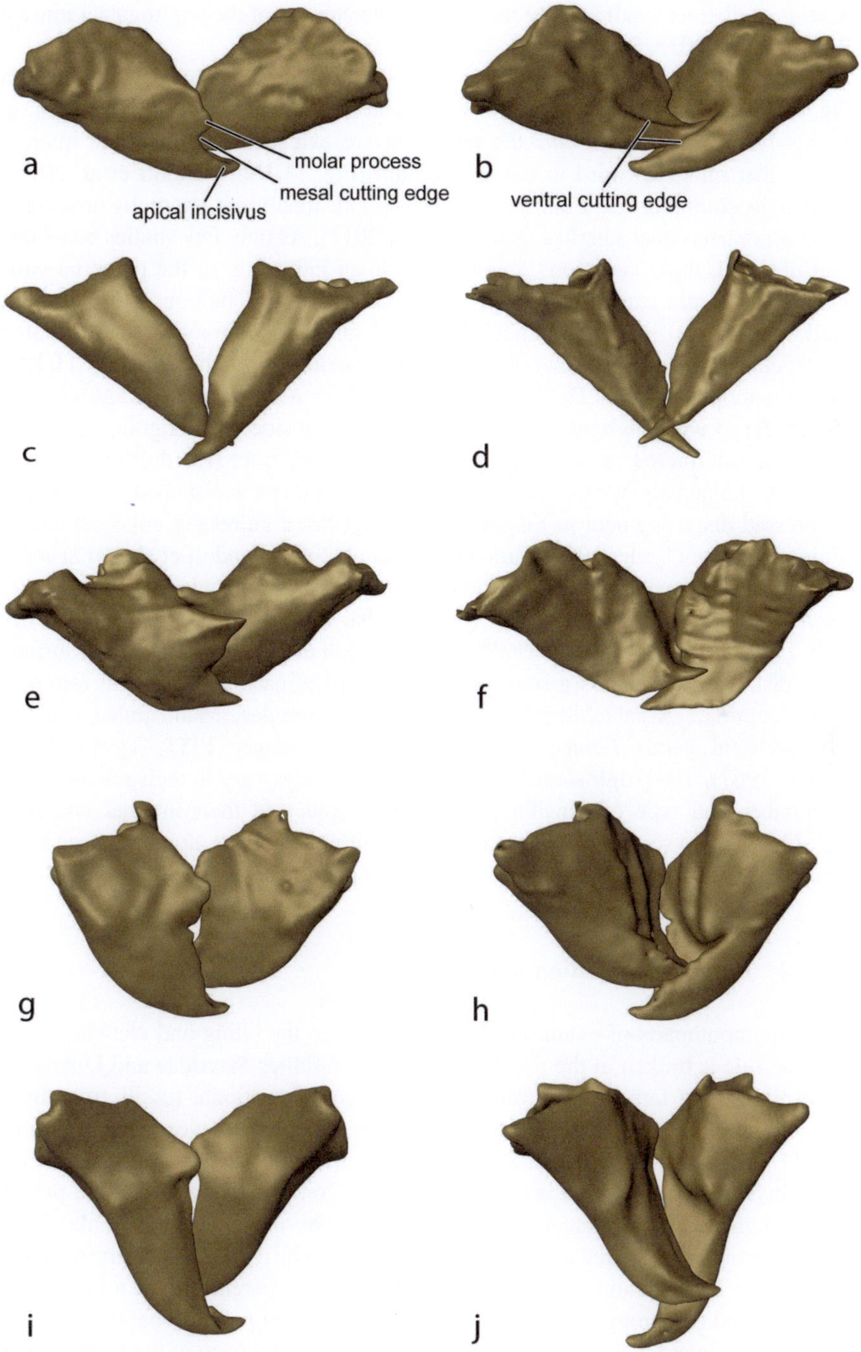


**Fig. 11.9** Diversity of mandibles in Neuroptera, 3D reconstructions (microCT data); **a**, **c**, **e**, **g**, **i** front view; **b**, **d**, **f**, **h**, **j** back view; (**a**, **b**) *Nevrorthus apatelios* (Nevrorthidae), (**c**, **d**) *Dilar turcicus* (Dilaridae), (**e**, **f**) *Podallea vasseana* (Berothidae), (**g**, **h**) *Nymphes* sp. (Nymphidae), (**i**, **j**) *Myrmeleon hyalinus* (Myrmeleontidae)

present—a distinct small sclerite basal of the attachment of the stipitogaleal muscle (Randolf et al. 2013, 2014).

The labium is tripartite, consisting of a submentum, mentum and prementum (Fig. 11.10). Even these sclerites, however, can harbour surprising peculiarities: in the Sisyridae *Sisyra terminalis* the palpigera are extended to an arcuate internal process that runs backward to the submentum (Fig. 11.10b; Randolf et al. 2013), while in the coniopterygid *Coniopteryx pygmaea* the mentum is internally prolonged to the hypopharyngeal sclerites (Randolf et al. 2017). As only few studies boast this level of detail, these variations merely provide an indication of the potential variability. The ligula can also take on different forms, e.g. bulbous (Fig. 11.10b; Randolf et al. 2013), quadrangular (Beutel et al. 2010b: *Osmylus*) or elongated (Krenn et al. 2008: *Nemoptera*); it is completely reduced in Dilaridae (Fig. 11.10c) and the nemopterid genus *Derhynchia* (Tjeder 1967). Membranous paraglossae are laterally fused with the ligula and fold onto the dorsal side of the ligula, creating a rim that is interpreted as an elongation of the salivary tract (Randolf et al. 2013, 2014). The labial palpi are usually three-segmented, with an axe-shaped and laterally compressed distal segment in Sisyridae (Fig. 11.10b), a bulbously enlarged one in Coniopterygidae (Tjeder 1957; Zimmermann et al. 2009; Randolf et al. 2017) and a pointed one in Nevrorthidae, Osmylidae, Berothidae, Chrysopidae, Hemerobiidae, Psychopsidae and Myrmeleontidae (Fig. 11.10a; Stitz 1931; Korn 1943; Tjeder 1959, 1960, 1961, 1966; Beutel et al. 2010b; Randolf et al. 2014). The lateral surface of the distal segment bears a sensory pit termed palpimaculae in several families: Myrmeleontidae, Ascalaphidae, Psychopsidae, Nymphidae, Mantispidae and the polystoechotid genus *Fontecilla* (Crampton 1921; Eisner 1953; Tjeder 1960; Shepard 1967). The palpimaculae of the different groups vary in their relative size, the number and type of sensilla present in the pits and their internal structure (Fig. 11.11). Chrysopidae and Osmylidae exhibit conspicuous, appressed sensillae in the same area that might be homologous to the palpimaculae (Beutel et al. 2010b).

#### 11.4.1.2 Eccentric Evolution in the Past

While the mouthparts of extant Neuroptera belong to the biting and chewing type, even this rule is broken in the fossil record: in the families Sisyridae and Dilaridae, and in the extinct family Kalligrammatidae, several exceptional fossils from Burmese amber with long siphonate mouthparts have been described recently (Fig. 11.12; Huang et al. 2015; Lu et al. 2016; Makarkin 2016, 2017; Liu et al. 2018). In Sisyridae the ligula is transformed into a long acute stylet, and the paired elongated galeae and laciniae may have formed a proboscis; the labial and maxillary palpi are strongly elongated and the mandibles seem to be reduced (Makarkin 2016). Similarly, in Kalligrammatidae the galeae and ligulae apparently form a functional proboscis, and the strongly elongated maxillary and labial palpi probably served to probe for nectar and pollen (Liu et al. 2018). The length and form of the proboscis in Kalligrammatidae is highly diverse (Labandeira et al. 2016; Liu et al. 2018). The mouthparts of the fossil Dilaridae differ in having a reduced ligula, which is a

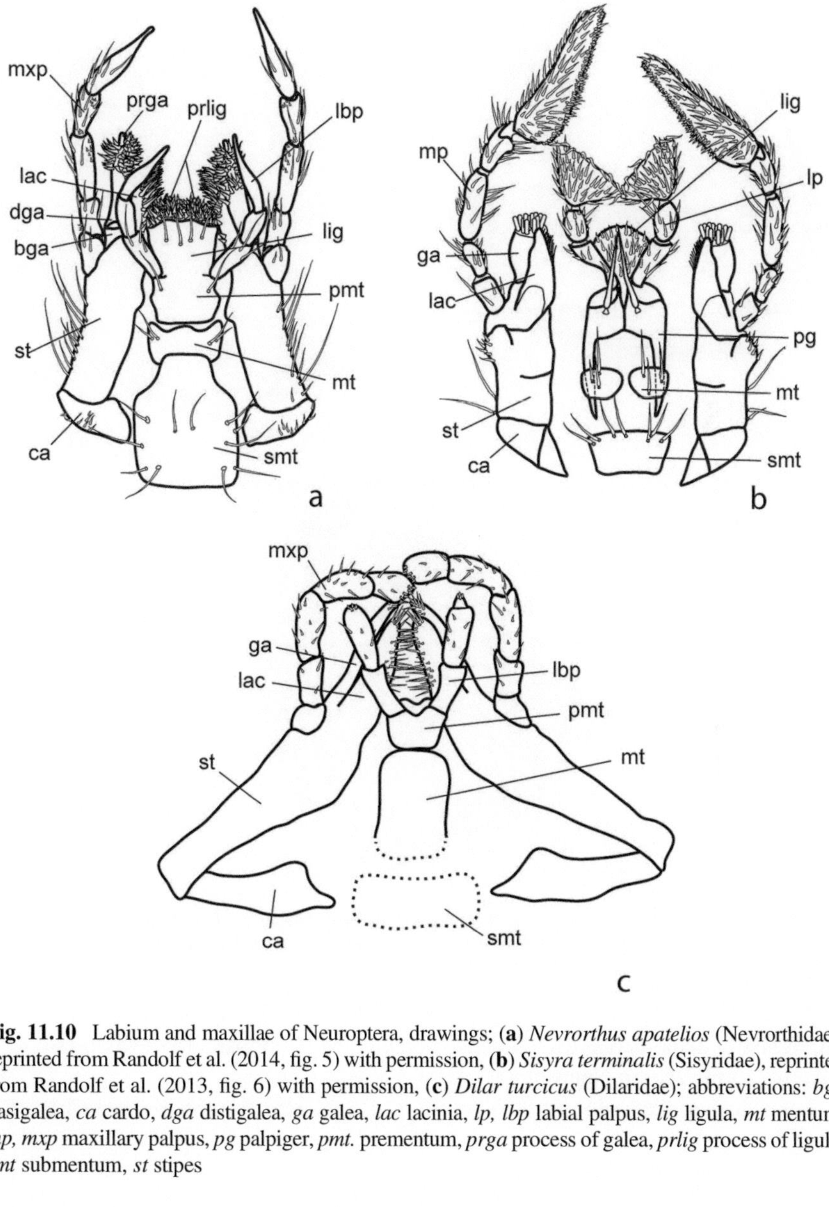


**Fig. 11.10** Labium and maxillae of Neuroptera, drawings; (**a**) *Nevrorthus apatelios* (Nevrorthidae), reprinted from Randolf et al. (2014, fig. 5) with permission, (**b**) *Sisyra terminalis* (Sisyridae), reprinted from Randolf et al. (2013, fig. 6) with permission, (**c**) *Dilar turcicus* (Dilaridae); abbreviations: *bga* basigalea, *ca* cardo, *dga* distigalea, *ga* galea, *lac* lacinia, *lp, lbp* labial palpus, *lig* ligula, *mt* mentum, *mp, mxp* maxillary palpus, *pg* palpiger, *pmt.* prementum, *prga* process of galea, *prlig* process of ligula, *smt* submentum, *st* stipes

synapomorphy of the entire family, and in having widely spaced galeae which probably could not form a probobscis (Fig. 11.12; Makarkin 2017). The mouthparts of all three groups seem well suited to enter the shallow calyx of many Burmese amber flowers. This would put these insects among the first forms specialized on exploiting the newly formed niche of flowers as a source of food (Labandeira 2010; Labandeira et al. 2016; Makarkin 2017; Liu et al. 2018). Altogether, these fossils

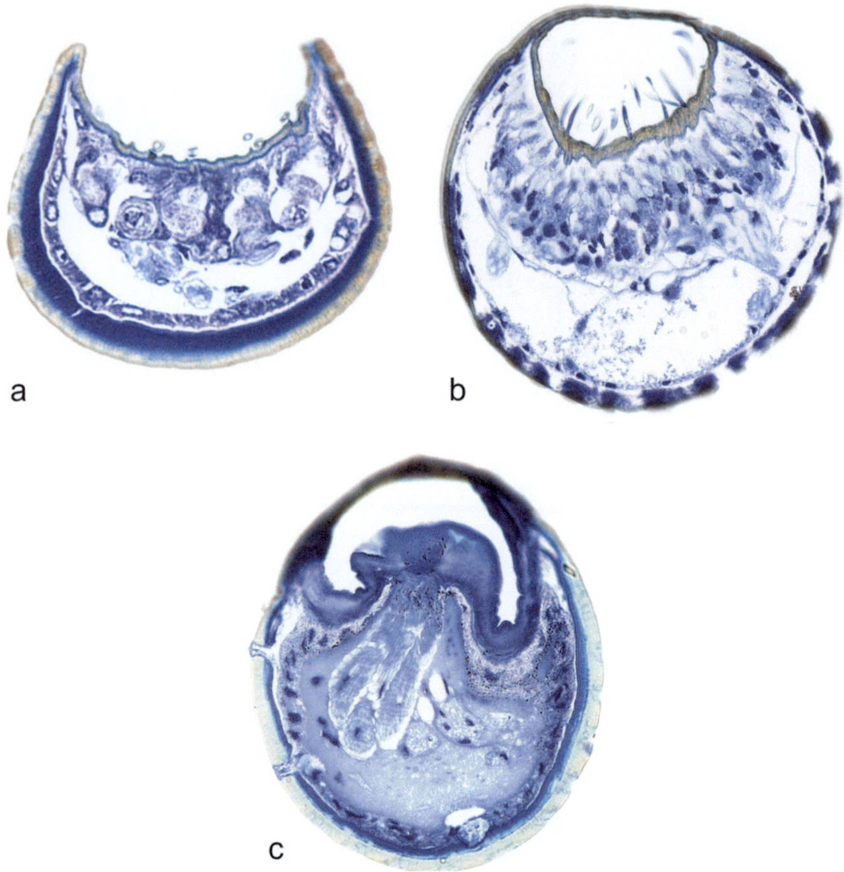


**Fig. 11.11** Histological cross sections through a sensory organ on the labial palpus, the so-called palpimacula; (**a**) *Mantispa styriaca* (Mantispidae), (**b**) *Myrmeleon hyalinus* (Myrmeleontidae), (**c**) *Nymphes* sp. (Nymphidae)

impressively demonstrate the immense potential of mouthparts to develop into new adaptive forms also in Neuroptera.

For a further discussion of the evolution of siphonate mouthparts, see Chap. 17.

## *11.4.2 Functional Correlations in the Mouthparts of Neuroptera*

This section puts the different mouthpart morphologies of Neuroptera in correlation to their feeding habits. Data on neuropteran feeding habits are scarce. The only broader study was done by Stelzl (1991) comprising data from 49 species and 8 families. Recently, Devetak and Klokočovnik (2016) summarized all available

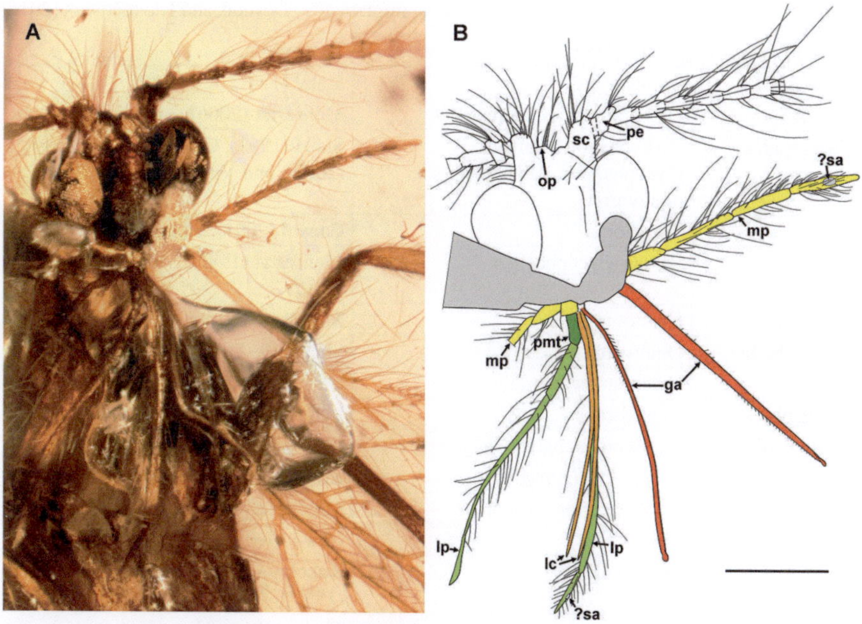


**Fig. 11.12** Siphonate mouthparts of the dilarid fossil *Burmopsychops groehni;* (**a**) photograph, frontal view, (**b**) line drawing; both reprinted from Makarkin (2017, fig. 6) with permission; *ga* galea, *lc* lacinia, *lp* labial palpus, *mp* maxillary palpus, *op* median ocular pulvinus, *pe* pedicellus, *pmt.* prementum, *sa* sensory area, *sc* scapus. Scale bar 0.5 mm (both to scale)

data in an extensive review (Table 11.1). Generally, adult Neuroptera variously feed on other arthropods, pollen, nectar and honeydew. Devetak and Klokočovnik (2016) distinguish six feeding types: carnivorous, omnivorous, carnivorous-glycophagous, glycophagous, palyno-glycophagous and palynophagous. For some families, data are completely lacking.

#### 11.4.2.1 Feeding on Honeydew

Honeydew, a sugar-rich, sticky substance excreted by aphids, is an important complementary food source for the adults of representatives of many neuropteran families (Stelzl 1991; Devetak and Klokočovnik 2016). Besides evidence from direct observation, information can also be gleaned from single Lepidoptera scales, pollen grains or spores in the digestive tract: these items indicate honeydew feeding because they easily get trapped in the sticky surface (Kokubu and Duelli 1983). In adult Neuroptera, honeydew is a reported food source of representatives of Chrysopidae, Sisyridae, Osmylidae, Hemerobiidae, Coniopterygidae, Myrmeleontidae and Ascalaphidae (Stelzl 1991; Devetak and Klokočovnik 2016). In Chrysopidae, honeydew has been shown to stimulate fecundity by nourishing a symbiotic yeast that resides in the crop of adults, providing them with essential amino acids (Finney 1948; Hagen and Tassan 1970).

**Table 11.1** Summary of feeding habits of adult Neuroptera after Devetak and Klokočovnik (2016)

| Family | Feeding habit | References |
|---|---|---|
| Ascalaphidae | Carnivorous | Stelzl (1991) |
| Berothidae | Omnivorous, palynophagous | Monserrat (2006) |
| Chrysopidae | Glycophagous, palyno-glycophagous, palynophagous, omnivorous | Stelzl (1991), Canard (2001), Bozsik (2000), and Villenave et al. (2006) |
| Coniopterygidae | Carnivorous | Canard (2001), Stelzl (1991) |
| Crocidae | Palyno-glycophagous | Monserrat (1983), Monserrat et al. (2012) |
| Dilaridae | Unknown | Monserrat (2014) |
| Hemerobiidae | Omnivorous | Stelzl (1991), Canard (2001) |
| Ithonidae | Carnivorous | De Jong (2011) |
| Mantispidae | Carnivorous | Redborg (1998), Tauber et al. (2002) |
| Myrmeleontidae | Carnivorous, palynophagous | Stange (1970), Stelzl and Gepp (1990) |
| Nemopteridae | Palyno-glycophagous | Krenn et al. (2008) |
| Nevrorthidae | Unknown (fungal spores?) | Monserrat (2005) |
| Nymphidae | Carnivorous | New (1986) |
| Osmylidae | Omnivorous | Kokubu and Duelli (1983), Stelzl (1991) |
| Psychopsidae | Unknown (carnivorous?) | Tjeder (1960), New (1988) |
| Sisyridae | Carnivorous, glycophagous | Kokubu and Duelli (1983), Stelzl (1991), and Monserrat (2005) |

Honeydew can be consumed in liquid form or dried on the leaf surface. Adaptations to this food source are primarily evident in the lacinia and the ligula. A spoon-shaped lacinia as found in many Hemerobiidae, Chrysopidae and Sisyridae is understood as an adaptation to the uptake of liquids such as honeydew drops or nectar (Stelzl 1992; Randolf et al. 2013). Desiccated honeydew is dissolved with saliva prior to its uptake (Bartlett 1962). A secondary prolongation of the salivary duct facilitates the application of saliva onto the leaf surface in the families Sisyridae, Chrysopidae, Hemerobiidae, Berothidae, Polystoechotidae and Nevrorthidae (Randolf et al. 2013, 2014). In these insects, the paraglossae are laterally fused with the ligula and folded onto its dorsal surface, so that the saliva is channelled to the tip of the ligula, enabling precise application. In Coniopterygidae, paraglossae are absent, but instead the hypopharynx reaches to the tip of the ligula, prolonging the salivary duct in a different way (Randolf et al. 2017).

#### 11.4.2.2 Carnivorous Feeding

Many Neuroptera feed entirely or partially on other arthropods. While the larger and more specialized adults of Mantispidae, Ascalaphidae and Myrmeleontidae (partim) feed on a variety of insect orders such as Diptera, Coleoptera, Hymenoptera and Heteroptera, the smaller representatives mainly feed on soft-bodied aphids and mites (Stelzl 1991; Devetak and Klokočovnik 2016). Aphids and mites are easy to disintegrate and can be located by simply palpating the leaf surface, waiving the need for hunting skills. Sex pheromones of the prey, as well as honeydew excreted by Homoptera, are known to serve as kairomones (Hagen 1986; Mendel et al. 1997). Characteristic for this food location strategy are long maxillary and labial palpi with sensilla patches on their tips. The mandibles of fully or partly carnivorous species are typically asymmetric (Ickert 1968). In most families, one mandible fits into a shovel-like extension of the other one (Fig. 11.9h), crushing and squashing the prey. The mandibles have sharply pointed incisors to capture the prey and convey it into the mouth opening.

#### 11.4.2.3 Pollen-Feeding

Pollen is a major food source for the adults of Nemopteridae (Fig. 11.7) and Crocidae, as well as some species of Chrysopidae, Berothidae, Sisyridae and Myrmeleontidae (Pupedis 1987; Devetak and Klokočovnik 2016). The process of pollen-feeding was studied in detail in *Nemoptera* (Krenn et al. 2008): in these flower-visiting lacewings the brush-shaped setaceous laciniae function as the main pollen-collecting organ. The maxillary structures are elongated, being folded beneath the head in resting position and extended forward during feeding. The mandibles and labrum are more or less stationary during feeding. They primarily serve to transfer the pollen from the lacinia into the mouth opening.

The use of mandibles for pollen manipulation, as reported by Krenn et al. (2005) for beetles, is not documented for Neuroptera. Generally, pollen-feeding Neuroptera are characterized by symmetrical mandibles (Stelzl 1992).

### *11.4.3 From Structure to Biology: What Might Dilaridae and Nevrorthidae Feed on?*

The food sources of only a few of the over 6000 described neuropteran species are known from direct observation or from gut content analyses (Devetak and Klokočovnik 2016). Nothing is known about the feeding habits of representatives of the families Dilaridae and Nevrorthidae. Nonetheless, studying the mouthparts enables drawing conclusions about their suitability to disintegrate and process certain kinds of food.

#### 11.4.3.1 The Diet of Dilaridae

The mouthparts of adult Dilaridae are characterized by a reduced labium lacking a ligula, a well-developed maxilla with an elongated cardo and an outward-directed cardo-stipital joint and sharply pointed mandibles without a grinding surface or cutting edge (Figs. 11.9c–d and 11.10c). In many neuropteran families, the ligula is extended with upfolded paraglossae and forms a prolongation of the salivary duct to the tip of the mouthparts. This formation mainly helps to dissolve desiccated honeydew from the surface of leaves (Randolf et al. 2013, 2014). The complete reduction of the ligula in Dilaridae indicates that honeydew-feeding does not play a role in these insects. In contrast to the labium, the maxillae are well developed and the cardo, stipes, lacinia and galea are elongated and folded sideways at the cardo-stipital joint (Fig. 11.10c), so that they can be extended forwards with the respective muscular contraction. A similar formation, though with a downward folding at the cardo-stipital joint, is present in adult Nemopteridae (Krenn et al. 2008), where the movement of the maxillae serves to collect nectar and pollen from flowers. Accordingly, a palyno-glycophagous feeding as in Nemopteridae seems to be the most plausible hypothesis.

#### 11.4.3.2 The Diet of Nevrorthidae

Malicky (1984) hypothesized that Nevrorthidae feed on honeydew based on the observation that adults are often found on sticky leaves. The presence of upfolded paraglossae, prolonging the salivary duct and facilitating the application of saliva onto the leaf surface to dissolve desiccated honeydew (see Sect. 11.4.2.1.), corroborates this observation. Nevertheless, based on the strong, asymmetric mandibles and well-developed labium and maxilla with a complete musculature (Randolf et al. 2014), they probably do not feed exclusively on honeydew. The mandibles resemble those of Sisyridae (Randolf et al. 2013), Osmylidae (Beutel et al. 2010b) and many Chrysopidae (Tjeder 1966) and indicate feeding on soft-bodied arthropods. The galea and lacinia are densely covered with setae, forming brush-like organs that might serve to collect pollen and also nectar through adhesion. Finally, Nevrorthidae also exhibit two interesting structures of unknown function: one is a paired, small, finger-like process on the distal margin of the ligula (Fig. 11.10a) that is not known from any other neuropteran family; the other is a paired submental gland with a multiporous opening which they share with Osmylidae (Randolf et al. 2014). Concluding from the mouthpart morphology, omnivorous feeding seems to be the most plausible hypothesis.

**Acknowledgements** We sincerely thank Brian Metscher (University of Vienna) for providing microCT scans which have proved to be helpful ever since. We are grateful to Peter Duelli (Eidgenössische Forschungsanstalt Birmensdorf), Heiko Bellmann (†), Harald Krenn (University of Vienna) and Vladimir Makarkin (Russian Academy of Sciences Vladivostok) for making their photographs and images available to us. The excellent histological sections in Fig. 11.11 were made by Barbara Schädl (Ludwig Boltzmann Institute Vienna).

## References

Acker TS (1958) The comparative morphology of *Stenorrhachus* (MacLachlan) and of *Nemopterella* sp. (Neuroptera; Nemopteridae). Microentomology 23:106–130

Anthony MH (1902) The metamorphosis of *Sisyra*. Am Nat 36:615–631

Aspöck U (1992) Crucial points in the phylogeny of the Neuroptera (Insecta). In: Canard M, Aspöck H, Mansell MW (eds) Current research in Neuropterology. Proceedings of the fourth international symposium on Neuropterology Toulouse, France, pp 63–73

Aspöck U (1993) Geklärtes und Ungeklärtes im System der Neuroptera (Insecta: Holometabola). Mitt Dtsch Ges Allg Angew Entomol 8:451–456

Aspöck U, Aspöck H (2007) Verbliebene Vielfalt vergangener Blüte. Zur Evolution, Phylogenie und Biodiversität der Neuropterida (Insecta: Endopterygota). Denisia 20:451–516

Aspöck U, Aspöck H (2008) Phylogenetic relevance of the genital sclerites of Neuropterida (Insecta: Holometabola). Syst Entomol 33:97–127

Aspöck H, Aspöck U, Hölzel H [In collaboration with Rausch H] (1980) Die Neuropteren Europas. Eine zusammenfassende Darstellung der Systematik, Ökologie und Chorologie der Neuropteroidea (Megaloptera, Raphidioptera, Planipennia) Europas. Goecke und Evers, Krefeld

Aspöck H, Hölzel H, Aspöck U (2001) Kommentierter Katalog der Neuropterida (Insecta: Raphidioptera, Megaloptera, Neuroptera) der Westpaläarktis. Denisia 2:1–606

Bartlett BR (1962) The ingestion of dry sugars by adult entomophagous insects and the use of this feeding habit for measuring the moisture needs. Ann Entomol Soc Am 55:749–753

Berland L, Grassé PP (1951) Super-ordre des Névroptéroïdes. In: Grassé PP (ed) Traité de zoologie, anatomie, systématique, biologie. Vol. 10, fasc. 1. Masson et Cie, Paris, pp 1–69

Beutel RG, Friedrich F (2008) Comparative study of larval head structures of Megaloptera (Hexapoda). Eur J Entomol 105:917–938

Beutel RG, Friedrich F, Aspöck U (2010a) The larval head of Nevrorthidae and the phylogeny of Neuroptera (Insecta). Zool J Linnean Soc 158:533–562

Beutel RG, Zimmermann D, Krauss M, Randolf S, Wipfler B (2010b) Head morphology of *Osmylus fulvicephalus* (Osmylidae, Neuroptera) and its phylogenetic implications. Org Divers Evol 10:311–329

Bozsik A (2000) Nahrungsanalytische Untersuchungen an einigen mitteleuropäischen Chrysopiden-Imagines (Neuroptera: Chrysopidae). Beitr Entomol [=Contrib Entomol] 50:237–246

Brauer F (1851a) Beschreibung und Beobachtung der österreichischen Arten der Gattung *Chrysopa*. In: Naturwissenschaftliche Abhandlungen, gesammelt und durch Subscription herausgegeben von Wilhelm Hadinger vol 4:1–12

Brauer F (1851b) Verwandlungsgeschichte des *Osmylus maculatus*. Archiv für Naturgeschichte 17:255–258

Brauer F (1852a) Verwandlungsgeschichte der *Mantispa pagana*. Archiv für Naturgeschichte 18:1–2

Brauer F (1852b) Versuch einer Gruppierung der Gattungen in der Zunft Planipennia mit besonderer Rücksicht auf die früheren Stände. Stett Ent Ztg 13:71–77

Brauer F, Löw F (1857) Neuroptera austriaca. Die im Erzherzogthum Oesterreich bis jetzt aufgefundenen Neuropteren nach der analytischen Methode zusammengestellt, nebst einer kurzen Charakteristik aller europäischen Neuropteren-Gattungen. C. Gerold's Sohn, Wien

Canard M (2001) Natural food and feeding habits of lacewings. In: McEwen PK, New TR, Whittington AE (eds) Lacewings in the crop environment. Cambridge University Press, Cambridge, pp 116–129

Crampton GC (1917) A phylogenetic study of the larval and adult head in Neuroptera, Mecoptera, Diptera, and Trichoptera. Ann Entomol Soc Am 10:337–344

Crampton GC (1921) The sclerites of the head and the mouthparts of certain immature and adult insects. Ann Entomol Soc Am 14:65–103

Crampton GC (1923) A phylogenetic comparison of the maxillae throughout the orders of insects. JNY Entomol Soc 31:77–107

De Jong GD (2011) Observations on the biology of *Polystoechotes punctatus* (Fabricius) (Neuroptera: Ithonidae): adult trophic status, description of the male reproductive system, and associations with mites. Proc Entomol Soc Wash 113:291–298

Devetak D, Klokočovnik V (2016) The feeding biology of adult lacewings (Neuroptera): a review. Trends Entomol 12:30–42

Eisner T (1953) The histology of a sense organ in the labial palps of Neuroptera. J Morphol 93:109–121

Engel MS, Winterton SL, Breitkreuz LCV (2018) Phylogeny and evolution of Neuropterida: where have wings of lace taken us? Annu Rev Entomol 63:531–551

Ferris GF (1940) The morphology of *Plega signata* (Hagen) (Neuroptera: Mantispidae). Microentomology 5:33–56

Finney GL (1948) Culturing *Chrysopa californica* and obtaining eggs for field distribution. J Econ Entomol 41:719–721

Gaumont J (1976) L'appareil digestif des larves de Planipennes. Ann Sci Nat Zool 18:145–250

Gillott C (1980) Entomology. Plenum, New York

Grebennikov VV (2004) Grub-like larvae of Neuroptera (Insecta): a morphological review of the families Ithonidae and Polystoechotidae and a description of *Oliarces clara*. Eur J Entomol 101:409–417

Hagen KS (1986) Ecosystem analysis: plant cultivars (HPR), entomophagous species and food supplements. In: Boethel DJ, Eikenbary RD (eds) Interactions of plant resistance and parasitoids and predators of insects. Wiley, New York, pp 151–197

Hagen KS, Tassan RL (1970) The influence of food wheast® and related *Saccharomyces fragilis* yeast products on the fecundity of *Chrysopa carnea* (Neuroptera: Chrysopidae). Can Entomol 102:806–811

Huang DY, Azar D, Cai CY, Garrouste R, Nel A (2015) The first Mesozoic pleasing lacewing (Neuroptera: Dilaridae). Cretac Res 56:274–277

Ickert G (1968) Beiträge zur Biologie einheimischer Chrysopiden (Planipennia, Chrysopidae). Entomol Abh 36:123–192

Jandausch K, Pohl H, Aspöck U, Winterton SL, Beutel RG (2018) Morphology of the primary larva of *Mantispa aphavexelte* Aspöck & Aspöck, 1994 (Neuroptera: Mantispidae) and phylogenetic implications to the order of Neuroptera. Arthropod Syst Phylo 76(3): 529–560

Kluge NJ (2005) Metamorphosis and homology of mouthparts in Neuropteroidea (Hexapoda: Metabola), with remarks on systematics and nomenclature. Russ Entomol J 14:87–100

Kokubu H, Duelli P (1983) Adult food of sponge flies: observations on the crop and gut content of *Sisyra terminalis* Curtis (Planipennia: Sisyridae). Neuroptera Int 2:157–162

Korn W (1943) Die Muskulatur des Kopfes und des Thorax von *Myrmeleon europaeus* und ihre Metamorphose. Zool Jahrb Abt Anat Ontog Tiere 68:273–330

Krenn HW, Plant JD, Szucsich NU (2005) Mouthparts of flower-visiting insects. Arthropod Struct Dev 34:1–40

Krenn HW, Gereben-Krenn BA, Steinwender BM, Popov A (2008) Flower visiting Neuroptera: mouthparts and feeding behaviour of *Nemoptera sinuata* (Nemopteridae). Eur J Entomol 105:267–277

Labandeira CC (2010) The pollination of mid Mesozoic seed plants and the early history of long-proboscid insects. Ann Missouri Bot Gard 97:469–513

Labandeira CC, Yang Q, Santiago-Blay JA, Hotton CL, Monteiro A, Wang YJ, Goreva Y, Shih CK, Siljeström S, Rose TR, Dilcher DL, Ren D (2016) The evolutionary convergence of mid-Mesoizoic lacewings and Cenozoic butterflies. Proc R Soc Lond [Biol] 283:20152893

Liu Q, Lu X, Zhang Q, Chen J, Zheng X, Zhang W, Liu X, Wang B (2018) High niche diversity in Mesozoic pollinating lacewings. Nat Commun 9:3793
Lu XM, Zhang WW, Liu XY (2016) New long-proboscid lacewings of the early cretaceous provide insights into ancient plant-pollinator interactions. Sci Rep 6:25382
Lucchese E (1956) Ricerche sulla *Mantispa perla* Pallas (Neuroptera Planipennia- Fam. Mantispidae). II. Contributo su nuovi reporti biologici e morfologici concernenti l'adulto, la larva della I. et de la completa evoluzione di questa nella sua sede definitiva. Ann Fac Agr Univ Perugia 12:83–213
MacLeod EG (1964) A comparative morphological study of the head capsule and cervix of larval Neuroptera (Insecta). PhD thesis. Harvard University, Cambridge, MA
Malicky H (1984) Ein Beitrag zur Autökologie und Bionomie der aquatischen Netzflüglergattung *Neurorthus* (Insecta, Neuroptera, Neurorthidae). Arch Hydrobiol 101:231–246
Makarkin VN (2016) Enormously long, siphonate mouthparts of a new, oldest known spongillafly (Neuroptera: Sisyridae) from Burmese amber imply nectarivory or hematophagy. Cretac Res 65:126–137
Makarkin VN (2017) New taxa of unusual Dilaridae (Neuroptera) with siphonate mouthparts from the mid-cretaceous Burmese amber. Cretac Res 74:11–22
Matsuda R (1965) Morphology and evolution of the insect head. Mem Am Entomol Inst 4:1–334
Meinander M (1972) A revision of the family Coniopterygidae (Planipennia). Acta Zool Fennica 136:1–357
Mendel Z, Adar K, Nestel D, Dunkelblum E (1997) Sex pheromone traps as a tool for the study of population trends of the predator of a scale insect and for the identification of potential predators for biological control. OBC/WPRS Bull 20:231–240
Minter LR (1990) A comparison of the eggs and first-instar larvae of *Mucroberotha vesicaria* Tjeder with those of other species in the families Berothidae and Mantispidae (Insecta: Neuroptera). In: Mansell MW, Aspöck H (eds) Advances in Neuropterology. Proceedings of the Third International Symposium on Neuropterology. South African Department of Agricultural Development, Pretoria, pp 115–129
Minter LR (1992) The egg and larval stages of *Nallachius krooni* Minter. In: Canard M, Aspöck H, Mansell MW (eds) Current research in Neuropterology. Proceedings of the fourth international symposium on Neuropterology. Toulouse, France
Möller A (2003) Aspects of the larval morphology and biology of south African *Podallea* species (Neuropterida: Neuroptera: Berothidae) MSc thesis. University of the North, South Africa
Monserrat VJ (1983) *Pterocroce capillaris* (Klug, 1836) en Europa (Neur., Plan., Nemopteridae). Neuroptera Int 2:109–128
Monserrat VJ (2005) Nuevos datos sobre algunas pequeñas familias de neurópteros (Insecta: Neuroptera: Nevrorthidae, Osmylidae, Sisyridae, Dilaridae). Heteropterus Rev Entomol 5:1–26
Monserrat VJ (2006) Nuevos datos sobre algunas especies de la familia Berothidae (Insecta: Neuroptera). Heteropterus Rev Entomol 6:173–207
Monserrat VJ (2014) Los dilaridos de la península Ibérica (Insecta, Neuropterida, Neuroptera, Dilaridae). Heteropterus Rev Entomol 14:187–214
Monserrat VJ, Acevedo F, Triviño V (2012) Los ascaláfidos de la Península Ibérica y Baleares (Insecta: Neuroptera: Ascalaphidae). Heteropterus Rev Entomol 12:33–58
New TR (1986) A review of the biology of Neuroptera Planipennia. Neuroptera Int Suppl Ser 1:1–57
New TR (1988) The Psychopsidae (Insecta: Neuroptera) of Australia and the Oriental region. Invertebr Taxon 2:841–883
New TR (1989) Planipennia, lacewings. Handbuch der Zoologie, (Arthropoda: Insecta), part 30, vol 4. Walter de Gruyter, Berlin
Oswald JD, Machado JP (2018) Biodiversity of the Neuropterida (Insecta: Neuroptera, Megaloptera and Raphidioptera). In: Footit RG, Adler PA (eds) Insect biodiversity: science and society, vol 2, pp 627–671
Poivre C (1978) Morphologie externe comparée de *Gerstaeckerella gigantea* Enderlein [Planipennia, Mantispidae]. Ann Soc Entomol France 14:191–206
Poivre C (1981) Mantispides du Cameroun I (Neuroptera, Planipennia). Neuroptera Int 1:77–89

Pupedis RJ (1980) Generic differences among New World spongilla-fly larvae and a description of the female of *Climacia striata* (Neuroptera: Sisyridae). Psyche 87:305–314

Pupedis RJ (1987) Foraging behavior and food of adult Spongila-flies (Neuroptera: Sisyridae). Ann Entomol Soc Am 80:758–760

Randolf S, Zimmermann D, Aspöck U (2013) Head anatomy of adult *Sisyra terminalis* (Insecta: Neuroptera: Sisyridae) – functional adaptations and phylogenetic implications. Arthropod Struct Dev 42:565–582

Randolf S, Zimmermann D, Aspöck U (2014) Head anatomy of adult *Nevrorthus apatelios* and basal splitting events in Neuroptera (Neuroptera: Nevrorthidae). Arthropod Syst Phylo 72:111–136

Randolf S, Zimmermann D, Aspöck U (2017) Head anatomy of adult *Coniopteryx pygmaea* Enderlein, 1906: effects of miniaturization and the systematic position of Coniopterygidae (Insecta: Neuroptera). Arthropod Struct Dev 46:304–322

Redborg KE (1998) Biology of the Mantispidae. Annu Rev Entomol 43:175–194

Rousset A (1966) Morphologie céphalique des larves de planipennes (Insectes Névroptéroïdes). Mem Mus Hist Nat Paris (A) 42:1–199

Shepard FD (1967) The head capsule and cervix of adult Neuroptera (Insecta). A comparative morphological study. PhD thesis, Cambridge, Massachusetts

Stange LA (1970) Revision of the ant-lion tribe Brachynemurini of North America (Neuroptera: Myrmeleontidae). Univ Calif Publ Entomol 55:1–192

Stelzl M (1991) Untersuchungen zu Nahrungsspektren mitteleuropäischer Neuropteren-Imagines (Neuropteroidea, Insecta) Mit einer Diskussion über deren Nützlichkeit als Opponenten von Pflanzenschädlingen. J Appl Entomol 111:469–477

Stelzl M (1992) Comparative studies on mouthparts and feeding habits of adult Raphidioptera and Neuroptera (Insecta: Neuropteroidea). In: Canard M, Aspöck H, Mansell MW (eds) Current research in Neuropterology. Proceedings of the Fourth International Symposium on Neuropterology. Bagnères-de-Luchon, Haute-Garonne, France, pp 341–347

Stelzl M, Gepp J (1990) Food-analysis of imagines of central European Myrmeleontidae (Insecta: Neuroptera). In: Mansell MW, Aspöck H (eds) Advances in neuropterology. Proceedings of the third international symposium on neuropterology. South African Department of Agricultural Development, Pretoria, pp 205–210

Stitz H (1931) Planipennia. In: Schulze P (ed) Biologie der Tiere Deutschlands, vol 35. Gebrüder Borntraeger, Berlin, pp 67–304

Sundermeier W (1940) Der Hautpanzer des Kopfes und des Thorax von *Myrmeleon europaeus* und seine Metamorphose. Zool Jahrb Abt Anat Onto Tiere 66:291–348

Svácha P (1992) What are and what are not imaginal discs: reevaluation of some basic concepts (Insecta, Holometabola). Dev Biol 154:101–117

Tauber CA, Tauber MJ, Albuquerque GS (2002) Neuroptera (lacewings, Antlions). In: Resh VH, Cardé R (eds) Encyclopedia of insects. Academic, San Diego, pp 785–798

Tillyard RJ (1922) The life-history of the Australian moth-lacewing, *Ithone fusca*, Newman (order Neuroptera Planipennia). Bull Entomol Res 13:205–223

Tjeder B (1957) Neuroptera-Planipennia. The lace-wings of southern Africa. 1. Introduction and families Coniopterygidae, Sisyridae, and Osmylidae. In: Hanström B, Brinck P, Rudebec G (eds) South African animal life, vol 4. Swedish Natural Science Research Council, Stockholm, pp 95–118

Tjeder B (1959) Neuroptera-Planipennia. The lace-wings of southern Africa. 2. Family Berothidae. In: Hanström B, Brinck P, Rudebec G (eds) South African animal life, vol 6. Swedish Natural Science Research Council, Stockholm, pp 256–314

Tjeder B (1960) Neuroptera-Planipennia. The lace-wings of southern Africa. 3. Family Psychopsidae. In: Hanström B, Brinck P, Rudebec G (eds) South African animal life, vol 7. Swedish Natural Science Research Council, Stockholm, pp 164–209

Tjeder B (1961) Neuroptera-Planipennia. The lace-wings of southern Africa. 4. Family Hemerobiidae. In: Hanström B, Brinck P, Rudebec G (eds) South African animal life, vol 8. Swedish Natural Science Research Council, Stockholm, pp 296–408

Tjeder B (1966) Neuroptera-Planipennia. The lace-wings of southern Africa. 5. Family Chrysopidae. In: Hanström B, Brinck P, Rudebec G (eds) South African animal life, vol 12. Swedish Natural Science Research Council, Stockholm, pp 228–534

Tjeder B (1967) Neuroptera-Planipennia. The lace-wings of southern Africa. 6. Family Nemopteridae. In: Hanström B, Brinck P, Rudebec G (eds) South African animal life, vol 13. Swedish Natural Science Research Council, Stockholm, pp 290–501

Tjeder B (1972) Coniopterygidae from the Snow Mountains, New Guinea (Neuroptera). Entomol Tidskr 93:186–209

Tjeder B (1992) The Ascalaphidae of the Afrotropical region (Neuroptera). 1. External morphology and bionomics of the family Ascalaphidae, and taxonomy of the subfamily Haplogleniinae including the tribes Proctolyrini n. tribe, Melambrotini n. tribe, Campylophlebini n. tribe, Tmesibasini n. tribe, Allocormodini n. tribe, and Ululomyiini n. tribe of Ascalaphidae. Entomol Scand Suppl 41:3–169

Villenave J, Deutsch B, Lodé T, Rat-Morris E (2006) Pollen preference of the *Chrysoperla* species (Neuroptera: Chrysopidae) occurring in the crop environment of western France. Eur J Entomol 103:771–777

Weber H (1933) Lehrbuch der Entomologie. Fischer, Jena

Winterton SL, Lemmon AR, Gilling JP, Garzon IJ, Badano D, Bakkes DK, Breitkreuz LCV, Engel MS, Lemmon EM, Liu X, Machado RJP, Skevington JH, Oswald JD (2018) Evolution of lacewings and allied orders using anchored phylogenomics (Neuroptera, Megaloptera, Raphidioptera). Syst Entomol 43:330–354

Withycombe CL (1924) Some aspects of the biology and morphology of the Neuroptera. With special reference to the immature stages and their possible phylogenetic significance. R Entomol Soc Lond 72:303–411

Wundt H (1961) Der Kopf der Larve von *Osmylus chrysops* L. (Neuroptera, Planipennia). Zool Jahrb Abt Anat Onto Tiere 79:557–662

Wundt H (1967) Die Metamorphose von Spezialisationsbildungen bei *Osmylus chrysops* L. (Neuroptera). Verh Dtsch Zool Ges 1967:734–741

Zimmermann D, Klepal W, Aspöck U (2009) The first SEM study on Coniopterygidae (Neuroptera) - structural evidence and phylogenetic implications. Eur J Entomol 106:651–662

Zwick P (1967) Beschreibung der aquatischen Larve von *Neurorthus fallax* (Rambur) und Errichtung der neuen Planipennierfamilie Neurorthidae fam. nov. Gewässer und Abwässer 44 (45):65–86

# Chapter 12
# Structure and Evolution of Mouthparts in Coleoptera

**Rolf Georg Beutel and Margarita Yavorskaya**

**Abstract** An unusually broad food spectrum has likely played an important role in the enormous diversification of Coleoptera (ca. 400,000 spp.). Nevertheless, the structure and configuration of the mouthparts are mostly conservative. Archostemata, often addressed as ancestral, are characterized by a trend towards limited food uptake in the adult stage, whereas the larval mouthparts are suitable for burrowing in wood. Adephaga are predacious as adults and larvae with few exceptions, with more or less advanced preoral digestion. Adults and larvae of Myxophaga feed on algae, and Polyphaga are primarily saprophagous. The ancestral adult feeding apparatus in Myxophaga and different groups of Polyphaga is characterized by large mandibular molae and epi- and hypopharyngeal bulges with microtrichia, a configuration suitable for processing soft and decaying plant tissues and also small particles. Polyphaga have evolved a broad spectrum of food preferences, with different modifications in immatures of carnivorous groups or groups specialized on faeces or fresh plant materials. Mouthparts as they are found in Myxophaga and basal groups of Polyphaga likely belong to the groundplan of Coleoptera. The major burst of diversification in Polyphaga was linked with the shift to feeding on fresh plant materials.

## 12.1 Introduction

With almost 400,000 described species (e.g. McKenna et al. 2015) and certainly many more to be discovered (Stork et al. 2015: ca. 1,500,000 expected species), Coleoptera clearly outnumber all other groups of insects. There is little doubt that the capacity to make use of a very broad spectrum of different food sources has

R. G. Beutel
Institut für Zoologie und Evolutionsforschung, FSU Jena, Jena, Germany
e-mail: rolf.beutel@uni-jena.de

M. Yavorskaya (✉)
Department of Biology II LMU München, Functional Morphology Group, Planegg-Martinsried, Germany



H. W. Krenn (ed.), *Insect Mouthparts*, Zoological Monographs 5,
https://doi.org/10.1007/978-3-030-29654-4_12

contributed to the diversification of beetles, especially in the megadiverse Polyphaga, which contain almost 90% of the known species.

A rich fundus of morphological studies on feeding structures of beetles is available. An earlier extensive article on the mouthparts of adults was published by Williams (1938). However, this study did not cover the mandibles, and no information on muscles and the configuration of the feeding apparatus was provided. A broad survey of mouthpart variation in insects was published by Chaudonneret (1990), also covering some group of beetles. Numerous characters including mouthparts of adult and larval Coleoptera were described and coded in Lawrence et al. (2011), however without anatomical data and without information on the general configuration and functional aspects. Extensive information on the morphology of adults and larvae of all groups of Coleoptera is provided in three volumes of the *Handbook of Zoology* series (Beutel and Leschen 2005a, 2016; Leschen et al. 2010; Leschen and Beutel 2014). Finally, mouthparts and other head structures are treated in a considerable number of studies dedicated to single species or few more or less closely related taxa (e.g. Dönges 1954; Noars 1956; Honomichl 1975; De Marzo 1978, 1979; Beutel 1986a, b; Belkaceme 1991). A time-honoured but outstanding example is a monograph on *Dytiscus marginalis* L. (Korschelt 1923–1924: "Der Gelbrand"). Extremely detailed and well-documented information on the adult and larval head structures and the digestive tract is provided in several chapters of this work. A study on the mouthparts of flower-visiting insects (Krenn et al. 2005) also contained information on various groups of beetles feeding on nectar or pollen or both.

The main purpose of the present chapter is to give a brief overview of the major types of mouthparts occurring in adults and larvae of the main subdivisions of Coleoptera: Archostemata (ca. 40 spp.), Myxophaga (ca. 120 spp.), Adephaga (ca. 50,000 spp.) and Polyphaga (ca. 350,000 spp.). This is mainly based on anatomical studies emerging from the Entomology Group of the Phyletisches Museum in the last two decades (e.g. Beutel and Haas 1998; Beutel et al. 1998, 2008; Anton and Beutel 2004, 2006, 2012; Dressler and Beutel 2010; Antunes-Carvalho et al. 2017; Yavorskaya et al. 2017, 2018a, b) but also on other contributions mentioned above. Head structures of adults and immature stages of representatives of all four suborders were examined with a broad spectrum of morphological techniques (e.g. Wipfler et al. 2016; Yavorskaya 2018a, b). The structural features are interpreted with respect to their functional, phylogenetic and evolutionary significance.

## 12.2 Mouthparts of Adults

### *12.2.1 General Features*

Despite of an extremely broad spectrum of different feeding habits, especially in the megadiverse Polyphaga (e.g. Crowson 1981; Leschen et al. 2010; Leschen and Beutel 2014; Beutel and Leschen 2016), the mouthparts of adult beetles are generally of a plesiomorphic biting type. Few derived features were pointed out by

Crowson (1981): four-segmented maxillary palps and the lack of glossae and paraglossae. The prognathous condition almost generally present in adults is likely primarily linked with the tendency to penetrate into narrow spaces, especially subcortical (under bark) microhabitats.

The labrum is usually free and retractile, with a short paired intrinsic muscle and external retractors (M. frontoepipharyngalis) attached to the tormae, but without a median M. frontolabralis. It is fused with the clypeus in few groups, for instance, *Omma* (Ommatidae), Micromalthidae (both Archostemata) (Beutel et al. 2008), *Lucanus* (Lucanidae), *Chauliognathus* (Cantharidae), *Chelonarium* (Chelonariidae) and Curculionoidea excluding Anthribiidae and Nemonychidae (Crowson 1981; Lawrence et al. 2011). An elongated and apically pointed labrum has evolved in the cerylonid *Cautomus* (Besuchet 1972) and few other taxa (Crowson 1981).

The mandibles articulate in a typical dicondylic manner, the primary ventral joint (posterior in orthognathous heads) usually with a mandibular condyle and cephalic socket, and the dorsal secondary joint vice versa. The mandibles vary considerably in shape and substructures, but are generally moved by two antagonistic muscles originating on the head capsule, a comparatively slender external extensor (M. craniomandibularis externus) and an internal flexor (M. craniomandibularis internus), almost always the largest cephalic muscle. A very thin tentorio-mandibular muscle, functioning as a proprioreceptor (Honomichl 1975, 1976, 1978), can be present or absent and may have been overlooked in some studies. A grinding mola is usually well-developed in beetles feeding on algae, fungi, spores or decaying plant substrates (e.g. Beutel et al. 2003; Yavorskaya et al. 2018a) and is likely part of the groundplan of Coleoptera (Yavorskaya 2018). Retinacula, mesally directed teeth of the middle mandibular region, brushes of hairs, and prosthecae (lacinia mobilis) occur in various groups.

The maxillae of most groups comprise a relatively small cardo, which articulates with the fossa maxillaris, a stipes subdivided into a basistipes and mediostipes, a triangular, fairly large lateral palpifer, a four-segmented palp, a two-segmented galea, and a lacinia, usually equipped with articulated spines and largely or completely fused with the mediostipes.

The labium is usually composed of a submentum fused with the gula posteriorly, a mentum, a retractile prementum, three-segmented palps inserted on a small palpiger, and often a more or less distinct unpaired ligula. The hypopharynx forms a structural and functional unit with the prelabium. A salivarium is always absent. Salivary ducts are almost generally missing but present in the archostematan species *Micromalthus debilis* LeConte (Micromalthidae) (Yavorskaya et al. 2018b).

### *12.2.2 Archostemata*

The feeding habits of adults are not well known. It is assumed that adults of Cupedidae feed on pollen (Crowson 1962; Hörnschemeyer 2005). Food uptake plays a minor role if at all in the "vestigial" miniaturized adults of *Micromalthus debilis* (Micromalthidae) (Yavorskaya et al. 2018b). The labrum is relatively small but free and movable in Cupedidae. In contrast, it is immobilized in the ommatid genus *Tetraphalerus* (Fig. 12.1; Beutel et al. 2008) and completely fused with the clypeus in *Omma* and in *Micromalthus*. The mandibles (Fig. 12.1) are elongated, widely separated at their bases and usually partly covered by cuticular scales like other parts of the body. They lack molae or basal protuberances and also retinacula or movable teeth. A longitudinal brush of short stout hairs is present on the mesal

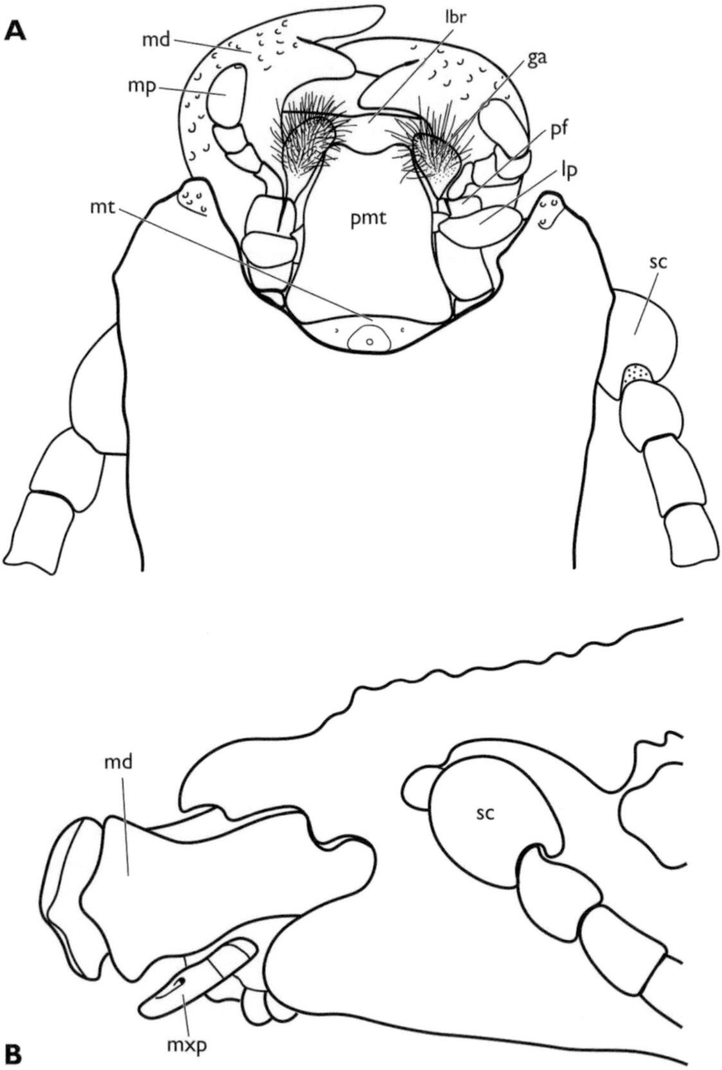


**Fig. 12.1** Mouthparts of adults of Archostemata: *Tetraphalerus bruchi* Heller; (**a**) ventral view; (**b**) lateral view. Abbreviations: *ga* galea, *lbr* labrum, *lp* labial palp, *md* mandible, *mt* mentum, *mp, mxp* maxillary palp, *pf* palpifer, *pmt* prementum, *sc* scapus. Redrawn from SEM micrographs in Beutel et al. (2008)

side in *Tetraphalerus* (Beutel et al. 2008). The apical and subapical teeth are arranged longitudinally in Cupedidae (Hörnschemeyer et al. 2002, 2006), as it is the case in adults of most groups of beetles, whereas three incisivi are arranged in a vertical row in Ommatidae and *Micromalthus debilis* (e.g. Beutel et al. 2008; Lawrence et al. 2011). The small maxillae (Fig. 12.1a) are inserted in a rather shallow fossa maxillaris. The cardo is small and the stipes subdivided into a basistipes and mediostipes. The four-segmented palp is inserted on a large palpifer, which is mesally connected with the basistipes. The lacinia is fused with the broad mediostipes, spade-shaped, and densely covered with hairs along its mesal edge. A prominent and pointed apical tooth is absent, and strongly developed articulated spines are also missing. In Cupedidae, the galea is composed of a smooth, sclerotized stalk, and a semimembranous onion-shaped galeomere 2 (distal subunit of galea), which is densely covered with fine setae. The proximal galeomere is less slender in *Tetraphalerus* (Fig. 12.1a) than in Cupedidae (Hörnschemeyer et al. 2002, 2006; Beutel et al. 2008). Both endite lobes are reduced in the miniaturized *Crowsoniella* and *Micromalthus*. Like in other groups of Coleoptera, the apical maxillary palpomeres of Archostemata bear two fields of campaniform sensilla, one apically and one apicolaterally (Beutel et al. 2008; Hörnschemeyer 2009). The apicolateral sensilla are countersunk in a groove. In contrast to almost all other beetles, digitiform sensilla are missing in Archostemata. The labium is composed of the usual subunits in the groundplan of Archostemata. The mentum is a distinctly delimited, short transverse sclerite in *Tetraphalerus* (Fig. 12.1a), but largely or completely fused to the submentum in Cupedidae and the other families (Beutel et al. 2008). The prementum is large and plate-like and equipped with a strong unpaired internal apodeme, visible as a deep groove externally (Hörnschemeyer et al. 2002: Fig. 2). The ligula is simple in the groundplan but can be modified as a complicated membranous structure in species of Cupedidae, divided into many digitiform appendages (Hörnschemeyer 2009), suggesting an association with flowers according to Crowson (1981).

### *12.2.3 Myxophaga*

Myxophaga mainly rely on algae as food as far as presently known (e.g. Reichardt 1973; Beutel and Arce-Pérez 2016). As a possible effect of miniaturization, they are characterized by a tendency to internalize the mouthparts in adults and larvae (Beutel et al. 1998; Yavorskaya et al. 2018a). The labrum (Fig. 12.2a) of adults is generally free and movable and covers considerable parts of the paired mouthparts. In contrast to Archostemata, the proximal part of the mandibles (Fig. 12.2b, c) is stout, and a grinding mola is present. A derived feature shared by Myxophaga (except for *Sphaerius*) is the presence of a movable tooth on the left mandible. A semimembranous hairy and rounded lobe is present proximad this structure (Reichardt 1973; Anton and Beutel 2006). The distal part of the mandible is more or less slender. The distinct reduction of the lacinia is a potential autapomorphy of

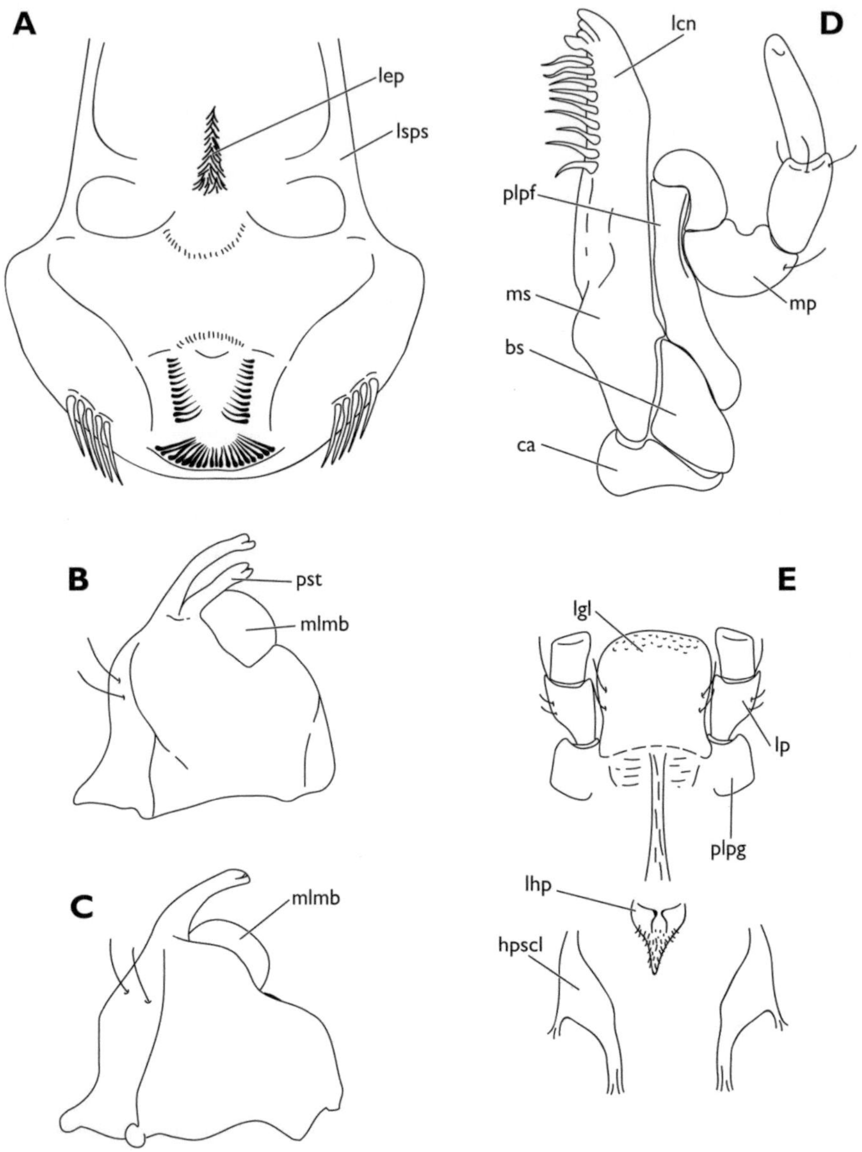


**Fig. 12.2** Mouthparts of adults of Myxophaga: *Lepicerus inaequalis* Sharp; (**a**) labrum, ventral view, with longitudinal epipharyngeal ridge; (**b**) the left mandible; (**c**) the left mandible; (**d**) maxilla; (**e**) labium. Abbreviations: *bs* basistipes, *ca* cardo, *hpscl* hypopharyngeal sclerite, *lcn* lacinia, *lep* longitudinal epipharyngeal process;, *lgl* ligula, *lhp* longitudinal hypopharyngeal process, *lp* labial palp, *lsps* labral suspensorium, *mlmb* median lobular mandibular brush, *mp* maxillary palp, *ms* mediostipes, *plpf* palpifer, *plpg* palpiger, *pst* prostheca. Redrawn from Anton and Beutel (2006)

the suborder (Fig. 12.2d). A vestigial lacinia is preserved in *Sphaerius* (Yavorskaya et al. 2018a) but lacking in the other groups (Reichardt 1973; Beutel 1999; Anton and Beutel 2006). A row of flattened teeth is present on the apical part of the maxillary mala of several genera of Torridincolidae (Beutel 1999; Anton and Beutel 2006). The distal maxillary palpomere is small and subulate in Myxophaga excluding Lepiceridae. The penultimate palpomere is strongly inflated in *Sphaerius* (Yavorskaya et al. 2018a). In the groundplan of Myxophaga, the submentum and mentum are separated, but they are fused in the hydroscaphid genera *Hydroscapha* and *Scaphydra* (Beutel 1999). The mentum forms a very large lid-like structure in *Lepicerus*, which encloses the paired mouthparts completely, together with the large labrum (Anton and Beutel 2006). An autapomorphic feature of larvae and adults of the suborder is a premental ligula more or less densely set with rounded papilla (Fig. 12.2e; Beutel 1999; Yavorskaya et al. 2018a).

The hairy semimembranous mandibular lobes of myxophagan adults interact with corresponding hairy longitudinal ridges of the epi- and hypopharynx (Fig. 12.2e; Anton and Beutel 2006; Yavorskaya et al. 2018a), as it is also the case in several lineages of Polyphaga (e.g. Anton and Beutel 2004; Antunes-Carvalho et al. 2017).

### *12.2.4 Adephaga*

The mouthparts of Adephaga (Fig. 12.3) are apparently adapted to the predacious habits of almost all subgroups of these suborders and specifically to liquid feeding with capillary forces in Carabidae (e.g. Forsythe 1982, 1983). The labrum is always separated from the anterior clypeal margin (Fig. 12.3a) but likely immobilized in most cases, as the retractor and levator (M. frontoepipharyngalis) is usually missing (Dressler and Beutel 2010). The mandible (Fig. 12.3c–f) is rather short in the aquatic groups and the terrestrial Trachypachidae (e.g. Beutel 1986a, 1989; Belkaceme 1991; Dressler and Beutel 2010; Beutel et al. 2017) but more or less elongated in Carabidae (e.g. Green 1956; Luff 1974; Forsythe 1982, 1983; Acorn and Ball 1991). Molae and movable teeth are always lacking (Figs. 12.3c–f and 12.4a), whereas mesal setal brushes are usually present, especially in the terrestrial groups (e.g. Acorn and Ball 1991). The presence of several strongly developed and pointed mesal teeth is likely an apomorphy of Cicindelinae (Ball et al. 2011). The maxillae are well-developed and composed of the usual components (Fig. 12.3g, h). The distal lacinia is usually hooklike and the mesal edge set with setae and spines. A characteristic feature is the galea resembling a small two-segmented palp (e.g. Williams 1938; Beutel 1989; Belkaceme 1991; Dressler and Beutel 2010). A derived feature of the labium is the presence of two anterolateral lobes on the mentum, which enclose the prementum (Fig. 12.3f). The prementum bears palpigers and two segmented labial palps. A ligula is usually present but inconspicuous, more or less integrated in the anterior premental margin (e.g. Williams 1938; Beutel 1986a, 1989; Belkaceme 1991; Dressler and Beutel 2010; Beutel et al. 2017).

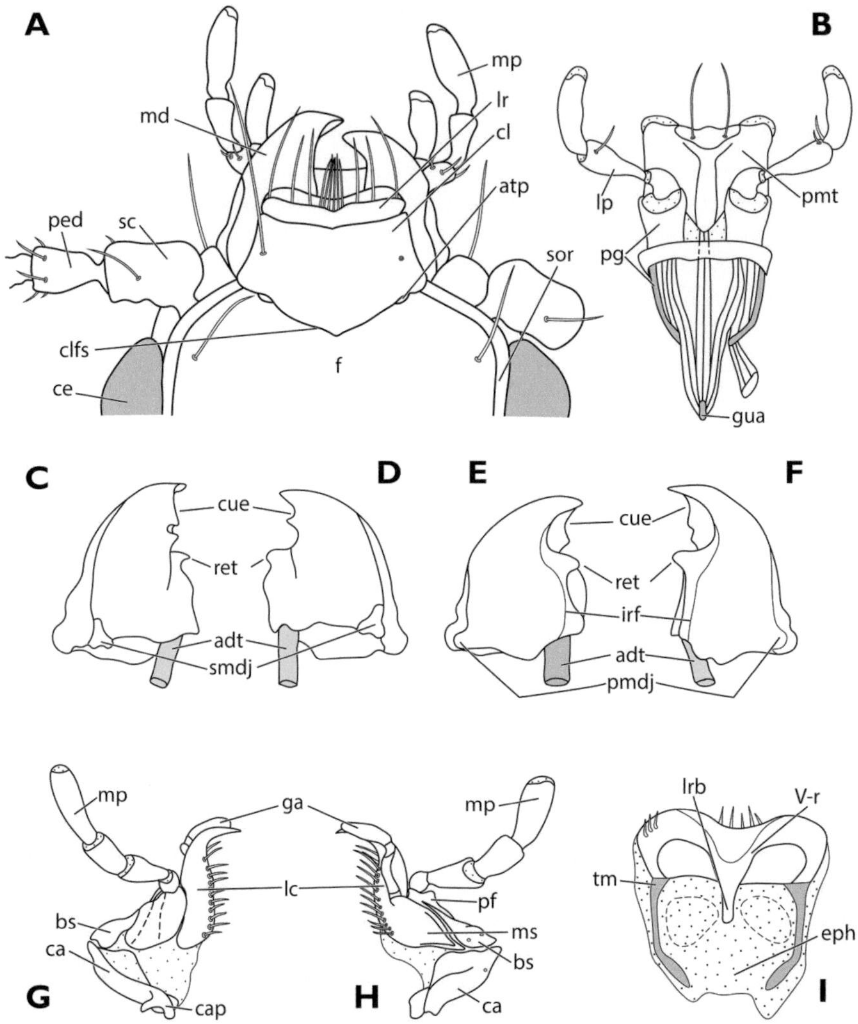


**Fig. 12.3** Mouthparts of adults of Adephaga: *Trachypachus holmbergi* Mannerheim; (**a**) ventral view of anterior part of head; (**b**) labium, ventral view; (**c–f**) mandibles; (**g, h**) maxillae; (**i**) labrum. Abbreviations: *adt* adductor tendon, *anc* circumantennal ridge with process, *atp* anterior tentorial pit, *bs* basistipes, *ca* cardo, *cap* cardo process, *ce* compound eye, *cl* clypeus, *clfs* clypeofrontal suture, *cue* cutting edge, *eph* epipharynx, *f* frons, *irf* insertion ridge of ventral fringe of hairs, *ga* galea, *gua* gular apodeme, *lc* lacinia, *lp* labial palp, *lrb* median bar of labrum, *lrr* transverse labral ridge, *md* mandible, *mp* maxillary palp, *ms* mediostipes, *ped* pedicellus, *pf* palpifer, *pg* palpiger, *pmdj* primary mandibular joint, *pmt* prementum, *ret* retinaculum, *sc* scapus, *smdj* secondary mandibular joint, *sor* supraocular ridge, *tm* torma, *V-r* v-shaped ventral labral ridge. Redrawn from Dressler and Beutel (2010)

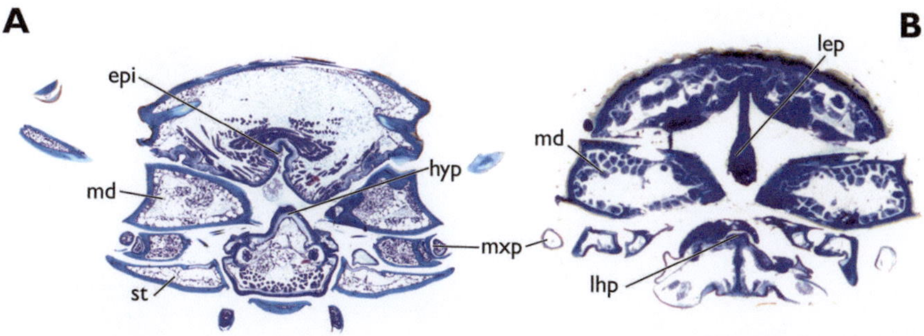


**Fig. 12.4** Light microscopic micrographs of transverse histological sections through anterior part of the head (molae, epi- and hypopharynx) of Adephaga and Polyphaga. (**a**) *Heterogyrus milloti* (Gyrinidae); (**b**) *Catops ventricosus* (Leiodidae). Abbreviations: *epi* epipharynx, *hyp* hypopharynx, *lep* longitudinal ridge of epipharynx, *lhp* lateral hypopharyngeal process, *md* mandible, *mxp* maxillary palp, *st* stipes. Modified from Antunes-Carvalho et al. (2017) and Beutel et al. (2017)

### 12.2.5 *Polyphaga Excluding Phytophaga*

The groundplan configuration of Polyphaga is largely preserved in most subgroups of Scirtoidea (Anton et al. 2016) and Staphyliniformia (Figs. 12.5 and 12.4b) (e.g. Beutel et al. 2003; Anton and Beutel 2004; Anton et al. 2016; Antunes-Carvalho et al. 2017). The labrum (Fig. 12.5a) is usually retractile, with a well-developed M. frontoepipharyngalis attached to the tormae. The mandibles are equipped with well-developed basal grinding molae and setal brushes (Fig. 12.5b, c). A ventral accessory process occurs in different groups (e.g. Lawrence et al. 2011). The maxillae comprise the typical components (Figs. 12.5f, g). A fimbriate galea with curved setae arranged in several rows is typical for groups with presumably ancestral microphagous or saprophagous feeding habits. The labium lacks anterolateral lobes of the mentum (Fig. 12.5d). The mandibular brushes interact with brushes of longitudinal epi- and hypopharyngeal ridges (Figs. 12.5a, e and 12.4b). Derived features of Staphyliniformia are a hypopharynx (Fig. 12.5e) which is hourglass shaped in cross section and a transverse cranial muscle inserted on the basal articulatory membrane of the maxilla (e.g. Anton and Beutel 2004: Mx).

### 12.2.6 *Phytophaga*

The phytophagous Chrysomelidae (e.g. Chamorro 2014; Nadein and Bezděk 2014; Reid 2014; Ge et al. 2015) possess a free labrum and short triangular mandibles, usually with a shovel-like distal region with two to six broad teeth. A membranous prostheca can be present (e.g. Chrysomelinae) or absent (e.g. Cryptocephalinae). The mola can be well-developed, partly reduced or absent (e.g. Chamorro 2014; Vencl and Leschen 2014; Reid 2014). The maxillae of Chrysomelinae bear a

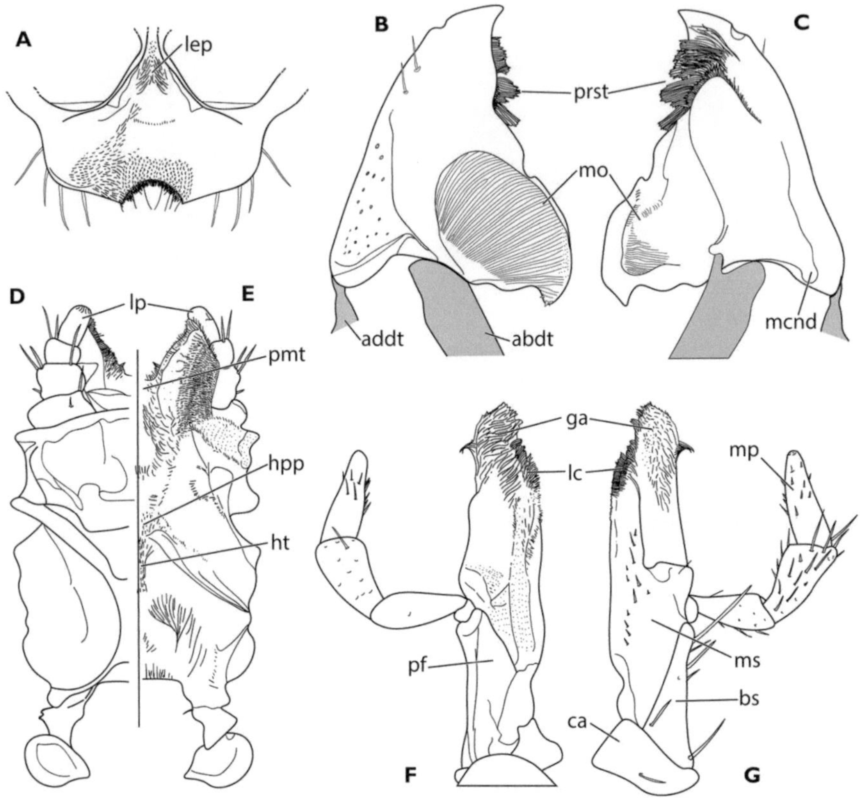


**Fig. 12.5** Mouthparts of adults of the ancestral type of Polyphaga: *Catops* sp.; (**a**) labrum, ventral view; (**b**) mandible, dorsal view; (**c**) mandible, ventral view; (**d**) labium, dorsal view; (**e**) labium, ventral view; (**f**) maxilla, dorsal view; (**g**) maxilla, ventral view. Abbreviations: *abdt* abductor tendon, *addt* adductor tendon, *bs* basistipes, *ca* cardo, *ga* galea, *hpp* hypopharynx, *ht* tuft of hairs of the hypopharynx (or longitudinal hypopharyngeal process), *lep* longitudinal epipharyngeal process, *lc* lacinia, *lp* labial palp, mcnd, mandibular condyle, *mo* mola, *mp* maxillary palp, *ms* mediostipes, *pf* palpifer, *pmt* prementum, *prst* prostheca. Redrawn form Antunes-Carvalho et al. (2017)

palpiform galea and membranized lacinia, both apically setose (Reid 2014). The mentum is usually strongly transverse in Chrysomelidae and closely related families (Chamorro 2014; Vencl and Leschen 2014; Reid 2014).

The adults of several ancestral groups of Cerambycidae do not feed or imbibe only fluids (Svacha and Lawrence 2014). In contrast to Chrysomelidae, the mandibles are usually unidentate. They are often short and broad but moderately elongated in some groups or even strongly enlarged (mainly males of some Prioninae). A prostheca is always absent. A flat platelike molar sclerite is present in Necydalinae and Lepturinae, whereas smaller molar protuberances can occur in some other subgroups (Svacha and Lawrence 2014).

A main characteristic of the extremely diverse Curculionoidea is a long rostrum (e.g. Dönges 1954). This feature is present in nearly all "ancestral" weevils (e.g. Nemonychidae, Belidae) and also in the earliest fossils and therefore very likely in the groundplan of the superfamily (Oberprieler 2014). It was emphasized by Oberprieler (2014) that the rostrum is primarily a tool for depositing eggs in plant tissue and a key adaptation in this extremely successful group. It is reduced in wood-feeding subgroups (e.g. Scolytinae) but also in Anthribiidae and some genera of other families (Oberprieler 2014). The labrum is fused in Curculionoidea with few exceptions (see above).

## 12.3 Specialized Mouthparts in Adult Polyphaga

### *12.3.1 Piercing-Sucking Mouthparts*

In contrast to a plesiomorphic configuration in adults of most members of Scirtoidea (e.g. Anton et al. 2016), mouthparts forming a piercing-sucking beak occur in some Eucinetidae. The modifications compared to unspecialized members of the family include a strongly acute labrum, very narrow and elongated mandibles and styliform maxillary endite lobes (Leschen 2016).

### *12.3.2 Falcate Mandibles of Predacious Adults*

Falcate mandibles without mola occur in predacious adults of several groups like Staphylinidae (e.g. Scydmaeninae), Omalisidae or Lampyridae (e.g. Crowson 1981; Bocak and Bocakova 2010: Fig. 4.10.1F; Bocak et al. 2010; Thayer 2016).

### *12.3.3 Elongated Mandibles*

Strongly elongated mandibles occur in *Prostomis* (Prostomidae, Tenebrionoidea) (Seago and Beutel 2010), in males of some Prioninae (Cerambycidae) and in males of Lucanidae (e.g. Scholtz and Grebennikov 2016: Fig. 15.14). They are used in mating fights by males but also for clutching food. Cherries are preferred by adults of *Lucanus cervus* (Krenn et al. 2002).

### *12.3.4 Sticky Labial Rod of Steninae*

A highly specialized case of mouthparts adapted to preying upon small agile arthropods, especially springtails, is the “sticky rod” of *Stenus*. Few coleopteran mouthparts have attracted as much attention as the protrusile elongated labium of members of this extremely species-rich staphylinid genus. The morphology, ultrastructure and function were investigated by Weinreich (1968), Bauer and Pfeiffer (1991), Betz (1996, 1998), Betz et al. (2018), Koerner et al. (2012, 2017) and others. The prey-catching device can be thrust forward rapidly (within 1–3 ms) towards the potential prey (e.g. Weinreich 1968; Betz 1996; Koerner et al. 2012). Attachment pads on the distal part of the prementum, commonly designated as paraglossae, bear numerous distally branching outgrowths. They create a maximum adhesive force when hitting the prey, which is seized by the mandible after the prementum is retracted by four pairs of extrinsic muscles. The submentum and mentum are largely unmodified (Betz et al. 2018: Fig. 11.23). The elongated membranous connecting tube functions like the finger of a glove when receiving the elongated proximal part of the prementum, which reaches the middle region of the prothorax in its retracted position (Weinreich 1968). The protrusion of the structure is effected by increased haemolymph pressure. The adhesion of the premental pads is supported by gland secretions (e.g. Betz et al. 2018).

### 12.3.5 Mandibles for Processing Soft Faeces

Largely membranous mandibles with only the outer edge sclerotized are typical for Scarabaeinae (Scholtz and Grebennikov 2016; Karolyi et al. 2016: Fig. 1J). The mandibles of the coprophagous *Canthon pilularius* (Scarabaeidae, Scarabaeinae) and related species were examined in detail by Hata and Edmonds (1983). The molar lobes are designed to finely grind particles in soft, pasty food. They function as a mortar-pestle system milling plant material contained in the faeces by a combination of squeezing and grinding actions (Hata and Edmonds 1983). The molar surface consists of series of ridges with rows of minute scrapers (“tritors”). The incisor lobes of the mandibles scrape food from the surfaces of the galeae and laciniae, which bring food into the preoral cavity (Hata and Edmonds 1983).

### 12.3.6 Mouthparts of Anthophilous Beetles

Specializations of mouthparts linked with nectar feeding were described by Krenn et al. (2005), who pointed out that the plesiomorphic orthopteroid configuration is largely maintained. Modifications are mainly restricted to specific arrangements of bristles on the paired mouthparts and also on labial substructures. A condition typical for floricolous beetles (e.g. *Dascillus*, Ptilodactylidae part, Mordellidae part) according to Crowson (1981) is elongated and hairy maxillary galeae. In some cases, the laciniae are also elongated and pubescent, and a ligula divided into paired structures with a hairy surface also occurs in some species (Crowson 1981). The elongated mouthparts of flower-visiting and nectar-feeding Nemagnathini

(Meloidae) were described in detail by Wilhelmi and Krenn (2012) (see also Chaudonneret 1990).

Beetles feeding on pollen are found in a number of families (e.g. Buprestidae, Elateridae, Nitidulidae, Mordellidae, Oedemeridae, Meloidae), and in some cases, their mouthparts are also suitable for the uptake of nectar (Krenn et al. 2005). As in the case of the nectar feeders, the modifications are limited. Characteristic features are pubescent areas, a soft lacinia mobilis and a postmola which kneads and conveys the pollen (Nel and Scholtz 1990; Krenn et al. 2005). Mandibles with a reduced biting capacity are mainly used for manipulating pollen in some cases (Krenn et al. 2005). Pollen-harvesting structures with spatulate or spoon-shaped and specifically arranged microtrichia occur in a number of families (e.g. Scarabaeidae, Cantharidae, Oedemeridae, Mordellidae, Cerambycidae) (Krenn et al. 2005; see Chap. 13).

### *12.3.7 Filter Feeding*

Filter feeding is a rare exception in Coleoptera. Presently, it is only known in adults of Spercheidae (Hydrophiloidea), aquatic beetles moving upside down along the surface film of water and ingesting small particles (Rothmeier and Jäch 1986). The apparatus and mechanism were described by Rothmeier and Jäch (1986) and Beutel et al. (2001: Figs. 1–3, 7–9, 13–15). Fringes of long setae on the clypeus, labrum, antennomere 2, submentum and prementum form an external food-collecting apparatus. The beetles gather particles from a current created by oscillating movements of the maxillary palps. The finger-shaped galea is used as a comb or cleaning device. Together with rows of short spines along the margin of the flattened lacinia, it removes particles from the external collecting apparatus (Rothmeier and Jäch 1986). The maxillary endite lobes are retracted into the preoral cavity and cleaned by microtrichia of the epi- and hypopharyngeal surfaces. A subapical semimembranous lobe of the mandibles removes particles from the epipharyngeal lobe and a ventral longitudinal brush from the median hypopharyngeal ridge. The food mass is then moved towards the large molae where it is intensively processed between surfaces with dense rows of toothlike asperities. The substrate is then likely diluted with secretions of tubular glands in the labrum, mandibles and labium and then sucked back into the pharynx by alternative contractions of dilators and ring muscles (Beutel et al. 2001: Fig. 4). Additional processing takes place in the proventriculus.

### *12.3.8 Reduced Mouthparts*

The mouthparts are almost completely reduced in some adults of Ripidiinae (Rhipiphoridae) (Crowson 1981), forming only a single unpaired tubercle in advanced forms (e.g. Batelka 2011).

## 12.4 Larval Mouthparts

The feeding habits of beetle larvae often differ distinctly from those of adults (e.g. Hydrophiloidea; Archangelsky et al. 2016), and the spectrum of variation of the mouthparts is much wider (e.g. Beutel 1993, 1995; Beutel and Molenda 1997; Gorb and Beutel 2000; Lawrence et al. 2011). Specialized conditions have evolved in different lineages: largely internalized paired mouthparts (Myxophaga part; Beutel et al. 1998), mandibular sucking channels (e.g. Gyrinidae, Haliplidae, Dytiscidae, *Graphipterus* [Carabidae; Brandmayr et al. 1993], Lampyridae, Brachypsectridae), protracted tactile maxillae (Adephaga major part; Hydrophiloidea partim; Beutel 1993, 1999), maxilla-labial complexes (Elateriformia major part, Cleroidea; Beutel 1995; Beutel and Pollock 2000), subdivided mandibles (Lycidae; e.g. Bocak and Matsuda 2003) or a snoutlike rostrum (*Holopsis*, Corylophidae; Yavorskaya et al. 2014).

### *12.4.1 Archostemata*

The larval mouthparts (Fig. 12.6) differ strongly from those of the adults. They are suitable for processing solid material like wood. The labrum is free (Fig. 12.6a), with a well-developed M. frontoepipharyngalis. Even a thin M. frontolabralis is present in the primary larvae of *Tenomerga* (Yavorskaya et al. 2015), which is missing in all other adult or immature beetles examined (e.g. Beutel 1986a, b, 1993, 1994, 1999;

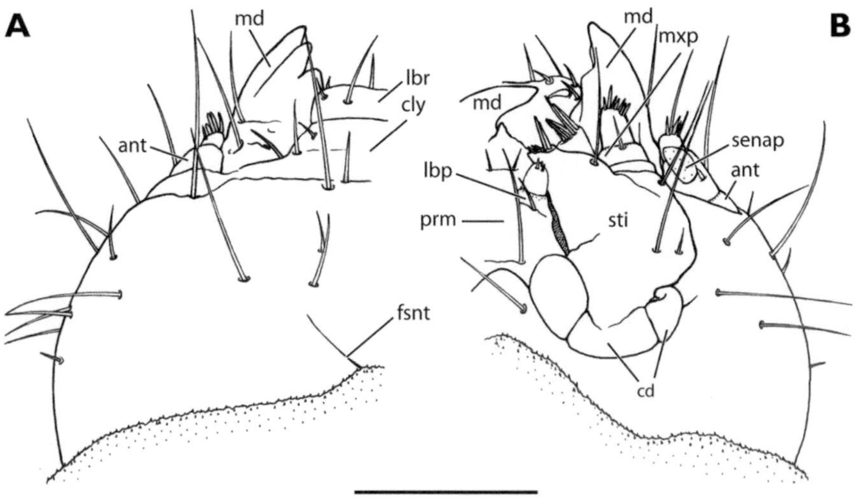


**Fig. 12.6** Mouthparts of larvae of Archostemata: *Tenomerga mucida*; (**a**) head, dorsal view; (**b**) head, ventral view. Abbreviations: *ant* antenna, *ca* cardo, *cl* clypeus, *fsnt* frontal suture, *lbr* labrum, *lbp* labial palp, *md* mandible, *mp* maxillary palp, *prm* prementum, *senap* sensorial appendage, *st* stipes. Scale bar: 0.1 mm. Modified from Yavorskaya et al. (2015)

Beutel et al. 2008; Dressler and Beutel 2010). The mandibles are compact, roughly triangular and articulated in a typical dicondylic manner (Figs. 12.6a, b). The grinding mola is well-developed and enclosed by a distinct quadrangular border in larvae of *Micromalthus* and later instars of Cupedidae (Beutel and Hörnschemeyer 2002a, b). A prostheca and mesal brushes of hairs are missing, but the apical teeth are strongly developed. The maxillae are inserted in deep maxillary grooves. The more or less transverse short cardo and the undivided stipes are connected by a hinge. The four-segmented palp is inserted on a small palpifer. The galea and the hook-shaped lacinia are distinctly separated (Fig. 12.6b). The proximal labial elements are fused into an undivided postmentum, which is strongly narrowed between the maxillary bases. The well-developed prementum bears two-segmented palps. A conspicuous, sclerotized wedge-shaped ligula is an autapomorphy of the suborder and obviously suitable for burrowing in wood (Beutel and Hörnschemeyer 2002a, b; Yavorskaya et al. 2015).

### *12.4.2 Myxophaga*

The larval and adult mouthparts are similar in their basic configuration. The labrum is large and like in the adults largely conceals the paired mouthparts, especially in Sphaeriusidae and Hydroscaphidae, which are characterized by a condition described as semi-entognathous (Beutel et al. 1998). The mandibles are broad at their base and equipped with a grinding mola. Distad to it, a lobe- or pad-like semimembranous prostheca is present. The distal mandibular part is slender. The maxillae are similar to those of the adults, with only a single sclerotized endite lobe, which is likely homologous with the lacinia. The premental ligula bears papillae as they are also present on the adult prementum.

### *12.4.3 Adephaga*

The larval mouthparts of Adephaga (Fig. 12.7) are distinctly modified compared to those of the adults, likely correlated with advanced predacious habits, distinct prognathism and extraoral digestion (e.g. Luff 1974; Tröster 1987; Beutel 1991a, 1993; Arndt and Beutel 1994, 1995). The labrum is always fused with the clypeus and forms a robust anterior edge of the head capsule (Fig. 12.7a). A nasale with four teeth as it is found in most Gyrinidae, Trachypachidae and some Carabidae (e.g. Beutel 1993) is likely ancestral for the suborder. Labral muscles are always absent. The mandibles are relatively stout in Trachypachidae, Haliplidae, Noteridae and Amphizoidae but transformed in long and sickle-shaped structures in most groups (Fig. 12.7a, b) (e.g. Bertrand 1972; Beutel 1986b, 1991b; Arndt and Beutel 1995). A mola is always absent, and a movable prostheca or lacinia mobilis does only occur in the specialized larvae of Paussinae (Arndt and Beutel 1994). A

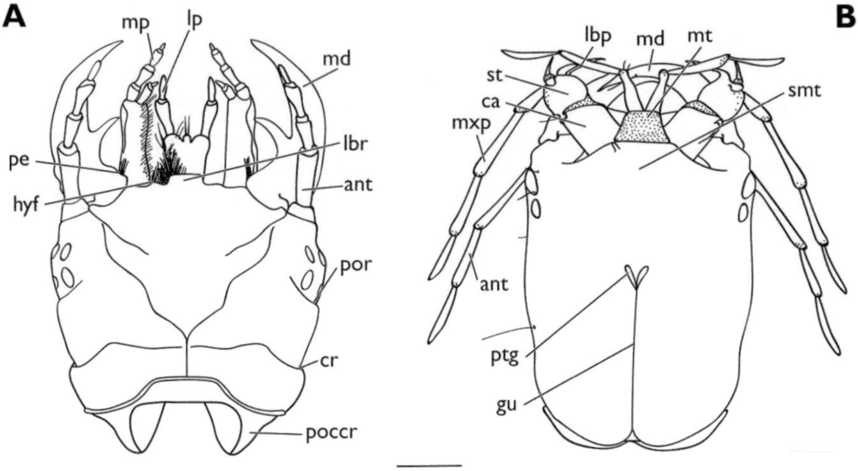


**Fig. 12.7** Mouthparts of larvae of Adephaga. (**a**) *Broscus cephalotes* (L3) (Carabidae), dorsal view; (**b**) *Orectogyrus* sp. (L3) (Gyrinidae), dorsal view. Abbreviations: *ant* antenna, *ca* cardo, *gu* gula, *hyf* hypopharyngeal fringe, *lp, lbp* labial palp, *md* mandible, *mp* maxillary palp, *pe* penicillium, *poccr* postoccipital ridge, *por* postocular ridge, *smt* submentum, *st* stipes. Scale bar: 1 mm. Redrawn from Beutel (1993)

sclerotized curved retinaculum is present in the groundplan. It is almost generally well-developed in Geadephaga (Fig. 12.7a) (Thompson 1979) but partly or completely reduced in the aquatic groups with the exception of Gyrinini (Noars 1956; Beutel 1993; Beutel and Roughley 1994). A basal brush-like group of hairs, the penicillium, is present in most groups of Carabidae (Fig. 12.7a) but missing in the groundplan of the family (Thompson 1979; Beutel 1991a) like in Trachypachidae and the aquatic groups. A single mesal cutting edge is present in Carabidae and the aquatic Hygrobiidae, whereas two cutting edges enclose a more or less deep mesal groove in Trachypachidae, Amphizoidae and most Noteridae (Ruhnau 1986; Beutel 1993; Alarie et al. 2004). A mandibular sucking channel has evolved independently in Gyrinidae (Fig. 12.7b), Haliplidae, Dytiscidae (excl. Copelatini and *Hydrotrupes*) and two genera of Noteridae (Beutel 1993; Beutel et al. 2006). In Adephaga excluding Gyrinidae, the maxillae of larvae are inserted at the anteroventral margin of the head capsule (protracted position) and movable in all directions. The maxillary fossa is largely or completely reduced. The maxilla is more or less transformed into an accessory ventral tactile organ. The cardo is small in all groups except for Gyrinidae and completely reduced or fused with the stipes in Hygrobiidae. It is subdivided into a small mesal and a transverse lateral sclerite in almost all groups of Carabidae (Beutel 1993). A hinge between cardo and stipes is only preserved in larvae of Gyrinidae (Fig. 12.7b). The stipes is relatively short in gyrinid larvae but more or less elongated in the remaining groups, often more or less tube-like. A cranial muscle dorsally attached to the stipital base, a levator of the maxilla, is likely a synapomorphy of Adephaga excl. Gyrinidae. The palp is four-segmented and inserted apicolaterally on the stipes. An elongated and movable hook-like lacinia with M. craniolacinialis attached to its base is only present in Gyrinidae (Fig. 12.7b) (Noars 1956; Beutel and Roughley 1994). It is still elongated and hook-like in Metriinae and Omophronini among Carabidae (Beutel 1991a, 1992a), but strongly reduced in the other groups and completely missing in many carabid larvae (Thompson 1979) and also in the aquatic groups, for instance, in Noteridae and Hygrobiidae (Beutel 1993; Alarie et al. 2004). The galea is usually peg-like. The submentum is completely fused with the ventral capsule and usually divided by a median ecdysial suture. The mentum is a transverse membranous area anteriorly connected with the prementum, which bears two-segmented labial palps. An anteromedian semimembranous ligula is usually present in Geadephaga (Thompson 1979). It can be elongated (e.g. Omophronini) but is usually short and apically rounded (Beutel 1991a, 1993).

### *12.4.4 Polyphaga*

The strong variation of the larval mouthparts in Polyphaga (Fig. 12.8) reflects the extremely broad food spectrum of the immature stages, ranging from saprophagous, sporophagous, fungivorous and algophagous feeding habits (e.g. Scirtoidea, most staphylinoid groups) (e.g. Beutel and Molenda 1997; Beutel and Haas 1998; Grebennikov and Beutel 2002) to specialized forms of carnivory (e.g. Hydrophilidae (Fig. 12.8a), Cantharoidea: Lampyridae (Fig. 12.8b), Lycidae, etc.) (e.g. Beutel 1995, 1999; Ge et al. 2012) or feeding on faeces (e.g. Geotrupidae, Aphodiinae), carrion (Silphidae, Trogidae, Dermestidae), wood (e.g. Buprestidae, Cerambycidae) or leaves or other living plant materials (especially Chrysomelidae and Curculionidae) (e.g. Beutel and Leschen 2016; Leschen et al. 2010; Leschen and Beutel 2014).

The ancestral condition in Polyphaga is likely similar to what is found in larvae of Myxophaga (Beutel and Haas 1998) and is mainly preserved in subgroups of Scirtoidea and Staphylinoidea (e.g. Hydraenidae, Leiodidae, basal groups of Staphylinidae; e.g. Newton 1991; Beutel and Molenda 1997; Beutel and Leschen 2005b). This includes a free labrum, and mandibles broad at base and equipped with a well-developed mola and often with a slender prostheca. The maxillae are inserted in a distinct maxillary fossa (retracted position), and cardo and stipes are connected by a hinge. Galea and a hook-shaped lacinia are present as distinctly separated endite lobes, the former often fimbriate and the latter usually equipped with mesal rows of spines. The posterior labium is subdivided into a submentum and mentum. The retractile prementum bears the two-segmented palp and a variously developed ligula (Newton 1991; Beutel and Molenda 1997; Beutel and Leschen 2005b).

A feature occurring in different polyphagan lineages is the presence of an accessory ventral process of the mandible, also referred to as crushing tubercle, or erroneously as accessory condyle (see Lawrence et al. 2011). It occurs in Nosodendridae (Beutel 1996), Derodontidae and Dascillidae, in most Scarabaeoidea and Scirtoidea, in a number of groups of Cucujoidea, and also in the myxophagan

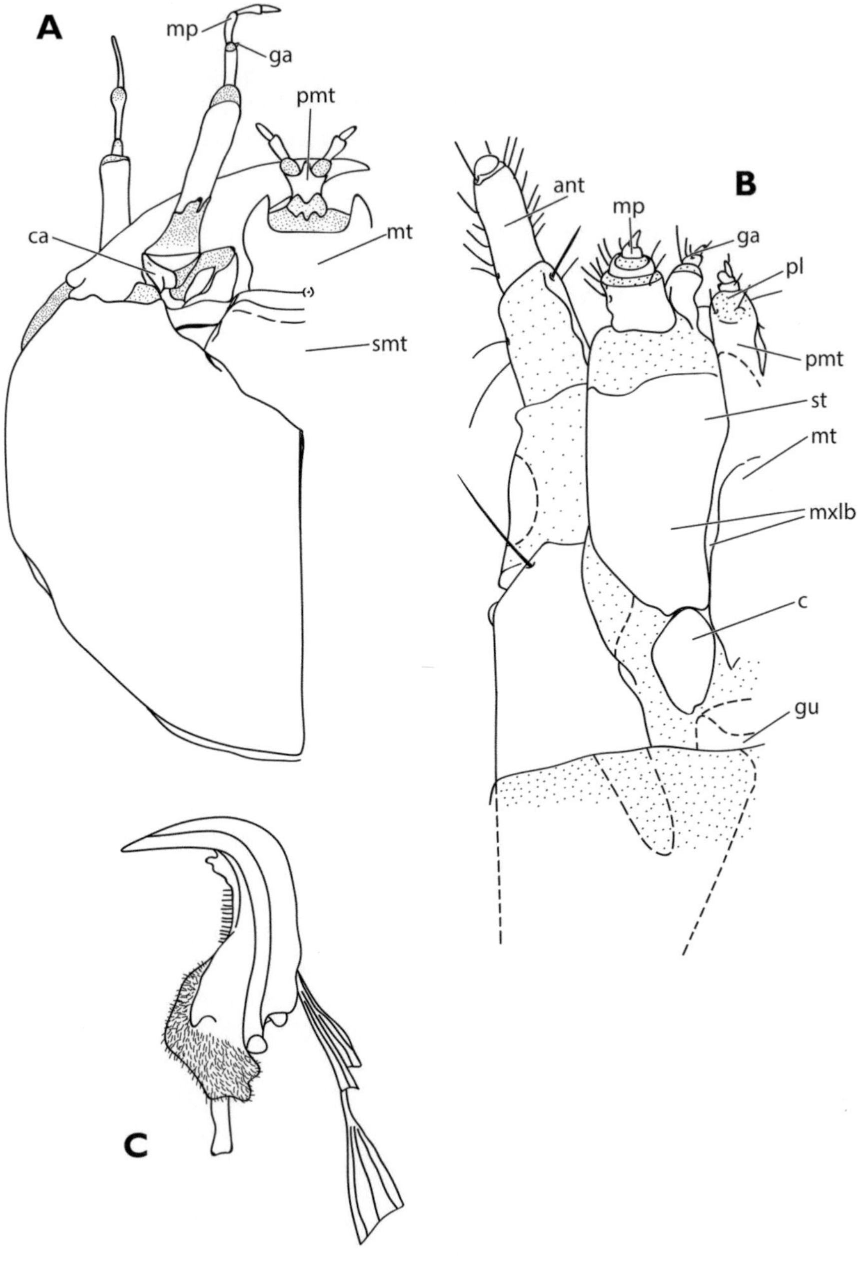


**Fig. 12.8** Mouthparts of specialized larva of Polyphaga. (**a**) *Hydrophilus piceus* (L3) (Hydrophilidae), ventral view of head; (**b**) *Phosphaena* sp. (L3) (Lampyridae), ventral view of head; (**c**) *Phosphaena* sp., mandible. Abbreviations: *ant* antenna, *c* cardo, *ga* galea, *gu* gula, *mp* maxillary palp, *mt* mentum, *mxlb* maxillolabial complex, *pl* labial palp, *pmt* prementum, *smt* submentum, *st* stipes. Redrawn from Beutel (1995) and Gorb and Beutel (2000)

genus *Hydroscapha* (Lawrence et al. 2011). This structure interacts with sclerotizations of the anterior hypopharynx in the process of crushing food particles.

## 12.4.5 *Specialized Mouthparts in Larval Polyphaga*

### 12.4.5.1 Grinding and Pressing Feeding Apparatus of Larval Scirtidae

The exceptionally complicated mouthparts of the larvae of Scirtidae were described in detail by Beier (1949) and Hannappel and Paulus (1987). The clypeolabrum, a large shield-shaped structure, is movably connected with the frons (Beier 1949). It bears a dense fringe of long setae along its deeply emarginated anterior margin. The epipharynx below the ventral labral lobe (Fig. 12.9) bears a very complex vestiture of different structural elements (Beier 1949). It comprises a transverse row of cuticular teeth, a transverse row of cupola-shaped sensilla, a dense field of long microtrichia, a plate with folded lamellae, and a structure referred to as pad plate by Beier (1949). A posteriorly directed median rod, a smooth and strongly sclerotized structure, mainly interacts with the molar parts of the mandible and the sclerotized tooth plate of the hypopharynx (Beier 1949). The distinctly asymmetric, roughly triangular mandibles bear very dense mesally directed fringes of long and distally split hairs, a sub-basal trapezoid prostheca with a row of stiff bristles forming a comb-like structure (Beier 1949), and a prominent sclerotized mola with a smooth ventral side and with very fine, serrated transverse lamellae on its dorsal surface. Additionally, an accessory ventral process is present proximad the mola, referred to as ventral hook by Beier (1949). The molae interact mechanically with the epipharyngeal rod and the tooth plate of the hypopharynx. The ventral hook stabilizes the feeding apparatus in a certain stage of the process of compacting food particles (Beier 1949). The maxillary lacinia and galea are largely fused. The large and rounded galea bears several dense rows of comb hairs (Beier 1949; Hannappel and Paulus 1987). The hypopharynx is even more complex than the epipharynx. In contrast to other groups of beetles, the hypopharynx of larvae (and adults) is not firmly fused with the prelabium but can be easily detached from it. A complex suspensorium connects it with the ventral wall of the anteriormost pharynx and with the posterior epipharynx (Fig. 12.9). The main part of the hypopharynx is formed by the median unpaired lingua and the paired lateral superlinguae (Beier 1949). The armature comprises paired oblique comb plates with stiff hairs, rows of teeth along rounded mesally directed plate-like structures, a claw apparatus (Beier 1949), an anteromedian pair of tooth bristles, a median pad, a median field of bristles and the posteromedian tooth plate of the pressing apparatus. Various parts are covered with fine microtrichia or rows of minute tooth-like structures, for instance, the surface of the superlinguae. The mentum and prementum form a large undivided prelabium. The anterolaterally inserted two-segmented palps are very small in relation to the entire structure. Well-developed salivary glands are present in the prelabial region. The ingestion of the aquatic detritophagous larvae takes place in four phases. This

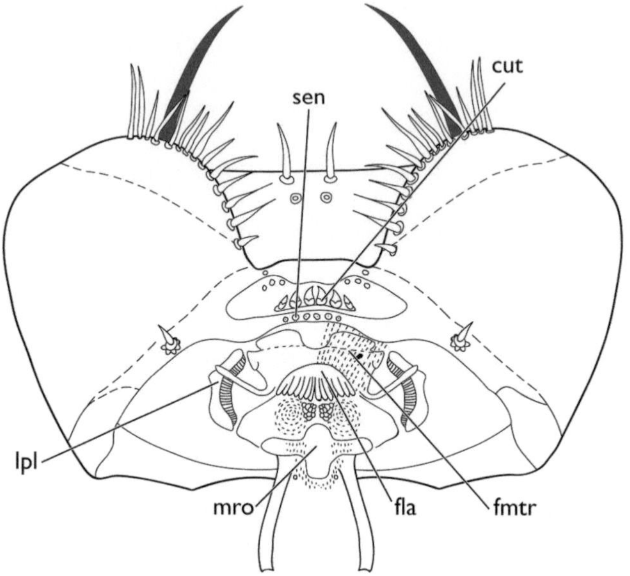


**Fig. 12.9** General structure of the clypeolabrum and epipharynx (Helodidae). Abbreviations: *cut* cuticular teeth, *fla* field of lamella, *fmtr* field of microtrichia, *lpl* lateral plate, *mro* median rod, *sen* sensilla. Redrawn from Hannappel and Paulus (1987)

process was described in detail by Beier (1949, 1952; see also Hannappel and Paulus [1987]): (1) collecting fine particles from plant surfaces and bottom substrate, (2) deposition on the collection apparatus, (3) transfer to the processing area and (4) compaction to form solid pellets for transport into the gut.

#### 12.4.5.2 Mouthparts Adapted for Predacious Habits

Modifications linked with predacious habits have evolved independently in different groups, for instance, in Hydrophiloidea (Fig. 12.8a), in "higher" Staphylinidae (e.g. Beutel 1994, 1999; Beutel and Molenda 1997; Archangelsky et al. 2016), in Cantharidae and related groups (Fig. 12.8b: Lampyridae) (Beutel 1995) and in the cucujiform Cleroidea (e.g. Kolibáč 2010; Beutel and Pollock 2000). Some of the derived features are similar to conditions found in larval Adephaga, like the fusion of the labrum with the clypeus, falcate mandibles without molae, the reduced lacinia, and the increased movability of the maxilla, which resembles a ventrally inserted accessory larval antenna (Beutel 1994, 1999). Like in Adephaga (excl. Gyrinidae), the intrinsic movability of the maxilla is restricted in most groups of Hydrophiloidea. A condition typical for Hydrophiloidea (excl. Spercheidae) is the subdivision of the cardo into several sclerites (Fig. 12.8a) (Beutel 1994, 1999; Archangelsky 1998, 1999). A characteristic feature of larvae of the family Hydrophilidae is hyperprognathism, with the mouthparts directed upwards and asymmetric mandibles (e.g. Gorb and Beutel 2000). These conditions are likely correlated with feeding on snails in an aquatic environment (Gorb and Beutel 2000; Archangelsky et al. 2016). A unique type of tripartite longitudinally split mandibles occurs in larvae of Lycidae (e.g. Beutel 1995; Kazantsev 2006), which may be partly carnivorous but probably mainly feed on fermenting juices (Bocak and Bocakova 2010).

#### 12.4.5.3 Maxillolabial Complex

A distinct functional and structural modification of the ventral mouthparts has evolved at least two times independently in Polyphaga, a maxillolabial complex (Fig. 12.8b). This functional unit formed by the maxillae and labium occurs in the majority of families of Elateriformia (Beutel 1995) and in Cleroidea (Beutel and Pollock 2000). The ventral mouthparts are closely connected and move like a shovel in vertical direction. The lateral movability of the maxillae is largely or completely restricted. The tentorio-cardinal and tentorio-stipitalis muscles are almost always vertically oriented in elateriform larvae and arise on strongly developed, modified posterior tentorial arms. The origin of the extrinsic maxillary muscles is shifted to the hypopharyngeal floor of the elongated prepharynx in Omalisidae, Lycidae, Drilidae and Lampyridae (and possibly in related elateroid groups), thus increasing the efficiency of the preoral pumping apparatus (Vogel 1915; Beutel 1995).

#### 12.4.5.4 Mouthparts Adapted for Feeding on Fresh Plant Material

The larval mouthparts of phytophagous groups like Chrysomelidae are similar to those of adults but appear slightly simplified (e.g. Oberprieler 2014; Reid 2014; Ge et al. 2015; Hua et al. 2015). The robust mandibles lack molae, and the distal part usually forms a palmate structure. Like in other cucujiform larvae, the endite lobes of the maxilla form an undivided mala (e.g. Lawrence et al. 2011).

#### 12.4.5.5 Larval Rostrum

A very unusual modification is the formation of a snoutlike rostrum in larvae of the corylophid genus *Holopsis*. Like in adults of Curculionoidea, small mouthparts are inserted on the apical part of the elongated part of the head. In the case of *Holopsis* larvae, the rostrum, which differs distinctly in length in both examined species, is apparently an adaptation to basidiomycete fungi (Yavorskaya et al. 2014).

## 12.5 Evolutionary Trends in Coleoptera

Even though the potential to process many different food substrates is likely linked with the enormous diversification of Coleoptera, the mouthparts are surprisingly conservative and almost always of the biting type in larvae and adults, with slightly more variation in the immature stages than in the adult beetles.

The basal branching events, seemingly crucial for tracing the evolutionary changes of larval and adult mouthparts, are not clarified yet (Fig. 12.10), with different data sets suggesting different phylogenetic concepts. Analyses of extensive morphological character sets suggest a branching pattern with Archostemata as sister to a clade comprising Adephaga, Myxophaga and Polyphaga, the latter two forming sister taxa (Beutel and Haas 2000; Beutel et al. 2008, 2018; Friedrich et al. 2009). In contrast to this, studies based on characters of the wing base and wing venation (Kukalová-Peck and Lawrence 1993, 2004) and recent phylogenetic results based on molecular data (e.g. McKenna et al. 2015; Zhang et al. 2018) support a sister group relationship between Polyphaga and the other three coleopteran suborders.

Even though the two main alternative phylogenies differ strongly, the distribution of the major types of mouthparts among Polyphaga and Myxophaga suggests that an adult feeding apparatus suitable for saprophagous habits, as it is found in myxophagans and basal polyphagan groups (e.g. Yavorskaya et al. 2018a), is ancestral for crown group Coleoptera (Figs. 12.5 and 12.11). The alternative interpretation that a very complex and specific configuration of the mouthparts has evolved independently in myxophagans and in polyphagan groups appears much less plausible. The entire apparatus includes mandibles of adults and larvae with well-developed molae and prosthecae and in the adults maxillary endite lobes

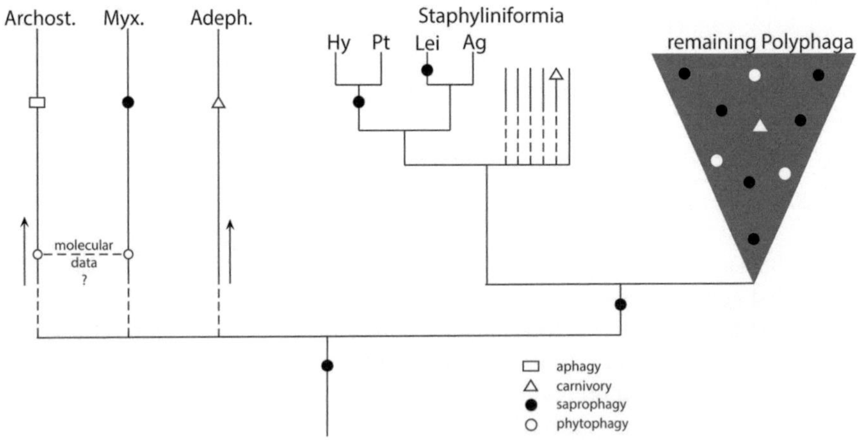


**Fig. 12.10** Phylogenetic tree showing evolutionary transformations of feeding habits. Interrelationships of suborder depicted as unresolved due to conflicting hypotheses based on morphological (e.g. Beutel and Haas 2000; Beutel et al. 2008, 2018) and molecular data sets (e.g. McKenna et al. 2015; Zhang et al. 2018). Abbreviations: *Adeph* Adephaga, *Ag* Agritidae, *Archost* Archostemata, *Hy* Hydraenidae, *Lei* Leiodidae, *Myx* Myxophaga, *Pt* Ptiliidae

suitable for detaching and handling small particles or soft substrates and hairy longitudinal epipharyngeal and hypopharyngeal lobes (Fig. 12.5). This pattern is found in all four families of Myxophaga with minor modifications and also in different polyphagan lineages, especially in Scirtoidea and Staphyliniformia (e.g. Anton and Beutel 2004, 2006, 2012; Anton et al. 2016; Antunes-Carvalho et al. 2017; Yavorskaya et al. 2018a), but also in groups of the elateriform and cucujiform series. The configuration of structures involved in feeding is usually slightly less complex in the larval stages (e.g. Beutel and Molenda 1997; Beutel and Haas 1998), even though an extremely complicated larval feeding apparatus has evolved in Scirtidae (Beier 1949).

The mouthparts of Permian stem group beetles (e.g. †Tshecardocoleidae, †Moravocoleidae, †Permocupedidae) are very insufficiently known (e.g. Ponomarenko 1969; Kukalová-Peck and Beutel 2012). However, the mandibles were definitely more compact than in extant Cupedidae and Ommatidae (Ponomarenko 1969). It is conceivable that they were close to the groundplan condition postulated for crown group Coleoptera in this chapter, even though any detailed evidence is presently lacking. A close link to narrow spaces under bark (e.g. Crowson 1975; Beutel 1997) tentatively suggests that the earliest beetles may have been feeding on fungal material, which would be consistent with a feeding apparatus similar to that of extant Myxophaga and different extant groups of Polyphaga (Figs. 12.2, 12.5 and 12.11).

The presumably ancestral feeding apparatus is largely maintained in the small suborder Myxophaga, with algophagous habits maintained in the vast majority of species (e.g. Reichardt 1973; Beutel and Arce-Pérez 2016). In contrast, shifts of diet took place in different groups of the megadiverse Polyphaga, for instance, carnivorous habits in subgroups of Staphylinidae (Thayer 2016), in Lampyridae (females) and in Cleridae with few exceptions (Kolibáč 2010). Highly specialized forms drilling holes in strongly armoured mites with their mandibles occur in Scydmaeninae, and Steninae have evolved an extrusible elongated labium as a catching device, analogous to the one of immatures of Odonata, but structurally distinctly different. Predacious habits have also evolved in larvae of staphyliniform subgroups. Adults of Hydrophiloidea have maintained saprophagous habits and the ancestral feeding apparatus (e.g. Anton and Beutel 2004), whereas the larvae are predators with strongly modified mouthparts (Fig. 12.8a), quite similar to those of immatures of Adephaga (Beutel 1993, 1999). Larvae of the primarily aquatic Hydrophilidae are specialized on snails as prey, with very robust asymmetric mandibles adapted to crush the shell (Archangelsky 1998, 1999; Beutel 1999; Gorb and Beutel 2000) and with a mandibular sucking channel in species of the genus *Berosus* (Böving and Henriksen 1938).

Aside from predacious habits, specializations on numerous different substrates have evolved in Polyphaga, including faeces (e.g. sphaeridiine Hydrophilidae [partim], Geotrupidae, Aphodiinae) or fresh (e.g. Silphidae) or dry cadavers (Dermestidae) (e.g. Beutel and Leschen 2016). The major shift occurred in the megadiverse Phytophaga (Chrysomeloidea + Curculionoidea), with fresh plant material as predominant or exclusive diet. The typical modifications in Chrysomelidae and closely related families are the formation a shovel-like distal part of the mandibles with several flattened teeth and the reduction of the mola

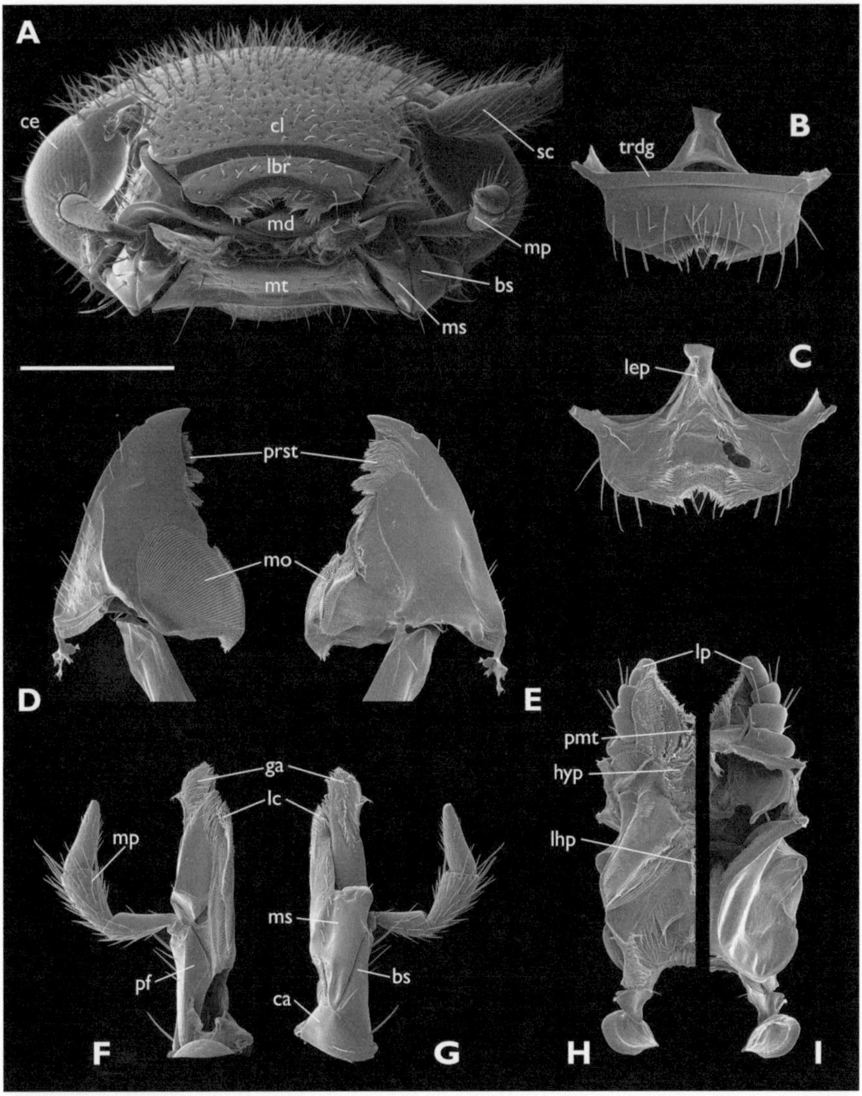


**Fig. 12.11** The head and mouthparts of *Catops* sp., scanning electron microscopic images; (**a**) head, frontal view; (**b**) dorsal surface of labrum; (**c**) ventral surface of labrum and epipharynx, ventral view; (**d**) dorsal view of the left mandible; (**e**) ventral view of left mandible; (**f**) dorsal view of the left maxilla; (**g**) ventral view of the left maxilla; (**h**) dorsal surface of hypopharynx-prementum complex; (**i**) ventral surface of hypopharynx-prementum complex. Abbreviations: *bs* basistipes, *ca* cardo, *ce* compound eye, *cl* clypeus, *ga* galea, *hyp* hypopharynx, *lbr* labrum, *lep* longitudinal epipharyngeal process, *lhp* longitudinal hypopharyngeal process, *lp* labial palp, *md* mandible, *mo* molae, *mp* maxillary palp, *ms* mediostipes, *mt* mentum, *pf* palpifer, *pmt* prementum, *prst* prostheca, *sc* scapus, *trdg* transverse dorsal ridge. Scale bar: 0.2 mm. Modified from Antunes-Carvalho et al. (2017)

(e.g. Reid 2014). The formation of a snout-like rostrum with small mouthparts at the apical part characterizes Curculionoidea, linked with the habit of inserting their eggs in live plant tissue.

Even though sometimes addressed as ancestral, Archostemata are clearly derived in terms of adult mouthparts (Fig. 12.1). The simplified structures involved in feeding in adults of the small relict taxon, especially in the monotypic families Crowsoniellidae and Micromalthidae (Pace 1975; Yavorskaya et al. 2018b), clearly reflect a tendency towards a reduced (or even absent) role of food uptake in the adult stage. This is apparently linked with a short adult life focused on reproduction in most cases and a prolonged postembryonic development (e.g. Hörnschemeyer 2005) with mouthparts and other structures of larvae modified for burrowing in rotting trees and processing wood infested with microorganisms (Fig. 12.6) (Beutel and Hörnschemeyer 2002a, b; Yavorskaya et al. 2015). Adults of Micromalthidae, which are characterized by an extremely complicated life cycle (e.g. Hörnschemeyer 2005; Perotti et al. 2016), are interpreted as relicts, occurring only sporadically and likely non-feeding. Their lacking involvement in reproduction was referred to as a ghost sex-life by Perotti et al. (2016).

Adephaga, the second largest suborder of beetles, have shifted to predacious habits, with only few secondary exceptions, for instance, in the aquatic Haliplidae, which mainly feed on algae as adults and exclusively as larvae (Seeger 1971; Beutel 1986b), or in Rhysodini (Carabidae or Rhysodidae), which are likely specialized on slime moulds (Myxomycetes) as larvae (Beutel 2016), with highly modified larval mouthparts (Beutel 1992b). To avoid loss or dilution of digestive fluids, the aquatic groups have evolved mandibular sucking channels several times independently (Fig. 12.7b) (e.g. Beutel 1986b, 1993, 1997). A conspicuous feature of almost all adephagan larvae is maxillae rather resembling accessory ventral antennae than mouthparts. Larvae of the extremely species-rich Carabidae are characterized by an increasing trend to preoral digestion, with an increasingly flattened hypopharynx and a dense preoral brush of long microtrichia, which create capillary forces and prevent particles from entering the narrow and elongated prepharynx (e.g. Beutel 1991a, 1992a, c). In adults of Adephaga, the mouthparts are largely conservative, with relatively compact mandibles in the aquatic groups and the very small family Trachypachidae (Fig. 12.3) (e.g. Beutel 1986a; Beutel et al. 2017; Dressler and Beutel 2010) and a tendency to elongation in Carabidae (e.g. Acorn and Ball 1991; Arndt et al. 2016). Secondary herbivory has also evolved in ground beetles (Acorn and Ball 1991). However, these shifts only played a minor role in this very successful family.

As a whole, the mouthparts of Coleoptera have largely maintained a plesiomorphic pattern, almost always with biting mandibles and maxillae for scraping food particles, manipulating them in the preoral space and pushing them toward the functional mouth opening. The ability to adapt to an immensely broad range of food substrates (e.g. Crowson 1981; Beutel and Leschen 2016; Leschen et al. 2010; Leschen and Beutel 2014) with minor variation of the feeding apparatus is likely one factor that contributed to the exceptionally successful diversification in insects.

## References


Acorn JH, Ball GE (1991) The mandibles of some adult ground beetles: structure, function, and the evolution of herbivory (Coleoptera: Carabidae). Can J Zool 69(3):638–650

Alarie Y, Beutel RG, Watts CHS (2004) Larval morphology of three species of Hygrobiidae (Coleoptera: Adephaga: Dytiscoidea) with phylogenetic considerations. Eur J Entomol 101:293–311

Anton E, Beutel RG (2004) On the head morphology and systematic position of *Helophorus* (Coleoptera: Hydrophiloidea: Helophoridae). Zool Anz 242:313–346

Anton E, Beutel RG (2006) On the head morphology of Lepiceridae (Coleoptera: Myxophaga) and the systematic position of the family and suborder. Eur J Entomol 103:85–95

Anton E, Beutel RG (2012) The head morphology of *Dascillus* (L.) (Dascilloidea: Dascillidae) and *Glaresis* Erichson (Scarabeoidea: Glaresidae) and its phylogenetic implications. Arthropod Syst Phylogeny 70(1):3–42

Anton E, Yavorskaya MI, Beutel RG (2016) The head morphology of adults of Clambidae (Coleoptera: Polyphaga: Scirtoidea). J Morphol 277:615–633

Antunes-Carvalho C, Yavorskaya M, Gnaspini P, Ribera I, Beutel RG (2017) Cephalic anatomy and three-dimensional reconstruction of the head of *Catops ventricosus* (Weise, 1877) (Coleoptera: Leiodidae: Cholevinae). Org Divers Evol 17(1):199–212

Archangelsky M (1998) Phylogeny of Hydrophiloidea (Coleoptera: Staphyliniformia) using characters from adult and preimaginal stages. Syst Entomol 23:9–24

Archangelsky M (1999) Adaptations of immature stages of Sphaeridiinae (Staphyliniformia, Hydrophiloidea: Hydrophilidae) and state of knowledge of preimaginal Hydrophilidae. Coleopt Bull 53(1):64–79

Archangelsky M, Beutel RG, Komarek A (2016) Hydrophiloidea. In: Beutel RG, Leschen RAB (eds) Handbook of zoology, Vol. IV Arthropoda: Insecta. Part 38. Coleoptera, Vol. 1, 2nd edn: Morphology and systematics (Archostemata, Adephaga, Myxophaga, Polyphaga (partim)). Walter De Gruyter, Berlin, pp 231–372

Arndt E, Beutel RG (1994) Descriptions of the larvae of *Paussus* (*Klugipaussus*) aff. *distinguendus* Reichensberger and *P.* (*Lineatopaussus*) *afzelii* Westwood (Carabidae) with phylogenetic considerations. Elytron 8:129–139

Arndt E, Beutel RG (1995) Larval morphology of *Systolosoma* Solier and *Trachypachus* Motschulsky (Coleoptera: Trachypachidae) with phylogenetic consideration. Entomol Scand 26:439–446

Arndt E, Beutel RG, Kipling W (2016) Carabidae Latreille, 1802. In: Beutel RG, Leschen RAB (eds) Handbook of zoology, Vol. IV Arthropoda: Insecta. Part 38. Coleoptera, Vol. 1, 2nd edn: Morphology and systematics (Archostemata, Adephaga, Myxophaga, Polyphaga (partim)). Walter De Gruyter, Berlin, pp 162–197

Ball GE, Acorn JH, Shpeley D (2011) Mandibles and labrum-epipharynx of tiger beetles: basic structure and evolution (Coleoptera, Carabidae, Cicindelitae). ZooKeys 147:39

Batelka J (2011) *Olemehlia krali*, a new genus and species of Ripidiinae from Baltic amber (Coleoptera: Ripiphoridae). Annl Soc Entomol Fr 47(3–4):361–364

Bauer T, Pfeiffer M (1991) 'Shooting' springtails with a sticky rod: the flexible hunting behaviour of *Stenus comma* (Coleoptera; Staphylinidae) and the counter-strategies of its prey. Anim Behav 41:819–828

Beier M (1949) Körperbau und Lebensweise der Larve von *Helodes hausmanni* Gredler (Col. Helodidae). Eos 25:49–100

Beier M (1952) Structure and function of the mouthparts in helodid larvae (Col.). Trans Int Congr Entomol 9(1):135–138

Belkaceme T (1991) Skelet und Muskulatur des Kopfes und Thorax von *Noterus laevis* Sturm. Ein Beitrag zur Morphologie und Phylogenie der Noteridae (Coleoptera: Adephaga). Stuttg Beitr Naturk (A) 462:1–94

Bertrand HPI (1972) Larves et Nymphes des Coléoptères Aquatiques du Globe. F. Paillart, Paris

Besuchet C (1972) Les Coléoptères Aculagnathides. Rev Suisse Zool 79:99–145
Betz O (1996) Function and evolution of the adhesion-capture apparatus of *Stenus* species (Coleoptera, Staphylinidae). Zoomorphology 116:15–34
Betz O (1998) Comparative studies on the predatory behaviour of *Stenus* spp. (Coleoptera: Staphylinidae): the significance of its specialized labial apparatus. J Zool 244(4):527–544
Betz O, Koerner L, Dettner K (2018) The biology of Steninae. In: Betz O, Irmler U, Klimaszewski J (eds) Biology of rove beetles (Staphylinidae). Springer, Cham, pp 229–283
Beutel R (1986a) Skelet und Muskulatur des Kopfes und Thorax von *Hygrobia tarda* (Herbst). Ein Beitrag zur Klärung der phylogenetischen Beziehungen der Hydradephaga (Insecta: Coleoptera). Stuttg Beitr Naturk (A) 388:1–54
Beutel R (1986b) Skelet und Muskulatur des Kopfes der Larve von *Haliplus lineatocollis* Mrsh. (Coleoptera). Stuttg Beitr Naturk (A) 390:1–15
Beutel RG (1989) The head of *Spanglerogyrus albiventris* Folkerts (Coleoptera: Gyrinidae). Contribution towards clarification of the phylogeny of Gyrinidae and Adephaga. Zool Jb Anat 118:431–461
Beutel RG (1991a) Larval head structures of *Omophron* and their implications for the relationships of Omophronini (Coleoptera: Carabidae). Entomol Scand 22:55–67
Beutel RG (1991b) Internal and external structures of the head of 3rd instar larvae of *Amphizoa lecontei* Matthews (Coleoptera: Amphizoidae). A contribution towards clarification of the systematic position of Amphizoidae. Stuttg Beitr Naturk (A) 469:1–24
Beutel RG (1992a) Study on the systematic position of Metriini (Coleoptera, Carabidae) based on characters of the larval head. Syst Entomol 17:207–218
Beutel RG (1992b) On the morphology of the head of the larva of *Omoglymmius hamatus* (LeConte) and the systematic position of Rhysodidae (Coleoptera, Adephaga). Entomol Scand 23:169–184
Beutel RG (1992c) External and internal structures of the head of 3rd instar larvae of *Carabus coriaceus* F. (Adephaga: Carabidae). Entomologica Basiliensia 15:147–168
Beutel RG (1993) Phylogenetic analysis of Adephaga (Coleoptera) based on characters of the larval head. Syst Entomol 18:127–147
Beutel RG (1994) Phylogenetic analysis of Hydrophiloidea (Coleoptera: Polyphaga: Staphyliniformia) based on characters of the head of adults and larvae. Koleopt Rdsch 64:103–131
Beutel RG (1995) Phylogenetic analysis of Elateriformia (Coleoptera: Polyphaga) based on larval characters. J Zool Syst Evol Res 33:145–171
Beutel RG (1996) Study of the larva of *Nosodendron fasciculare* (Olivier 1790) (Coleoptera: Nosodendridae) with implications to the phylogeny of Bostrichiformia. J Zool Syst Evol Res 34:121–134
Beutel RG (1997) Über Phylogenese und Evolution der Coleoptera (Insecta), insbesondere der Adephaga. Verh Naturwiss Ver Hamburg NF 31:1–164
Beutel RG (1999) Morphology and evolution of the larval head structures of Hydrophiloidea and Histeroidea (Coleoptera: Staphylinidae). Tijdschr Entomol 142:9–30
Beutel RG (2016) Family Rhysodidae Laporte, 1840 (=Rhysodini S. Bell). In: Beutel RG, Leschen RAB (eds) Handbook of zoology, Vol. IV Arthropoda: Insecta. Part 38. Coleoptera, Vol. 1, 2nd edn: Morphology and systematics (Archostemata, Adephaga, Myxophaga, Polyphaga (partim)). Walter De Gruyter, Berlin, pp 191–197
Beutel RG, Arce-Pérez R (2016) Sphaeriusidae Erichson, 1845 (Jäch 1999) (=Microsporidae). In: Beutel RG, Leschen RAB (eds) Handbook of zoology, Vol. IV Arthropoda: Insecta. Part 38. Coleoptera, Vol. 1, 2nd edn: Morphology and systematics (Archostemata, Adephaga, Myxophaga, Polyphaga (partim)). Walter De Gruyter, Berlin, pp 70–71
Beutel RG, Haas A (1998) Larval head morphology of *Hydroscapha natans* LeConte, 1874 (Coleoptera, Myxophaga, Hydroscaphidae) with special reference to miniaturization. Zoomorphology 18:103–116

Beutel RG, Haas F (2000) Phylogenetic relationships of the suborders of Coleoptera (Insecta). Cladistics 16:103–141

Beutel RG, Hörnschemeyer T (2002a) Larval morphology and phylogenetic position of *Micromalthus debilis* LeConte (Coleoptera: Micromalthidae). Syst Entomol 27(2):169–190

Beutel RG, Hörnschemeyer T (2002b) Description of the larva of *Rhipsideigma raffrayi* (Coleoptera, Archostemata), with phylogenetic and functional implications. Eur J Entomol 99:53–66

Beutel RG, Leschen RAB (eds) (2005a) Handbook of zoology, Vol. IV Arthropoda: Insecta. Part 38. Coleoptera, Vol. 1: Morphology and systematics (Archostemata, Adephaga, Myxophaga, Polyphaga (partim)), 1st edn. Walter De Gruyter, Berlin

Beutel RG, Leschen RAB (2005b) Phylogenetic analysis of Staphyliniformia (Coleoptera) based on characters of larvae and adults. Syst Entomol 30:510–548

Beutel RG, Leschen RAB (eds) (2016) Handbook of zoology, Vol. IV Arthropoda: Insecta. Part 38. Coleoptera, Vol. 1: Morphology and systematics (Archostemata, Adephaga, Myxophaga, Polyphaga (partim)), 2nd edn. Walter De Gruyter, Berlin

Beutel R, Molenda R (1997) Comparative morphological study of larvae of Staphylinoidea (Coleoptera, Polyphaga) with phylogenetic implications. Zool Anz 236:37–67

Beutel RG, Pollock DA (2000) Larval head morphology of *Phycosecis litoralis* (Pascoe) (Coleoptera, Phycosecidae) with phylogenetic implications. Invertebr Tax 14:825–835

Beutel RG, Roughley RE (1994) Phylogenetic analysis of Gyrinidae based on characters of the larval head (Coleoptera: Adephaga). Entomol Scand 24:459–468

Beutel RG, Maddison DR, Haas A (1998) Phylogenetic analysis of Myxophaga (Coleoptera) using larval characters. Syst Entomol 24:1–23

Beutel RG, Anton E, Bernhard D (2001) Head structures of adult *Spercheus*: their function and possible significance to staphyliniform phylogeny. Ann Zool 51(4):473–484

Beutel RG, Anton E, Jäch MA (2003) On the evolution of adult head structures and the phylogeny of Hydraenidae (Coleoptera, Staphyliniformia). J Zool Syst Evol Res 41(4):256–275

Beutel RG, Balke M, Steiner WE (2006) On the systematic position of Meruidae (Coleoptera, Adephaga) and the phylogeny of the smaller hydradephagan families. Cladistics 22:102–131

Beutel RG, Ge S-Q, Hörnschemeyer T (2008) On the head morphology *Tetraphalerus*, the phylogeny of Archostemata and the basal branching events in Coleoptera. Cladistics 23:1–29

Beutel RG, Yan E, Richter A, Büsse S, Miller KB, Yavorskaya M, Wipfler B (2017) The head of *Heterogyrus milloti* (Coleoptera: Gyrinidae) and its phylogenetic implications. Arthr Syst Phyl 75(2):261–280

Beutel RG, Pohl H, Yan EV, Anton E, Liu SP, Ślipiński SA, McKenna D, Friedrich F (2018) The phylogeny of Coleopterida (Hexapoda) – morphological characters and molecular phylogenies. Syst Entomol 44(1):75–102. https://doi.org/10.1111/syen.12316

Bocak L, Bocakova M (2010) Family Omalisidae Lacordaire, 1857. Lycidae Laporte, 1836. In: Leschen RAB, Beutel RG, Lawrence JF (eds) Handbook of zoology, Vol. IV Arthropoda: Insecta. Part 39. Coleoptera, Vol. 2: Morphology and systematics (Elateroidea, Cucujifomia excl. Phytophaga). Walter De Gruyter, Berlin, pp 110–123

Bocak L, Matsuda K (2003) Review of the immature stages of the family Lycidae (Insecta: Coleoptera). J Nat Hist 37(12):1463–1507

Bocak L, Branham MA, Kundrata R (2010) Family Drilidae Blanchard, 1845. In: Leschen RAB, Beutel RG, Lawrence JF (eds) Handbook of zoology, Vol. IV Arthropoda: Insecta. Part 39. Coleoptera, Vol. 2: Morphology and systematics (Elateroidea, Cucujifomia excl. Phytophaga). Walter De Gruyter, Berlin, pp 104–110

Böving AG, Henriksen KL (1938) The developmental stages of the Danish Hydrophilidae (Ins., Coleoptera). Vidensk Medd Dansk naturh Foren 102:27–162

Brandmayr TZ, Brandmayr P, Marano I, Paarmann W (1993) The larva of *Graphipterus serrator* (Forskål 1775) (Coleoptera Carabidae Graphipterini): description and functional morphology. Trop Zool 6(2):299–312

Chamorro ML (2014) Lamprosomatinae Lacordaire, 1848. In: Leschen RAB, Beutel RG (eds) Handbook of zoology, Vol. IV Arthropoda: Insecta. Part 38. Coleoptera, Vol. 3: Morphology and systematics (Phytophaga). Walter De Gruyter, Berlin, pp 226–230

Chaudonneret J (1990) Les pièces buccales des insects. Thèmes et variations. Edition hors-serie du Bulletin Scientifique de Bourgogne, Dijon

Crowson RA (1962) Observations on the beetle family Cupedidae, with the descriptions of two new fossil forms and a key to the recent genera. Ann Mag Nat Hist 5(13):147–157

Crowson RA (1975) The systematic position and implications of *Crowsoniella*. Boll Mus Civ St Nat Verona 2:459–463

Crowson RA (1981) The biology of Coleoptera. John Murray, London

De Marzo L (1978) Studi sulle larve dei Coleotteri Ditiscidi. IX. Morfologia delle mandibole delle larve in alcune specie delle subff. Laccophilinae e Hydroporinae. Ipotesi sulla derivazione morfogenetica dei relativi modelli di mandibola. Entomologica (Bari) 14:137–157

De Marzo L (1979) Studi sulle larve dei Coleotteri Ditiscidi. X. Anatomia e funzionamento dell' apparato succhiante cibario-faringeo in alcune forme larvali delle subff. Dytiscinae, Colymbetinae, Laccophilinae e Hydroporinae. Entomologica (Bari) 15:5–72

Dönges J (1954) Der Kopf von *Cionus scrophulariae* L. Curculionidae. Zool Jahrb Anat 74:1–76

Dressler C, Beutel RG (2010) The morphology and evolution of the adult head of Adephaga (Insecta, Coleoptera). Arthropod Syst Phylogeny 68:239–287

Forsythe TG (1982) Feeding mechanisms of certain ground beetles (Coleoptera: Carabidae). Coleopt Bull 36:26–73

Forsythe TG (1983) Mouthparts and feeding of certain ground beetles (Coleoptera: Carabidae). Zool J Linnean Soc 79(4):319–376

Friedrich F, Farrell BD, Beutel RG (2009) The thoracic morphology of Archostemata and the relationships of the extant suborders of Coleoptera (Hexapoda). Cladistics 25:1–37

Ge S-Q, Wipfler B, Pohl H, Hua Y, Ślipiński A, Yang X-K, Beutel RG (2012) The first complete 3D reconstruction of a Spanish Fly primary larva (*Lytta vesicatoria*, Meloidae, Coleoptera). PLoS One 7(12):e52511

Ge S-Q, Hua Y, Ren J, Ślipiński A, Heming B, Beutel RG, Yang X-K, Wipfler B (2015) Transformation of head structures during the metamorphosis of *Chrysomela populi* (Coleoptera, Chrysomelidae). Arthropod Syst Phylogeny 73(1):129–152

Gorb S, Beutel RG (2000) Head-capsule design and mandible control in beetle larvae: a three-dimensional approach. J Morphol 244:1–14

Grebennikov VV, Beutel RG (2002) Morphology of the minute larva of *Ptinella tenella*, with special reference to effects of miniaturisation and the systematic position of Ptiliidae (Coleoptera: Staphylinoidea). Arthropod Struct Dev 31:157–172

Green J (1956) The mouthparts of *Eurynebria complanata* (L.) and *Bembidion laterale* (Sam.) (Col., Carabidae). Entomol Mon Mag 92:110–113

Hannappel U, Paulus HF (1987) Arbeiten zu einem phylogenetischen System der Helodidae (Coleoptera) – Feinstrukturuntersuchungen an europäischen Larven. Zool Beitr NF 31:77–150

Hata K, Edmonds WD (1983) Structure and function of the mandibles of adult dung beetles (Coleoptera: Scarabaeidae). Int J Insect Morphol Embryol 12(1):1–12

Honomichl K (1975) Beitrag zur Morphologie des Kopfes der Imago von *Gyrinus substriatus* Stephens, 1829 (Coleoptera, Insecta). Zool Jb Anat 94:218–295

Honomichl K (1976) Feinstruktur eines Muskelrezeptors im Kopf von *Dermestes maculatus* De Geer (Insecta, Coleoptera). Zoomorphologie 85:59–71

Honomichl K (1978) Feinstruktur zweier Propriorezeptoren im Kopf von *Oryzaephilus surinamensis* (L) (Insecta, Coleoptera). Zoomorphology 90:213–226

Hörnschemeyer T (2005) Archostemata Kolbe, 1908. In: Leschen RAB, Beutel RG (eds) Handbook of zoology, Vol. IV Arthropoda: Insecta. Part 38. Coleoptera, Vol. 1, 2nd edn: Morphology and systematics (Archostemata, Adephaga, Myxophaga, Polyphaga (partim)), Walter De Gruyter, Berlin, pp 29–42

Hörnschemeyer T (2009) The species-level phylogeny of archostematan beetles—where do *Micromalthus debilis* and *Crowsoniella relicta* belong? Syst Entomol 34(3):533–558

Hörnschemeyer T, Beutel RG, Pasop F (2002) Head structures of *Priacma serrata* Leconte (Coleoptera, Archostemata) inferred from x-ray tomography. J Morphol 252(3):298–314

Hörnschemeyer T, Goebbels J, Weidemann G, Faber C, Haase A (2006) The head morphology of *Ascioplaga mimeta* (Coleoptera: Archostemata) and the phylogeny of Archostemata. Eur J Entomol 103(2):409–423

Hua Y, Beutel RG, Ge S-Q, Yang X-K (2015) The larval head structures of *Podagricomela shirahatai* (Chûjô) (Chrysomelidae, Galerucinae, Alticini) and morphological effects of leaf mining. J Morphol 276(4):446–457

Karolyi F, Hansal T, Krenn HW, Colville J (2016) Comparative morphology of the mouthparts of the megadiverse South African monkey beetles (Scarabaeidae: Hopliini): feeding adaptations and guild structure. PeerJ 4:e1597

Kazantsev SV (2006) Comparative morphology of mandibular structures in lycid larvae and its phylogenetic implications (Polyphaga, Hexapoda). Acta Zool 87(3):229–238

Koerner L, Gorb SN, Betz O (2012) Functional morphology and adhesive performance of the stick-capture apparatus of the rove beetles *Stenus* spp. (Coleoptera, Staphylinidae). Zoology 115:117–127

Koerner L, Garamszegi LZ, Heethoff M, Betz O (2017) Divergent morphologies of adhesive predatory mouthparts of *Stenus* species (Coleoptera: Staphylinidae) explain differences in adhesive performance and resulting prey-capture success. Zool J Linnean Soc 181(3):500–518

Kolibáč J (2010) Cleridae Latreille, 1802. In: Leschen RAB, Beutel RG, Lawrence JF (eds) Handbook of zoology, Vol. IV Arthropoda: Insecta. Part 39. Coleoptera, Vol. 2: Morphology and systematics (Elateroidea, Cucujifomia excl. Phytophaga). Walter De Gruyter, Berlin, pp 257–261

Korschelt E (1923–1924) Bearbeitung einheimischer Tiere. Erste Monographie: Der Gelbrand *Dytiscus marginalis* L. Bd. 1 (1923), Bd. 2 (1924). W. Engelmann, Leipzig

Krenn HW, Pernstich A, Messner T, Hannappel U, Paulus HF (2002) Kirschen als Nahrung des männlichen Hirschkäfers, *Lucanus cervus* (Linnaeus 1758) (Lucanidae: Coleoptera). Entomologische Zeitschrift 112:165–170

Krenn HW, Plant JD, Szucsich NU (2005) Mouthparts of flower-visiting insects. Arthropod Struct Dev 34(1):1–40

Kukalová-Peck J, Beutel RG (2012) Is the Carboniferous *Adiphlebia lacoana* really the "oldest beetle"? Critical reassessment and description of a new Permian beetle family. Eur J Entomol 109(4):633–645

Kukalova-Peck J, Lawrence JF (1993) Evolution of the hind wing in Coleoptera. Can J Zool 125:181–258

Kukalova-Peck J, Lawrence JF (2004) Relationships among coleopteran suborders and major endoneopteran lineages: evidence from hind wing characters. Eur J Entomol 101:95–144

Lawrence JF, Ślipiński A, Seago AE, Thayer MK, Newton AF, Marvaldi AE (2011) Phylogeny of the Coleoptera based on morphological characters of adults and larvae. Ann Zool 61:1–217

Leschen RAB (2016) Eucinetidae Lacordaire, 1857. Clambidae Fischer von Waldheim, 1821. In: Beutel RG, Leschen RAB (eds) Handbook of zoology, Vol. IV Arthropoda: Insecta. Part 38. Coleoptera, Vol. 1, 2nd edn: Morphology and systematics (Archostemata, Adephaga, Myxophaga, Polyphaga (partim)). Walter De Gruyter, Berlin, pp 206–201

Leschen RAB, Beutel RG (eds) (2014) Handbook of zoology, Vol. IV Arthropoda: Insecta. Part 39. Coleoptera, Vol. 3: Morphology and systematics (Phytophaga). Walter De Gruyter, Berlin

Leschen RAB, Beutel RG, Lawrence JF (eds) (2010) Handbook of zoology, Vol. IV Arthropoda: Insecta. Part 39. Coleoptera, Vol. 2: Morphology and Systematics (Elateroidea, Cucujifomia excl. Phytophaga). Walter De Gruyter, Berlin

Luff ML (1974) Adult and larval feeding habits of *Pterostichus madidus* (F.) (Coleoptera: Carabidae). J Nat Hist 8(4):403–409

McKenna DD, Wild AL, Kanda K, Bellamy CL, Beutel RG, Caterino MS, Farnum CW, Hawks DC, Ivie MA, Jameson ML, Leschen RAB, Marvaldi AE, Mchugh JV, Newton AF, Robertson JA, Thayer MK, Whiting MF, Lawernce JF, Ślipiński A, Maddison DR, Farrell BD (2015) The beetle tree of life reveals Coleoptera survived end Permian mass extinction to diversify during the Cretaceous terrestrial revolution. Syst Entomol 40(4):835–880

Nadein KS, Bezdek J (2014) Galerucinae Latreille, 1802. In: Leschen RAB, Beutel RG (eds) Handbook of zoology, Vol. IV Arthropoda: Insecta. Coleoptera, Vol. 3: Morphology and systematics (Phytophaga). Walter De Gruyter, Berlin, pp 251–259

Nel A, Scholtz CH (1990) Comparative morphology of the mouthparts of adult Scarabaeoidea (Coleoptera). Entomol Mem 80:1–85

Newton AF Jr (1991) Leiodidae (Staphylinoidea). Leptinidae (Staphylinoidea). Silphidae (Staphylinoidea). Carrion Beetles (Silphinae) and Burying Beetles (Nicrophorinae). In: Stehr F (ed) Immature insects, Vol. 2. Kendall/Hunt, Dubuque, Iowa, pp 327–331, 339–341

Noars R (1956) Contribution à la connaissance de la larve d'*Orectochilus villosus* Müll. Coléoptères Gyrinides. Trav Lab Zool Fac Sci Dijon 17:1–32

Oberprieler R (2014) Caridae Thompson, 1992. Brentidae Billberg, 1820. Eurhynchinnae Lacordaire, 1863, Ithycerinae Schoenherr, 1823 and Microcerinae Lacordaire, 1863. In: Leschen RAB, Beutel RG (eds) Handbook of zoology, Vol. IV Arthropoda: Insecta. Coleoptera, Vol. 3: Morphology and systematics (Phytophaga). Walter De Gruyter, Berlin, pp 356–384

Pace R (1975) An exceptional endogeous beetle: *Crowsoniella relicta* n. gen. n. sp. of Archostemata Tetraphaleridae from Central Italy. Boll Mus Civico Storia Nat Verona 2:445–458

Perotti MA, Young DK, Braig HR (2016) The ghost sex-life of the paedogenetic beetle *Micromalthus debilis*. Sci Rep 6:27364

Ponomarenko AG (1969) The historical development of archostematan beetles (in Russian). Trudy Paleont Inst Akad Nauk SSSR 125:1–238

Reichardt H (1973) A critical study of the suborder Myxophaga, with a taxonomic revision of the Brazilian Torridincolidae and Hydroscaphidae (Coleoptera). Arq Zool 24(2):73–162

Reid C (2014) Chrysomelinae Latreille, 1802. In: Leschen RAB, Beutel RG (eds) Handbook of zoology, Vol. IV Arthropoda: Insecta. Coleoptera, Vol. 3: Morphology and systematics (Phytophaga). Walter De Gruyter, Berlin, pp 243–251

Rothmeier G, Jäch MA (1986) Spercheidae, the only filter-feeders among Coleoptera. In: Proceedings of the IV. European congress of entomology (Amsterdam, 1986), pp 133–137

Ruhnau S (1986) Phylogenetic relations within the Hydradephaga (Coleoptera) using larval and pupal characters. Entomologica Basiliensia 11:231–271

Scholtz CH, Grebennikov VV (2016) Scarabaeoidea Latreille, 1802. In: Beutel RG, Leschen RAB (eds) 2016. Handbook of zoology, Vol. IV Arthropoda: Insecta. Part 38. Coleoptera, Vol. 1: Morphology and systematics (Archostemata, Adephaga, Myxophaga, Polyphaga (partim)), 2nd edn. Walter De Gruyter, Berlin, pp 443–525

Seago A, Beutel RG (2010) Prostomidae G.G. Thomson, 1859 Spangler and Steiner 2005. In: Leschen RAB, Beutel RG, Lawrence JF (eds.) Handbook of zoology, Vol. IV Arthropoda: Insecta. Part 39. Coleoptera, Vol. 2: Morphology and systematics (Elateroidea, Cucujifomia excl. Phytophaga). Walter De Gruyter, Berlin, pp 659–667

Seeger W (1971) Morphologie, Bionomie und Ethologie von Halipliden, unter besonderer Berücksichtigung funktionsmorphologischer Gesichtspunkte (Haliplidae; Coleoptera). Arch Hydrobiol 68(9):400–435

Stork NE, McBroom J, Gely C, Hamilton AJ (2015) New approaches narrow global species estimates for beetles, insects, and terrestrial arthropods. Proc Natl Acad Sci 112:7519–7523

Svacha P, Lawrence JF (2014) Cerambycidae Latreille, 1802. In: Leschen RAB, Beutel RG, Lawrence JF (eds) Handbook of zoology, Vol. IV Arthropoda: Insecta. Part 39. Coleoptera, Vol. 3: Morphology and systematics (Phytophaga). Walter De Gruyter, Berlin, pp 659–667

Thayer MK (2016) Staphylinidae Latreille, 1802. In: Leschen RAB, Beutel RG, Lawrence JF (eds) Handbook of zoology, Vol. IV Arthropoda: Insecta. Part 39. Vol. 1: Morphology and systematics (Archostemata, Adephaga, Myxophaga, Polyphaga (partim)), 2nd edn. Walter De Gruyter, Berlin, pp 394–442

Thompson RG (1979) Larvae of North American Carabidae with a key to the tribes. In: Erwin TL, Ball GE, Whitehead GE, Halpern AL (eds) Carabid beetles: their evolution, natural history, and classification. W. Junk, The Hague, pp 209–291

Tröster G (1987) Skelett und Muskulatur des Kopfes der Larve von *Pterostichus nigrita* (Paykull) (Coleoptera: Carabidae). Stuttg Beitr Naturk (A) 399:1–23

Vencl FV, Leschen RAB (2014) Criocerinae Latreille, 1807. In: Leschen RAB, Beutel RG (eds) Handbook of zoology, Vol. IV Arthropoda: Insecta. Coleoptera, Vol. 3: Morphology and systematics (Phytophaga). Walter De Gruyter, Berlin, pp 237–242

Vogel R (1915) Beitrag zur Kenntnis des Baues und der Lebensweise der Larve von *Lampyris noctiluca*. Z Wiss Zool 112:291–432

Weinreich E (1968) Über den Klebfangapparat der Imagines von *Stenus* Latr. (Coleopt., Staphylinidae) mit einem Beitrag zur Kenntnis der Jugendstadien dieser Gattung. Z Morph Tiere 62(2):162–210

Wilhelmi AP, Krenn HW (2012) Elongated mouthparts of nectar-feeding Meloidae (Coleoptera). Zoomorphology 131(4):325–337

Williams IW (1938) The comparative morphology of the mouthparts of the order Coleoptera treated from the standpoint of phylogeny. J NY Entomol Soc 46:245–289

Wipfler B, Pohl H, Yavorskaya MI, Beutel RG (2016) A review of methods for analysing insect structures - the role of morphology in the age of phylogenomics. Curr Opin Ins Sci 18:60–68

Yavorskaya MI (2018) Evolution of head structures in Coleoptera with special emphasis on the feeding apparatus and miniaturized forms. PhD dissertation, FSU, Jena

Yavorskaya MI, Leschen RAB, Polilov AA, Beutel RG (2014) Unique rostrate larvae and basidiomycophagy in the beetle family Corylophidae. Arthropod Struct Dev 43(2):153–162

Yavorskaya MI, Kojima K, Machida R, Beutel RG (2015) Morphology of the 1st instar larva of *Tenomerga mucida* (Chevrolat, 1829) (Coleoptera: Archostemata: Cupedidae). Arthropod Syst Phylogeny 73(2):239–258

Yavorskaya M, Beutel RG, Polilov A (2017) Head morphology of the smallest beetles (Coleoptera: Ptiliidae) and the evolution of sporophagy within Staphyliniformia. Arthropod Syst Phylogeny 75(3):417–434

Yavorskaya MI, Anton E, Jałoszyński P, Polilov A, Beutel RG (2018a) Cephalic anatomy of Sphaeriusidae and a morphology-based phylogeny of the suborder Myxophaga (Coleoptera). Syst Entomol 43(4):777–797. https://doi.org/10.1111/syen.12304

Yavorskaya MI, Hörnschemeyer T, Beutel RG (2018b) The head morphology of *Micromalthus debilis* (Coleoptera: Archostemata: Micromalthidae) – an "archaic" beetle with an unusual morphology and a unique life cycle. Arthropod Syst Phylogeny 76(3):475–486

Zhang SQ, Che LH, Li Y, Pang H, Ślipiński A, Zhang P (2018) Evolutionary history of Coleoptera revealed by extensive sampling of genes and species. Nat Commun 9(1):205

# Chapter 13
# What's on the Menu: Floral Tissue, Pollen or Nectar? Mouthpart Adaptations of Anthophilous Beetles to Floral Food Sources

**Florian Karolyi**

**Abstract** Anthophilous Coleoptera are frequent flower visitors, and many have specialised mouthparts to feed on floral tissue, pollen grains and/or nectar. Although not all of these flower visitors are pollinators, many beetles pollinate various plant families. The present chapter deals with the mouthpart morphology and feeding preferences of beetles. The evolutionary pathways to flower visiting and pollination by beetles are discussed in the context of mouthpart specialisations. Compared to destructive floral tissue feeding, many pollen-feeding beetles have lost cutting or biting structures on the mandibles and maxillae. Instead, their characteristic mouthparts include characteristic types of hairs, bristles or setae on the maxillae, responsible for collecting and transporting food inside the preoral cavity. Massive, paired mandibles with a lacinia mobilis and a prominent molar crush pollen like mortar and pestle. Adaptations to nectar feeding include either elongated maxillary palps or an elongation of the hairy galea. Although Coleoptera are considered rather unspecialised pollinators, various plant species have evolved floral syndromes, such as flower odour, colour and morphology, to attract anthophilous beetles. As such, specialised beetle pollination systems are found across various plant families.

## 13.1 Introduction

Anthophilous Coleoptera are common and quite charismatic, often striking, flower visitors (Fig. 13.1). Many species spend extended periods of time on a single flower, feeding on pollen and nectar with biting-chewing mouthparts (Krenn et al. 2005; Karolyi et al. 2016). Although many beetles are unspecialised and visit numerous plant species, there are many cantharophilous plants from many different families that are specialised on beetle pollination and rely on beetles for pollen transfer. Such beetle

F. Karolyi (✉)
Department of Integrative Zoology, University of Vienna, Vienna, Austria
e-mail: florian.karolyi@univie.ac.at



H. W. Krenn (ed.), *Insect Mouthparts*, Zoological Monographs 5,
https://doi.org/10.1007/978-3-030-29654-4_13

**Fig. 13.1** Flower-visiting beetles on host flowers. (**a**) *Ctenicera cuprea* (Elateridae). (**b**) *Clania glenlyonensis* (Scarabaeidae, Hopliini) on *Gazania rigida* (Asteraceae). (**c**) *Cetonia aurata* (Scarabaeidae). (**d**) *Cerambyx scopolii* (Cerambycidae)

pollination systems are particularly important in moist tropical and arid areas (Proctor et al. 1996). However, beetles are generally considered as "mess-and-soil" pollinators (Kevan and Baker 1983; Johnson and Midgley 2001; Steenhuisen and Johnson 2012), feeding on both floral rewards and reproductive floral parts (Harmon et al. 2011). Beetles are known to be important pollinators of various plant families in certain regions of the world, e.g. the Cape Floristic Region (Goldblatt and Manning 2011) and the East Mediterranean region (Dafni et al. 1990). Indeed, the numbers of plant species pollinated by beetles may currently still be underestimated (Wardhaugh 2015).

Approximately 180 angiosperm plant species in 34 families are specialised for beetle pollination (Bernhardt 2000). Beetles are considered one of the "big four" of insect orders that contribute to the bulk of flower-visiting and pollinating insects (Wardhaugh 2015). Additionally, anthophilous beetle species appear to be associated with a wide range of plant species, mainly within angiosperms (Dafni et al. 1990; Steiner 1998b; Ratnayake et al. 2007; Procheş and Johnson 2009; Steenhuisen et al. 2010). Although morphological adaptations and specialisations towards flower food sources are often inconspicuous in beetles, they may be more prominent than originally thought. This suggests that pollen and/or nectar feeding and associated adaptations are an important and a relatively unexplored driving force in explaining patterns of anthophilous beetle diversity (Karolyi et al. 2016).

Beetles have been found to be important elements of the insect pollination fauna in certain ecosystems and play a major role in specialised pollination systems. Pollinating beetle species also appear to be an important driver in plant speciation, as seen across several plant families that show flower shape and colour adaptations specifically towards beetle pollination. For example, South African monkey beetles (Hopliini) play a vital role in the pollination of numerous plant species across various families within the Greater Cape Floristic Region (Picker and Midgley 1996; Goldblatt et al. 1998, 2000; Steiner 1998b; Schnitzler et al. 2011). In the Mediterranean region, a group of Glaphyridae is responsible for pollinating various red-flowered plant species across several families (Dafni et al. 1990). Thermophile scarab beetles of the genus *Cyclocephala* pollinate heat-producing flowers of *Philodendron* (Araceae) in the Neotropics. In this way, the beetles save energy for activity by spending most of their time inside thermogenic flowers (Seymour et al. 2003). Numerous other examples could be added to this list, with a wide range of pollinating beetles with high levels of host specialisation (Wardhaugh et al. 2013; Wardhaugh 2015).

Considering the mouthparts of highly specialised flower-visiting insects, like the foldable proboscis of Apidae, the coilable galea of Lepidoptera or the elongated, straw-like proboscis of some Diptera (Szucsich and Krenn 2000, 2002; Krenn et al. 2005; Bauder et al. 2011; Karolyi et al. 2012, 2014), the morphological adaptations to floral food sources found in Coleoptera are usually less conspicuous. In general, beetles display prognathous mouthparts that consist of the labrum, paired mandibles and maxilla as well as the labium. Usually, both the labrum and labium, with the hairy epipharynx and hypopharynx, respectively, serve as the dorsal and ventral cover of the cibarium and seal the preoral cavity. While the galeae of the maxillae represent the primary food uptake organs, the mandibles are responsible for the processing of food within the cibarium (Karolyi et al. 2009). Although conspicuous deviations of this pattern can only be found in a few highly specialised groups, most anthophilous beetle species show characteristic adaptations of their mouthparts to preferred floral food sources. Within this group, species have evolved specialised mouthpart structures to harvest pollen grains or to take up nectar from flowers (Karolyi et al. 2009, 2016; Wilhelmi and Krenn 2012).

Detailed studies devoted to the functional mouthpart morphology of flower-visiting beetles are generally scarce. This chapter gives an overview of the morphological adaptations to different floral food sources found in specialised flower-visiting beetle species. In addition, the chapter will present new results in modern functional morphology to draw a broader picture of the adaptations found in the mouthparts of flower-visiting beetles.

## 13.2 From Florivory to Feeding on Pollen and Nectar: The Evolutionary Road to Specialised Flower Feeding

After insect diversification followed in the wake of the angiosperm radiation, the earliest beetles that visited Cretaceous flowers were most likely generalised flower feeders (Grimaldi 1999). With the radiation of flowering angiosperms in the

Cretaceous, new food sources appeared, and new feeding strategies evolved in insects (Grimaldi 1999; Smith et al. 2010). Ancestral adult Coleoptera were most likely zoophagous or saprophagous (Lovell 1915; Yavorskaya 2018; see Chap. 12). Over the past 230 million years, they evolved various patterns of ecological food partitioning (Grimaldi and Engel 2005). Pollen and nectar consumption, as well as pollination mutualisms, have originated numerous times within the Coleoptera (Labandeira 2010). Primitive cantharophilous pollination syndromes found from the Late Jurassic and Early Cretaceous suggest that plant-beetle associations are an evolutionary ancient relationship. Primitive beetles most likely have fed on plant material, floral tissue and pollen, with some plants relying on large-bodied beetles for "mess-and-soil" pollination (Labandeira 1997; Nel et al. 2018).

Today, anthophilous beetles are found amongst various taxa of Polyphaga (Table 13.1). Species characterised as destructive floral or foliage tissue feeders have teeth and cutting edges on the maxillae and mandibles (Fuchs 1974; Karolyi et al. 2016). Some of the earliest beetle fossil records from the Jurassic displayed such toothed structures suggesting a phytophagous lifestyle (Grimaldi and Engel 2005).

Considering the mouthpart morphology, pollen-feeding species of Scarabaeidae most likely derived from flower- and foliage-feeding ancestors that already existed in the Early Cretaceous (130 Mya) (Grimaldi and Engel 2005). Deposits from the Eocene revealed beetle fossils (Buprestidae) with well-preserved pollen in the gut (Labandeira 1997). Scarabaeidae mouthparts are typically characterised by powerful mandibles that are used to grind food particles; several lineages of Scarabaeidae have evolved flexible mouthparts modified for pollen consumption (Fig. 13.2). Adaptations towards pollen feeding led to the reduction of toothed structures on the maxilla and the development of characteristic bristles on the galeae. While facultative pollen feeders have numerous long, curled bristles, obligatory pollen feeders display fewer spatulate galea bristles (Fuchs 1974).

In Cantheridae and Cleridae, phyllophages and pollen feeders derived from zoophagous species that used to prey on insects that occur on flowers (Fuchs 1974), while pollen-feeding Staphylinidae evolved from saprophagous ancestors (Betz et al. 2003). As representatives of these groups still feed on small insects, some species switched to pollen, which represents an easily accessible and richly available protein source. Recent representatives of these groups are known to be zoophagous but have also been observed feeding on pollen. Dissections of specimens revealed pollen grains in their guts (Fuchs 1974).

Considering their successful evolutionary pathway since the Early Permian (Grimaldi and Engel 2005), Coleoptera went from ancestral saprophagy or mycophagy (Betz et al. 2003) to a predacious lifestyle, before pollen feeders emerged. Various taxa evolved adaptations to use nectar as either an additional or a primary food source.

### *13.2.1 Feeding on Floral Tissues: Rather Devastate than Pollinate*

Early reports described anthophilous beetles as destructive flower visitors and rather unsuitable as pollinators (Kirchner 1911). However, florivorous Coleoptera can be

**Table 13.1** Anthophilous beetle taxa which include flower-visiting species and plant families which include cantharophilous species, as well as food sources for beetles. Bold letters indicate taxa with well-investigated mouthpart morphology in flower-visiting, anthophilous beetle species

| Taxa | Plant families | Food source | References |
|---|---|---|---|
| Anthicidae | Myristicaceae | Pollen | Bernhardt (2000) |
| Cantharidae | Apiaceae, Asteraceae, Orchidaceae | Floral tissue, pollen | Fuchs (1974) |
| **Cerambycidae** | Various | Pollen, nectar | Goldman (1933), Kevan and Baker (1983), Frame (2003) |
| Chrysomelidae | *Ranunculus* spp., Asteraceae: Lactuceae | Pollen | Kevan and Baker (1983), Frame (2003) |
| Cleroidea | | Floral tissue, pollen | Fuchs (1974), Kevan and Baker (1983) |
| Melyridae | | | Kevan and Baker (1983) |
| **Malachiinae** | Poaceae, Cyperaceae: *Carex* | Pollen | Schicha (1967) |
| Curculionidae | Magnoliaceae | Floral tissue | Armstrong and Irvine (1990), Frame (2003) |
| Elateridae | Orchidaceae: *Eulophia, Listera* | Pollen | Kevan and Baker (1983), Peter and Johnson (2009) |
| **Glaphyridae** | Ranunculaceae, Liliaceae, Papaveraceae | Pollen | Dafni et al. (1990), Keasar et al. (2010) |
| Hydrophilidae<br>Hydraenidae | Cyclanthaceae, Monimiaceae | Floral tissue, pollen | Bernhardt (2000) |
| Lycidae | Orchidaceae | | Bernhardt (2000) |
| Meloidae | | | |
| **Nemognathinae** | Asteraceae | Nectar | Bologna and Pinto (2001), Wilhelmi and Krenn (2012) |
| **Mordellidae** | Magnoliids | Floral tissue | Kevan and Baker (1983), Thien et al. (2009) |
| Nitidulidae | Ranunculaceae, Brassicaceae, Annonaceae | Floral tissue, pollen | Kevan and Baker (1983), Bernhardt (2000) |
| **Oedemeridae** | Apiaceae: *Eryngium* | Nectar | Fuchs (1974), Kevan and Baker (1983) |
| Ptiliidae | Cyclanthaceae | Floral tissue, pollen | Bernhardt (2000) |
| Rhizophagidae | Monimiaceae | Pollen | Bernhardt (2000) |
| Scraptiidae | Magnoliids | Pollen | Bernhardt (2000) |
| Scarabaeidae | | | Nel and Scholtz (1990) |
| **Cetoniinae** | Apocynaceae, Asclepiadoideae, Orchidaceae: Eulophia | Pollen, nectar | Ollerton et al. (2003), Peter and Johnson (2009) |
| ***Cetonia*** | Rosaceae | Pollen | Karolyi et al. (2009) |

(continued)

**Table 13.1** (continued)

| Taxa | Plant families | Food source | References |
|---|---|---|---|
| *Cyclocephala* | Arecaceae: *Bactris gasipaes*; Annonaceae | Floral tissue, pollen | Rickson et al. (1990), Costa et al. (2017) |
| **Dynastinae** | Araceae | Floral tissue, pollen | Nel and Scholtz (1990), Gottsberger et al. (2013) |
| **Rutelinae** | | | |
| **Hopliini** | Iridaceae: *Morea*, *Romulea*, *Hesperantha*, *Sparaxis*, Asteraceae, Aizoaceae, Proteaceae and Orchidaceae | Pollen, nectar, floral tissue | Steiner (1998a, b), Goldblatt et al. (2002), Van Kleunen et al. (2007), Karolyi et al. (2016), Johnson and Nicolson (2001) |
| **Melolonthinae** | Arecaceae | Pollen, nectar, floral tissue | Nel and Scholtz (1990), Küchmeister et al. (1997) |
| Scraptiidae | Magnoliids | Floral tissue | Bernhardt (2000) |
| **Staphylinidae** | Various | Floral tissue, pollen | Betz et al. (2003), Frame (2003) |

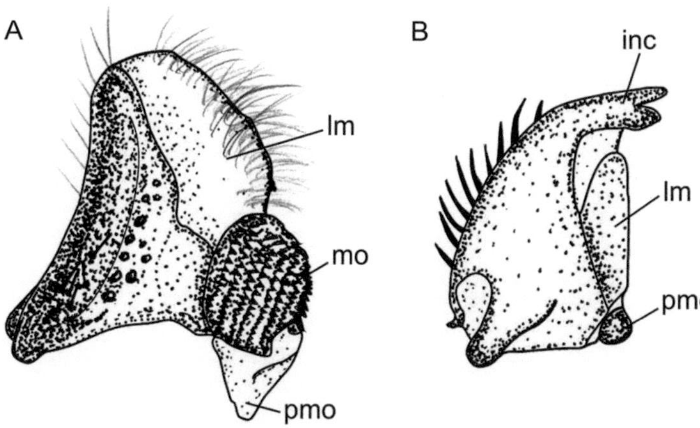


**Fig. 13.2** Mandibles of beetles with different feeding habits. (**a**) *Pachycnema flavolineata*, mandible with prominent, setose lacinia mobilis and molar part displaying numerous rows of pointy teeth. (**b**) *Malachius bipustulatus*, ventral view of the left mandible with lacinia mobilis. Abbreviations: *inc* incisivus; *lm* lacinia mobilis; *mo* mola; *pmo* postmola

considered to play a significant role in insect-plant systems. Basal angiosperms in particular display floral modifications to reward floral visitors with a nutritive tissue, a sheltered breeding location and a larval brood place (Gottsberger 1974, 1977; Beach 1982; Thien et al. 2009). Part of the angiosperm's success was the evolution of a flower that is meant to be eaten. Since flower visitors are attracted to the flower, plants are able to control the consumption of plant tissue and enhance pollination (Frame 2003). Today, many herbivorous beetles still feed on floral tissue, including petals and green flower tissue (McCall and Irwin 2006; McCall 2008, 2010; Soper Gorden and Adler 2016). Flowers can be highly nutritious and are an abundant food source for anthophilous beetles (Higginson et al. 2015).

When it comes to studies dealing with insect-plant interactions, research has focussed mostly on pollination aspects, and only a few studies have dealt with the phenomenon of florivory. The most speciose beetle families, e.g. Cerambycidae, Chrysomelidae, Curculionidae, Scarabaeidae and Staphylinidae, include flower visitors that feed on floral tissue (Frame 2003). Some genera of Glaphyridae are also known to feed on whole flowers, including ovaries and petals (Keasar et al. 2010). Beetles characterised as floral or foliage tissue feeders display a mouthpart morphology characterised by teeth and cutting edges on the maxillae and mandibles and are therefore regarded as being destructive to their host flowers (Midgley 1993; Picker and Midgley 1996; Karolyi et al. 2016).

Several adaptations of the morphology of the mandibles, especially in the shape of the incisivi, are seen in hopliines (Fig. 13.4a–d). The sclerotized cutting edges on the mandibles, together with the toothed galea, are considered as adaptations towards floral tissue feeding. The rows of short, tooth-shaped hairs, as well as the single, lateral tooth found in some species on the lacinia mobilis of the mandible, are possibly used to grate floral or foliage tissue, whereas the sclerotized molar is most likely used to grind floral particles (Karolyi et al. 2016).

Magnoliids (e.g. Annonaceae, Eupomatiaceae, Magnoliaceae, Monimiaceae and Myristicaceae) are typically beetle-pollinated, and some species possess a highly specialised beetle pollination system (Thien et al. 2009). The Mexican *Magnolia tamaulipana* (Magnoliaceae) uses thermogenetic flowers that spread floral odour to attract nocturnal *Cyclocephala* species (Scarabaeidae: Dynastinae) and reward them with carbohydrate-rich petals (Dieringer et al. 1999). The Australian Magnoliales *Eupomatia laurina* is pollinated by weevils that feed on special odour-emitting staminodes. In addition, beetles also use magnoliid flowers as mating sites and brood substrate (Armstrong and Irvine 1990). During mating and feeding, the body of a beetle becomes covered in pollen, which they then transport to the next flower. A similar system was described for *Annona coriacea* (Annonaceae) (Costa et al. 2017) and the species of the *Philodendron bipinnatifidum* complex (Araceae) in Brazil (Gottsberger et al. 2013). These plants are pollinated by dynastid scarab beetles from the genus *Cyclocephala*, which have also been recorded to feed on the floral and inflorescence parts.

## 13.2.2 *Pollen Feeding in Flower-Visiting Beetles*

### 13.2.2.1 Hairs, Bristles and Setae: Hairy Mouthparts Are the Key to Success

Pollen is a nutritious food source for many flower-visiting insects, containing various lipids, carbohydrates, vitamins, minerals and different types of protein (Kevan and Baker 1983; Roulston and Cane 2000; Johnson and Nicolson 2001; Cook et al. 2004). It has also been recorded to have a positive effect on survival and reproductive fitness of beetles (Cook et al. 2004). Consequently, pollen feeding has been recorded for numerous species across several families of beetles (Johnson and Nicolson 2001; Karolyi et al. 2009). Nevertheless, pollen grains are highly resistant, durable particles, encased by a hard and highly resistant outer exine (Johnson and Nicolson 2001) that challenges pollen feeders to gain access to the encapsulated nutrition.

The mouthparts of pollen-feeding beetles are characteristically equipped with hairs, bristles or setae (Fig. 13.3). These structures are responsible for collecting and transporting food inside the preoral cavity, which is enclosed by the mouthparts. A

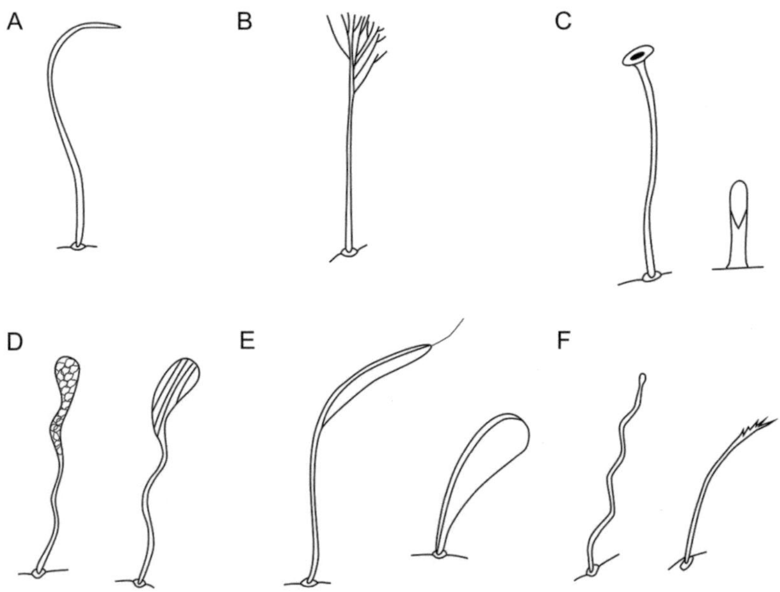


**Fig. 13.3** Different maxilla bristles of various pollen-feeding beetles. (**a**) Hooked bristle of *Cteniopus flavus* (Alleculidae). (**b**) Branched bristles as can be found on the galea tip of *Nemognatha chrysomelina* (Meloidae). (**c**) *Malachius bipustulatus* (Melyridae), trumpet- and spoon-shaped bristles on maxilla and labium. (**d**) Adhesion surfaces with sculptured cuticle. (**e**) Variously broadened tips without sculptured surfaces of *Rutpela maculata* (Cerambycidae) and *Nemognatha* sp. (Cantharidae). (**f**) *Cetonia aurata*, wavy bristles on the galea and serrated lacinia bristles (modified after Schicha 1967; Fuchs 1974; Karolyi et al. 2009)

special feature of pollen feeders is that, instead of the mandibles, the paired maxillae are responsible for taking up pollen and for transporting it into the cibarium. Therefore, they represent the primary structures for pollen uptake. Specialised flower-visiting Scarabaeidae generally show a conspicuous relationship between the number and shape of galea bristles, where the number of bristles is reduced with an increasing differentiation of the shape and surface structure. A study comparing seven genera of anthophilous Scarabaeidae has shown a correlation between feeding type and the development of the maxillae (Fuchs 1974). The increasing specialisation towards pollen feeding resulted in a reduction of the maxillary teeth and the development of highly specialised galea bristles to take up pollen and nectar from flowers (Fuchs 1974). Each maxilla possesses a distal galea with numerous long hairs, bristles or setae to brush up pollen grains from flowers in a semi-circular action of the maxillae. In addition, a bristled lacinia transports food inside the cibarium. The hairs and bristles on the galea are often conspicuously shaped. For example, spatula-shaped, hooked or curled bristles provide structures to which pollen grains adhere. Further, the common pollen-feeding scarab beetle *Trichius fasciatus* possesses galea bristles with a corrugated surface that interact with the sculptured pollen exine. The scarab *Valgus hemipterus* has a knoblike surface structure on the galea bristles that enhance the attachment of sticky pollen and adhesive nectar (Fuchs 1974).

The rather small mandibles of pollen-feeding beetles are characterised by a large membranous lacinia mobilis to transport pollen grains within the preoral cavity. In addition, large molar parts, which are equipped with cuticular knops, teeth or spikes, are able to crush and grind pollen (Schremmer 1961; Schicha 1967; Matthes 1991; Karolyi et al. 2009, 2016).

#### 13.2.2.2 Monkey Beetles: Megadiverse Pollen Harvesters in South Africa

The highly specious monkey beetles (Scarabaeidae: Hopliini) are part of one of the characteristic insect pollination systems in South Africa's floristic hotspot of the Greater Cape Region (Goldblatt et al. 2000; Colville et al. 2014). Their global centre of diversity and adaptive radiation is centred within South Africa with roughly 63% of the world's species and 38% of the genera concentrated here (Colville 2009). Based on flower colour preference, feeding behaviour and mouthpart morphology, they can be classified into feeding guilds, including specialised pollen and/or nectar feeders (Picker and Midgley 1996; Karolyi et al. 2016).

Adaptations for pollen feeding in monkey beetles can be seen in several modifications of the mandible and the maxilla (Fig. 13.4e–h). They are characterised by a prominent, toothed mola and a large, densely bristled and sometimes toothed lacinia mobilis as well as a bristled galea on the maxilla. The galeae act as the primary organs for pollen uptake, with the sclerotized hairs functioning as an ideal pollen brush. Although the lacinia is reduced in specialised hopliine pollen feeders, the remaining hairy comb of the lacinia conveys pollen grains to the mandibles. On the

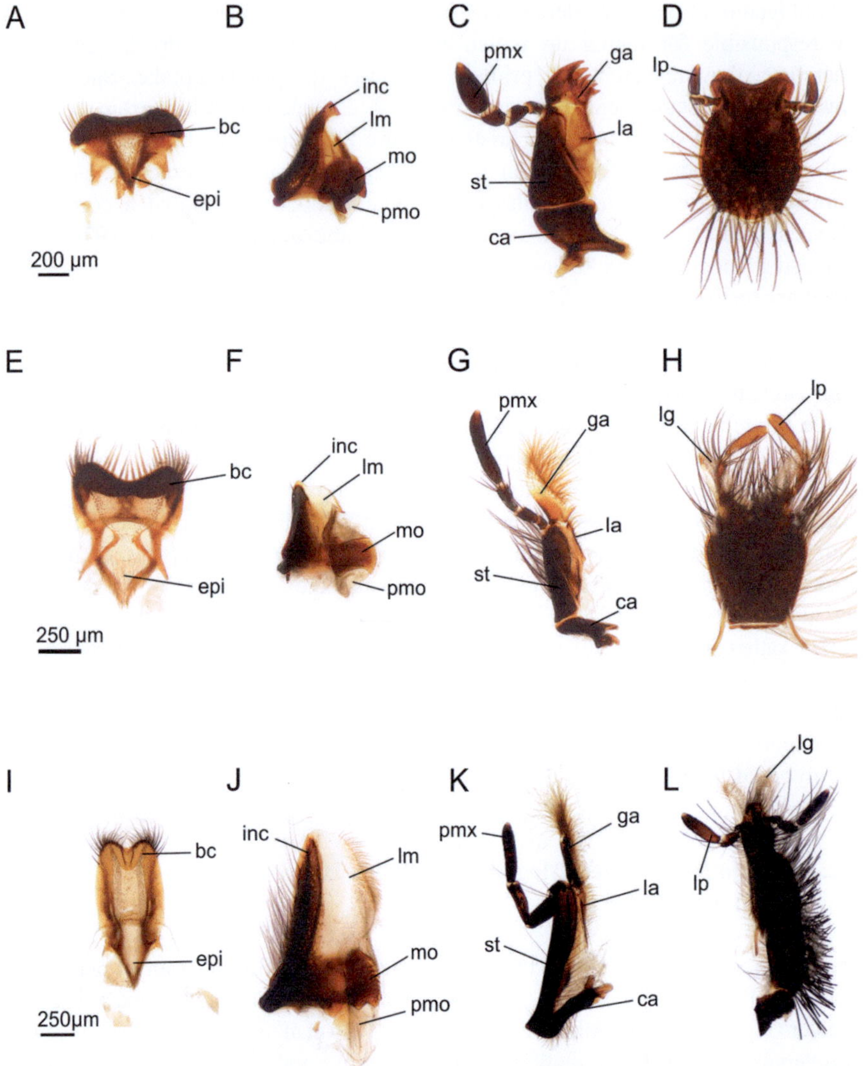


**Fig. 13.4** Mouthparts of three different hopliine (Rutelinae, Scarabaeidae) according to their designated feeding guilds. (**a–d**) Heavy sclerotized mouthparts of flower-feeding *Mauromecistoplia nieuwoudtvillensis*: (**a**) Labrum with hair crest and epipharynx. (**b**) Mandible with conspicuously formed incisive and prominent molar part. (**c**) Maxilla with sclerotized teeth on the galea. (**d**) Rounded labium. (**e–h**) Mouthparts of pollen-feeding *Lepithrix* sp. (**e**) Labrum with hair crest at epipharynx. (**f**) Mandible displays a membranous lacinia mobilis with a proximal, sclerotized tooth and well-developed molar region. (**g**) Maxilla with haired, membranous galea. (**h**) Labium. (**i–l**) Elongated mouthparts of nectar-feeding *Anisonyx ursus*. (**i**) Weaker sclerotized labrum and epipharynx. (**j**) Mandible with prominent, membranous lacinia mobilis and relatively small molar part. (**k**) Maxilla with conspicuously haired galea. (**l**) Elongated labium. Abbreviations: *bc* hair crest; *epi* epipharynx; *ca* cardo; *ga* galea; *inc* incisivus; *la* lacinia; *lg* ligula; *lm* lacinia mobilis; *lp* labial palpus; *mo* mola; *pmo* postmola; *pmx* palpus maxillaris; *st* stipes (modified from Karolyi et al. 2016)

mandible, the lobe-like lacinia mobilis, with its hooked tooth, functions as a pollen-collecting device inside the cibarium, where during movement of the mandibles the lacinia mobilis acts as a scraper, wiping pollen grains to the molar part. In addition, specialised pollen-feeding hopliine species possess a prominent mola with numerous rows of teeth that may be used for perforating pollen grains (Karolyi et al. 2016).

#### 13.2.2.3 *Cetonia aurata*: Shiny Examples for a Pollen Mop

The European rose beetle *C. aurata* (Scarabaeidae) is mainly known as regular visitor of cultivated and wild Rosaceae. These green shiny flower visitors are often believed to be destructive to their host flowers, although they are obligatory pollen feeders (Schremmer 1961; Bürgis 1986, 1987, 1989; Matthes 1991). Using cryo-SEM, Karolyi et al. (2009) investigated the bristles and brushes found on the mouthparts of *C. aurata*. Similarly to other pollen-feeding beetles, the mouthparts of *C. aurata* function together as an effective pollen-harvesting and pollen-processing device (Fig. 13.5a–d). The maxilla possesses numerous long hairs on the soft galea and a comb of bristles on the lacinia. The mandible lacks any cutting edges or biting structures but possesses a large, membranous lacinia mobilis densely covered with hairs. However, the hairy mouthparts of *C. aurata* are covered with a liquid layer of nectar, mixed with saliva to a grainy flower cocktail (Fig. 13.5e–g). The liquid improves the adhesion of pollen grains and transforms the brushlike structures into moist pollen mops. The lacinia mobilis functions as a pollen conveying structure by sweeping pollen between the mola surfaces that work as a pollen mill, crushing pollen grains before they are swallowed. A moist fringe prevents pollen from slipping off the grinding surface. These moistened bristles represent a unique adaptation in pollen-feeding beetles and represent a highly effective method of taking up powdery and dry pollen from flowers.

#### 13.2.2.4 Malachiinae: Highly Effective Pollen-Harvesting Beetles

European malachid beetles are highly adapted, efficient pollen feeders that prefer grass pollen (Poaceae) as their main host plants. These are typically wind pollinated and therefore have dry and powdery pollen grains with a smooth surface. They are not covered with pollen kit, but with a thin layer of an oily substance (Pohl 1929). Schicha (1967) investigated the functional mouthpart morphology and feeding behaviour of several different Malachiinae species in a comparative study. In order to feed on this pollen, malachid beetles evolved unique morphological adaptations of their mouthparts (Schicha 1967). *Malachius bipustulatus* (Melyridae, Malachiinae) represents an especially well-investigated species concerning the mouthpart morphology and mode of nutrition (Fig. 13.6).

The pyramidal mandibles bear two heavily sclerotized apical teeth that are used as pincers for opening the anthers of a flower and a rather small lacinia mobilis and a sclerotized molar part (Fig. 13.2c). The mandibles appear to have undergone a functional change from biting to chewing. The incisivi can be regarded as a remnant

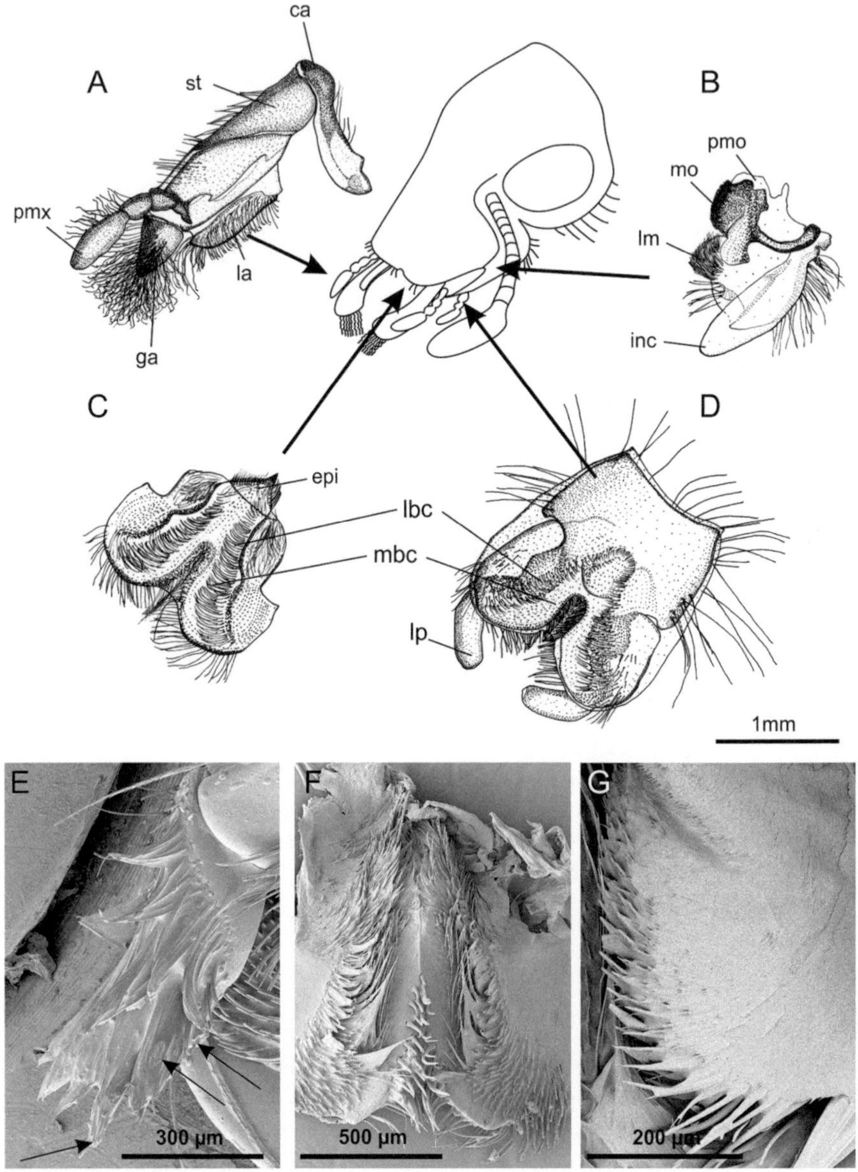


**Fig. 13.5** The head and mouthparts of *Cetonia aurata* (Scarabaeidae); (**a**–**d**) modified and redrawn from Karolyi et al. (2009). (**a**) Maxilla swivels out laterally and extends forward. During its retraction, the galea turns inward sweeping pollen to the lacinia. (**b**) Mandible with nonbiting incisivus, lacinia mobilis and pollen-crushing mola. (**c**) Labrum, buccal view with lateral and median bristle crests. The smooth dorsal side lies directly under the heavily sclerotized clypeus. (**d**) Buccal side of labium with bristle crests. (**e**–**g**) Cryo-SEM micrographs. (**e**) Galea tip, bristles forming a wet brush. Arrows indicate pollen grains that adhere to the liquid layer. (**f**) Ventral view of labrum with lateral and median bristle crest and epipharynx. All bristles are covered with a fluid

of their basal carnivore lifestyle. The maxillae, however, show no biting structures. Instead, distinctive flexible hairs or setae can be found on the galea and lacinia of the maxilla, as well as on the ligula of the labium (Fig. 13.6c, d). These bristles are not rigid, but flexible with a special abutment in the soft cuticula. The soft, shovel-shaped galea of the maxilla is rounded and bears numerous trumpet-shaped bristles. In addition, the dolabriform, soft lacinia is densely equipped with conspicuous spoon-shaped bristles, which can also be found on the ligula of the labium. Similar to the trumpet-shaped bristles on the galea, these labial bristles are flexible and are

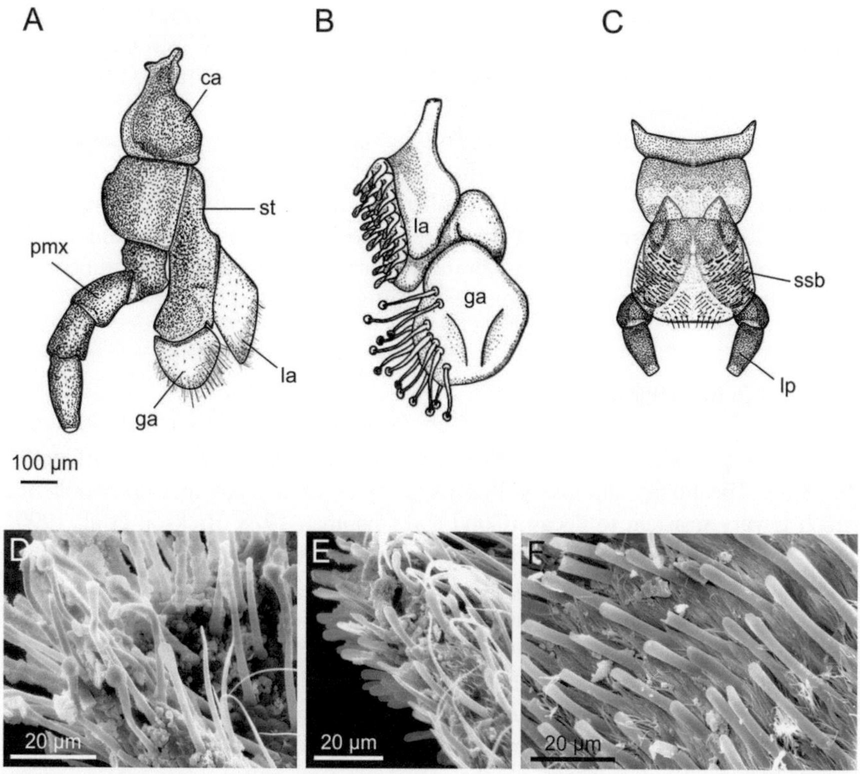


**Fig. 13.6** *Malachius bipustulatus* (Melyridae) mouthparts (mandible see Fig. 13.2b). (**a**) Maxilla. (**b**) Details of the maxilla: galea with trumpet-shaped bristles, lacinia with spoon-shaped bristles. (**c**) Dorsal view of the labium with spoon-shaped bristles (**d**) Trumpet-shaped bristles on the galea (SEM). (**e**) Spoon-shaped bristles on the lacinia and (**f**) on the lateral labium (SEM). Abbreviations: *ca* cardo; *ga* galea; *la* lacinia; *lp* labial palpus; *pmx* palpus maxillaris; *ssb* spoon-shaped bristles; *st* stipes

**Fig. 13.5** (continued) layer. (**g**) Lacinia mobilis of the mandible with liquid-covered bristles that form a pollen-trapping and transport device. Abbreviations: *ca* cardo; *epi* epipharynx; *inc* incisivus; *ga* galea; *la* lacinia; *lbc* lateral bristle crest; *lm* lacinia mobilis; *lp* labial palpus; *mbc* median bristle crest; *mo* mola; *pmo* postmola; *pmx* palpus maxillaris; *st* stipes

responsible for transporting pollen within the cibarium. Both laciniae and ligula form a functional unit within the cibarium. Laterally inbounded bristles on the lacinia, together with dorsally inbound setae on the ligula, transport pollen grains inside the cibarium.

Pollen uptake and transport within the cibarium is accomplished by the maxillae. By stretching them forward, the galeae get in contact with the flower, and pollen is gathered between the trumpet-shaped bristles. During retraction, the maxillae proceed in a median gathering-type movement, and pollen is transported to the labium. Here, the spoon-shaped setae on the labium sweep through the galea bristles and remove the pollen. Finally, during the next movement of the maxilla, the laciniae perform a complex twisting movement, and the spoons on the laciniae brush the pollen between the mandibles where they are kneaded and transported to the pharynx. The membranous lacinia mobilis forms the ventral closure of the cibarium and prevents pollen grains from falling off the maxillary feeder band. It also acts as an additional pollen transport area during feeding. In addition, epipharyngeal bristles on the ventral side of the labrum retain pollen grains that are moved to the mandibles by back and forth movements of the labrum. Together with the tonguelike hypopharynx on the dorsal labium, the labrum and labium have a supporting function in holding the pollen between the molar parts of the mandibles.

#### 13.2.2.5 Pollen Digestion

In addition to pollen uptake, beetles have to deal with the challenge of pollen digestion. The highly durable pollen exine includes the polymer sporopollenin, which is very resistant to decay (Good and Chapman 1978; Rickson et al. 1990). Therefore, beetles have evolved various strategies to overcome the resistant exine of pollen grains. The most common strategy is to crush pollen grains with the heavy sclerotized and sometimes toothed molar part of the mandibles (Schicha 1967; Johnson and Nicolson 2001; Karolyi et al. 2009, 2016).

*Cyclocephala amazona* (Scarabaeidae) from Costa Rica uses gastroliths provided by its host plant, the peach palm *Bactris gasipaes*, to grind pollen grains in the gut. The plant cells of the peach palm have thick, highly lignified cell walls that are ingested by the beetles. These are then used to crush pollen gains inside the gut (Rickson et al. 1990). In protea beetles (Cetoniini) and monkey beetles (Hopliini), weakening of the exine due to osmotic concentration differences in the gut, combined with digestive enzyme penetration, has been described (Johnson and Nicolson 2001).

### *13.2.3 Nectar Feeding*

Some flower-visiting beetles are known to use nectar as a high-energy source. For example, nectar has been shown to be an important source for longevity of cerambycid beetles (Wang and Chen 2005). Nectar is mainly a sugar solution with concentrations ranging between 15 and 60%, depending on the plant species (Daniel et al. 1989; Parachnowitsch et al. 2018). Nectar may also contain small amounts of amino acids, proteins, organic acids, phosphates, vitamins and enzymes (Baker and Baker 1973; Kevan and Baker 1983; Nicolson 2007). Most plants attract fast-flying flower visitors, such as bees and flies. Some plant species however, have adapted their flower morphology towards attracting nectar-feeding Coleoptera (Goldblatt and Manning 2011). Nectar feeding can be regarded as a non-destructive type of flower visiting and requires special mouthpart structures for directing the capillary flow of nectar towards the mouth (Labandeira 1997). Such elongated proboscis-like structures that take up small amount of nectar per flower are realised in only a few taxa of anthophilous Coleoptera.

#### 13.2.3.1 Elongated Mouthparts in Nectar-Feeding Meloidae

Although the vast majority of Meloidae possess prognathous, unmodified biting-chewing mouthparts, some species, evolved elongated mouthparts that are adapted to take up concealed nectar from flowers (Schremmer 1961). Nectar feeding appears to have evolved at least two times within Meloidae, and mouthparts adapted to concealed nectar are found in at least three taxa in the Nemognathini (Handschin 1929; Kaszab 1963; Bologna and Pinto 2001; Bologna et al. 2008; Wilhelmi and Krenn 2012). Adaptations resulted in either an elongation of the maxillary palps, as seen in species of *Leptopalpus*, or an elongation of the hairy galea, as seen in representatives of *Nemognatha* and *Gnathium* (Fig. 13.7). In contrast to nectar-feeding Lepidoptera, Apidae and Diptera, the proboscis-like mouthparts of Meloidae do not form a median food canal. Instead, nectar is most likely transported via capillary action along bristles on the galea or the maxillary palps, supported by a suction pump in the head (Wilhelmi and Krenn 2012). For example, *L. rostratus* inserts its elongated maxillary palps into florets of Asteraceae, and nectar is imbibed via capillary action of the hairs on the inner side of the palps, galea and lacinia (Fig. 13.7a–c). In *N. chrysomelina*, branched bristles on tip of the galea (Fig. 13.7d–f) act as an effective brushlike tongue to take up nectar from thistle flowers.

Nectar-feeding Meloidae also consume pollen, using their mandibles. In contrast to other pollen-feeding Coleoptera, their mandibles consist of a large, sclerotized incisivi part but only a small, bristled lacinia mobilis and molar part. The hairy galea and lacinia of *Leptopalpus* and *Nemognatha* suggest that they are also used as organs for pollen uptake, possibly taking up pollen at the same time as the proboscis-like maxillary palps are inserted into the inflorescence of the host flower to take up nectar (Wilhelmi and Krenn 2012). Unfortunately, studies of their flower-visiting behaviour are lacking. These would be important to estimate the pollination potential of this group, as specialised flower visitors are not necessarily pollinators of their host plants.

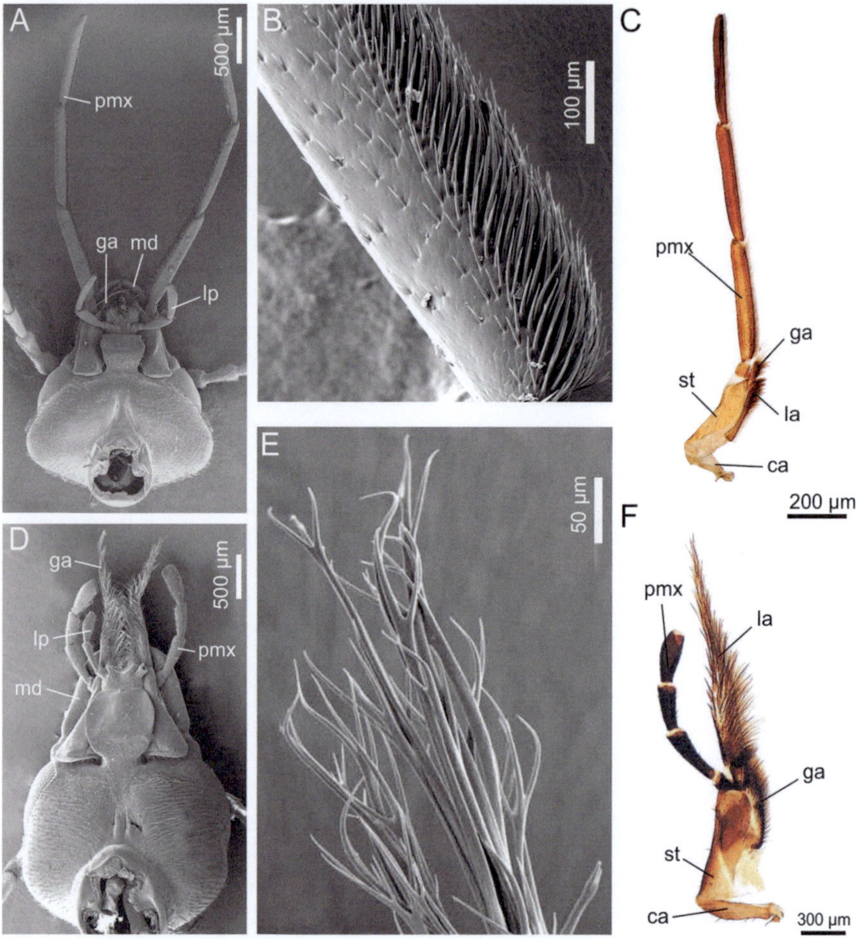


**Fig. 13.7** Elongated mouthparts of Meloidae (new arrangement of micrographs from Wilhelmi and Krenn 2012). (**a**–**c**) *Leptopalpus rostratus*. (**a**) Ventral view of the head (SEM). (**b**) Second palp segment of the maxilla with median bristle row (SEM). (**c**) Ventral view of right maxilla with elongated palpus. (**d**–**f**) *Nemognatha chrysomelina*. (**d**) Ventral view of the head (SEM). (**e**) Branched bristles at the galea tip (SEM). (**f**) Ventromesal view of the right maxilla with elongated, hairy maxilla. Abbreviations: *ca* cardo; *ga* galea; *la* lacinia; *lp* labial palp; *md* mandible; *pmx* maxillary palp; *st* stipes

### 13.2.3.2 Nectar-Feeding Monkey Beetles

Several monkey beetle species of the Greater Cape Floristic Region appear to take up nectar as their primary food source. Monkey beetle species characterised as nectar feeders possess long and slender mouthparts (Fig. 13.4i–l), with the labrum rather weakly sclerotized, compared to other flower- and pollen-feeding monkey beetle species. The broad shape and soft texture of the lacinia mobilis also suggests the ability to sweep nectar into the mouth. The long and slender maxillae, with the hirsute and sclerotized galea, are most likely adapted for probing into flowers and for mopping-up nectar. The long and slender cardo found in nectar-feeding monkey beetle species presumably allows protrusion into the flower and a higher movability of the distal maxilla parts during food uptake. The elongated labium, together with the elongated, hirsute ligulae, can also be considered as an adaptation towards nectar feeding (Karolyi et al. 2016). As in Meloidae, more detailed studies on the flower-visiting behaviour would be important to investigate the fluid-feeding technique and the role that nectar-feeding species play as pollinators.

### 13.2.4 Pollen or Nectar Feeding or a Mixed Cocktail?

Investigating the functional mouthpart morphology of flower-visiting beetles usually gives indications about their feeding preferences. However, several species are difficult to categorise, as some anthophilous beetle species show mouthpart adaptations for a dual diet, suggesting that they consume a nutritious mixed cocktail of pollen and nectar (Karolyi et al. 2016). For example, preliminary results of gut examinations in monkey beetles revealed beetles to be rather generalists than specialists when it comes to floral food sources (Karolyi et al. 2016). In addition, various species of Scarabaeidae, Curculionidae, Cerambycidae, Chrysomelidae, Staphylinidae and Elateridae that visit tropical palm trees (Arecaceae) have been observed to feed on both floral tissue, nectar and pollen (Küchmeister et al. 1997). Without detailed experimental studies, the specialisation to certain food types remains unclear in many species of anthophilous Coleoptera.

## 13.3 How Flowers Attract Anthophilous Beetles: When Beetles See Red

Although Coleoptera are generally messy pollinators, numerous plant species from various families rely on beetle pollination. Many plants pollinated by flower-visiting insects have evolved common characters regardless of taxonomic relationships. These patterns of common pollination syndromes (Proctor et al. 1996), such as floral odour, shape, size and colour, stimulate the visual and olfactory organs of their particular pollinators. Flower-visiting insects use these floral cues to locate flowers and to discriminate between flower species that offer different food sources (Chittka and Raine 2006). Since many diurnal insects are believed to have a well-developed colour vision, colour is regarded to play a key role in flower-pollinator interactions and is used by insects to locate flowers (Steiner 1998a; Martínez-Harms et al. 2012; Reverté et al. 2016). However, angiosperms like Magnoliaceae, Arecaceae, Araceae and Annonaceae rely on odour to attract beetle pollinators from various families, especially when they are reliant on nocturnal beetle species (Armstrong and Irvine

1990; Dieringer et al. 1999; Gibernau et al. 1999; Thien et al. 2009; Gottsberger et al. 2013; Costa et al. 2017). Nonetheless, many flower-visiting Coleoptera shifted to colour as primary cue (Dafni et al. 1990; Picker and Midgley 1996; Steiner 1998a; Goldblatt et al. 2005; Van Kleunen et al. 2007; Martínez-Harms et al. 2012). In beetles where adult activities on flowers are associated with feeding and mating, flower odour (Steiner 1998b; Van Kleunen et al. 2007) and colour play an important role in the location of flowers and brightly coloured mates (Wang and Chen 2005; Keasar et al. 2010; Martínez-Harms et al. 2012). Red sensitivity can be regarded as an important key stimulus in anthophilous beetles and might have evolved independently more than once within Coleoptera (Martínez-Harms et al. 2012). Convergent evolution amongst phylogenetically distant plant species resulted in a range of floral traits to attract beetles. Indeed, specialised beetle pollination systems are found in various plant species, especially in the Cape and KwaZulu-Natal grasslands of South Africa, as well as in the Middle East region (Dafni and Potts 2004).

Within plant-insect interactions, Coleoptera represent a frequent group of flower visitors, and some groups show certain colour preferences (Picker and Midgley 1996; Steiner 1998b; Peter and Johnson 2009; Keasar et al. 2010). Typically, beetles are attracted to bowl- or cup-shaped flowers or dense, capitate inflorescences. Cantharophilous flowers are generally scentless, display red to orange-yellow colours without any UV reflection and often have a dark contrasting centre (Dafni et al. 1990; Steiner 1998b; Goldblatt et al. 2002; Ollerton et al. 2003; Van Kleunen et al. 2007). For example, Mediterranean scarabs from the genus *Amphicoma* and *Pygopleurus* (Glaphyridae) show an affinity towards red bowl-shaped flowers from different families, including Liliaceae, Papaveraceae and Ranunculaceae (Dafni et al. 1990; Martínez-Harms et al. 2012). Flowers of this "poppy guild" are odourless and nectarless and display a dark centre which results in an attractive red-black contrast for beetles. It is worth noting that a geographic colour distribution exists within the genera of the "poppy guild". This is especially true for the plant genus *Ranunculus*: only 3 out of 400 species worldwide display red flowers, and all occur in the Mediterranean region which also represents the centre of speciation for the beetle genus *Amphicoma*. Species of this genus have mouthparts that are partly adapted to pollen feeding (Nel and Scholtz 1990). In addition, all members of the guild usually reflect UV light and are pollinated by bees. However, in the Eastern Mediterranean, flowers lack UV reflection, indicating a convergence to beetle pollination (Dafni et al. 1990; Dafni and Potts 2004). Scentless, red bowl-shaped flowers can therefore be considered as typical of beetle pollination syndrome in this pollination system.

Furthermore, South African asclepiads (Apocynaceae: Asclepiadoideae), amongst others, rely on pollination by fruit chafer beetles (Cetoniini) where they display a highly evolved floral morphology (Ollerton et al. 2003). In South Africa, several orchid species of the genus *Eulophia* are known to be pollinated by cetoniid and elaterid beetles (Peter and Johnson 2006, 2009). Although non-rewarding, these orchids display dense inflorescences that are usually associated with beetle pollination. Further, they co-occur with rewarding *Helichrysum nudifolium* and attract beetles by colour mimicry of other co-occurring, rewarding Asteraceae. The attraction with colour by orchids has already been demonstrated for the South African genus *Ceratandra*, which shifted from an oil-collecting bee pollination system to a non-rewarding beetle pollinating system (Steiner 1998a).

Dark markings in Iridaceae, as well as yellow or red flower colour in iris and in many daisy species, play an important role in the attraction of monkey beetles in the Greater Cape Floristic Region (Steiner 1998b; Goldblatt et al. 2002; Ollerton et al. 2003; Van Kleunen et al. 2007). For example, of the four pollination groups that are found in *Romulea*, one of them is exclusively pollinated by monkey beetles (Goldblatt et al. 2002). Further, monkey beetles are responsible for pollinating three species of spring-flowering peacock *Moraea* and a group of *Romulea* (both Iridaceae) in the southwestern Cape Region (Steiner 1998b; Goldblatt et al. 2002). These originally bee-pollinated *Moraea* species shifted to beetle pollination and therefore display flowers with iridescent spots on the broad outer tepals that are thought to attract monkey beetles.

## 13.4 Conclusion

The history of flower-visiting Coleoptera dates back to the Early Cretaceous, and numerous studies have altered our understanding of the role of beetles as flower visitors and pollinators. Of particular note is the fact that pollinating Coleoptera often have rather unspecialised mouthparts, while species with mouthparts adapted to pollen and nectar are not necessarily pollinating their host plant. Species that specialised on pollen feeding evolved a conspicuous mouthpart type which can be characterised by a loss of biting structures on the maxillae and mandibles. Instead, the membranous galea is equipped with various differentiated setae and responsible for food uptake, whereas an increasing specialisation towards pollen feeding has led to a decreased number of setae on the galea. In return, these bristles are typically modified for pollen uptake. The pollen-feeding mandible has a rounded apex, a soft membranous lacinia mobilis for transporting pollen in the preoral cavity and a prominent molar part for crushing or grinding pollen grains.

Broadly conceived studies of the mouthpart morphology in addition to investigations of the flower-visiting and flower-feeding behaviour in the field are required to fully understand food preferences and feeding ecology of cantharophilous Coleoptera. Further, detailed studies about the alimentary tract are missing to answer the question of how beetles are actually able to digest pollen.

## References

Armstrong JE, Irvine AK (1990) Functions of staminodia in the beetle-pollinated flowers of *Eupomatia laurina*. Biotropica 22:429–431. https://doi.org/10.2307/2388563

Baker HG, Baker I (1973) Amino-acids in nectar and their evolutionary significance. Nature 241:543–545. https://doi.org/10.1038/241543b0

Bauder JAS, Lieskonig NR, Krenn HW (2011) The extremely long-tongued Neotropical butterfly *Eurybia lycisca* (Riodinidae): proboscis morphology and flower handling. Arthropod Struct Dev 40:122–127. https://doi.org/10.1016/j.asd.2010.11.002

Beach JH (1982) Beetle pollination of *Cyclanthus bipartitus* (Cyclanthaceae). Am J Bot 69:1074–1081. https://doi.org/10.1002/j.1537-2197.1982.tb13352.x

Bernhardt P (2000) Convergent evolution and adaptive radiation of beetle-pollinated angiosperms. Plant Syst Evol 222:293–320. https://doi.org/10.1007/BF00984108

Betz O, Thayer MK, Newton AF (2003) Comparative morphology and evolutionary pathways of the mouthparts in spore-feeding Staphylinoidea (Coleoptera). Acta Zool 84:179–238. https://doi.org/10.1046/j.1463-6395.2003.00147.x

Bologna MA, Pinto JD (2001) Phylogenetic studies of Meloidae (Coleoptera), with emphasis on the evolution of phoresy. Syst Entomol 26:33–72. https://doi.org/10.1046/j.1365-3113.2001.00132.x

Bologna MA, Oliverio M, Pitzalis M, Mariottini P (2008) Phylogeny and evolutionary history of the blister beetles (Coleoptera, Meloidae). Mol Phylogenet Evol 48:679–693. https://doi.org/10.1016/j.ympev.2008.04.019

Bürgis H (1986) Blütenfressende Blatthornkäfer (Anthophaga), Der Rosenkäfer *Cetonia aurata* und seine Verwandten. 2. Das Zusammenspiel der Mundwerkzeuge. Mikrokosmos 75:327–330

Bürgis H (1987) Blütenfressende Blatthornkäfer (Anthophaga), Der Rosenkäfer *Cetonia aurata* und seine Verwandten. 3. Der Kaubezirk. Mikrokosmos 76:346–366

Bürgis H (1989) Blütenfressende Blatthornkäfer (Anthophaga), Der Rosenkäfer *Cetonia aurata* und seine Verwandten. 4. Die Nahrung des Rosenkäfers *Cetonia*. Mikrokosmos 78:50–53

Chittka L, Raine NE (2006) Recognition of flowers by pollinators. Curr Opin Plant Biol 9:428–435. https://doi.org/10.1016/j.pbi.2006.05.002

Colville JF (2009) Understanding the evolutionary radiation of the mega-diverse monkey beetle fauna (Scarabaeidae: Hopliini) of South Africa. Front Biogeogr 1:56–61

Colville JF, Potts A, Bradshaw PL, Measey GJ, Snijman D, Picker MD (2014) Floristic and faunal Cape biochoria: do they exist? In: Allsopp N, Colville JF, Verboom A (eds) Fynbos: ecology, evolution, and conservation of a megadiverse region, first. Oxford University Press, Oxford

Cook SM, Murray DA, Williams IH (2004) Do pollen beetles need pollen? The effect of pollen on oviposition, survival, and development of a flower-feeding herbivore. Ecol Entomol 29:164–173. https://doi.org/10.1111/j.0307-6946.2004.00589.x

Costa MS, Silva RJ, Paulino-Neto HF, Pereira MJB (2017) Beetle pollination and flowering rhythm of *Annona coriacea* Mart. (Annonaceae) in Brazilian cerrado: behavioral features of its principal pollinators. PLoS One 12:e0171092. https://doi.org/10.1371/journal.pone.0171092

Dafni A, Potts SG (2004) The role of flower inclination, depth, and height in the preferences of a pollinating beetle (Coleoptera: Glaphyridae). J Insect Behav 17:823–834. https://doi.org/10.1023/B:JOIR.0000048991.45453.73

Dafni A, Bernhardt P, Shmida A, Ivri BY, Greenbaum S, O'Toole C, Losito L (1990) Red bowl-shaped flowers: convergence for beetle pollination in the Mediterranean region. Isr J Bot 39:81–92. https://doi.org/10.1080/0021213X.1990.10677134

Daniel TL, Kingsolver JG, Meyhöfer E (1989) Mechanical determinants of nectar-feeding energetics in butterflies: muscle mechanics, feeding geometry, and functional equivalence. Oecologia 79:66–75. https://doi.org/10.1007/BF00378241

Dieringer G, Cabrera RL, Lara M, Loya L, Reyes-Castillo P (1999) Beetle pollination and floral thermogenecity in *Magnolia tamaulipana* (Magnoliaceae). Int J Plant Sci 160:64–71. https://doi.org/10.1086/314099

Frame D (2003) Generalist flowers, biodiversity and florivory: implications for angiosperm origins. Taxon 52:681–685. https://doi.org/10.2307/3647343

Fuchs G-V (1974) Die Gewinnung von Pollen und Nektar bei Käfern. Nat Mus 104:45–54

Gibernau M, Barabé D, Cerdan P, Dejean A (1999) Beetle pollination of *Philodendron solimoesense* (Araceae) in French Guiana. Int J Plant Sci 160:1135–1143

Goldblatt P, Manning JC (2011) Hopliine beetles (Scarabaeidae: Rutelinae: Hopliini), specialized pollinators of the Southern African flora. Curtis's Bot Mag 28:238–259. https://doi.org/10.1111/j.1467-8748.2011.01752.x

Goldblatt P, Manning JC, Bernhardt P (1998) Pollination of petaloid geophytes by monkey beetles (Scarabaeidae: Rutelinae: Hopliini) in Southern Africa. Ann Mo Bot Gard 85:215–230
Goldblatt P, Manning JC, Bernhardt P (2000) Adaptive radiation of pollination mechanisms in *Sparaxis* (Iridaceae: Ixioideae). Direct 22:57–70
Goldblatt P, Bernhardt P, Manning JC (2002) Floral biology of *Romulea* (Iridaceae: Crocoideae): a progression from a generalist to a specialist pollination system. Adansonia 24:243–262
Goldblatt P, Bernhardt P, Manning JC (2005) Pollination mechanisms in the African genus *Moraea* (Iridaceae, Iridoideae): floral divergence and adaptation for pollinators. Adansonia 27:21–46
Goldman EH (1933) Comparisons of the mouth-parts of adult longhorn beetles with reference to their food (Coleoptera: Cerambycidae). Trans Am Entomol Soc 59(2):85–102
Good BH, Chapman RL (1978) The ultrastructure of *Phycopeltis* (Chroolepidaceae: Chlorophyta). I. Sporopollenin in the cell walls. Am J Bot 65:27–33. https://doi.org/10.1002/j.1537-2197.1978.tb10830.x
Gottsberger G (1974) The structure and function of the primitive angiosperm flower - a discussion. Acta Bot Neerl 23:461–471. https://doi.org/10.1111/j.1438-8677.1974.tb00962.x
Gottsberger G (1977) Some aspects of beetle pollination in the evolution of flowering plants. In: Flowering plants. Springer, Vienna, pp 211–226
Gottsberger G, Silberbauer-Gottsberger I, Dötterl S (2013) Pollination and floral scent differentiation in species of the *Philodendron bipinnatifidum* complex (Araceae). Plant Syst Evol 299:793–809. https://doi.org/10.1007/s00606-013-0763-4
Grimaldi D (1999) The co-radiations of pollinating insects and angiosperms in the Cretaceous. Ann Mo Bot Gard 86:373–406. https://doi.org/10.2307/2666181
Grimaldi D, Engel M (2005) Evolution of the insects. Cambridge University Press, Cambridge, NY
Handschin E (1929) Ein neuer Rüsseltypus bei einem Käfer - Biologische und Morphologische Beobachtungen an *Leptopalpus rostratus* F. Zeitschrift für Morphol und Ökologie der Tiere 14:513–521. https://doi.org/10.1007/BF00419307
Harmon JP, Ganguli AC, Solga MJ (2011) An overview of pollination in rangelands: who, why, and how. Rangelands 33:4–8. https://doi.org/10.2111/1551-501X-33.3.4
Higginson AD, Speed MP, Ruxton GD (2015) Florivory as an opportunity benefit of Aposematism. Am Nat 186:728–741. https://doi.org/10.1086/683463
Johnson SD, Midgley JJ (2001) Pollination by monkey beetles (Scarabaeidae: Hopliini): do color and dark centers of flowers influence alighting behavior? Environ Entomol 30
Johnson SA, Nicolson SW (2001) Pollen digestion by flower-feeding scarabaeidae: protea beetles (Cetoniini) and monkey beetles (Hopliini). J Insect Physiol 47:725–733. https://doi.org/10.1016/S0022-1910(00)00166-9
Karolyi F, Gorb SN, Krenn HW (2009) Pollen grains adhere to the moist mouthparts in the flower visiting beetle *Cetonia aurata* (Scarabaeidae, Coleoptera). Arthropod Plant Interact 3:1–8. https://doi.org/10.1007/s11829-008-9052-5
Karolyi F, Szucsich NU, Colville JF, Krenn HW (2012) Adaptations for nectar-feeding in the mouthparts of long-proboscid flies (Nemestrinidae: *Prosoeca*). Biol J Linn Soc 107:414–424. https://doi.org/10.1111/j.1095-8312.2012.01945.x
Karolyi F, Colville JF, Handschuh S, Metscher BD, Krenn HW (2014) One proboscis, two tasks: adaptations to blood-feeding and nectar-extracting in long-proboscid horse flies (Tabanidae, Philoliche). Arthropod Struct Dev 43:403–413. https://doi.org/10.1016/j.asd.2014.07.003
Karolyi F, Hansal T, Krenn HW, Colville JF (2016) Comparative morphology of the mouthparts of the megadiverse South African monkey beetles (Scarabaeidae: Hopliini): feeding adaptations and guild structure. PeerJ 4:e1597. https://doi.org/10.7717/peerj.1597
Kaszab Z (1963) Merkmale der Adaption, Spezialisation, Konvergenz, Korrelation und Progression bei den Meloiden (Coleoptera). Acta Zool Acad Sci Hung 9:135–175
Keasar T, Harari AR, Sabatinelli G, Keith D, Dafni A, Shavit O, Zylbertal A, Shmida A (2010) Red anemone guild flowers as focal places for mating and feeding by Levant glaphyrid beetles. Biol J Linn Soc 99:808–817. https://doi.org/10.1111/j.1095-8312.2010.01384.x

Kevan PG, Baker HG (1983) Insects as flower visitors and pollinators. Annu Rev Entomol 28:407–453. https://doi.org/10.1146/annurev.en.28.010183.002203
Kirchner O (1911) Blumen und Insekten. Ihre Anpassungen aneinander und ihre Gegenseitige Abhängigkeit. B.G. Teubner, Leipzig
Krenn HW, Plant JD, Szucsich NU (2005) Mouthparts of flower-visiting insects. Arthropod Struct Dev 34:1–40. https://doi.org/10.1016/j.asd.2004.10.002
Küchmeister H, Silberbauer-Gottsberger I, Gottsberger G (1997) Flowering, pollination, nectar standing crop, and nectaries of *Euterpe precatoria* (Arecaceae), an Amazonian rain forest palm. Plant Syst Evol 206:71–97. https://doi.org/10.1007/BF00987942
Labandeira CC (1997) Insect mouthparts: ascertaining the paleobiology of insect feeding strategies. Annu Rev Ecol Syst 28:153–193. https://doi.org/10.1146/annurev.ecolsys.28.1.153
Labandeira CC (2010) The pollination of mid Mesozoic seed plants and the early history of long-proboscid insects. Ann Mo Bot Gard 97:469–513. https://doi.org/10.3417/2010037
Lovell JH (1915) The origin of anthophily among the Coleoptera. Psyche 22:67–84
Martínez-Harms J, Vorobyev M, Schorn J et al (2012) Evidence of red sensitive photoreceptors in *Pygopleurus israelitus* (Glaphyridae: Coleoptera) and its implications for beetle pollination in the southeast Mediterranean. J Comp Physiol A 198:451–463. https://doi.org/10.1007/s00359-012-0722-5
Matthes D (1991) Die Mundwerkzeuge Nektar und Blütenstaub fressender Käfer. Mikrokosmos 80:330–333
McCall AC (2008) Florivory affects pollinator visitation and female fitness in *Nemophila menziesii*. Oecologia 155:729–737. https://doi.org/10.1007/s00442-007-0934-5
McCall AC (2010) Does dose-dependent petal damage affect pollen limitation in an annual plant? Botany 88:601–606. https://doi.org/10.1139/B10-032
McCall AC, Irwin RE (2006) Florivory: the intersection of pollination and herbivory. Ecol Lett 9:1351–1365
Midgley J (1993) Asteraceae - an evaluation of hutchinsons beetle-daisy hypothesis. Bothalia 23 (1):70–72
Nel A, Scholtz CH (1990) Comparative morphology of the mouthparts of adult Scarabaeoidea (Coleoptera). Entomol Mem Dep Agric Dev 80:1–84
Nel P, Bertrand S, Nel A (2018) Diversification of insects since the Devonian: a new approach based on morphological disparity of mouthparts. Sci Rep 8:3516. https://doi.org/10.1038/s41598-018-21938-1
Nicolson SW (2007) Nectaries and nectar. In: Nepi M, Pacini E (eds) Nectaries and nectar. Springer, Dordrecht, p XVIII + 396
Ollerton J, Johnson SD, Cranmer L, Kellie S (2003) The pollination ecology of an assemblage of grassland asclepiads in South Africa. Ann Bot 92:807–834. https://doi.org/10.1093/aob/mcg206
Parachnowitsch AL, Manson JS, Sletvold N (2018) Evolutionary ecology of nectar. Ann Bot 123 (2):247–261. https://doi.org/10.1093/aob/mcy132
Peter CI, Johnson SD (2006) Anther cap retention prevents self-pollination by elaterid beetles in the South African orchid *Eulophia foliosa*. Ann Bot 97:345–355. https://doi.org/10.1093/aob/mcj041
Peter CI, Johnson SD (2009) Pollination by flower chafer beetles in *Eulophia ensata* and *Eulophia welwitschii* (Orchidaceae). S Afr J Bot 75:762–770. https://doi.org/10.1016/j.sajb.2009.07.008
Picker MD, Midgley JJ (1996) Pollination by monkey beetles (Coleoptera: Scarabaeidae: Hopliini): flower and colour preferences. Afr Entomol 4:7–14
Pohl F (1929) Beziehungen zwischen Pollenbeschaffenheit, Bestäubungsart und Fruchtknotenbau. Bot Zbl Beih:247–285
Procheş Ş, Johnson SD (2009) Beetle pollination of the fruit-scented cones of the South African cycad *Stangeria eriopus*. Am J Bot 96:1722–1730. https://doi.org/10.3732/ajb.0800377
Proctor M, Yeo P, Lack A (1996) The natural history of pollination. Harper Collins, London

Ratnayake RMCS, Gunatilleke UN, Wijesundara DSA, Saunders RMK (2007) (Annonaceae): curculionid beetle pollination, promoted by floral scents and elevated floral temperatures. Int J Plant Sci 168:1255–1268. https://doi.org/10.1086/521689

Reverté S, Retana J, Gómez JM, Bosch J (2016) Pollinators show flower colour preferences but flowers with similar colours do not attract similar pollinators. Ann Bot 118:249–257. https://doi.org/10.1093/aob/mcw103

Rickson FR, Cresti M, Beach JH (1990) Plant cells which aid in pollen digestion within a beetle's gut. Oecologia 82:424–426. https://doi.org/10.1007/BF00317493

Roulston TH, Cane JH (2000) Pollen nutritional content and digestibility for animals. Plant Syst Evol 222:187–209. https://doi.org/10.1007/BF00984102

Schicha E (1967) Morphologie und Funktion der Malachiidenmundwerkzeuge unter besonderer Berücksichtigung von *Malachtus bipustulatus* L. (Coleopt., Malacodermata). Z Morphol Okol Tiere 60:376–433. https://doi.org/10.1007/BF00424639

Schnitzler J, Barraclough TG, Boatwright JS, Goldblatt P, Manning JC, Powell MP, Rebelo T, Savolainen V (2011) Causes of plant diversification in the cape biodiversity hotspot of South Africa. Syst Biol 60:343–357. https://doi.org/10.1093/sysbio/syr006

Schremmer F (1961) Morphologische Anpassungen von Tieren - insbesondere Insekten - an die Gewinnung von Blumennahrung. Ver Dtsch Z 25:375–401

Seymour RS, White CR, Gibernau M (2003) Heat reward for insect pollinators. Nature 426:243–244. https://doi.org/10.1038/426243a

Smith SA, Beaulieu JM, Donoghue MJ (2010) An uncorrelated relaxed-clock analysis suggests an earlier origin for flowering plants. Proc Natl Acad Sci U S A 107:5897–5902. https://doi.org/10.1073/pnas.1001225107

Soper Gorden NL, Adler LS (2016) Florivory shapes both leaf and floral interactions. Ecosphere 7: e01326. https://doi.org/10.1002/ecs2.1326

Steenhuisen SL, Johnson SD (2012) Evidence for beetle pollination in the African grassland sugarbushes (Protea: Proteaceae). Plant Syst Evol 298:857–869. https://doi.org/10.1007/s00606-012-0589-5

Steenhuisen SL, Raguso RA, Jürgens A, Johnson SD (2010) Variation in scent emission among floral parts and inflorescence developmental stages in beetle-pollinated *Protea* species (Proteaceae). S Afr J Bot 76:779–787. https://doi.org/10.1016/j.sajb.2010.08.008

Steiner KE (1998a) The evolution of beetle pollination in a South African orchid. Am J Bot 85:1180–1193. https://doi.org/10.2307/2446351

Steiner KE (1998b) Beetle pollination of peacock moraeas (Iridaceae) in South Africa. Plant Syst Evol 209:47–65. https://doi.org/10.1007/BF00991523

Szucsich NU, Krenn HW (2000) Morphology and function of the proboscis in Bombyliidae (Diptera, Brachycera) and implications for proboscis evolution in Brachycera. Zoomorphology 120:79–90. https://doi.org/10.1007/s004350000025

Szucsich NU, Krenn HW (2002) Flies and concealed nectar sources: morphological innovations in the proboscis of Bombyliidae (Diptera). Acta Zool 83:183–192. https://doi.org/10.1046/j.1463-6395.2002.00111.x

Thien LB, Bernhardt P, Devall MS, Chen Z-D, Luo Y-B, Fan J-H, Yuan L-C, Williams JH (2009) Pollination biology of basal angiosperms (ANITA grade). Am J Bot 96:166–182. https://doi.org/10.3732/ajb.0800016

Van Kleunen M, Nänni I, Donaldson JS, Manning JC (2007) The role of beetle marks and flower colour on visitation by monkey beetles (hopliini) in the greater cape floral region, South Africa. Ann Bot 100:1483–1489. https://doi.org/10.1093/aob/mcm256

Wang Q, Chen LY (2005) Mating behavior of a flower-visiting longhorn beetle *Zorion guttigerum* (Westwood) (Coleoptera: Cerambycidae: Cerambycinae). Naturwissenschaften 92:237–241. https://doi.org/10.1007/s00114-005-0615-3

Wardhaugh CW (2015) How many species of arthropods visit flowers? Arthropod Plant Interact 9:547–565. https://doi.org/10.1007/s11829-015-9398-4

Wardhaugh CW, Stork NE, Edwards W, Grimbacher PS (2013) Insects on flowers: the unexpectedly high biodiversity of flower-visiting beetles in a tropical rainforest canopy. Commun Integr Biol 6:1–3

Wilhelmi AP, Krenn HW (2012) Elongated mouthparts of nectar-feeding Meloidae (Coleoptera). Zoomorphology 131:325–337. https://doi.org/10.1007/s00435-012-0162-3

Yavorskaya MI (2018) Evolution of head structures in Coleoptera with special emphasis on the feeding apparatus and miniaturized forms. PhD dissertation, Friedrich-Schiller Universität Jena

# Chapter 14
# Evolution of the Multifunctional Mouthparts of Adult Vespidae

**Volker Mauss, Kenneth Kuba, and Harald W. Krenn**

**Abstract** The morphology of mouthparts and their use in context with nutrition, nesting behaviour and brood provisioning are compared across various lineages of Vespidae. In the plesiomorphic state, adult wasps take up nectar and pollen with a short labiomaxillary complex; females construct subterranean nests with the mandibles, moisten hard soil with water collected using the glossa, and larvae are mass-provisioned with paralysed insect larvae which are gathered and transported by the mandibles. The evolution of nectar-feeding proboscises occurred multiple times in the Vespidae. The Masarinae evolved a bee-like life form and collect pollen to supply their larvae. Two lineages have evolved a proboscis composed of the glossa and paraglossa, or only the elongated glossa forms a food canal to imbibe nectar or water. The females of the taxon [Eumeninae + Raphiglossinae + Zethinae + Polistinae + Vespinae] feed on malaxated prey in addition to sugary fluids. Few representatives of the Eumeninae, the Raphiglossinae and some Zethinae evolved a proboscis composed of the elongated galeae, glossa and paraglossae. Tusk-like mandibles evolved in some male Eumeninae in the context of male-male competition. In the representatives of the [Raphiglossinae + Zethinae + Polistinae + Vespinae] the mandibles have adapted to processing plant material for nest construction. The short mouthparts of the social [Polistinae + Vespinae] also build paper nests and are used for trophallaxis.

## 14.1 Introduction

The mandibulate mouthparts of Hymenoptera are characterized by the labiomaxillary complex which is formed by the functional interaction of basal components of the labium and the maxillae (e.g. Snodgrass 1935; Seifert 1995; Beutel et al. 2014). This

V. Mauss
Staatliches Museum für Naturkunde, Stuttgart, Germany

K. Kuba · H. W. Krenn (✉)
Department of Integrative Zoology, University of Vienna, Vienna, Austria
e-mail: harald.krenn@univie.ac.at



H. W. Krenn (ed.), *Insect Mouthparts*, Zoological Monographs 5,
https://doi.org/10.1007/978-3-030-29654-4_14

unique functional unit of the mouthparts as well as the four-segmented labial palpus are regarded as autapomorphies of the Hymenoptera (Krenn 2007). The mouthparts of adult vespid wasps (Vespidae) have the same general but complex organization, as in many other representatives of Hymenoptera (Bischoff 1927; Duncan 1939; Osten 1982; Jervis 1998; Jervis and Vilhelmsen 2000). They consist of well-developed biting mandibles and a more or less elongated unit composed of the maxillae and labium that are moved as a unit. The mandibles are used for various purposes in the different taxa, like hunt and nest construction, but the labiomaxillary complex is usually engaged in the uptake of carbohydrate-rich fluids, like nectar, water or suspended food (Fig. 14.1) (Spradbery 1973; Baranek et al. 2018).

Vespidae have been extensively studied (e.g. Schremmer 1962; Duncan 1939; Spradbery 1973; Carpenter 1982; Gess 1996; Turillazzi 2012; Archer 2012; Gess and Gess 2014) and comprise more than 5000 species worldwide (Pickett and Carpenter 2010). They represent a diverse range of life forms and use a variety of different food sources to nourish themselves and to provision their brood (Mauss 2007). The oldest known fossils have been dated approximately 129 million years ago with *Curiosivespa antiqua* being placed in the stem lineage of the Euparagiinae and *Priorvespa longiceps* in the stem lineage of the remaining Vespidae (Perrard et al. 2017). A fossil nest, from the stem lineage of the Polistinae and Vespinae, provides evidence for the existence of a social life form within the Vespidae dating back 63 million years (Wenzel 1990).

The phylogenetic position of the Vespidae within the Aculeata s.str. is unresolved, since recent molecular studies (Peters et al. 2017; Branstetter et al. 2017; Bank et al. 2017) offered incongruent results which were also in conflict with older morphological studies that placed the Scoliidae as the sister group to the Vespidae (Brothers and Carpenter 1993; Brothers 1999). Despite these conflicting phylogenetic hypotheses, it is likely that the Vespidae evolved from an ancestor with a scoliid-like life form as it occurs in the representatives of Scoliidae and Tiphiidae (Mauss 2007). The adults of these solitary wasps are flower visitors which use various nectar-rich flowers with easily accessible nectaries (Osten 1982; Landeck 2002). Pollen consumption has been demonstrated in some taxa (e.g. *Proscolia*, some Campsomerini), and therefore pollen has been considered to be a protein source in the diet of at least some adult Scoliidae (Osten 1988; Mauss 2000, 2007). Their larvae are ectoparasitoids of beetle larvae living in friable soil. The female scoliid wasps use their strong shovel-shaped mandibles for digging in the ground in search of host larvae, whilst the legs move the soil particles backwards. The transport of particles is done with additional help of a brush on the ventral margin of the mandibles, the labrum and the maxillae (Osten 1982). When a digging female scoliid finds a beetle larva, she paralyses it with her sting and lays an egg on it. After hatching the wasp larva feeds on this host.

Comparison of mouthpart morphology of representatives of Scoliidae/Tiphiidae and Vespidae allows for the reconstruction of the ground pattern and plesiomorphic character state of mouthpart use. This provides the basis to discuss the evolution of form and function of the mouthparts in context of feeding behaviour and lifestyle in Vespidae, which may range from solitary to social insects and includes prey hunting, nest construction and brood care (Mauss 2007).

**Fig. 14.1** The mouthparts of vespid wasps (Vespidae) are used for taking up fluids (e.g. nectar, water), for hunting, for brood provisioning and for nest construction; (**a**) *Polistes dominula* (Polistinae) queen feeding on open nectary of an ivy (*Hedera helix*) blossom; (**b**) *Celonites abbreviatus* (Masarinae) male, retraction of the proboscis after drinking from the concealed nectaries of *Micromeria juliana* (Lamiaceae); (**c**) *Delta unguiculatum* (Eumeninae) male taking up water with short mouthparts; (**d**) *Delta unguiculatum* (Eumeninae) female transporting a caterpillar into the nest (photo by courtesy of Albert Krebs/ETH Zürich); (**e**) *Tropidodynerus interruptus* forms a mud pellet to seal the nest; (**f**) worker of *Polistes dominula* (Polistinae) scrapes off wood fibres for building a paper nest

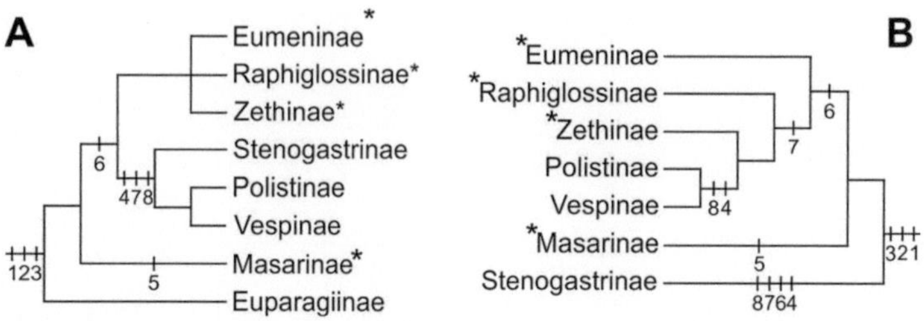


**Fig. 14.2** Conflicting phylogenies of Vespidae with key innovations of mouthparts and life history; redrawn and modified from Carpenter (1982) on the left side (the study included representatives of Raphiglossinae and Zethinae; however both were regarded to be members of the Eumeninae s.l.) and Bank et al. (2017) on the right side (Euparagiinae not studied). Evolution of a long proboscis occurred in some species in several taxa independently (indicated with an asterisk). The stem lineage of the Vespidae evolved three apomorphic traits: (1) the construction of a multicellular ground nest which is built after the soil is moistened by regurgitated fluid; water is collected with the labiomaxillary complex and stored in the crop; soil particles are formed to mud pellets and transported using the mouthparts; (2) oviposition before provisioning the nest cell; (3) mass provisioning of larval food; epigaeic hunt for insect larvae; prey is paralysed with the sting and stored in the brood cell; insects are hunted and transported using the mouthparts (Mauss 2007); (4) eusocial lifestyle; (5) pollen provision for larvae; (6) malaxated prey as protein source for adults; (7) use of plant fibres for nest construction; (8) adult-adult trophallaxis between nest mates

Despite the phylogeny of the Vespidae being intensively studied (Carpenter 1982; Pickett and Carpenter 2010; Perrard et al. 2017; Bank et al. 2017; Piekarski et al. 2018), the position of the Stenogastrinae (hover wasps) is still uncertain (Fig. 14.2). Based on extensive molecular studies, the Stenogastrinae were placed as sister taxon to the rest of the Vespidae (Bank et al. 2017; Piekarski et al. 2018), but when using combined morphological and molecular data, it was recognized as the sister group of the [Polistinae + Vespinae] (Pickett and Carpenter 2010; see also Perrard et al. 2017). The Eumeninae sensu Carpenter (1982) seem to be paraphyletic with Zethinae and Raphiglossinae being more closely related to the [Polistinae + Vespinae] than to the remaining Eumeninae (Fig. 14.2) (Bank et al. 2017; Piekarski et al. 2018).

## 14.2 Life History and Mouthparts in the Stem Lineage of the Vespidae

Compared to the out-group taxa, three apomorphic behavioural complexes evolved within the stem group of the Vespidae in which mouthparts are involved: (1) Using mainly the mandibles, the females construct a subterraneous multicellular nest in the hard ground before searching for prey insect larvae. (2) The excavation of the nest is facilitated by softening the soil with water which was previously collected at water sites using the labiomaxillary complex of the mouthparts. (3) The brood cells are mass-provisioned with several small insect larvae after egg laying. Adults feed themselves by the uptake of nectar as a source of carbohydrates and pollen as the main source of protein. Mouthparts are involved in nectar feeding, pollen uptake, nest construction and brood cell provisioning with prey insect larvae (Mauss 2007). A precondition of the evolution of mass provisioning is epigaeic hunt and the transportation of the prey insect larvae to a previously constructed nest (Mauss 2007).

In the ground pattern of the mouthparts, the mandibles are decussate and equipped with several apical teeth, typically found in females (Carpenter 1982; Carpenter and Kimsey 2009). The maxillae are short; the laciniae and the galeae are flat, rounded and unmodified in comparison to other Aculeata; the maxillary palpus has six palpomers of equal length. Similarly, the labium is short and bears a four-segmented palpus. The glossa is bifid and equipped with transverse cuticular lamella; it is flanked by the paraglossae of the same length (Richards 1962).

## 14.3 Stenogastrinae (Hover Wasps)

Based on recent molecular phylogenetic studies, the Stenogastrinae are the sister group to the remaining Vespidae (Hines et al. 2007; Bank et al. 2017; Piekarski et al. 2018). They appear as an early diverging clade distantly related to Polistinae and Vespinae indicating two independent origins of eusocial behaviour within the Vespidae (Fig. 14.2) (Hines et al. 2007).

All Stenogastrinae studied so far are primitively eusocial, living in temporary matrifilial societies with a small number of individuals (Turillazzi 2012). The complexity of the social structure within colonies varies between the genera. The females are reproductively plastic (i.e. retain the ability to reproduce) in all species, and there are clear behavioural castes, but morphophysiological caste differences have not been observed (Turillazzi 2012; Piekarski et al. 2018). The existing temporal division of labour is regulated by dominance hierarchies between female nestmates (Turillazzi 2012).

The Dufour's gland of Stenogastrinae is much larger than in other Vespidae. It produces large amounts of a secretion called the abdominal substance (Keegans et al. 1992) utilized for attaching eggs and larvae inside the brood cells, and in some taxa, it is also used for the construction of ant guards close to or on the nest. During ant guard construction the secretion is often indirectly applied to the substrate with the mouthparts (Turillazzi 2012). The mouthparts are also used for indirect egg deposition: An egg-laying female bends her gaster ventrally towards her head attaching her egg as it emerges to a pad of Dufour's gland secretion held between her mouthparts. Then, she places the secretion together with the egg inside an empty cell with her mandibles (Turillazzi 2012). The Stenogastrinae exhibit a progressive form of provisioning whereby the adults deposit food with their mouthparts on the secretion next to small larvae or inside the coils of more mature larvae, where the larvae consume it little by little (Turillazzi 1991). Thus, in contrast to other social Vespidae, the food is not directly placed onto the larval mouthparts (Turillazzi 2012). The mouthparts are also involved in adult-adult trophallaxis, e.g. the sharing of solid food or exchange of regurgitated fluid (Turillazzi 2012).

**Fig. 14.3** Head and mouthparts of a female worker of *Eustenogaster hauxwelli* (Stenogastrinae, Vespidae) (microphotos from a dried museum specimen); (**a**) head with spread mouthparts; (**b**) mouthparts in lateral view; abbreviations: *gl* glossa; *ma* mandible; *mx* maxilla; *mxp* maxillary palpus; *lmc* labiomaxillary complex

Adult wasps have been recorded to feed on plant sap, licking honeydew from leaves and nectar from flowers (Turillazzi 2012). Various small arthropods are used as protein source. In a unique behaviour, females of some genera seize small prey items in flight, whilst hovering, from spider webs but also from vertical surfaces (Turillazzi 2012). If the prey is too big, the mandibles are used to cut it up. The prey is intensively malaxated with the mouthparts resulting in a food bolus that might be shared with nestmates (Turillazzi 2012).

Nest architecture is species-specific and differentiates the Stenogastrinae from other social wasps, firstly due to the lack of a peduncle (Carpenter 1988a; Turillazzi 2012) and secondly based on the quality of the nest material. The material consists of a mixture of mud and plant fibres, but the proportion of the components used varies strongly between different species. The material is processed with the mouthparts and mixed with a salivary secretion in order to produce a composite suitable for nest construction (Turillazzi 2012).

Mouthparts are composed of slender mandibles and a short, unspecialized labiomaxillary complex (Fig. 14.3). Females possess tridentate, decussate mandibles; male mandibles are elongate and are equipped with three teeth in the ground pattern but in some species reduced to two or one (Carpenter 1988a). In females of *Liostenogaster*, the shape of the second tooth on the mandible is squared and sharp in species using mainly plant material for nest construction, whilst it is pointed and blunt in species that use mud (Turillazzi 2012). The maxillary palpus consists of six palpomers; in some genera the second palpomer is longer than the third (Carpenter 1988a). The function of the unequal length of palpomers is unknown. The labium bears a four-segmented palpus where the first palpus segment is elongate and longer than the rest (Vecht 1975). Acroglossal buttons are lacking in Stenogastrinae (Vecht 1977; Carpenter 1988a). The mouthparts have not been studied in great detail, and therefore no complete illustration is available.

## 14.4 Euparagiinae: Relict Group of Mesozoic Vespidae

Mesozoic compression fossils and amber inclusions indicate that the Euparagiinae evolved in the early Cretaceous (Carpenter and Rasnitsyn 1990; Brothers 1992; Perrard et al. 2017). The position of the Euparagiinae within the basal clades of the Vespidae is unsolved as it has been placed as the sister group to all recent Vespidae (Pickett and Carpenter 2010; Perrard et al. 2017) or as the sister group to the Masarinae (Hines et al. 2007) or to the Gayellini within the Masarinae (Piekarski et al. 2018).

Adult Euparagiinae possess short and unmodified mouthparts (two species were illustrated by Bradley 1922 and Richards 1962). The mandibles show two apical teeth, reduced to a single tooth in *Euparagia unidentata* (Carpenter and Kimsey 2009). The maxillae are short and bear a six-segmented palpus. The labium has a short distal bifid glossa similar in length or may slightly exceed the paraglossae. Acroglossal buttons are lacking. The labial palpi are composed of four segments. The mouthpart morphology and feeding behaviour is similar to the ground pattern of the Vespidae.

The adult life form of representatives of the Euparagiinae corresponds with the plesiomorphic state in the ground pattern of the Vespidae. To nourish themselves the adults visit flowers with open nectaries (Bohart 1948, 1989; Clement and Grissell 1968). Moderate amounts of pollen have been found in the alimentary tract of female *Euparagia* indicating that pollen was actively consumed (Mauss 2000). During nest excavation the females moisten the hard ground with a regurgitated fluid which is probably water, since females have been regularly observed taking up water with their mouthparts whilst standing on the surface of small water sites (Longair 1985). The mouthparts are used to remove the moist soil by forming little mud pellets, which are carried out of the burrow with the mouthparts and discarded during short circular pellet dropping flights (Clement and Grissell 1968). A turret is built over the nest entrance by continually adding moist mud pellets excavated from the burrow to the circumference of the opening with the mouthparts, legs and tip of the metasoma (Clement and Grissell 1968). After egg laying the brood cells are provisioned with curculionid larvae that are brought in flight, one by one, to the nest and carried primarily using the mandibles (Williams 1927; Clement and Grissell 1968; Moore 1975; Trostle and Torchio 1986).

## 14.5 Masarinae (Pollen Wasps): Wasps with a Bee-Like Life Form

The Masarinae are comprised of two monophyletic clades, the Gayellini and the Masarini, which are considered to be sister groups (Carpenter 1982, Carpenter 1988b). The females of these taxa provision their larvae with pollen instead of paralysed insect prey as a protein source (Claude-Joseph 1930; Gess 1996; Mauss

2007; Gess and Gess 2010). This bee-like life form has probably evolved within the stem lineage of Masarinae (Carpenter 1982, 1988b), though a recent molecular phylogenetic analysis indicates that pollen provisioning might have evolved independently within Gayellini and Masarini (Piekarski et al. 2018).

In the ground pattern of the Masarinae, the plesiomorphic nesting condition is a multicellular sub-vertical burrow in hard horizontal ground excavated by the nester, with an entrance turret constructed from mud pellets extracted from within the burrow (Mauss 2007). For nest excavation the females moisten the ground with regurgitated water. The moistened soil is removed with the mandibles supported by the labiomaxillary complex and in some species of the genus *Ceramius* by scratching movements of the fore and mid legs (Mauss and Müller 2000; Mauss 2007). During this process the soil particles accumulate behind the mandibles resulting in a little mud pellet. Pellets of soft, clayey soil are carried behind the mandibles whereas little clumps or stones are held between them (Mauss 2007; Mauss et al. 2010). The mud pellets are removed from the burrow by walking backwards out of the nest entrance. The required water is ingested with the labiomaxillary complex at water collection sites. In the most basal clades of the Masarini, i.e. *Priscomasaris*, *Paragia* and some species groups of *Ceramius*, the females stand on the water surface during water uptake as in *Euparagia* (Naumann and Cardale 1987; Gess 1996, 1999, 2001; Gess and Gess 2010) indicating a plesiomorphic condition adopted from the ground pattern of the Vespidae. Water uptake from the water edge or from damp soil is probably derived and has only been observed in clades in which the elongated glossa forms a concealed food tube (Mauss 2007; Mauss et al. 2010; Gess and Gess 2010). In these species the glossa is extended and held in slightly curved position during water uptake (Fig. 14.4a). Water is imbibed with the bifurcated tip. During water uptake the outer surface of the glossa appears to be wet (Krenn et al. 2002). Use of nectar instead of water for soil moistening is an apomorphic condition and has been observed in *Masarina strucki*, *Celonites*, *Quartinia* and *Pseudomasaris* (Torchio 1970; Gess 1996; Gess and Gess 2010; Mauss 2007; Mauss and Müller 2014). In species of *Quartinia*, the walls of the turret, the shaft and the brood cells of the hypogaeic burrow consist of sand grains bonded together with a dense, continuous lining of silk-like fibres on the inside (Gess and Gess 1992, 2010; Mauss and Müller 2016; Mauss et al. 2018). The silk is released from the oral region of the nest-building female and applied to the walls with the mouthparts. In this process the mandibles are moderately spread and the labrum is erected. Moreover, moist shining, protruded parts of the epi- and hypopharynx are visible as well as distal parts of the maxillae performing some movements (Mauss and Müller 2016). Epigaeic nests consisting of a group of mud cells attached to plant stems or stones evolved independently in the Gayellini (Claude-Joseph 1930) and the Masarini, in which aerial nesting has been recorded in *Celonites* and *Pseudomasaris* (Torchio 1970; Gess and Gess 1992; Gess 1996; Mauss 2007). The aerial brood cells are built from little mud pellets gathered in a pre-existing quarry site. The behaviour during removal, forming and transport of mud pellets with the mouthparts and the way they are placed onto the wall of the cell during construction are very similar to the behaviour of the hypogaeic nesting species during turret building (Torchio 1970; Bellmann 1984; Mauss 2007; Mauss and Müller 2014; Gess and Gess 1992, 2010). The mouthparts are used for nectar feeding and pollen consumption. The more or less elongated labiomaxillary complex is the main organ for nectar uptake (Fig. 14.4b).

**Fig. 14.4** Fluid uptake using a proboscis in pollen wasps (Masarinae, Vespidae); (**a**) water uptake with proboscis composed of the elongated glossa in *Ceramius maroccanus* female; (**b**) *Celonites abbreviatus* male leaving flower after nectar feeding, with partly extended proboscis in the process of retracting the glossa; glossal sac indicated by an arrow

In the ground pattern, mouthparts of the Masarinae resemble the ground pattern of the Vespidae with the exception of the presence of acroglossal buttons that are lacking in the basal most clades (Bradley 1922; Carpenter 1988b; Krenn et al. 2002). The clypeus of the Masarinae projects over the labrum which bears long bristles. The closed mandibles obscure the frontal view of the labiomaxillary complex which is retracted under the head and visible only from the posterior in resting position (Krenn et al. 2002). Cardo and stipes lie between the labial base and head capsule. Distally the stipes bears the flat lobe-shaped lacinia and the galea which is composed of several plates and bears rows of bristles which function as pollen combs. In several species the infrabuccal pouch has been recorded as being full of pollen. The prementum bears the 4-segmented labial palpus and the bifid glossa which is flanked by the paraglossae. Both labial endits are shorter than the prementum in the plesiomorphic state. The glossa and paraglossae are extendible and retractable and form the principle organ of fluid uptake. They bear rows of lamella-shaped cuticle structures at the anterior side; acroglossal buttons are present at the apex. Capillarity is assumed to be the main mechanism for fluid adhesion to the large surface that is formed by the numerous cuticle structures (Krenn et al. 2002).

The basal taxa of pollen wasps, e.g. *Gayella*, *Priscomasaris* and *Paragia*, possess a relatively short glossa which has cuticular structures that allow uptake of nectar from easily accessible flowers and water, probably by adhesion to the cuticular lamellae on the anterior sides of the extendable glossa and paraglossae (Carpenter 1988b; Gess 2001; Krenn et al. 2002). This passive mode of fluid uptake is similar to that of other wasps like Euparagiinae (Bradley 1922), Eumeninae (Richards 1962; Osten 1982) and Vespinae (Duncan 1939).

The evolution of a concealed nectar extraction apparatus (sensu Jervis 1998) is an apomorphic character state that allows for nectar drinking from long-tubed flowers (Figs. 14.1b and 14.4b). The comparison of mouthparts in various representatives of

Masarinae indicates that a proboscis has independently evolved twice, namely, in *Metaparagia* of the Paragiina and in the stem lineage of the Masarina (encompassing the genera *Ceramius*, *Ceramiopsis*, *Trimeria*, *Jugurtia*, *Masarina*, *Quartinia*, *Celonites*, *Masaris*, *Pseudomasaris*) (Carpenter 1996; Krenn et al. 2002). In the Paragiina, the proximal region of the glossa is elongated, and the paraglossae are long and reach the bifid apex of the glossa. Cuticle lamella on the surface of the glossa and paraglossa take up fluid probably by adhesion. The glossal structures form an open canal on the anterior side where liquid can be transported over the glossa to the mouth. This proboscis-like organ is shorter than the head. In resting position, the labium is folded under the head (Krenn et al. 2002).

In the members of the taxon Masarina, the distal glossa forms a closed feeding tube, whilst the paraglossae are not elongated (Fig. 14.5). This proboscis is composed of the glossa only. The glossa may reach double the length of the body and is stored in a particular sac in a backward flipped loop inside the voluminous prementum (Fig. 14.6). The cuticle lamellae of the glossa are large, flat and curved to the mid-line where they overlap each other and shape the food canal on the anterior side of the glossa (Krenn et al. 2002). These glossa lamellae compose the food canal from the bifid tip of the glossa where drinking slits occur to the base of the proboscis (Fig. 14.5). The proboscis functions according to the siphoning feeding principle which is primarily based on a pressure gradient. Most likely fluid is transported from the bifid glossa tip into the preoral cavity by the action of the pharyngeal sucking pump (Fig. 14.6) (Krenn et al. 2002). The glossa tips are equipped with an acroglossal button which probably has sensory functions (Fig. 14.5d).

In this way, many representatives of the Masarina are able to feed nectar from concealed nectaries in spurred and resupinate flowers using their elongated proboscis formed by the glossa (Gess and Gess 1989; Gess 1996; Mauss and Müller 2000, 2014, 2016; Krenn et al. 2002; Mauss et al. 2010). The long and thin glossa is extended after the pollen wasp has landed on the flower, whilst the mandibles are slightly open (Figs. 14.1b and 14.4b). Proboscis extension is due to the stiff rod inside the glossa. The compression of the space in the prementum by the action of labial muscles is responsible for the glossa extension. It immediately propels the glossa forwards and out of the prementum where it is stored in a loop-shaped resting position (Fig. 14.6). The retraction is achieved similarly by the labial muscles. They draw the glossa back and invaginate the anterior side of the proximal glossa (Fig. 14.5b) in a way that the glossal rods fold back due to the elastic properties (Fig. 14.6). In some species with a particularly long proboscis, the backwardly looped glossa extends over the prementum lying in a membranous sac between the cardines, for example, in *Celonites* (Fig. 14.4b) (Schremmer 1961), or in several loops inside the prementum (Krenn et al. 2002).

The adults feed not only on nectar but also on pollen which is used as protein source for their own nourishment (Hunt et al. 1991; Mauss 2000, 2007, unpubl.). Moreover, female masarine wasps transport ingested pollen in their crop and regurgitate it along with nectar when provisioning a brood cell after egg laying (Gess 1996; Mauss 2007; Gess and Gess 2010). During pollen ingestion, the mandibles and maxillae are moved, whilst the proboscis stays completely retracted (Mauss and

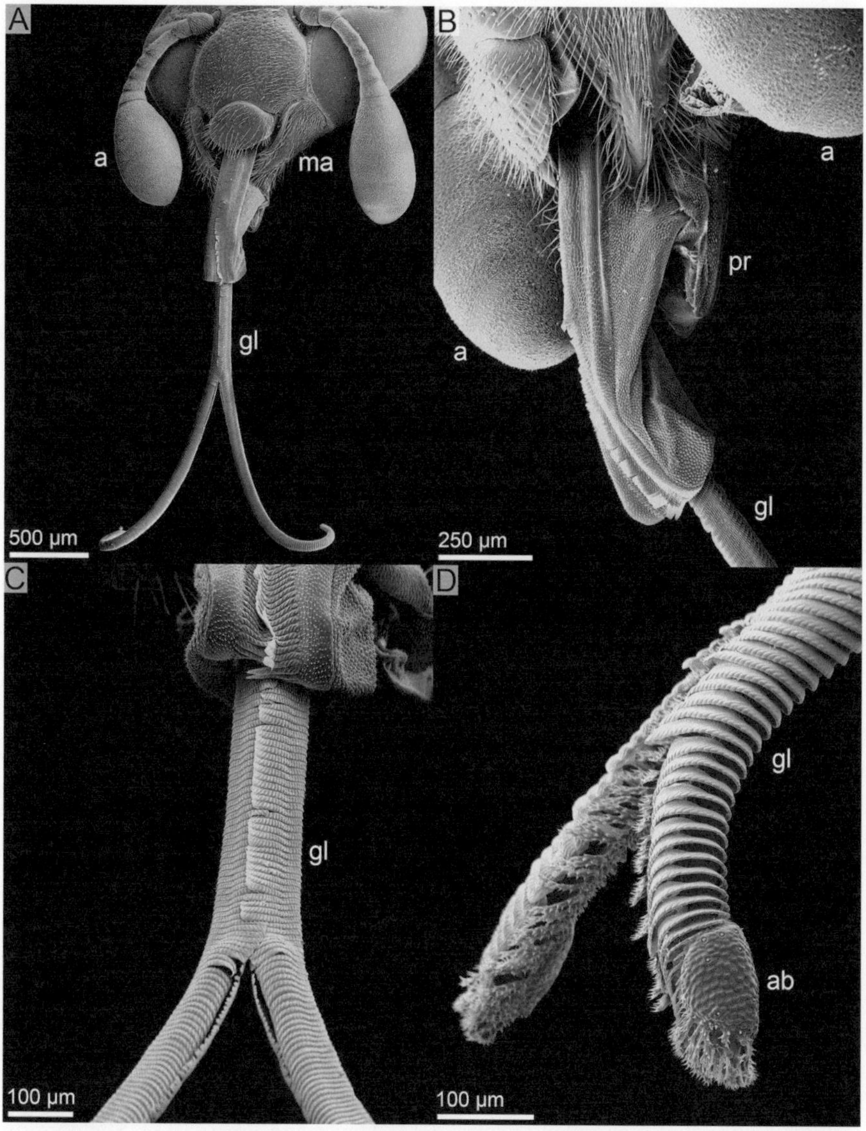


**Fig. 14.5** Head and mouthparts of *Celonites fischeri* (Masarinae, Vespidae) (scanning electron microscopical images); (**a**) head in frontal view, glossa partly extended; (**b**) proboscis proximally invaginated and flexed inside the prementum (lateral view of head); (**c**) bifurcation of the glossa, cuticle lamellae form the food canal; food canal divides distally; (**d**) glossa tip bears acroglossal button; abbreviations: *a* antenna; *ab* acroglossal button; *gl* glossa; *ma* mandible; *pr* prementum

Müller 2014, 2016). Pollen is often directly removed from the pollen sacs by clasping and nibbling the anthers with the mouthparts (Gess 1996; Gess and Gess 2010; Mauss 2007; Mauss and Müller 2014, 2016; Mauss and Mauss 2016). This is frequently accompanied by alternating brushing movements of the forelegs from the

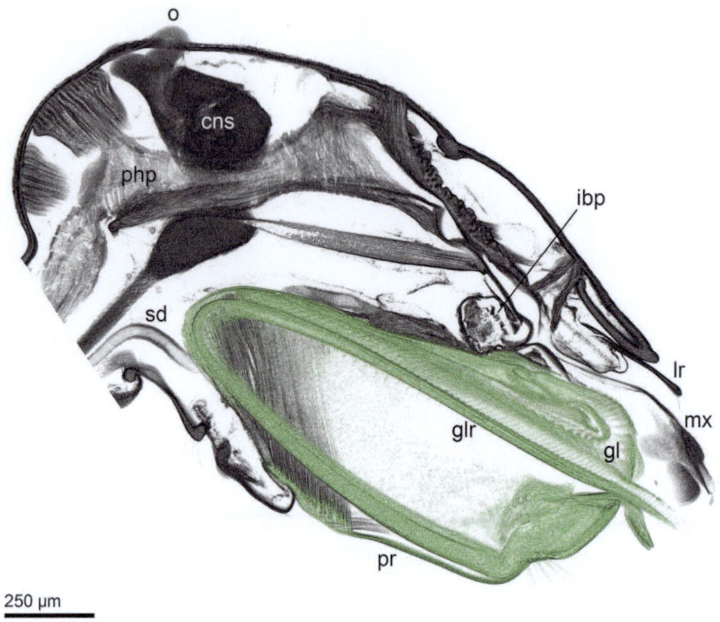


**Fig. 14.6** Proboscis of *Ceramius hispanicus* (Masarinae, Vespidae) in resting position, retracted into the head (micro-CT image); glossal rod flexed backwards into the prementum, proximal region of the proboscis invaginated below the preoral cavity and infrabuccal pouch; abbreviations: *cns* central nervous system; *gl* glossa; *glr* glossal rod; *ibp* infrabuccal pouch; *lr* labrum; *mx* maxillary components; *o* ocellus; *pr* prementum; *php* pharyngeal sucking pump; *sd* salivary duct

anthers towards the mouth bringing pollen grains to the mouthparts. Facultative indirect pollen uptake has been observed in various species of *Ceramius*, *Masarina*, *Quartinia* and *Celonites*, when pollen accumulates on parts of their exoskeleton as a result of contact with the anthers during flower visits. In this case, the wasps remain for a short while on the flower or nearby on the ground brushing pollen grains from the exoskeleton to the mouthparts by alternating movements of their forelegs (Gess 1996; Mauss 1996; Gess and Gess 2010; Mauss et al. 2006; Mauss and Müller 2000, 2014, 2016). Obligate indirect pollen uptake from nototribic flowers of Lamiaceae with specialized stiff 'knobbed' setae on the frons evolved in the stem line of the *Celonites abbreviatus*-complex (Mauss 2013; Mauss and Prosi 2018). Standing on the lower lip of the flowers, the females of this species-complex rub their head over the nototribic anthers. In this manner pollen grains are removed from the pollen sacs, accumulating on the frons and clypeus from where they are brushed to the mouthparts at regular intervals with the forelegs (Schremmer 1959; Müller 1996; Mauss et al. 2016). Different morphological adaptations to indirect pollen uptake from flowers with anthers enclosed in narrow corolla tubes exist in *Trimeria* and *Ceramiopsis*. Females of *Trimeria buyssoni* remove pollen from flowers of Verbenaceae and probably also Boraginaceae by inserting their modified foretarsi with hooked bristle-like setae into the narrow corolla tubes (Neff and Simpson

1985). In *Ceramiopsis* females the proboscis bears distinct spine-like ventral processes over the proximal two-thirds of its length (Richards 1962). Both species of the genus are restricted to flowers of Pontederiaceae (Garcete-Barrett and Klassen Dück 2010). Pollen collection from these plants by *Ceramiopsis* begins with a short stationary flight of a female in front of an inflorescence in which she explosively extends her proboscis; once alighting on a flower, she immediately positions her proboscis in the corolla, followed by repeated backward and forward shaking of her body, thereby rubbing her proboscis over the anthers in the narrow corolla tube. Finally, when leaving the flower, she remains in a stationary flight in front of it during which the extended glossa is quickly pulled down a couple of times with the forelegs probably to transfer the pollen from the proboscis towards the mouthparts. Then the whole behavioural sequence starts anew on another flower of the same inflorescence. The proboscis is not retracted until the female finally leaves the inflorescence in normal horizontal flight (Garcete-Barrett and Klassen Dück 2010).

The key innovation in the evolution of the pollen wasps is the transition of provisioning the larvae with pollen. It is unknown if the larval mouthparts are different from other Vespidae. The starting point was the feeding behaviour of female masarine wasps which already consumed pollen as protein source and carried it inside the crop (Mauss 2007). It is postulated that pollen from the crop was placed inside the brood cells when the females started to regurgitate crop content during the provisioning phase of the brood cells. This might have been the result of continued regurgitation behaviour that was primarily only performed during the previous phase of brood cell construction to moisten the soil by regurgitating water. It is assumed that a transitional stage existed where pollen and paralysed insects were concurrently used to provision larvae. In a second step, only pollen was used to provision larvae (Mauss 2007). Independent evolution of a proboscis is explained by the regular flower visits which started with pollen requirements and is an adaptation to using concealed nectaries in long-tubed flowers.

## 14.6 The Taxon [Eumeninae + Raphiglossinae + Zethinae + Polistinae + Vespinae]

An innovation in the stem line of this lineage is the nutrition of adult females with malaxated insect prey. The insects captured for provisioning brood cells are intensively chewed using the mandibles that are decussate and 4-toothed in the ground pattern (Carpenter 1982, Carpenter and Cumming 1985). However, sometimes the prey insect is completely consumed by the female offering an important source of protein (Spradbery 1973; Chilcutt and Cowan 1992; Hunt 1994). As in the ground pattern of the Vespidae, nectar serves as a source of carbohydrates and is taken up by using a short and broad labiomaxillary complex (Mauss 2007).

## 14.7 Eumeninae (Potter Wasps): Insect Hunters with Various Nesting Behaviours

The mouthparts of the representatives of the Eumeninae are primarily designed for biting and prey mastication as well as licking for fluid feeding with the labiomaxillary complex (Spradbery 1973). The morphology of the mouthparts corresponds with the ground pattern of the Vespidae except that the bilobed glossa is slightly elongated (Figs. 14.7 and 14.8) and distinctly narrower than in Euparagiinae, Gayellini, Polistinae and Vespinae (cf. Bradley 1922, Richards 1962, Carpenter 1988b). The glossa is equipped with numerous spatula-shaped cuticle structures which form a large surface for fluid adhesion (Fig. 14.7b). The paraglossae are as long as the glossa, conspicuously slender and much more palpate in form (Fig. 14.7c). The galeae are slightly longer and narrower (Spradbery 1973; Parker 1966; Gereys 2016).

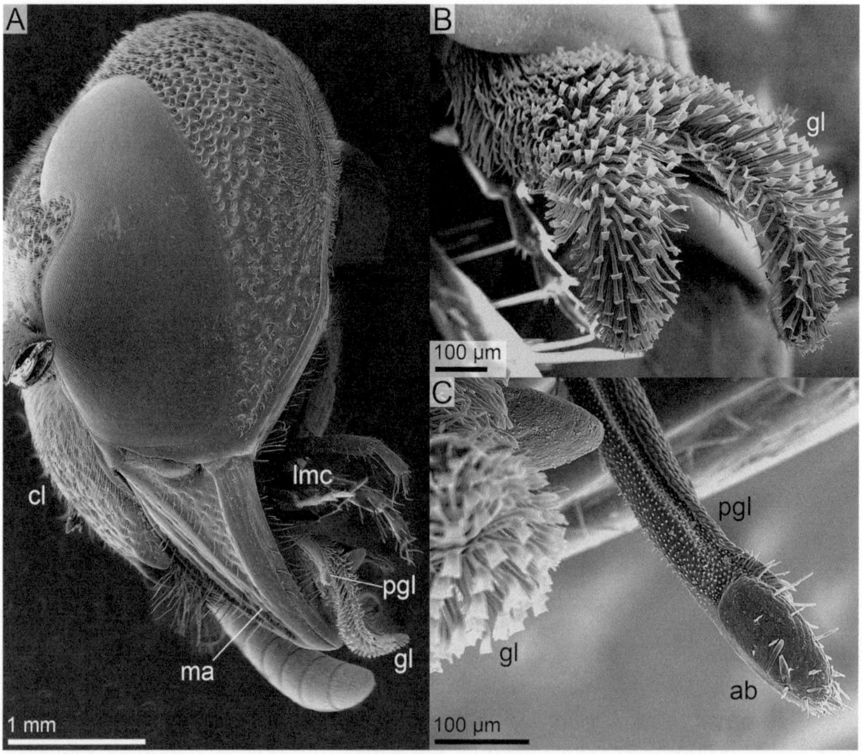


**Fig. 14.7** Head and mouthparts of *Tricarinodyneus guerinii* (Eumeninae, Vespidae) (scanning electron microscopical images); (**a**) mandibles and short labiomaxillary complex in lateral view bears maxillary palpi and labial palpi; (**b**) short, bifid glossa bears curved, spatula-shaped microtrichia on the anterior surface; (**c**) slender paraglossa bears an acroglossal button at the tip; abbreviations: *ab* acroglossal button; *cl* clypeus; *gl* glossa; *lmc* labiomaxillary complex; *ma* mandible; *pgl* paraglossa

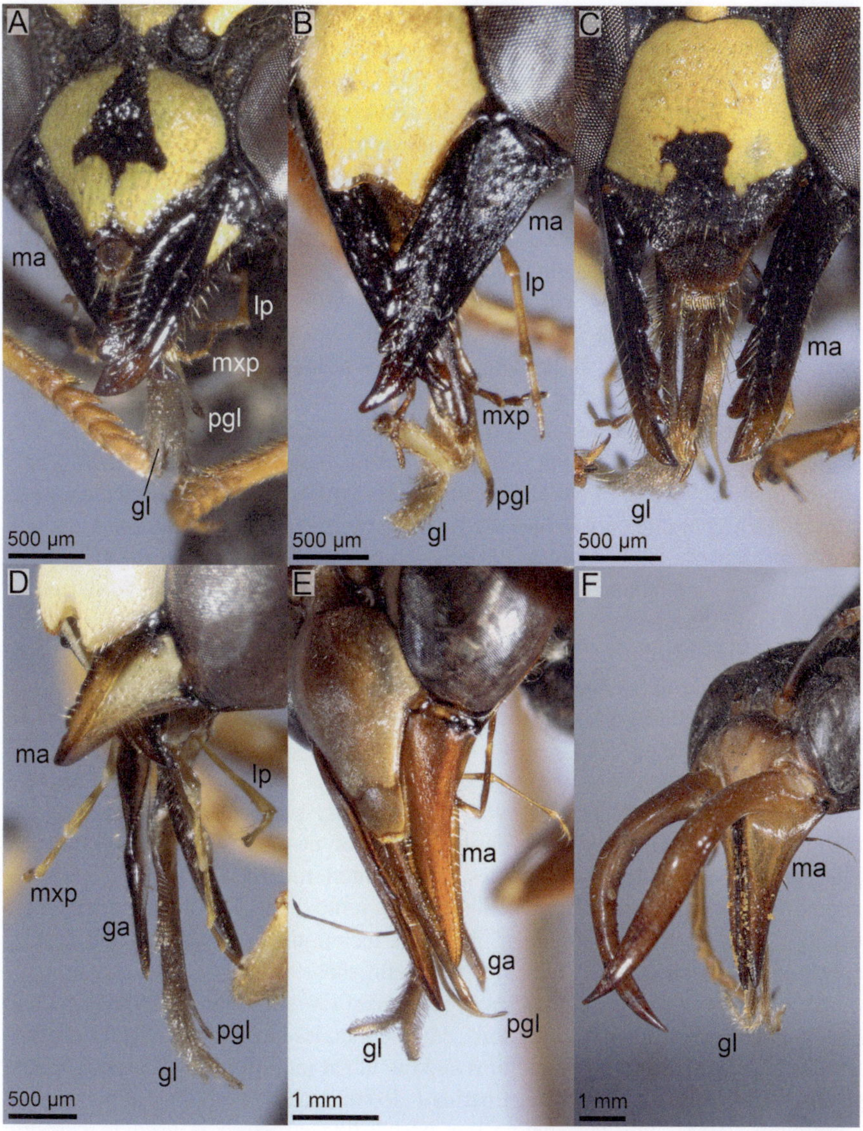


**Fig. 14.8** Examples of short and long mouthparts in Eumeninae (Vespidae) (microphotos of dried museum specimens); (**a**) *Ancistrocerus parietinus*, (**b**) *Delta unguiculatum* and (**c**) *Eumenes papillarius* short unspecialized mouthparts with slender mandibles for hunting prey and nest construction; the short labiomaxillary complex is distally composed of a bifid glossa, slender paraglossae and galeae, all the same length; (**d**) *Pterocheilus pedicellatus* proboscis formed by elongated galeae and glossa plus paraglossa; (**e**) *Zeta canaliculatum* elongated mandibles, galeae, paraglossae; bifid hairy glossa extending beyond the mandibles; (**f**) Head of a male *Synagris cornuta*, mandibles bear large tusks; abbreviations: *ga* galea; *gl* glossa; *lp* labial palpus; *ma* mandible; *mxp* maxillary palpus; *pgl* paraglossa

Eumeninae use their mouthparts for feeding on freely accessible nectar and honey dew, for water collection (Fig. 14.1c) and nest construction as well as for hunting and transporting prey to provision brood cells (Fig. 14.1d) (Mauss 2007).

In several genera of the Eumeninae, the nest consists of a burrow which has been excavated in compacted soil by the female wasp. The entrance is surmounted by a turret, built with mud pellets removed from within the burrow (Evans 1956; Lith 1956; Móczár 1960, 1962; Miotk 1979; Haeseler 1997; Arens 1999; Fateryga and Ivanov 2013; Fateryga 2013, 2018; Fateryga and Amolin 2014; Gess and Gess 2014). As in Euparagiinae and Masarinae, the females soften the soil with regurgitated fluids and remove the material with their mandibles and the labiomaxillary complex forming mud pellets that are held with the mouthparts. The pellets are removed from the burrow moving backwards out of the nest entrance. Smooth pellets are carried behind the mandibles attached to the posterior surface with the aid of the labiomaxillary complex, i.e. the labial palpi, and larger compact particles or little stones are carried between the mandibles. Turret building behaviour is also very similar to the behaviour of Masarinae and Euparagiinae, except that eumenine wasps construct a higher diversity of turret forms. Therefore, nesting in the ground in a self-excavated burrow is probably a plesiomorphic condition belonging to the ground pattern of the Eumeninae (Parker 1966; Iwata 1976; Mauss 2007). The fluid used for soil softening during nest construction is likely to be mainly water, since the females of many species regularly visit water sites, where they take up water with the aid of the labiomaxillary complex standing on the water surface or on the edge. In addition, a secretion of the labial glands may be added to the regurgitated liquid that stabilizes the mud (Cowan 1991).

Within the Eumeninae the transition to nesting in pre-existing cavities or self-constructed aerial mud nests evolved independently several times (Carpenter and Cumming 1985). Even in these derived cases of nest construction the building material is always mud or soil particles. The required mud is collected by the females at quarry sites in the form of mud pellets that are removed, transported and applied to the nest with the mouthparts using the same techniques as in the ground nesting species (e.g. documented by Olberg 1959; Bellmann 1995).

Ground nesting eumenine species of the genera *Pterocheilus*, *Leptochiloides* and *Pseudopterocheilus* inhabit sandy soils, and the females have independently evolved fringes of long setae on the mandibles and the labial palpi forming a psammophore (Bohart 1940; Evans and West-Eberhard 1970; Haeseler 1975; Carpenter and Cummings 1985). In *Pterocheilus* this basket-like structure is used in carrying sandy soil while excavating the nest (Grissell 1975; Witt 2009). The sand grains are carried between the posterior surface of the mandibles and the labial palpi enclosed by the long setae of the psammophore that minimizes the loss of loose sand grains during transport. However, when the sand is very dry, females have been observed loosing part of their load because the sand would simply run through the palpal setae (Grissell 1975). The behaviour during nest excavation and the position of the removed soil behind the mandibles are very similar to the ground pattern of Eumeninae; however, the derived morphology of the psammophore evolved as an adaptation for transporting loose soil particles of a higher grain size in sandy habitats.

All eumenine wasps are predaceous. The females hunt mainly caterpillars for brood cell provisioning (Iwata 1976; Krombein 1979), which is probably part of the ground pattern of the Eumeninae, while the use of sawfly- or beetle-larvae is derived (Carpenter and Cumming 1985). During the transport back to the nest, the paralysed prey is held between the mandibles and each pair of legs in a longitudinal direction under the body of the wasp (Olberg 1959). Mass provisioning of the brood cells is part of the ground pattern of the Eumeninae (Carpenter and Cumming 1985) whereas progressive provisioning evolved within the Afrotropical genus *Synagris* (Roubaud 1911; Bequaert 1918). In *Synagris cornuta* each female rears only one larva at a time in an open brood cell of her aerial mud nest. She nourishes her larva from day to day with a paste of malaxated caterpillars placed directly on the ventral side of the thorax close to the mouth of the larva with her mandibles (Roubaud 1911; Bequaert 1918). When the larva is fully grown, the female seals the cell and immediately starts building a new one, alongside the sealed cell, for her next offspring.

Prey larvae are also the main protein source for the adult females. The captured insect larvae are often masticated with the 4–5 toothed mandibles (Chilcutt and Cowan 1992). In addition, pollen may be consumed in smaller quantities (Hunt et al. 1991). The most important source of carbohydrate is nectar collected from flowers with open nectaries and honey dew (Evans 1966; Kugler 1970; Spradbery 1973; Fateryga 2010). For the uptake of these sugar-rich fluids, the adult wasps use their labiomaxillary complex. The evolution of an elongated proboscis for nectar feeding from flowers with deep corollae occurred independently at least four times within the Eumeninae (Fig. 14.8), specifically in *Leptochilus* (*L. bellulus* group, cf. Parker 1966), *Eumenes* (Osten 1982), *Synagris* (Maidl 1914; Bequaert 1918; Richards 1962) and *Pterocheilus* (Haeseler 1975). In these taxa the proboscis consists of a distinctly narrowed, elongated and distally bifid glossa enclosed dorso-laterally by the elongated galeae and ventro-laterally by the similarly narrowed and elongated paraglossae. In resting position, the proboscis is folded backwards along the ventral side of the mesosoma reaching the mid-coxae or, in the case of *Pterocheilus phalaeratus,* even the hind-coxae. In this species the proboscis is nearly 5 mm long, i.e. about two-thirds of the body length, and is mainly used for stealing nectar from the papilionoid flowers of *Lotus corniculatus* (Haeseler 1975).

Striking sexual dimorphism of the mandible has evolved independently in several eumenine genera (Carpenter and Cummings 1985). For example, in several genera, a deep incision exists between the second and third tooth of the male mandible that often correlates with a semicircular emargination of the ventral margin of the clypeus (Carpenter and Cumming 1985). These structures might be associated with the mating system, i.e. specific male behaviour during copulation or in male-male competition. In *Synagris cornuta* and some other species within the genus, many males have remarkable elongate tusks on the mandible. Unarmed males exhibit only a short, tooth-like protuberance on the mandibles, whereas armed males bear tusks of up to 17 mm in length extending forwards from the anterior portion of the base of each mandible (Fig. 14.8f) (Maidl 1914; Longair 2004). The tusked males spend some time guarding aerial mud nests containing females that are close to emergence. They are able to keep other, usually smaller males off the nest using the tusks in threat display and as weapons when grappling with any individual landing on the nest. Thus larger, guarding males with large tusks probably obtain a high proportion of matings, while smaller and unarmed males function as satellites (Longair 2004).

## 14.8 Raphiglossinae: Nesting in Plants and Nectar Feeding with a Proboscis

The Raphiglossinae contain only a small number of species in three genera, i.e. *Raphiglossa*, *Psiliglossa* and *Elisella* (Giordani Soika 1974). As in the following taxon [Zethinae + Polistinae + Vespinae], the short 4-toothed mandibles of the representatives of the Raphiglossinae are used for the removal and mastication of plant material used for nest construction, in addition to the uptake of soil particles. The use of plant material for nest building is a key innovation and a novel function of the mouthparts (Bank et al. 2017).

In all species of *Raphiglossa* and *Psiliglossa,* for which nesting behaviour has been recorded, the nest consists of a burrow excavated in the pith of dry stems of herbaceous plants or shrubs (Brauns in Meade-Waldo 1913; Ferton 1911, 1920; Gess and Gess 2014). Scratchmarks observed on the inner surface of the unbranched burrow indicate that the pith is scraped off with the mandibles (Ferton 1920). Within the burrow are one to three cells in linear series with the cell partitions constructed only from sand as in *Raphiglossa natalensis*, or partitions are constructed with a combination of sand, fine dust, balls of pith or plant fibres, discs of powdered masticated pith and small stones as in *Raphiglossa zethoides* and *Psiliglossa algeriensis*. While the pith or fibres are derived from within the burrow, sand pellets and small stones are brought into the nest from outside carried in with the mouthparts. The brood cells are provisioned with paralysed larvae of Coleoptera or Lepidoptera. During emergence the adults do not leave the stem in the ordinary way through the cell partitions, but each one gnaws a circular emergence hole in the lateral wall of its cell with the mandibles (Brauns in Meade-Waldo 1913; Ferton 1920).

Afrotropical species of *Raphiglossa* have been recorded visiting flowers of Asteraceae, Aizoaceae, Fabaceae and Scrophulariaceae (Gess and Gess 2003). During flower visits to *Berkheya*, the adults of *Raphiglossa* stand on the capitula taking up nectar with their extended proboscis (Mauss unpubl.), which is as long as the body (Krenn et al. 2005). The food tube is formed by the long and narrow galeae curving over the linear glossa, which is apically bifid (Fig. 14.9a). The lateral sides of the glossa are rolled in; the apical region of the glossa is equipped with many cuticle structures and looks like a brush. The wide but not long stipes covers the fossa where the proboscis is retracted with the galea over the glossa (Plant unpubl.). The thin paraglossae are almost as long as the glossa and lack acroglossal buttons (Richards 1962). In resting position, the long proboscis cannot be retracted fully under the head and lies for the most part under the mesosoma extending beyond the hind coxae (Fig. 14.9a). The extremely elongated proboscis composed of glossa, galeae and paraglossae is an apomorphic trait of the ground pattern

**Fig. 14.9** Head and mouthparts of representatives of Raphiglossinae (Vespidae) and Zethinae (Vespidae) (microphotos of dried museum specimens); (**a**) *Raphiglossa eumenoides* (Raphiglossinae), the labiomaxillary complex forms a long and thin proboscis which is flexed under the body in resting position; the glossa and galeae are greatly elongated; (**b**) *Zethus arietis* (Zethinae) possesses a short labiomaxillary complex; (**c**) *Zethus hilarianus* (Zethinae) with long distal components of the labiomaxillary complex forming a proboscis composed of galeae, glossa and paraglossae; abbreviations: *ga* galea; *gl* glossa; *ma* mandible; *mxp* maxillary palpus; *lp* labial palpus; *pgl* paraglossa

of the Raphiglossinae. Its general structure is very similar to that in all Eumeninae with elongated proboscis, which is probably caused by convergent adaptation to concealed nectaries.

## 14.9 Zethinae

The Zethinae are similar to the Raphiglossinae in that the females bear short 4-toothed mandibles that are used for processing plant material during nest building, but they differ distinctly in that their labiomaxillary complex is not elongated as in the ground pattern (Fig. 14.9b) (Richards 1962). However, as in the ground pattern of

Eumeninae, the bilobed glossa, the paraglossae and the galeae are distinctly narrower and slightly longer than in the plesiomorphic condition represented by basal groups of Vespidae, like *Euparagia* or *Gayella* (cf. Bradley 1922; Bohart and Stange 1965; Carpenter 1988b).

Members of the zethine genus *Discoelius* and some species of *Zethus* utilize old insect burrows in twigs, in wood or in the ground for nesting (Claude-Joseph 1930; Blüthgen 1961; Bohart and Stange 1965; Tischendorf et al. 2015). Such cavity nesting is probably the plesiomorphic type of nest-building behaviour within the Zethinae (Bohart and Stange 1965). The females separate the brood cells within the pre-existing cavity with cell partitions composed of leaf fragments cemented together and fixed to the wall with a paste of masticated plant material mixed with saliva (Claude-Joseph 1930; Blüthgen 1961). The leaf fragments come from trees, but the degree of specialization of the wasps to certain tree species is unknown. During leaf cutting a female stands on a leaf and starts cutting it from the edge using its mandibles like a pair of scissors, until a more or less rounded or longitudinal piece is finally cut off (Claude-Joseph 1930; Blüthgen 1961). Leaf-cutting is facilitated by the saw-like configuration of the teeth along the inner distal margin of the mandibles (Claude-Joseph 1930). The brood cells of cavity nesters are mass-provisioned with paralysed lepidopteran larvae after egg laying and sealed with a cell partition before the larva hatches.

Pre-existing cavities are sometimes cleaned or maybe even enlarged by the nester in a way that resembles nest excavation behaviour in the ground pattern of the Vespidae. For example, females of *Zethus dicomboda* were observed removing fine compacted sawdust that made up a solid filling in the abandoned burrows of wood-boring beetles (Claude-Joseph 1930). The material was formed into a little pellet and held between the mandibles while it was carried out of the nest and dropped, whereupon the female immediately returned into the nest and started the whole process anew (Claude-Joseph 1930). *Zethus ceylonicus* also excavates tightly packed powdery frass from abandoned buprestid tunnels. Moreover, this species builds a trumpet-shaped turret of 15 mm in length at the tunnel opening which is composed of the tips of plant stems bearing tiny leaves lined with dried resin (Krombein 1991).

Building aerial nests, as it occurs in *Calligaster* and the subgenus *Zethoides* of *Zethus*, is probably derived within the Zethinae (Bohart and Stange 1965). The nests are constructed from leaf fragments by cutting leaves with the mandibles in the same way as cavity nesters (Williams 1919). The nest material consists of leaf fragments of specific form and size, glued together with a paste composed of highly masticated leaf material, and *Zethoides* may use an additional resin-like substance (Ducke 1914; Williams 1919; Bohart and Stange 1965; Nugroho et al. 2016). The nests are constructed by attaching the first cell to little twigs, aerial roots or palm leaflets, with successive cells being added to the first (Williams 1919). The number of brood cells per nest varies between only a few cells in *Calligaster* to more than 30 cells in the polygynous nests of *Zethus miniatus* (Ducke 1914; Williams 1919; Bohart and Stange 1965; Nugroho et al. 2016).

Aerial nesting is accompanied by progressive provisioning of the larvae. As in the eumenine wasp *Synagris cornuta*, a female rears only one larva at a time that she feeds directly with freshly killed caterpillars, though these are not malaxated (Williams 1919; Ducke 1914). When the larva has finished its development and is ready to spin a cocoon, the female seals the cell with a cap composed of leaf fragments. This derived brood care in *Calligaster williamsi* probably led to a change in hunting behaviour in that the prey caterpillar is killed by chewing off its head with the mandibles instead of paralyzing it with the sting (Williams 1919). This is similar to the hunting behaviour of eusocial wasps belonging to Polistinae and Vespinae that also provision their larvae progressively. Primitively social behaviour exists in *Zethus miniatus*, in which polygynous groups of sometimes more than fifteen females share large aerial nests but without reproductive division of labour (Ducke 1914; West-Eberhard 1978, 1987).

The adults of the Central European species of the genus *Discoelius* receive carbohydrates in the form of floral nectar mainly from Apiaceae with open nectaries (Tischendorf et al. 2015). Neo- and Afrotropical species of *Zethus* have also been recorded visiting flowers with easily accessible nectar such as Asteraceae and Euphorbiaceae, but also from Fabaceae, Portulacaceae, Sapindaceae, Rubiaceae, Celastraceae and even Lamiaceae (Bohart and Stange 1965; Gess and Gess 2003). As in representatives of Eumeninae and Raphiglossinae, the glossa, paraglossae and galeae are distinctly narrowed and more or less elongated (Fig. 14.9b, c). In several species of *Zethus* as well as species of *Macrocalymma*, *Ischnocoelia* and *Discoelius*, the distal components of the labiomaxillary complex form a proboscis (Bohart and Stange 1965) indicating convergent elongation of the proboscis (Fig. 14.9c), which is probably an adaptation to flowers with concealed nectaries. Similar to Eumeninae and Raphiglossinae, the elongated proboscis in all Zethinae cannot be retracted and is folded under the mesosoma when in resting position.

## 14.10 The Taxon [Polistinae + Vespinae]: Mouthpart Use in Eusocial Wasps

A transition from a solitary to a primitively eusocial life form evolved within the stem line of the [Polistinae + Vespinae] (Bank et al. 2017; Piekarski et al. 2018). Correspondingly, the adults of these taxa live in colonies with a reproductive division of labour between two female castes: the fertile queens lay eggs thereby producing the offspring, while the sterile workers provide the colony with resources and care for the offspring. Typical worker tasks are nest building, thermoregulation, sanitation, foraging for plant pulp, flesh or carbohydrate-rich fluid, feeding flesh to larvae, passing fluid to adults and larvae, receiving fluid from larvae, defence against intruders and guarding (Archer 2012). Most of these tasks are associated with the use of the mouthparts, which are in principle similar to the ground pattern of the representatives of the [Eumeninae + Raphiglossinae + Zethinae + Polistinae + Vespinae] consisting of short 4-toothed mandibles and a short, broad, lobate glossa and paraglossae of the labiomaxillary complex (Fig. 14.10).

**Fig. 14.10** Mouthparts of Vespinae and Polistinae (Vespidae); (**a**) worker of *Dolichovespula saxonica* (Vespinae) builds a paper nest, arrow indicates freshly applied pulp; (**b**) *Vespa orientalis* (Vespinae) licks up water with the extended hairy glossa while the mandibles are slightly opened; (**c**) *Polistes biglumis* (Polistinae) with malaxated prey; (**d**) worker of *Vespula vulgaris* (Vespinae) cuts open a killed butterfly using the mandibles and cuts out the insect body parts for brood supply and takes up body fluid of the butterfly using the glossa; (**e**) mouthparts of *Vespula germanica* (Vespinae) comprise short broad 4-toothed mandibles and short labiomaxillary complex (frontal view); (**f**) labiomaxillary complex folded behind the mandibles in resting position (lateral view); abbreviations: *gl* glossa; *lp* labial palpus; *ma* mandible; *mxp* maxillary palpus; *pgl* paraglossa

In the ground pattern, colonies are founded by a single queen and the first workers that are reared are all daughters of the foundress (Carpenter 1991; Archer 2012). Independent nest founding by a group of queens as well as swarm founding by a group of queens and workers evolved independently within some tropical taxa of Polistinae and Vespinae (Carpenter 1991; Archer 2012). During the worker phase of the independent nest-founding group, more workers are increasingly reared until the colony switches over to the production of queens and males. The maximum colony size varies depending on the species from less than 10 to more than 20,000 individuals due to significant differences in social organization between various taxa (Jeanne 1991; Spradbery 1991). A convergent evolution of inter-specific obligatory social parasitism from primitively eusocial ancestors including a secondary loss of the worker caste took place at least once within the Polistinae (Carpenter 1997) and twice within the Vespinae (Carpenter and Perera 2006).

The last common ancestor of the [Polistinae + Vespinae] built an aerial nest that consisted of a simple comb of brood cells without an envelope, attached to a substrate by a pedicel (Wenzel 1991). From this rather simple form, a large variety of highly modified nest architectures evolved especially within the Polistinae (Wenzel 1991). The nest is made of a paper-like material, mainly produced from malaxated woody fibres or plant fibres mixed with saliva from the thoracic glands and sometimes also secretions of other glands associated with the oral region (Downing and Jeanne 1983; Schremmer et al. 1985; Hansell 1987; Wenzel 1991; Kirshboim and Ishay 1998; Archer 2012). In *Vespa orientalis*, a polymer is incorporated that is secreted by labial exocrine glands. The outlets of these glands are situated in the upper portion of the acroglossal buttons of the glossa (Kirshboim and Ishay 1998). In derived cases, social wasps may also add mud to the plant pulp in variable proportions (Wenzel 1991).

During pulp collection a social wasp queen or worker moistens the wood or plant surface with saliva and scrapes off wood fibres with the mandibles moving slowly backwards in parallel with the grain of the wood (Fig. 14.1f) (Edwards 1980; Archer 2012). The accumulating fibres are manipulated and formed into a pellet with the aid of the mandibles and labial palpi, which may be supported by the forelegs. During transport to the nest, the pellet is held behind the mandibles with the labial palpi (Edwards 1980). Altogether, pulp collection is quite similar to the formation and transport of mud pellets by ground-nesting solitary Euparagiinae, Masarinae and Eumeninae and is therefore probably homologous to excavating behaviour in the ground pattern of the Vespidae (Mauss 2007). Prior to use, the pulp pellet is malaxated with the mandibles and mixed with saliva (Spradbery 1973) whilst often held and rotated by the forelegs (Archer 2012). Then the pellet is applied to the free edge of a paper wall under construction, and the wasp moves backwards unrolling the pulp pellet into a thick strip which readily adheres to the edge. This is followed by the wasp moving forwards to the beginning of the strip where she starts thinning it (Fig. 14.10a) moving her head up and down with the mandibles opening on the down movement and closing around the strip on the up movement while she is slowly moving backwards (Archer 2012).

In all representatives of Polistinae and Vespinae, the brood cells are provisioned progressively (Carpenter 1982). The larvae are always reared in open brood cells and fed with masticated pieces of prey that are directly placed in front of the open larval mandibles by the feeding females with their mouthparts (Archer 2012). Even during the solitary founding phase, several larvae are reared simultaneously by a single female (Carpenter 1982). The cells are not sealed by the females when the larvae are fully developed, but each larva spins a firm cocoon by itself forming a functional cap over the cell opening (Edwards 1980).

Female foragers provide the colony with carbohydrates obtained from flowers with open nectaries (Fig. 14.1a), honeydew, ripe fruits and tree sap (Spradbery 1973; Edwards 1980; Matsuura and Yamane 1990; Hunt 1994; Raveret Richter 2000, Mauss 2008). The major protein source is arthropod prey (Hunt 1994; Raveret Richter 2000; Mauss 2008). In addition, many species opportunistically scavenge vertebrate or invertebrate carrion (Kemper and Döhring 1962; Raveret Richter 2000). Hunting females generally kill the prey with their strong mandibles (Fig. 14.10d), whilst the sting is rarely used in this situation (Raveret Richter 2000). The mouthparts are also required to cut up and malaxate prey items (Fig. 14.10c) and to transport the flesh load to the nest and feed it to the brood.

Sharing nutrients between all members of the colony is essential for its success (Spradbery 1973; Edwards 1980). In larva-adult trophallaxis, the adults take up droplets of larval saliva with their labiomaxillary complex either in exchange for food or after squeezing the larva with the mandibles (Maschwitz 1966; Ishay and Ikan 1968; Edwards 1980). The larval saliva contain sugars, amino acids and proteins in considerable quantities; the larvae therefore serve as a food reserve for the colony during unfavourable foraging conditions (Maschwitz 1966; Hunt et al. 1982, 1987). Furthermore, trophallaxis also occurs between adults: In this case a donor wasp gives some regurgitated liquid from her crop to a receptor wasp, which may be a worker, queen or male (Edwards 1980). Adult-adult trophallaxis is initiated by a soliciting wasp through stroking the glossa of the prospective donor with one antenna between the donor's mandibles, whilst the other antenna is used to stimulate the donor's maxillary and labial palps (Spradbery 1973; Hunt et al. 1987). Moreover, the donor is also stimulated with the palps of the soliciting wasp (Spradbery 1973). The correct signals elicit contact between the glossae of both wasps, whilst the mandibles are opened, in which liquid food is transferred from the donor to the soliciting wasp (Edwards 1980; Hunt et al. 1987). Biased trophallactic food transfer is part of the development of dominance hierarchies between females (Edwards 1980), with dominant individuals receiving considerably more nutrients than subordinates (Pardi 1948; Edwards 1980; Reeve 1991; Gadagkar 1991).

The general morphology of the mouthparts of some *Vespula* and *Vespa* species have been studied in great detail (Kirmayer 1909; Brocher 1922; Duncan 1939; Spradbery 1973; Edwards 1980; Kirshboim and Ishay 1998; Baranek et al. 2018). They consist of the large biting mandibles and the short labiomaxillary complex which is fully flexed behind the mandibles in resting position (Fig. 14.10e, f). No proboscis formations are reported in any representative of the Polistinae and Vespinae.

The labrum (termed as epipharynx in Seifert 1995) is situated under the frontal rim of the clypeus (Fig. 14.10b, termed as labrum in Seifert 1995). The mandibles are stout and heavily sclerotized, since they primarily serve as cutting tools. In female wasps they are used to catch insects and to scrape fibres from woody surfaces (Duncan 1939). In addition, the mandibles are the main tools for nest building. They are used for scraping loose earth and debris in underground nests (Spradbery 1973) and to help shape the paper nests. Furthermore, the mandibles are applied by emerging wasps to cut through the pupal cocoon (Schremmer 1962). The mandibles bear three frontal teeth and a cutting edge. In repose, one mandible is folded over the other, which indicates a scissor-like function in action (Fig. 14.10e). Each maxilla consists of a slender cardo and the broad, flat stipes bearing the small lacinia and lateral galea as well as the 6-segmented maxillary palpus. The labium is divided into the submentum, forming the movable lateral connection to the cardines, the roof-shaped mentum and the distal prementum with the 4-segmented labial palpi as well as the apical ligula which is composed of the united glossae and paired paraglossae each bearing an apical acroglossal button (Duncan 1939; Kirshboim and Ishay 1998). The ligula has four lobes which are all equipped with numerous small cuticle structures on the anterior sides and functions like a tongue for fluid feeding (Baranek et al. 2018).

In adult wasps of both sexes, the short labiomaxillary complex is extensible and the major organ for uptake of liquid food or water (Figs. 14.1a and 14.10b). The glossa and paraglossa are used for fluid uptake, characterized by the dense endowment of microtrichia on the dorsal/frontal side. The high number of spatula-shaped microtrichia and the particular shape of the microtrichia create a space underneath and between the cuticle structures. Based on the assumption that the cuticle is wettable, it can be supposed that this arrangement and the specific micromorphology of the glossal microtrichia is crucial to the uptake of fluids into the mouthparts (López-Cubillos and Sarmiento 2013; Baranek et al. 2018). In addition, the microtrichia of the ligula could serve as a rasping device, since wasps often feed from soft ripe fruits by rasping over the surface to obtain more fruit juice (Duncan 1939; Baranek et al. 2018).

Since social wasps are not equipped with a closed tubular proboscis, feeding of fluid has to be accomplished in a different way than suction. A short food canal for the uptake of liquid food is temporally formed by parts of the labiomaxillary complex and the labrum/epipharynx. The labium constitutes the bottom, the stipites including the lacinia and galea form the sides and the labrum/epipharynx composes the roof of this temporarily constructed food tube (Duncan 1939) which was termed "Wespenrüssel" (Kirmayer 1909) or "wasps proboscis" (Fig. 14.10b, d). The glossae and paraglossae produce lapping motions, in extended position of the ligula; the spatula-shaped microtrichia adhere liquids that are thus transported towards the mouth when the glossae and paraglossae are retracted (Seifert 1995).

Female Vespinae use their mouthparts not only for the uptake of carbohydrate-rich solutions (Fig. 14.1a) but also for water collection (Fig. 14.10b), killing and consuming prey (Fig. 14.10c, d), wood fibre harvesting (Fig. 14.1f), paper nest building (Fig. 14.10a) and picking up soil and stones when colonizing underground cavities, as well as in defensive behaviour. Hence, the female mouthparts can be compared to a set of microtools for different kinds of tasks. In contrast, male wasps mainly feed on nutrients obtained from freely accessible sugary fluids and larval saliva (Maschwitz 1966; Baranek et al. 2018). Spradbery (1973) reported that male mouthparts are similar although smaller than the mouthparts of female workers and queens. The smaller size of the mandibles in male *Vespula* wasps is associated with inconspicuous median teeth and mola as well as smaller musculature (Baranek et al. 2018). It was concluded that the biting force is weaker in males than in females, since males do not use their mandibles for hunting or nest building. In male wasps, the mandibles are mainly used to open the cocoon cap during emergence (Schremmer 1962). Males may also utilize their mandibles for prey malaxation, which frequently occurs in *Polistes* and in some other Polistinae and has also been reported for some Vespinae including *Vespula* (Hunt and Noonan 1979).

## 14.11 Evolution of the Mouthparts in Vespid Wasps

In the plesiomorphic state, the adult mouthparts of the Vespidae consist of short mandibles and a short labiomaxillary complex with a broad bilobed glossa and broad paraglossae of equal length. The primary functions of these mouthparts are (1) to open the cocoon and the sealed brood cell during emergence, (2) for adult nourishment and (3) in females to take part in various behavioural tasks associated with brood care. Larval mouthpart morphology has not been studied in detail for most taxa. Therefore, it is unknown if different larval food (insect prey vs. pollen) is reflected in the various mouthpart morphologies of vespid immatures.

Biting open a hole into the cocoon or the brood cell for emergence is an inevitable constraint in all species and as one would therefore expect both sexes have always had completely functionable mandibles, despite the fact that sexual dimorphism frequently occurs.

Adult nourishment requires the consumption of carbohydrates and proteins. In the ground pattern carbohydrates are obtained from flowers with open nectaries that can be easily exploited with the short labium (Figs. 14.1a and 14.11a). There is clear evidence within several clades of solitary Vespidae for the evolution of elongated mouthparts forming proboscises adapted for nectar uptake from concealed nectaries (Fig. 14.2). However, there are two main types of elongation: (1) In Masarinae two independent events led to an elongated proboscis composed only of elements of the labium without participation of the maxillae. In both cases the proboscis can be fully retracted into the prementum so that only the tip is visible at rest (Figs. 14.6 and 14.11b); (2) In contrast, proboscis formation in Eumeninae, Raphiglossinae and Zethinae always includes the elongation from parts of the labium together with parts of the maxillae, i.e. the galeae. Elongated proboscises of this type are folded under the metasoma in resting position because they cannot be retracted (Figs. 14.8d and 14.9a, c). This very different type of proboscis indicates that there might be a constraint that prevented the elongation of the maxillae in Masarinae. The maxillae of Masarinae take part in pollen ingestion. Since pollen became the only protein source for larval provisioning in the stem line of the Masarinae and continued to be

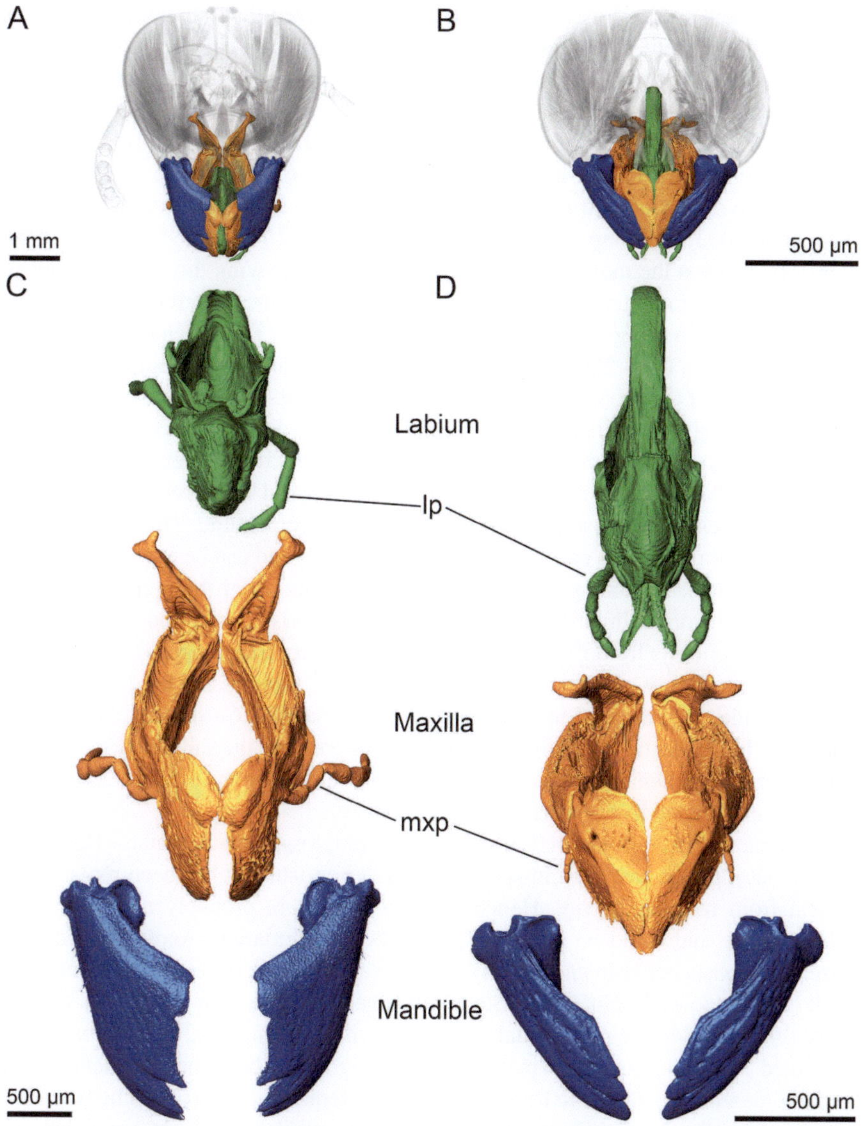


**Fig. 14.11** Anatomy of the head and mouthparts of Vespidae with short and long labium (reconstruction of micro-computer tomography images); (**a**, **c**) *Vespula germanica* (Vespinae), labium with short glossa, maxilla with long palpi, large powerful mandibles with big sharp teeth; (**b**, **d**) *Ceramius hispanicus* (Masarinae), labium with long glossa retracted into the prementum, glossa forms the proboscis; maxilla engaged in pollen collection and has short palpi; mandibles with blunt tips; abbreviations: *mxp* maxillary palpus; *lp* labial palpus

the sole protein source for adults, it can be hypothesized that the elongation of the maxillae was prevented by their essential function in pollen consumption.

To date, the functional mechanism of a wasp proboscis has only been studied in some species of the Masarinae (Krenn et al. 2002). In the proboscis of long-proboscid pollen wasps, only the labium is elongated and the glossa was modified to extend and retract. The labiomaxillary complex of all aculeate Hymenoptera allows some extension and retraction. The mechanism has been described for ants (Paul et al. 2003; see Chap. 10), Scoliidae (Osten 1982, 1988) and *Vespula* (Duncan 1939). It involves the movements of the cardines which swing the proboscis out of the proboscidial fossa under the head, and at the same time, it unfolds the z-shape-folded prementum and glossa, paraglossae and the galeae. This mechanism is similar to Apidae (Krenn et al. 2005) and has also been hypothesized as being involved in the movement of the proboscis of Eumeninae, Raphiglossinae and Zethinae. In case of the elongated proboscis of the Masarina, the glossa lies in one or several loops inside the enlarged prementum. Labial muscles compress the prementum and extend the flexible glossa which protracts due to its glossal rod that probably lies under elastic compression in resting position inside the prementum. This proboscis is particularly thin since it is composed of the glossa only and can be extended directly into a floral tube. The invagination of the proximal region of the proboscis allows extension in a frontal direction, and therefore no space is needed under the body to unfold the proboscis from the resting position, as is necessary in bees. The comparison of the anatomy of various components of the mouthparts in short-tongued Vespidae, like *Vespula* and long-proboscid Masarina, like *Ceramius*, clearly indicates that fundamental anatomical modifications concern the labium (Fig. 14.11). The other components of the mouthparts have only minor differences that are probably not related to movement of the proboscis.

In the members of the taxon [Eumeninae + Raphiglossinae + Zethinae + Polistinae + Vespinae], pollen as a protein source for adult nourishment has been replaced by protein from malaxated prey. Hence, the maxillae were probably not crucial for pollen uptake and could be included into proboscis formation. Another noticeable correlation exists between an eusocial life form and the lack of proboscis elongation in all [Polistinae + Vespinae] and Stenogastrinae, despite a high number of species (Figs. 14.3 and 14.10). A potential constraint arising from sociality that may have prevented proboscis elongation in these groups might have been the important role of the labiomaxillary complex in trophallaxis, which requires close contact between the mouthparts of the donor and recipient. However, eusocial Apidae (e.g. honey bees or bumble bees) do have a proboscis. However, ecological factors may also be at play, since the short mouthparts of the social Vespinae are possibly associated with being able to access a broader nutrional niche, which might be of importance for a long-lived wasp colony, at the same nest site, in a seasonal changing environment with variable food sources becoming available.

Substantial brood care was already performed by the solitary stem species of the Vespidae: the females nest in hard ground, in a self-excavated burrow surmounted by a turret made of mud pellets obtained from inside the burrow; during nest excavation the soil is softened with regurgitated water previously collected at water sites; after egg laying the brood cells are mass-provisioned with paralysed phytophagic lepidopteran or coleopteran larvae and finally sealed with mud. From this ground pattern, preserved in the basal groups of the Euparagiinae and Eumeninae, several different adaptations have evolved. Nesting in pre-existing cavities and aerial, free-standing mud nests evolved several times. Besides the changes in location of nest site or nest architecture, the behavioural sequences such as the removal, transport and application of mud using the mouthparts are similar and appear to be highly conserved. In the stem line of the [Polistinae + Vespinae], these plesiomorphic techniques of processing the building material with the mouthparts were also applied to new substance, i.e. the regurgitation of water to soften the plant fibres was replaced by secretion of large amounts of saliva. The representatives of the Raphiglossinae and Zethinae also use plant matter for nest construction, and the mandibles of Raphiglossinae, Zethinae and [Polistinae + Vespinae] are fundamentally similar forming strong scissor-like cutting tools for plant material. However, although recent phylogenetic studies based on molecular data suggest homology of the morphological adaptations of the mandibles to the use of plant material, it should be highlighted that there are distinct behavioural differences between the taxa. The Raphiglossinae are tube-nesting wasps that excavate pith from dry stems. Nevertheless, the cell partitions are always made from soil particles originating from outside of the nest carried into the tunnel by the female with her mouthparts, while removed pith may only be used in addition to the soil particles. Moreover, the cell partitions of Zethinae are built from leaf pieces obtained by the females by cutting the leaves with their mandibles, which is a highly derived apomorphic trait. Cut leaves are also the basic material in producing masticated plant paste which is used for gluing the plant fragments together. The derived behavioural pattern underlying the utilization of plant material in aerial nest building of the Zethinae and the [Polistinae + Vespinae] is therefore analogous.

The modification of certain morphological characters in mouthpart evolution within Vespidae can be interpreted as an adaptation from the plesiomorphic pattern of excavation behaviour to specific environmental constraints. For example, colonization of sand habitats led to the convergent evolution of a psammophore behind the mandibles formed primarily by the labial palpi. This structure makes it possible to remove loose sand grains during nest excavation in the plesiomorphic way behind the mandibles. In female *Quartinia*, the apomorphic production of a silk-like lining with their mouthparts enables them to stabilize their self-excavated burrow in loose sandy soil thus adapting the plesiomorphic ground nesting behaviour to sandy habitats.

In the ground pattern the brood cells are mass-provisioned with paralysed lepidopteran or coleopteran larvae. In the stem line of the Masarinae, mass provisioning persisted, but the provisions changed from prey to pollen. This change provided a new source of protein for larval development. Flower visiting and pollen consumption by adults was a plesiomorphic behavioural preadaption adopted from the ground pattern of the Vespidae that enabled the Masarinae to evolve into their new bee-like life form. Progressive provisioning evolved from mass provisioning independently in Stenogastrinae, Eumeninae, Zethinae and [Polistinae + Vespinae]. It is closely associated with prey malaxation by the females as a preadaptation that evolved separately within the stem line of the Stenogastrinae and the [Eumeninae + Raphiglossinae + Zethinae + Polistinae + Vespinae].

**Acknowledgements** Gabriele Hellenschmidt kindly helped with the translation of French literature. Dominique Zimmermann provided access to the collection of the Natural History Museum in Vienna. We are grateful to Daniela Gruber from the electron microscopical laboratory (CIUS) and Brian Metscher (Department of Theoretical Biology) at the Faculty of Life Sciences (University of Vienna) who helped with CT imaging. We thank Andrea Spiegel for helpful comments and Annalie Melin for linguistic help.

## References

Archer ME (2012) Vespine wasps of the world. Siri Scientific Press, Manchester

Arens W (1999) Zum Verhalten von *Tropidodynerus interruptus* (Brullé 1832) (Vespoidea, Eumenidae) und seines Brutparasits *Chrysis jaxartis* Sem. am Nest. Linzer Biologische Beiträge 31:147–158

Bank S, Sann M, Mayer C, Meusemann K, Donath A, Podsiadlowski L, Kozlov A, Petersen M, Krogmann L, Meier R, Rosa P, Schmitt T, Wurdack M, Liu S, Zhou X, Misof B, Peters RS, Niehuis O (2017) Transcriptome and target DNA enrichment sequence data provide new insights into the phylogeny of vespid wasps (Hymenoptera: Aculeata: Vespidae). Mol Phylogenet Evol 116:213–226

Baranek B, Kuba K, Bauder JA-S, Krenn HW (2018) Mouthpart dimorphism in male and female wasps of *Vespula vulgaris* and *Vespula germanica* (Vespidae, Hymenoptera). Deutsche Entomologische Zeitschrift 65:65–74

Bellmann H (1984) Beobachtungen zum Brutverhalten von *Celonites abbreviatus* Villers (Hymenoptera, Masaridae). Zool Anz 212:321–328

Bellmann H (1995) Bienen, Wespen, Ameisen – Hautflügler Mitteleuropas. Kosmos Naturführer. Franckh-Kosmos, Stuttgart

Bequaert J (1918) A revision of the Vespoidea of the Belgian Congo based on the collection of the American Museum Congo expedition, with a list of Ethiopian *Diplopterous* wasps. Bull Am Mus Nat Hist 39:1–384

Beutel RG, Friedrich F, Ge S-Q, Yang X-K (2014) Insect morphology and phylogeny – a textbook for students of entomology. De Gruyter Graduate, Berlin, Boston

Bischoff H (1927) Biologie der Hymenopteren. Springer, Berlin

Blüthgen P (1961) Die Faltenwespen Mitteleuropas (Hymenoptera, Diploptera). Abhandlungen der Deutschen Akademie der Wissenschaften zu Berlin Klasse für Chemie, Geologie und Biologie 2:1–249

Bohart RM (1940) A revision of the North American species of *Pterocheilus* and notes on related genera (Hymenoptera, Vespidae). Ann Entomol Soc Am 33:162–208

Bohart RM (1948) The genus *Euparagia* in North America (Hymenoptera, Vespidae, Euparagiinae). Pan Pac Entomol 24:149–154

Bohart RM (1989) A review of the genus *Euparagia* (Hymenoptera: Masaridae). J Kansas Entomol Soc 62:462–467

Bohart RM, Stange LA (1965) A revision of the genus *Zethus* in the western hemisphere (Hymenoptera, Eumenidae). University of California Publications in Entomology 40, Berkeley, pp 1–208

Bradley JC (1922) The taxonomy of the masarid wasps, including a monograph of the North American species. Univ Calif Publ Technical Bulletins College of Agriculture, Agricultural Experiments Station Entomology 1. University of California Press, Berkeley, pp 369–464

Branstetter MG, Danforth BN, Pitts JP, Faircloth BC, Ward PS, Buffington ML, Gates MW, Kula RR, Brady SG (2017) Phylogenomic insights into the evolution of stinging wasps and the origins of ants and bees. Curr Biol 27:1019–1025

Brocher F (1922) La tête du Frelon (*Vespa crabro*) – Étude anatomique. Bulletin de l 'Institut National Genevois 45:223–256

Brothers DJ (1992) The first Mesozoic Vespidae (Hymenoptera) from southern hemisphere. J Hymenopt Res 1:119–124

Brothers DJ (1999) Phylogeny and evolution of wasps, ants and bees (Hymenoptera, Chrysoidea, Vespoidea and Apoidea). Zool Scr 28:233–250

Brothers DJ, Carpenter JM (1993) Phylogeny of Aculeata: Chrysidoidea and Vespoidea. J Hymenopt Res 2:227–302

Carpenter JM (1982) The phylogenetic relationships and natural classification of the Vespoidea (Hymenoptera). Syst Entomol 7:11–38

Carpenter JM (1988a) The phylogenetic system of the Stenogastrinae (Hymenoptera: Vespidae). J N Y Entomol Soc 96:140–175

Carpenter JM (1988b) The phylogenetic system of the Gayellini (Hymenoptera: Vespidae, Masarinae). Psyche 95:211–241

Carpenter JM (1991) Phylogenetic relationships and the origin of social behavior in the Vespidae. In: Ross KG, Matthews RW (eds) The social biology of wasps. Cornell University, Ithaca, pp 7–32

Carpenter JM (1996) Generic classification of the Australian pollen wasps (Hymenoptera: Vespidae; Masarinae). J Kansas Entomol Soc 69:384–400

Carpenter JM (1997) Phylogenetic relationships among European *Polistes* and the evolution of social parasitism (Hymenoptera: Vespidae, Polistinae). Mémoires du Muséum National d'Histoire Naturelle 173:135–161

Carpenter JM, Cumming JM (1985) A character analysis of the North American potter wasps (Hymenoptera; Vespidae; Eumeninae). J Nat Hist 19:877–916

Carpenter JM, Kimsey LS (2009) The genus *Euparagia* Cresson (Hymenoptera: Vespidae; Euparagiinae). Am Mus Novit 3643:1–11

Carpenter JM, Perera EP (2006) Phylogenetic relationships among yellowjackets and the evolution of social parasitism (Hymenoptera: Vespidae, Vespinae). Am Mus Novit 3507:1–19

Carpenter JM, Rasnitsyn PA (1990) Mesozoic Vespidae. Psyche 97:1–20

Chilcutt CF, Cowan DP (1992) Carnivory in adult female euminid wasps (Hymenoptera: Vespidae: Eumeninae) and its effect on egg production. Great Lakes Entomol 25:297–301

Claude-Joseph F (1930) Recherches biologiques sur les prédateurs du Chili. Annales des sciences naturelles Zoologie (10e série) 13:289–354

Clement SL, Grissell EE (1968) Observations on the nesting habits of *Euparagia scutellaris* Cresson (Hymenoptera: Masaridae). Pan Pac Entomol 44:34–37

Cowan DP (1991) The solitary and presocial Vespidae. In: Ross KG, Matthews RW (eds) The social biology of wasps. Cornell University, Ithaca, pp 33–73

Downing HA, Jeanne RL (1983) Correlation of season and dominance status with activity of exocrine glands in *Polistes fuscatus* (Hymenoptera: Vespidae). J Kansas Entomol Soc 56:387–397

Ducke A (1914) Über Phylogenie und Klassifikation der sozialen Vespiden. Zoologisches Jahrbuch für Systematik 36:303–330

Duncan DC (1939) A contribution to the biology of North American vespine wasps. Stanford University Publications, University Series, Biological Sciences 8. Stanford University, California University Press, Redwood City, CA, pp 1–257

Edwards R (1980) Social wasps. Rentokil Limited, East Grinstead

Evans HE (1956) Notes on the biology of four spcies of ground-nesting Vespidae, (Hymenoptera). Proc Entomol Soc Wash 58:265–270

Evans HE (1966) The behaviour patterns of solitary wasps. Annu Rev Entomol 11:123–154

Evans HE, West-Eberhard MJ (1970) The wasps. University of Michigan Press, Ann Arbor, 265 pp

Fateryga AV (2010) Trophic relations between vespid wasps (Hymenoptera, Vespidae) and flowering plants in the Crimea. Entomological Review 90:698–705
Fateryga AV (2013) Nesting biology of *Odynerus albopictus calcaratus* (Morawitz, 1885) and *Odynerus femoratus* de Saussure, 1856 (Hymenoptera: Vespidae: Eumeninae). J Insects 13:1–8
Fateryga AV (2018) Nesting biology of *Paragymnomerus signaticollis tauricus* (Kostylev, 1940) (Hymenoptera: Vespidae: Eumeninae). Zootaxa 4378:429–441
Fateryga AV, Amolin AV (2014) Nesting and biology of *Jucancistrocerus caspicus* (Hymenoptera, Vespidae, Eumeninae). Entomol Rev 94:73–78
Fateryga AV, Ivanov S (2013) Nesting biology of *Paravespa rex* (von Schulthess 1924) (Hymenoptera: Vespidae: Eumeninae) in the Crimea, Ukraine. Zootaxa 3721:589–600
Ferton C (1911) Notes détachées sur l'instinct des Hyménoptères melliféres et ravisseurs (7e Série) avec la description de quatre espèces nouvelles. Ann Soc Entomol Fr 80:383–387
Ferton C (1920) Notes détachées sur l'instinct des Hyménoptères mellifères et ravisseurs avec la description deux espèces nouvelles. (9e Série). Ann Soc Entomol Fr 89:329–375
Gadagkar R (1991) *Belonogaster*, *Mischocyttarus*, *Parapolybia*, and independent-founding *Ropalidia*. In: Ross KG, Matthews RW (eds) The social biology of wasps. Cornell University, Ithaca, pp 149–190
Garcete-Barrett BR, Klassen Dück CH (2010) Biological observations on the genus *Ceramiopsis* Zavattari (Insecta: Hymenoptera: Vespidae: Masarinae). Boletin del Museo Nacional de Historia Natural del Paraguay 16:8–17
Gereys B (2016) Vespidae solitaires de France métropolitaine (Hymenoptera: Eumeninae, Masarinae). Faune de France 98:1–330
Gess SK (1996) The pollen wasps – ecology and natural history of the Masarinae. Harvard University Press, Cambridge, MA
Gess SK (1999) Distribution and ethology of *Ceramius damarinus* Turner (Hymenoptera: Vespidae: Masarinae) in Namibia. University of Kansas Natural History Museum Special Publication 24:25–32
Gess SK (2001) Distribution and ethology of *Priscomasaris* Gess (Hymenoptera: Vespidae: Masarinae: Priscomasarina) in Namibia. J Hymenopt Res 10:16–28
Gess SK, Gess FW (1989) Flower visiting by masarid wasps in southern Africa (Hymenoptera: Vespoidea: Masaridae). Annals of the Cape Provincial Museums (Natural History) 18:95–134
Gess FW, Gess SK (1992) Ethology of three southern african ground nesting Masarinae, two *Celonites* species and a silk-spinning *Quartinia* species, with a discussion of nesting by the subfamily as a whole (Hymenoptera: Vespidae). J Hymenopt Res 1:145–155
Gess SK, Gess FW (2003) A catalogue of flower visiting records for aculeate wasps and bees in the semi-arid to arid areas of southern Africa. Albany Museum, Grahamstown
Gess SK, Gess FW (2010) Pollen wasps and flowers in southern Africa, SANBI Biodiversity Series, vol 18, pp 1–147
Gess SK, Gess FW (2014) Wasps and bees in southern Africa, SANBI Biodiversity Series, vol 24, pp 1–320
Giordani Soika A (1974) Revisione della sottofamiglia Raphiglossinae (Hym. Vesp.). Bolletino della Museo Civico di Storia Naturale di Venezia 25:107–146
Grissell EE (1975) Ethology and larvae of *Pterocheilus texanus* (Hymenoptera: Eumenidae). J Kansas Entomol Soc 48:244–253
Haeseler V (1975) *Pterocheilus phaleratus* (Hymenoptera: Vespoidea), ein Nektardieb an den Blüten von *Lotus corniculatus* (Fabales: Fabaceae). Entomologica Germanica 1:213–221
Haeseler V (1997) Die *Odynerus*-Arten *O. eburneofasciatus* Dusmet 1903, *O. antigai* Dusmet 1903 sowie *O. annulicornis* Blüthgen 1956 und zur Biologie von *O. eburneofasciatus* Dusmet 1903 (Vespoidea: Eumenidae). Linzer Biologische Beiträge 29:151–166
Hansell MH (1987) Nest building as a facilitating and limiting factor in the evolution of eusociality in the Hymenoptera. Oxf Surv Evol Biol 4:155–181
Hines HM, Hines JH, O'Connor TK, Gillespie JJ, Cameron SA (2007) Multigene phylogeny reveals eusociality evolved twice in vespid wasps. Proc Natl Acad Sci U S A 104:3295–3299

Hunt JH (1994) Nourishment and social evolution in wasps sensu lato. In: Hunt JH, Nalepa CA (eds) Nourishment and evolution in insect societies. Westview's Studies in Insect Biology Westview Press, Oxford, pp 211–244

Hunt JH, Noonan KC (1979) Larval feeding by male *Polistes fuscatus* and *Polistes metricus* (Hymenoptera: Vespidae). Insect Soc 26:247–251

Hunt JH, Baker HG, Baker I (1982) Similarity of amino acids in nectar and larval saliva: the nutritional basis for trophallaxis in social wasps. Evolution 36:1318–1322

Hunt JH, Jeanne RL, Baker I, Grogan DE (1987) Nutrient dynamics of swarm founding social wasp species, *Polybia occidentalis* (Hymenoptera: Vespidae). Ethology 75:291–305

Hunt JH, Brown PA, Sago KM, Kerker JA (1991) Vespid wasps eat pollen (Hymenoptera: Vespidae). J Kansas Entomol Soc 64:127–130

Ishay J, Ikan R (1968) Food exchange between adults and larvae in *Vespa orientalis* F. Anim Behav 16:298–303

Iwata K (1976) Evolution of instinct – comparative ethology of Hymenoptera. Smithsonian Institution and the National Science Foundation, New Dehli, 535 pp

Jeanne RL (1991) The swarm-founding Polistinae. In: Ross KG, Matthews RW (eds) The social biology of wasps. Cornell University, Ithaca, pp 191–231

Jervis M (1998) Functional and evolutionary aspects of mouthpart structure in parasitoid wasps. Biol J Linn Soc 63:461–493

Jervis M, Vilhelmsen L (2000) Mouthpart evolution in adults of the basal, "symphytan", hymenopteran lineages. Biol J Linn Soc 70:121–146

Keegans SJ, Morgan ED, Jackson BD, Berresford G, Turillazzi S, Billen J (1992) The chemical nature and origin of the abdominal substance of the social wasp *Liostenogaster flavolineata* (Vespidae, Stenogastrinae). In: Billen J (ed) Biology and evolution of social insects. Leuven University Press, Leuven

Kemper H, Döhring E (1962) Untersuchungen zur Ernährung sozialer Faltenwespen Deutschlands, insbesondere von *P. germanica* und *P. vulgaris*. Zeitschrift für angewandte Zoologie 49:227–280

Kirmayer R (1909) Bau und Entwicklung der Mundteile bei *Vespa vulgaris*. Gegenbaurs Morphologisches Jahrbuch Zeitschrift für Anatomie und Entwicklungsgeschichte 39:1–30

Kirshboim S, Ishay J (1998) The labium of the oriental hornet: micromorphology and functions. J Electron Microsc 47:505–515

Krenn HW (2007) Evidence from mouthpart structure on internodial relationships in Endopterygota? Arthropod Systematics & Phylogeny 65(1):7–14

Krenn HW, Mauss V, Plant J (2002) Evolution of the suctorial proboscis in pollen wasps (Masarinae, Vespidae). Arthropod Structure and Development 31:103–120

Krenn HW, Plant JD, Szucsich NU (2005) Mouthparts of flower-visiting insects. Arthropod Structure and Development 34:1–40

Krombein KV (1979) Vespoidea. In: Krombein KV, Hurd PD, Smith DR, Burks BD (eds) Catalog of Hymenoptera in America North of Mexico 2. Smithsonian Institution Press, Washington, pp 1469–1522

Krombein KV (1991) Biosystematic studies of Ceylonese wasps XIX: natural history notes in several families (Hymenoptera: Eumenidae, Vespidae, Pompilidae and Crabronidae). Smithsonian Contributions to Zoology 515(i–iv):1–41

Kugler H (1970) Blütenökologie. Gustav Fischer, Stuttgart

Landeck I (2002) Nektarpflanzen der Borstigen Dolchwespe *Scolia hirta* in der Lausitz (Mitteleuropa) bei Berücksichtigung von Blütenfarbe, Blüten- und Blütenstandsmorphologie (Hymenoptera: Scoliidae). Entomologica Generalis 26:107–120

Longair RW (1985) Male behavior in *Euparagia richardsi* Bohart (Hymenoptera: Vespidae). Pan Pac Entomol 61:318–320

Longair RW (2004) Tusked males, male dimorphism and nesting behavior in a subsocial afrotropical wasp, *Synagris cornuta*, and weapons and dimorphism in the genus (Hymenoptera: Vespidae: Eumeninae). J Kansas Entomol Soc 77:528–557

López-Cubillos S, Sarmiento CE (2013) A mandible arresting system in neotropical social wasps (Vespidae, Polistinae): structural diversity within homogeneous functionality. Naturwissenschaften 100:429–435

Maidl F (1914) Monographie der Gattung *Synagris* Latreille (Hymen.-Vespidae). Denkschriften der Akademie der Wissenschaften Mathematisch Naturwissenschaftliche Klasse 91 (1–119):215–333. 5 pls

Maschwitz U (1966) Das Speichelsekret der Wespenlarven und seine biologische Bedeutung. Z Vgl Physiol 53:228–252

Matsuura M, Yamane S (1990) Biology of the vespine wasps. Springer, Berlin

Mauss V (1996) Contribution to the bionomics of *Ceramius tuberculifer* Saussure (Hymenoptera, Vespidae, Masarinae). J Hymenopt Res 5:22–37

Mauss V (2000) Anatomie des Darmtraktes von Pollenwespen (Hymenoptera, Vespidae, Masarinae): Anpassungen an Pollentransport und -konsum. Mitteilungen der Deutschen Gesellschaft für allgemeine und angewandte Entomologie 12:571–574

Mauss V (2007) Evolution verschiedener Lebensformtypen innerhalb basaler Teilgruppen der Faltenwespen (Hymenoptera, Vespidae). Denisia 20, zugleich Kataloge der oberösterreichischen Landesmuseen Neue Serie 66:701–722

Mauss V (2008) Bionomie und Abwehrverhalten der in Deutschland vorkommenden allergologisch bedeutsamen Bienen und Faltenwespen. (Hymenoptera, Apidae, Vespidae). Hautarzt 59:184–193

Mauss V (2013) Description of *Celonites andreasmuelleri* sp. n. (Hymenoptera, Vespidae, Masarinae) from the Middle East with a key to the Palaearctic species of the *C. abbreviatus*-complex of the subgenus *Celonites* s.str. J Hymenopt Res 31:79–95

Mauss V, Mauss H (2016) Field observations of the behaviour of the pollen wasp *Quartinia tenerifina* Richards, 1969 (Hymenoptera, Vespidae, Masarinae) at flowers in Tenerife (Canary Islands, Spain). Linzer Biologische Beiträge 48:1327–1332

Mauss V, Müller A (2000) A study of the bionomy of the Spanish pollen wasp *Ceramius hispanicus* Dusmet (Hymenoptera, Vespidae, Masarinae): nesting, mating, and flower association. J Hymenopt Res 9:1–17

Mauss V, Müller A (2014) First contribution to the bionomics of the pollen wasp *Celonites fischeri* Spinola, 1838 (Hymenoptera, Vespidae, Masarinae) in Cyprus. J Hymenopt Res 39:119–153

Mauss V, Müller A (2016) Contribution to the bionomics of the pollen wasp *Quartinia canariensis* Blüthgen, 1958 (Hymenoptera, Vespidae, Masarinae) in Fuerteventura (Canary Islands, Spain). J Hymenopt Res 50:1–24

Mauss V, Prosi R (2018) Identity and distribution of *Celonites hermon* Gusenleitner, 2002 (Hymenoptera, Vespidae, Masarinae) from the Middle East with a description of the hitherto unknown male. J Hymenopt Res 66:55–70

Mauss V, Müller A, Prosi R (2006) Mating, nesting and flower association of the east Mediterranean pollen wasp *Ceramius bureschi* in Greece (Hymenoptera: Vespidae: Masarinae). Entomologia Generalis 29:1–26

Mauss V, Müller A, Yildirim E (2010) First contribution to the bionomics of the pollen wasp *Ceramius palestinensis* (Giordani Soika, 1957) (Hymenoptera: Vespidae: Masarinae) in Turkey. Entomological Science 13:42–59

Mauss V, Fateryga A, Prosi R (2016) Taxonomy, distribution and bionomics of *Celonites tauricus* Kostylev, 1935, stat. n. (Hymenoptera, Vespidae, Masarinae). J Hymenopt Res 48:33–66

Mauss V, Müller A, Prosi R (2018) Flower associations and nesting of the pollen wasp *Quartinia major* Kohl, 1898 (Hymenoptera, Vespidae, Masarinae) in Morocco. J Hymenopt Res 62:15–31

Meade-Waldo G (1913) New species of *Diploptera* in the collection of the Bristish Museum Part IV. Ann Mag Nat Hist 8(11):44–54

Miotk P (1979) Zur Biologie und Ökologie von *Odynerus spinipes* (L.) und *O. reniformis* (Gmel.) an den Lößwänden des Kaiserstuhls (Hymenoptera: Eumenidae). Zoologisches Jahrbuch für Systematik 106:374–405

Móczár L (1960) The loess wall of Tihany and the nesting of *Odynerus spiricornis* Spin. (Hymenoptera, Eumenidae). Annales Historico-Naturales Musei Nationalis Hungarici. Pars Zoologica 52:383–409

Móczár L (1962) Data on the development of *Odynerus spiricornis* Spin. (Hymeoptera, Eumenidae). Annales Historico-Naturales Musei Nationalis Hungarici. Pars Zoologica 54:339–351

Moore WS (1975) Observations on the egg laying and sleeping habits of *Euparagia scutellaris* Cresson (Hymenoptera, Masaridae). Pan-Pac Entomol 51:286

Müller A (1996) Convergent evolution of morphological specializations in Central European bee and honey wasp species as an adaptation to the uptake of pollen from nototribic flowers (Hymenoptera, Apoidea and Masaridae). Biol J Linn Soc 57:235–252

Naumann ID, Cardale JC (1987) Notes on the behaviour and nests of an Australian masarid wasp *Paragia* (*Paragia*) *decipiens decipiens* Shuckard (Hymenopter: Vespoidea: Masaridae). Aust Entomol Mag 13:59–65

Neff JL, Simpson BB (1985) Hooked setae and narrow tubes: foretarsal pollen collection by *Trimeria buyssoni* (Hymenoptera: Masaridae). J Kansas Entomol Soc 58:730–732

Nugroho H, Ubaidillah R, Kojima J (2016) Taxonomy of the Indo-Malayan presocial potter wasp genus *Calligaster* de Saussure (Hymenoptera, Vespidae, Eumeninae). J Hymenopt Res 48:19–32

Olberg G (1959) Das Verhalten der solitären Wespen Mitteleuropas (Vespidae, Pompilidae, Sphecidae). VEB Deutscher Verlag der Wissenschaften, Berlin

Osten T (1982) Vergleichend-funktionsmorphologische Untersuchungen der Kopfkapsel und der Mundwerkzeuge ausgewählter "Scolioidea" (Hymenoptera, Aculeata). Stuttgarter Beiträge zur Naturkunde Serie A (Biologie) 354:1–60

Osten T (1988) Die Mundwerkzeuge von *Proscolia spectator* Day (Hymenoptera: Aculeata). Ein Beitrag zur Phylogenie der "Scolioidea". Stuttgarter Beiträge zur Naturkunde Serie A (Biologie) 414:1–30

Pardi L (1948) Dominance order in *Polistes* wasps. Physiol Zool 21:1–13

Parker FD (1966) A revision of the North American species in the genus *Leptochilus* (Hymenoptera: Eumenidae). Misc Publ Entomol Soc Am 5:151–229

Paul J, Roces F, Hölldobler B (2003) How do ants stick out their tongues? J Morphol 254:39–52

Perrard A, Grimaldi D, Carpenter JM (2017) Early lineages of Vespidae (Hymenoptera) in Cretaceous amber. Syst Entomol 42:379–386

Peters RS, Krogmann L, Mayer C, Donath A, Gunkel S, Meusemann K, Kozlov A, Podsiadlowski L, Petersen M, Lanfear R, Diez PA, Heraty J, Kjer KM, Klopfstein S, Meier R, Polidori C, Schmitt T, Liu S, Zhou X, Wappler T, Rust J, Misof B, Niehuis O (2017) Evolutionary history of the Hymenoptera. Curr Biol 27:1–6

Pickett KM, Carpenter JM (2010) Simultaneous analysis and the origin of eusociality in the Vespidae (Insecta: Hymenoptera). Arthropod Syst Phylog 68:3–33

Piekarski PK, Carpenter JM, Lemmon AR, Moriarty Lemmon E, Sharanowski BJ (2018) Phylogenetic evidence overturns current conceptions of social evolution in wasps (Vespidae). Mol Biol Evol 35:2097–2109

Raveret Richter M (2000) Social wasps (Hymenoptera: Vespidae) foraging behaviour. Annu Rev Entomol 45:121–150

Reeve HK (1991) Polistes. In: Ross KG, Matthews RW (eds) The social biology of wasps. Cornell University, Ithaca, pp 99–148

Richards OW (1962) A revisional study of the masarid wasps (Hymenoptera, Vespoidea). British Museum (Natural History), London, 294 pp

Roubaud E (1911) The natural history of the solitary wasps of the genus *Synagris*. Annual report of the Smithsonian Institution 1910:507–525

Schremmer F (1959) Der bisher unbekannte Pollensammelappart der Honigwespe *Celonites abbreviatus* Vill. (Vespidae, Masarinae). Z Morphol Ökol Tiere 48:424–438

Schremmer F (1961) Morphologische Anpassungen von Tieren – insbesondere Insekten – an die Gewinnung von Blumennahrung. Verhandlungen der Deutschen Zoologischen Gesellschaft (Zoologischer Anzeiger Supplement) 25:375–401
Schremmer F (1962) Wespen und Hornissen. Die einheimischen sozialen Faltenwespen. A. Ziemsen Verlag, Wittenberg Lutherstadt
Schremmer F, März L, Simonsberger P (1985) Chitin im Speichel der Papierwespen (Soziale Faltenwespen, Vespidae): Biologie, Chemismus, Feinstruktur. Mikroskopie 42:52–56
Seifert G (1995) Entomologisches Praktikum, 3. Auflage. G. Thieme Verlag, Stuttgart, New York
Snodgrass RE (1935) Principles of insect morphology. McGraw-Hill Book Company, New York. (Reprint with a new foreword by Goerge C Eickwort, Cornell University Press 1993)
Spradbery JP (1973) An account of the biology and natural history of social and solitary wasps. University of Washington Press, Seattle
Spradbery JP (1991) Evolution of queen number and queen control. In: Ross KG, Matthews RW (eds) The social biology of wasps. Cornell University, Ithaca, pp 336–388
Tischendorf S, Engel M, Flügel H-J, Frommer U, Geske C, Schmalz K-H (2015) Atlas der Faltenwespen Hessens. FENA Wissen 3. Hessen-Forst FENA, Gießen
Torchio PF (1970) The ethology of the wasp *Pseudomasaris edwardsii* (Cresson), and a description of its immature forms. Contrib Sci Nat Hist Mus Los Angeles County 202:1–32
Trostle GE, Torchio PF (1986) Notes on the nesting biology and immature development of *Euparagia scutellaris* Cresson (Hymenoptera: Masaridae). J Kansas Entomol Soc 59:641–647
Turillazzi S (1991) The Stenogastrinae. In: Ross KG, Matthews RW (eds) The social biology of wasps. Cornell University, Ithaca, pp 4–98
Turillazzi S (2012) The biology of hover wasps. Springer, Heidelberg
van der Vecht J (1975) A review of the genus *Stenogaster* Guérin (Hymenoptera: Vespoidea). J Aust Entomol Soc 14:283–308
van der Vecht J (1977) Studies of oriental Stenogastrinae (Hymenoptera, Vespoidea). Tijdschrift voor Entomologie 120:55–75
van Lith HP (1956) *Hoplomerus* (*Hoplomerus*) *spinipes* (L.) en *Hoplomerus* (*Spinicoxa*) *reniformis* (Gmel.). Entomologische Berichten 16:259–263
Wenzel JW (1990) A social wasp's nest from the cretaceous period, Utah, USA, and its biogeographical significance. Psyche 97:21–29
Wenzel JW (1991) Evolution of nest architecture. In: Ross KG, Matthews RW (eds) The social biology of wasps. Cornell University, Ithaca, pp 480–519
West-Eberhard MJ (1978) Polygyny and the evolution of social behaviour in wasps. J Kansas Entomol Soc 51:832–856
West-Eberhard MJ (1987) Flexible strategy and social evolution. In: Ito Y, Brown JL, Kikkawa J (eds) Animal societies: theories and facts. Japan Scientific Societies Press, Tokyo, pp 35–51
Williams FX (1919) Philippine wasp studies. Part 2. Descriptions of new species and life history studies. Bull Exp Stn Hawaiian Sugar Plant Assoc 14:19–187
Williams FX (1927) *Euparagia scutellaris* Cresson, a Masarid wasp that stores its cells with the young of a curculionid beetle. Pan Pac Entomol 4:38–39
Witt R (2009) Wespen. 2. neubearbeitete Auflage. Vademecum-Verlag, Oldenburg

# Chapter 15
# Superlong Proboscises as Co-adaptations to Flowers

**Julia A.-S. Bauder and Florian Karolyi**

**Abstract** Extraordinarily long mouthparts evolved as co-adaptations to drink nectar from deep-tubed, highly rewarding flowers. Although superlong-proboscid taxa represent only a minority of flower-visiting insects, some play an important role as pollinators of many plant species, while others profit from consuming floral nectar without performing pollination services. Superlong proboscises evolved several times independently in euglossine bees (Hymenoptera: Apidae), butterflies (Lepidoptera: Riodinidae, Hesperiidae), hawk moths (Lepidoptera: Sphingidae), tangle-veined flies (Diptera: Nemestrinidae) and horse flies (Diptera: Tabanidae). Novel techniques in micromorphology allow for new insights into the ecomorphology of superlong mouthparts, organs for proboscis movement and fluid uptake inside the insect's head which are associated with fluid feeding. Common features of long-proboscid insect taxa concern the elongation of rather simple structures, such as only the prementum of the labium in flies or the galeae in butterflies and hawk moths. Morphological adaptations such as larger food canals help to overcome biophysical problems resulting from mouthpart elongation such as increasing nectar flow resistance. Functional costs in terms of increased flower manipulation times arise from a reduced sensory equipment of superlong proboscises. Furthermore, reinforcements of the organs for proboscis movement and nectar uptake evolved to maintain the performance of the feeding apparatus. Finally, evolutionary trade-offs and behavioural adaptations as well as limits to further proboscis elongation in various superlong-proboscid insect taxa are discussed.

## 15.1 General Introduction

Floral nectar is commonly regarded as the world's most ubiquitous food source (Nicolson 2007). Species all over the animal kingdom – mainly those capable of flight – such as birds, bats and insects, have independently evolved morphological

J. A.-S. Bauder (✉) · F. Karolyi
Department of Integrative Zoology, University of Vienna, Vienna, Austria
e-mail: julia.bauder@univie.ac.at; florian.karolyi@univie.ac.at

 
H. W. Krenn (ed.), *Insect Mouthparts*, Zoological Monographs 5,
https://doi.org/10.1007/978-3-030-29654-4_15

structures which enable its uptake (Pellmyr 2002; Muchhala and Thomson 2009; Johnson and Anderson 2010; Krenn 2010; Karolyi et al. 2012, 2014).

Countless insects rely on flower nectar as an energy source to fuel their daily activities and take up these carbohydrate-rich fluids with a tubular, slender proboscis which frequently measures about the length of the head or the body (Krenn et al. 2005; Borrell and Krenn 2006). Nectar-extracting mouthparts are composed of various elongated mouthpart structures and have attained impressive extremes in five different taxa among insects, such as euglossine bees (Hymenoptera: Euglossini), butterflies and hawk moths (Lepidoptera: Papilionoidea, Sphingidae), tangle-veined flies (Diptera: Nemestrinidae) and horse flies (Diptera: Tabanidae), and often exceed twice the body length (Amsel 1938; Borrell 2005; Borrell and Krenn 2006; Kunte 2007; Pauw et al. 2009; Karolyi et al. 2012, 2014; Bauder et al. 2014).

The function of these structures is governed by the physical laws of fluid dynamics (Kingsolver and Daniel 1979; Pivnick and McNeil 1985; Daniel et al. 1989; Kim et al. 2011b; Lee et al. 2014). Biophysical models describe the various factors influencing the speed of fluid feeding and predict that the rate of nectar intake declines with escalating proboscis length (Kingsolver and Daniel 1979, 1995; Lee et al. 2014). Since the rate of energy intake during feeding influences foraging efficiency and reproductive fitness (Wolf et al. 1972; Heinrich 1975; Whitham 1977; May 1988; Hainsworth et al. 1991), rapid feeding should be favoured by natural selection (Emlen 1966; Schoener 1971; Pyke et al. 1977), and insects with extreme proboscis lengths either evolved morphological adaptations of the feeding apparatus (Bauder et al. 2013, 2015a; Karolyi et al. 2014; Düster et al. 2018; Krenn and Bauder 2018) or have to bear functional costs in terms of decreased nectar intake rates and longer flower handling times (Bauder et al. 2011; Karolyi et al. 2013). Other factors that influence the rate of nectar intake include the radius of the proboscis food tube, the mechanics and the size of the suction pump which creates a pressure gradient as well as the nectar viscosity (Daniel et al. 1989). Furthermore, compensating for the biophysical disadvantages of a superlong proboscis by morphological or biometrical adaptations of the feeding apparatus, such as changes in the size of the food tube or the suction pump, could also give rise to material costs for sustaining the functionality of the slender and long proboscis.

Despite the obvious advantages of having a superlong proboscis, such as exclusive access to highly rewarding deep-tubed flowers which cannot be used by insects with short proboscises (Nilsson 1988; May 1992; Bauder et al. 2015b) and a broadening of the food spectrum (Haber and Frankie 1989; Miller 1997), superlong proboscises are rather the exception than the rule. Examining the functional consequences of superlong proboscises in consideration of the whole feeding apparatus combined with ecomorphological analyses is inevitable to complete our understanding of the evolution of flower-visiting insects and their role in pollinator-plant networks. The morphological and functional comparison of superlong mouthparts, i.e. proboscises which exceed body length by at least 1.5 times, allows for novel conclusions on the costs and benefits of these extraordinary feeding organs. Detailed morphological and sometimes even biometrical analyses of the proboscis components including its musculature and sensilla equipment, organs for proboscis movements and the uptake of nectar provide new insights into evolutionary constraints and co-adaptations of extremely long-proboscid flower visitors. Furthermore, superlong-proboscid insects display a fascinating feeding ecology and flower-visiting behaviour, which is astonishingly revealed to vary with respect to the impact on their nectar host plants' fitness and assigns these insects different roles in plant-pollinator networks.

## 15.2 Jewels of the Rainforests: Euglossine Bees (Apidae: Euglossini)

### *15.2.1 Feeding Ecology of Long-Proboscid Orchid Bees*

The earliest euglossine bee fossil with a superlong proboscis (*Euglossa moronei*) dates back to 20 million years ago, and their main radiation probably happened in the mid-Cretaceous, which coincides with the appearance of other long-proboscid pollinators such as brachyceran flies (Grimaldi 1999; Grimaldi and Engel 2005; Peñalver et al. 2015). All fossil and extant euglossines are restricted to the Neotropics, which suggests that they originated there after the supercontinent Gondwanaland was split apart forming our present-day continents. Combined fossil and biogeographic evidence suggests that euglossines originated sometime between 100 and 20 million years ago. Orchid bees are known to be the pollinators of at least one genus of every major orchid group in the Neotropics, which appeared 65 million years ago. Eugossines likely coevolved with orchid flowers repeatedly since then and promoted an increase of orchid species richness (Roubik and Hanson 2004).

Euglossini is one of four highly distinct taxa (Meliponini, Bombini, Apini) composing the monophyletic corbiculate clade within Apinae, which all share the synapomorphy of a concave, pollen-collecting depression on the hindlegs, the corbicula. The extant Euglossini contain approximately 190 species among five genera (Cameron 2004). Also known as orchid bees, euglossines are a group of brilliantly iridescently coloured Neotropical bees which are famous among tropical flower-visiting insects due to their extraordinary relationship with orchids: They are the exclusive pollinators of approximately 700 species of orchids (Dressler 1982; Roubik and Hanson 2004). Male euglossine bees visit the flowers of orchids to collect fragrances using their modified legs—not the proboscis—to emit these volatile components at mating sites to attract females (Dressler 1982; Eltz et al. 2005). Nevertheless, both female and male euglossines use their superlong proboscises to extract nectar of a wide array of deep-tubed flowers (Roubik and Hanson 2004).

Within euglossine bees, proboscis length varies extremely depending on the species, but some possess superlong mouthparts which can measure up to 41.3 mm and 2.6 times the body length (Table 15.1). The shortest proboscis length within euglossines measures 12.2 mm which corresponds to 0.8 times the body length (Roubik and Hanson 2004). The bees forage for nectar on flowers belonging to

**Table 15.1** Euglossine bees (Hymenoptera: Apidae) with superlong proboscises. Given are absolute length [mm] and relative proboscis length, i.e. proboscis length divided by body length, and known nectar host plants

| Species | Absolute proboscis length | Relative proboscis length | Nectar host plants |
|---|---|---|---|
| *Eufriesea ornata* (Mocsáry, 1896) | 41.3[a] | 1.7[b] | *Costus*[c], *Eschweilera*[d] |
| *Eufriesea schmidtiana* (Friese, 1925) | 32.2[a] | 1.7[b] | *Calathea*[e], *Costus*[c, e], *Dimerocostus*[e] |
| *Euglossa allosticta* Moure, 1969 | 23.4[a] | 1.8[b] | *Calathea*[e], *Costus*[e], *Dimerocostus*[e], *Drymonia*[c], *Odontadenia*[c], *Stemmadenia*[c] |
| *Euglossa asarophora* Moure & Sakagami, 1969 | 38.9[a] | 2.6[b] | *Calathea*[e], *Costus*[e], *Dimerocostus*[e] |
| *Euglossa bursigera* Moure, 1970 | 19.8[a] | 1.8[b] | *Drymonia*[f] |
| *Euglossa cybelia* Moure, 1968 | 18.9[a] | 1.7[b] | *Calathea*[c] |
| *Euglossa despecta* Moure, 1968 | 19.0[a] | 1.7[b] | N/A |
| *Euglossa dodsoni* Moure, 1965 | 18.7[a] | 1.9[b] | *Calathea*[c] |
| *Euglossa flammea* Moure, 1969 | 35.1[a] | 2.5[b] | N/A |
| *Euglossa gorgonensis* Cheesman, 1929 | 17.4[a] | 1.7[b] | *Calathea*[c], *Costus*[c] |
| *Euglossa imperialis* Cockerell, 1922 | 32.1[a] | 2.1[b] | *Calathea*[c], *Costus*[c], *Mandevilla*[c], *Pachyptera*[c], *Stemmadenia*[c] |
| *Euglossa purpurea* Friese, 1899 | 18.8[a] | 1.7[b] | N/A |
| *Euglossa variabilis* Friese, 1899 | 17.0[a] | 1.7[b] | *Arrabidaea*[c], *Drymonia*[c] |

[a]Roubik and Hanson (2004), [b]Calculated from values given in Roubik and Hanson (2004), [c]Ackerman (1985), [d]Prance et al. (1983), [e]Borrell (2005), [f]Steiner (1985)

51 different plant families, e.g. Orchidaceae, Costaceae, Marantaceae, Melastomataceae, Solanaceae, Lecithidaceae, Bixaceae, Heliconiaceae, Zingiberaceae, Rubiaceae, Apocynaceae, Bignoniaceae, Convolvulaceae, Polygalaceae and Verbenaceae (Table 15.1). The capacity of orchid bees for flying long distances, combined with their orientation ability, allows them to repeatedly visit the same, widely separated flowers over several weeks or even months. This so-called trap-lining behaviour contributes to their important role as effective pollinators of many steady-state understory plants (Janzen 1971; Roubik and Hanson 2004). However, euglossine bee species generally forage at any flowers from which they can physically extract nectar. Euglossine bees with longer proboscises visit a greater

number of nectar host plants than bees with shorter proboscises. On the other hand, flowers with longer tubes have fewer visitor species than flowers with shorter tubes. Euglossine bees with superlong proboscises are generalistic flower visitors, but flowers with long tubes have specialized on them as pollinators (Borrell 2005). Since these steady-state understory plants with long flower tubes are small in size and occur in low densities, they are unable to support the energetic needs of an entire pollinator population (Schemske 1981). Increasing nectar production would be energetically costly (Southwick 1984), but by excluding short-proboscid bees as consumers, long-tubed plants can provide sufficient nectar rewards to superlong-proboscid euglossines, which learn to include rewarding nectar sources on their daily traplines (Janzen 1971; Kay and Schemske 2003). In this case, long flower tubes evolved by competition among sympatric flowers for inclusion on an individual bee's foraging route (Rathcke 1992; Garrison and Gass 1999) and not by directional selection exercised by specialized pollinators, which is one of the most widespread explanations for the existence of long flower tubes and long proboscises of pollinators (Darwin 1862). The need for pollinator sharing because of problems to provide rewarding nectar sources for efficient bee pollinators may be intensified due to superlong-proboscid skipper and riodinid butterflies that also take nectar from these flowers but do not pollinate them (Bauder et al. 2015b).

## *15.2.2 Morphology of a Superlong Bee Proboscis*

### 15.2.2.1 Proboscis Components and Movement

The principle mouthpart composition of extremely long-proboscid euglossine bees is similar to that of other long-proboscid bees such as *Apis mellifera* and *Bombus* (Winston 1979; Plant and Paulus 2015; Düster et al. 2018). The mouthparts consist of the short unpaired labrum, paired biting mandibles and the labiomaxillary complex, which is designed for fluid feeding (Plant and Paulus 2015) (Fig. 15.1a, b). The labiomaxillary complex is composed of the basally linked pair of maxillae and the unpaired labium (Snodgrass 1956; Winston 1991). The proboscis consists of the enlongated maxillary galeae and labial palps which together form the food canal that encompasses the glossa. The components which form the food canal are laterally curved to fit tightly into each other and are held secure by interlocking cuticle structures to form a closed tube enveloping the central glossa. These structural adaptations account for a permanent food tube formation in orchid bees in contrast to other Apidae, where the components of the food tube separate from each other in resting position (Düster et al. 2018). The permanent connection of the mouthparts composing the food tube probably represents an adaptation to avoid problems with proboscis assembly, which is more complex with a superlong proboscis.

The first two segments of the labial palps are elongated and as long as the galeae but shorter than the glossa. The third and fourth segment of the labial palps are short

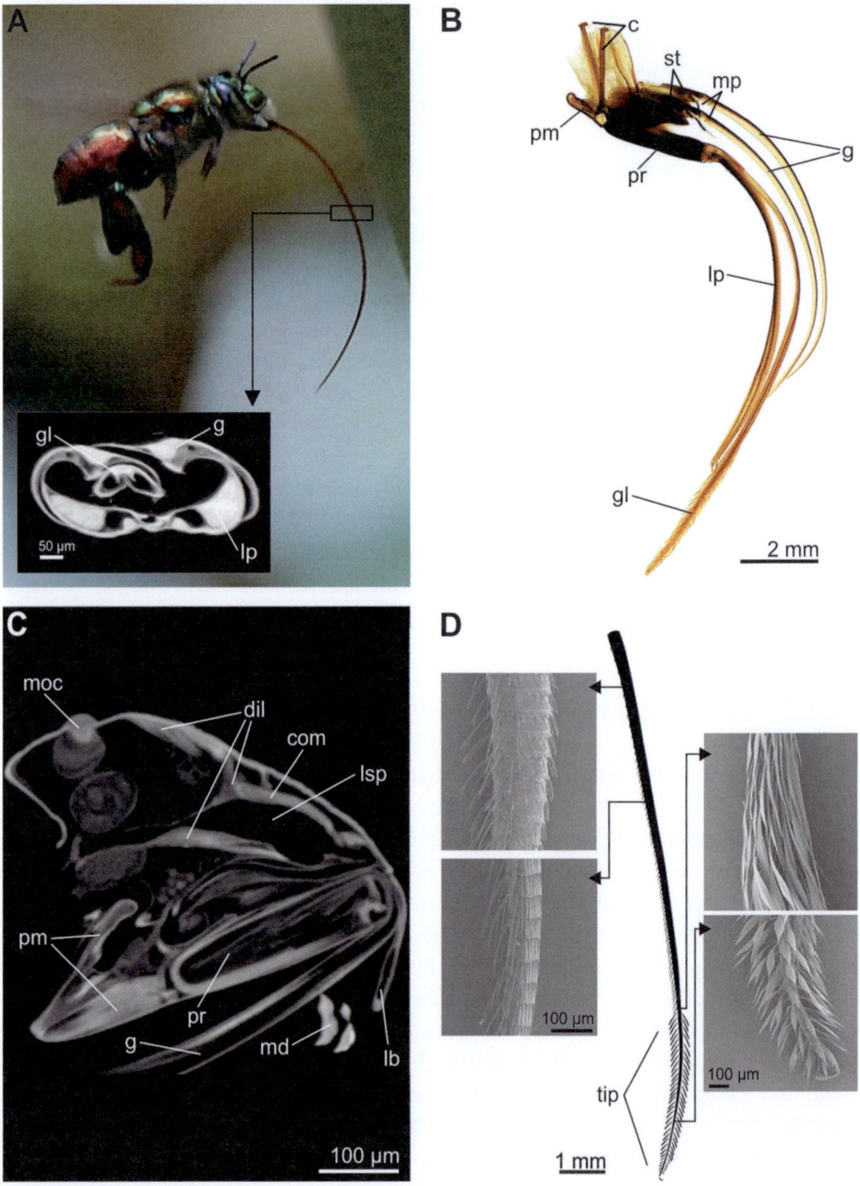


**Fig. 15.1** Euglossine bee (Apidae: Euglossini) (**a**) Orchid bee approaching a flower with unfolded proboscis. Inserted detail: Micro-CT scan of a proboscis cross section showing the food tube in resting position. Ensheathing structures that form the food canal are the interlocked galeae and labial palps that enclose the glossa (Düster et al. 2018). (**b**) Light microscopical image of the proboscis of a male *Euglossa championi* (Düster et al. 2018). (**c**) Micro-CT scan of a sagittal section through the head of *Euglossa*. Dilator and compressor muscles attach at the lumen of the suction pump. (**d**) Glossa of a male *Euglossa championi*. Schematic illustration of the brush-like tip region combined with SEM images along the glossa from the proximal region (top left), mid-region, distal

and diverge laterally from the first labial palp segment. The short, two-segmented maxillary palpus is located at the joint between stipes and galea base on each side (Fig. 15.1a, b). The glossa is protractible and covered with sensilla and spatula-shaped microtrichia, which are longer distally (Fig. 15.1b, d). These long and lanceolate microtrichia function as a brush by transferring liquids into the food canal via capillarity. Proximally, the food canal empties into the functional mouth, which leads into the cibariopharyngeal cavity (Düster et al. 2018).

A comparison of euglossine bees with different proboscis lengths showed that superlong proboscises come from the elongation of the parts which compose the food tube, i.e. galeae and labial palps, and the glossa, whereas the basal components of the mouthparts, i.e. cardo, stipes and prementum, are similar in size (Düster et al. 2018). Compared to other long-proboscid insects such as Lepidoptera, the bee proboscis is more complex in its composition and functionality. In contrast to long-proboscid Lepidoptera and Diptera, the mouthparts of the bee proboscis contain no muscles inside. However, this is true for all bees and thus cannot be interpreted to be an adaptation of euglossine bees with superlong proboscises.

The mechanism of proboscis extension in orchid bees is the same as in the honey bee (Düster et al. 2018). Proboscis unfolding is achieved as soon as the galeae and the labial palpi swing forward and the basal components of the labiomaxillary complex are extended frontally: As the labium moves forward, the folded postmentum turns from a vertical to a horizontal position and the glossa is extended. The protracted proboscis is longer than the folded proboscis since the unfolding of the postmentum enables an enormous enhancement of length. Additionally, the previously contracted and curved glossa base in the prementum protracts, which increases proboscis length up to 21%. The capability of the prementum of sliding back and forth combined with the considerable postmentum extension is unique in euglossine bees and has never been observed in other bees. The morphological adaptations of the proboscis components, such as movements of prementum, postmentum and glossa, account for an increased functional proboscis length in extremely long-proboscid *Euglossa* species, which has been observed to be essential for taking up nectar from much deeper flowers than would be expected when the food tube length alone is considered (Düster et al. 2018).

**Fig. 15.1** (continued) region and the tip region with the flabellum (bottom right). The shape and size of microtrichia changes along the glossa from short, bristle-shaped in the proximal region to long, lanceolate at the proboscis tip. Reproduced with permission of Elsevier Ltd. from Düster et al. (2018). Abbreviations: *c* cardo, *com* compressor muscle, *dil* dilator muscle, *g* galea, *gl* glossa, *lb* labrum, *lp* labial palp, *lsp* lumen of suction pump, *md* mandible, *moc* median ocellus, *mp* maxillary palp, *pm* postmentum, *pr* prementum, *st* stipes

#### 15.2.2.2 Sensory Equipment of the Proboscis

Euglossine bees are equipped with different types of bristle-shaped sensilla throughout the length of the glossa, galeae, labial palpi and maxillary palpi. The glossa and the distal labial palpi are covered with sensilla chaetica, which serve as mechanosensitive sensilla and/or gustatory sensilla and are similar to those of the honey bee (Galic 1971; Whitehead and Larsen 1976; Düster et al. 2018). At the tip of the glossa, these sensilla help to detect nectar inside the flower, whereas the sensilla on the labial palpi serve to locate the flower opening. Sensilla chaetica are also located along the inner surface of the galeae, the proximal labial palpi, the glossa and inside the food canal. They probably function as gustatory sensilla to detect nectar flow as in the food canal of the butterfly proboscis (Inoue et al. 2009). The number of these sensilla does not correlate with proboscis length. As in butterflies, euglossine bees with superlong proboscises are characterized by a reduction of the sensilla equipment inside the food canal. Bristle-shaped sensilla of the maxillary palpi protrude laterally at the basis of the proboscis and probably provide information about the maximal proboscis insertion into a floral tube. Proboscis movements can also be monitored by sensilla chaetica on the lateral galeae and the proximal labial palp segments (Düster et al. 2018).

#### 15.2.2.3 Suction Pump for Fluid Uptake

Unlike most other bees, extremely long-proboscid euglossines primarily use a suction feeding technique to ingest large amounts of nectar from deep-tubed flowers, while the glossa remains motionless and usually does not perform lapping movements (Borrell 2004; Düster et al. 2018). Like all other superlong-proboscid insects, suction feeding functions with the action of a suction pump inside the head (Fig. 15.1c). Assuming that the suction pump creates a pressure gradient from the proboscis tip to the preoral and cibariopharyngeal cavity, this could be achieved by the contraction of dilator muscles, which extend the lumen of the suction pump. However, detailed analyses of the suction pump morphology are still missing for euglossine bees.

### *15.2.3 Flexible Feeding Behaviour: Lapping-Sucking Mouthparts*

Orchid bees were recently shown to have flexible fluid-feeding techniques depending on the amount of liquid available (Düster et al. 2018). In addition to pure suction feeding from deep-tubed flowers, euglossines can take up residual amounts of liquids by lapping, i.e. repeated glossal protraction and retraction movements. Microtrichia covering the distal glossa are longer, flat, lanceolate and widely spaced, which creates a large volume between them that can be filled with nectar via capillarity (Düster et al. 2018). Extremely long-proboscid orchid bees such as *Euglossa imperialis* are able to take up more than twice the liquid volume by lapping compared to shorter-proboscid orchid bees like *Euglossa championi*. This is simply due to the two times longer proboscis of *Euglossa imperialis* and indicates that a similar fluid amount can be loaded on the glossa per mm length (Düster et al. 2018).

### 15.2.4 Behavioural Adaptations as Consequences of Suction Feeding

The purely suctorial feeding technique employed by long-proboscid euglossines to drink from deep-tubed flowers comes along with problems regarding nectar viscosity: As viscosity rises exponentially with sugar concentration (Baker 2016), the suction feeding style is constrained by nectar concentration: The optimal energy intake rate is achieved for approximately 10–20% lower sugar concentrations ranging between 30 and 40% compared to the lapping feeding style employed by bees with shorter proboscises (Borrell 2004). Such dilute nectars are characteristic for flowers with deep tubes visited by euglossines (Schemske 1981; Roubik et al. 1995; Borrell 2007). However, euglossines are strong flyers which cover extremely long distances daily and therefore have a highly active metabolism. The problem that comes with a long proboscis is how to cover high-energy expenditures on predominantly dilute nectars (Pokorny et al. 2014).

Further, sugar concentration does not only have a severe impact on an adult's energy budget but plays an important role in brood provisions as well (Nicolson 2011). Euglossine bees provision their brood with pollen and added nectar after the egg has been laid, like most other bees (Rozen 1984; Velthuis et al. 1984; Neff and Simpson 1992; Cane et al. 2011). A higher sugar content of the provision correlates positively with higher larval weight, which has a beneficial effect on adult weight and fecundity (Elliott et al. 2008; Burkle and Irwin 2009).

A common technique employed by short-proboscid bees is to dehydrate nectar. The bees stretch out their proboscis and regurgitate nectar by forming a droplet at the proboscis tip or around the whole proboscis. After some time, dehydrated nectar with a higher sugar concentration is taken up for a second time from the proboscis tip through the whole length of the proboscis (Pokorny et al. 2014).

For euglossine bees with proboscises much longer than the body, stretching out the proboscis while perching on a substrate is difficult, even more without wiping off the droplet at the proboscis tip. Apart from that, the re-uptake of highly concentrated nectar droplets from the tip would be constrained in the same way as drinking nectar with high sugar contents in the first place. However, superlong-proboscid euglossine bees have evolved a derived behaviour called tongue flicking, which differs from nectar dehydration practiced by other bees and helps to offset the disadvantageous constraints of suction feeding by increasing the sugar content of nectar (Pokorny et al. 2014): After feeding, euglossine bees perch motionless with the proximal proboscis parts split and spread apart into galeae, glossa and labial palps to draw a fluid film between them for water evaporation. Drops of liquids form around the mandibles and the basal proboscis, which are moved back and forth rhythmically. Exposing regurgitated nectar droplets at the proboscis base and splitting the food canal into its parts circumvents viscosity constraints that would otherwise impede nectar dehydration performed with a long proboscis. Crop sugar concentration after tongue flicking was up to 34.8% higher than the sugar content of the imbibed flower nectar.

Aside from tongue flicking, female euglossines display a derived, possibly unique behaviour leading to a higher sugar content of aliments for their brood. Female euglossines also visit nectar flowers during pollen collection trips and add regurgitated nectar to the pollen loads in the corbiculae while hovering. By moving the basitarsi of the forelegs along the extended proboscis, they swipe droplets of regurgitated nectar off the proboscis tip and transfer them onto the corbiculae via middle and hind tarsi. Later, they spread the pollen loads across their whole ventral body surface and four legs to create a large surface, move and knead the mass while continuously adding more nectar to it. By this extensive manipulation process, water evaporates and the sugar content of the mix rises. Finally, the pollen-nectar mix is scraped off the bee's body and transferred back to the corbiculae (Pokorny et al. 2014). This unique behaviour circumvents problems owed to long proboscises, such as the inability to regurgitate dehydrated crop content onto pollen loads already positioned in the brood cell as short-proboscid bees do (Velthuis et al. 1984).

### *15.2.5 Limits to Proboscis Elongation?*

During hovering flight, which is employed by euglossine bees during a flower visit, insects reach rates of energy production that are among the highest measured in the animal kingdom (Darveau 2004). Furthermore, the Euglossini have relatively high body temperatures when active (Evoy and Jones 1971) and cover larger distances daily than other bees (Janzen 1971). It is obvious that they have to consume relatively large quantities of nectar (Dressler 1982). For example, the superlong-proboscid bee *Euglossa imperialis* requires 40 mg of sugar per day only to maintain body mass (Borrell 2005). Additionally, the rate of nectar intake was shown to decline with increasing proboscis length in euglossine bees, in contrast to long-proboscid Lepidoptera (Borrell 2004, 2007; Bauder et al. 2015a).

Further, the shift of the fluid-feeding technique from capillary-based lapping to suction feeding during drinking from deep-tubed flowers (Düster et al. 2018) led to a decrease in the nectar sugar concentration that maximizes the rate of energy intake (Borrell 2004). Long-proboscid orchid bees drink dilute nectars between 30 and 40% sugar concentration as do long-proboscid butterflies, hawk moths and flies (Pivnick and McNeil 1985; Kingsolver and Daniel 1995; Josens and Farina 2001; Borrell

2004). At some point, drinking relatively dilute nectars from deep-tubed flowers that occur in low densities might result in insufficient energy supply despite having exclusive access to large nectar amounts. Taking into account that these bees fly to widely spaced nectar host plants via their daily traplines while also having to spend time for nectar regurgitation and water evaporation by tongue flicking, the combination of these time-consuming tasks might constrain the evolution of even longer proboscises. A so far untouched aspect of evolutionary limits to mouthpart length concerns the enclosed confinements of the euglossine bee larvae inside the brood cell.

## 15.3 Of Nectar Thieves, Pollinators and the Longest Proboscises Among Insects: Butterflies and Hawk Moths (Lepidoptera: Papilionoidea and Sphingidae)

### *15.3.1 Occurrence of Superlong Proboscises in Lepidoptera*

Among the glossatan Lepidoptera, both Papilionoidea and Sphingidae comprise representatives with particularly long proboscises (Fig. 15.2).

The butterflies (Papilionoidea) comprise nearly 20,000 species in more than 1800 genera worldwide (Van Nieukerken et al. 2011). Although the majority of adult butterflies feeds on floral nectar (Krenn et al. 2010), their role as effective pollinators remains doubtful in many examples (Wiklund et al. 1979; Wiklund 1981; Courtney et al. 1982). A mutualistic coevolutionary relationship between particular butterfly species and their preferred nectar host plants has only been demonstrated in few cases (Grant and Grant 1965; Gilbert 1972, 1975; Levin and Berube 1972). In fact, some authors regard butterflies as opportunistic flower visitors that use resources as they become available during the season (Shreeve 1992; Tudor et al. 2004; Stefanescu and Traveset 2009). Proboscis length ranges from 2.8 mm in *Sarota gyas* (Riodinidae) (Bauder et al. 2013) to 52.7 mm in *Damas clavus* (Hesperiidae), which is the current world record holder in proboscis length among butterflies (Bauder et al. 2014). In European butterflies, proboscis length ranges between 4.9 and 17 mm (Paulus and Krenn 1996).

The hawk moths (Sphingidae) comprise about 1400 species in approximately 200 genera (Rothschild and Jordan 1903; Kitching and Cadiou 2000; Kawahara et al. 2009). Hawk moths are best known as agile nectar foragers and are considered as important pollinators (e.g. Grant and Grant 1983; Nilsson 1983; Nilsson et al. 1985, 1987; Haber and Frankie 1989; Miller 1997; Martins and Johnson 2013; Johnson and Raguso 2016). However, as many as one-fifth of hawk moth species have extremely short proboscises and do not feed on nectar (Miller 1997). Proboscis length variation within Sphingidae is tremendous (Miller 1997; Kitching and Cadiou 2000), ranging from merely 1 mm in the Palaearctic species *Andriasa* and *Marumba* as well as the Australian species *Hopliocnema* (Kawahara 2007) to 280 mm in the

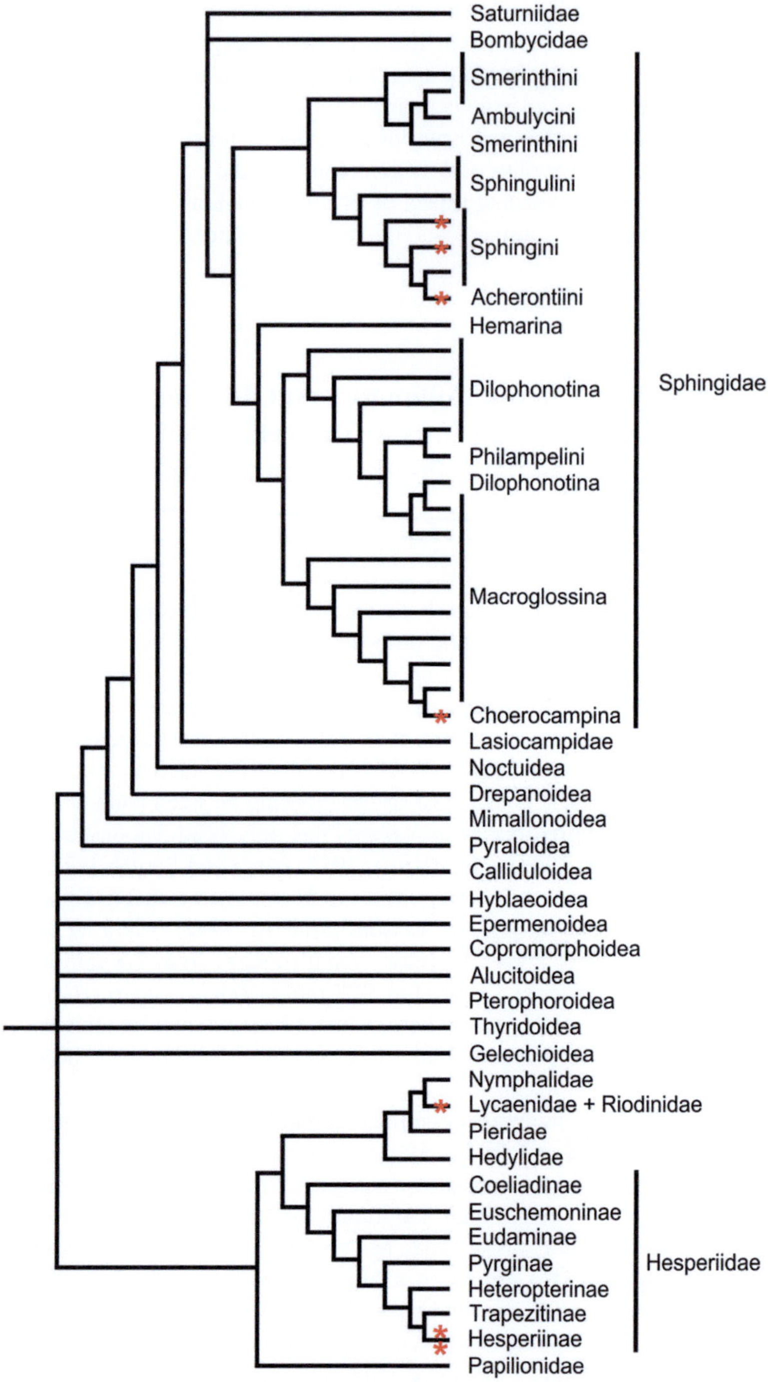


**Fig. 15.2** Phylogenetic tree of Macroplepidoptera based on Regier et al. (2013), Mitter et al. (2017), Kawahara et al. (2009) and Warren et al. (2009). Red stars indicate the existence of species with superlong proboscises within the clade

Neotropical *Amphimoea walkeri*, which corresponds to the fourfold of body length (Amsel 1938). The most famous example of the whole insect world in context with outstanding proboscis length represents the legendary Palaeotropical species *Xanthopan morganii praedicta* (Rothschild and Jordan 1903): Charles Darwin predicted its existence (Darwin 1862) based on the extremely long nectar spur (up to 290 mm) of the Malagasy star orchid *Angraecum sesquipedale* (Arditti et al. 2012). Decades later, the "predicted" hawk moth species with a matching superlong proboscis was described by Rothschild and Jordan (1903) as the potential pollinator of the star orchid.

## *15.3.2 Evolution and Feeding Ecology of Superlong-Proboscid Lepidoptera*

### 15.3.2.1 Papilionoidea

Superlong proboscises evolved multiple times independently in Papilionoidea in two different butterfly families, the skipper butterflies (Hesperiidae) and the metalmark butterflies (Riodinidae) (Table 15.2). Among Neotropical Hesperiidae superlong proboscises presumably evolved twice within groups of Hesperiinae, i.e. Calpodini and Clade 113 (Warren et al. 2009; Bauder et al. 2014). Extremely long proboscises in butterflies outside of the Hesperiidae occur only within a single genus of Riodinidae, i.e. *Eurybia* (Kunte 2007, Bauder et al. 2011, 2013). Extremely long-proboscid skipper and metalmark butterflies were found to take nectar from highly rewarding, deep-tubed flowers, where they benefit from exclusive access to high amounts of nectar. This confirms the hypothesis of competition avoidance stating that food resource partitioning is a driving force for the coexistence of animals (Hespenheide 1973; Schoener 1974; Inouye 1980; Ranta and Lundberg 1980) and contradicts other studies, where butterflies are regarded as generalist flower visitors (Nilsson et al. 1985; Nilsson 1988; Corbet 2000) and the number of visited plant species correlates with proboscis length (Agosta and Janzen 2005). Extremely long-proboscid metalmarks and skippers, such as the calpodines (Hesperiinae), are known to live in shady, forested habitats (Warren et al. 2009). *Calathea* (Marantaceae) plants, which they use as their primary nectar plants, usually grow in the understory of the forest (Weber et al. 2001) and thus probably have similar habitat requirements as extremely long-proboscid Riodinidae and Hesperiidae. These extremely long-proboscid butterflies probably make the best of their situation by using nectar plants flowering in their proximity and harvesting nectar out of deep-tubed flowers. Furthermore, the larvae of many long-proboscid skipper species, such as representatives of the Calpodini, feed on monocotyledons (Janzen and Hallwachs 2009) that occur in the understory of the forest (Weber et al. 2001), including Marantaceae, Costaceae, Heliconiaceae and Zingiberaceae (Janzen and Hallwachs 2009). It was argued that adult butterflies would search for these plants to lay their eggs anyway and would simultaneously have the opportunity to take nectar from the deep-tubed

**Table 15.2** Hawk moths (Lepidoptera: Sphingidae) and butterflies (Lepidoptera: Riodinidae, Hesperiidae) with superlong proboscises. Given are absolute proboscis length [mm] and relative proboscis length, i.e. proboscis length divided by body length, and known nectar host plants. Values of relative proboscis length labelled with an asterisk are given as proboscis length divided by forewing length (not body length) as a surrogate for body size

| Taxon | Absolute proboscis length | Relative proboscis length | Nectar host plants |
|---|---|---|---|
| **Sphingidae** | | | |
| *Amphimoea walkeri* (Boisduval, 1875) | 280[a] | N/A | N/A |
| *Cocytius antaeus* (Drury, 1773) | 139[b] | N/A | *Cereus*[c], *Dendrophylax*[d], *Habenaria*[e] |
| *Dolbogene igualana* (Schaus, 1932) | 50 ± 1.8[f] | *1.8[g] | N/A |
| *Euryglottis aper* (Walker, 1856) | 93.2 ± 14.3[f] | 1.8 ± 0.4[h] | *Nicotiana*[i] |
| *Manduca ochus* (Klug, 1836) | 136 ± 3.4[f] | *2.5[g] | N/A |
| *Manduca quinquemaculata* (Haworth, 1803) | 122.0 ± 6.2[j] | *2.2[k] | *Acleisanthes*[l], *Agave*[m], *Cucurbita*[m], *Datura*[l], *Kallstroemia*[m], *Mimosa*[m], *Mirabilis*[l], *Oenothera*[l], *Platanthera*[n] |
| *Manduca rustica* (Fabricius, 1775) | 144 ± 9.8[f] | *2.4[g] | *Acleisanthes*[l], *Agave*[m], *Datura*[m], *Hylocereus*[f], *Inga*[o] |
| *Manduca scutata* (Rothschild & Jordan, 1903) | 94.0 ± 7.6[h] | 1.7 ± 0.1[h] | N/A |
| *Manduca sexta* (Linnaeus, 1763) | 95 ± 9.8[f] | *1.7[g] | *Agave*[m], *Caesalpinia*[p], *Cucurbita*[m], *Datura*[m], *Mimosa*[m], *Mirabilis*[m], *Petunia*[q], *Tocoyena*[r] |
| *Nannoparce balsa* Schaus, 1932 | 50.7[s] | *1.7[t] | N/A |
| *Neococytius cluentius* (Cramer, 1775) | 198.2 ± 11.4[h] | 2.7 ± 0.3[h] | N/A |
| *Panogena lingens* (Butler, 1877) | 115 ± 7.2[u] | *2.5 ± 0.1[u] | *Aerangis*[v], *Angraecum*[w], *Jumellea*[v], *Neobathiea*[v] |
| *Xanthopan morganii praedicta* (Walker, 1856) | 217 ± 4.2[x] | *2.8 ± 0.3[x] | *Angraecum*[w] |
| *Agrius cingulata* (Fabricius, 1775) | 106 ± 1.2[f] | *2.2[g] | *Bauhinia*[r], *Caesalpinia*[p], *Nicotiana*[i], *Petunia*[q], *Platanthera*[n] |
| *Agrius convolvuli* (Linnaeus, 1758) | 73.9 ± 17.0[h] | 1.8[h] | *Aerangis*[y], *Ammocharis*[y], *Angraecum*[w, u], *Cladostemon*[z], *Conostomium*[y], *Crinum*[z], *Datura*[y], *Gardenia*[z], *Gladiolus*[aa,z], *Gynandropsis*[o], *Harveya*[z], *Hedychium*[z], *Ipomoea*[z], *Lilium*[z], *Lippia*[y], *Rangaeris*[z], *Sesamothamnus*[z] |

(continued)

**Table 15.2** (continued)

| Taxon | Absolute proboscis length | Relative proboscis length | Nectar host plants |
|---|---|---|---|
| *Coelonia fulvinotata* (Butler, 1875) | 98.5 ± 16.8[j] | *2.1[k] | *Aerangis*[y], *Angraecum*[w], *Cladostemon*[z], *Oxyanthus*[z], *Rangaeris*[y] |
| *Coelonia solani* (Boisduval, 1833) | 192 ± 0.7[u] | *3.7 ± 0.11[u] | *Angraecum*[w] |
| *Megacorma obliqua* (Walker, 1856) | 135.0 ± 7.1[j] | *2.6[k] | *Hymenocallis*[ab] |
| *Pergesa acteus* (Cramer, 1779) | 80.3 ± 3.2[j] | *2.6[k] | N/A |
| **Riodinidae** | | | |
| *Eurybia elvina* Stichel, 1910 | 35.8 ± 3.2[h] | 2.0 ± 0.1[h] | *Calathea*[ac], *Centropogon*[ac], *Cephaelis*[ac], *Ischnosiphon*[ac], *Psychotria*[ac], *Stachytarpheta*[ac] |
| *Eurybia lycisca* Westwood, 1851 | 35.4 ± 3.8[h] | 2.1 ± 0.2[h] | *Calathea*[ac], *Cephaelis*[ac], *Ischnosiphon*[ac], *Lantana*[ac], *Psychotria*[ac], *Stachytarpheta*[ac] |
| *Eurybia unxia* Godman & Salvin, 1885 | 28.2 ± 3.4[h] | 1.7 ± 0.2[h] | *Calathea*[ac], *Cephaelis*[ac], *Ischnosiphon*[ac] |
| **Hesperiidae** | | | |
| *Lycas godart boisduvalii* (Ehrmann, 1909) | 45.7[ae] | 1.8[ae] | *Calathea*[ad] |
| *Perichares adela* (Hewitson, 1867) | 44.5 ± 4.9[ae] | 1.9 ± 0.1[ae] | *Calathea*[ad] |
| *Perichares lotus* (A. Butler, 1870) | 48.3[ae] | 2.1[ae] | *Calathea*[ad] |
| *Damas clavus* (Herrich-Schäffer, 1869) | 49.5 ± 2.1[ae] | 2.1 ± 0.1[ae] | *Calathea*[ad], *Lantana*[ad] |
| *Damas immaculata* Nicolay, 1973 | 52.0 ± 1.0[ae] | 2.4 ± 0.2[ae] | *Stachytarpheta*[ad] |
| *Saliana esperi esperi* Evans, 1955 | 36.5[ae] | 2.0[ae] | *Calathea*[ad] |
| *Saliana salius* (Cramer, 1775) | 47.2 ± 5.7[ae] | 2.0 ± 0.2[ae] | *Calathea*[ad] |
| *Saliana severus* (Mabille, 1895) | 51.8[ae] | 1.8[ae] | *Calathea*[ad] |
| *Saliana triangularis* (Kaye, 1914) | 41.1 ± 2.1[ae] | 2.0 ± 0.1[ae] | *Calathea*[ad] |
| *Talides hispa* Evans, 1955 | 45.0 ± 0.7[ae] | 1.8 ± 0.1[ae] | *Calathea*[ad] |
| *Talides sergestus* (Cramer, 1775) | 36.6[ae] | 1.7[ae] | N/A |

(continued)

**Table 15.2** (continued)

| Taxon | Absolute proboscis length | Relative proboscis length | Nectar host plants |
|---|---|---|---|
| *Tromba xanthura* (Godman, 1901) | 48.2[ae] | 2.3[ae] | *Stachytarpheta*[ad] |

[a]Amsel (1938), [b]Miller (1997), [c]Locatelli and Machado (1999), [d]Luer (1972), [e]Stewart and Kane (2007), [f]Haber and Frankie (1989), [g]calculated from values given in Haber and Frankie (1989), [h]Bauder unpublished, [i]Nattero et al. (2003), [j]Kawahara (2007), [k]calculated from values given in Kawahara (2007), [l]Grant and Grant (1983), [m]Alarcón et al. (2008), [n]Fox et al. (2015), [o]Johnson et al. (2017), [p]Moré et al. (2006), [q]Brandenburg and Bshary (2011), [r]Sazatornil et al. (2016), [s]Bullock and Pescador (1983), [t]calculated from values given in Bullock and Pescador (1983), [u]Wasserthal (1997), [v]Nilsson et al. (1987), [w]Nilsson et al. (1985), [x]Kritsky (2001), [y]Martins and Johnson (2013), [z]Johnson and Raguso (2016), [aa]Alexandersson and Johnson (2002), [ab]Dvořák (2017), [ac]DeVries (1997), [ad]Bauder et al. (2015b), [ae]Bauder et al. (2014)

flowers of the larval host plants and would thus be candidates for a mutualistic coevolutionary relationship if they pollinated these flowers. However, the proboscis of long-proboscid skippers and metalmarks is longer than the corolla tube of their preferred nectar plants (Bauder et al. 2011, 2014). Since they are not forced to plunge their head into the corolla entrance (Fig. 15.3a), a successful pollination is very unlikely. Neither skippers nor metalmark butterflies were able to release the trigger mechanism of *Calathea* flowers effectively, which is a precondition for successful pollen transfer (Schemske and Horvitz 1984; Bauder et al. 2011, 2014) and thus must be regarded as nectar thieves, not pollinators. Furthermore, these plants suffer from flower predation exerted by the larvae of long-proboscid *Eurybia* butterflies, which feed on *Calathea* flowers and not on foliage (Schemske and Horvitz 1984; DeVries 1997). Convergent evolution of superlong proboscises in Neotropical deep-forest butterflies would provide them exclusive access to deep-tubed flowers, which occur in their microhabitat and cannot be exploited by the vast majority of other short-proboscid flower visitors (Bauder et al. 2015a).

#### 15.3.2.2 Sphingidae

Recent phylogenetic analyses based on morphological and molecular data confirmed the existence of three monophyletic clades within Sphingidae: the Smerinthinae, the Sphinginae and the Macroglossinae (Kawahara et al. 2009). Generally, short proboscises are found in most Smerinthinae, which do not feed as adults at all. Species with superlong proboscises measuring more than 200 mm are found within the Sphinginae in two subtribes (Sphingini, Acherontiini) as well as within the Macroglossinae in the subtribe Choreocampina (Table 15.2, Fig. 15.2). The first lineage to branch off within Sphingini contains *Xanthopan* and a Neotropical group of species including *Amphimoea*, *Cocytius* and *Neococytius* (Kitching 2002; Kawahara et al. 2009; Johnson et al. 2017). These Old and New World long-

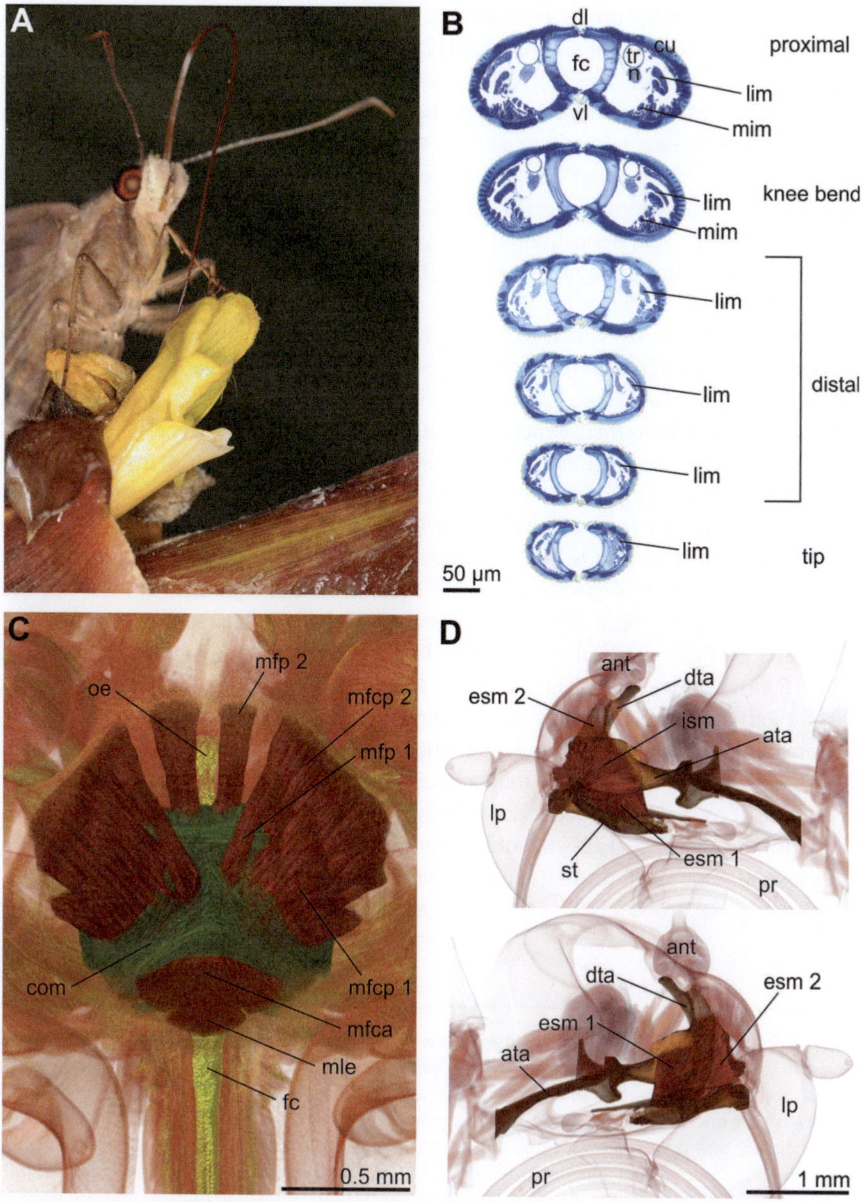


**Fig. 15.3** Superlong-proboscid Hesperiidae of the genera *Carystoides* and *Damas*. (**a**) The extremely long-proboscid skipper *Carystoides escalantei* (Hesperiidae) drinks nectar from a deep-tubed *Calathea lutea* flower (Marantaceae). (**b**) Semithin sections of the extremely long-proboscid skipper *Damas clavus* (Hesperiidae) from the proboscis base to the distal tip. Lateral and median muscles occur in the proximal region and the knee bend, from the distal region to the proboscis tip only lateral muscle sets are present (Krenn and Bauder 2018). (**c**) 3D reconstruction based on micro-CT images of the suction pump of the extremely long-proboscid *D. clavus*

proboscid hawk moths probably share this trait through common descent rather than independent evolution (Johnson et al. 2017). Since hawk moths likely originate from the Old World, chances are that some hawk moths in this group that colonized the New World already possessed long proboscises (Kawahara et al. 2009; Johnson et al. 2017). However, taking into account that other extremely long-proboscid hawk moths are found in another branch of the Sphingini (*Manduca*) as well as in the Acherontiini (*Megacorma*, *Agrius* and *Coelonia*) and even in another subfamily (Macroglossinae, Choreocampina: *Pergesa*), superlong proboscises evolved convergently within Sphingidae, as it has been demonstrated for butterflies (Bauder et al. 2011, 2013, 2014).

The mechanisms by which extreme proboscis lengths in hawk moths came to existence have been discussed fervently ever since Darwin's theory of coevolution (Darwin 1862): "As certain moths of Madagascar became larger through natural selection in relation to their general conditions of life, either in the larval or mature state, or as the proboscis alone was lengthened to obtain honey from the *Angraecum* and other deep tubular flowers, those individual plants of the *Angraecum* which had the longest nectaries (and the nectary varies much in length in some orchids) and which, consequently, compelled the moths to insert their probosces up to the very base, would be fertilised". Darwin thus did not picture a one-on-one interdependency between one moth species and one orchid species, which was supported by field observations of orchid-sphingid interactions (Wasserthal 1997) (Fig. 15.4a). Despite some suggestions to the contrary, e.g. that long proboscises in hawk moths evolved primarily as an adaptation to evade attacks from predatory spiders lying in ambush for them in non-specialized, short-tubed flowers (Wasserthal 1997), today it is almost certain that coevolution with plants was responsible for the evolution of very long proboscises in hawk moths (Darwin 1862; Nilsson 1988, 1998; Johnson et al. 2017; Netz and Renner 2017). Recent studies support the idea that several, not one, deep tubular flowers might be involved in selection for extremely long proboscises: The Madagascan subspecies *X. morganii praedicta*

---

**Fig. 15.3** (continued) (Hesperiidae); frontal view. Various dilator muscles (red) expand the cibarium to suck nectar through the food canal (yellow) into the cibarium, and compressor muscles (green) compress the cibarium to swallow nectar into the oesophagus (yellow). (**d**) 3D reconstruction based on micro-CT images of the stipes pump of the extremely long-proboscid *D. clavus* (Hesperiidae); top: right side of head as seen from within, bottom: right side of head as seen from outside. Two external and one internal stipes muscle (red) originate at different sites of the tentorium (brown) and insert at the stipes (brown). Alternating contraction of these muscles uncoils the proboscis. Reproduced with permission of John Wiley and Sons from Krenn and Bauder (2018). Abbreviations: *ant* antenna, *ata* anterior tentorial arm, *cu* cuticula, *dl* dorsal linkage, *dta* dorsal tentorial arm, *esm 1* extrinsic stipes muscle 1, *esm 2* extrinsic stipes muscle 2, *fc* food canal, *ism* intrinsic stipes muscle, *lim* lateral intrinsic muscles, *lp* labial palpus, *mim* median intrinsic muscles, *mfca* musculus frontoclypeo-cibarialis anterior, *mfcp 1* musculus frontoclypeo-cibarialis posterior 1, *mfcp 2* musculus frontoclypeo-cibarialis posterior 2, *mfp 1* musculus frontoclypeo-pharyngealis 1, *mfp 2* musculus frontoclypeo-pharyngealis 2, *mle* musculus labro-epipharyngealis, *n* nerve, *oe* oesophagus, *pr* proboscis, *st* stipes, *tr* trachea, *vl* ventral linkage

and the African subspecies *morganii* diverged 7.4 million years ago, which overlaps the divergence of *A. sesquipedale* from its sister, *A. sororium*, at 7.5 million years ago, and since both have extremely long spurs, those probably existed before that split. Moreover, phylogenies showed that several long-proboscid moths and long-spurred orchids coevolved in Madagascar since the mid-Miocene (Netz and Renner 2017). Therefore, trait matching in Madagascan hawk moths with extremely long proboscises and *Angraecum* orchids with extremely long nectar spurs is best explained by guild coevolution and the pollinator shift model (Wasserthal 1997; Whittall and Hodges 2007), which posits repeated partner switching and not one-on-one coevolution (Netz and Renner 2017). Existing studies suggest that adult hawk moths are highly polyphagous nectar feeders and readily drink from flowers which have tubes much shorter than their proboscis lengths, while plants adapted to hawk moth visitors are more specialized, particularly when long-tubed (Nilsson et al. 1987; Haber and Frankie 1989; Martins and Johnson 2013; Amorim et al. 2014). This trend for longer-proboscid hawk moths to be generalist nectar foragers seems to contradict coevolutionary theories since this process would involve a certain degree of reciprocal specialization between plants and their pollinators (Johnson et al. 2017). However, larger nectar rewards of longer-tubed flowers which can only be exploited by long-proboscid hawk moths would promote temporal constancy for flower visits to deep-tubed flowers (Johnson et al. 2017). Long-proboscid hawk moths concentrate their foraging efforts on longer-tubed plant species although they are polyphagous (Sazatornil et al. 2016). Longer proboscis lengths are correlated with larger body size and lower abundance, so that long-proboscid species may be at a disadvantage in scramble competition with the more species-rich and individually abundant short-proboscid species visiting shorter-tubed flowers (Rodríguez-Gironés and Llandres 2008). Miller (1997) noted that hawk moths with superlong proboscises mostly utilize larval host plants with inconspicuous or ephemeral growth forms, whereas those with shorter proboscises or non-functional mouthparts use more apparent, long-lived plants (shrubs or trees) as hosts. Further, long-proboscid hawk moths are larger in size and do not mature all of their eggs at eclosion (Miller 1997), indicating that adult nectar meals are needed both to fuel long-distance dispersal flights for oviposition and to increase fitness by maturing more eggs (O'Brien 1999). Because of that, long-proboscid hawk moths are generalist foragers that should visit flowers opportunistically including those that are not primarily adapted to them as pollinators, as they disperse over long distances (Miller 1997; Amorim et al. 2014; Johnson et al. 2017).

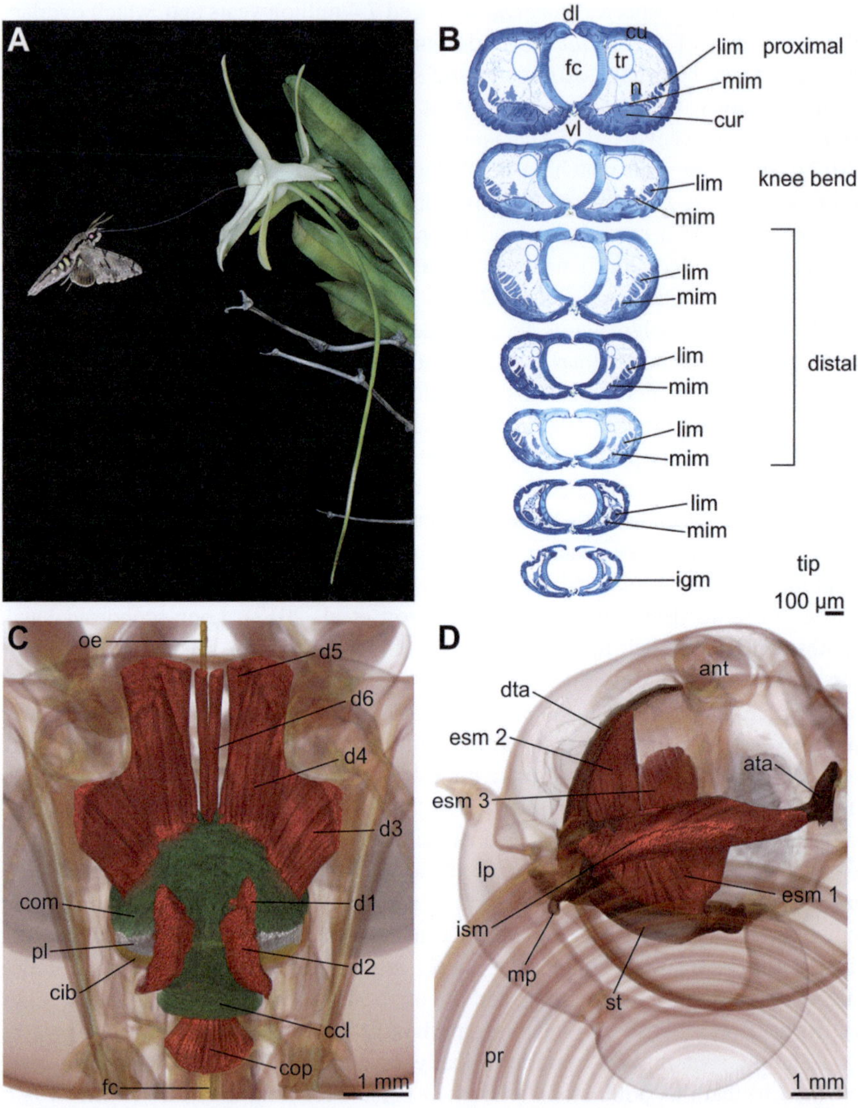


**Fig. 15.4** Superlong-proboscid Sphingidae of the genera *Xanthopan* and *Neococytius*. (**a**) The extremely long-proboscid hawk moth *Xanthopan morganii praedicta* (Sphingidae) drinks nectar from a deep-tubed *Angraecum sesquipedale* flower (Orchidaceae). Reproduced with permission of John Wiley and Sons from Wasserthal (1997). (**b**) Semithin sections of the extremely long-proboscid hawk moth *Neococytius cluentius* (Sphingidae) from the proboscis base to the distal tip. Lateral and median muscles occur throughout the whole proboscis length, only in the apical tip lateral or median muscle sets cannot be safely differentiated. (**c**) 3D reconstruction based on micro-CT images of the suction pump of the extremely long-proboscid *N. cluentius* (Sphingidae); frontal view. Various dilator muscles (red) expand the cibarium to suck nectar through the food canal (yellow) into the cibarium. All dilators pass through a flexible plunger (white) to insert on the cibarial roof. Compressor muscles (green) compress the cibarium to swallow nectar into the

## 15.3.3 *Fluid Feeding in Lepidopterans: Proboscis, Its Movement and Nectar Uptake*

### 15.3.3.1 Proboscis

The proboscis of all Lepidoptera is coiled up and stored under the head during rest, while it is uncoiled in a split of a second for feeding (Krenn 1990, 2008). Before landing on the flower, the proboscis is unwound from its tightly coiled position, flexed upwards and uncoiled. In most cases, this results in a characteristic bending of the proboscis at one third of its length called knee bend (Fig. 15.3a). The knee bend separates the proximal region at the proboscis base of the head, which is positioned horizontally, from the distal region, which constitutes the remaining two thirds of the proboscis which aims downwards. The distal region terminates at the apical proboscis drinking region, which is equipped with many sensilla and slits for nectar intake (Krenn 2008, Lehnert et al. 2016).

In general, the proboscis morphology in all long-proboscid Lepidoptera is similar and does not show differences compared to short-proboscid Lepidoptera (Bauder et al. 2013). Each galea contains one nerve, a vertical septum and one trachea both extending longitudinally from the basis to the tip. There are two muscle sets within the galeal lumen (Table 15.3): Lateral intrinsic muscles comprise a series of intrinsic galeal muscles which are stacked one on the other and originate from the dorso-lateral side and extend to the ventral bottom of the galea in an oblique angle. The median intrinsic muscles form a series of small galeal muscles which extend along the ventral wall in a nearly longitudinal course. Both muscle sets contract during proboscis coiling for tightening the proboscis coil (Wannenmacher and Wasserthal 2003; Krenn 2010).

The musculature arrangement throughout the proboscis varies between superlong-proboscid species: The long-proboscid butterfly *Damas clavus* (Hesperiidae) is characterized by the presence of lateral and median sets of intrinsic galeal musculature throughout the proximal region and the knee bend. In the distal region, the median muscles disappear, and only lateral intrinsic muscles occur until the drinking region (Krenn and Bauder 2018) (Fig. 15.3b). The long-proboscid butterflies of the genus *Eurybia* (Riodinidae) are characterized by the occurrence

**Fig. 15.4** (continued) oesophagus (yellow). (**d**) 3D reconstruction based on micro-CT images of the stipes pump of the extremely long-proboscid *N. cluentius* (Sphingidae); right side of head as seen from within. Three external and one internal stipes muscle (red) originate at different sites of the tentorium (brown) and insert at the stipes (brown). Alternating contraction of these muscles uncoil the proboscis. Abbreviations: *ant* antenna, *ata* anterior tentorial arm, *ccl* cibarial closer, *cib* cibarium, *com* compressor muscles, *cop* cibarial closer, *cur* cuticle ridge, *cu* cuticula, *d1–d6* dilator muscles 1–6, *dl* dorsal linkage, *dta* dorsal tentorial arm, *esm 1–3* extrinsic stipes muscle 1–3, *fc* food canal, *igm* intrinsic galeal muscles, *ism* intrinsic stipes muscle, *lim* lateral intrinsic muscles, *lp* labial palpus, *mim* median intrinsic muscle, *mp* maxillary palpus, *n* nerve, *oe* oesophagus, *pl* plunger, *pr* proboscis, *st* stipes, *tr* trachea, *vl* ventral linkage

**Table 15.3** Musculature of the feeding apparatus (mouthparts and suction pump) in Lepidoptera: muscles names in various studies, their proposed functions and progressions; nomenclature in bold letters is used in this review

| Muscle name and synonyms | Function | Progression |
|---|---|---|
| **Proboscis** | | |
| **Proximal basal muscle** | Elevation | |
| Dorsal elevator muscle[a] | | Galea base—dorsal apodeme of proximal galea inflexion |
| Galea extensor muscle[b] | | Proximal galea base—anterior galea surface |
| Proximal basal muscle[c] | | Proximal end of galea—dorsal galea fold |
| Basal galeal muscles[d] | | N/A |
| **Distal basal muscle** | Elevation | |
| Ventral elevator muscle[a] | | Galea base—dorsal apodeme of distal galea inflexion |
| Distal basal muscle[c] | | Ventral galea fold—dorsal median galea |
| **Lateral intrinsic muscles** | Coiling | |
| Primary oblique muscles[a, c] | | Proximal lateral galea cuticle—distal medio-ventral galea cuticle |
| Galea retractor muscles[b] | | Lateral galea—posterior surface of galea |
| Lateral galeal muscles[d] | | N/A |
| Lateral intrinsic muscles[e, f, g] | | Dorso-lateral galea wall—ventro-lateral galea wall |
| **Median intrinsic muscles** | Coiling | |
| Secondary oblique muscles[a, c] | | Ventral median galea cuticle—ventral lateral galea close to septum |
| Ventral galeal muscles[d] | | N/A |
| Median intrinsic muscles[e, f, g] | | Proximal ventral galeal wall—distal ventral galeal wall |
| **Stipes pump** | | |
| **Internal stipes muscle** | Galea retraction | |
| Posterior tentorial proboscis extensor[h] | | Mesal surface of anterior tentorial arm—distal stipes (paired) |
| Posterior tentorial stipital adductor muscle[a] | | Inner face of posterior tentorial arm—anterior half of stipital apodeme (paired) |
| Stipes retractor[b] | | Medial edge of anterior tentorial arm—via tendon on distal stipes (paired) |
| Posterior stipital muscle[d] | | Posterior medial surface of anterior tentorial arm—distal stipes near galea base (paired) |
| Internal stipes muscle[f] | | Posterior tentorium—distal stipes near galea base (paired) |
| Musculus tentorio-stipitalis internalis[g] | | Posterior tentorium—median part of stipes (paired) |
| **External stipes muscle 1** | Compression | |
| Anterior tentorial proboscis extensor[h] | | Lateral surface of anterior tentorial arm—median distal flat stipes (paired) |
| Anterior stipital adductor muscle[a] | | Inner and outer face of anterior tentorial arm—posterior stipital apodeme (paired) |

(continued)

**Table 15.3** (continued)

| Muscle name and synonyms | Function | Progression |
|---|---|---|
| Stipes extensor 1[b] | | Ventral anterior tentorial arm—medial stipes (paired) |
| Tentorial stipital muscle[d] | | Lateral surface of anterior tentorial arm—flat part of stipes (paired) |
| External stipes muscle 2[f] | | Anterior tentorium—flat part of stipes (paired) |
| Musculus tentorio-stipitalis externalis *2*[g] | | Ventral tentorium—median flat stipes sclerite (paired) |
| **External stipes muscle 2** | Compression | |
| Cranial proboscis extensor[h] | | Anterior part of gena—flat mesal sclerite of stipes (paired) |
| Cranial adductor muscle[a] | | Gena and clypeus—stipes (paired) |
| Genal stipital muscle[d] | | Gena—flat part of stipes (paired) |
| External stipes muscle 1[f] | | Gena—flat part of stipes (paired) |
| Musculus clypeo-stipitalis[g] | | Clypeus lateral from anterior tentorial pit—via tendon on stipes (paired) |
| **External stipes muscle 3** | Compression | Dorsal tentorial arm—distal stipes (paired) |
| Stipes extensor 2[b] | | Frontoclypeus lateral to suction pump dilators—stipes lateral to *stie 1* (paired) |
| **Suction pump** | | |
| **Compressor musculature** [h, a, i, j, f, g] | Compression | Several layers on dorsal side of cibarium |
| **Musculus epipharyngeo-transversalis**[i, f, g] | Compression | Epipharyngeal fold |
| Transverse sphincter of cibarium[a] | | Cibarial wall |
| Cibarial closer[j] | | Roof of cibarium, inserted on each side of floor of valve |
| **Musculus labro-epipharyngealis**[i, f, g] | Dilation | Labrum—epipharynx (unpaired) |
| Dilator of cibarium[h] | | Clypeus—roof of pump (unpaired) |
| Labral compressor[a] | | Labrum—cibarium (unpaired) |
| Cibarial dilator 1[b] | | Ventral frontoclypeus—anteroventral wall of pump (unpaired) |
| Cibarial opener[j] | | Clypeus—roof of pump (unpaired) |
| **Musculus frontoclypeo-cibarialis anterior**[i, f, g] | Dilation | Frontoclypeus—anterior roof of cibarium (unpaired) |
| Dilator of buccal cavity[h] | | Clypeus—roof of pump (unpaired) |
| Anterior cibarial dilator[a] | | Clypeus—dorsal roof of pump (unpaired) |
| Cibarial dilator 2[b] | | Anterolateral frontoclypeus—dorsoanterior pump (paired) |
| Dilator 1[j] | | Lateral frontoclypeus—roof of cibarial valve and anterior roof of buccal chamber (paired) |
| Dilator 2[j] | | Median frontoclypeus—roof of cibarial valve and anterior roof of buccal chamber (paired) |

(continued)

**Table 15.3** (continued)

| Muscle name and synonyms | Function | Progression |
|---|---|---|
| **Musculus frontoclypeo-cibarialis posterior**[i, f] | Dilation | Lateral posterior frontoclypeus—roof of cibarium in front of frontal ganglion (paired) |
| Dilator of buccal cavity[h] | | Clypeus—roof of pump (paired) |
| Lateral cibarial dilator[a] | | Lateral clypeus—roof of pump (paired) |
| Cibarial dilator 3[b] | | Lateral frontoclypeus below antennae, medial to compound eyes—lateral dorsal pump wall (paired) |
| Cibarial dilator 4[b] | | Lateral frontoclypeus near antennal base—dorsal pump wall (paired) |
| Cibarial dilator 5[b] | | Boundary of frontoclypeus and vertex—dorsal pump wall anterior to frontal connectives (paired) |
| Dilator 3[j] | | Anterior medial antennal socket, frontal head capsule—roof of buccal chamber (paired) |
| Dilator 4[j] | | Top of head capsule—roof of buccal chamber (paired) |
| Musculus frontoclypeo-cibarialis posterior[g] | | Lateral posterior frontoclypeus—lateral median dorsal roof of suction pump (paired) |
| Musculus frontoclypeo-cibarialis posterior[g] | | Lateral posterior frontoclypeus—lateral posterior dorsal roof of suction pump (paired) |
| **Musculus frontoclypeo-pharyngealis**[i, f] | Dilation | Lateral frontoclypeus—posterior roof of cibarium behind frontal ganglion (paired) |
| Dilator of pharynx[h] | | Frons—roof of pump (paired) |
| Posterior cibarial dilator[a] | | Middle of clypeus—postero-dorsal roof of pump (unpaired) |
| Pharyngeal dilator 1[b] | | Medial anterior edge of vertex—dorsal pump wall posterior to frontal connectives (paired) |
| Pharyngeal dilator 2[b] | | Medial to *dlph 1* on anterior edge of vertex—dorsal pump wall medial to *dlph 1* (paired) |
| Dilator 5[j] | | Median top of head capsule—posterior roof of buccal chamber (paired) |
| Dilator 6[j] | | Median top of head capsule—posterior roof of buccal chamber (paired) |
| Musculus frontoclypeo-pharyngealis 1[g] | | Median frontoclypeus—median dorsal roof of suction pump (paired) |
| Musculus frontoclypeo-pharyngealis 2[g] | | Posterior median frontoclypeus—posterior median dorsal roof of suction pump (paired) |

[a]Eastham and Eassa (1955), [b]Eaton (1988), [c]Krenn (2000), [d]Wannenmacher and Wasserthal (2003), [e]Bauder et al. (2011), [f]Bauder et al. (2013), [g]Krenn and Bauder (2018), [h]Schmitt (1938), [i]Eberhard and Krenn (2005), [j]Davis and Hildebrand (2006)

of lateral intrinsic muscles throughout the whole galea and the presence of median intrinsic muscles only in the knee bend (Bauder et al. 2013). Lateral and median intrinsic muscles are present in the proximal region, the knee bend and throughout the whole distal region to the tip in the long-proboscid hawk moth *Neococytius cluentius* (Sphingidae) (Fig. 15.4b).

Studies on the number or volume of intragaleal muscles in relation to proboscis length show the same pattern in both Hesperiidae and Riodinidae: The number of single lateral intrinsic muscles is estimated to be 25 per mm galea length in the long-proboscid *D. clavus* and 22 per mm galea length in the short-proboscid *P. ocola* (Hesperiidae) (Krenn and Bauder 2018). Within Riodinidae, a comparison of the percentage of the intragaleal muscle volume per galeal volume showed that muscle volume is similar in long-proboscid and short-proboscid metalmark species (Bauder et al. 2013). These studies showed consistently that although absolutely longer proboscises require absolutely more muscles, the number and volume of intragaleal muscles per mm of proboscis length and of galeal volume, respectively, is constant regardless of proboscis length. Therefore, there is no indication for additional investments for intragaleal muscles in a superlong proboscis to ensure the coiling process.

By contrast, anatomical extra-investment in the cuticular proboscis wall occurs in butterflies and sphingids with superlong proboscises: Long-proboscid riodinids have a two times thicker cuticle wall than short-proboscid riodinids (Bauder et al. 2013). Furthermore, a very conspicuous feature of the long-proboscid hawk moth *Neococytius cluentius* is a large cuticle ridge running along the ventral side of each galea from the proboscis base to the distal region (Fig. 15.4b). This extremely reinforced cuticle ridge probably serves to enforce the superlong proboscis. These morphological adaptations of the proboscis wall found in butterflies and hawk moths may be necessary to stabilize the proboscis and retain its manoeuvrability to take up nectar from deep and narrow flower tubes (Bauder et al. 2011, 2013).

Another morphological adaptation of superlong proboscises to increase the nectar intake rate in butterflies is a large food canal: Long-proboscid riodinids have a significantly larger food canal volume in relation to the galeal volume than short-proboscid species (Bauder et al. 2013). The same holds true for long-proboscid hesperiids—proboscis length correlates positively with food canal cross-sectional area (Bauder et al. 2015a). Since physical laws predict that nectar intake rate declines linearly with increasing proboscis length (Kingsolver and Daniel 1979, 1995), the enlargement of the food canal is a very effective way to compensate for the negative influence of a superlong proboscis.

#### 15.3.3.2 Sensory Equipment

The external surface of the lepidopteran proboscis is covered with various types of sensilla with different sensory modalities: Bristle-shaped sensilla trichodea or chaetica (Faucheux 2013) function as mechanoreceptors, cover the whole proboscis and are the most numerous in butterflies. In skipper butterflies, proboscis length does not correlate significantly with the number of bristle-shaped sensilla (Krenn and Bauder 2018). A similar pattern occurs in metalmark butterflies, where long-proboscid species even have a significantly lower number of bristle-shaped sensilla per mm proboscis length than short-proboscid metalmarks (Bauder et al. 2013).

Sensilla styloconica function as contact chemo-mechanoreceptors to find the corolla entrance of the flower and taste nectar (Krenn 1998). They are distributed only near the drinking region. The number of sensilla styloconica is very low and does not correlate with proboscis length in skippers. Their proboscis tip is uniformly equipped with only two sensilla styloconica that are interspaced by two sensilla basiconica on the dorsal side and two sensilla styloconica which are arranged in a pair located in a cuticle depression on the lateral side and a bristle-shaped sensillum followed by a third sensillum styloconicum. In contrast to many other butterflies, hesperiid sensilla styloconica have a uniform and simple appearance regardless of proboscis length. They are composed of a cylindrical stylus and a pointed sensory cone on top (Krenn and Bauder 2018).

As in long-proboscid skippers, the long-proboscid *Eurybia* species have a significantly lower number of sensilla styloconica per mm tip length than short-proboscid metalmark species. Four sensilla styloconica form a characteristic cluster at the tip. Again, the morphology of these sensilla is very simple: The stylus is very short; it resembles a collared socket and bears a long sensory peg (Bauder et al. 2013).

Sensilla basiconica are chemoreceptors and occur inside the food canal as well as on the external lateral surface of the galea. In skippers, the number of external sensilla basiconica does not correlate significantly with proboscis length. By contrast, the number of sensilla basiconica covering the food canal increased significantly with increasing proboscis length. They are very small and distributed in a single row from the base of the proboscis to the drinking region (Krenn and Bauder 2018). In Sphingidae, short and smooth sensilla styloconica are located in pits and never extend beyond the surface of the proboscis, which is interpreted to be an adaptation to penetrate narrow flower tubes without problems. Furthermore, these sensilla are spread out over the whole length of the proboscis and not restricted to the drinking region in *Macroglossum stellatarum*, which is interpreted to be an adaptation for gathering nectar in flight. They are often arranged in three longitudinal rows on each galea (Faucheux 2013).

On the whole, long proboscises of skippers and metalmarks are very similar with regard to their sensory equipment: The remarkably slender proboscises possess only a few short sensilla styloconica. Neither the number of sensilla trichodea nor sensilla basiconica on the lateral sides of the galea correlates with proboscis length. Only the number of sensilla basiconica of the median food canal is positively correlated with proboscis length. Sensilla styloconica on the proboscis tip are uniform in their shape, length, and number in skippers regardless of proboscis length differences, whereas long-proboscid riodinids have even fewer sensilla styloconica than short-proboscid metalmarks. Given the great diversity in the shape of the proboscis tip sensilla of Nymphalidae, Lycaenidae and some moths (Krenn 1998; Krenn and Penz 1998; Krenn et al. 2001; Petr and Stewart 2004; Molleman et al. 2005; Faucheux 2013;

Lieskonig 2013), the uniform sensory equipment of the proboscis tip in long-proboscid Hesperiidae and Riodinidae is astonishing. In all long-proboscid hesperiid and riodinid species, the sensilla styloconica are short, cylindriformous and without special microsculpture. By contrast, butterflies that use other energy sources than nectar, such as juice from rotten fruit, are known to have brushy proboscises formed of dense rows of numerous, elongated sensilla styloconica (Lehnert et al. 2016). Since sensilla styloconica are chemo-mechanosensilla and have been functionally interpreted to be responsible for detecting the opening of flowers and the presence of nectar inside the corolla tube (Krenn 1998), the remarkably low number of tip sensilla—even when compared to other flower-visiting butterflies (Paulus and Krenn 1996; Krenn et al. 2001)—on a long and difficult to manoeuvre proboscis can explain longer flower manipulation times of long-proboscid butterflies (Bauder et al. 2015a).

#### 15.3.3.3 Proboscis Movement by the Stipes Pump

The stipes pump of Lepidoptera uncoils the proboscis before nectar feeding. The uncoiling process most likely relies on a hydraulic mechanism. The contraction of external stipes muscles moves the median sclerotized part of the stipes and compresses the lateral stipital tube, which forces haemolymph into the attached galea and results in proboscis uncoiling. Because the structures of the stipes form a valve, the proboscis can be uncoiled stepwise when haemolymph pressure inside the galeae increases. Proboscis coiling is achieved by the elasticity of the cuticle as well as by the contraction of intragaleal muscles after the stipes valve opens. Contraction of the internal stipes muscles flexes the proboscis back into its resting position (Wannenmacher and Wasserthal 2003; Krenn 2010).

The number of stipes muscles differs in long-proboscid butterflies and hawk moths: Three muscles, two paired external stipes muscles and one paired internal stipes muscle, were found to be attached to the stipes pump of long-proboscid hesperiids (Table 15.3, Fig. 15.3d). The stipital pump of the long-proboscid hawk moth *N. cluentius* comprises even four prominent muscles (Table 15.3, Fig. 15.4d).

A biometrical comparison of long-proboscid and short-proboscid hesperiids indicates conspicuous differences in size of the stipes sclerites and various volumes of the stipes musculature. The stipes of the long-proboscid *D. clavus* has approximately twice the length of the stipes of the short-proboscid *P. ocola*, and the area of origin of the stipes muscles in the long-proboscid *D. clavus* is much larger than in the species with short proboscis (Krenn and Bauder 2018).

Within Riodinidae, haemolymph pump morphology also correlates with proboscis length. In long-proboscid *Eurybia* species, the stipes musculature is significantly larger in relation to the head capsule than in short-proboscid metalmarks (Bauder et al. 2013). So far, only two external stipes muscles and one internal stipes muscle were found to be present in long-proboscid butterflies and hawk moths (Eaton 1988; Wannenmacher and Wasserthal 2003; Bauder et al. 2013; Krenn and Bauder 2018). However, the long-proboscid *N. cluentius* possesses even four stipes muscles

(Fig. 15.4d). Future studies focussing on the comparison of the stipes musculature and its biometry between hawk moths with differently sized proboscises would help to clarify if there were morphometrical specializations for uncoiling an extremely long proboscis in hawk moths.

#### 15.3.3.4 Suction Pump for Nectar Ingestion

The suction pump for nectar uptake is the largest muscular organ in the head and lies in the anterior half of the head. Its principle composition is similar in all Lepidoptera regardless of proboscis length. A tracheal air sac is located above the pump to make possible rapid volume displacements within the head as the pump is dilated and compressed. At the same time, the action of the pump on this air sac provides ventilation of the tracheal system (Davis and Hildebrand 2006). Its ventral side is sclerotized, and the dorsal roof is covered with layers of transversal compressor muscles (Table 15.3). Another smaller compressor muscle, the musculus epipharyngeo-transversalis, is located in the epipharyngeal fold at the transition of the suction pump lumen to the proximal food canal of the proboscis (Eberhard and Krenn 2005; Bauder et al. 2013; Krenn and Bauder 2018).

Dilator muscles attach on different sites of the suction pump (Table 15.3). Long-proboscid hesperiids have an additional pair of large dilator muscles compared to short-proboscid skippers. Six dilators, two unpaired and four paired muscles, were found in hesperiids with superlong proboscises (Fig. 15.3c) (Krenn and Bauder 2018). In metalmarks, two pairs of dilator muscles and two unpaired dilators were found in both long-proboscid and short-proboscid species (Bauder et al. 2013). The long-proboscid hawk moth *N. cluentius* is characterized by six pairs of dilator muscles and an unpaired cibarial opener (Fig. 15.4c).

Biometrical comparisons of the proportions of the dilator and compressor muscles in relation to the head capsule showed that both long-proboscid metalmark and skipper butterflies have larger dilators than those with shorter proboscises (Bauder et al. 2013; Krenn and Bauder 2018). Such biometrical comparisons are lacking for hawk moths yet. Since these muscles account for the creation of a pressure drop to transport fluid throughout the proboscis, enlarged dilators allow for an efficient nectar uptake (Eberhard and Krenn 2005; Bauder et al. 2015a): Long-proboscid skipper butterflies can take more nectar in a given time than short-proboscid skippers (Bauder et al. 2015a).

### *15.3.4 Superlong Nectar-Extracting Organs: A Reasonable Compromise?*

Benefits of a long proboscis are obvious since flower-visiting insects possessing a particularly long proboscis can monopolize deep-tubed flowers as an exclusive food source for nectar feeding (Haber and Frankie 1989; Miller 1997; Bauder et al. 2015b). However, functional costs coming from longer flower handling times (Kunte 2007; Bauder et al. 2015a) or anatomical costs for maintaining the performance of the feeding apparatus and the functionality of the proboscis also play a role in the interface of biophysical constraints and adaptability.

Superlong butterfly proboscises are characterized by a reduction of the sensilla equipment, which probably explains the longer flower manipulation times proven for long-proboscid Neotropical butterflies (Bauder et al. 2011, 2015a). This may be caused by the difficulty of inserting the long proboscis into a narrow floral tube due to a poor supply of mechano- or chemosensory information (Bauder et al. 2011, 2013). Detailed knowledge on the sensory equipment of long-proboscid hawk moths is lacking yet. Since the hovering flight of hawk moths is an energetically expensive activity, the flower handling efficiency should be subject to selective pressure. In addition, mechanoreception of tactile cues on the floral surface is known to play an important part of successful flower handling in hawk moths (Goyret and Raguso 2006). However, hawk moths possess very inconspicuous sensilla at the proboscis tip (Faucheux 2013) and probably suffer from longer flower manipulation times.

Taking a longer time to find nectar in the first place can lower the energy intake rate by decreasing the proportion of foraging time devoted to actually imbibing nectar (Heinrich 1983; May 1985) and may constitute functional costs of long proboscises. Alternatively, longer manipulation times of long-proboscid butterflies could also be due to differences in flower morphology and length. It has been shown that bumble bees require more time to learn complex flowers, such as long floral tubes with concealed nectar (Laverty 1994). Several studies on the foraging behaviour of butterflies showed that individual experience gained by successive attempts to forage on a flower can shorten flower manipulation time (Lewis 1986; Kandori and Ohsaki 1996; Goulson et al. 1997). This is also true for hawk moths, which can decrease their nectar discovery time during a single foraging bout (Goyret and Raguso 2006). Therefore, learning the flower morphology could serve as an adaptive strategy for increasing the efficiency of nectar collection (Kandori and Ohsaki 1996). Further, long-proboscid Lepidoptera could compensate for long manipulation times by visiting fewer but nectar-rich flowers instead of many flowers with tiny nectar volumes.

By contrast to the reduced sensory equipment, anatomical extra-investments of long butterfly proboscises include very thick proboscis cuticle walls for stabilizing the long and slender proboscis, as well as large food canals (Bauder et al. 2013, 2015a). Enlargement of the food canal is a very effective way to increase the intake of nectar, since according to biophysical laws, nectar intake rate increases with the food canal radius to the exponent four (Kingsolver and Daniel 1979, 1995; see Chap. 8).

Although longer proboscises contain absolutely more intragaleal muscles for coiling the proboscis, the volume of muscles per galeal length is similar regardless of proboscis length (Bauder et al. 2013; Krenn and Bauder 2018). It seems that a defined set of intragaleal muscles without extra-investments is essential for successful proboscis coiling.

Adaptations of the feeding apparatus concern both the stipital pumps for proboscis movements as well as the cibarial pump for nectar uptake. Long-proboscid butterflies have larger stipes muscles and a larger suction pump lumen as well as larger suction pump dilator muscles, indicating reinforced organs for proboscis movements and nectar uptake, respectively (Bauder et al. 2013; Krenn and Bauder 2018). Higher nectar intake rates of long-proboscid species might be due to a more powerful suction pump (aside from enlarged food canals), as long-proboscid species have larger pump volumes, larger musculature and additional muscles of the suction pump (Bauder et al. 2015a).

In conclusion, the evolution of a superlong proboscis comes along with "hidden" anatomical costs that have to be considered in addition to sole increase of length when proboscis adaptations to long nectar spurs are discussed: Additional anatomical costs, such as cuticle investments in the proboscis wall, reinforced stipes pumps and suction pumps probably balance the benefits of a particularly long proboscis. Further, functional costs in terms of prolonged flower manipulation times, the time necessary to enter and leave a flower, add up to anatomical costs of long proboscises. This might be a reason why most nectar-feeding insects possess medium-sized mouthparts which normally do not exceed the body length although a much longer proboscis would give beneficial access to more flowers and exclusive access to highly rewarding deep-tubed flowers. Conversely, these considerations call for comparative investigations of the feeding apparatus and the measurement of nectar intake rates of hawk moths.

### 15.3.5 What Precludes Further Elongation?

The evolution of extreme absolute proboscis lengths in Lepidoptera is closely linked to extreme relative proboscis lengths, since body size and absolute proboscis length scale allometrically (Bauder et al. 2014). In hawk moths, the extreme proboscis length of *Amphimoea walkeri*, 280 mm, corresponds to the fourfold of body length (Amsel 1938), whereas relative proboscis length in butterflies never exceeds 2.5 (Kunte 2007; Bauder et al. 2011, 2013). Proboscis length in hawk moths can exceed that of butterflies not only because hawk moths are larger but also because of a steeper scaling relationship between body size and proboscis length. Explanations for what keeps butterflies from evolving equally long mouthparts in relation to body size as hawk moths could be found in differences regarding the flower-visiting behaviour and/or metamorphosis.

A crucial difference between butterflies and hawk moths regards their flower-visiting behaviour: Hawk moths typically hover over or in front of flowers during nectar uptake (Farina et al. 1994), whereas nearly all butterflies need to sit on the flower to feed, except for Troidini (Papilionidae) (Krenn 2008). In butterflies, uncoiling a very long proboscis is limited by how far a butterfly can bend back its head and stretch its legs to allow for straightening of the proboscis spiral while perching on the flower. In addition, superlong-proboscid butterflies probably are constrained by the need to take up fluids, such as water, from the ground in the same way as they are during uncoiling their proboscis to probe a flower. None of these problems applies to hawk moths, which can modulate the space needed for uncoiling by hovering at an acceptable distance in front of or over the flower. Although absolute proboscis length determines access to nectar in flowers with deep tubes, relative proboscis length plays a crucial role during the uncoiling process and might constrain butterflies from evolving even longer mouthparts.

Further, developmental constraints could limit the evolution of proboscis length in butterflies since proboscis formation takes place in a developmental sheath on the ventral side of the pupa (Lowe et al. 2013), where the galeae are straight, separate from each other and arranged parallel. Since the developmental sheath contains the full length of the unfolded proboscis, this organ grows accordingly to accommodate the extreme length of the adult proboscis and may extend a full body length beyond the last abdominal segment (DeVries 1997). Further elongation of this fragile and thin pupal organ might constrain proboscis length evolution in butterflies. By contrast, the pupae of long-proboscid hawk moths during metamorphosis develop a heavily sclerotized, hook-shaped external outgrowth that contains a loop of the developing proboscis that allows for the formation of a proboscis of much greater length (Patočka 1993; Kitching 2002).

## 15.4 Superlong-Proboscid Flies: Nectar-Feeding Nemestrinidae and Blood-Feeding Tabanidae

### *15.4.1 The Long-Proboscid Fly Pollination System in South Africa*

Southern Africa, in particular the Greater Cape Floristic Region, represents one of the global hot spots for plant diversity and endemism, harbouring approximately 10% of the world's total plant species (Johnson and Steiner 2003). Similar to the exceptional plant diversity (Johnson and Steiner 2003; Van der Niet et al. 2006; Johnson 2010; Schnitzler et al. 2011), the pollinator fauna shows high levels of adaptive radiation, and the Cape Floristic Region is the centre of global diversity for several insect pollinator groups (Colville et al. 2014). Consequently, Southern Africa supports a wide range of highly specialized pollination systems, including the unique long-proboscid fly pollination system (Johnson and Steiner 2003; Goldblatt and Manning 2006).

Several species of South African long-proboscid Nemestrinidae and Tabanidae are considered as keystone species pollinating almost 200 plant species across various families (including Iridaceae, Geraniaceae, Orchidaceae and Proteaceae) characterized by similar floral traits (Goldblatt et al. 1995; Johnson and Steiner 1997; Goldblatt and Manning 1999, 2000; Potgieter and Edwards 2005; Anderson and Johnson 2008; Pauw et al. 2009; Johnson 2010). Typically, flowers visited by long-proboscid flies show an elongated and narrow, straight or slightly curved floral tube (Goldblatt and Manning 2000). However, a particularly high variation in spur length among populations of plant communities has been noticed, which matches the high variability of proboscis length of a single, geographically variable, pollinating fly species (Anderson and Johnson 2008, 2009). In an in-depth study, Pauw et al. (2009) highlighted the role of reciprocal selection leading to the eye-catching intraspecific variation of proboscis lengths on certain nemestrinid species. Following the predictions of the geographic mosaic theory of coevolution (Thompson 1994, 2005), selective pressure of corolla tube length appears to have led to adaptations of proboscis lengths in flies.

Four separate long-proboscid fly pollination systems have evolved independently in South Africa (Goldblatt and Manning 2006). Within Nemestrinidae, the genera *Prosoeca*, *Moegistorhynchus* and *Stenobasipteron* include species that have developed a disproportionally long proboscis (Table 15.4). Diptera from the family Nemestrinidae are the world record holders in terms of relative proboscis length with proboscises exceeding four times the body length (Borrell and Krenn 2006). Linked into a geographical network of coevolutionary interactions, they have become model organisms for reciprocal adaptation and plant speciation in several floristically rich regions of South Africa (Goldblatt et al. 1995; Johnson and Steiner 1997; Goldblatt and Manning 1999, 2000; Potgieter and Edwards 2005; Anderson and Johnson 2008; Pauw et al. 2009).

In addition, some species of the tabanid genus *Philoliche* also display remarkably elongated proboscises (Barraclough 2006; Morita 2008) (Table 15.4). Despite their blood-sucking behaviour, these long-proboscid Tabanidae are important flower visitors and pollinators (Goldblatt and Manning 2000) and have proboscises that measure three times the body length to extract nectar from deep-tubed flowers (Morita 2008). In contrast to strictly nectarivorous Nemestrinidae, female *Philoliche* are anautogenous and require blood for nourishing their developing eggs (Lehane 2005). In addition, nectar represents an important source of energy for both male and female flies and is necessary to sustain daily activities for both sexes (Downes 1958; Johnson and Johnson 1993). Compared to nectar, vertebrate blood is a heterogeneous suspension and its viscosity varies with the diameter of the food tube (Kingsolver and Daniel 1979; Kim et al. 2011a, b). Consequently, the feeding apparatus of insects with such a twofold diet imposes particular adaptations. The elongated proboscis of Tabanidae is a unique structure that is adapted to perform a dual task: blood sucking and nectar feeding within the same female individual.

## *15.4.2 Morphological Adaptations of the Feeding Apparatus*

### 15.4.2.1 General Mouthpart Morphology

The proboscis of long-proboscid species of *Prosoeca* (Fig. 15.5a) and *Philoliche* (Fig. 15.6a) is at least twice as long as the body (Table 15.4). In the resting position,

**Table 15.4** Horse flies and tangle-veined flies (Diptera: Tabanidae, Nemestrinidae) with superlong proboscises. Given are absolute proboscis length [mm] and relative proboscis length, i.e. proboscis length divided by body length, and known nectar host plants

| Taxon | Absolute proboscis length | Relative proboscis length | Nectar host plants |
|---|---|---|---|
| **Tabanidae** | | | |
| *Philoliche gulosa* | 24.2 ± 2.2[a] | N/A | *Aristea*[b], *Geissorhiza*[b], *Gladiolus*[b], *Ixia*[b], *Lapeirousia*[b], *Lobelia*[b], *Pelargonium*[b], *Tritonia*[b], *Watsonia*[b] |
| *Philoliche rostrata* | 21.0 ± 8.4[a] | N/A | *Disa*[b], *Geissorhiza*[b], *Gladiolus*[b], *Ixia*[b], *Pelargonium*[b], *Tritonia*[b], *Watsonia*[b] |
| **Nemestrinidae** | | | |
| *Moegistorhynchus longirostris* | 40–100[c] | N/A | *Babiana*[c], *Disa*[c], *Gladiolus*[c], *Ixia*[c], *Lapeirousia*[c], *Pelargonium*[c], *Tritonia*[c] |
| *Prosoeca ganglbaueri* | 20–45[d] | N/A | *Babiana*[c], *Brunsvigia*[c], *Hesperantha*[c], *Lapeirousia*[c], *Nerine*[c], *Pelargonium*[c], *Spari*[c], *Zaluziankya*[c] |
| *Prosoeca marinusi* | 36.3 ± 3.9[e] | 1.8[e] | *Babiana*[e], *Lapeirousia*[e] |
| *Prosoeca nitidula* | 18–28[b] | N/A | *Gladiolus*[b], *Nerine*[b], *Watsonia*[b] |
| *Prosoeca longipennis* | 38–40[b] | N/A | *Gladiolus*[b], *Pelargonium*[b] |
| *Prosoeca robusta* | 33[b] | N/A | *Disa*[b], *Gladiolus*[b] |
| *Stenobasipteron wiedmanni* | 18–30[f] | N/A | *Barleria*[f], *Bownleea*[f], *Gladiolus*[f], *Hesperantha*[f], *Hypoestes*[f], *Impatiens*[f], *Isoglossa*[f], *Orthosiphon*[f], *Plectranthus*[f], *Streptocarpus*[f] |

[a]Barraclough et al. (2018), [b]Goldblatt and Manning (2000), [c]Karolyi et al. (2012), [d]Karolyi et al. (2014), [e]Manning and Goldblatt (1997), [f]Potgieter and Edwards (2005)

the proboscis of *Prosoeca* lies ventrally along the mid-line of the body between the fly's legs and projects beyond the abdomen (Karolyi et al. 2012; Barraclough et al. 2018). During feeding from flowers, the proboscis is projected in a slightly forward downward position, up to 100° from the resting position. Overall, the proboscis of *Prosoeca* and *Philoliche* species consists of the proximal, strongly sclerotized labrum-epipharynx unit, paired maxilla structures, lance-shaped hypopharynx and labium, with mandibles missing in *Prosoeca*. The proboscis is divided into a membranous basal joint region, a proximal and an elongated distal region that comprises more than half of the proboscis length. The basal region articulates with the head, allowing the proboscis to flex backwards from feeding into a resting position (Table 15.5).

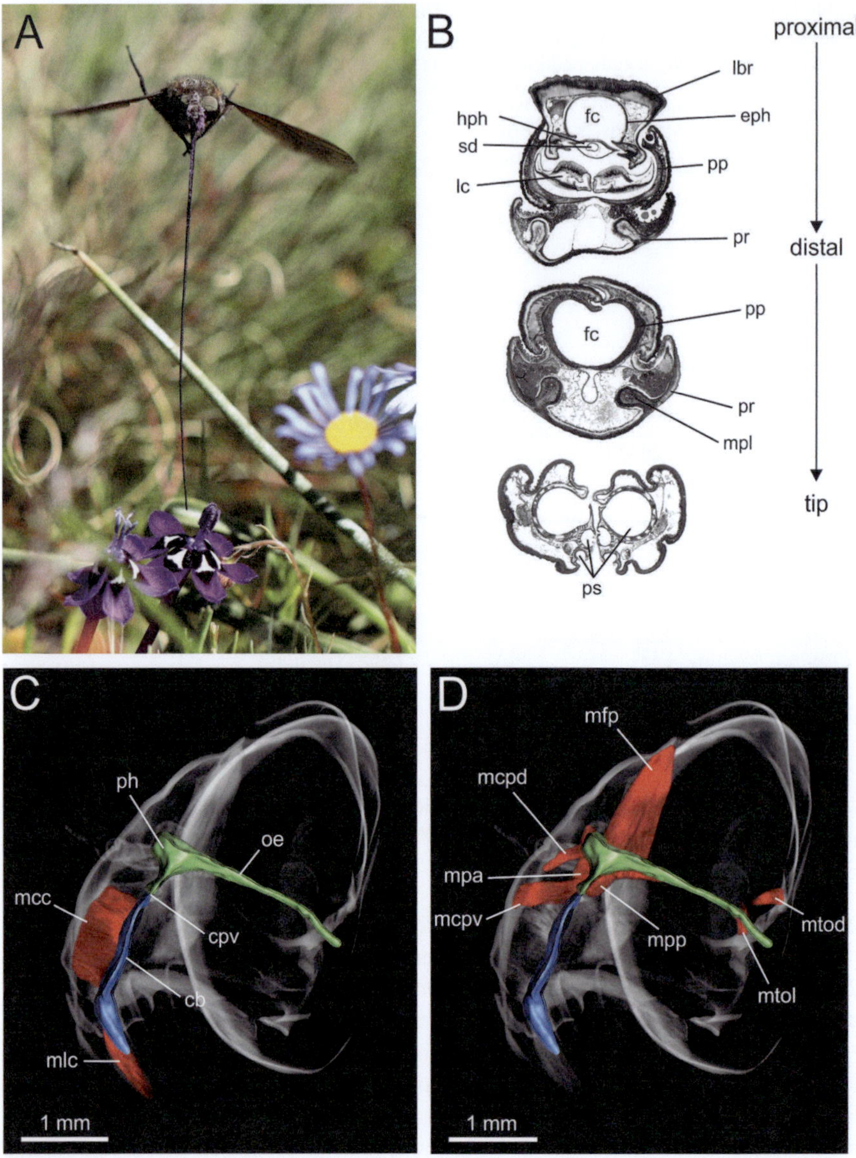


**Fig. 15.5** Long-proboscid Nemestrinidae of the genus *Prosoeca*. (**a**) *Prosoeca marinusi* approaching a flower of *Lapeirousia oreogena*. During foraging the proboscis is swung forward into a vertical position. (**b**) *Prosoeca ganglbaueri*, proboscis cross sections. Proximal part; labium sheath, labrum-epipharynx complex with epipharyngeal food canal, hypopharynx with salivary duct and laciniae. Distal part; the labial food canal with lateral prementum walls bend upwards and dorsally interlocking in a tongue and groove. Labella with paired, nearly closed headers of the pseudotracheae. Ventrally the labella fold apart to allow nectar uptake. Reproduced with permission of John Wiley and Sons from Karolyi et al. (2012). (**c**) Micro-CT head scan of *Prosoeca marinusi* displaying a lateral view of the cibarial pump (blue) with cibarial valve, cibarial dilator and cibarial-

#### 15.4.2.2 Proximal Proboscis

The labial base articulates with the head capsule via a rostral membrane on the ventral side, and dorsally it is connected with the hypopharynx. This articulation and the foldable cuticle of the rostral membrane ensure the movability of the proboscis from the resting to the feeding position. The proximal region is composed of the interlocked, labrum-epipharynx unit sheathed in the labial prementum that also contains the slender laciniae, based on a fused, heavy sclerotized cardo and stipes with a two-segmented maxillary palp. In cross section, the labrum contains a half-pipe lined with the sclerotized epipharynx which forms the food canal and is connected to the dorsal cibarium, joining the epipharyngeal food canal with the suction pump the head (Figs. 15.5b and 15.6b). The sclerotized hypopharynx is connected to the bottom of the cibarium and traversed by the salivary duct. It is engaged into a longitudinal fold of the epipharynx in a key and slot joint.

In *Philoliche*, the proximal proboscis represents the rather short piercing structures for blood sucking. However, the sclerotized mandible stylets are only present in females. The ribbed mandible blades have a tooth-shaped and serrated tip. The overlapping mandible blades lie beneath the labrum-epipharynx complex, closing the proximal food canal ventrally (Fig. 15.6b). The sclerotized lacinia stylets reside left and right of the hypopharynx below the mandible blades. The last third of each stylet is laterally equipped with a toothed edge ending in a strongly toothed tip. In cross section, the sclerotized, lance-shaped hypopharynx is dorsally flattened and positioned between the laciniae and below the mandible blades.

Since nemestrinids are obligatory nectar feeders, the maxillae of *Prosoeca* function considerably different from *Philoliche*. The stipes supports elongated, needle-shaped laciniae which extend into the proximal region of the proboscis. They attain one third of the proboscis length and rest between the paraphyses of the labium. In the process of moving the proboscis forward, contraction of the lacinia promotor pulls the stipes up, moving the laciniae forward, which therefore act as a lever for the elongated prementum (Table 15.5). According to their position within the proboscis, it appears that the elongated laciniae of long-proboscid *Prosoeca* underwent a functional change from a piercing structure to a lever which functions as a lifting structure.

**Fig. 15.5** (continued) pharyngeal valve. Reproduced with permission of Springer Karolyi et al. (2013). (**d**) Micro-CT head scan of *Prosoeca marinusi* displaying a lateral view of the pharyngeal pump with oesopharynx (both green), pharyngeal dilator and compressor muscles together with oesopharyngeal dilators. Reproduced with permission of Springer from Karolyi et al. (2013). Abbreviations: *cb* cibarium, *cpv* cibarial-pharyngeal valve, *eph* epipharynx, *fc* food canal, *hph* hypopharynx, *lc* lacinia, *lbr* labrum, *mcc* musculus clypeo-cibarialis, *mcpd* m. clypeo-pharyngealis dorsalis, *mcpv* m. clypeo-pharyngealis ventralis, *mfp* m. fronto-pharyngealis, *mlc* m. labro-cibarialis, *mpa* m. pharyngealis anterior, *mpl* mediolateral prementum ledges, *mpp* musculus pharyngealis posterior, *mtod* m. tentorio-oesopharyngealis dorsalis, *mtol* m. tentorio-oesopharyngealis lateralis, *oe* oesopharynx, *ph* pharynx, *pp* paraphyses, *pr* prementum, *ps* pseudotracheae, *sd* salivary duct

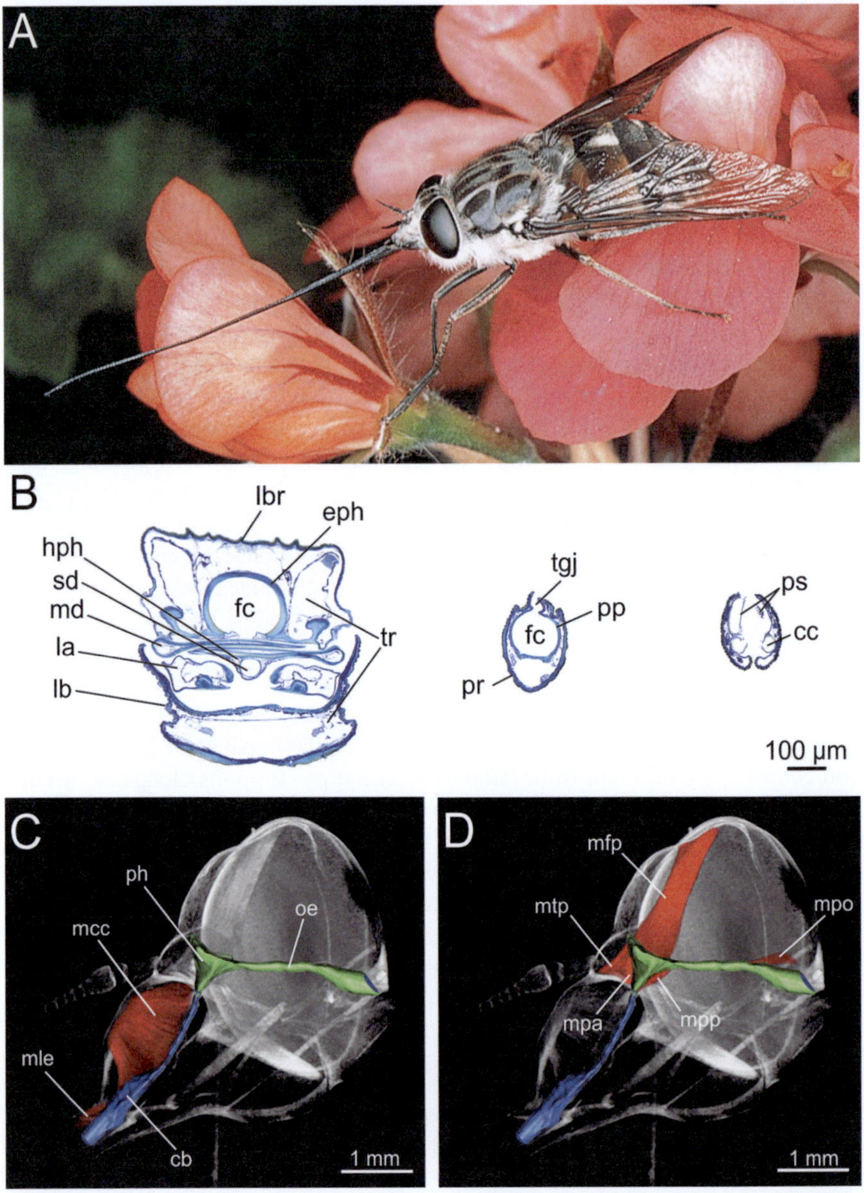


**Fig. 15.6** Long-proboscid Tabanidae *Philoliche rostrata*. (**a**) Proboscis divided into the proximal piercing part and the distal part responsible for nectar intake. Image provided by M. Picker & C. Griffith. (**b**) Proboscis cross sections. Proximal part (left): labrum (*lbr*) with epipharyngeal (eph) food canal (fc), mandibles (md) and laciniae stylet (la), in between hypopharynx (hph) with salivary duct (sd); trachea (tr). The piercing structures are sheathed in the labium (lb). Distal part (middle): prementum (pr) containing the labial food canal (fc) lined by the paraphyses (pp) and closed by a dorsal tongue and groove joint (tgj). Apical labella (right) with pseudotracheal system (ps) and collecting canals (cc). Reproduced with permission of Elsevier Ltd. from Karolyi et al. (2014). (**c**)

**Table 15.5** Head muscles responsible for proboscis movements in Nemestrinidae (*Prosoeca*) and Tabanidae (*Philoliche*). Functions marked with a question mark remain uncertain (after Karolyi et al. 2012, 2014)

| Muscle | | Function |
|---|---|---|
| ***Prosoeca marinusi*** | | |
| 1 | M. clypeo-labralis | Promotor of the labrum |
| 2 | M. clypeo-maxillaris | Retractor of the lacinia |
| 3 | M. tentorio-lacinialis | Promotor of the lacinia |
| 4 | M. tentorio-praementalis | Retractor of the labium |
| 5 | M. premento-kappalis | Depressor of the labella? |
| 6 | M. praemento-labellaris | Abductor of the labella? |
| 7 | M. praemento-paraphysalis | Flexor of the paraphyses? |
| ***Philoliche rostrata*** | | |
| 1 | M. clypeo-labralis | Retractor of the labrum |
| 2 | M. postgeno-mandibularis d. | Abductor of the mandible |
| 3 | M. postgeno-mandibularis s. | Outer adductor of the mandible |
| 3 | M. tentorio-mandibularis | Tentorial adductor of the mandible |
| 4 | M. tentorio-cardinalis | Promotor of the cardo |
| 5 | M. tentorio-stipitalis | Tentorial retractor of the lacinia |
| 6 | M. tentorio-lacinialis | Tentorial retractor of the lacinia |
| 7 | M. postgeno-stipitalis | Outer retractor of the lacinia |
| 8 | M. tentorio-labialis | Retractor of the labium |

### 15.4.2.3 Distal Proboscis Region

In both families, the lance-shaped, labial prementum is the longest, most conspicuous part of the proboscis. In the distal part, the prementum alone composes the food canal with the paired and short apical labella. The labial food canal is formed by the longitudinally rolled up prementum, strengthen by paired lateral paraphyses. Dorsally a tongue and groove joint of the lateral prementum walls close the proboscis. The labium is heavily sclerotized and reinforced by premental ledges. The food canal leads into the paired collecting canals of the short labella, which are connected to the

**Fig. 15.6** (continued) Micro-CT scan of the head displaying the cibarial pump (blue) in a lateral view with epipharyngeal compressor and cibarial dilator. Reproduced with permission of Elsevier Ltd. from Karolyi et al. (2014). (**d**) Micro-CT scan of the head displaying the pharyngeal pump with the oesophagus (both green) in a lateral view, main and secondary pharyngeal dilators, pharyngeal compressors and oesopharyngeal dilator. Reproduced with permission of Elsevier Ltd. from Karolyi et al. (2014). Abbreviations: *cb* cibarium, *mcc* musculus clypeo-cibarialis, *mtp* m. clypeo-pharyngealis, *mfp* m. fronto-pharyngealis, *mle* m. labro-epipharyngealis, *mpo* m. postoccipitalis-oesopharyngealis, *oe* oesopharynx, *ph* pharynx

pseudotracheal system. These canals are the only opening of the otherwise sealed proboscis (Figs. 15.5b and 15.6b).

The similar mouthpart morphology and composition in both families highlight the convergence in mouthpart adaptations associated with nectar feeding from long-tubed flowers using a superlong proboscis. This is especially evident in the distal section of the food canal, which is formed by a single component, making this proboscid structure unique among nectar-feeding insects (Krenn et al. 2005).

#### 15.4.2.4 The Suction Pump of Long-Proboscid Flies

Diptera are characterized by fluid-feeding mouthparts and elongated proboscises evolved independently among many taxa specialized on nectar or blood feeding (Krenn et al. 2005; Krenn and Aspöck 2012). The morphology of the suction pump of nectar- and blood-feeding *Philoliche* and nectar-feeding *Prosoeca* species appears to be comparable (Morita 2011; Karolyi et al. 2012). The pumping organ is composed of two consecutive parts, with corresponding inlet valves (Figs. 15.5c, d and 15.6c, d, Table 15.6). The cibarial pump is located behind the clypeus and the frons and separated from the pharyngeal pump by a cibarial-pharyngeal valve. However, the muscle which controls the cibarial-pharyngeal valve in *Prosoeca* is missing in *Philoliche*. Both pumping organs consist of a set of massive, paired dilator muscles, and the pharynx is additionally equipped with secondary dilators as well as enclosed by two unpaired compressor muscles. The consecutive pumps form a right angle with the oesophagus, which proceeds from the pharynx through the brain to the back of the head. Further, in *Prosoeca*, two small muscles reside between the posterior tentorial arms and the oesophagus; in *Philoliche* however, only one paired muscle resides between the post-occipital ridge and the oesophagus (Karolyi et al. 2014).

Both long-proboscid *Philoliche* and *Prosoeca* are highly active flyers reliant on ample nectar amounts for energy intake. Their feeding apparatus represents a two-level fluidic system with a strawlike proboscis and the systaltic motion of a cibarial and a pharyngeal suction pump to transport nectar out of long, narrow flowers to the mouth. The suction pumps work in three functional phases. First, the cibarial lumen is extended by massive paired dilator muscles, and due to the emerging negative pressure, nectar is sucked from the food canal into the cibarium. In the second phase, contractions of the primary and secondary, paired pharyngeal dilators draw nectar into the pharynx. In the last phase, unpaired compressor muscles push the nectar into the oesopharynx. Finally, dorsal and lateral dilators open the distal oesophagus valve to the midgut. During this procedure, paired cibarial retractor and protractor muscles hold the pump stationary by working antagonistically against the massive dilator muscles.

Similar pumping systems have been described for short-proboscid Tabanidae (Bonhag 1951), Bombyliidae (Szucsich and Krenn 2000) and mosquitoes (Snodgrass 1944; Kim et al. 2012). In long-proboscid flies, the morphological adaptation of a two-pump system is inevitable to connect the food canal within the proboscis with the midgut (Kim et al. 2011a). In addition, since the nectar flow

**Table 15.6** Suction pump muscles of cibarial and pharyngeal pump of *Prosoeca marinusi* (Nemestrinidae) and *Philoliche rostrata* (Tabanidae) including muscle name and function (after Karolyi et al. 2013, 2014)

| Muscle | Function |
|---|---|
| ***Prosoeca marinusi*** | |
| M. labro-epipharyngealis | Cibarial valve dilator |
| M. clypeo-cibarialis | Cibarial dilator |
| M. clypeo-pharyngealis ventralis | Pharyngeal dilator |
| M. clypeo-pharyngealis dorsalis | Pharyngeal dilator |
| M. pharyngealis anterior | Pharyngeal compressor |
| M. pharyngealis posterior | Pharyngeal compressor |
| M. fronto-pharyngealis | Pharyngeal dilator |
| M. geno-cibarialis | Cibarial retractor |
| M. fronto-cibarialis | Cibarial protector |
| ***Philoliche rostrata*** | |
| M. labro-epipharyngealis | Cibarial valve dilator |
| M. clypeo-cibarialis | Cibarial dilator |
| M. tentorio-pharyngealis | Pharyngeal dilator |
| M. fronto-pharyngealis | Pharyngeal dilator |
| M. pharyngealis anterior | Anterior pharynx compressor |
| M. pharyngealis posterior | Posterior pharynx compressor |
| M. postoccipitalis-oesopharyngealis | Dorsal dilator of oesopharynx |

suffers from energy loss along the curved path, an additional pump is necessary to regulate the flow effectively. In this way, long-proboscid flies are able to suck liquid efficiently using the phase-shifted motion of cibarial and pharyngeal pump. Considering the intraspecific proboscis length variations found in South African long-proboscid flies, increasing pump dimensions which correlate with increasing proboscis lengths underline the importance of both pumps and indicate that a longer proboscis demands larger pumping organs (Kunte 2007; Karolyi et al. 2013, 2014).

### *15.4.3 Flower-Visiting Behaviour of* Prosoeca

The flower-visiting behaviour of *Prosoeca marinusi* has been described for its main host plant *Lapeirousia oreogena* (Iridaceae) (Karolyi et al. 2013). The flies approach a flower in horizontal flight, and the proboscis is swung forward into feeding position while hovering above a flower for a few seconds. After contact with the flower, the fly alights down, inserting the proboscis into the opening of the nectar spur and pushes down as deep as possible. Finally, after drinking, the proboscis is removed with a rapid upward movement. It has been shown that flower handling time correlates with proboscis length. In particular, the drinking time significantly increased with an increasing proboscis length (Karolyi et al. 2013). Previous studies of the flower-visiting behaviour of *Prosoeca marinusi* on *L. oreogena* revealed the importance of the white arrow markings as functional nectar guides to minimize flower handling times. Flies were no longer able to insert their proboscis into flowers with nectar markings which have been covered with black marker ink (Hansen et al. 2011).

### *15.4.4 Blood Feeding and Nectar Sucking in* Philoliche

While male Tabanidae only feed on nectar, females are nectar and blood feeders (Bonhag 1951). The dual task of female long-proboscid *Philoliche* is accomplished by separately bending the labium backwards to expose the piercing structures. Hence, long-proboscid *Philoliche* either use their elongated mouthparts to take up nectar from long-spurred flowers, or females suck blood with the relatively short piercing proximal proboscis part (Goodier 1962; Morita 2011; Karolyi et al. 2014). Depending on the feeding habit, two different proboscis positions can be distinguished. While feeding from flowers, the labium is inserted straight forward into the corolla (Johnson and Morita 2006). During blood feeding however, the labium is flapped backwards, and only the mandibles and maxillary structures are used to pierce the skin and suck up blood through the epipharyngeal food canal (Tetley 1917; Goodier 1962). In order to maintain a pressure drop in the extraordinarily long food canal, both the proximal and distal food canal form a concealed, strawlike nectar-extracting apparatus, which leads to the pseudotracheal system of the labella (Karolyi et al. 2012).

During blood feeding, the initial laceration of the host skin is achieved by scissor-like movement of the apical serrated mandible blades, while the maxilla stylets are responsible for puncturing the skin, representing the piercing structures that are actively inserted into the host skin (Bonhag 1951; Matsuda 1965; Chaudonneret 1986; Krenn and Aspöck 2012; Karolyi et al. 2014). Retractor muscles and their antagonists bend and stretch the maxillary base, resulting in protraction and retraction of the stylets that work like a push drill to penetrate the host skin with the toothed tip region, and the sclerotized inner edge widens the puncture wound. In addition, the hypopharynx acts as a syringe that injects saliva into the wound (Karolyi et al. 2014).

Coevolutionary relationships between flowers and long-proboscid flies appear to have been long associated with long-proboscid Diptera. Fossils of Nemestrinidae from the Jurassic period found in China show well-developed long proboscises that suggest nectar feeding from long tubular flowers (Ren 1998). The elongated prementum of *Philoliche* and *Prosoeca* probably evolved by “diffuse coevolution” (Janzen 1980) along with long-tubed flower guilds (Anderson and Johnson 2009; Pauw et al. 2009), while the female piercing structures retain a length that is adapted to piercing the host skin. Therefore, the evolution of a distal nectar uptaking proboscis in long-proboscid flies is regarded as an adaptation that gave exclusive access to nectar-rewarding long-tubed flowers, while the proximal piercing part in female *Philoliche* remained unmodified in comparison to other Tabanidae (Bonhag 1951).

## References

Ackerman JD (1985) Euglossine bees and their nectar hosts. In: D'Arcy WG, Correa MD (eds) The botany and natural history of Panama. Missouri Botanical Garden, St. Louis, pp 225–233
Agosta SJ, Janzen DH (2005) Body size distributions of large Costa Rican dry forest moths and the underlying relationship between plant and pollinator morphology. Oikos 108(1):183–193
Alarcón R, Davidowitz G, Bronstein JL (2008) Nectar usage in a southern Arizone hawkmoth community. Ecol Entomol 33(4):503–509
Alexandersson R, Johnson SD (2002) Pollinator–mediated selection on flower–tube length in a hawkmoth–pollinated *Gladiolus* (Iridaceae). Proc R Soc Lond B 269:631–636
Amorim FW, Wyatt GE, Sazima M (2014) Low abundance of long-tongued pollinators leads to pollen limitation in four specialized hawkmoth-pollinated plants in the Atlantic Rain Forest, Brazil. Naturwissenschaften 101:893–905
Amsel HG (1938) *Amphimoea walkeri* Bsd., der Schwärmer mit dem längsten Rüssel! Entomol Rundschau 55:165–167
Anderson B, Johnson SD (2008) The geographical mosaic of coevolution in a plant-pollinator mutualism. Evolution 62:220–225. https://doi.org/10.1111/j.1558-5646.2007.00275.x
Anderson B, Johnson SD (2009) Geographical covariation and local convergence of flower depth in a guild of fly-pollinated plants. New Phytol 182:533–540. https://doi.org/10.1111/j.1469-8137.2009.02764.x
Arditti J, Elliott J, Kitching IJ, Wasserthal LT (2012) "Good Heavens what insect can suck it" – Charles Darwin, *Angraecum sesquipedale* and *Xanthopan morganii praedicta*. Bot J Linn Soc 169(3):403–432
Baker HG (2016) Sugar concentrations in nectars from hummingbird flowers. Assoc Trop Biol Conserv 7:37–41
Barraclough DA (2006) An overview of the South African tangle-veined flies (Diptera: Nemestrinidae), with an annotated key to the genera and a checklist of species. Zootaxa 1277:39–63
Barraclough DA, Colville JF, Karolyi F, Krenn HW (2018) A striking new species of *Prosoeca* Schiner, 1867 (Diptera: Nemestrinidae): an important pollinator from the Bokkeveld Plateau, Northern Cape Province, South Africa. Zootaxa 4497:411–421. https://doi.org/10.11646/zootaxa.4497.3.5
Bauder JA-S, Lieskonig NR, Krenn HW (2011) The extremely long-tongued Neotropical butterfly *Eurybia lycisca* (Riodinidae): proboscis morphology and flower handling. Arthropod Struc Dev 40:122–127
Bauder JA-S, Handschuh S, Metscher BD, Krenn HW (2013) Functional morphology of the feeding apparatus and evolution of proboscis length in metalmark butterflies (Lepidoptera: Riodinidae). Biol J Linn Soc 110:291–304
Bauder JA-S, Warren AD, Krenn HW (2014) Evolution of extreme proboscis lengths in Neotropical Hesperiidae (Lepidoptera). J Res Lepidoptera 47:65–67
Bauder JA-S, Morawetz L, Warren AD, Krenn HW (2015a) Functional constraints on the evolution of long butterfly proboscides: lessons from Neotropical skippers (Lepidoptera: Hesperiidae). J Evol Biol 28:678–687
Bauder JA-S, Warren AD, Krenn HW (2015b) The ecological role of extremely long-proboscid Neotropical butterflies (Lepidoptera: Hesperiidae) in plant-pollinator networks. Arthropod-Plant Inte 9(4):415–421

Bonhag PF (1951) The skeleto-muscular mechanism of the head and abdomen of the adult horsefly (Diptera: Tabanidae). Trans Am Entomol Soc 77:131–202
Borrell BJ (2004) Suction feeding in orchid bees (Apidae: Euglossini). Proc R Soc B 271(Suppl 4):164–166
Borrell BJ (2005) Long tongues and loose niches: evolution of euglossine bees and their nectar flowers. Biotropica 37(4):664–669
Borrell BJ (2007) Scaling of nectar foraging in orchid bees. Am Nat 169(5):569–580
Borrell BJ, Krenn HW (2006) Nectar feeding in long-proboscid insects. In: Herrel A, Speck T, Rowe NP (eds) Ecology and biomechanics: a mechanical approach to the ecology of animals and plants. CRC, Boca Raton, pp 185–212
Brandenburg A, Bshary R (2011) Variable responses of hawkmoths to nectar-depleted plants in two native *Pentunia axillaris* (Solanaceae) populations. Arthr-Plant Inte 5:141–148
Bullock SH, Pescador A (1983) Wing and proboscis dimensions in a sphingid fauna from western Mexico. Biotropica 15(4):292–294
Burkle LA, Irwin R (2009) Nectar sugar limits larval growth of solitary bees (Hymenoptera: Megachilidae). Envirol Entomol 38(4):1293–1300
Cameron SA (2004) Phylogeny and biology of Neotropical orchid bees (Euglossini). Annu Rev Entomol 49:377–404. https://doi.org/10.1146/annurev.ento.49.072103.115855
Cane JH, Gardner DR, Harrison PA (2011) Nectar and pollen sugars constituting larval provisions of the alfalfa leaf-cutting bee (*Megachile rotundata*) (Hymenoptera: Apiformes: Megachilidae). Apidologie 42:401–408. https://doi.org/10.1007/s13592-011-0005-0
Chaudonneret J (1986) Les pieces buccales des insectes: theme et variations. 5. Bull Sci Bourgogne 39:6–45. https://doi.org/10.1093/aesa/85.3.358
Colville JF, Potts A, Bradshaw PL, Measey GJ, Snijman D, Picker MD (2014) Floristic and faunal Cape biochoria: do they exist? In: Allsopp N, Colville JF, Verboom A (eds) Fynbos: ecology, evolution and conservation of a megadiverse region. Oxford University Press, Oxford
Corbet SA (2000) Butterfly nectaring flowers: butterfly morphology and flower form. Entomol Exp Appl 96(3):289–298
Courtney SP, Hill CJ, Westerman A (1982) Pollen carried for long periods by butterflies. Oikos 38 (2):260–263
Daniel TL, Kingsolver JG, Meyhöfer E (1989) Mechanical determinants of
ing energetics in butterflies: muscle mechanics, feeding geometry, and functional equivalence. Oecologia 79(1):66–75
Darveau C-A (2004) Evolution of energy metabolism in a lineage of orchid bees (Apidea: Euglossini): from morphology to molecules. PhD-Thesis, University of British Columbia, Vancouver
Darwin C (1862) On the various contrivances by which British and foreign orchids are fertilised by insects and on the good effects of intercrossing. John Murray, London
Davis NT, Hildebrand JG (2006) Neuroanatomy of the sucking pump of the moth *Manduca sexta* (Sphingidae, Lepidoptera). Arthropod Struct Dev 35:15–39
DeVries PJ (1997) The butterflies of Costa Rica and their natural history, vol II: Riodinidae. Princeton University Press, Chichester, p 288
Downes JA (1958) The feeding habits of biting flies and their significance in classification. Annu Rev Entomol 3:249–266. https://doi.org/10.1146/annurev.en.03.010158.001341
Dressler RL (1982) Biology of the orchid bees (Euglossini). Ann Rev Ecol Sys 13:373–394
Düster JV, Gruber MH, Karolyi F, Plant JD, Krenn HW (2018) Drinking with a very long proboscis: functional morphology of orchid bee mouthparts (Euglossini, Apidae, Hymenoptera). Arthropod Struct Dev 47:25–35. https://doi.org/10.1016/j.asd.2017.12.004
Dvořák B (2017) Evidence of *Hymenocallis speciosa* (Amaryllidaceae) as a nectar source for *Megacorma obliqua* (Walker, 1856) by C. W. Gan in Malaysia (Lepidoptera: Sphingidae). Entomologische Zeitschrift mit Insekten-Börse 127(3):187–189
Eastham LES, Eassa YEE (1955) The feeding mechanism of the butterfly *Pieris brassicae* L. Philos Trans R Soc Lond B Biol Sci 239:1–43

Eaton JL (1988) Lepidopteran anatomy. Wiley, New York

Eberhard SH, Krenn HW (2005) Anatomy of the oral valve in nymphalid butterflies and a functional model for fluid uptake in Lepidoptera. Zool Anz 243:305–312

Elliott SE, Irwin RE, Adler LS, Williams NM (2008) The nectar alkaloid, gelsemine, does not affect offspring performance of a native solitary bee, *Osmia lignaria* (Megachilidae). Ecol Entomol 33:298–304

Eltz T, Roubik DW, Lunau K (2005) Experience-dependent choices ensure species-specific fragrance accumulation in male orchid bees. Behav Ecol Sociobiol 59:149–156

Emlen JM (1966) The role of time and energy in food preference. Am Nat 100:611–617

Evoy WH, Jones BP (1971) Motor patterns of male euglossine bees evoked by floral fragrances. Anim Behav 19:583–588

Farina WM, Varjú D, Zhou Y (1994) The regulation of distance to dummy flowers during hovering flight in the hawk moth *Macroglossum stellatarum*. J Comp Physiol A 174:239–247. https://doi.org/10.1007/BF00193790

Faucheux MJ (2013) Sensillum types on the proboscis of the Lepidoptera: a review. Ann Soc Entomol Fr 49(1):73–90

Fox K, Anderson KM, Andres R, Foster MC, Foster CE, Vik D, Vitt P, Harris MO (2015) Nectar robbery and thievery in the hawk moth (Lepidoptera: Sphingidae)-pollinated western prairie fringed orchid *Platanthera praeclara*. Ann Entomol Soc Am 108(6):1000–1013

Galic M (1971) Die Sinnesorgane an der Glossa, dem Epipharynx und dem Hypopharynx der Arbeiterin von *Apis mellifera* L. (Insecta, Hymenoptera). Zeitschrift für Morphol der Tiere 70:201–228

Garrison JSE, Gass CL (1999) Response of a traplining hummingbird to changes in nectar availability. Behav Ecol 10(6):714–725

Gilbert LE (1972) Pollen feeding and reproductive biology of *Heliconius* butterflies. Proc Natl Acad Sci U S A 69(6):1403–1407

Gilbert LE (1975) Ecological consequences of a coevolved mutualism between butterflies and plants. In: Gilbert LE, Raven PH (eds) Coevolution of animals and plants. University of Texas Press, Austin, London, pp 210–240

Goldblatt P, Manning JC (1999) The long-proboscid fly pollination system in *Gladiolus* (Iridaceae). Ann Missouri Bot Gard 86:758–774

Goldblatt P, Manning JC (2000) The long-proboscid fly pollination system in Southern Africa. Ann Missouri Bot Gard 87:146–170. https://doi.org/10.2307/2666158

Goldblatt P, Manning JC (2006) Radiation of pollination systems in the Iridaceae of sub-Saharan Africa. Ann Bot 97:317–344. https://doi.org/10.1093/aob/mcj040

Goldblatt P, Manning JC, Bernhardt P (1995) Pollination biology of *Lapeirousia* subgenus *Lapeirousia* (Iridaceae) in Southern Africa; floral divergence and adaptation for long-tongued fly-pollination. Ann Missouri Bot Gard 82:517–534

Goodier R (1962) Blood feeding by *Philoliche* (*Dorcaloemus*) *silverlocki* Austen (Diptera, Tabanidae). Nature 193:1003

Goulson D, Stout JC, Hawson SA (1997) Can flower constancy in nectaring butterflies be explained by Darwin's interference hypothesis? Oecologia 112:225–231

Goyret J, Raguso RA (2006) The role of mechanosensory input in flower handling efficiency and learning by *Manduca sexta*. J Exp Biol 209:1585–1593

Grant V, Grant KA (1965) Flower pollination in the *Phlox* family. Columbia University Press, New York

Grant V, Grant KA (1983) Hawkmoth pollination of *Mirabilis longiflora* (Nyctaginaceae). Proc Natl Acad Sci U S A 80:1298–1299

Grimaldi DA (1999) The co-radiations of pollinating insects and angiosperms in the Cretaceous. Ann Missouri Bot Gard 86:373–406

Grimaldi DA, Engel MS (2005) Evolution of the insects. Cambridge University Press, New York

Haber WA, Frankie GW (1989) A tropical hawkmoth community: Costa Rican dry forest Sphingidae. Biotropica 21:155–172

Hainsworth FR, Precup E, Hamill T (1991) Feeding, energy processing rates and egg production in painted lady butterflies. J Exp Biol 156(1):249–265

Hansen DM, Van der Niet T, Johnson SD (2011) Floral signposts: testing the significance of visual "nectar guides" for pollinator behaviour and plant fitness. Proc R Soc B 279:634–639. https://doi.org/10.1098/rspb.2011.1349

Heinrich B (1975) Energetics of pollination. Ann Rev Ecol Sys 6:139–170

Heinrich B (1983) Insect foraging energetics. In: Jones CE, Little RJ (eds) Handbook of experimental pollination biology. Scientific and Academic Editions, New York, pp 187–214

Hespenheide HA (1973) Ecological inferences from morphological data. Ann Rev Ecol Syst 4:213–229

Inoue T, Asaoka K, Seta K, Imaeda D, Ozaki M (2009) Sugar receptor response of the food-canal taste sensilla in a nectar-feeding swallowtail butterfly, *Papilio xuthus*. Naturwissenschaften 96 (3):355–363

Inouye DW (1980) The effect of proboscis and corolla tube lengths on patterns and rates of flower visitation by bumblebees. Oecologia 45(2):197–201

Janzen DH (1971) Euglossine bees as long-distance pollinators of tropical plants. Science 171:203–205

Janzen DH (1980) When is it coevolution? Evolution 34(3):611–612

Janzen DH, Hallwachs W (2009) Dynamic database for an inventory of the macrocaterpillar fauna, and its food plants and parasitoids, of Area de Conservación Guanacaste (ACG), northwestern Costa Rica (nn-SRNP-nnnnn voucher codes). http://janzen.sas.upenn.edu

Johnson SD (2010) The pollination niche and its role in the diversification and maintenance of the Southern African flora. Philos Trans R Soc Lond B Biol Sci 365:499–516. https://doi.org/10.1098/rstb.2009.0243

Johnson SD, Anderson B (2010) Coevolution between food-rewarding flowers and their pollinators. Evo Ed Outreach 3:32–39

Johnson SD, Johnson K (1993) Beauty and beast: a Cape orchid pollinated by horse flies. Veld Flora 79:38–39

Johnson SD, Morita S (2006) Lying to Pinocchio: floral deception in an orchid pollinated by long-proboscid flies. Bot J Linn Soc 152:271–278. https://doi.org/10.1111/j.1095-8339.2006.00571.x

Johnson SD, Raguso RA (2016) The long-tongued hawkmoth pollinator niche for native and invasive plants in Africa. Ann Bot 117(1):25–36

Johnson SD, Steiner KE (1997) Long-tongued fly pollination and evolution of floral spur length in the *Disa draconis* complex (Orchidaceae). Evolution 51(1):45–53

Johnson SD, Steiner KE (2003) Specialized pollination systems in Southern Africa. S Afr J Sci 99:345–348

Johnson SD, Moré M, Amorim FW, Haber WA, Frankie GW, Stanley DA, Cocucci AA, Raguso RA (2017) The long and the short of it: a global analysis of hawkmoth pollination niches and interaction networks. Funct Ecol 31(1):101–115

Josens RB, Farina WM (2001) Nectar feeding by the hovering hawk moth *Macroglossum stellatarum*: intake rate as a function of viscosity and concentration of sucrose solutions. J Comp Physiol A 187:661–665

Kandori I, Ohsaki N (1996) The learning abilities of the white cabbage butterfly, *Pieris rapae*, foraging for flowers. Res Popul Ecol 38:111–117

Karolyi F, Szucsich NU, Colville JF, Krenn HW (2012) Adaptations for nectar-feeding in the mouthparts of long-proboscid flies (Nemestrinidae: *Prosoeca*). Biol J Linn Soc 107:414–424. https://doi.org/10.1111/j.1095-8312.2012.01945.x

Karolyi F, Morawetz L, Colville JF, Handschuh S, Metscher BD, Krenn HW (2013) Time management and nectar flow: flower handling and suction feeding in long-proboscid flies (Nemestrinidae: *Prosoeca*). Naturwissenschaften 100(11):1083–1093. https://doi.org/10.1007/s00114-013-1114-6

Karolyi F, Colville JF, Handschuh S, Metscher BD, Krenn HW (2014) One proboscis, two tasks: adaptations to blood-feeding and nectar-extracting in long-proboscid horse flies (Tabanidae, *Philoliche*). Arthropod Struc Dev 43(5):403–413

Kawahara AY (2007) Molecular phylogenetic analysis of the hawkmoths (Lepidoptera: Bombycoidea: Sphingidae) and the evolution of the sphingid proboscis. PhD-thesis, University of Maryland, p 159

Kawahara AY, Mignault AA, Regier JC, Kitching IJ, Mitter C (2009) Phylogeny and biogeography of hawkmoths (Lepidoptera: Sphingidae): evidence from five nuclear genes. PLoS One 4(5): e5719

Kay KM, Schemske DW (2003) Pollinator assemblages and visitation rates for 11 species of Neotropical *Costus* (Costaceae). Biotropica 35(2):198–207. https://doi.org/10.1646/02159

Kim BH, Kim HK, Lee SJ (2011a) Experimental analysis of the blood-sucking mechanism of female mosquitoes. J Exp Biol 214:1163–1169. https://doi.org/10.1242/jeb.048793

Kim W, Gilet T, Bush JWM (2011b) Optimal concentrations in nectar feeding. PNAS 108 (40):16618–16621

Kim BH, Seo ES, Lim JH, Lee SJ (2012) Synchrotron X-ray microscopic computed tomography of the pump system of a female mosquito. Microsc Res Tech 75:1051–1058. https://doi.org/10.1002/jemt.22030

Kingsolver JG, Daniel TL (1979) On the mechanics and energetics of nectar feeding in butterflies. J Theor Biol 76:167–179

Kingsolver JG, Daniel TL (1995) Mechanics of food handling by fluid-feeding insects. In: Chapman RF, De Boer G (eds) Regulatory mechanisms in insect feeding. Chapman and Hall, New York, pp 32–73

Kitching IJ (2002) The phylogenetic relationships of Morgan's Sphinx, *Xanthopan morganii* (Walker), the tribe Acherontiini, and allied long-tongued hawkmoths (Lepidoptera: Sphingidae, Sphinginae). Zool J Linnean Soc 135:471–527

Kitching IJ, Cadiou J-M (2000) Hawkmoths of the world: an annotated and illustrated revisionary checklist (Lepidoptera: Sphingidae). Ann Entomol Soc Am 93:1195–1196

Krenn HW (1990) Functional morphology and movements of the proboscis of Lepidoptera (Insecta). Zoomorphology 110:105–114

Krenn HW (1998) Proboscis sensilla in *Vanessa cardui* (Nymphalidae, Lepidoptera): functional morphology and significance in flower-probing. Zoomorphology 118:23–30

Krenn HW (2008) Feeding behaviours of Neotropical butterflies (Lepidoptera, Papilionoidea). Stapfia 88, zugleich Kataloge der oberösterreichischen Landesmuseen Neue Serie. Biologiezentrum Linz/Austria 80:295–304

Krenn HW (2010) Feeding mechanisms of adult Lepidoptera: structure, function, and evolution of the mouthparts. Annu Rev Entomol 55:307–327

Krenn HW, Aspöck H (2012) Form, function and evolution of the mouthparts of blood-feeding Arthropoda. Arthropod Struct Dev 41:101–118. https://doi.org/10.1016/j.asd.2011.12.001

Krenn HW, Bauder JA-S (2018) Morphological fine tuning of the feeding apparatus to proboscis length in Hesperiidae (Lepidoptera). J Morphol 279(3):396–408

Krenn HW, Penz CM (1998) Mouthparts of *Heliconius* butterflies (Lepidoptera: Nymphalidae): a search for anatomical adaptations to pollen-feeding behavior. Int J Insect Morphol 27 (4):301–309

Krenn HW, Zulka KP, Gatschnegg T (2001) Proboscis morphology and food preferences in nymphalid butterflies (Lepidoptera: Nymphalidae). J Zool (Lond) 254:17–26

Krenn HW, Plant JD, Szucsich NU (2005) Mouthparts of flower-visiting insects. Arthropod Struct Dev 34:1–40. https://doi.org/10.1016/j.asd.2004.10.002

Krenn HW, Wiemers M, Maurer L, Pemmer V, Huber W, Weissenhofer A (2010) Butterflies of the Golfo Dulce region, Costa Rica. Verein zur Förderung der Tropenstation La Gamba, Vienna, p 76

Kritsky G (2001) Darwin's Madagascan hawkmoth prediction. Am Entomol 37:206–210

Kunte K (2007) Allometry and functional constraints on proboscis lengths in butterflies. Funct Ecol 21:982–987

Laverty TM (1994) Bumblebee learning and flower morphology. Anim Behav 47(3):531–545
Lee SC, Kim BH, Lee SJ (2014) Experimental analysis of the liquid-feeding mechanism of the butterfly *Pieris rapae*. J Exp Biol 217:2013–2019
Lehane MJ (2005) Gonotrophic concordance. In: The biology of blood-sucking insects. Cambridge University Press, New York, pp 96–97
Lehnert MS, Beard CE, Gerard PD, Kornev KG, Adler PH (2016) Structure of the lepidopteran proboscis in relation to feeding guild. J Morphol 277(2):167–182
Levin DA, Berube DE (1972) *Phlox* and *Colias*: the efficiency of a pollination system. Evolution 26 (2):242–250
Lewis AC (1986) Memory constraints and flower choice in *Pieris rapae*. Science 232:863–865
Lieskonig NR (2013) Comparative morphology of proboscides in Nymphalidae (Lepidoptera) - phylogenetic and functional conclusions. Master thesis, University of Vienna, Vienna
Locatelli E, Machado ICS (1999) Floral biology of *Cereus fernambucensis*: a sphingophilous cactus of restinga. Bradleya 17:86–95
Lowe T, Garwood RJ, Simonsen TJ, Bradley RS, Whiters PJ (2013) Metamorphosis revealed: time-lapse three-dimensional imaging inside a living chrysalis. J R Soc Interface 10:20130304
Luer CA (1972) The native orchids of Florida. The New York Botanical Garden, New York
Manning JC, Goldblatt P (1997) The *Moegistorhynchus longirostris* (Diptera: Nemestrinidae) pollination guild: long-tubed flowers and a specialized long-proboscid fly pollination system in southern Africa. Plant Syst Evol 206(1–4):51–69
Martins DJ, Johnson SD (2013) Interactions between hawkmoths and flowering plants in East Africa: polyphagy and evolutionary specialization in an ecological context. Biol J Linn Soc 110 (1):199–213
Matsuda R (1965) Morphology and evolution of the insect head. Mem Am Entomol Inst 4:1–334
May PG (1985) Nectar uptake rates and optimal nectar concentrations of two butterfly species. Oecologia 66(3):381–386
May PG (1988) Determinants of foraging profitability in two nectarivorous butterflies. Ecol Entomol 13:171–184. https://doi.org/10.1111/j.1365-2311.1988.tb00345.x
May PG (1992) Flower selection and the dynamics of lipid reserves in two nectarivorous butterflies. Ecology 73(6):2181–2192
Miller WE (1997) Diversity and evolution of tongue length in hawkmoths (Sphingidae). J Lepid Soc 51:9–31
Mitter C, Davis DR, Cummings MP (2017) Phylogeny and evolution of Lepidoptera. Annu Rev Entomol 62:265–283
Molleman F, Krenn HW, Van Alphen ME, Brakefield PM, DeVries PJ, Zwaan BJ (2005) Food intake of fruit-feeding butterflies: evidence for adaptive variation in proboscis morphology. Biol J Linn Soc 86:333–343
Moré M, Sérsic AN, Cocucci AA (2006) Specialized use of pollen vectors by *Caesalpinia gilliesii*, a legume species with brush-type flowers. Biol J Linn Soc 88:579–592
Morita SI (2008) A phylogeny of long-tongued horse flies (Diptera: Tabanidae: *Philoliche*) with the first cladistic review of higher relationships within the family. Invertebr Syst 22:311–327. https://doi.org/10.1071/IS07005
Morita SI (2011) Repeatability and precision in proboscis length measurements for long proboscid flies. Zootaxa 3112:49–58
Muchhala N, Thomson JD (2009) Going to great lengths: selection for long corolla tubes in an extremely specialized bat-flower mutualism. Proc R Soc B 276:2147–2152
Nattero J, Moré M, Sérsic AN, Cocucci AA (2003) Possible tobacco progenitors share long-tongued hawkmoths as pollen vectors. Plant Syst Evol 241:47–54
Neff L, Simpson BB (1992) Partial bivoltinism in a ground-nesting bee: the biology of *Diadasia rinconis* in texas. J Kansas Entomol Soc 65:377–392
Netz C, Renner SS (2017) Long-spurred *Angraecum* orchids and long-tongued sphingid moths on Madagascar: a time frame for Darwin's predicted *Xanthopan*/*Angraecum* coevolution. Biol J Linn Soc 122(2):469–478

Nicolson SW (2007) Nectar consumers. In: Nicolson SW, Nepi M, Pacini E (eds) Nectaries and nectar. Springer, Dordrecht, pp 298–342

Nicolson SW (2011) Bee food: the chemistry and nutritional value of nectar, pollen and mixtures of the two. African Zool 46:197–204. https://doi.org/10.3377/004.046.0201

Nilsson LA (1983) Processes of isolation and introgressive interplay between *Platanthera bifolia* (L.) Rich and *P. chlorantha* (Custer) Reichb. (Orchidaceae). Bot J Linn Soc 87(4):325–350

Nilsson LA (1988) The evolution of flowers with deep corolla tubes. Nature 334:147–149

Nilsson LA (1998) Deep flowers for long tongues. TREE 13(7):259–260

Nilsson LA, Jonsson L, Rason L, Randrianjohany E (1985) Monophily and pollination mechanisms in *Angraecum arachnites* Schltr. (Orchidaceae) in a guild of long-tongued hawk-moths (Sphingidae) in Madagascar. Biol J Linn Soc 26:1–19

Nilsson LA, Jonsson L, Ralison L, Randrianjohany E (1987) Angraecoid orchids and hawkmoths in Central Madagascar: specialized pollination systems and generalist foragers. Biotropica 19:310–318

O'Brien D (1999) Fuel use in flight and its dependence on nectar feeding in the hawkmoth *Amphion floridensis*. J Exp Biol 202:441–451

Patočka J (1993) Die Puppen der mitteleuropäischen Schmetterlinge aus den Familien Endromidae, Lemoniidae, Saturniidae und Sphingidae (Lepidoptera). Ann Naturhist Mus Wien 94/95 B:503–545

Paulus HF, Krenn HW (1996) Vergleichende Morphologie des Schmetterlingsrüssels und seiner Sensillen - Ein Beitrag zur phylogenetischen Systematik der Papilionoidea (Insecta, Lepidoptera). J Zool Syst Evol Res 34(4):203–216

Pauw A, Stofberg J, Waterman RJ (2009) Flies and flowers in Darwin's race. Evolution 63 (1):268–279. https://doi.org/10.1111/j.1558-5646.2008.00547.x

Pellmyr O (2002) Pollination by animals. In: Herrera CM, Pellmyr O (eds) Plant-animal interactions: an evolutionary approach. Blackwell Science Ltd, Oxford, pp 157–184

Peñalver E, Arillo A, Pérez-De La Fuente R, Riccio ML, Delclòs X, Barrón E, Grimaldi DA (2015) Long-proboscid flies as pollinators of cretaceous gymnosperms. Curr Biol 25:1917–1923. https://doi.org/10.1016/j.cub.2015.05.062

Petr D, Stewart KW (2004) Comparative morphology of sensilla styloconica on the proboscis of North American Nymphalidae and other selected taxa (Lepidoptera): systematic and ecological considerations. Trans Am Entomol Soc 130(4):293–409

Pivnick KA, McNeil JN (1985) Effects of nectar concentration on butterfly feeding: measured feeding rates for *Thymelicus lineola* (Lepidoptera: Hesperiidae) and a general feeding model for adult Lepidoptera. Oecologia 66(2):226–237

Plant JD, Paulus HF (2015) Evolution and phylogeny of bees: review and cladistic analysis in light of morphological evidence (Hymenoptera, Apoidea). Zoologica 161:1–368

Pokorny T, Lunau K, Eltz T (2014) Raising the sugar content - orchid bees overcome the constraints of suction feeding through manipulation of nectar and pollen provisions. PLoS One 9:1–16. https://doi.org/10.1371/journal.pone.0113823

Potgieter CJ, Edwards TJ (2005) The *Stenobasipteron wiedmanni* (Diptera, Nemestrinidae) pollination guild in Eastern Southern Africa. Ann Missouri Bot Gard 92:254–267

Prance GT, Idrobo JM, Castaño OV (1983) Mecanismos de polinización de *Eschweilera garagarae* Pittier en El Chocó, Colombia. Mutisia (Acta Botânica Colombiana) 60:1–7

Pyke GH, Pulliam HR, Charnov EL (1977) Optimal foraging: a selective review of theory and tests. Q Rev Biol 52(2):137–154

Ranta E, Lundberg H (1980) Resource partitioning in bumblebees: the signifcance of differences in proboscis length. Oikos 35:298–302

Rathcke BJ (1992) Nectar distributions, pollinator behavior, and plant reproductive success. In: Hunter MD, Ohgushi T, Price PW (eds) Effects of resource distribution on animal-plant interactions. Academic, San Diego, pp 113–138

Regier JC, Mitter C, Zwick A, Bazinet AL, Cummings MP, Kawahara AY, Sohn J-C, Zwickl DJ, Cho S, Davis DR, Baixeras J, Brown J, Parr C, Weller S, Lees DC, Mitter KT (2013) A large-

scale, higher-level, molecular phylogenetic study of the insect order Lepidoptera (moths and butterflies). PLoS One 8(3):e58568. https://doi.org/10.1371/journal.pone.0058568
Ren D (1998) Flower-associated Brachycera flies as fossil evidence for jurassic angiosperm origins. Science 280:85–88. https://doi.org/10.1126/science.280.5360.85
Rodríguez-Gironés MA, Llandres AL (2008) Resource competition triggers the co-evolution of long tongues and deep corolla tubes. PLoS One 3(8):e2992
Rothschild L, Jordan K (1903) A revision of the lepidopterous family Sphingidae. Novit Zool 9:1–972
Roubik DW, Hanson PE (2004) Orchid bees of tropical America. Biology and fieldguide. Inbio, Santo Domingo de Heredia, Costa Rica
Roubik DW, Yanega D, Aluja S, Buchmann SL, Inouye DW (1995) On optimal nectar foraging by some tropical bees (Hymenoptera: Apidae). Apidologie 26(3):197–211
Rozen JG Jr (1984) Comparative nesting biology of the bee tribe Exomalopsini, Apoidea, Anthophoridae. Am Museum Novit 2798:1–37
Sazatornil FD, Moré M, Benitez-Vieyra S, Cocucci AA, Kitching IJ, Schlumpberger BO, Oliveira PE, Sazima M, Amorim FW (2016) Beyond neutral and forbidden links: morphological matches and the assembly of mutualistic hawkmoth–plant networks. J Anim Ecol 85(6):1586–1594
Schemske DW (1981) Floral convergence and pollinator sharing in two bee-pollinated tropical herbs. Ecology 62(4):946–954
Schemske DW, Horvitz CC (1984) Variation among floral visitors in pollination ability: a precondition for mutualism specializiation. Science 225:519–521
Schmitt JB (1938) The feeding mechanism of adult Lepidoptera. Smithson Misc Coll 97:1–28
Schnitzler J, Barraclough TG, Boatwright JS, Goldblatt P, Manning JC, Powell MP, Rebelo T, Savolainen V (2011) Causes of plant diversification in the cape biodiversity hotspot of South Africa. Syst Biol 60:343–357. https://doi.org/10.1093/sysbio/syr006
Schoener TW (1971) Theory of feeding strategies. Annu Rev Ecol Syst 2:369–404. https://doi.org/10.1146/annurev.es.02.110171.002101
Schoener TW (1974) Resource partitioning in ecological communities. Science 185:27–39
Shreeve TG (1992) Adult behaviour. In: RLH D (ed) The ecology of butterflies in Britain. Oxford University Press, Oxford, pp 22–45
Snodgrass RE (1944) The feeding apparatus of biting and sucking insects affecting man and animals. Smithson Misc Collect 104:1–133
Snodgrass RE (1956) The anatomy of the honey bee. Cornell University Press, Ithaca
Southwick EE (1984) Photosynthate allocation to floral nectar: a neglected energy investment. Ecology 65:1775–1779
Stefanescu C, Traveset A (2009) Factors influencing the degree of generalization in flower use by Mediterranean butterflies. Oikos 118(7):1109–1117
Steiner KE (1985) The role of nectar and oil in the pollination of *Drymonia serrulata* (Gesneriaceae) by *Epicharis* bees (Anthophoridae) in Panama. Biotropica 17(3):217–229
Stewart SL, Kane ME (2007) Orchid conservation in the Americas – lessons learned in Florida. Lankesteriana 7(1–2):382–387
Szucsich NU, Krenn HW (2000) Morphology and function of the proboscis in Bombyliidae (Diptera, Brachycera) and implications for proboscis evolution in Brachycera. Zoomorphology 120:79–90. https://doi.org/10.1007/s004350000025
Tetley H (1917) The structure of the mouth-parts of *Pangonia longirostris* in relation to the probable feeding-habits of the species. Bull Entomol Res 8(3):253–267. https://doi.org/10.1017/S0007485300037652
Thompson JN (1994) The coevolutionary process. University of Chicago Press, Chicago
Thompson JN (2005) The geographic mosaic of coevolution. University of Chicago Press, Chicago
Tudor O, Dennis RLH, Greatorex-Davies JN, Sparks TH (2004) Flower preferences of woodland butterflies in the UK: nectaring specialists are species of conservation concern. Biol Conserv 119(3):397–403

Van der Niet T, Johnson SD, Linder HP (2006) Macroevolutionary data suggest a role for reinforcement in pollination system shifts. Evolution 60:1596–1601. https://doi.org/10.1111/j.0014-3820.2006.tb00504.x

Van Nieukerken EJ, Kaila L, Kitching IJ, Kristensen NP, Lees DC, Minet J, Mitter C, Mutanen M, Regier JC, Simonsen TJ, Wahlberg N, Yen S-H, Zahiri R, Adamski D, Baixeras J, Bartsch D, Bengtsson BÅ, Brown JW, Bucheli SR, Davis D, De Prins J, De Prins W, Epstein M, Gentili-Poole P, Gielis C, Hättenschwiler P, Hausmann A, Holloway JD, Kallies A, Karsholt O, Kawahara A, Koster SJC, Kozlov MV, Lafontaine JD, Lamas G, Landry J-F, Lee S, Nuss M, Penz C, Rota J, Schmidt BC, Schintlmeister A, Sohn J, Solis MA, Tarmann G, Warren AD, Weller SJ, Yakovlev R, Zolotuhin VV, Zwick A (2011) Order Lepidoptera Linnaeus, 1758. In: Zhang Z-Q (ed) Animal biodiversity : an outline of higher classification and survey of taxonomic richness. Zootaxa 3148:212–221

Velthuis HHW, Wolf Y, Gerling D (1984) Provisioning and preparation of the brood cell in two carpenter bees, *Xylocopa sulcatipes* Maa and *Xylocopa pubescens* Spinola (Hymenoptera: Anthophorodae). Isr J Entomol 18:39–51

Wannenmacher G, Wasserthal LT (2003) Contribution of the maxillary muscles to proboscis movement in hawkmoths (Lepidoptera: Sphingidae) - an electrophysiological study. J Insect Physiol 49:765–776

Warren AD, Ogawa JR, Brower AVZ (2009) Revised classification of the family Hesperiidae (Lepidoptera: Hesperioidea) based on combined molecular and morphological data. System Entom 34(3):467–523

Wasserthal LT (1997) The pollinators of the Malagasy star orchids *Angraecum sesquipedale*, *A. sororium* and *A. compactum* and the evolution of extremely long spurs by pollinator shift. Bot Acta 110(5):343–430

Weber A, Huber W, Weissenhofer N, Zamora N, Zimmermann G (2001) An introductory field guide to the flowering plants of the Golfo Dulce rain forests, Costa Rica. Corcovado National Park and Piedras Blancas National Park ("Regenwald der Österreicher"). Stapfia 78, Oberösterreichisches Landesmuseum, Linz

Whitehead AT, Larsen JR (1976) Ultrastructure of the contact chemoreceptors of *Apis mellifera* L. (Hymenoptera: Apidae). Int J Insect Morphol Embryol 5:301–315

Whitham TG (1977) Coevolution of foraging in *Bombus* and nectar dispensing in *Chilopsis*: a last dreg theory. Science 197:593–596. https://doi.org/10.1126/science.197.4303.593

Whittall JB, Hodges SA (2007) Pollinator shifts drive increasingly long nectar spurs in columbine flowers. Nature Letters 447:706–710

Wiklund C (1981) On the pollination efficiency of butterflies: a reply to Courtney et al. Oikos 38 (2):263

Wiklund C, Eriksson T, Lundberg H (1979) The wood white butterfly *Leptidea sinapis* and its nectar plants: a case of mutualism or parasitism? Oikos 33:358–362

Winston ML (1979) The proboscis of the long-tongued bees: a comparative study. Univ Kansas Sci Bull 51:631–667

Winston ML (1991) The biology of the honeybee. Harvard University Press, Cambridge

Wolf LL, Hainsworth FR, Stiles FG (1972) Energetics of foraging: rate and efficiency of nectar extraction by hummingbirds. Science 176:1351–1352

# Chapter 16
# Morphogenesis of Piercing Stylets in Hemiptera

**Roman Rakitov**

**Abstract** This chapter describes how next-instar piercing stylets of Hemiptera develop between molts and how they replace the old stylets, lost during molting. Because the stylets contain epithelium and bodies of neurons in their widened bases and only sensory dendrites in their narrow long shafts, the new stylet cannot grow inside the old one. Instead, next-instar stylets develop within two-layered "retortiform organs" formed by the basal epithelium invaginating into the head cavity. Their development, styletogenesis, generally resembles the development of holometabolan imaginal discs but has unique features. In particular, the cuticle of the new stylet is secreted by parallel long filamentous projections of styligenic cells. Longitudinal sculpture for interlocking between stylets develops along the lines of contact between these projections. Usually the new stylets are passively extracted during ecdysis because their tips are attached to bases of the old stylets. Some Sternorrhyncha with extra-long stylets extract them actively, probably with the aid of the labium. Newly extracted stylets become interlocked into a bundle by pressure from the maxillary plates, labrum, and labium. Hemipteran stylets are not attached to muscles directly and slide into coupling cuticular structures during ecdysis. Stylet development in Thysanoptera may be similar, but its details remain unknown.

## 16.1 Introduction

Hemipterans have two pairs of setiform stylets (Fig. 16.1a) forming together a piercing-sucking bundle (Fig. 16.1b) encased at rest by the labium. Penetration of the substrate is achieved by in-and-out movements of individual stylets within the bundle and is facilitated by anchoring or cutting serrations of their apices (Fig. 16.1b–d). The adjacent surfaces of the central (maxillary) pair of stylets form two separate conduits for ingestion of liquid food and for ejection of saliva (Figs. 16.2a–c and 16.3b–f). Additionally, most stylets contain inner sensory equipment monitoring their load and deformation during piercing (Fig. 16.3h). The

R. Rakitov (✉)
Borissiak Paleontological Institute, Russian Academy of Sciences, Moscow, Russia



H. W. Krenn (ed.), *Insect Mouthparts*, Zoological Monographs 5,
https://doi.org/10.1007/978-3-030-29654-4_16

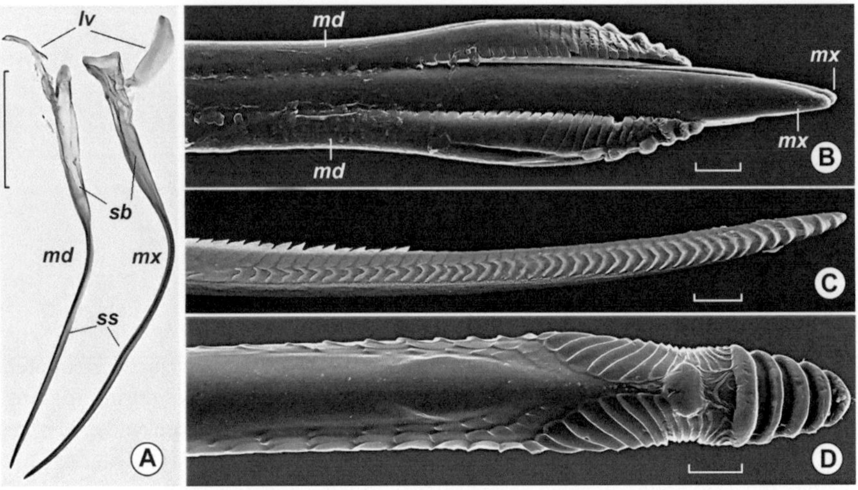


**Fig. 16.1** General structure of stylets and preapical serration of their lateral surfaces. (**a**) A maxillary and a mandibular stylets of an adult *Homalodisca vitripennis* (Germar), dissected out together with the braces (not distinct from stylet bases) and levers; (**b**) apical part of the stylet bundle of *Aphrophora* sp., fifth-instar nymph; note that the maxillary stylets are rotated with respect to the mandibular ones; compare with the cross section in Fig. 16.3d; (**c**) apical part of a mandibular stylet of an adult *H. vitripennis*, dorsolateral view; (**d**) same, fifth-instar nymph of *Tibicen plebejus* (L.), lateral view. Abbreviations: *md* mandibular stylet, *mx* maxillary stylet, *sb* stylet bases, *lv* stylet levers, *ss* stylet shafts. Scale bars: (**a**) 500 μm; (**b–d**) 25 μm

piercing function of hemipteran stylets requires them to have small cross sections and sufficiently strong walls, leaving inside no room for epithelium except in the widened bases.

Unlike mosquitoes and fleas, which acquire piercing-sucking mouthparts only at the final stage of metamorphosis, all feeding stages of Hemiptera and related Thysanoptera replace their old stylets, lost with the exuviae, after each molt. Because the main length of a stylet contains no live epithelium, these insects have evolved a mechanism of stylet development—styletogenesis—which is unlike that of any other body part. As Snodgrass (1927) succinctly put it: “Ordinarily, when a new appendage is to be similar in form and size to the one it replaces, it is formed immediately within the cuticle of the old. With the mouth setae of the Hemiptera, however, the new setae obviously cannot grow inside the old ones; for this reason they are produced within invaginations of the hypodermis immediately above the bases of the old setae, their tips alone projecting into the latter. The setal sacs of the Hemiptera are thus analogous to the peripodial pouches of the imaginal appendages in holometabolous insects.”

Understanding styletogenesis is important to better understand the morphology and function of the piercing-sucking mouthparts in Hemiptera and Thysanoptera, as well as their evolutionary origin. Below I will briefly review the organization of hemipteran stylets and then describe their development, ecdysis, and assembly based on literature and original observations.

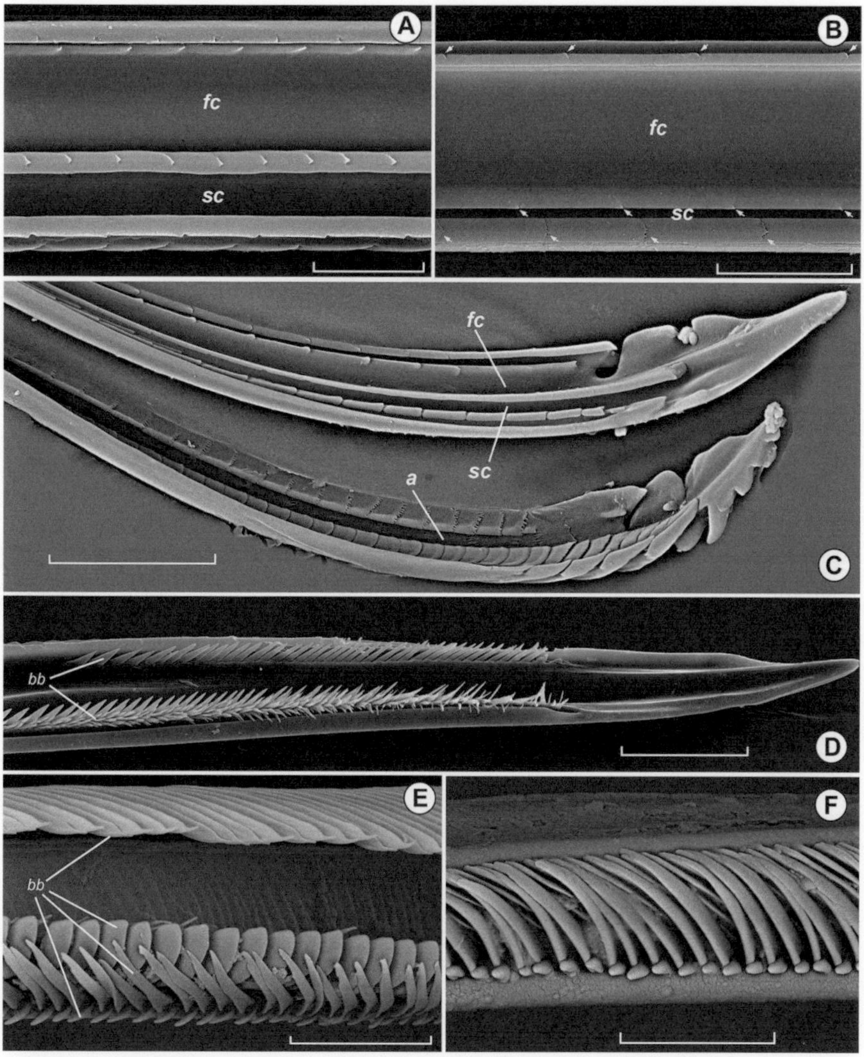


**Fig. 16.2** Sculpture of the mesal surfaces of stylets, SEM images. (**a**) Pentatomidae, *Palomena prasina* (L.), fifth-instar nymph, maxillary stylet; (**b**) Cicadidae, *Tibicen plebejus* (L.), fifth-instar nymph, maxillary stylet, arrows indicate olistheter scales; (**c**) Pyrrhocoridae, *Pyrrhocoris apterus* (L.), fifth-instar nymph, left maxillary and mandibular stylets, apices; (**d**) Reduviidae, *Platymeris biguttatus* (L.), adult, right maxillary stylet, apex; (**e**) Gerridae, *Gerris argentatus* Schumm., adult, apical region of a maxillary stylet, note three different types of "barbs"; (**f**) same, pair of maxillary stylets, apical regions interlocked by barbs. Abbreviations: *a* olistheter aulax on the mesal surface of the mandibular stylet, *bb* barbs, *fc* food canal, *sc* salivary canal. Scale bars: (**a**) 20 μm; (**b**, **c**) 50 μm; (**d**) 100 μm; (**e**, **f**) 10 μm

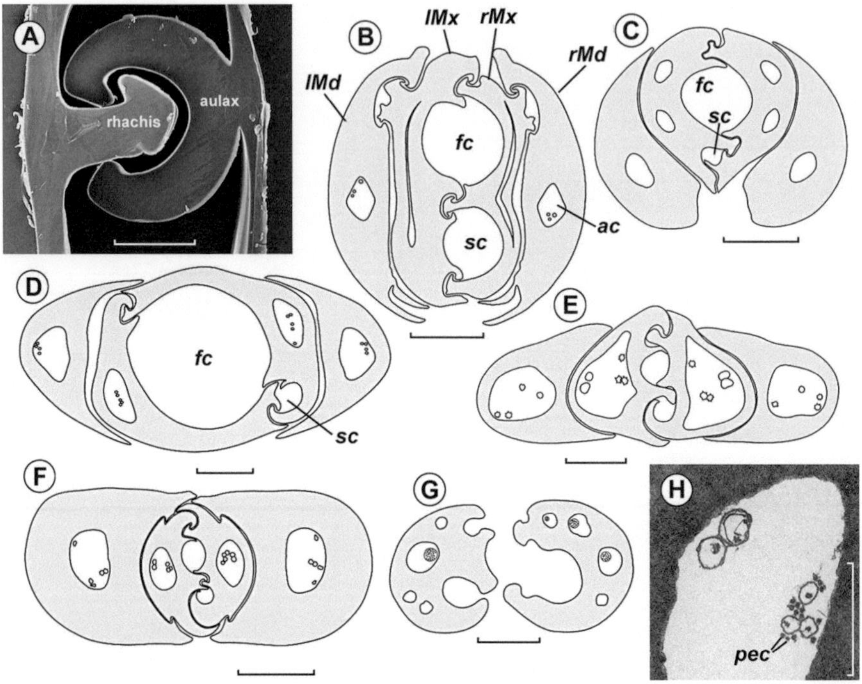


**Fig. 16.3** Olistheter interlocking and cross sections of stylet bundles. (**a**) Olistheter interlocking illustrated by a "ziplock" of a reclosable plastic bag, shown in cross section, SEM image; (**b–f**) stylet bundles of Hemiptera (for all species the orientation is as in (**b**)): (**b**) Pyrrhocoridae, *Pyrrhocoris apterus* (L.), adult; (**c**) Delphacidae, *Stiroma* sp., adult, sensory dendrites not shown; (**d**) Aphrophoridae, *Aphrophora* sp., fifth-instar nymph; (**e**) Cicadellidae, *Scenergates viridis* (Vilb.), first-instar nymph; (**f**) Peloridiidae, *Peloridium* sp., adult; (**g**) two maxillary stylets of *Adrothrips intermedius* (Bianchi) (Thysanoptera: Phlaeothripidae), adult; (**h**) cross section of the axial cavity of the left maxillary stylet of an adult *Peloridium* sp. showing five sensory dendrites (two n-type and three p-type, as classified by Backus 1985; dendrites of the latter type are accompanied with projections of their enveloping cells, here dissociated from the dendrites possibly due to inadequate fixation). Abbreviations: *ac* axial cavity; *fc* food canal; *lMd, rMd, lMx, rMx* left and right mandibular and maxillary stylets; *pec* projections of the enveloping cell; *sc* salivary canal. Scale bars: (**a**) 500 μm; (**b**, **d**) 10 μm; (**c**, **e**) 2 μm; (**f**) 5 μm; (**g**, **h**) 1 μm

## 16.2 Morphology

Despite their presumed derivation from dissimilar chewing mouthparts—the mandibles and lacinial endites of the maxillae—the mandibular and the maxillary stylets have a remarkably similar organization (Fig. 16.1a). It is convenient to recognize three areas along the stylet length: a funnel-shaped base and a narrow parallel-sided shaft with a tapered apex. Because the stylet shafts are generally flattened, it is sufficient to recognize their two surfaces: mesal (facing the axis of the bundle) and lateral. The stylet base contains live epithelium and bodies of sensory neurons, while the shaft and the apex contain no live tissue other than dendrites of the abovementioned neurons. The dendrites run in a liquid-filled axial cavity (Fig. 16.3b, d–f, h), which often branches into two or more parallel cavities (Fig. 16.3c, dendrites not shown), each containing one or several dendrites. Some stylets lack dendrites, and some also lack an axial cavity (e.g., the maxillary stylets of most Sternorrhyncha). The maxillary stylets are slightly to significantly longer than the mandibular stylets. In relative terms, the length of stylets varies from a fraction to several times the body length. Their width (the maximum dimension of the cross section) varies from almost 100 μm in large cicadas and belostomatids down to less than 2 μm in first-instar nymphs of scale insects.

### *16.2.1 Interlocking Devices*

The shafts of the maxillary stylets bear on their mesal surfaces parallel longitudinal grooves and ridges, which allow the stylets to interlock and form between them the food and the salivary canals (Figs. 16.2a–c and 16.3b–f). The grooves and ridges are absent from the base and usually disappear a short distance from the stylet tip.

The basic interlocking unit consists of a channel-like groove with a narrow entrance on one stylet, which embraces a bolster-like ridge on the opposite stylet. In literature this is usually referred to as tongue-and-groove or dovetail interlocking. This principle is implemented in "ziplock" fasteners of reclosable plastic bags (Fig. 16.3a). Similar grooves and ridges also interlock the valves of primary insect ovipositors. Because, while preventing separation of its components, this mechanism allows, at least in theory, their independent longitudinal sliding, Smith (1969) introduced for it the term olistheter (Greek, olistheros = slippery); he called the groove part the aulax and the tongue part the rhachis. These terms have so far been applied only to ovipositors, but they merit wider use. The term "olistheter" is applied here in a wide sense to include the typical structure described above and both its simplified and more complicated variants; the latter are formed, for example, when one or both walls of an aulax serve as rhachises.

The mandibular stylets usually tightly adhere to the maxillary stylets without any coupling structures (Fig. 16.3c–e). It is assumed that the interface between the adjacent mandibular and maxillary stylets is filled with some liquid; the mandibular stylets readily dissociate when the stylet bundle is placed in KOH solution or a detergent (Pollard 1970). However, in some heteropterans, including most Pentatomomorpha, Miridae, Tingidae, and Ochteridae (Rieger 1976; Cobben 1978), the mesal surface of each mandibular stylet forms in its dorsal part an aulax, which accepts the rhachis formed by the lateral surface of the adjacent maxillary stylet (Fig. 16.3b). An unusual type of coupling is observed in *Peloridium* (Coleorrhyncha: Peloridiidae): two opposed sharp ridges near the dorsal and ventral margins of the mandibular stylet fit into coadapted grooves on the maxillary stylet, which can be interpreted as a wide olistheter (Fig. 16.3f).

Because of the olistheters, pairs of hemipteran maxillary stylets are never exactly bilaterally symmetrical. While the mandibular stylets are usually bilaterally symmetrical, although not always (for examples of asymmetrical mandibular stylets in Reduviidae and Lygaeidae, see Cobben 1978 and Cohen 1990), the maxillary stylets vary from almost bilaterally symmetrical (except the interlocking sculpture) to strongly asymmetrical, and to altogether dissimilar (e.g., Brożek 2013). The asymmetry is directional: the left and right stylets have their specific morphologies.

Interlocking between the maxillary stylets is assumed to be inherited by Hemiptera from their common ancestors with Thysanoptera (Emeljanov 2002). The maxillary stylets of Thysanoptera are interlocked by what looks like primitive olistheters, which permit independent protraction of each stylet (Chisholm and Lewis 1984; Hunter and Ullman 1989) and form between them the food canal, also used for ejecting saliva (Heming 1993). Pairs of thysanopteran maxillary stylets are bilaterally asymmetrical, but unlike in Hemiptera, one stylet forms two rhachises and another two aulaces (Fig. 16.3g). An example of convergent evolution present styletiform movable cheliceral digits of plant-sucking mites, interlocked by asymmetrical primitive olistheters into a tube, used for puncturing the plant but not for sucking (Andre and Remacle 1984; Lillo et al. 2001; Beard et al. 2012).

### *16.2.2 Serrations, Scales, and Barbs*

Lateral surfaces of stylets bear preapical serrations for cutting the substrate and for anchoring. They may occur on the maxillary stylets but are more profound on the mandibular stylets. They vary from a few denticles to elaborate devices (Fig. 16.1b–d) and are amply described in literature (e.g., Pollard 1970, 1972; Cobben 1978; Cohen 1990; Leopold et al. 2003; Brożek 2013; Wang et al. 2015; Hao et al. 2016), so they will not receive more attention here.

The mesal surfaces of the maxillary stylets of Auchenorrhyncha, Cicadomorpha, and Heteroptera bear minute dents, scales, or notches arranged metamerically along rhachises and aulaces (Fig. 16.2a, b). These structures closely resemble the so-called olistheter scales of hymenopteran ovipositors, which do not have a known function (Smith 1969, introduced the term; Quicke et al. 1994). Such structures appear less well developed in Auchenorrhyncha Fulgoromorpha and may be absent in Sternorrhyncha and Coleorrhyncha. At least in some studied species (*Homalodisca vitripennis* (Germar), *Palomena prasina* (L.)), the dents on one stylet are aligned with the notches on the opposite stylet, but their placement is such that they can neither form an additional interlocking mechanism nor impede independent longitudinal displacement of the stylets. Toward the stylet apex, the olistheter scales become more closely spaced and in some species grow into conspicuous projections discussed below. In the heteropterans with interlocked mandibular stylets, their mesal surfaces also bear such scales (Fig. 16.2c). Being small and hidden from observation, these structures have rarely been discussed. However, Pollard (1972, 1973) noticed them in leafhoppers, spittlebugs, psyllids, and whiteflies and commented that “The function of these is problematical; they may assist the interlocking process, but are more likely used for filtering or cleaning.” Because the scales barely protrude into the food and the salivary canals, they cannot play a role in filtering and cleaning (both purely hypothetical). In both hymenopteran ovipositors and hemipteran stylets, these minute projections and notches may in some way facilitate the assembly of olistheters after molts (Sect. 16.3.4). In the case of hemipteran stylets, it seems also possible that the notches allow the new stylet to straighten from its curved state during ecdysis without rupturing.

In many predaceous heteropterans, preapical mesal surfaces of the maxillary stylets bear longitudinal series of longer, setiform projections (Fig. 16.2d; Cobben 1978), which apparently assist in lacerating soft tissues of the prey. In several families of Nepomorpha and Gerromorpha, the maxillary stylets bear on their mesal surfaces multiple longitudinal series of long setiform projections, variably called setae, spines, bristles, or barbs, which form filtering or rupturing apparatuses resembling a baleen (Parsons 1959; Cobben 1978; Swart and Felgenhauer 2003; Brożek 2013). While more basally the shafts of the maxillary stylets are interlocked by olistheters (Brożek and Herczek 2004; Brożek 2013), in their apical parts the barbs of the opposing stylets interpenetrate and form another interlocking mechanism. Strongly modified, flattened and curved barbs of some gerromorphans closely resemble the legulae of lepidopteran proboscises (see Chap. 3) and interlock in a similar manner (Fig. 16.2e, f). In some species the barbs appear to emerge from basal sockets (Cobben 1978), but their true structure remains to be studied.

### *16.2.3 Food and Salivary Canals*

Cross sections of interlocked stylet bundles vary between taxa and can inform phylogenetic studies (Brożek and Bourgoin 2013); their evolution has been discussed by Cobben (1978) and Emeljanov (1987, 2002). In order to make comparisons, it should be kept in mind that (1) the structure of the bundle can vary along its length (Pollard 1972), thus potentially impeding comparisons based on isolated sections, (2) it can also vary between stages of development (Hao et al. 2016), and (3) the transmission electron microscopic (TEM) sections (Parrish 1967; Forbes 1966, 1969, 1972, 1976, 1977; Forbes and Mullick 1970; Forbes and Raine 1973; Cobben 1978; Brożek 2006) are far superior in preserving details of cross sections compared to the scanning electron microscopic (SEM) sections (Cobben 1978; Brożek 2007, 2013; Brożek and Herczek 2001, 2004; Brożek et al. 2006; Brożek and Bourgoin 2013).

In Heteroptera and Sternorrhyncha, a typical pair of interlocked maxillary stylets forms three olistheters: dorsal, ventral, and middle. The middle olistheter separates the food canal in the dorsal part of the bundle from the salivary canal in its ventral part (Fig. 16.3b). This condition has been interpreted as plesiomorphic (Brożek 2007). In contrast, in Auchenorrhyncha and Coleorrhyncha, there are only two olistheters, and the salivary canal is essentially a diverticulum of the ventral olistheter (Fig. 16.3c–f; Brożek and Herczek 2001; Brożek and Bourgoin 2013).

Basally the integrity of the two liquid flows is maintained by the hypopharynx tightly wedging between diverging bases of the maxillary stylets, so that the salivary duct at the tip of the hypopharynx inserts directly into the salivary canal between the stylets, while the cibarial space between the hypopharynx and the epipharynx continues as the food canal. The stylet bases, hypopharynx, epipharynx, and maxillary plates form together an intricately coadapted liquid-tight mechanism (e.g., Pesson 1944, 1951), which allows the stylets to slide during protraction and retraction past the immobile hypopharynx without compromising liquid tightness of the canals (Wenk et al. 2010). A salivary canal is reportedly absent in Hydrometridae, in which case the single wide canal is assumed to serve alternately for food ingestion and saliva ejection (Cobben 1978).

### 16.2.4 Braces, Levers, and Muscles

The bases of stylets lie within the head in intergnathal spaces referred to as stylet sacs (Parsons 1959). The sac of the mandibular stylet is formed by the inner walls of the mandibular and maxillary plates, and that of the maxillary stylet is formed by the inner wall of the maxillary plate and the outer wall of the hypopharyngeal wing; anteriorly the two sacs at each side of the head fuse into a single channel, which contains both stylets. The basalmost part of the sac snugly embraces the stylet base and serves for attachment of stylet muscles; it has been referred to by Pesson (1944, 1951) as the "manchon," translated here as the "brace." The braces can be variably sclerotized and equipped with apodemes (Fig. 16.4a, b). They often appear as integral parts of stylet bases (Fig. 16.1a). Their true nature becomes clear during molts: like other cuticular structures of the head, the braces molt in situ and retain their connection to muscles, while the stylet base develops as part of the progressively invaginating retortiform organ (Sect. 16.3.3) and is disconnected from muscles until inserted into the brace during ecdysis (Fig. 16.4c). The structures traditionally described as outgrowths of the stylet base, such as the apodemes connecting the base of the mandibular stylet of Auchenorrhyncha to its retractors and the lever (e.g., Snodgrass 1927), are instead parts of the brace.

The bases of stylets are connected to the inner side of the cranium with rod-shaped or plate-shaped stylet levers (Figs. 16.1a and 16.4a–c). In fact, the levers are reinforced parts of the mostly membranous folds between the mandibular and the maxillary plates (mandibular lever) and between the maxillary plate and the hypopharyngeal wing (maxillary lever) (Benwitz 1956; Parsons 1959, 1964; Rieger 1976). The cuticle of the lever is continuous with that of the brace. Usually the levers also serve as surfaces for attachment of stylet muscles, but the maxillary levers of Heteroptera do not, and in predaceous aquatic families and Reduvioidea they are lost to maximize the range of stylet motion (Quadri 1959; Cobben 1978; Rieger 1976).

The muscles allow stylets to move in the axial direction, i.e., to protract and retract; in some species all four stylets are capable of independent movement, while in others the interlocked maxillary stylets work as a unit (Pollard 1969, 1973). No

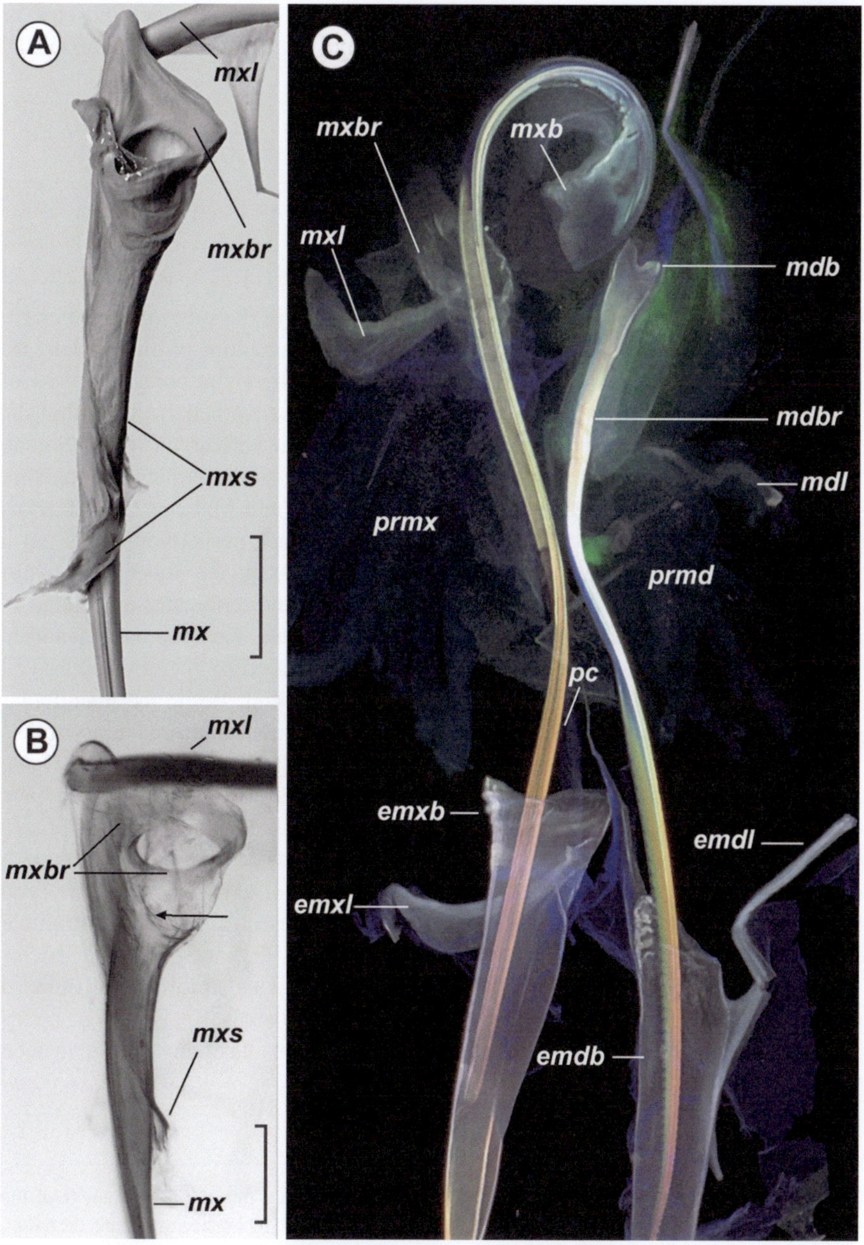


**Fig. 16.4** Connection of stylet bases to braces and levers. (**a**) Base of a maxillary stylet of an adult *Palomena prasina* (L.) (Pentatomidae) with the stylet brace, sac, and lever (muscles and other soft tissues removed), SEM image; (**b**) same, a light microscope image; the basal margin of the stylet (arrow) is visible inside a funnel-shaped brace; compare to Fig. 16.5g showing new stylet bases of this species prior to molting; (**c**) confocal laser scanning microscopy image of a mandibular and a maxillary stylets dissected from a molting adult of *Graphocephala fennahi* Young (Cicadellidae);

muscles are attached directly to the stylets. Details on the stylet muscles of Auchenorrhyncha, Sternorrhyncha, and Coleorrhyncha have been summarized by Singh (1971) and of Heteroptera by Rieger (1976).

### *16.2.5 Sensory Equipment*

The maxillary and the mandibular stylets are innervated each by a branch of, correspondingly, the maxillary or the mandibular nerve of the subesophageal origin (Benwitz 1956; Pinet 1963, 1970; for alternative interpretations of innervation, see Pesson 1944 and Ponsen 1972). The nerve entering the stylet contains axons of bipolar neurons, whose bodies (perikarya) lie among the epithelial cells within the stylet base. One stylet sensillum comprises either a single perikaryon wrapped in an enveloping cell or a cluster of two perikarya wrapped in an enveloping cell together (Pinet 1963; Wensler 1974; Foster et al. 1983; Backus and McLean 1982; Backus 1985). One stylet usually contains several sensilla. Each perikaryon sends out a single dendrite, whose outer segment either terminates in the stylet base (short dendrites) or runs down the axial cavity of the stylet and terminates embedded in the cuticular wall of the stylet near its apex (long dendrites). The long outer dendritic segments are surrounded with a sheath of cuticle produced by the enveloping cell and often are accompanied by thin projections of that cell. While the mandibular stylets appear to always have sensory dendrites in axial cavities, the maxillary stylets of most Sternorrhyncha (except some Coccoidea, see figures in Brożek 2006) have neither dendrites nor axial cavities; the maxillary stylets of some Heteroptera have axial cavities but no dendrites in them (Fig. 16.3b; Forbes 1976; Cobben 1978; also Benwitz 1956 found no nerves entering the maxillary stylets of *Corixa*). Although numerous cross sections of stylets showing dendrites in axial cavities have been published, detailed reconstructions of stylet sensilla, based on sections at multiple levels, are limited to one reduviid species (Pinet 1963, 1970) and one aphid species (Wensler 1974). The sensilla perceive stylet displacement and deformation (Bernard et al. 1970; Bernard and Pinet 1973; Wensler 1974; Backus 1985). No gustatory sensilla are associated with hemipteran stylets, and no sensilla of any kind have been found on their surface.

---

**Fig. 16.4** (continued) note that the apices of the new stylets enter into the basal funnels of the exuvial stylets and their bases are not yet completely inserted into stylet braces, which are, therefore, clearly seen (especially the brace of the maxillary stylet); in contrast, the exuvial braces are indistinguishable from the bases of exuvial stylets. Abbreviations: *emdb* exuvial mandibular stylet base, *emdl* exuvial mandibular lever, *emxb* exuvial maxillary stylet base, *emxl* exuvial maxillary lever, *mdb* mandibular stylet base, *mdl* mandibular lever, *mdbr* mandibular stylet brace, *mx* maxillary stylet, *mxb* maxillary stylet base, *mxbr* maxillary stylet brace, *mxl* maxillary lever, *mxs* maxillary sac, *pc* peripodial cuticle, *prmx* protractor muscle of the maxillary stylet, *prmd* protractor muscle of the mandibular stylet. Scale bars: (**a**, **b**) 200 μm

## 16.3 Development, Molting, and Assembly

Early development of mouthparts during embryogenesis is beyond the scope of this chapter. It has been studied in selected species of Heteroptera (Heymons 1899; Muir and Kershaw 1911; Newcomer 1948), Auchenorrhyncha (Heymons 1899; Muir and Kershaw 1912; Singh 1971), and Sternorrhyncha (Mecznikow 1866; Witlaczil 1882; Pesson 1944). Each maxillary stylet develops from the internal lobe of a two-lobed primordial maxilla, the rest of which develops into the maxillary plate (a flattened lobe immovably connected to the cranium and covering the bases of stylets laterally and posteriorly). By the time the second embryonic (pronymphal) cuticle is deposited, both pairs of stylets appear as narrowly conical short processes emerging between the clypeolabrum and the maxillary plates (Fig. 16.5c).

Later on, the epithelium at each stylet base apolyses from the cuticle and forms a circular invagination, which grows into the head as a two-layered tube, inside which the first-instar nymphal stylet begins to form. Mecznikow (1866) described the shape of these structures in aphids and scale insects as retortiform (Fig. 16.5a), and subsequent authors referred to them as "retortiform organs." Although these are not organs in the strict sense (Cicero 2017) and seldom resemble a retort (Fig. 16.5b), for lack of a better term, the invaginated structure inside of which the stylet is formed will be referred here to as the retortiform organ. Its inner layer, referred here to as the styligenic epithelium, forms the styligenic cone eventually developing into the stylet (Fig. 16.6b). The outer layer, which forms a sac around the styligenic cone, is referred to as the peripodial epithelium, the term commonly applied to invaginated epithelia enclosing imaginal discs of holometabolans. The two epithelia are continuous with one another across the fold at the free end of the retortiform organ; the peripodial epithelium is also continuous with the epidermis of the stylet sac. The lumen of the styligenic cone is an extension of the body cavity, while the cavity between the cone and the peripodial epithelium is continuous with the molting space, which forms by delamination of epidermis from the old cuticle. A fully formed stylet is pulled out into the working position during hatching. Prior to each subsequent molt, the epithelium inside the stylet base expands into a new retortiform organ (Figs. 16.5d–g, 16.6a–c, and 16.7). Its development is similar between the embryonic and the nymphal stages and will be described below in detail based on studies of nymphs.

### 16.3.1 *History of Study*

Retortiform organs with developing stylets have been briefly described in early papers on the embryology and immature morphology of Hemiptera (Mecznikow 1866; Witlaczil 1882; Heymons 1899; Bugnion and Popoff 1911; Muir and Kershaw 1911, 1912). Snodgrass (1927) concisely described the retortiform organs and the process of stylet extraction in molting cicada nymphs and commented that "the

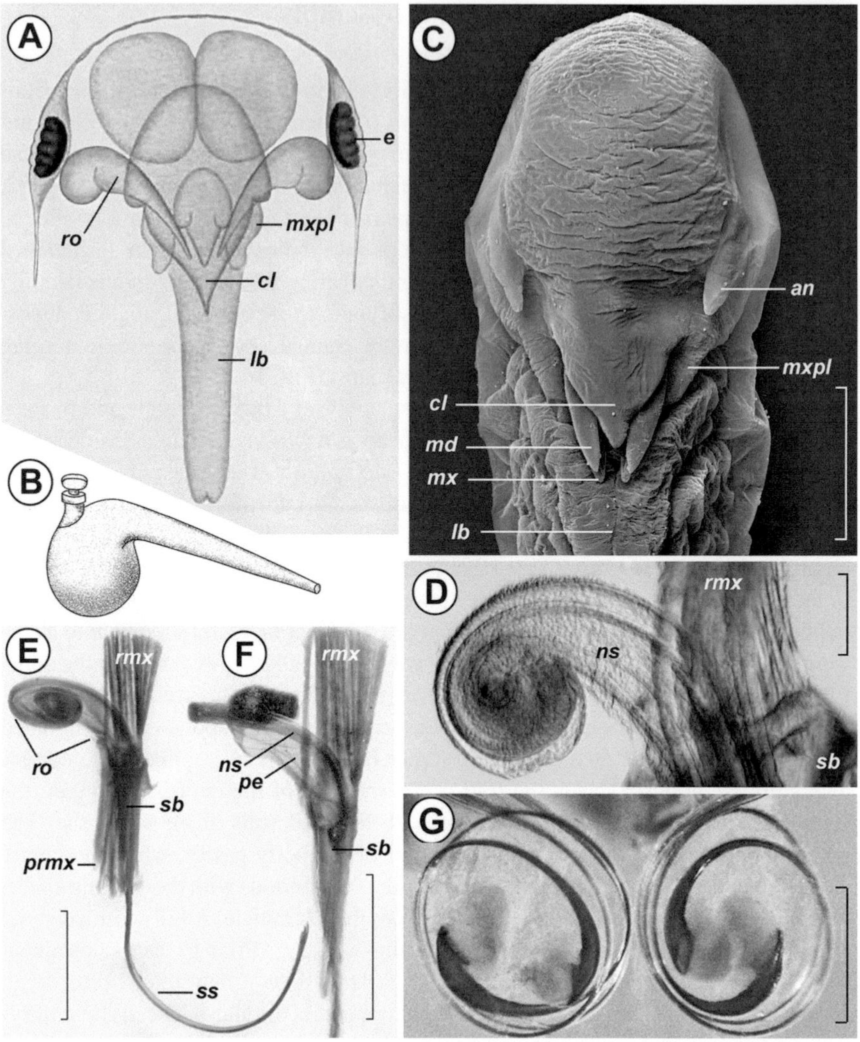


**Fig. 16.5** Retortiform organs. (**a**) Mecznikow's (1866) drawing of a pronymph of *Aphis rosae* Koch (Aphididae) showing a pair of retortiform organs (the second pair is omitted), with new labeling added; (**b**) glass retort; (**c**) head of a pronymph (late embryo coated with the second embryonic cuticle) of *Aphrophora pectoralis* Mats. (Aphrophoridae), note the exposed stylets; (**d**) *Poophilus nebulosus* (Leth.) (Aphrophoridae), fifth-instar nymph, maxillary stylet base with a retortiform organ, early stage of development; (**e**, **f**) *A. pectoralis*, same, late stage of development, different views; (**g**) *Palomena prasina* (L.) (Pentatomidae), ready-to-molt fifth-instar nymph, four coils of fully developed adult stylets (at each side the maxillary and mandibular stylet coils are close-set), note ball-shaped masses of styligenic epithelial cells protruding from stylet bases. Abbreviations: *an* antenna, *e* eye, *cl* clypeolabrum, *lb* labium, *md* mandibular stylet, *mx* maxillary stylet, *mxpl* maxillary plate, *ns* nascent stylet, *pe* peripodial epithelium, *prmx* protractor of the maxillary stylet, *rmx* retractor of the maxillary stylet, *ro* retortiform organ, *sb* stylet base, *ss* stylet shaft. Scale bars: (**c**, **d**) 200 μm; (**e**, **f**, **g**) 500 μm

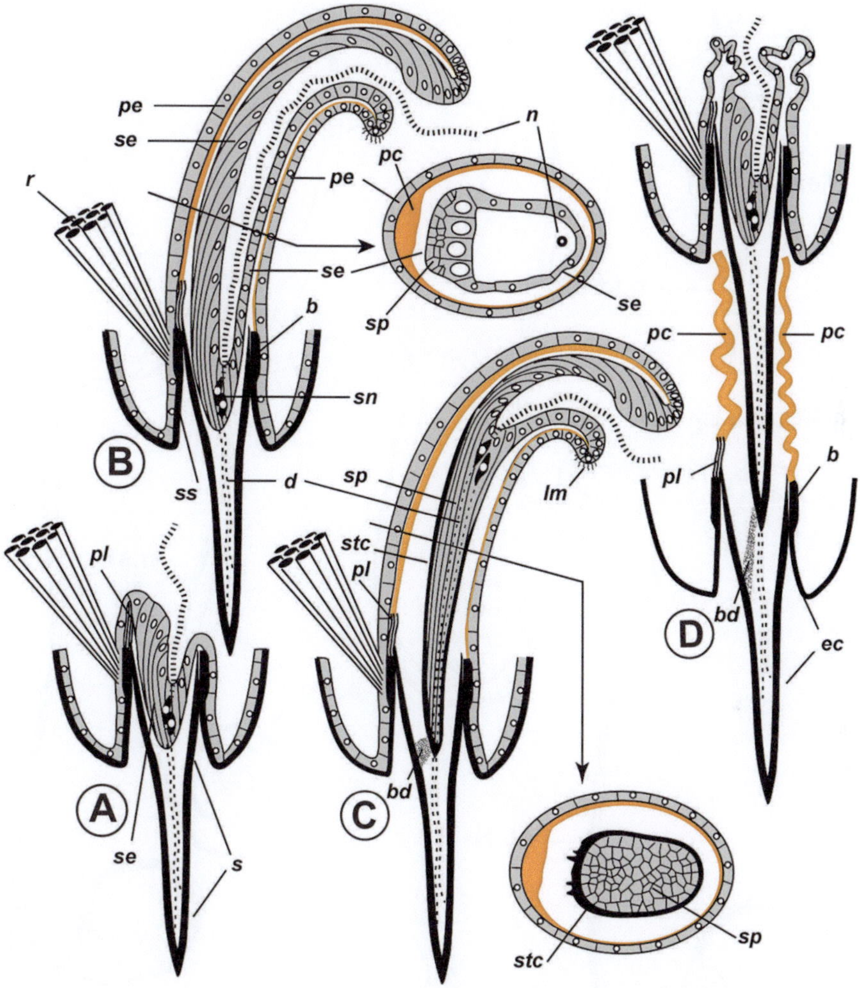


**Fig. 16.6** Styletogenesis; different parts are not drawn to scale, the lever and protractor muscles not shown. (**a**) Intermolt condition; (**b**) stage 1 of styletogenesis, with a cross section; (**c**) stage 2 of styletogenesis, with a cross section; (**d**) ecdysis. See text for details. Abbreviations: *b* brace, *bd* bond between the old and the new stylets, *d* sensory dendrites, *ec* ecdysial cuticle, *lm* branched outgrowths of the basal lamina of the peripodial epithelium, *n* stylet nerve, *pe* peripodial epithelium, *pc* peripodial cuticle, *pl* pleated cuticle, *r* retractor muscle, *s* stylet, *se* styligenic epithelium, *sn* stylet sensilla, *sp* styligenic cellular projections, *ss* stylet sac, *stc* stylet cuticle

details apparently have never been closely studied." Soon afterward, Weber (1929, 1930) proposed a model of stylet development and molting, which was only partially correct because it assumed that stylets were produced by secretion of cuticle into the central lumen of the retortiform organ. Based on serial sections of the coccid *Icerya purchasi* Mask, Pesson (1944) established that the stylet is produced by deposition of cuticle on top of a conical epithelial process and proposed a detailed model of

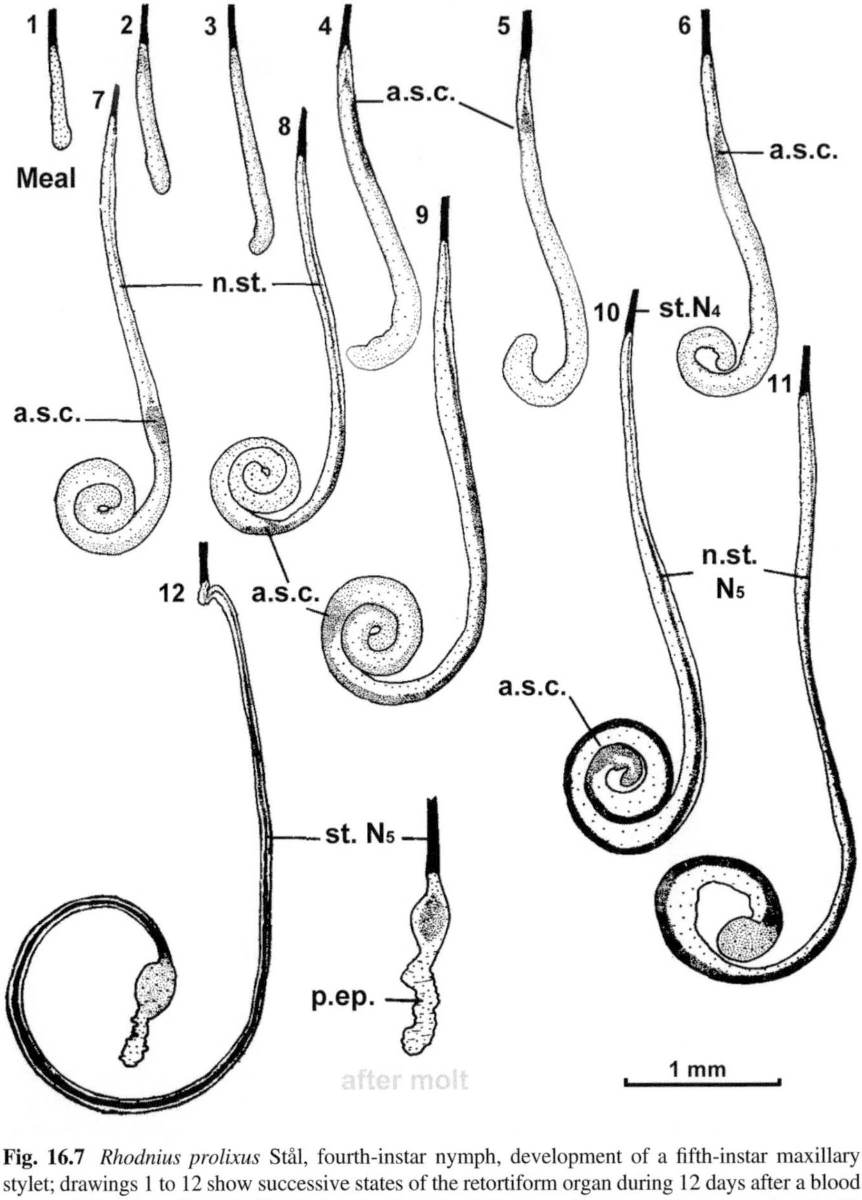


**Fig. 16.7** *Rhodnius prolixus* Stål, fourth-instar nymph, development of a fifth-instar maxillary stylet; drawings 1 to 12 show successive states of the retortiform organ during 12 days after a blood meal at 29 °C (from Pinet 1968a, © Annales de la Société entomologique de France, reprinted with permission from Taylor & Francis Ltd., http://www.tandfonline.com on behalf of Annales de la Société entomologique de France). Abbreviations: *a.s.c.* apex of the styligenic cone (Pinet's "bourgeon de l'organe rétortiforme," this boundary marks the apical extent of the cell bodies of the styligenic epithelium excluding filamentous styligenic projections), *n.st.* (same as n.st.N5 and st.N5), developing new stylet, *p.ep.* peripodial epithelium (Pinet's "gaine atrial"). See text and Fig. 16.6 for details

stylet development, which remains accurate today. Newcomer (1948) described an essentially the same process during the embryonic development of the true bug *Oncopeltus fasciatus* (Dall.), and Benwitz (1956) found that the retortiform organs of *Corixa punctata* (Illig.) had a similar structure. Later, Pesson's student Pinet studied styletogenesis in nymphs of the triatomine bug *Rhodnius prolixus* Stål. The large size of that species allowed for the first time to study the structure of stylet sensilla (Pinet 1963) and their fate during styletogenesis, which he thoroughly described (Pinet 1968a). Additionally, Pinet studied developmental mechanics of retortiform organs implanted into the bug's abdomen, either intact (and then producing a normal stylet) or upon various surgical treatments; he found the disparate developmental potentials of the styligenic and the peripodial epithelia to be predetermined (Pinet 1968b). All of these results have been summarized in the thesis (Pinet 1970), which contains the most comprehensive analysis of hemipteran styletogenesis produced so far. This superb study remains little known, which is especially unfortunate because it contains results of Pinet's ultrastructural studies of every aspect of stylet development and innervation—not incorporated in journal publications. In particular, the use of transmission electron microscopy allowed Pinet to demonstrate that the maxillary stylet is formed by filamentous projections of styligenic cells, and these projections form intricate profiles on the prospective interlocking mesal surfaces of the stylets prior to secretion of the cuticle. Observations on developing stylets of another true bug, *Dysdercus andreae* (L.), have been included in an unpublished student thesis produced under Pesson and Pinet's supervision (Atachi 1976). Recently, late stages of styletogenesis in last-instar nymphs of two psyllid species have been described in detail using transmission electron microscopy (Ammar et al. 2015; Cicero 2017; Cicero et al. 2018).

### *16.3.2 New Observations*

The coccid studied by Pesson (1944) and the triatomine studied by Pinet (1963, 1968a, 1968b, 1970) are in most respects as dissimilar as two hemipterans can be, yet their stylets develop similarly. It is, therefore, not surprising that my observations on developing stylets of *Aphrophora* sp. (Auchenorrhyncha: Aphrophoridae), *Stomaphis quercus* (L.) (Sternorrhyncha: Aphididae), and *Pyrrhocoris apterus* (L.) (Heteroptera: Pyrrhocoridae) agree well with observations of these authors. Specimens for the study were preserved in phosphate-buffered 2.5% glutaraldehyde (pH 7.4), stained with osmium tetroxide, embedded in Epon, and sectioned according to conventional protocols; ultrathin sections were examined on a JEOL-1011 transmission electron microscope.

### 16.3.3 Styletogenesis

Between consecutive ecdyses different areas of insect epidermis apolyse from the cuticle and begin mitotic growth at different times (Wigglesworth 1973). The growth of retortiform organs begins very early. In the fourth-instar *Rhodnius prolixus* nymphs, it is already noticeable the next day after a blood meal (Pinet 1968a), which is earlier than in most other body parts (Wigglesworth 1973). Pinet (1968a, 1970) described daily changes through 12 days of this process (Fig. 16.7); a similar development was observed in *Dysdercus* nymphs (Atachi 1976). Styletogenesis is diagrammatically represented in Fig. 16.6a–d. For convenience, two stages can be recognized as described below. Because the zone of cell division is situated at the fold at the basal (free) end of the retortiform organ, one can often observe stage 1 in its basal and stage 2 in its apical parts. Ultrastructural details of stylet development were similar between *Pyrrhocoris apterus* (Figs. 16.8, 16.9, 16.10 and 16.11), *Aphrophora* sp. (Fig. 16.12), and *Stomaphis quercus* (Figs. 16.13 and 16.14).

During stage 1 epidermis around the stylet base invaginates as a fold, circular at first and then tubular; the latter structure is the two-layered retortiform organ already described (Fig. 16.6b). Until the end of this stage, the styligenic epithelial cells with embedded among them perikarya of sensory neurons, which prior to the onset of styletogenesis resided in the stylet base, remain there. This epithelium, now forming the apex of the growing styligenic cone, separates from the stylet's cuticle, except at one point where the connection is preserved (see below). The stylet nerve passes through the lumen of the styligenic cone (Figs. 16.8a, c and 16.12a).

In the area of the basal fold at the free end of the retortiform organ, the cells of both styligenic and peripodial epithelia undergo mitotic divisions. As the retortiform organ becomes longer, its free end bends and usually winds into a coil. Depending on the stylet length, the coil of a fully grown retortiform organ can lie in the head or in the prothorax (typical of heteropterans), or even further back. The coils of long-styleted species contain numerous turns. For example, in the examined nymphs of *Stomaphis quercus*, an aphid with stylets well exceeding the body length (Pesson 1951; Brożek et al. 2015), each maxillary and mandibular retortiform organ formed five turns (partially visible in Fig. 16.13a). The cross section of a coiled retortiform organ can be circular (Fig. 16.8a) or narrowly triangular, with the inner angle stretched out toward the coil's axis (Fig. 16.13a). Shortly before ecdysis the retortiform organs of *Rhodnius* and *Dysdercus* partially uncoil (Pinet 1968a, 1970; Atachi 1976).

The convex side of the turns of a coiled styligenic cone will form the mesal surface of the stylet. The styligenic epithelium here is thickened, and its outer contour is flat rather than convex (Figs. 16.6b, 16.9a, b, 16.10a, 16.11, 16.12a, 16.13a, and 16.14a). The cells comprising this area send out long filamentous projections—referred here to as the styligenic cellular projections—along the outer surface of the styligenic cone toward its apex (e.g., Fig. 16.8b, e). Several tiers of such projections form together what is visible on light microscopic sections as a nucleus-free hyaline area (Newcomer 1948), which eventually forms the bulk of the

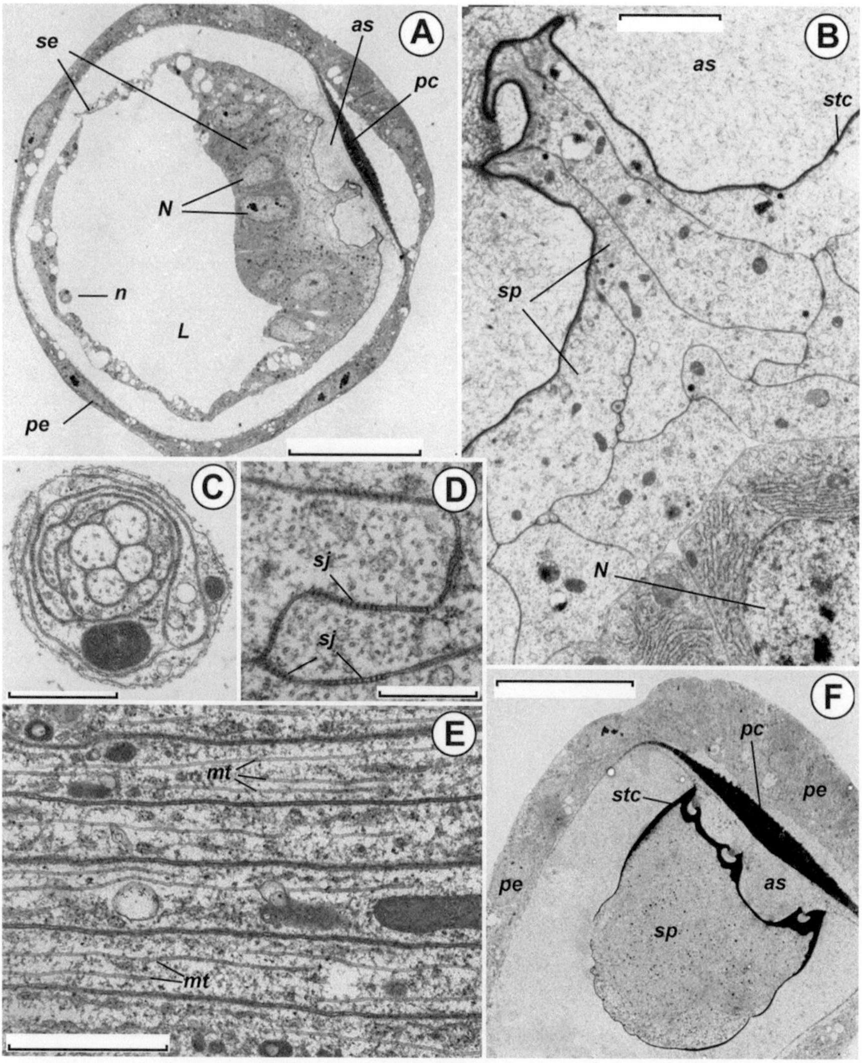


**Fig. 16.8** *Pyrrhocoris apterus* (L.), fourth-instar nymph, development of fifth-instar maxillary stylets. (**a**) Cross section of a retortiform organ of a right maxillary stylet during stage 1 of styletogenesis; note that the styligenic epithelium is strongly thickened at the prospective mesal side of the stylet and forms here a nucleus-free zone; (**b**) same, detail of the nucleus-free zone consisting of cellular projections forming the rhachis separating the food and the salivary canals (compare to Fig. 16.3b); (**c**) same, maxillary stylet nerve in the lumen of the styligenic cone; (**d**) detail of the nucleus-free zone, showing numerous septate junctions between styligenic cellular projections; (**e**) tangential section showing parallel styligenic projections, note numerous microtubules; (**f**) cross section of a retortiform organ of a left maxillary stylet during stage 2 of styletogenesis; the entire volume of the developing stylet is filled with styligenic cellular projections (the axial cavity not yet cleared); note that the cuticle is much thinner on the lateral side of the stylet compared to its mesal side. Abbreviations: *as* amorphous substance, *L* lumen of the styligenic cone, *mt* microtubules, *n* maxillary stylet nerve, *N* nuclei of styligenic epithelial cells, *pc* peripodial cuticle, *pe* peripodial epithelium, *se* styligenic epithelium, *sj* septate cell junctions, *sp* styligenic cellular projections, *stc* stylet cuticle. Scale bars: (**a, f**) 20 μm; (**b**) 2 μm; (**c, e**) 1 μm; (**d**) 0.5 μm

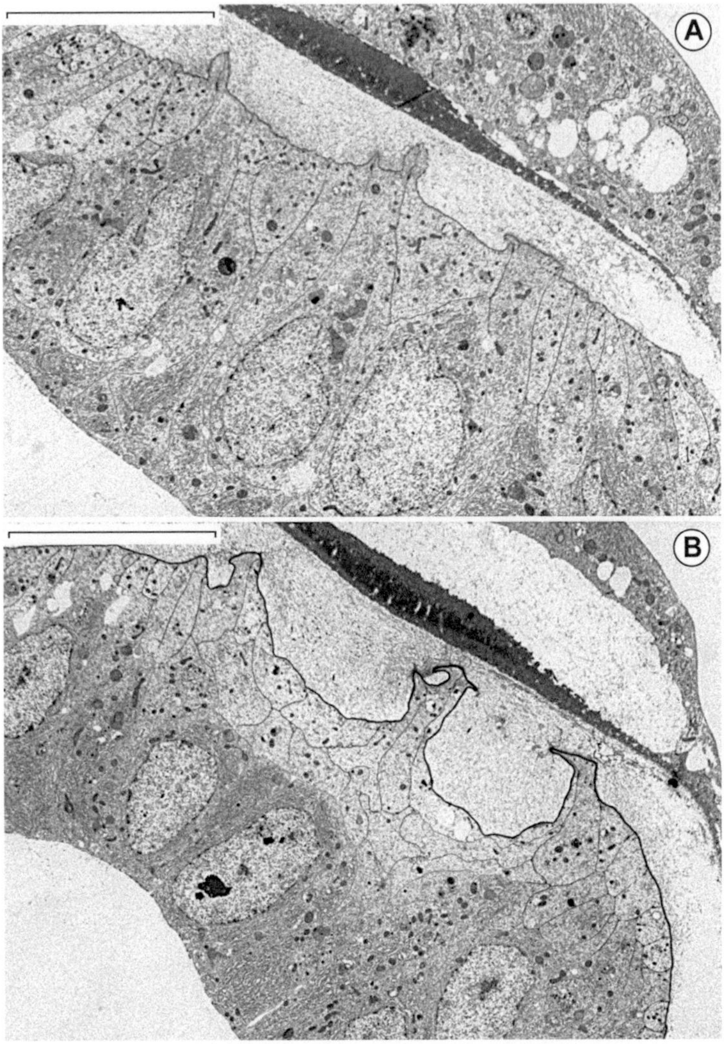


**Fig. 16.9** *Pyrrhocoris apterus* (L.), fourth-instar nymph, development of interlocking sculpture on the prospective mesal surface of a fifth-instar right maxillary stylet during stage 1 of styletogenesis: cross sections of the same retortiform organ. The section (**a**) is located more proximally and shows an earlier stage of development than the section (**b**). See Fig. 16.11a for tracing of the cells and cellular projections. Scale bars: (**a**, **b**) 10 μm

stylet. The projections contain numerous microtubules (Figs. 16.8d, e and 16.13e) and are held together by adherence junctions near the exposed surface (Fig. 16.14a) and by septate junctions deeper below (Fig. 16.8d); this arrangement of junctions is typical of insect epidermis (e.g., Chapman 2013: Fig. 16.1). Some pairs of adjacent projections form ridges along the line of contact (Fig. 16.9a). These parallel ridges and grooves between them grow and acquire more complex shapes until the

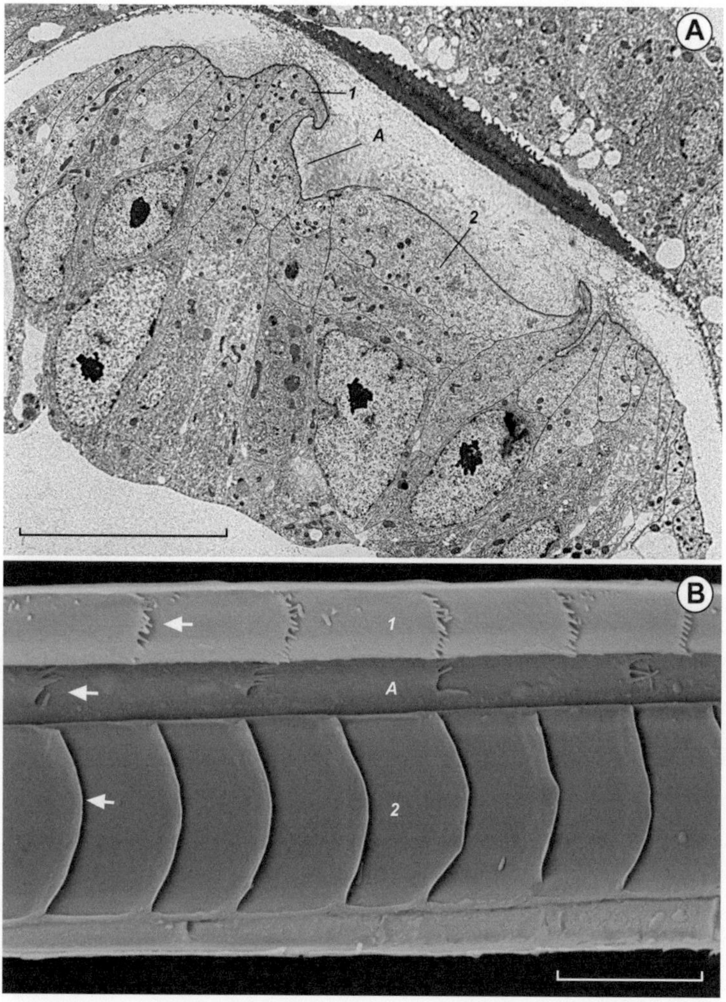


**Fig. 16.10** *Pyrrhocoris apterus* (L.), development of microsculpture on the mesal surface of a fifth-instar nymphal mandibular stylet. (**a**) Cross section through a retortiform organ of a fourth-instar nymph during stage 1 of styletogenesis showing how the mesal surface is patterned by styligenic projections; see Fig. 16.11 for tracing of the cells and cellular projections; (**b**) microsculpture of the same surface in a fifth-instar nymph. Note that the aulax ("A") and the raised parts at its both sides (labeled as "1" and "2") are each formed by a single row of styligenic projections. The metamerically arranged "olistheter scales" (arrows) obviously demarcate the termini of successive overlapping styligenic projections within each row. Scale bars: (**a**, **b**) 10 μm

characteristic mesal profile of each stylet develops before the onset of cuticle secretion (Figs. 16.8a, b, 16.9b, 16.10a, 16.11a, b, 16.13a, and 16.14a, b). The continuity of ridges and grooves of developing olistheters results from partially overlapping projections lying precisely on top of one another, with the projections of more basal origin reaching the surface sooner. Regularly spaced "olistheter

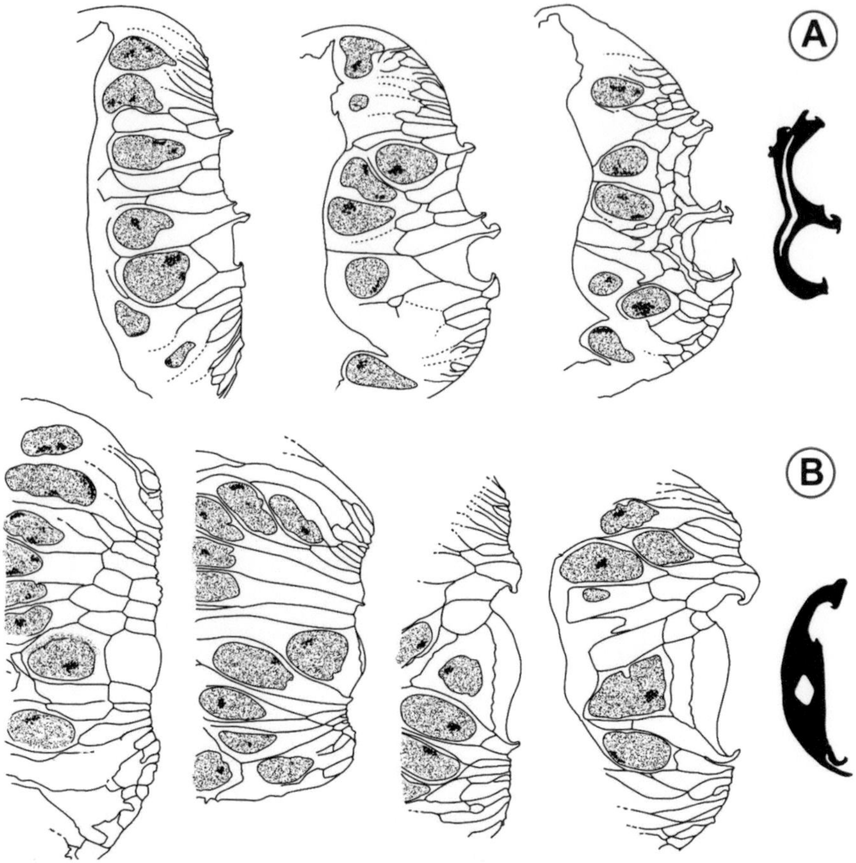


**Fig. 16.11** *Pyrrhocoris apterus* (L.), development of interlocking profiles of fifth-instar nymphal stylets; cross sections of the styligenic epithelium at the prospective mesal side of coiled retortiform organs of the same fourth-instar nymph taken at different levels, from more basal to more distal, which corresponds to successive stages of development; compare to profiles of the mature stylets shown on the right. Cell boundaries in the areas farther from the apical surface are poorly visible, traced incompletely (dashed lines). (**a**) Right maxillary stylet; (**b**) right mandibular stylet

scales" (Figs. 16.2a, b and 16.10b) obviously correspond to the points where the upper projection terminates and the next one comes to the surface, which explains their metameric linear arrangement (compare Fig. 16.10a and Fig. 16.10b). The long barbs on the mesal stylet surfaces of Reduvioidea, Nepomorpha, and Gerromorpha (Fig. 16.2d–f) have not yet been closely studied but are probably formed by free terminal parts of such projections. Olistheter scales of hymenopteran ovipositors (Smith 1969) probably have a similar origin. The formation of the rhachis on the lateral surface of the maxillary stylet of *Pyrrhocoris* (Figs. 16.2c and 16.3b) was not observed, and its origin is unclear. It also remains unknown how the apical serrations of the stylets develop.

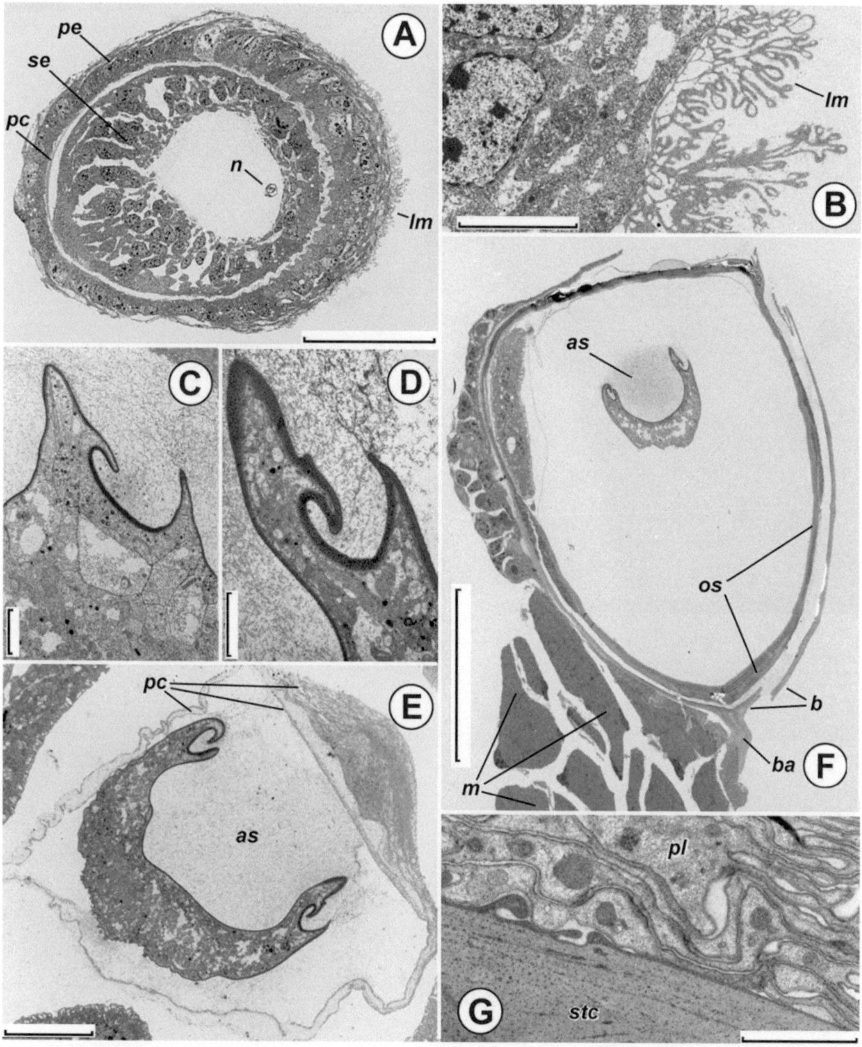


**Fig. 16.12** *Aphrophora* sp., fourth- and fifth-instar nymphs, development of next-instar stylets. (**a**) Cross section through a retortiform organ of a mandibular stylet near its basal end showing an early stage 1 of styletogenesis; note a thin layer of the peripodial cuticle and a flattened styligenic epithelium on the prospective mesal (here left) side of the developing stylet; (**b**) same, detail of the peripodial epithelium on the opposite side with its basal lamina forming branching outgrowths; (**c**, **d**) stages of development of the olistheter aulax enclosing the salivary canal of the left maxillary stylet; (**e**) left maxillary stylet within a retortiform organ at stage 2 of styletogenesis, compare with the profile in Fig. 16.3d, also note the complex structure of the peripodial cuticle; (**f**) same, at a more distal level within the base of the old maxillary stylet, note the stylet brace and the absence of the peripodial cuticle; (**g**) pleated cuticle of the brace in contact with the stylet cuticle. Abbreviations: *as* amorphous substance, *b* brace (broken at one point), *ba* apodeme of the brace, *lm* outgrowths of the basal lamina of the peripodial epithelium, *m* maxillary protractor muscle, *n* stylet nerve, *os* old stylet, *pc* peripodial cuticle, *pe* peripodial epithelium, *pl* pleated cuticle of the brace, *se* styligenic epithelium, *stc* stylet cuticle. Scale bars: (**a**, **f**) 50 μm; (**b**) 5 μm; (**c**, **d**, **g**) 2 μm; (**e**) 10 μm

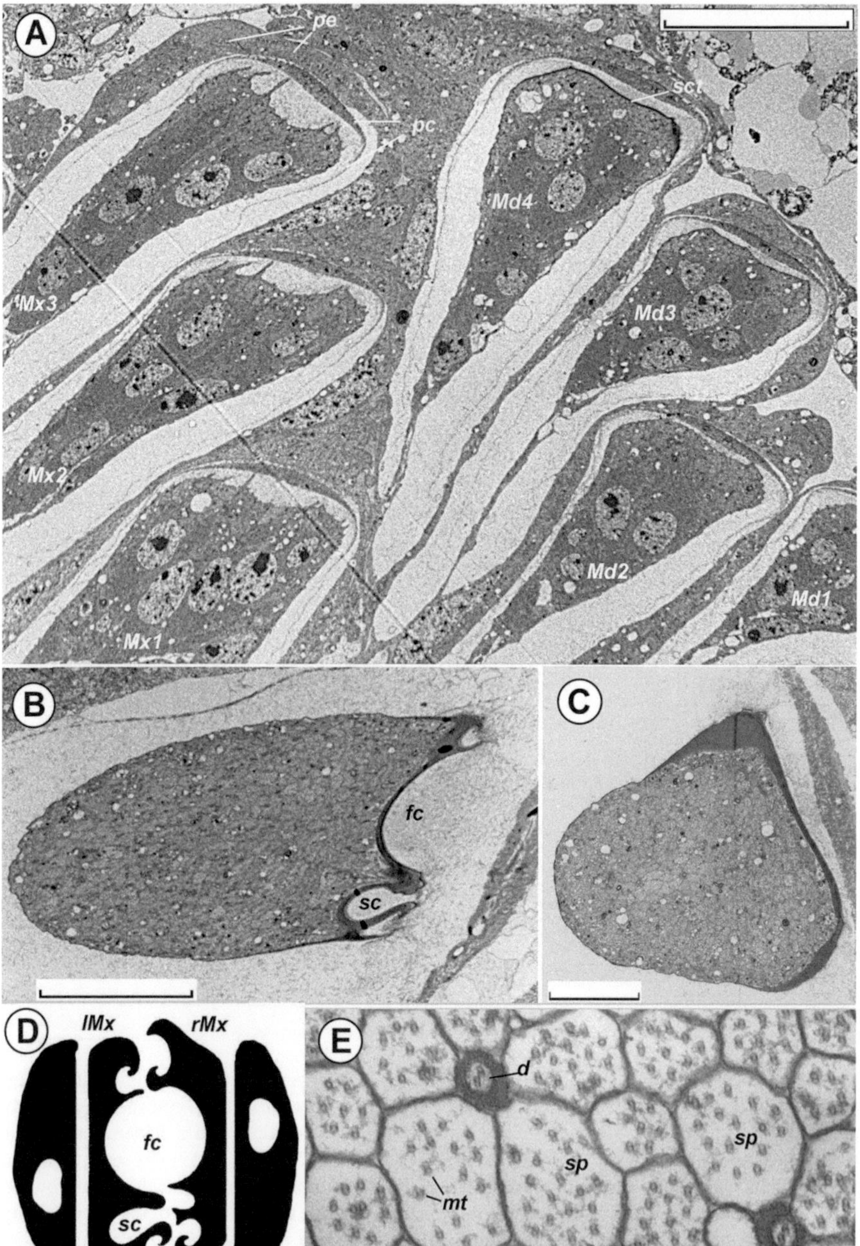


**Fig. 16.13** *Stomaphis quercus* (L.), first-instar nymph, development of second-instar stylets. (**a**) Cross section through three turns of a retortiform organ with a developing maxillary stylet and four turns of an adjacent retortiform organ with a developing mandibular stylet; note goblet-shaped cross sections of the thickened mesal part of the styligenic cone; the cone's lumen is obliterated; (**b**) left

The peripodial epithelium consists of flattened cells, which have similar structure around the circumference of the retortiform organ (Fig. 16.8a). At the basal end of the latter, the basal membrane underlying the peripodial epithelium forms a brush of branching outgrowths (Fig. 16.12a, b; also observed in *Rhodnius*, see Pinet 1970).

Already at this stage, a thin layer of stylet cuticle begins to be deposited over the exposed surfaces of styligenic projections (Figs. 16.8b and 16.14c). But even before the stylet cuticle becomes noticeable, a conspicuous layer of extracellular amorphous-looking material forms between the styligenic and the peripodial epithelia. It appears to be produced by the peripodial cells. Unlike the stylet cuticle, this peripodial cuticle remains mostly or entirely unsclerotized. It is not digested during ecdysis and is discarded together with the exuviae in the form of peripodial sheaths attached to the old stylet bases (Sect. 16.3.4). Such discarded dry sheaths are not soluble in aqueous KOH solution, suggesting the presence of chitin. Above the flattened presumptive mesal surface of the developing stylet, the peripodial cuticle is particularly thick and is also flattened, appearing as a lamina (in some cases strongly osmiophilic) detached from both the peripodial and the styligenic epithelia; elsewhere it is thin and usually attached to the peripodial epithelium (Figs. 16.8a, f, 16.9a, b, 16.10a, 16.12a, e, 16.13a, and 16.14a–d).

During stage 2 the styligenic epithelium undergoes restructuring (Fig. 16.6c). Starting from the apex of the cone, the cell bodies retreat to its base, leaving behind a conical bunch of parallel nucleus-free styligenic projections enveloped in a layer of cuticle they continue to deposit—a nascent stylet. The perikarya of sensory neurons migrate together with the surrounding epidermal cell bodies. This mass of cell bodies is visible under a stereomicroscope as a bulge gradually relocating toward the base of the retortiform organ. In *Rhodnius*, its migration takes at least 4 days (Fig. 16.7, stages 6 through 10; Pinet 1968a). It appears that the cell bodies slide up past the layer of styligenic projections, which have become firmly interconnected via cell junctions and have begun depositing cuticle during the previous stage. The driving force behind this movement is enigmatic. The nascent stylet is completely filled with styligenic projections and has no cavities (Figs. 16.8f, 16.12e, 16.13b, c, and 16.14d). The sensory dendrites run between styligenic projections (Fig. 16.13e). As the stylet matures, the projections disappear and the axial cavity forms. At first, the cuticle deposited at the lateral side of the stylet is much thinner in comparison to

**Fig. 16.13** (continued) maxillary stylet during stage 2 of development; (**c**) mandibular stylet at the same stage; (**d**) reconstruction of the stylet bundle cross section of *S. quercus* based on Fig. 16.13b, c and bundle cross sections of *Rhopalosiphum* (Parrish 1967) and *Acyrthosiphon* (Uzest et al. 2010); (**e**) detail of (**c**), showing sensory dendrites between styligenic projections. Abbreviations: *d* outer dendritic segment of a stylet sensillum, *fc* food canal, *lMx, rMx, lMd, rMd* left and right maxillary and mandibular stylets, *mt* microtubules, *Mx1–3, Md1–4* successive turns of the maxillary and the mandibular coils, respectively, numbered from basal to apical, *pc* peripodial cuticle, *pe* peripodial epithelium, *sc* salivary canal, *sct* stylet cuticle, *sp* styligenic cellular projections. Scale bars: (**a**) 20 μm; (**b**) 10 μm; (**c**) 5 μm; (**e**) 0.5 μm

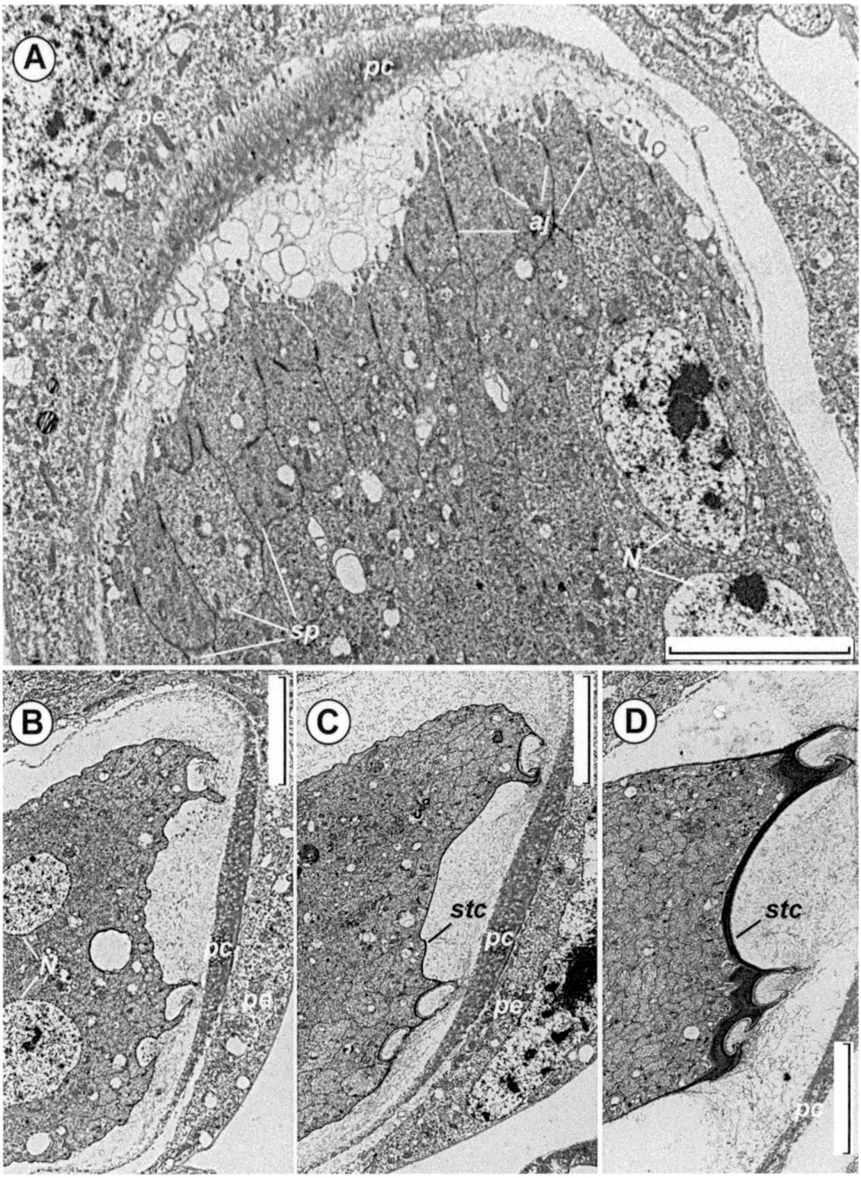


**Fig. 16.14** *Stomaphis quercus* (L.), first-instar nymph, development of a second-instar right maxillary stylet. (**a–d**) Cross sections at different levels of the same retortiform organ showing successive stages of development of the stylet's mesal surface. Note the adherence cell junctions between the styligenic projections in (**a**). For the profile of the mature stylet, see Fig. 16.13d. Abbreviations: *aj* adherence cell junctions, *N* nuclei of styligenic epithelial cells, *pc* peripodial cuticle, *pe* peripodial epithelium, *sp* styligenic cellular projections, *stc* stylet cuticle. Scale bars: (**a–d**) 5 μm

the mesal side (Figs. 16.8f, 16.13b, c, and 16.14d), but eventually it becomes almost as thick (Fig. 16.3b–f). Pinet (1970) suggested that the microtubules contained in the projections serve as a scaffold for deposition of the stylet cuticle, into which they become incorporated. However, incorporation of styligenic projections into the cuticle has, so far, not been confirmed by ultrastructural data.

Prior to ecdysis the apex of a mature new stylet lies within the old stylet's basal funnel (Fig. 16.4c) or is situated outside of it (in heteropterans, see Cobben 1978: Fig. 5). In both cases, it is connected to the inner wall of the old stylet's funnel by an amorphous-looking or finely flocculent substance. This anchoring is crucial because it allows a passive extraction of the stylet during ecdysis (see below). The assertion by Benwitz (1956) that the old and the new stylets are connected by fibers is probably erroneous. According to Pinet (1970), the amorphous substance is produced by the styligenic epithelium. Apparently the same material is visible along the entire developing stylet between its mesal surface and the peripodial cuticle (Figs.16.8a, b, f and 16.12e, f).

## *16.3.4 Ecdysis and Assembly of Stylets into a Bundle*

During ecdysis the newly formed stylets are pulled out of the head into the working position. Until this process is complete, the new stylets are disconnected from stylet protractors and retractors and, therefore, cannot be moved by them. Two methods of stylet extraction can be recognized, called here passive and active.

The passive method has been described by Weber (1929, 1930) and appears to be employed by most hemipterans: the new stylets slide out as the insect extricates itself from the exuviae because their tips are anchored to the old stylet bases (Figs. 16.6d and 16.15a–d). The new stylets are pulled out while being coated with sheaths of unsclerotized peripodial cuticle. Once the stylets are extracted to their full length, their basal funnels become locked in stylet braces (Fig. 16.4a–c), while the peripodial sheaths continue to be pulled off until completely removed (Fig. 16.16a–c, g). In the words of Snodgrass (1927): "... As the imago extracts its head from the nymphal cuticula, four long white threads, similar in appearance to the tracheal linings, but attached within [in fact, outside – R.R.] the hollow bases of the nymphal setae, are seen to pull out from the setal pouches of the imago. ... The sheathes are evidently cuticular invaginations formed about the setae, but an explanation of their true morphology must be left for a histological study of the early developmental stages of the setae." Some aspects of the peripodial cuticle still remain poorly understood; in particular, it remains unclear why, although secreted simultaneously with the new stylet cuticle, the sheaths are attached to the old cuticle of the stylet sacs and are discarded together with the exuviae.

The peripodial sheaths are sticky and, once pulled off the stylets, form together a tether, which also includes cuticular lining of the cibarium and foregut. In molting adults of *Pyrrhocoris apterus*, this tether has a gooey appearance (Fig. 16.16a, b) and is easily stretchable. In order to get rid of the sheaths, the bug repeatedly pulls and

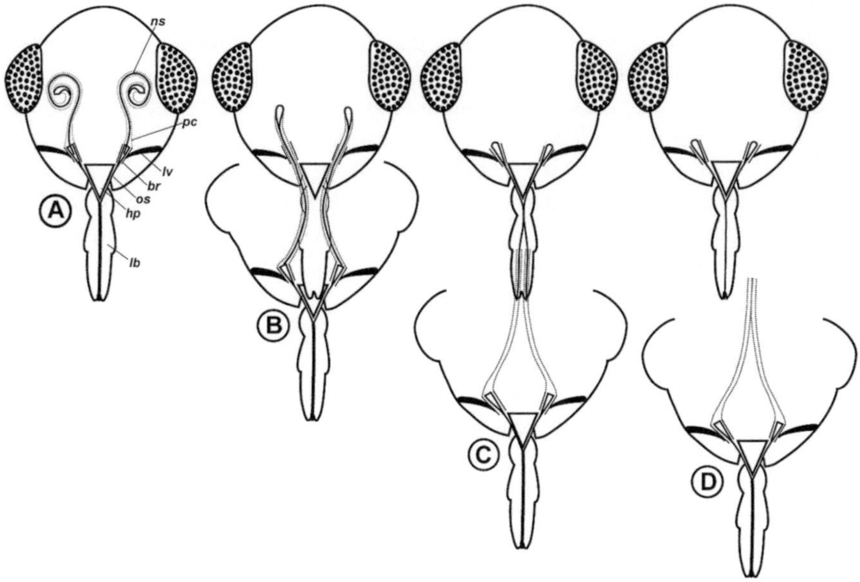


**Fig. 16.15** Passive extraction of new stylets during ecdysis. The antennae, ocelli, epithelia, muscles, and most of other head structures are omitted for clarity. (**a–d**) Successive stages, see text for details. Abbreviations: *br* stylet brace, *hp* hypopharynx, *lb* labium, *lv* lever, *ns* new stylet, *os* old stylet, *pc* peripodial cuticular sheaths

releases it. Every time the tether is stretched out, a short length of the sheaths comes off the stylet apices. The sharp bent-in apices of the maxillary stylets are anchored in the sheaths, which facilitates extraction of the stylets but makes it harder to remove the sheaths, with the resisting stylet apices cutting through the sheath's matrix (Fig. 16.16b). The dry thread consisting of the four peripodial sheaths and the foregut cuticle stuck together remains attached to the exuviae (Fig. 16.15d).

Simultaneously with extraction of stylets, the dorsal surface of the labium invaginates to form the labial groove and the new stylets gradually sink into it (Fig. 16.16 a–b) until concealed completely. Labial sculpture, such as the presumably non-sensory spines at the labial apex of various Heteroptera and Auchenorrhyncha (Cobben 1978; Brożek and Zettel 2014; Parveen et al. 2015; Wang et al. 2015), which were suggested to clean hardened saliva and other contaminants off the stylet bundle after feeding (e.g. Leopold et al. 2003), may assist in removal of the peripodial sheaths during ecdysis.

The active method of stylet extraction has been observed in hatching first-instar nymphs of adelgids, scale insects, and psyllids (Sternorrhyncha: Aphidoidea, Coccoidea, and Psylloidea), all of which have extra-long stylet bundles exceeding the length of the labium and stowed away at rest in the crumena, a membranous invagination unique to Sternorrhyncha, which extends from the base of the labium into the thorax and, in some cases, further into the abdomen, which accommodates the stylet bundle bent into a loop (Weber 1928). During ecdysis the first-instar stylets

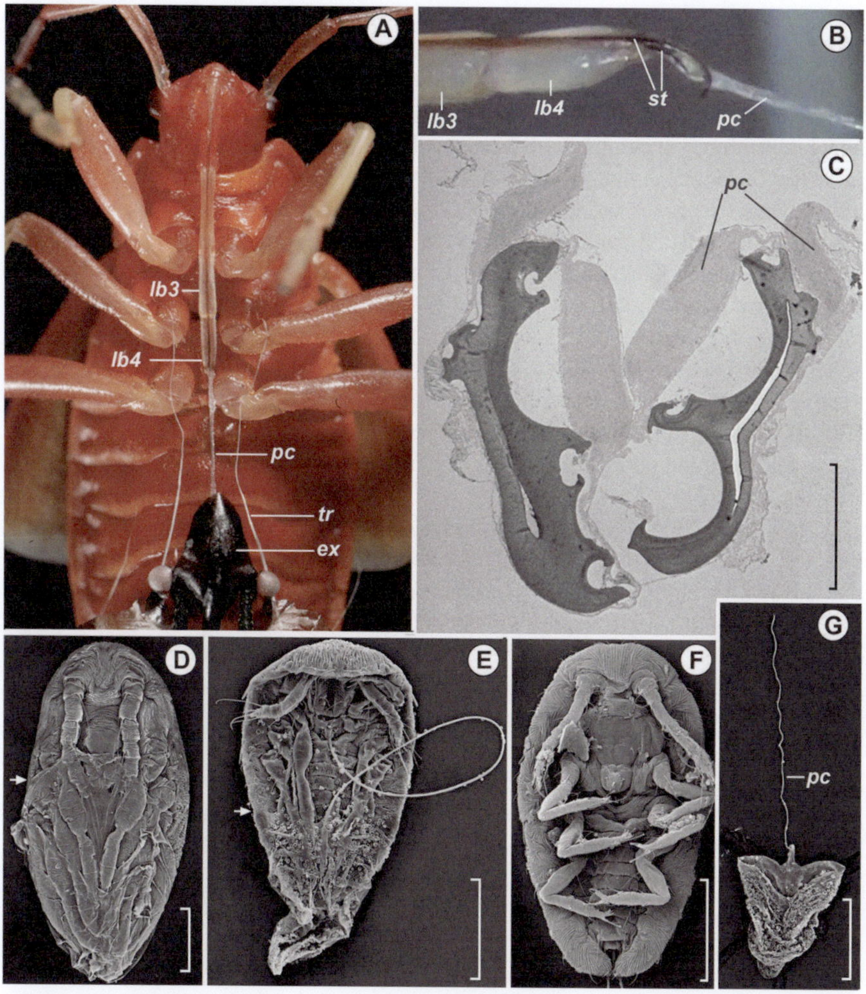


**Fig. 16.16** Extraction of stylets and removal of peripodial cuticle during ecdysis. (**a**) Freshly molted adult *Pyrrhocoris apterus* (L.) during removal of peripodial sheaths; (**b**) same, enlarged lateral view of peripodial cuticular sheaths being stripped off the stylets; (**c**) cross section of a pair of maxillary stylets of adult *P. apterus* drawn out of the head but still coated with peripodial cuticular sheaths; note the thickened mesal areas of the sheaths (compare to Fig. 16.8a, f); (**d–e**) emerging first-instar nymphs of *Coccus hesperidum* L., successive stages of shedding of the pronymphal cuticle: (**d**) the nymphal head begins emerging; in this and the next photo the arrow shows the anterior extent of the pronymphal cuticle; (**e**) the stylets drawn out of the retortiform organs appear as a single bundle (although not yet interlocked) forming a large loop, while their tips are still attached to the exuvial pronymphal cuticle coating the posterior one-third of the body; (**f**) freshly emerged first-instar nymph with the stylets retracted into the crumena, only the tip of the bundle protrudes from the labium; (**g**) cast-off pronymphal cuticle of *C. hesperidum* with a long filament formed by peripodial cuticular sheaths of stylets stuck together. Abbreviations: *ex* exuvial head, *lb3, lb4* third and fourth segments of the labium, *pc* peripodial cuticular sheaths, *st* stylets, *tr* exuvial tracheae. Scale bars: (**c**) 10 μm; (**d**) 50 μm; (**e–g**) 100 μm

are incrementally protruded out of the head, apparently propelled by alternating gripping and pulling movements of the labium. Once the stylets have been drawn out to their full length, they are retracted into the crumena, probably also with the labium. During retraction the stylets slide out of the peripodial sheaths, which remain attached to the exuviae, as in all hemipterans (Fig. 16.16g). The process has been observed by Heriot in first-instar crawlers of *Adelges abietis* (L.) (Adelgidae), which have stylets five times the length of the body and hatch openly on spruce needles (Heriot 1936). It is the same for crawlers of the scale insect *Coccus hesperidum* (L.) (Coccidae), which hatch and then stay for a while among their siblings in a narrow space under their mother's abdomen (Weber 1930: Fig. 212), where the passive method of stylet extraction would be hard to practice given the great length of the stylets. The hatchlings are able to draw their stylets out of the retortiform organs and then retract them into the crumena while being still attached to the exuviae (Fig. 16.16d–f). In hatching first-instar nymphs of *Psylla buxi* (L.), withdrawal of the stylets from the retortiform organs begins even before the egg chorion is broken (Wilcke 1941: Fig. 12).

It has long been established that such sternorrhynchans protrude their stylet bundles out of the crumena during and retract them after the act of feeding with the aid of a clamping mechanism in the penultimate labial segment. Both Weber (1928) and Pollard (1970) came to the conclusion that in feeding *Psylla* nymphs this clamp is used only to alternately arrest and release the stylets but not to propel them, which is done by contraction and retraction of the stylet muscles. Other authors instead believed that the labium actively propels the clamped stylet bundle using muscles of its clamping mechanism (Pesson 1944: Fig. 77; for additional references, see Pollard 1970). Although Pollard's (1970) model of stylet movements during feeding has become widely accepted, and even extrapolated onto all plant-sucking hemipterans, it cannot explain the extraction and retraction of newly formed stylets, not yet attached to stylet muscles, during ecdysis of *Adelges* and *Coccus*. In contrast, the labial mechanism suggested by Pesson (1944) is their most likely explanation and, if so, may play a role also during feeding. Heriot (1934, 1936) stated that during ecdysis of the scale *Lepidosaphes ulmi* (L.) (Diaspididae), the stylets pass from their coiled positions inside the head directly into the crumena, believed by him to be formed by peripodial sheaths; this account does not seem plausible.

As the new stylets are pulled out of the head, they become straight and parallel. Even before the peripodial sheaths are removed, the extracted stylets are already partially aligned. They are held closely together due to adhesiveness of their peripodial sheaths and appear to stick together even closer as the sheaths are pulled off. The presence of the peripodial sheaths prevents interlocking of olistheters until the stylets have been fully extracted and the sheaths at least partly removed. A study of hatching leafhoppers and froghoppers suggested that coalescence of their maxillary stylets begins basally, near the tip of the hypopharynx, and propagates toward their apices in the wake of receding peripodial sheaths (Fig. 16.15c–d; Rakitov 2018). However, removal of the sheaths is not by itself sufficient to make the two halves of an olistheter to coalesce. In a typical olistheter, the apical expansion of the rhachis will prevent it from going through the narrow entrance of an aulax unless the two are pressed together. Everyday experience of using plastic ziplock bags (Fig. 16.3a) also suggests that interlocking of stylet olistheters requires lateral pressure from other parts of the body. The source of such lateral pressure acting on the stylets at their basal point of coalescence is hard to establish. Soon after ecdysis the maxillary plates, hypopharynx, anteclypeus, and labrum assemble into a coadapted liquid-tight mechanism, inside of which the stylet bases are aligned with high precision. In hatching leafhoppers, a pair of lobes at the transition between the anteclypeus and labrum unfold and embrace the stylets at the coalescence point, possibly squeezing them together (Rakitov 2018). Once the olistheters are interlocked at one point, their closure is easier to propagate, which is done by pressure from the sides of the labial groove (Fig. 16.15c–d). Adult *Pyrrhocoris apterus* are capable of reclosing their maxillary stylets, artificially separated from their apices to the point of their exit from the labrum (i.e., with more basal interlocking left intact) within a few hours with the aid of the labium. In scales and other Sternorrhyncha with looped extra-long stylets, the labial clamp probably interlocks the maxillary stylets when it grips and pushes the stylets into the crumena.

Interestingly, the assembly of the feeding tube in lepidopterans emerging from the pupa also requires a special mechanism. Unlike hemipteran stylets, the two lepidopteran galeae have intrinsic musculature and are able to gradually interlock—from base to apex—mostly as a result of their own specific movements (Krenn 1997).

### *16.3.5 Assembly of Stylet Bases*

The development of new stylets is incompatible with direct attachment of muscles to stylet bases and requires their attachment to coupling cuticular structures. Like with all other cuticular body parts, except the stylets, the epidermis of stylet levers and stylet braces apolyses and secretes new cuticle in situ, so that their muscle attachments are preserved. When, during ecdysis, the new stylet is being extracted, it slides through the brace (Fig. 16.4c) until its basal funnel becomes locked there like a boot in a stirrup. Upon sclerotization of the cuticle, the stylet and the brace become firmly connected and cannot be dissected out separately. In *Palomena* (Pentatomidae), the edge of the stylet base remains visible through a semitransparent wall of the brace (Fig. 16.4b, arrow, compare to Fig. 16.5g showing bases of new stylets of this species prior to ecdysis), but in most cases these are not discernible as separate structures (Fig. 16.1a).

Pesson's (1944, 1951) description of styletogenesis does not specify the fate of the peripodial epithelium during ecdysis. On his diagrams it appears shrinking out of existence, so that nothing remains between the stylet brace epidermis and the residual cone of styligenic epithelium inside the base of the new stylet. Both the brace and the stylet base continue producing cuticle after ecdysis, the two cuticles merging proximally, thus closing the stylet sac. At least in the examined Cicadidae and Aphrophoridae, the basalmost part of the stylet cuticle indeed appears to be continuous with the brace cuticle. However, when and how this transition forms is unclear because it is hard to imagine the peripodial epithelium disappearing without a trace. Immediately prior to ecdysis, the peripodial epithelium forms a tubular sac enclosing the entire new stylet, and its cells show no signs of degradation. During ecdysis this structure shrinks to a minute vestige surrounding the mass of styligenic cells at the base of the new stylet. The mechanism behind this dramatic size change is unknown; in theory, it may involve both apoptosis and cell volume reduction. In freshly molted adults of *Pyrrhocoris*, this vestige is two-layered because the basal half of the shrunk epithelium becomes infolded into its apical half; the same condition has been observed in a 1-day-old aphid nymph (Ponsen 1972: Fig. 4). Therefore, at least immediately after molt, a stretch of peripodial epithelium remains between the stylet base and the stylet brace. It is possible that once the unsclerotized peripodial cuticle is discarded with the exuviae, the shrunk peripodial epithelium deposits another, thin cuticular layer, which becomes sclerotized.

In several examined species, approximately at the place where the retractor muscles are attached, the brace cuticle forms a cushion consisting of intricately convoluted layers with some residual material between them (Fig. 16.12g). This so-called pleated cuticle ("cuticule plissée," Pinet 1968a, 1970) is situated at the mesal side of the stylet brace (both maxillary and mandibular); therefore, during styletogenesis it is continuous with the thickened side of the peripodial cuticle. According to Pinet (1968a, 1970), who observed it in *Rhodnius*, the pleated cuticle is secreted immediately before ecdysis, and its characteristic structure is due to exfoliation of layers. Cicero (2017) and Cicero et al. (2018) found the same structure in psyllids and interpreted it as bolus of thin sclerotized cuticle produced by the peripodial epithelium during the preceding cycle of styletogenesis and collected during ecdysis at the stylet base instead of being discarded with the exuviae. New observations on *Pyrrhocoris* do not confirm this theory because the pleated cuticle appears on its periphery as a normal cuticle in contact with the underlying epidermis, which is not expected for a previous-instar cuticle after ecdysis. Moreover, the exfoliated brace cuticle of *Aphrophora* (Fig. 16.12g) shows intermediate stages between a normal, layered endocuticle and a stack of convoluted loose layers.

In the retortiform organs discussed so far, seemingly representing the most common type, the cuticle of the developing stylet base is connected to the cuticle of the next-instar lever and brace only through the unsclerotized peripodial cuticle, eventually discarded. In ready-to-molt last-instar nymphs of *Magicicada* (Cicadidae) Snodgrass described weakly chitinous prolongations of the mandibular and the maxillary levers connecting each of these to the base of the corresponding new stylet (Snodgrass 1927: Figs. 6 and 11; also see Lew 1959). Snodgrass suggested that these bars maintain the continuity between developing new stylets, as they invaginate into the head, and developing new levers; once the mature new stylet has been pulled out into the working position, the connecting bar obliterates, which brings the new stylet and lever into immediate contact. I observed these structures in pharate adults of *Magicicada* and *Poophilus* (Aphrophoridae). The bar runs inside the retortiform organ along its concave margin, i.e., opposite to where the stylet lies. At one end it is continuous with the edge of the new stylet's basal funnel, and at the other, it appears connected to the new lever. The absence of any signs of these structures from cross sections of aphrophorid retortiform organs (*Aphrophora*) during early and intermediate stages of styletogenesis suggests that they appear late in the development. The role of these structures and their fate after ecdysis are unknown.

### 16.3.6 Stylet Sensilla During Styletogenesis and Ecdysis

Development of new stylets takes multiple days, during which most hemipterans continue feeding. There is little doubt that stylet sensilla remain functional during this period, but how this is accomplished is not entirely clear. In particular, it is not known whether the length of the stylet nerve, which passes through the lumen of the retortiform organ, increases when the latter grows, or the nerve originally has extra length to accommodate for that growth. When later the perikarya ascend toward the base of the retortiform organ, the nerve must become shorter again, while the outer dendritic segments, together with their cuticular sheaths and accompanying processes of the enveloping cell, must grow in order to retain their attachment inside the old stylet's apex. It seems most likely that the old dendrites are retained at the next instar in the same manner the sensory dendrites of various arthropod sensilla are retained during molting: the growing dendrite remains attached to the old cuticle while the new cuticle is deposited around it, so that eventually the dendrite penetrates the newly secreted cuticle through a hole, referred to as the molting pore; during ecdysis the part of the dendrite distal of the molting pore is discarded with the exuviae (Geiselbrecht and Melzer 2014 and references therein). Examination of hemipteran stylet apices in SEM so far failed to reveal molting pores, probably because they become occluded. Discarded cuticular sheaths of dendrites were observed inside the axial cavities of exuvial stylets of Aphrophoridae and Cicadidae.

### 16.3.7 Similarities with Styletogenesis in Thysanoptera

Development of stylets in Thysanoptera has so far been known only from light microscopic observations, which nevertheless revealed similarity with hemipteran styletogenesis. During embryonic development, the right mandible degenerates, while the prospective left mandible and the median lobe (lacinia) of the bilobed maxilla invaginate into the head as "stylet-secreting organs" producing stylets only, while the levers have a separate origin (Heming 1980). During intermolt periods, stylet-secreting organs develop at stylet bases, each appearing as a globular mass of cells sending out thin cytoplasmic projections, together forming a nucleus-free cone, around which the new stylet cuticle is deposited (Reyne 1927). It remains unknown whether the stylet-forming organ invaginates with formation of a peripodial sac as in Hemiptera. This must be so as there is no room inside the old stylet to accommodate the new one, especially in the species with long maxillary stylets (Mound 1970). More similarities with Hemiptera will undoubtedly be found when the development of thysanopteran stylets is studied using TEM.

### 16.3.8 Similarities with Morphogenesis of Holometabolan Antennae

Deep invagination of apolysed epidermis to accommodate for the growth of the next-instar structure occurs during setogenesis in some crustaceans (e.g., Guse 1983; Espeel 1986) but is best known during the development of holometabolan imaginal discs. Švácha (1992) has argued that formation of classical holometabolan imaginal discs is merely a modified form of a common mechanism of molting of epidermal structures. This is also true about hemipteran styletogenesis. Additionally, Švácha's (1992) ultrastructural study of the development of antennae in prepupal caterpillars of *Bombyx mori* L. revealed two specific features not commonly discussed in literature and strongly reminiscent of hemipteran styletogenesis: (1) secretion of the last-instar larval antennal cuticle by filamentous cellular projections and (2) secretion of unsclerotized cuticle by the peripodial epithelium.

At the end of the fourth instar, when the antennal cuticle of the fifth (last prepupal) instar is being secreted, the antennal epidermis has an unusual structure. The cell bodies are separated from the cuticle by a nucleus-free layer of long, parallel cellular projections running along the outer surface of the antenna toward the apex. These projections are identical in their dimensions and ultrastructure to the styligenic cellular projections of hemipterans. In particular, they contain numerous longitudinal microtubules and are interconnected by cell junctions (Švácha 1992: Figs. 8, 9, 11–14). The projections form oblique tiers under the cuticle, so that the projections lying deeper come to the surface more apically. The exposed apical parts of the projections deposit the antennal cuticle. The lines of contact between the projections produce ridges of reticulate microsculpture on the surface of the antenna (Švácha 1992: Figs. 1–3).

During the fifth instar, the epidermal cells apolyse from the cuticle, retract the projections, and form a two-layered tubular invagination, the inner layer of which is referred to as the antennal disc. It is unknown if the cells also form such projections to deposit the cuticle of the pupal and later the imaginal antennae of *Bombyx* and if such projections are formed during earlier instars. It is possible that, like in hemipteran styletogenesis, sending out the projections allows the epidermal cells to retreat toward the base of the invaginating organ while continuing to secrete the cuticle.

Unlike hemipteran styligenic cellular projections, produced only by the apical cell membranes, the projections described by Švácha in *Bombyx* extend from both apical and basal cell membranes (the role of the latter type of processes is unclear). Among diverse kinds of eukaryotic cellular projections, these structures most resemble the "epidermal feet" produced during molting cycles by epidermal cells of various insects; the epidermal feet are formed strictly by the basal cell membranes, but they have similar dimensions, contain bundles of microtubules, and are interconnected by cell junctions (Locke 1985).

According to Švácha (1992), during the fifth larval instar of *Bombyx*, the epidermis of the basal antennal segment apolyses from the cuticle and sinks into the head cavity as a tubular invagination consisting of the outer peripodial layer and the inner cone (antennal disc). The two layers are separated by an amorphous-looking cuticular lamella, continuous with the ecdysial membrane formed by delamination of the inner layers of the larval head cuticle. Švácha believed that this lamella was the cuticle produced by both the peripodial and the disc epithelia and that during molting it became digested together with the endocuticle of the larval head. It appears more likely that this cuticle is produced mostly or exclusively by the peripodial cells.

The above similarities suggest that some of the most peculiar features of hemipteran styletogenesis also occur during the development of more typical insect appendages, such as antennae. This is consistent with the common notion of hemipteran stylets being modified gnathal appendages or their lobes, i.e., mandibles and maxillary laciniae. However, an alternative hypothesis that the stylets had evolved from spines on the ancestral mandibles and maxillae (e.g., Puchkova 1980; Emeljanov 2002) cannot be refuted at this time due to the lack of data. While the development of insect mechanosensory setae and bristles is known to be rather different from the development of appendages (Tilney and DeRosier 2005 and references therein), which appears to be true also of the crustacean sensory setae developing between ecdyses in an invaginated state (Guse 1983; Espeel 1986), I am unaware of any detailed studies of morphogenesis of spines, i.e., multicellular protuberances (Richards and Richards 1979). When such spines are large enough, like, for example, leg spurs, the next-instar structure forms inside the old one, but one can imagine a transition to a morphogenesis with invagination when such a structure narrows beyond a certain limit.

In *Bombyx*, the filamentous cellular projections secreting the antennal cuticle leave on it imprints separated by ridges (Švácha 1992). The ridges of olistheters and olistheter scales of hemipteran stylets also arise along the lines of contact between styligenic projections and, therefore, can be viewed as specialized types of cuticular microsculpture, related to those far more common insect microsculptures where ridges are formed along boundaries between epidermal cells.

## 16.4 Conclusion

This review shows that the model of styletogenesis developed by Pesson (1944, 1951) and Pinet (1968a, 1970) remains accurate for newly examined representatives of Sternorrhyncha, Heteroptera, and Auchenorrhyncha. Nevertheless, many aspects of the process have been pointed above that remain poorly understood or unknown.

In this chapter I have emphasized the role of filamentous styligenic cellular projections in the formation of stylets and, in particular, in the patterning of their interlocking surfaces, which has been discovered by Pinet but remains unpublished except in his thesis (Pinet 1970). Similar nucleus-free cellular projections participate in the development of lepidopteran larval antennae (Švácha 1992), and there is little doubt that they will be discovered during the development of other elongate structures. It is unfortunate and surprising that apparently no ultrastructural study yet has examined the development of other arthropod stylets (e.g., in mosquitoes, fleas, mites) and ovipositor valves (e.g., in hymenopterans). The styletogenesis in Thysanoptera also remains unstudied beyond light microscopic data, which indicated that the stylet is formed by thin cellular projections (Reyne 1927). The interlocking cuticular grooves and ridges of hemipteran olistheters are formed at the boundaries between styligenic projections, and the olistheter scales of unknown function are formed at their termini, which explains the strict parallelism of the first structures and the metameric arrangement of the second. The development of asymmetrical interlocking profiles on the left and right maxillary stylets provides a striking example of precisely coadapted structures developing in the absence of physical contact between them (Sahuc 1968, 1969). The development of apical serrations of stylets remains unstudied.

The absence of epithelium from stylet shafts is a consequence of their function, which requires a combination of thinness with strength and flexibility. Thus, a special mechanism is necessary for generation of next-instar stylets. This mechanism, called styletogenesis, is incompatible with a direct attachment of muscles to stylets because the base of a developing stylet grows away from the old base (Figs. 16.6b, c and 16.7). Pesson (1944: Fig. 65) suggested a transition between ancestral mandible muscles and stylet muscles, in which the mandible divides into the basal brace, retaining its attachment to muscles through molts, and the setiform apical part, which develops inside a retortiform organ. Because the next-instar stylets develop virtually independently from the old ones, this mechanism can generate remarkably long stylets, even several times exceeding the body length (in some Sternorrhyncha and Heteroptera), which are not known among other arthropods with piercing mouthparts. In association with this peculiar morphogenesis, Hemiptera and probably also Thysanoptera have evolved mechanisms to draw the new stylets out of the head, to connect their bases to braces and levers, to discard the peripodial cuticle, and finally to interlock the new stylets into a functional bundle.

**Acknowledgments** I am grateful to Harald Krenn for inviting this contribution, to Esther Appel and Stanislav Gorb for making possible my work on a confocal laser scanning microscope at the University of Kiel, and to Wayne Hunter, Lawrence Mound, and Chris Simon for kindly providing important insect material. The study was supported in part by the Russian Foundation for Basic Research Grant No. 18-04-00322.

## References

Ammar ED, Hall DG, Shatters RG (2015) Ultrastructure and development of the new stylets inside pre-molting first instar nymphs of the Asian citrus psyllid *Diaphorina citri* (Hemiptera: Liviidae). Florida Entomol 98:373–376

Andre HM, Remacle C (1984) Comparative and functional morphology of the gnathosoma of *Tetranychus urticae* (Acari: Tetranychidae). Acarologia 25:179–190

Atachi P (1976) Etude de la tete, des pieces buccales et des glandes salivaires de *Dysdercus andreae* L. (Heteroptera, Pyrrhocoridae). These présentée à l'universite Pierre et Marie Curie de Paris VI pour l'obtention du grade de docteur de 3e cycle

Backus EA (1985) Anatomical and sensory mechanisms of leafhopper and planthopper feeding behavior. In: Nault LR, Rodriguez JG (eds) The leafhoppers and planthoppers. John Wiley & Sons, New York, Chichester, Brisbane, Toronto, Singapore, pp 163–194

Backus EA, McLean DL (1982) The sensory systems and feeding behavior of leafhoppers. 1. The aster leafhopper, *Macrosteles fascifrons* Stål (Homoptera, Cicadellidae). J Morphol 172:361–379

Beard JJ, Ochoa R, Bauchan GR, Welbourn WC, Pooley C, Dowling APG (2012) External mouthpart morphology in the Tenuipalpidae (Tetranychoidea): *Raoiella* a case study. Exp Appl Acarol 57:227–255

Benwitz G (1956) Der Kopf von *Corixa punctata* Ill. (*geoffroyi* Leach) (Hemiptera-Heteroptera). Zool Jahrb Abt Anat Ontogenie Tiere 75:311–378

Bernard J, Pinet JM (1973) Origine des potentiels propagés dans un mécanorécepteur à dendrite longue. J Comp Physiol 87:361–377

Bernard J, Pinet JM, Boistel J (1970) Électrophysiologie des récepteurs des stylets maxillaires de *Triatoma infestans*—Action de la température et de la teneur en eau de l'air. J Insect Physiol 16:2157–2180

Brożek J (2006) Internal structures of the mouthparts of Coccinea (Hemiptera: Sternorrhyncha). Pol Pismo Entomol 75:255–265

Brożek J (2007) Labial sensillae and the internal structure of the mouthparts of *Xenophyes cascus* (Bergroth 1924) (Peloridiidae: Coleorrhyncha: Hemiptera) and their significance in evolutionary studies on the Hemiptera. Aphids and other Hemipterous Insects 13:35–42

Brożek J (2013) A comparison of external and internal maxilla and mandible morphology of water bugs (Hemiptera: Heteroptera: Nepomorpha). Zootaxa 3635:340–378

Brożek J, Bourgoin T (2013) The phylogenetic information carried by a new set of morphological characters in planthoppers: the internal mouthpart structures and test in the Cixiidae model (Hemiptera: Fulgoromorpha). Zoomorphology 132:403–420

Brożek J, Herczek A (2001) Modifications in the mouth parts structure in selected species of Cicadellidae (Hemiptera: Cicadomorpha). Acta Entomol Silesiana 7-8:19–25

Brożek J, Herczek A (2004) Internal structure of the mouthparts of true bugs (Hemiptera: Heteroptera). Polish J Entomol 73:79–106

Brożek J, Zettel H (2014) A comparison of the external morphology and functions of labial tip sensilla in semiaquatic bugs (Hemiptera: Heteroptera: Gerromorpha). Eur J Entomol 111:275–297

Brożek J, Bourgoin T, Szwedo J (2006) The interlocking mechanism of maxillae and mandibles in Fulgoroidea (Insecta: Hemiptera: Fulgoromorpha). Pol Pismo Entomol 75:239–253

Brożek J, Mróz E, Wylężek D, Depa Ł, Węgierek P (2015) The structure of extremely long mouthparts in the aphid genus *Stomaphis* Walker (Hemiptera: Sternorrhyncha: Aphididae). Zoomorphology 134:431–445

Bugnion E, Popoff N (1911) Les pieces buccales des Hemipteres (premiere partie). Arch Zool expérimentale générale 5 Serie 7:643–674

Chapman RF (2013) In: Simpson SJ, Douglas AE (eds) The insects - structure and function, 5th edn. Cambridge University Press, Cambridge

Chisholm IF, Lewis T (1984) A new look at thrips (Thysanoptera) mouthparts, their action and effects of feeding on plant tissue. Bull Entomol Res 74:663–675

Cicero JM (2017) Stylet biogenesis in *Bactericera cockerelli* (Hemiptera: Triozidae). Arthropod Struct Dev 46:644–661

Cicero JM, Alba-Tercedor J, Hunter WB, Cano LM, Saha S, Mueller LA, Brown SJ (2018) Asian citrus psyllid stylet morphology and applicability to the model for inter-instar stylet replacement in the potato psyllid. Arthropod Struct Dev 47:542–551

Cobben RH (1978) Evolutionary trends in Heteroptera: part II. Mouthpart-structures and feeding strategies. Meded Landbouwhogesch Wageningen 78/5:1–407

Cohen AC (1990) Feeding adaptations of some predaceous Hemiptera. Ann Entomol Soc Am 83:1215–1223

Emeljanov AF (1987) The phylogeny of the Cicadina (Homoptera, Cicadina) based on comparative morphological data. Trans All-Union Entomol Soc 69:19–109. [In Russian]

Emeljanov AF (2002) Evolutionary scenario of rostrum formation in the Rhynchota. Entomol Rev 82:1197–1206

Espeel M (1986) Morphogenesis during molting of the setae in the statocyst sensilla of the mysid shrimp *Neomysis integer* (Leach, 1814) (Crustacea, Mysidacea). J Morphol 187:61–68

Forbes AR (1966) Electron microscope evidence for nerves in the mandibular stylets of the green peach aphid. Nature 212:726

Forbes AR (1969) The stylets of the green peach aphid, *Myzus persicae* (Homoptera: Aphididae). Can Entomol 101:31–41

Forbes AR (1972) Innervation of the stylets of the pear psylla, *Psylla pyricola* (Homoptera: Psyllidae), and the greenhouse whitefly, *Trialeurodes vaporariorum* (Homoptera: Aleyrodidae). J Entomol Soc Br Columbia 69:27–30

Forbes AR (1976) The stylets of the large milkweed bug, *Oncopeltus fasciatus* (Hemiptera: Lygaeidae), and their innervation. J Entomol Soc Br Columbia 73:29–32

Forbes AR (1977) The mouthparts and feeding mechanism of aphids. In: Harris KF, Maramorosch K (eds) Aphids as virus vectors. Academic, New York, pp 83–103

Forbes AR, Mullick DB (1970) The stylets of the balsam woolly aphid, *Adelges piceae* (Homoptera: Adelgidae). Can Entomol 102:1074–1082

Forbes AR, Raine J (1973) The stylets of the six-spotted leafhopper, *Macrosteles fascifrons* (Homoptera: Cicadellidae). Can Entomol 105:559–567

Foster S, Goodman LJ, Duckett JG (1983) Sensory receptors associated with the stylets and cibarium of the rice brown planthopper, *Nilaparvata lugens*. Cell Tissue Res 232:111–119

Geiselbrecht H, Melzer RR (2014) Fine structure and ecdysis of mandibular sensilla associated with the lacinia mobilis in *Neomysis integer* (Leach, 1814) (Crustacea, Malacostraca, Peracarida). Arthropod Struct Dev 43:221–230

Guse GW (1983) Ultrastructure, development, and moulting of the aesthetascs of *Neomysis integer* and *Idotea baltica* (Crustacea, Malacostraca). Zoomorphology 103:121–133

Hao Y, Dietrich CH, Dai W (2016) Development of mouthparts in the cicada *Meimuna mongolica* (Distant): successive morphological patterning and sensilla differentiation from nymph to adult. Sci Rep 6:38151

Heming BS (1980) Development of the mouthparts in embryos of *Haplothrips verbasci* (Osborn) (Insecta, Thysanoptera, Phlaeothripidae). J Morphol 164:235–263

Heming BS (1993) Structure, function, ontogeny, and evolution of feeding in thrips (Thysanoptera). In: Schaefer CW, Leschen RAB (eds) Functional morphology of insect feeding. Thomas Say Publ Entomol, Entomol Soc Am, Lanham, MD, pp 3–41

Heriot AD (1934) The renewal and replacement of the stylets of sucking insects during each stadium, and the method of penetration. Can J Res 11:602–612

Heriot AD (1936) The crumena of the Coccidae and the Adelges. J Entomol Soc Br Columbia 33:22–24

Heymons R (1899) Beiträge zur Morphologie und Entwicklungsgeschichte der Rhynchoten. Nova Acta Acad Leopold 74:351–456

Hunter WB, Ullman DE (1989) Analysis of mouthpart movements during feeding of *Frankliniella occidentalis* (Pergande) and *F. schultzei* Trybom (Thysanoptera: Thripidae). Int J Insect Morphol Embryol 18:161–171

Krenn HW (1997) Proboscis assembly in butterflies (Lepidoptera) – a once in a lifetime sequence of events. Eur J Entomol 94:495–501

Leopold RA, Freeman TP, Buckner JS, Nelson DR (2003) Mouthpart morphology and stylet penetration of host plants by the glassy-winged sharpshooter, *Homalodisca coagulata* (Homoptera: Cicadellidae). Arthropod Struct Dev 32:189–199

Lew GL (1959) The external morphology of the last-instar nymph of *Magicicada septendecim* (L.) (Hemiptera: Homoptera). Entomol Am NS 39:101–182

Lillo E, Palma A, Nuzzaci G (2001) Morphological adaptations of mite chelicerae to different trophic activities (Acari). Entomologica Bari 35:125–180

Locke M (1985) The structure of epidermal feet during their development. Tissue Cell 17:901–921

Mecznikow E (1866) Embryologische Studien an Insekten. Die Enwicklung der viviparen Aphiden. Z Wiss Zool 16:437–478

Mound L (1970) Convoluted maxillary stylets and the systematics of some phlaeothripine Thysanoptera from *Casuarina* trees in Australia. Aust J Zool 18:439–463

Muir F, Kershaw JC (1911) On the later embryological stages of the head of *Pristhesancus papuensis* (Reduviidae). Psyche 18-19:75–79

Muir F, Kershaw JC (1912) The development of the mouthparts in the Homoptera, with observations on the embryo of *Siphanta*. Psyche 19:77–89

Newcomer WS (1948) Embryological development of the mouth parts and related structures of the milkweed bug, *Oncopeltus fasciatus* (Dallas). J Morphol 82:365–411

Parrish WB (1967) The origin, morphology, and innervation of aphid stylets (Homoptera). Ann Entomol Soc Am 60:273–276

Parsons MC (1959) Skeleton and musculature of the head of *Gelastocoris oculatus* (Fabricius) (Hemiptera-Heteroptera). Bull Mus Comp Zool 122:1–53

Parsons MC (1964) The origin and development of the hemipteran cranium. Can J Zool 42:409–432

Parveen S, Ahmad A, Broek J, Ramamurthy V (2015) Morphological diversity of the labial sensilla of phytophagous and predatory Pentatomidae (Hemiptera: Heteroptera), with reference to their possible functions. Zootaxa 4039:359–372

Pesson P (1944) Contribution à l'étude morphologique et fonctionnelle de la tête, de l'appareil buccal et du tube digestif des femelles de Coccides. Thèse, Paris, p 206

Pesson P (1951) Ordre des Homoptères (Homoptera Leach, 1815). In: Grasse PP (ed) Insectes supérieurs et Hémiptéroïdes, Traité de Zoologie X(II). Masson, Paris, pp 1390–1656

Pinet JM (1963) L'innervation sensorielle des stylets mandibulaires et maxillaires de *Rhodnius prolixus* Stål. (Insecte Hémiptère Hétéroptère ). C R Hebd Seances Acad Sci 257:3666–3668

Pinet JM (1968a) Structure et formation des coaptations des stylets maxillaires de *Rhodnius prolixus* [Heter. Reduviidae]. Ann Société Entomol Fr 4:455–475

Pinet JM (1968b) Etude de la morphogenese du stylet maxillaire de *Rhodnius prolixus* par des transplantations d'organes styligenes incomplets. Arch Zool expérimentale générale 109:613–641

Pinet JM (1970) Recherches sur les stylets mandibulaires at maxillaires de *Rhodnius prolixus*. Thèse d'Etat, Paris, p 253

Pollard DG (1969) Directional control of the stylets in phytophagous Hemiptera. Proc R Entomol Soc London Ser A, Gen Entomol 44:173–185

Pollard DG (1970) The mechanism of stylet movement in *Psylla mali* Schmidberger (Homoptera: Psyllidae). Zool J Linnean Soc 49:295–307

Pollard DG (1972) The stylet structure of a leafhopper (*Eupteryx melissae* Curtis: Homoptera, Cicadellidae). J Nat Hist 6:261–271

Pollard DG (1973) Plant penetration by feeding aphids (Hemiptera, Aphidoidea): a review. Bull Entomol Res 62:631–714

Ponsen MB (1972) The site of potato leafroll virus multiplication in its vector, *Myzus persicae*: an anatomical study. Meded Landbouwhogesch Wageningen 72-16:1–147
Puchkova LV (1980) Skeletomuscular, sensory and glandular apparatus of the head and thorax of aquatic Heteroptera. Trans All-Union Entomol Soc 62:1–234. [In Russian]
Quadri MAH (1959) Mechanism of feeding in Hemiptera. Smithson Misc Collect 137:237–245
Quicke DLJ, Fitton MG, Tunstead JR, Ingram SN, Gaitens PV (1994) Ovipositor structure and relationships within the hymenoptera, with special reference to the Ichneumonoidea. J Nat Hist 28:635–682
Rakitov R (2018) Pronymphs, hatching, and proboscis assembly in leafhoppers and froghoppers (Hemiptera, Cicadellidae and Aphrophoridae). Arthropod Struct Dev 47:529–541
Reyne A (1927) Untersuchungen über die Mundteile der Thysanopteren. Zool Jahrb Abt Anat Ontogenie Tiere 49:391–500
Richards GA, Richards PA (1979) The cuticular protuberances of insects. Int J Insect Morphol Embryol 8:143–157
Rieger C (1976) Skelett und Muskulatur des Kopfes and Prothorax von *Ochterus marginatus* Latreille. Zoomorphologie 83:109–191
Sahuc LJ-M (1968) Etude historique des coaptations chez les Invertébrés. Rev Hist Sci (Paris) 21:245–264
Sahuc LJ-M (1969) Coaptations chez le grillon domestique. Bull Biol Fr Belg 103:9–130
Singh S (1971) Morphology of the head of Homoptera. Res Bull Panjab Univ 22:261–316
Smith EL (1969) Evolutionary morphology of external insect genitalia. 1. Origin and relationships to other appendages. Ann Entomol Soc Am 62:1051–1079
Snodgrass R (1927) The head and mouth parts of the cicada. Proc Entomol Soc Wash 29:1–16
Švácha P (1992) What are and what are not imaginal discs: reevaluation of some basic concepts (Insecta, Holometabola). Dev Biol 154:101–117
Swart C, Felgenhauer B (2003) Structure and function of the mouthparts and salivary gland complex of the giant waterbug, *Belostoma lutarium* (Stål) (Hemiptera: Belostomatidae). Ann Entomol Soc Am 96:870–882
Tilney LG, DeRosier DJ (2005) How to make a curved *Drosophila* bristle using straight actin bundles. Proc Natl Acad Sci USA 102:18785–18792
Uzest M, Gargani D, Dombrovsky A, Cazevieille C, Cot D, Blanc S (2010) The 'acrostyle': a newly described anatomical structure in aphid stylets. Arthropod Struct Dev 39:221–229
Wang T, Pan L, Zhang Y, Dai W (2015) Morphology of the mouthparts of the spittlebug *Philagra albinotata* Uhler (Hemiptera: Cercopoidea: Aphrophoridae). Arthropod Struct Dev 44:121–130
Weber H (1928) Zur vergleichenden Physiologie der Saugorgane der Hemipteren. Mit besonderer Berücksichtigung der Pflanzenläuse. J Comp Physiol A Neuroethol Sensory, Neural, Behav Physiol 8:145–186
Weber H (1929) Kopf und thorax von *Psylla mali* Schmidb. (Hemiptera-Homoptera). Eine morphogenetische Studie. Z Morphol Okol Tiere 14:59–165
Weber H (1930) Biologie der Hemipteren. Verlag von Julius Springer, Berlin, p 543
Wenk P, Lucic S, Betz O (2010) Functional anatomy of the hypopharynx and the salivary pump in the feeding apparatus of the assassin bug *Rhodnius prolixus* (Reduviidae, Heteroptera). Zoomorphology 129:225–234
Wensler RJ (1974) Sensory innervation monitoring movement and position in the mandibular stylets of the aphid, *Brevicoryne brassicae*. J Morphol 143:349–363
Wigglesworth VB (1973) The significance of 'apolysis' in the moulting of insects. J Entomol Ser A, Gen Entomol 47:141–149
Wilcke J (1941) Biologie en morphologie van *Psylla buxi* L. Tijdschrift over Plantenziekten 47:41–89
Witlaczil E (1882) Zur Anatomie der Aphiden. Arbeit aus dem Zool Inst d Univ Wien IV 397:441

# Chapter 17
# The Fossil Record of Insect Mouthparts: Innovation, Functional Convergence, and Associations with Other Organisms

**Conrad C. Labandeira**

**Abstract** The mouthparts of insects are a phenomenal example of a multi-element, modular, feeding apparatus that repeatedly has been modified structurally to perform every feeding function imaginable in the terrestrial and freshwater realms, a process that began in the Early Devonian. Insect mouthparts have been structured to chew, pierce and suck, siphon, lap, sponge, bore, and mine on and within a wide variety of tissues, as well as filter, sieve, and collect particulate food such as plankton and pollen. Thirty-seven fundamental mouthpart classes perform these roles in the modern and fossil record, a result that has been expanded somewhat from earlier, phenetic cluster analyses of modern insect mouthparts. A broad survey of fossil insect mouthparts, in conjunction with the phenetic mouthpart analysis, revealed patterns of mouthpart innovation occurring in bursts of cladogenesis separated from intervals of rather static mouthpart morphology. For the Paleozoic Era, based on direct (body fossil) and indirect (trace fossil) evidence, and commencing during the Devonian Period, the four earliest mouthpart classes were present, accounting for 11.4% of all mouthpart classes in the fossil record. In the succeeding Mississippian Subperiod, no mouthparts are documented; the four mouthparts from the Devonian continue into the succeeding Pennsylvanian Subperiod. During Pennsylvanian time, there was a spectacular burst of new mouthpart classes, coincident with the appearance of approximately 15 major insect lineages. By the end of the period, 29.7% of all insect classes had appeared. The following Permian Period added another seven



C. C. Labandeira (✉)
Department of Paleobiology, National Museum of Natural History, Smithsonian Institution, Washington, DC, USA

Department of Entomology and BEES Program, University of Maryland, College Park, MD, USA

College of Life Sciences, Capital Normal University, Beijing, China
e-mail: labandec@si.edu



H. W. Krenn (ed.), *Insect Mouthparts*, Zoological Monographs 5,
https://doi.org/10.1007/978-3-030-29654-4_17

mouthpart classes, particularly those from early hemimetabolous and holometabolous lineages, resulting in 48.6% of all mouthpart classes present. The profound ecological crisis at the end of the Permian notably saw the near extirpation of only one mouthpart class, the Robust Beak of piercing-and-sucking paleodictyopteroid insects, which eventually was extinguished sometime during the ensuing Triassic Period. For the Mesozoic Era, the Triassic Period added another seven mouthpart classes, particularly involving aquatic naiads and larvae, and early dipteran mouthparts, resulting in 67.6% of all mouthpart classes at the end of the period. During the Jurassic, the Mesozoic Lacustrine Revolution had begun, reaching a peak in the invasion of freshwater ecosystems that commenced during the Late Triassic, but undergoing a major diversification of mouthparts in terrestrial lineages, resulting in 83.3% of all mouthpart classes present, notably before the ecological expansion of angiosperms in the subsequent Early Cretaceous. The Jurassic also was a time for the origin and initial innovation of mouthpart design in early Siphonaptera, and a largely parallel diversification event among hematophagous Diptera; both processes continued into the Early Cretaceous. The Cretaceous Period exhibits a considerable diversity in compression deposits and especially amber deposits, preserving relict lineages that bore mouthparts at a Permian and Triassic stage of evolution as well as new lineages with bizarre mouthpart structures that are difficult to place among existing mouthpart classes. During the Cretaceous, three new mouthparts classes are added, yielding 97.1% of all mouthparts at the end of the period. For the Cenozoic Era, no mouthpart classes are added during the Paleogene Period, and only one mouthpart class, lacking a fossil record, is added during the Neogene Period. During this time, there is modification and expansion of mouthpart classes established during the mid Mesozoic and the development of special mouthpart elements involved in leaf mining, blood feeding, and pollination.

> Ces possibilités se manifestent avec un exuberance toute particulière, tant sur le plan anatomique que sur plan fonctionnel, dans le cas des pieces buccales des Insectes qui sont susceptibles de s'adapter aux régimes alimentaires les plus divers, aux modalities es plus diverses de prise de la nourriture . . .
>
> J. Chaudonneret (1990)

## 17.1 Introduction

No other biological structure possessed by multicellular organisms is responsible for the wholesale transfer of energy from one trophic level to a superjacent trophic level than insect mouthparts. Put another way, insect mouthparts are the dominant mode by which the trophic pyramid in terrestrial and freshwater ecosystems is functionally realized. This condition likely has dominated terrestrial ecosystems from Pennsylvanian time from 323 to 299 million years ago (Ma), although direct and indirect

**Table 17.1** The geologic timescale[a]

| Era | Period | Epoch | Upper boundary date (Ma)[b] |
|---|---|---|---|
| CENOZOIC | Neogene | Holocene | 0.0 |
| | | Pleistocene | 0.011 |
| | | Pliocene | 2.6 |
| | | Miocene | 5.3 |
| | Paleogene | Oligocene | 23.0 |
| | | Eocene | 33.9 |
| | | Paleocene | 56.0 |
| MESOZOIC | Cretaceous | Late | 66.0[c] |
| | | Early | 100.0 |
| | Jurassic | Late | 145.0 |
| | | Middle | 164.0 |
| | | Early | 174.0 |
| | Triassic | Late | 201.0 |
| | | Middle | 237.0 |
| | | Early | 247.0 |
| PALEOZOIC | Permian | Lopingian | 252.0[d] |
| | | Guadalupian | 260.0 |
| | | Cisuralian | 272.0 |
| | Pennsylvanian[e] | Late | 299.0 |
| | | Middle | 307.0 |
| | | Early | 315.0 |
| | Mississippian[e] | Late | 323.0 |
| | | Middle | 331.0 |
| | | Early | 347.0 |
| | Devonian | Late | 359.0 |
| | | Middle | 383.0 |
| | | Early[f] | 411.0 |
| | Silurian | Prídolí[f] | 419.0 |

[a]The source of this timescale is Walker et al. (2013)
[b]The upper boundary age for the epoch at left is given in millions of years, designated as "Ma"
[c]This is the Mesozoic—Cenozoic boundary, 66 Ma, equivalent to a major ecological crisis and extinction event
[d]This is the Paleozoic—Mesozoic boundary, 252 Ma, equivalent to a major ecological crisis and extinction event
[e]The Pennsylvanian and Mississippian are subperiods of the Carboniferous Period (not shown), and are used here to achieve greater geochronologic resolution
[f]For the Paleozoic Era, this table only includes the Permian, Pennsylvanian, Mississippian, and Devonian Periods, and latest Silurian, that are relevant to insect mouthpart history. The Paleozoic Era also includes the Silurian, Ordovician, and Cambrian Periods, in successively older intervals of time

evidence indicates that several modes of insect feeding extend to the Early Devonian (Table 17.1) from 419 to 408 million years ago (Ma), at least near hot-spring environments. The great terrestrial expansion of major insect mouthpart types that began during Pennsylvanian times and had a resurgence during the mid Mesozoic from 174 to 100 Ma, and lessened considerably into modern times (Labandeira 1997). For freshwater ecosystems, the evidence indicates that the earliest aquatic insects, albeit with few mouthpart types, occurred during the Middle Pennsylvanian at approximately 311 Ma. However, aquatic mouthpart diversity did not dramatically expand until the mid Mesozoic during the Mesozoic Lacustrine Revolution (Buatois et al. 2016), unlike the pattern for terrestrial insects. This increase in insect mouthpart types—as well as associated functional feeding groups (the fundamental ways that insect access and process food) and dietary guilds—has been propelled by an increase in insect taxonomic diversity (Labandeira 2005a). Although this taxonomic diversity is characterized historically by phases of expansion and diminution, mouthpart diversity follows a trend in which the variety or disparity of mouthpart types originated considerably earlier than the diversity of insect taxa bearing those mouthpart types (Labandeira 1997). Parallel with this pattern is the hypothesis that the multi-element and modular nature of the insect feeding apparatus is a major driver of insect diversity (Popadić et al. 1998; Yang 2001; Rogers et al. 2002; Mayhew 2007).

Because of the vast number of studies on fossil insects and their feeding structures, this review is not a comprehensive compendium of mouthpart types in time and space. Rather, I provide an account of well-documented fossil insect species with a well-preserved ensemble of a prothorax, head, and feeding features that collectively offer a wealth of mouthpart diversity from the fossil record. Occasionally, I focus on particular examples of unique mouthpart structures in the fossil record, some of which may be key innovations. The kinds of documentation for this account come from the direct evidence of often older compression-impression (or adpression) and younger amber body fossils. Indirect evidence also is brought to bear, such as mouthpart-induced damage on fossil plants, sedimentary structures with mouthpart imprint impressions, and clades with inferred mouthpart types based on their sister-group relationship with a coexisting clade bearing a known mouthpart type. For this chapter, there is a focus on Late Paleozoic (Pennsylvanian, Permian) and Mesozoic (Triassic, Jurassic, Early Cretaceous) taxa. Less of a concentration has been placed on Late Cretaceous and Cenozoic mouthparts, many of which occur in amber deposits, particularly as insects possessing these mouthparts often are very similar to their modern descendant taxa described in other chapters of this volume. Nevertheless, several unique, modern insect mouthpart ensembles and individual elements from the Cenozoic, lacking or having a poor fossil record, are discussed, providing a view that mouthpart evolution is an ongoing process. In this review, mouthpart types are linked, where appropriate, with major global events associated with increases or decreases in insect diversity and other parallel, clade specific processes, such as mouthpart innovation, structural convergence, and the relationship of mouthpart types to the emergence of new food sources.

The initial motivation for this review was to better extend to the fossil record the results of an updated version of the author's dissertation at The University of Chicago (Labandeira 1990). That work, subsequently updated (Labandeira 1997), constituted the first comprehensive phenetic analysis of modern insect mouthparts, and is the basis for this chapter. The assignments of fossil insects discussed in this chapter and illustrated in the figures reference various modern (and fossil) mouthpart classes described in Table 17.2 and borrow heavily from the original analysis of 1990. As a postscript to the 1990 work and the current contribution, it is clear that new major fossil and modern insect mouthpart types will be discovered in the near future. Such discoveries will occur particularly in amber deposits such as Myanmar amber (e.g., Bai et al. 2016; Mey et al. 2017) and in the lesser-known interstices of the modern world (e.g., Besuchet 1972; Dajoz 1976; Vit 1981).

## 17.2 Methods and Conventions

### *17.2.1 The Original Phenetic Analysis*

A phenetic classification of modern insect mouthparts (Labandeira 1990), subsequently revised (Labandeira 1997), and updated (this report), is the basis for categorization of fossil mouthpart classes in this review. Approximately 1200 book chapters, monographs, and especially journal articles, describing the head and mouthpart morphology of particular insect species, were used for the original analysis. Multiple sources of data were used in the 72% of all species characterized. The original phenetic analysis, forming the basis of this report, consisted of 49 head and mouthpart characters that were analyzed across 1365 insect taxa. These results produced 34 mouthpart classes, to which two fossil mouthpart classes were added that were not part of the original analysis (Labandeira 1997), and one additional mouthpart class is added in this report as a consequence of dividing post hoc a former mouthpart class into two separate mouthpart classes (Table 17.2).

The 49 phenetic variables for the analysis sampled nine regions of the insect head and mouthparts. General head characters consisted of (1) mouthpart-related features of the head, such as descriptions of head shape, genae, gula, and lorum (eight characters); (2) sense organs (three variables); (3) general features of the mouthparts such as position, type, symmetry, and protractability (six characters); and (4) features of the clypeus (two characters). Mouthpart-related aspects were (5) labrum features (four characters), aspects of the pharyngeal region (three characters), (6) mandible features (five characters), (7) maxillae features (nine characters), and (8) labium features (nine characters). The types of characters were quality (27 characters), number consisting of an integer (eight characters), mouthpart element co-optation (six characters), shape (five characters), and aspect ratio (three characters). A subset of the data consisted of four diet-related characters that were associated with the mouthpart class dendrogram, but were not part of the mouthpart analysis.

The 1365 examined species sampled all 33 orders and 70.1% of all family-level taxa based on the classification of Parker (1982) and updates as of 1990. (Since the analysis, the "orders" Homoptera and Heteroptera have been combined into a single order Hemiptera.) Subadult taxa (nymph, naiad, and larva) represented 37.7% of the dataset. Coverage of the number of species per order as a percentage of the total number of dataset species is, in rank order for the top five: Coleoptera, 25.4%; Diptera, 17.3%; Lepidoptera, 10.3%; Hymenoptera, 10.0%; and Hemiptera, 9.7%,

**Table 17.2** Insect mouthpart morphology series and classes[a]

| | Mouthpart class number studied, figured material[b] | Description of mouthpart morphology[c] | Traditionally included clades[d] | Life stages[e] | Principal diets[f] | Geochronologic range[g] | Key references[h] |
|---|---|---|---|---|---|---|---|
| **Mandibulate Series** | | | | | | | |
| 1 | **Entognathate** (*n* = 10) (Fig. 17.1a) | Head capsule ovoidal; mouthparts not protractile or projecting; gena and gula regions well to minimally developed. Mouthpart ensemble mandibulate, entognathous, mostly prognathous, symmetrical, nonprotractile. Clypeus and labrum not subdivided. Mandibles monocondylic, incisiform, moved by adduction/abduction. Interregional co-optation of mouthparts absent | Collembola; Diplura | Nymphs, adults | Dead matter, insects, dead plants, wood rots, fungi, carrion, detritus | E. Devonian to recent | Wolter (1963), Goto (1972) |
| 2 | **Maxillopalpate**[i] (*n* = 3) (Fig. 17.2a) | Head quadrate in lateral profile, mostly enveloped by prothorax, with short, projecting mouthparts; gena area well developed, gula area absent. Mouthpart ensemble mandibulate, entognathous, hypognathous, symmetrical, nonprotractile. Clypeus and labium quadrate, each undivided. Mandibles weakly dicondylic, with molariform and incisiform regions; maxillary palpus leg-like of six articles. No interregional mouthpart co-optation | Archaeognatha | Nymphs, adults | Dead plants, detritus | E. Devonian to recent | Bitsch (1963) |
| 3 | **Adult Ectognathate** (*n* = 29) (Figs. 17.1b, 17.2b, 17.4a–f, 17.7d, 17.8b, c, 17.9a, c, d and 17.13b, c) | Head capsule spheroidal to ovoidal to ellipsoidal, typically free of prothorax. Mouthparts mostly compact gena region prominent to almost absent, gula region well developed. Clypeus mostly transverse, some subdivided or fused to frons; labrum shape variable. Mandibles strongly dicondylic, ectognathate, mostly tetragonal to falcate, with incisor and molar teeth or brustiate; maxilla and labium of generalized form and palp article number. No interregional mouthpart co-optation | Thysanura; Dictyoptera; Phasmatodea; Grylloblattodea; Orthoptera; Isoptera; Dermaptera; Embioptera; Plecoptera; Zoraptera; Megaloptera; Raphidioptera; Neuroptera; Coleoptera; Lepidoptera (Zeugloptera); Hymenoptera (Symphyta, Apocrita-Parasitica) | Adults, some larvae | Insects, plants, no food, dead matter, flowers, omnivory, sporangia, wood[j], pollen, unknown, wood rot, dead plants, nectar, fungi, ectoparasitic, spores, carrion, roots, seeds, mycelia | E. Devonian to recent | Crampton (1930), Walker (1933), Strenger (1949), Butt (1951), Kelsey (1954), Matsuda (1957), Popham (1961), Achtelig (1967), Honomichl (1975) |

| | | | | | | | |
|---|---|---|---|---|---|---|---|
| 4 | **Larval Ectognathate**<br>($n = 163$)<br>(Figs. 17.4g, 17.7e and 17.9b) | Head capsule spheroidal to ellipsoidal in shape, with mouthparts retractile and not projecting. Mouthparts variously positioned, mandibulate, generally symmetrical; gena region expansive, often forming most of the head capsule; gula mostly poorly developed. Frontal region wide, bearing an inverted Y-shaped epistomal suture; clypeus typically transverse, occasionally subsegmented; labrum vertically oriented. Mandibles dicondylic, ectognathate; tetragonal, falcate, palmate, or pectinate; dentition variable. Maxilla and labium typically reduced from corresponding adult stage. Interregional mouthpart co-optation absent | Coleoptera; Mecoptera; Siphonaptera; Diptera; (Anisopodidae); Trichoptera (Phryganopsychidae); Hymenoptera (Symphyta, Apocrita-Parasitica) | Larvae | Insects, wood rots, live fungi, carrion, foliage, roots, wood[j], detritus, dead plants, dead matter, plants, mycelia, animals, omnivory, grain, seeds | E. Pennsylvanian to recent | Sikes (1930), Bierbrodt (1943), Hiznay and Krause (1955), Jaboulet (1960), Quennedey (1965), Baker and Ellsbury (1988) |
| 5 | **Maxillolabiate**<br>($n = 24$)<br>(Figs. 17.12b and 17.14f) | Head capsule spheroidal, often dorsoventrally or laterally flattened; with conspicuous mandibles. Mouthparts mandibulate, symmetrical, nonprotractile, prognathous or hypognathous; gena and gula regions well developed. Clypeus mostly transversely oriented; labrum single or absent. Mandibles dicondylic, typically falcate and incisiform, occasionally with powerful adduction/abduction. Maxillae joined to labium in a distinctive maxillolabial apparatus, indicating interregional co-optation of mouthpart elements | Hymenoptera (Apocrita) | Adults | Insects, nectar, unknown, blood, sporangia, omnivory, seeds, flowers, foliage, fruits, endoparasitic, nectar and pollen | E. Jurassic to recent | Grandi (1924), Bucher (1948), Seghal (1963), Whiting et al. (1989) |
| 6 | **Raptorial Ectognathate**<br>($n = 17$)<br>(Fig. 17.2c, d) | Head capsule spheroidal to transversely elongate, with nonprotractile, compact mouthparts. Mouthparts mandibulate, hypognathous, and symmetrical. Clypeus transversely elongate, with an extensive cibarial pump bulge; labrum transversely elongate, lower margin setate. Mandibles robust, raptorial, dicondylic, strongly incisiform, powered by adduction/abduction. Maxillary lacinia fused to galea | Odonatoptera | Adults | Insects | E. Pennsylvanian to recent | Chao (1953), Short (1955), Mathur and Mathur (1961), Popham and Bevan (1979) |

(continued)

**Table 17.2** (continued)

| | Mouthpart class number studied, figured material[b] | Description of mouthpart morphology[c] | Traditionally included clades[d] | Life stages[e] | Principal diets[f] | Geochronologic range[g] | Key references[h] |
|---|---|---|---|---|---|---|---|
| | | forming a raptorial mala. Interregional co-optation of mouthparts absent | | | | | |
| 7 | **Labial Mask**[k]<br>(*n* = 5)<br>(Fig. 17.2e) | Head spheroidal to moderately transversely elongate; mouthparts mandibulate, symmetrical, hypognathous, and nonprotractile except labium. Clypeus equant, supporting a modest cibarial pump; labrum transverse, mobile; genae and gula well developed. Hypopharynx not evident. Mandibles dicondylic, quadrate, strongly incisiform, powerful adduction/abduction. Maxillae with palps and elongate galeolaciniae, each with several acuminate incisors. Labium highly modified into a protractile, rapid strike organ (labial mask), with an elongate postmentum and prementum bearing a pair of acuminate, hook-like paraglossae for impalement. No interregional co-optation of mouthpart features | Odonatoptera | Naiads | Insects, small aquatic vertebrates | E. Pennsylvanian to recent | Marshall (1914), Munscheid (1933), Asahina (1954), Snodgrass (1954), Watson (1955), Popham and Bevans (1979) |
| 8 | **Mortar and Pestle**<br>(*n* = 27)<br>(Figs. 17.7a, 17.5e (i) and 17.8a) | Head capsule spheroidal, with nonprotractile, compact mouthparts. Mouthparts mandibulate, overwhelmingly hypognathous, symmetrical, or with single element asymmetry; gena prominent, gula variably developed. Clypeus prominent, bearing a bulging cibarial pump; labrum inner surface modified into a projecting process (pestle) received by the sitophore depression (mortar) of the hypopharynx. Mandibles dicondylic and variable in shape. Maxilla and labium typical, except lacinia modified into a pick. Limited co-optation of mouthpart elements | Psocoptera; Phthiraptera (Amblycera, Ishnocera) | Nymphs, adults | Ectoparasitic, omnivory, fungi, dead matter, dead plants | E. Permian to recent | Badonnel (1931), Weber (1938), Cope (1940), Buckup (1959), Haub (1967) |

| | | | | | | | |
|---|---|---|---|---|---|---|---|
| 9 | **Rostrate**<br>($n = 13$)<br>(not figured) | Head capsule prolonged vertically into a rostrum, with nonprotractile, hypognathous mouthparts located at tip. Mouthparts mandibulate, symmetrical; gena expansive and gula typically well developed. Clypeus broad, quadrate, housing a slightly bulging food pump, occasionally fused to labrum. Mandibles dicondylic, serrated, or incisiform, with adduction/abduction. Maxillary stipes rod-like; labium usually lacking a glossa or paraglossa. Interregional co-optation of mouthparts absent | Mecoptera; Neuroptera (Nemopteridae) | Adults | Insects, foliage, dead matter, carrion, unknown | E. Permian to recent | Otanes (1922), Heddergott (1938), Acker (1958), Hepburn (1969), Picker (1987) |
| 10 | **Rhynchophorate**<br>($n = 18$)<br>(Fig. 17.9f) | Head capsule generally anteriorly prolonged, often considerably so, into a nonprotractile, rhynchophorous beak, with mouthparts located at the tip. Mouthparts mandibulate, with single element asymmetry; gena well developed, often expansive; gula well developed, elongated into a pregula. Clypeus quadrate to uncommonly elongate; labrum absent. Mandible ratchet shaped or tetragonal, with endodont or exodont dentition, moved by rope muscles on a ball-and-socket condyle. Maxilla and labium often abbreviated, with stubby palps. Interregional mouthpart co-optation absent | Coleoptera (Curculionoidea) | Adults | Wood[i], seeds, foliage, grain, plants, fungi, pollen, flowers | L. Triassic to recent | Dennell (1942), Donges (1954), Morimoto (1962), Muñiz and Barrera (1969), Brack-Egg (1973) |
| 11 | **Pectinate**<br>($n = 26$)<br>(not figured) | Head capsule spheroidal, often dorsoventrally flat, with nonprotractile, mandibulate, hypognathous mouthparts, some with large mandibular, forwardly projecting tusks, others modified into pectines. | Ephemeroptera | Naiads | Detritus, insects, plants | E. Permian to recent | Strenger (1954), Brown (1961), Froehlich (1964), Schonmann (1981) |

(continued)

**Table 17.2** (continued)

| | Mouthpart class number studied, figured material[b] | Description of mouthpart morphology[c] | Traditionally included clades[d] | Life stages[e] | Principal diets[f] | Geochronologic range[g] | Key references[h] |
|---|---|---|---|---|---|---|---|
| | | Clypeus and labrum variously shaped, undivided. Hypopharynx trilobed, with extensive setation. Mandibles dicondylic, distally expanded, planar, with pectines and prosthecae; musculation of adduction/abduction. Maxillae and labium generalized, but palpi either reduced and penicillate or formed into pectines. Mouthpart interregional co-optation absent | | | | | |
| 12 | **Mouthbrush** (*n* = 34) (Fig. 17.9e) | Head capsule spheroidal but often with projecting pectinate fan, rake, or leaf-mining structures. Mouthparts mandibulate, prognathous, symmetrical, often retractile; gena comprising much of the head capsule; gula typically prominent. Clypeus often fused to frons, labrum vertical with an expanded margin often a retractile pectinate fan. Hypopharynx of long, robust setae. Mandibles dicondylic, falcate, pectinate, incisiform, and with a prominent mesial brush or leaf-mining features; muscles of adduction/abduction. Maxillae and labium reduced, palps stubby or absent. Interregional mouthpart co-optation of mouthparts absent, but functional coordination important in filtering, leaf mining, and other mouthpart processes | Diptera (Nematocera); Coleoptera (Buprestidae); Lepidoptera (Phyllocnistidae, Gracillariidae) | Larvae | Detritus, insects, dead plants, omnivory, wood[j], wood rots, dead matter, mycelia, blood | L. Triassic to recent | Tragardh (1913), Jaywickreme (1940), Shalaby (1956), Gardner et al. (1973), Schremmer (1950a, b), Felix (1962), Craig (1977), Müh (1985), Borkent and Wood (1986) |

| | | | | | | | |
|---|---|---|---|---|---|---|---|
| 13 | **Mandibulobrustiate**<br>($n = 28$)<br>(not figured) | Head capsule spheroidal to broadly ellipsoidal, anteriorly prolonged. Mouthparts compact, mandibulate, symmetrical, nonprotractile, and prognathous or hypognathous. Clypeus shape various, fused to frons; labrum vertical, margin often pectinate. Hypopharynx bilobed, with lateral brushes (brustia) associated with a labiomaxillary spinneret; mandibles dicondylic, tetragonal, incisiform with a brush or hair tufts; movement by adduction/abduction. Maxilla integrated with labium into a labiomaxillary complex; palps typically stubby or mucronate. Interregional mouthpart co-optation present | Trichoptera | Larvae | Detritus, dead plants, omnivory, crustaceans, insects, wood rots | L. Triassic to recent | Botosaneanu (1956), Sattler (1958), Frochot (1962), Fotius-Jaboulet (1964), Cowley (1978) |
| 14 | **Sericterate**<br>($n = 125$)<br>(not figured) | Head capsule spheroidal, often withdrawn into prothorax; mouthparts nonproboscate, generally protractile. Gena expanded dorsoventrally forming most of cranium; gula variable. Mouthparts mandibulate, prognathous, or hypognathous, symmetrical. Clypeus quadrate or transversely elongate, often fused to frons; labrum vertical, undivided, with projecting setae. Hypopharynx bilobed, associated with labial spinneret. Mandibles dicondylic, triangular, palmate, with adduction/abduction. Maxilla and labium well developed but with small, stubby, papillate, or mucronate palps. Interregional co-optation of mouthparts present | Lepidoptera (Glossata); Hymenoptera | Larvae | Foliage, insects, ectoparasitic, regurgitate, endoparasitic, wood[j], roots, nectar and pollen, fruits, seeds, flowers, pollen | L. Pennsylvanian to recent | Grandi (1924), Lopez (1932), Dumbleton (1952), Arora (1959), Briolini (1960, 1965), Randriamamonj (1963), Wong (1963), Ippolito (1977) |

(continued)

**Table 17.2** (continued)

| | Mouthpart class number studied, figured material[b] | Description of mouthpart morphology[c] | Traditionally included clades[d] | Life stages[e] | Principal diets[f] | Geochronologic range[g] | Key references[h] |
|---|---|---|---|---|---|---|---|
| **Mandibulocanaliculate Series** | | | | | | | |
| 15 | **Fossate Complex** (*n* = 15) (not figured) | Head capsule ellipsoidal and anteriorly prolonged, nonretractile; mouthparts projecting, mandibulosuctorial, prognathous, and symmetrical. Clypeus shape variable, fused to labrum; genae expansive; gula mostly absent. Hypopharynx lobate, adjoined to labrum. Mandible dicondylic, grooved and falcate, lacking teeth, forming a complementary interlocking structure when paired with similarly shaped maxillary lacinia, both forming a gutter (fossa) for uptake of food. Labium reduced, bearing only palpi. Interregional co-optation of mouthparts present | Neuroptera | Larvae | Insects, sponges | E. Permian to recent | Lozinski (1908), Tillyard (1923), Rousset (1936), Korn (1943), Gurney (1947), Principi (1954), Wundt (1961), Zwick (1967) |
| 16 | **Tubulomandibulate** (*n* = 20) (not figured) | Head capsule spheroidal to broadly ellipsoidal, often partly hidden by prothorax; mouthparts often projecting, prognathous or hypognathous, symmetrical, and nonprojectile. Clypeus quadrate or transversely elongate, fused to frons, labrum indistinct; genae much expanded dorsoventrally; gula mostly well developed. Hypopharynx lingulate. Mandible dicondylic, falcate, internally tubulate to rarely forming an external gutter; edentate; muscles of adduction/abduction. Maxillae and labium reduced, often bearing only palpi. Interregional co-optation of mouthpart elements absent | Coleoptera (Hygrobiidae, Dytiscidae, Gyrinidae, Haliplidae, Brachyspectridae, Lycidae, Drilidae, Phenogodidae, Telegeusidae, Lampyridae, Cantharidae) | Larvae | Insects, soft invertebrates | L. Triassic to recent | Vogel (1915), Blunck (1918), Raj (1943), Noars (1956), Jaboulet (1960), DeMarzo (1976a, b, 1977) |

| | | | | | | | |
|---|---|---|---|---|---|---|---|
| 17 | **Mouthhook** (*n* = 55) (Fig. 17.14a) | Head inconspicuous to hidden, sunk into thorax, protractile; mouthparts reduced, developmentally remodeled. Clypeus variously shaped, difficult to discern, fused to frons; labrum vertical, with modified margin. Hypopharynx withdrawn internally, forming upper branch (cornua) of the endoskeletal cephalopharyngeal apparatus; the lower branch formed by labial elements. Mandibles range from dicondylic and distinct, moving in the lateral plane, to vertically oriented and fused to maxillary elements into hook-like to falcate mouthhooks, with adduction/abduction muscles in the sagittal plane. Extensive interregional co-optation of mouthparts present | Diptera (Cecidomyiidae, Brachycera) | Larvae | Insects, dead matter, carrion, dead plants, detritus, endoparasitic, excreta, roots, plant fluids, soft invertebrates, fruit, seeds, live plants | E. Jurassic to recent | Greene (1929), Ludwig (1949), Schremmer (1952), Wirth (1952), Valliant and Delhom (1956), McAlpine (1966), Foote (1970), Lobanov (1970), Sluss and Foote (1971) |
| **Labellate Series** | | | | | | | |
| 18 | **Haustoriate** (*n* = 30) (not figured) | Head capsule squarose, spheroidal, dorsoventrally flattened, with a retractile proboscis; mouthparts hypognathous, symmetrical. Clypeus quadrate and fused to the frons; labrum typically vertically oriented; genal region prominent, gular region moderately developed to absent. Hypopharynx conspicuous, multilobate, fleshy, bearing a network of pseudotracheae-like canals, forming the prominent haustorium. Maxillae and labium generalized, lacking some elements but palps prominent. Interregional co-optation of mouthpart elements absent | Trichoptera | Adults | Nectar, omnivory | M. Triassic to recent | Durand (1955), Crichton (1957), Klemm (1966), Korboot (1964), Riek (1968) |
| 19 | **Labellate** (*n* = 88) (Fig. 17.11b) | Head capsule spheroidal to ellipsoidal, dorsoventrally compressed, with a variously deployed, retractile proboscis; mouthparts labellate, hypognathous and protractile. Clypeus quadrate, fused to frons; labrum vertical; gena and gula typically well developed. Hypopharynx chitinized, often a salivarium. Mandibles | Diptera (Tipuloidea, Tanyderidae, Ptychopteridae, Bibionoidea, Brachycera); Mecoptera (Nannochoristidae) | Adults | Nectar, dead matter, insects, live plants, unknown, dead plants, grain, blood, nontrophic, leaves, excreta, plant fluids, fungi | M. Triassic to recent | Williams (1933), Cregan (1941), Hoyt (1952), Bletchly (1954), Gouin (1958), Chassagnard and Tsacas (1974), Zaitland and Larsen (1984) |

(continued)

**Table 17.2** (continued)

| | Mouthpart class number studied, figured material[b] | Description of mouthpart morphology[c] | Traditionally included clades[d] | Life stages[e] | Principal diets[f] | Geochronologic range[g] | Key references[h] |
|---|---|---|---|---|---|---|---|
| | | absent. Maxilla bearing filiform to clavate, two-segmented palps. Labial palps conjoined to form a fleshy, distally expanded organ (labellum) for sponging food through a branching network of pseudotracheae, the distal surface of which may be invested with small teeth for rasping surfaces. Interregional co-optation of mouthpart elements present | | | | | |
| **Siphonate Series** | | | | | | | |
| 20 | **Siphonate**<br>(*n* = 53)<br>(Fig. 17.14e) | Head capsule spheroidal to transversely ellipsoidal, nonprotractile, bearing a short to very long proboscis. Mouthparts a siphonate proboscis, hypognathous, capable of being recoiled. Clypeus quadrate, rarely transversely elongate, a sclerite fused to frons; labrum quadrate to broadly triangular, margin a pilifer. Hypopharynx diminutive. Mandibles absent, rarely vestigial. Each maxilla bears an elongate galea conjoined to its opposite by an interlocking system to form a double tubular siphon with a central food canal; lacinia absent; palps stubby or a brush. Labium bearing a pair of large palps. Co-optation of mouthparts absent | Lepidoptera (Glossata) | Adults | Nectar, unknown, pollen, nontrophy, fluidized mud, plant fluids, foliage, omnivory, fruit | E. Permian to recent | Eastham and Eassa (1955), Vasudeva (1956), Puri (1958), Rouchy (1964), Kristensen (1968a, b), Miller (1971), Common (1973) |
| 21 | **Siphonomandibulate**<br>(*n* = 5)<br>(not figured) | Head capsule spheroidal to ellipsoidal in the vertical position; with a long nonretractile or recoilable proboscis. Mouthparts hypognathous, rarely prognathous, siphonate, mandibulate. Clypeus quadrate; labrum quadrate, with margin variable; gena modestly and gula well developed. Hypopharynx various. Mandibles functional; dicondylic, triangular to falcate, dentition variable, adduction/ | Coleoptera (Hydrophilidae, Meloidae, Rhipiphoridae) | Adults | Nectar, dead plants | E. Cretaceous to recent | Handschin (1929), Rivnay (1929), Grinfel'd (1975), Barth (1985) |

| | | | | | | | |
|---|---|---|---|---|---|---|---|
| | | abduction movement. Maxillae with conjoined galeae in some forms and conjoined palps in others that form the siphon. Labium normal. Interregional co-optation of mouthparts absent | | | | | |
| **Glossate Series** | | | | | | | |
| 22 | **Glossate** (*n* = 29) (Fig. 17.12c) | Head capsule spheroidal to ellipsoidal, dorsoventrally flattened, often transversely elongate; mouthparts hypognathous, protractile, formed into a maxillolabial complex. Clypeus quadrate to transversely elongate, labrum vertically elongate; gena and gula regions small. Hypopharynx a lobe, inconspicuous. Mandibles dicondylic, flattened, excavate, incisiform and a basal brush; with adduction/abduction movement. Maxilla fused to labium by a membrane into a labiomaxillary complex whose basal sclerites are modified into a lever-based protraction/retraction mechanism; the labium modified into a medial glossa with hairs, transverse ridges and a brush. Interregional mouthpart co-optation present | Hymenoptera (Torymidae, Sapygidae, Scoliidae, Vespoidea, Sphecoidea, Apoidea); Neuroptera (Nemopteridae) | Adults | Nectar, pollen and nectar, omnivory, pollen, insects | L. Jurassic to recent | Demoll (1908), Ulrich (1924), Strenger (1952), Winston (1979), Michener (2000) |
| **Stylate Nonhaustellate Series** | | | | | | | |
| 23 | **Entognathous Stylate** (*n* = 4) (not figured) | Head capsule spheroidal to ellipsoidal; mouthparts projecting, often canted in a position intermediate between prognathous and hypognathous. Clypeus transverse, subsegmented; labrum quadrate or transversely elongate, distal margin acute to acuminate; gena prominent, gula moderately developed. Hypopharynx often styliform. Mandibles monocondylous, stylate, and edentate; musculation protraction/retraction. Maxilla always bearing stylate laciniae and frequently stylate | Protura; Collembola (Neanuridae) | Nymphs, adults | Mycelia, fungi | E. Devonian to recent | Tuxen (1959), François (1959, 1969), Wolter (1963), Eisenbeis and Wichard (1987) |

(continued)

**Table 17.2** (continued)

| | Mouthpart class number studied, figured material[b] | Description of mouthpart morphology[c] | Traditionally included clades[d] | Life stages[e] | Principal diets[f] | Geochronologic range[g] | Key references[h] |
|---|---|---|---|---|---|---|---|
| | | galeae; palps pectinate, short, or absent. Labium with glossae and paraglossae joined to form an oral cone containing up to nine stylets; palps small. Interregional co-optation of mouthparts absent | | | | | |
| 24 | **Ectognathous Stylate**<br>($n = 13$)<br>(not figured) | Head capsule spheroidal, anteriorly prolonged; visible mouthparts exserted protractile, mostly a prognathous fascicle of stylets two, four, or six in number. Clypeus separate or fused to frons, shape various; labrum also variable, often a brace to stylets. Hypopharynx variable. Mandibles secondarily monocondylic, rarely absent, tetragonal or stylate edentate; moved by protraction/retraction. Maxillae with stylate laciniae and often stylate galeae; palps normal, four-segmented. Labium normal, three-segmented palps and glossae forming a labial cone housing the stylets. Interregional co-optation of elements absent | Orthoptera (Cooloolidae); Coleoptera (Sphaeriidae, Rhysodidae, Leiodidae, Cerylonidae, Corylophidae, Eucinetidae) | Adults, one larva | Mycelia, fungi, spores, roots, wood rots, insects, unknown | L. Triassic to recent | Besuchet (1972), Sen Gupta and Crowson (1973), Lawrence and Stephan (1975), Dajoz (1976), Vit (1977, 1981), Rentz (1980), Pakaluk (1987) |
| 25 | **Laciniate**[l]<br>($n = 7$)<br>(Figs. 17.5a–d, e (ii, iii), 17.12a and 17.13a) | Head capsule spheroidal to ellipsoidal, flattened; mouthparts hypognathous (without rostrum) or prognathous (with rostrum), transitional between mandibulate and stylate conditions. Clypeus quadrate to triangular, often fused to frons, subsegmented, absent, a pronounced swelling; labrum triangular, small to elongate, or massive; gena narrow, expanded laterally, or broadly developed; gular region prominent in prognathous forms. Mandibles dicondylic, but highly elongate with small | Acercaria (Cacurgidae, Paleuthygrammatidae, Synomaloptilidae); Permopsocida (Archipsyllidae, Permopsocidae, Protopsyllidiidae, Psocidiidae) | Nymphs, adults | Inferred: insects, fungi, plants | E. Pennsylvanian to L. Cretaceous | Rasnitsyn (1980), Shear and Kukalová-Peck (1990), Grimaldi (2003), Huang et al. (2016), Yoshizawa and Lienhard (2016) |

| | | | | | | | |
|---|---|---|---|---|---|---|---|
| | | terminal teeth, or stylet-like. Maxilla elongate, long palps, and stylet-like laciniae with margin serrations or terminal teeth; galeae elongate. Labium prolonged, often forming a cover for styliform mouthparts. Interregional co-optation of elements not evident | | | | | |
| **Stylate-Haustellate Series** | | | | | | | |
| 26 | **Robust Beak**[k]<br>($n = 4$)<br>(Fig. 17.3a–d) | Head capsule spheroidal, often elongated transversely along compound eyes. Mouthparts hypognathous to slightly prognathous, piercing and sucking, typically of five stylets, two mandibular, one hypopharyngeal, two lacinial, each internally hollow, connected by an interlocking mechanism among stylets enclosing incoming food and outgoing salivary tubes; forming a beak from short-truncate to long-gracile. Clypeus domed, subsegmented; labrum subsegmented, broad to narrow triangular brace; gena narrow and gula region variably developed. Mandibles strongly sclerotized, robust, engaged by protractor–retractor muscles. Maxilla bearing two lacinial stylets less robust than the similar muscle-engaged mandibular stylets; five-segmented palps support the beak, extending to its tip. Labium reduced, lacking palps, with a pair of lobes in some. Integration of mouthpart elements pronounced | Paleodictyopteroidea | Adults, nymphs | Inferred: plant fluids, seeds, sporangia, nectar, pollen, live plants | E. Pennsylvanian to L. Triassic | Kukalová-Peck (1972, 1990), Carpenter (1992), Shear and Kukalová-Peck (1990), Pecharová et al. (2015), Prokop et al. (2016), Pecharová and Prokop (2017) |

(continued)

**Table 17.2** (continued)

| | Mouthpart class number studied, figured material[b] | Description of mouthpart morphology[c] | Traditionally included clades[d] | Life stages[e] | Principal diets[f] | Geochronologic range[g] | Key references[h] |
|---|---|---|---|---|---|---|---|
| 27 | **Segmented Beak** (*n* = 118) (Figs. 17.5e (iv), 17.6a–f and 17.12a, d) | Head capsule highly variable, spheroidal, ellipsoidal, or anteriorly prolonged cylindrically; mouthparts prognathous, hypognathous, or opisthognathous, two pairs of protractile stylets. Clypeus prominent, subdivided, typically a bulging cibarial pump with rows of compressor muscles; genae and gula developed to absent. Hypopharynx nonstylate, but a salivary syringe with piston pump. Mandibles acondylic, stylate, with serrate, barbate, or edentulous lateral margins, attached to lever system of protractor/retractor muscles. Maxilla with blade-like lacinial stylets working in tandem with mandibles for tissue penetration; galea absent. The only evident structure of the labium is the segmented beak ensheathing the stylets, likely consisting of fused palpi | Hemiptera (Sternorrhyncha, Fulgoromorpha, Cicadomorpha, Coelorrhyncha, Heteroptera) | Nymphs, adults | Foliage, insects, roots, plant sap, live plants, blood, mycelia, seeds, carrion, detritus, other foods | L. Pennsylvanian to R recent | Tower (1914), Davidson (1914), Snodgrass (1921), Neiswander (1926), Weber (1929), Willis (1949), Barth (1952, 1953), Benwitz (1956), Rathore (1961), Arora and Singh (1962), Parsons (1966), Verma et al. (1973), Rieger (1976), Cobben (1978) |
| 28 | **Buccal Cone** (*n* = 6) (not figured) | Head ellipsoidal, held porrect; mouthparts short to rarely long; gena absent; gula prominent. Mouthparts prognathous, protractile, with two stylets enveloped in a haustellum, rarely a single stylet. Clypeus often quadrate, above a cibarial pump; labrum various but with an aperturate or notched apex. Hypopharynx long, stylate, powered by a salivary syringe. Mandibles monocondylous, diminutive, edentulous, excavate, with protractor/retractor muscles. Maxillae reduced, of basal sclerites and tubulate galeae forming the buccal cone. Labium reduced, bearing a ligula | Phthiraptera (Anoplura, Rhynchophthirina) | Nymphs, adults | Blood, lymph, ectoparasitic | E. Cretaceous to recent | Peacock (1919), Stojanovich (1945), Mukerji and Sen Sarma (1955), Ramcke (1965), Hirsch (1986) |

| | | | | | | | |
|---|---|---|---|---|---|---|---|
| | | modified into a stylet. Interregional mouthpart co-optation present | | | | | |
| 29 | **Mouthcone** (*n* = 9) (Figs. 17.7a–c and 17.8a) | Head capsule spheroidal and downwardly prolonged; mouthparts opisthognathous, stylate-haustellate, consisting of three stylets, housed in an asymmetric mouthcone. Clypeus equant or transverse elongate, undivided with an aperturate terminus. Hypopharynx with super-lingulae, two lateral ovoidal sclerites, and basal sitophore. Left mandible stylate, edentulous, protractile, with lever-based protraction-retraction; right mandible absent. Maxillae contributing to mouthcone sidewall and two lacinial stylets; palp abbreviated. Labium also part of mouthcone wall elements and stubby, one-segmented palps. Interregional co-optation of mouthparts present | Thysanoptera | Nymphs, adults | Foliage, flowers, pollen, fungi, insects, unknown | E. Permian to recent (includes Lophoneurida) | Reyne (1927), Jones (1954), Risler (1957), Mickoleit (1963), Heming (1978), Moritz (1989) |
| 30 | **Monostylate/ Distylate** (*n* = 8) (not figured) | Head capsule equant to prolonged downwardly, dorsoventrally flattened; proboscis long to very long, mouthparts prognathous, stylate-haustellate, consisting of one or two nonretractile stylets. Clypeus equant, rarely transversely elongate, fused to frons; labrum very elongate; genae and gula well to minimally developed. Hypopharynx long, stylate (the "dagger"); epipharynx membranous. Mandibles absent. Maxilla reduced, bearing a palpus, often a brush, and a lacinia elongate and fleshy or an acuminate stylet. Labium mostly complete, with two-segmented palps. Interregional co-optation of mouthparts present | Diptera (Asilidae, Acroceridae, Empididae, Bombyliidae) | Adults | Insects, nectar | M. Jurassic to recent | Wesché (1902, 1904, 1909), Crampton (1942), Snodgrass (1943), Hoyt (1952), Bletchley (1954) |

(continued)

**Table 17.2** (continued)

| | Mouthpart class number studied, figured material[b] | Description of mouthpart morphology[c] | Traditionally included clades[d] | Life stages[e] | Principal diets[f] | Geochronologic range[g] | Key references[h] |
|---|---|---|---|---|---|---|---|
| 31 | **Distylate/ Tetrastylate** ($n = 9$) (Fig. 17.11b, c, e) | Head capsule ellipsoidal, prolonged downward, and dorsoventrally flattened; mouthparts hypognathous, retractile, with two or four stylets; labellum large. Clypeus shape various, fused to frons; labrum transversely elongate or equant; genae and gula prominent to small. Hypopharynx long, stylate; epipharynx stylate, adjoined to labrum. Mandibles absent. Maxillae with a basal sclerite, palpi of variable size, shape, and ornamentation; when present, laciniae stylate, acuminate, long. Labium reduced, with two-segmented palps and a prominent labellum. Interregional mouthpart co-optation present | Diptera (Anisopodidae, Rhagionidae, Mydidae, Therevidae, Ironomyiidae) | Adults | Insects, nectar, plant sap, unknown | E. Jurassic to recent | Bletchley (1955), McAlpine (1966), Chassagnard and Tsacas (1974), Irwin and Lyneborg (1981), Zaitlin and Larsen (1984) |
| 32 | **Hexastylate** ($n = 27$) (Fig. 17.11a, d) | Head capsule spheroidal, some dorsoventrally compressed; proboscis short to very long, mouthparts hypognathous, stylate-haustellate, hexastylate to very rarely tetrastylate, nonretractile. Clypeus equant, undivided, rarely vertically elongate, fused to frons; labrum long, margin spinose, often an acute tip. Epipharynx stylate, adjoined to labrum. Hypopharynx stylate, with a salivary syringe. Mandibles monocondylic, stylate (piercing) or blades (slashing), serrated, rarely edentulous, musculated by lever-based protraction/ retraction. Maxillary sclerites variable; palps variable; galea mostly absent; lacinia acuminate stylets. Labium reduced; glossae and paraglossae often absent; palpus two-segmented, a haustellum encompassing the six stylets. Interregional co-optation of mouthparts present | Diptera (Nematocera) | Adults | Blood and lymph, insects, nectar, seeds, nontrophic, dead animals | L. Triassic to recent | Adler and Theodor (1926), Gibbins (1938), Robinson (1939), Gad (1951), Bonhag (1951), Hoyt (1952), Schiemenz (1957), Wenk (1962), Stoffolano and Yin (1983), Sutcliffe and Deepan (1988) |

| | | | | | | | |
|---|---|---|---|---|---|---|---|
| 33 | **Tristylate**<br>($n = 15$)<br>(Fig. 17.10a–h) | Head capsule spheroidal; proboscis long, mouthparts hypognathous, protractile, stylate-haustellate, housing three stylets. Clypeus fused to frons; labrum median, elongate and bracing a stylet ensemble; genae enveloping most of cranium; gula inconspicuous. Hypopharynx a salivary syringe with a piston pump; epipharyngeal stylet adjoined to labrum. Mandibles absent. Maxillae with four-segmented palpi, lacinial stylets, and galeae, which are modified into a large, robust lever for lacinial stylet protraction. Labium reduced, palpi housing three stylets during engagement. Interregional co-optation of mouthparts present | Siphonaptera | Adults | Blood, lymph | M. Jurassic to recent | Lass (1905), Rothschild (1906), Wenk (1953), Hopkins and Traub (1955), Kluge (2002) |
| 34 | **Tubulostylate**<br>($n = 10$)<br>(Fig. 17.14c) | Head capsule spheroidal, often dorsoventrally flattened, proboscate; mouthparts prognathous to hypognathous, protractile, stylate-haustellate with a single stylet. Clypeus mostly equant in shape, fused to frons; labrum vertical and very elongate, divided or undivided, fused to epipharynx; genae and gula well developed. Hypopharynx a bulbous reservoir ending in aperturate lobe. Mandibles absent. Maxillae reduced, bearing one segment, pectinate palpi. Labium reduced, lacking glossae and paraglossae, with three-segmented palps modified into a long, tubular stylet with cutting, terminal denticles. Interregional co-optation of mouthparts present | Diptera (Glossinidae, Hippoboscidae, Streblidae, Nycteribiidae) | Adults | Blood, lymph | L. Paleocene to recent | Hansen (1903), Jobling (1926, 1928, 1929, 1933), Snodgrass (1943), Elzinga and Broce (1986) |
| 35 | **Siphonostylate**<br>($n = 2$)<br>(Fig. 17.14d) | Head capsule spheroidal or transversely elongated; proboscis typically very long, mouthparts hypognathous, retractile, stylate-haustellate, consisting of one stylet. Clypeus equant, fused to frons; labrum vertically elongate; genae and gula well to minimally developed. Hypopharynx | Lepidoptera (Erebidae) | Adults | Blood, lymph, fruit | E. Pleistocene to recent | Srivastava and Bogawat (1969), Bänziger (1970, 1980, 1986), Chaudonneret (1986), Zaspel et al. (2012) |

(continued)

**Table 17.2** (continued)

| | Mouthpart class number studied, figured material[b] | Description of mouthpart morphology[c] | Traditionally included clades[d] | Life stages[e] | Principal diets[f] | Geochronologic range[g] | Key references[h] |
|---|---|---|---|---|---|---|---|
| | | absent; epipharynx membranous. Mandibles absent. Maxillae with palpus and lacinia absent; galeae conjoined into a specialized, piercing, stylate siphon. Labium reduced, only three segmented, penicillate palps present. No interregional co-optation of mouthpart elements present | | | | | |
| **Agnathate Series** | | | | | | | |
| 36 | **Reduced Trophic** ($n = 52$) (Fig. 17.14b) | Head capsule shape variable, mostly free or commonly a cephalothorax, retractile or nonretractile; proboscis absent except rarely a short process. Clypeus generally absent, when present, equant to transversely elongate, undivided or fused; labrum generally absent, when present fused to clypeus. Hypopharynx and epipharynx mostly absent. Mandibles often absent, when present, hooked, ensiform, falcate; edentulous and weakly musculated by adduction/abduction. Maxillae mostly absent, when present a main body and small palps present. Labium frequently absent, when present, palpi or a vestigial labellum present. Interregional co-optation of mouthpart elements absent | Hemiptera (Coccoidea); Coleoptera (Ripiphoridae); Strepsiptera; Diptera (Canthyloscelidae, Nymphomyiidae, Oestridae, Acroceridae); Hymenoptera (Apocrita) | Nymphs, larvae, adults | Nontrophic, endoparasitic, ectoparasitic, wood rots, seeds, fruits, nectar, insects | E. Permian to recent | Silvestri (1924), Balachowsky (1937), Russell (1941), Bohart (1943), Clancy (1946), Jancke (1955), Theron (1958), Kinzelbach (1966), Giliomee (1968) |
| 37 | **Nontrophic** ($n = 9$) (not figured) | Head capsule transversely elongate, nonproboscate; mouthparts, when present, hypognathous, typically nonfunctional and asymmetrical. Clypeus equant, lacking evident food pump; labrum variously shaped, undivided; genae modestly to minimally developed, gula small. Hypopharynx absent, rarely a bulbous | Ephemeroptera; Lepidoptera (Hepialoidea, Lasiocampidae, Saturniidae) | Adults | Nontrophic, insects | L. Jurassic to recent | Murphy (1922), Schmitt (1938), Taylor (1957), Rouchy (1964), Nielson and Scoble (1986) |

| | | | | | | | |
|---|---|---|---|---|---|---|---|
| | | reservoir; epipharynx membranous or absent. Mandibles absent or vestigial. Maxillae reduced, laciniae absent, less commonly galeae; palps short, stubby, or vestigial. Labium medially elongate; rudimentary palps present but glossae and paraglossae absent. Inter-regional co-optation of mouthparts absent | | | | | |

[a]This table is recently upgraded from its two precursor studies (Labandeira 1990, 1997); in particular, three, additional mouthpart classes now are included. The constituent taxa and geochronologic ranges of each mouthpart class also was updated, based on Walker et al. (2013)

[b]The number of examined taxa for each mouthpart class, ranging from 2 to 291, is the basis for the data at right. About two-thirds of the mouthpart classes are illustrated in Figs. 17.1–17.14. Additional exemplars for each mouthpart class may improve the accuracy of these data. See Labandeira (1990) for additional details

[c]Based on the number of taxa examined, indicated at left

[d]Taxa with unusual mouthparts from the larger taxon in which they occur may have been assigned to other mouthpart classes. Conversely, the cited, more inclusive taxa may not include all members of the stated mouthpart class

[e]The terms nymph, naiad, larva, and adult are used in the North American and British sense—and not the continental European sense—in which a larva pertains to an immature endopterygote insect with holometabolous metamorphosis. A nymph is a terrestrial, nonholometabolous, preadult insect; by contrast, a naiad is an aquatic, nonholometabolous, preadult insect

[f]These are commonly understood translations of diet-type terms used in Labandeira (1990), and are taken directly from the literature pertaining to the diets (when recorded) of insects whose mouthparts and heads were examined. The list is presented in rank order of decreasing frequency

[g]Includes indirect and direct evidence for the presence of a mouthpart class. The standard geologic timescale used in this table is Walker et al. (2013). The geochronology is restricted to the epoch level in which Early (E), Middle (M), and Late (L) are used for the Paleozoic and Mesozoic Eras and the standard epochs of Paleocene, Eocene, Oligocene, Miocene, Pliocene, Pleistocene, and Holocene are used for the Cenozoic Era. (See Table 17.1 for additional details)

[h]Most of these references predate 1990. Line drawings are more useful for interpreting general head and mouthpart structure, than less utilitarian micro-CT scans, SEM micrographs, and other modern techniques

[i]After this phenetic mouthpart analysis was published in 1990, it was found that Archaeognatha, indeed, have dicondylous mandibles, although the anterior articulation is a weak one (Blanke et al. 2015). For this reason, the name of this mouthpart class has been changed from the erroneous Monocondylate mouthpart class to the Maxillopalpate mouthpart class, reflecting the other distinguishing mouthpart feature of these insects—their long, massive, and leg-like maxillary palps

[j]The diet "wood" is here interpreted as insects feeding on associated fungi and microorganisms that degrade wood. This is particularly important, as wood, essentially cellulose, is too refractory in chemical nutrients as a viable source of food in insects

[k]The Labial Mask Mouthpart class is an addition. Labial Mask mouthparts of odonatan naiads are distinctive but formerly were assigned to the Raptorial Ectognathate mouthpart class of odonatan adults. Labial Mask mouthparts represent a distinct cluster in the phenetic analysis of Labandeira (1990) that was not previously recognized

[l]The Laciniate and Robust Beak mouthpart classes are added, as indicated by Labandeira (1997), and were not part of the original analyses of Labandeira (1990) as as that study did not involve extinct mouthpart classes

equivalent to 72.8% of the total dataset. Percent coverage of families within each of these same orders was Coleoptera, 83.3%; Diptera, 90.2%; Lepidoptera, 52.2%; Hymenoptera, 77.5%; and Hemiptera, 76.7%.

Because of limitations in the similarity indices of commonly used programs at the time, such as BMDP, SAS, and SPSS, the cluster analysis initially was implemented by a specially coded in FORTRAN Jaccard similarity index equipped to handle an exceptionally large 1365 × 1365 matrix. The similarity matrix then was inputted into a BMDP1M program (Hartigan 1988) that used an unweighted pair-group clustering method, specifically an unweighted average linkage algorithm for joining clusters. The output of the analysis was a dendrogram, shown in Fig. 28 of Labandeira (1990) or Fig. 1 of Labandeira (1997). Criteria for interpreting the dendrogram and determination of mouthpart classes included (1) subcluster discreteness, (2) overall dendrogram topology, (3) cluster branch lengths, (4) cluster similarity levels, (5) chaining among larger subclusters, and (6) taxon membership in the focal subclusters. The analysis was run on a large mainframe computer at The University of Chicago (Labandeira 1990). Specific details of this analysis can be found in Labandeira (1990).

### *17.2.2 Conventions Used in This Report*

Several conventions will be followed in this contribution. The formal taxon, Insecta, is used in its more comprehensive sense, to include the primitively wingless taxa of Zygentoma, Archaeognatha, Diplura, Protura, and Collembola. Consequently, the taxon Insecta is used as a synonym of Hexapoda (Misof et al. 2014), and consistent with the theme of this volume. The term "larva" is used in the North American and British restrictive sense to refer to the immature holometabolan (endopterygote) stage sandwiched between the egg and the pupa. Accordingly, nonholometabolous immature stages are termed nymphs if they are terrestrial or naiads if they are aquatic. The abbreviation "Ma" designates millions of years ago, and refers to dates based on the standard geologic timescale (Walker et al. 2013), provided in Table 17.1. All figures consist of drawings by the author of insects and their mouthparts from original sources cited in the figure captions. These drawings were rendered as accurately as possible and were made to achieve consistency throughout this contribution. All figures are original. Abbreviations of museum repositories for drawn specimens in the figures are the following:

AMNH, American Museum of Natural History in New York, New York, USA
CAS, Canadian Agricultural Survey in Ottawa, Canada
CNU, Capital Normal University paleontological collections in Beijing, China
FMNH, Field Museum of Natural History in Chicago, Illinois, USA
FMW, Fuhlrott Museum in Wuppertal, Germany
ISM, Illinois State Museum in Springfield, Illinois, USA

MCZ, Museum of Comparative Zoology of Harvard University in Cambridge, Massachusetts, USA
MNHN, Muséum National d'Histoire Naturelle in Paris, France
NHM, Natural History Museum in London, UK
NIGP, Nanjing Institute of Geology and Palaeontology, in Nanjing, China
NMS, National Museums of Scotland in Edinburgh, United Kingdom
OSU, Oregon State University entomological collections in Corvallis, Oregon, USA
PICU, Paleontological Institute of Charles University, Prague, Czech Republic
PIN, Paleontological Institute in Moscow, Russia
SMNS, State Museum of Natural History in Stuttgart, Germany
YU, Yunnan University Institute of Deep Time Terrestrial Ecology in Kunming, China

One observation became evident during an examination of the primary literature of fossil insect mouthparts from deposits particularly from Asia. Poor camera-lucida drawings contextualized much of the illustrations of mouthparts. The reason for this is unclear, but the poor quality of drawings typically represented two-dimensional compression-impression (or adpression) specimens as well as the three-dimensional drawings of amber specimens. A higher quality of light illumination, better microscope optics, and greater attention to mouthpart detail may have allowed a better camera-lucida representation of such specimens.

## 17.3 The Fossil Record of Insect Mouthparts

The fossil record of insect mouthparts represents an archive documenting the first appearances of new mouthpart morphologies from insect lineages that evolved new ways of encountering, accessing, processing, and digesting existing and new food sources. Two important features characterize the insect mouthpart record. First, although the archive of mouthpart types is ultimately driven by taxonomic diversity, it also represents the immediate opportunism of new feeding modes that typically precede later taxonomic expansions of the lineages that are the bearers of those mouthparts (Labandeira and Sepkoski 1993). A second aspect is the frequent, rampant level of convergences of mouthpart types (Peris et al. 2017) that likely respond to the limited number of tissue types that plants, fungi, and animals offer to their insect consumers (e.g., Labandeira 2013). These two macroevolutionary features of the insect mouthpart record indicate that there are probably around 40 basic mouthpart types in time and space (Labandeira 1997, and subsequent discoveries) that are determined by a fixed number of basic food resources and by structural restrictions on the mouthpart assemblies designed to access those resources.

### *17.3.1 Direct Versus Indirect Evidence for Insect Mouthparts*

The fossil record of insect mouthparts consists of two separate and complementary records: one direct and the other indirect. The direct fossil record consists of mouthpart assemblies and individual elements found on insects as part of their body-fossil record. The direct record of insects will be emphasized in this contribution. By contrast, the indirect fossil record consists of two categories of evidence. The first category are activities resulting from the foods that insects consume, consisting of distorted plant tissues and the ensuing physical responses of host plants. These host–plant responses include features such as surface abrasion, hole and margin feeding excisions, lesions, stylet paths, and leaf mines as well as resulting scar tissue of cellular hypertrophy and hyperplasia that form callus and other response tissues. This spectrum of evidence can reveal the presence of insects with mandibulate, piercing-and-sucking, scraping, leaf-mining, and other mouthpart types in the absence of mouthpart structures from the body fossils co-occurring in the same fossil deposits. In particular, the patterns of excision on foliage can reveal the direction of mandibular movement on the leaf during chewing, resulting in the creation of major excision arcs and minor cuspules indicating the type of molar and incisor purchase on leaf tissue in the absence of body fossils (Gangwere 1966; Iannuzzi and Labandeira 2008). Similarly, damage to plant vascular tissues can reveal the type of tissues consumed and feeding strategies by piercing-and-sucking insects with stylate mouthparts (Labandeira and Phillips 1996a). Fossil leaf-mining damage to typically flattened plant organs and tissues frequently indicates a culprit with specialized, prognathous mouthparts housed in a dorsoventrally compressed head capsule (Ding et al. 2014).

A second category of indirect evidence is the consequences of feeding resulting in fossilized fecal pellets preserved as coprolites (Scott and Taylor 1983; Labandeira 2002a). Such coprolites occur in abundance during certain time intervals, unveiling digested and often anatomically identifiable contents (Labandeira 2001). Occasionally, evidence is available at other trophic levels, such as predation, and coprolites of mandibulate insects that reveal mouthparts, ovipositors, and other resistant, cuticular structures of prey items (Labandeira 2002b). Seed predation often documents specialized larval or adult mouthparts based on modern analogy and has a distinctive fossil record of seed punctures extending into the Paleozoic (Sharov 1973). A much poorer and recent record commencing during the Early Jurassic is parasitoid attack on insect hosts. Parasitoids are recorded as exit holes from fossil bee cells (Houston 1987) or the emergence of a hymenopteran parasitoid larva from a host adult ant entombed in Dominican amber (Poinar and Miller 2002).

### 17.3.2 *Quality of the Fossil Insect Mouthpart Record*

The fossil record of insect mouthparts has strengths and weaknesses. The direct record of insect body fossils displaying recognizable mouthparts begins with the Rhynie Chert of the Early Devonian in Scotland, at about 415 Ma, consisting of silica-permineralized material. However, the subsequent record of insect mouthparts undergoes an approximate 60–70 million-year-long hiatus (Schachat et al. 2018), ending in a broad diversity of mouthpart types that emerge during the 323–299 Ma interval, corresponding to the Pennsylvanian Period (Table 17.1). At this time, the fossil record becomes almost entirely adpression (Cleal et al. 1990) in nature, corresponding typically to flattened insect fossils preserved between thin layers of fine-grained strata. During this Pennsylvanian to recent time interval, hiatuses between successive fossil-insect occurrences decrease considerably, consisting of time intervals maximally of several million years. The most prolonged of these intervals is the approximately five million-year absence of insects with definable mouthparts during the Early Triassic. Notably, beginning at approximately 130–135 Ma with the appearance of Lebanese amber and other major Early Cretaceous amber deposits shortly thereafter, the amber fossil record kicks in, continuing to the present where it is represented by imperfectly devolatilized copal deposits (Labandeira 2014a). Although fossil ambers of the Early Cretaceous and Early–Late Cretaceous boundary typically preserve insect mouthparts minimally different from their modern descendants, nevertheless they offer a rare glimpse into unique mouthpart types that were present during the late Paleozoic and early Mesozoic, and became extinct shortly after this Cretaceous interval. The preservation of these ancient lineages in Early Cretaceous amber provided details of mouthpart morphology that were not available in earlier fossils of the same lineages preserved as adpressions.

### 17.3.3 *Results of a Phenetic Classification of Modern Insect Mouthparts*

The original phenetic analysis produced a large dendrogram with 1365 terminal, modern insect taxa that were clustered into 34 basic clusters, each of which were assigned to a mouthpart class (Table 17.2). These mouthpart classes are described in terms of standard mouthpart terminology (Snodgrass 1928; Chaudonneret 1990) and were grouped into mouthpart series (e.g., Metcalf 1929), after the analysis, based on functional similarities involved in feeding. However, a few minor changes have been made. In this report, one of the mouthpart classes, the Monocondylate mouthpart class, is renamed the Maxillopalpate mouthpart class. This reattribution resulted from the key feature of a monocondylous mandible which was judged inaccurate from a recent assessment of the Archaeognatha mandible as being dicondylous instead of monocondylous (Blanke et al. 2015). Additionally, the Raptorial

Ectognathate mouthpart class now is divided into two mouthpart classes—the Raptorial Ectognathate mouthpart class proper consisting of odonatopteran adults and the new Labial Mask mouthpart class that encompasses odonatopteran naiads, with their distinctive labial mouthparts. In a previous report (Labandeira 1997), two additional mouthpart classes from mostly the Pennsylvanian and Permian fossil record, the Robust Beak and Laciniate mouthpart class, were added. These changes are currently are reflected in Table 17.2.

## 17.4 The Paleozoic: Establishment of Major Mouthpart Classes, Functional Feeding Groups, and Dietary Guilds

The earliest evidence for insect taxa and their mouthparts appears rather late in the colonization of land, and is preceded by other terrestrialized arthropod groups, such as myriapods and scorpions, nonvascular and vascular land plants, and fungi or fungal-like forms (Taylor and Osborn 1996). The feeding on plants and other arthropods took a two-stage process in which there was insect consumption of stems and sporangia probably by forms with Entognathate and Adult Ectognathate mouthparts (Labandeira 2007a), early in the Devonian (Kevan et al. 1975; Habgood et al. 2004). This was followed by a long absence of evidence for consumption of other plant tissues throughout the Late Devonian and Mississippian time, except for two examples of foliar damage in the Mississippian Period (Iannuzzi and Labandeira 2008; Donovan, pers. comm.). This evidence indicates a considerable hiatus of a 60-million-year-long interval in the origin of major mouthpart classes.

### *17.4.1 The Earliest Insect Mouthpart Classes During the Devonian*

The Early Devonian Rhynie Chert from Scotland is a rare and atypical deposit representing a hot-spring environment in which terrestrial and aquatic organisms were entombed three dimensionally within a silica matrix (Anderson and Trewin 2003). Importantly, the Rhynie Chert provides a rare glimpse into the earliest documented body fossils of insects and other arthropods. The paleocontinent of Euramerica that extended from the Ural Mountains to the American Southwest before the second opening of the Atlantic Ocean also contains two localities, Gaspé in Québec, Canada (Labandeira et al. 1988), and Gilboa in New York state, USA (Shear et al. 1984), which are important for documenting the earliest presence of these mouthpart classes. By the end of the Devonian, four mouthpart classes, or 11.4% of the total, had originated: Entognathate, Maxillopalpate, Adult Ectognathate, and Entognathous Stylate (Table 17.2).

#### 17.4.1.1 Collembola: Early Mouthpart Specializations

Prominent among the Rhynie Chert insects is the collembolan *Rhyniella praecursor* (Hirst and Maulik 1926; Scourfield 1940) that contains distinctive, likely specialized mouthparts (Fig. 17.1a), similar to modern isotomid collembolans (Greenslade and Whalley 1986), typical of the Entognathate mouthpart class. As isotomid Collembola are phylogenetically nested within other collembolan taxa that bear Entognathous Stylate mouthparts, indirect evidence strongly suggests the presence of this mouthpart class as well (Labandeira 1997). Given the common presence of fungi in the Rhynie ecosystem (Taylor et al. 2004) and the ubiquitous associations between modern Collembola and fungi (Hopkin 1997), it is likely that *Rhyniella* mouthparts were used for consuming mycelia, fruiting bodies, and other fungal structures of the Rhynie hot-spring system (Labandeira 2005b).

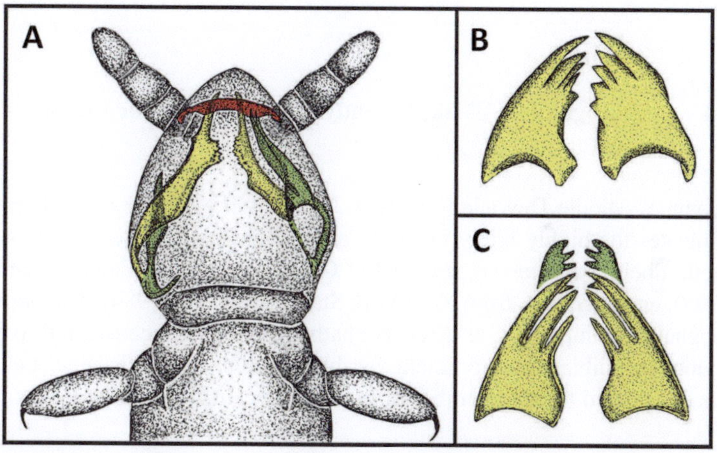


**Fig. 17.1** The earliest insect mouthparts. (**a**) The Entognathate mouthpart class: *Rhyniella praecursor* Hirst and Maulik 1926 (Collembola: Isotomidae), from the Rhynie Chert, Early Devonian of Scotland, UK. Redrawn from a reconstruction by Engel and Grimaldi (2004), from Fig. 1 on page 628. In (b) and (c) are two interpretations of the affinities of the earliest insect or insect-like member of the Adult Ectognathate mouthpart class, also from the Rhynie Chert. Redrawn from Haug and Haug (2017), from Fig. 3 on page 9. In (**b**) is the mandible pair *Rhyniognatha hirsti* Tillyard (1928), originally interpreted as a larval insect (Hirst and Maulik 1926) or later an adult neopterous insect (Engel and Grimaldi 2004); specimen NHML in 38234. In (**c**) is another major interpretation of *R. hirsti*, with possibly associated maxillae belonging to an early scutigeromorph centipede, as proposed by Haug and Haug (2017). Mouthpart color scheme: red, clypeus; yellow, mandibles; and green, maxilla

#### 17.4.1.2 *Rhyniognatha's* Enigmatic Jaws

While *Rhyniella* is validated as a Devonian collembolan with distinctive mouthparts (Whalley and Jarzembowski 1981), the other Rhynie taxon, also historically attributed to Insecta, *Rhyniognatha hirsti*, has its taxonomic affiliation recently challenged (Haug and Haug 2017). *Rhyniognatha hirsti* consists of a pair of mandibles, each of which is apparently dicondylous, flattened, lacking a molar region, and displaying either five (Tillyard 1928) or three (Haug and Haug 2017) incisors (Fig. 17.1b). Originally described as an insect, a closer examination based on modern three-dimensional imaging techniques has revealed detailed structures of likely maxillary elements, including palps, a clypeo-labrum, and a partial head capsule (Haug and Haug 2017). These structures collectively have been interpreted as myriapodan in origin (Rasnitsyn and Quicke 2002), and in particular having a possible affiliation to Chilopoda (centipedes). An assignment to an insect is preferred here, and the mandible pair accordingly indicates the earliest example of the Adult Ectognathate mouthpart class.

#### 17.4.1.3 Archaeognatha: Weak, Dicondylous Mandibles for Litter Feeding

From younger Middle Devonian deposits, body-fossil evidence other than mouthparts indicates the highly likely presence of Archaeognatha based on the superbly preserved, albeit controversial, material at Gaspé, in Québec, Canada (Labandeira et al. 1988), and at Gilboa, from New York State (Shear et al. 1984). The presence of Archaeognatha is important, as Archaeognatha currently are considered to possess dicondylous mandibles and prominent, leg-like maxillary palps (Blanke et al. 2015), and thus considered to be a member of the Maxillopalpate mouthpart class.

#### 17.4.1.4 Putative Devonian Insects

Other presumptive Devonian insect discoveries, such as *Eopterum devonicum*, *Devonohexapodus boksbergensis*, *Wingertshellicus backesi*, *Leverhulmia mariae*, and *Strudiella devonica* (Haug and Haug 2017), were once considered as Insecta or having relationships close to or paraphyletic to Insecta. These taxa now are considered as lacking features for inclusion in Insecta (Willmann 2005; Kühl and Rust 2009; Hörnschemeyer et al. 2013). In summary, current evidence about the variety of Early and Middle Devonian insect mouthpart types indicates that four types were directly or indirectly present—two different collembolan mouthparts referred to the Entognathate and Ectognathate Stylate classes, the archaeognathan Maxillopalpate mouthpart class, and *Rhyniognatha* indicating presence of the Adult Ectognathate mouthpart class, although the latter has been considered more myriapodan than insectan.

### 17.4.2 *Few Clues Regarding Mouthpart Classes During the Mississippian*

There is a prolonged hiatus from the Middle Devonian to the Mississippian–Pennsylvanian boundary interval (Table 17.3), representing approximately 60 million years of lapsed time, during which the body-fossil record of insect mouthparts is absent (Ward et al. 2006; Schachat et al. 2018). However, two rare occurrences of cuspate excisions on seed ferns (pteridosperms) indicate the presence of an externally feeding folivore with mandibulate mouthparts, presumably an insect, during the Early Mississippian (Tournasian Stage) at about 353 Ma, and a second occurrence, probably an orthopteroid insect ancestor, from the Late Mississippian (Visean Stage) at about 327 Ma. The older occurrence consists of small cuspate excisions with reaction rims on pinnules of two seed-plant genera from Eastern North America, indicative of a small insect herbivore with dicondylous, incisiform mandibles (M. Donovan, pers. comm.).

The younger occurrence consists of complex excisions on *Triphyllopteris austrina*, a probable lyginopterid seed fern, from the Sydney Basin of Australia (Iannuzzi and Labandeira 2008). The damage, replete with veinal stringers and dark reaction rims occurring along the chewed edge, is similar to modern orthopteran damage. The *Triphyllopteris* damage pattern paralleled the way that modern acridid grasshoppers feed on foliage, in which there is overall forward progression of the head and mouthparts that create a broad cusps but backward movement of individual mouthpart elements at each feeding event, resulting in subsets of small, semicircular cuspules (Gangwere 1966). The two Mississippian occurrences of plant damage suggest the presence of an orthopteroid insect or another folivorous, mandibulate insect in lieu of a body-fossil record. A member of the Adult Ectognathate mouthpart class is inferred to have made the two examples of margin folivory during the Mississippian. An alternative hypothesis would maintain that these cuspate excisions were made by myriapods, particularly millipedes, but it is unclear which millipede group would be responsible for active herbivory, particularly as millipedes are nearly entirely detritivorous, and the few herbivores among the group are root feeders (Hopkin and Read 1992).

No new mouthpart classes are recorded for the Mississippian. By the end of Mississippian time, the throughput of four earlier originating mouthpart classes continued, with four mouthpart classes present. By the end of the Mississippian, 11.4% of all mouthpart classes were still present (Table 17.3). Within the first few million years after the Mississippian–Pennsylvanian boundary interval at 323 Ma, many major order-level lineages of insects make their earliest documented appearances in the fossil record (Brauckmann et al. 1985; Labandeira 2001). Included in this geochronologically sudden increase in insect diversity were the rapid appearances of major, distinctive mouthpart types representing a variety of dietary guilds, functional feeding groups, and in particular mouthpart classes (Labandeira 2006b, 2007a).

**Table 17.3** The number and percentage of insect mouthpart classes through time

| Geologic period | Additional and removed mouthpart classes | Number of mouthpart classes for the period[a] | | |
|---|---|---|---|---|
| | | Additional | Cumulative | Percent of running total |
| DEVONIAN | *Added*: Entognathate, Entognathous Stylate, Maxillopalpate, Adult Ectognathate | 4 | 4 | 11.4 |
| MISSISSIPPIAN | [none] | 0 | 4 | 11.4 |
| PENNSYLVANIAN | *Added*: Larval Ectognathate, Raptorial Ectognathate, Labial Mask, Sericterate, Laciniate, Robust Beak, Segmented Beak | 7 | 11 | 29.7 |
| PERMIAN | *Added*: Mortar and Pestle, Rostrate, Pectinate, Fossate Complex, Siphonate, Mouthcone, Ectognathous Stylate, Reduced Trophic | 7 | 18 | 48.6 |
| TRIASSIC | *Added*: Rhynchophorate, Mandibulobrustiate, Tubulomandibulate, Haustoriate, Labellate, Distylate/Tetrastylate, Hexastylate | 7 | 25 | 67.6 |
| JURASSIC | *Added*: Maxillolabiate, Mouthhook, Glossate, Monostylate/Distylate, Distylate/Tetrastylate, Tristylate, Nontrophic. *Removed*: Robust Beak[b] | 7 | 32 | 88.9 |
| CRETACEOUS | *Added*: Siphonomandibulate, Buccal Cone | 2 | 34 | 94.4 |
| PALEOGENE | *Added*: Tubulostylate. *Removed*: Laciniate[b] | 0 | 34 | 97.1 |
| NEOGENE | *Added*: Siphonostylate | 1 | 35 | 100.0 |

[a]At the end of the said time interval

[b]These two mouthpart classes were extirpated, reducing the standing total of mouthpart classes from 37 to 35

### 17.4.3 *Major Expansion of New Mouthpart Classes During the Pennsylvanian*

During the Pennsylvanian Period, 15 primary lineages of insects are newly recorded or are a range-through continuation from an earlier fossil record. The range-through designation implies that a fossil lineage is present between earlier and later occurrences even though it is not recorded in the interval of interest. These ordinal-level lineages, including those unassigned to order, are Collembola; Diplura; Archaeognatha; Monura; Zygentoma; the paleodictyopteroid orders of Paleodictyoptera, Megasecoptera, Diaphanopterodea, and Dicliptera; Ephemeroptera; Odonatoptera; Archaeorthoptera; Dictyoptera; Acercaria (including Paraneoptera); and Holometabola (Shear and Kukalová-Peck 1990; Labandeira and Sepkoski 1993; Labandeira 1999; Grimaldi and Engel 2005; Haug et al. 2015). These mouthpart classes represented four mouthpart classes present during the Early and Middle Devonian (Entognathate, Maxillopalpate, Entognathous stylate, and Adult ectognathate) and seven additional, newly emerging, mouthpart classes (Larval Ectognathate, Raptorial Ectognathate, Labial Mask, Pectinate, Sericterate, Laciniate, and Robust Beak), two of which are based on indirect evidence (Labandeira 1997). Matching this expansion of mouthpart classes were their associated functional feeding groups that included external foliage feeding, piercing and sucking, galling, seed predation, and stem boring in the terrestrial realm, and filtering and collecting in the aquatic realm (Labandeira 1998, 2006a, b). The major dietary guilds of detritivory, fungivory, herbivory, palynivory, and carnivory were present by the end of the Pennsylvanian, based on evidence from gut contents, coprolites, and plant damage from two-dimensionally, well-preserved compression deposits (Scott and Taylor 1983; Labandeira 2006a; Xu et al. 2018) and three-dimensionally, exquisitely permineralized coal-ball deposits (Labandeira and Phillips 1996a, b, 2002; Labandeira 1998, 2001, 2006a). By the end of Pennsylvanian time, seven new mouthpart classes originated, and including the throughput of four earlier originating mouthpart classes, there were 11 mouthpart classes present. By the end of the Pennsylvanian, 29.7% of the total number of mouthpart classes had originated (Table 17.3).

#### 17.4.3.1 Monura: Evidence for Large Maxillary Palps in the Sedimentological Record

Monurans historically were known from the Early Pennsylvanian to the Late Permian, although recently the iconic genus *Dasyleptus* has been found in Middle Triassic deposits of Switzerland (Bechly and Stockar 2011). The most conspicuous feature of monurans was their very large, leg-like maxillary palps that were probably tactile and sensory in function and gave rise to the designation of Maxillopalpate mouthparts typical of the clade. Closely related to Archaeognatha (Bechly and Stockar 2011), which has, albeit controversially, an Early Devonian origin (Shear et al. 1984; Labandeira et al. 1988), monurans apparently had dicondylous, weakly articulating, and probably weakly adducting-abducting, milling mandibles that were edentulous or had diminutive teeth. Recent evidence indicates that modern Archaeognatha possess a previously unrecognized, weak anterior mandibular articulation that renders the group as essentially dicondylous (Blanke et al. 2015), and this feature has been identified in Monura. Notably, monuran mandibles may have been sufficiently prominent that they produced a distinctive soft-sediment trace fossil, *Tonganoxichnus buildexensis* (Mángano et al. 1997), which included a paired

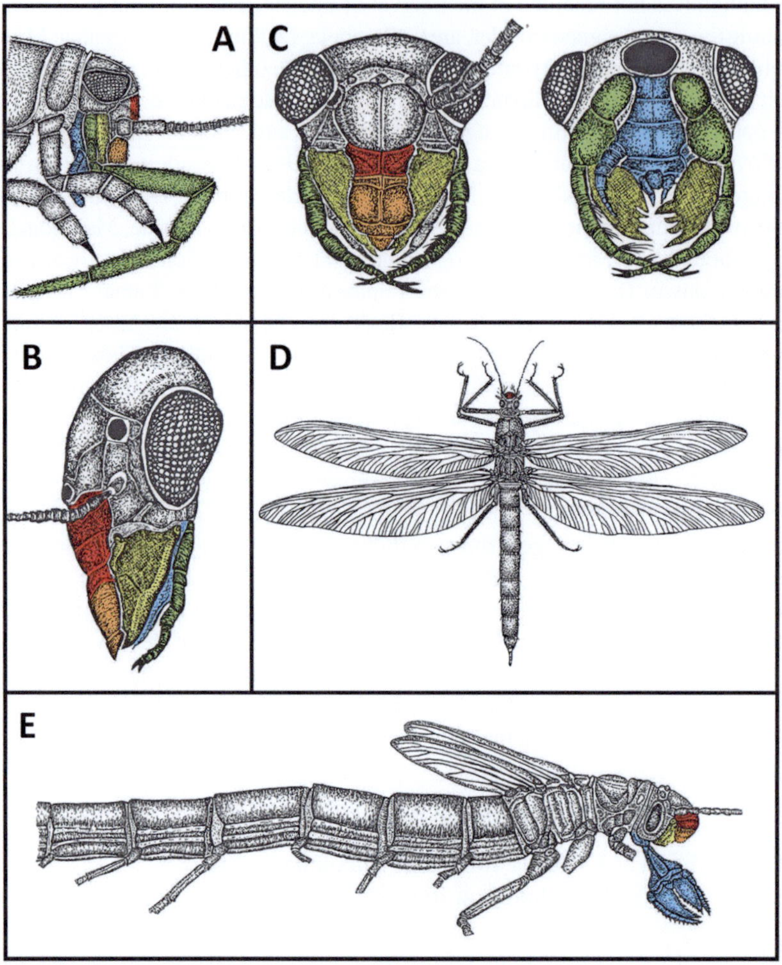


**Fig. 17.2** Initial expansion of mouthpart morphologies during the Pennsylvanian Period (Late Carboniferous) to early Permian Period from 323 to 272 million years ago. Part 1: Monura, Ephemeroptera, and Odonatoptera. (**a**) A representative of the Adult Ectognathate Class is *Dasyleptus brongniarti* Sharov (1957) (Monura: Dasyleptidae), from the late Pennsylvanian to early Permian Kuznetsk Basin of Russia; based mostly on specimen PIN 1197/594. Redrawn from Sharov (1957), Fig. 1 on page 796. (**b**) Representing the Adult Ectognathate Class is an adult mayfly (Ephemeroptera: Protereismatidae), from the early Permian of Kansas, USA, based on a composite of specimens reconstructed by Kukalová-Peck (1990). Redrawn from Kukalová-Peck (1990) from Fig. 6.14c on page 160. (**c**) The Raptorial Ectognathate Class is represented by the griffinfly *Meganeura monyi* (Brongniart) (1884) (Protodonata: Meganeuridae), with a wingspan of about 28 cm, from the late Pennsylvanian of Commentry, France, whose head and mouthparts are shown here in anterior (left) and posterior (right) aspect; from a reconstruction in Shear and Kukalová-Peck on page 1824. (**d**) The adult griffinfly *Meganeurula selysii* Brongniart (1893) (Protodonata: Meganeuridae), possessing Raptorial Ectognathate mouthparts and a wingspan of about 28 cm, from the late Pennsylvanian of Commentry, France. Redrawn from Kukalová-Peck (2009) from Fig. 4 on page 183; specimen MNHN 1422. (**e**) The earliest documented example of

"frontal impression" anterior to thoracic paired walking-limb impressions. This frontal impression (FI) likely represents sharp indentations of paired mandibles of a saltatorial *Dasyleptus* as it landed on the soft sediment of an estuarine valley adjacent to a shoreface from the Late Pennsylvanian of Kansas (Mángano et al. 1997). The same trace fossil contains broad, paired impressions designated as the anteriormost appendage (A1), separated by some distance from the FI impressions, and likely representing the bend of the articulation between the second and third articles of the massive, leg-like maxillary palps. The two FI and A1 features, plus their head-capsule feeding cavity formed by surrounding mouthparts (Fig. 17.2a), would be consistent with landing impressions of an individual with Maxillopalpate mouthparts. It should be noted that in fine-grained sediments, such as mudstones, the impressions of mouthpart elements of some fossil insects could provide data that may be difficult to glean from anatomical studies alone. Maxillopalpate mouthparts are consistent with a detritivorous, externally feeding insect (Shear and Kukalová-Peck 1990).

#### 17.4.3.2 Ephemeroptera: Adults with Functioning Adult Ectognathate Mouthparts

Ephemeropteran adults of the Paleozoic had functioning, mandibulate mouthparts that are assigned to the Adult Ectognathate mouthpart class (Fig. 17.2b), unlike their successors, which possessed nonfunctional, highly reduced mouthparts assigned typically to the Nontrophic mouthpart class (Labandeira 1990). The nonfunctional nature of modern ephemeropteran mouthparts befits their status of existing from a few hours to a few days as ephemeral, nonfeeding adults (Grimaldi and Engel 2005). Unlike their adults, the mouthparts of modern immature Ephemeroptera are highly functional, assigned to the Pectinate mouthpart class, and are noted for their distally expanded mandibles, each bearing an articulating brush-like process, the prostheca. Notably, the labium is well developed and has a pair of distinctive, large, plate-like lobes forming the often transversely elongate paraglossae (Labandeira 1990). This characterization of Pectinate mouthparts is consistent with naiad mouthparts of the ephemeropteran Protereismatidae from the early Permian of Kansas (Hubbard and Kukalová-Peck 1980), a lineage that extends into the latest Pennsylvanian based on sister-group relationships with other earliest ephemeropteran lineages (Grimaldi and Engel 2005). Early members of the Pectinate mouthpart class likely belonged to the scraper and filterer functional feeding groups, and had diets of plant litter and dead

**Fig. 17.2** (continued) mouthparts belonging to the Labial Mask Class, a large naiad of *Dragonympha sroka* Kukalová-Peck (2009) (Protodonata) of uncertain family relationships, but likely affiliated to Meganeuridae, from the Middle Pennsylvanian of Mazon Creek, Illinois, USA; specimen ISM 004ab. Redrawn from Kukalová-Peck (2009), Fig. 1A on page 177. Mouthpart color scheme: red, clypeus; orange, labrum; yellow, mandibles; brown, hypopharynx; green, maxilla; and blue, labium

animals as detritivores and possibly aquatic vegetation as herbivores, as does their modern descendants (Edmunds 1984).

#### 17.4.3.3 Odonatoptera Adults: Raptorial Ectognathate Mouthparts and Predation

Odonatoptera are the preeminent representatives of the Raptorial Ectognathate and Labial Mask mouthpart classes as adults and naiads, respectively. Gigantic dragonfly adults, some species 60–70 cm in wingspan (Carpenter 1960; Wootton 1981), comprised the subclade Protodonata that have been known for over a century (Brongniart 1884). Nevertheless, very little has been revealed about their mouthparts until recently, principally from well-preserved and recently reexamined specimens of *Meganeura monyi* (Meganeuridae), from the Late Pennsylvanian coal-mine deposits in north-central France (Shear and Kukalová-Peck 1990). From this material, anterior and posterior views of the *M. monyi* head have been reconstructed (Fig. 17.2c), indicating a distinctive branchial basket in which mouthpart elements were able to disassemble relatively large prey items. The robust clypeus–labrum complex is subsegmented and offered a substantial upper lip to secure prey from the top, matched by thick maxillary palps from the side that apparently allowed confinement of prey items entering from below (Shear and Kukalová-Peck 1990). From each side, the mandible and associated maxillary galea and lacinia bore sharp, incisiform teeth to impale and macerate prey items. Notably, prothoracic and mesothoracic legs were canted forwardly to assist the capture and feeding process as auxiliary elements to the mouthparts. These features represent a more massive mouthpart construction than modern Odonata, even allowing for differences related to overall body size (Fig. 17.2d). Members of the Raptorial Ectognathate mouthpart class were the primary insect carnivore-feeding guild and the functional feeding group that pursued ambush predation from stationary perches, rather than pursuit predation on the wing among modern dragonflies (Riek and Kukalová-Peck 1984).

#### 17.4.3.4 Odonatoptera Naiads: Labial Mask Mouthparts in Meganeurid Dragonflies

Until recently, mouthparts of naiads of Paleozoic Protodonata were not known. Previously, it was unclear if Protodonatan naiad mouthparts would be composed of the labial mask found in Cretaceous and Cenozoic nymphs of the Odonata (Fleck et al. 2002), or whether their mouthparts would be entirely different, given the considerable phylogenetic separation of these two lineages since the Paleozoic. However, recent discovery of a Protodonatan naiad from the Middle Pennsylvanian of Mazon Creek site in northern Illinois, USA (Fig. 17.2e), indicates that the naiad labial mask is a fundamental feature of the broader clade Odonatoptera (Kukalová-Peck 2009). The relevant specimen, *Dragonympha srokai*, is incomplete, with a preserved body length (excluding antennae) of 3.81 cm, and probably a total body length of 5 cm—rather small for a later instar of a large meganeurid adult (Fig. 17.2e). The clypeus and labrum are frontally positioned, transversely divided into sclerites, large and expanded, indicating a significant cibarial pump underneath. A gena and basal maxillary segment of the head are present, and the mandible base indicates a wide, dicondylous articulation with the underside of the genal margin. Maxillary appendages are absent. In contrast, all features of the labium are present, except for a brief break of one mentum hinge at the labial mask base. The labial mask itself exhibits considerable detail including the claw-like paraglossae, the rimmed and laterally attached labial palps with outwardly directed teeth, and almost all of the articulatory sclerites of the mentum, suggesting a modern-aspect strike apparatus. Kukalová-Peck (2009) suggests that *Dragonympha* was semiaquatic and that the Labial Mask mouthparts originated as a capture device for terrestrial prey, later co-opted to secure prey in aquatic environments. Interestingly, a similar extrudable labial mask device has independently arisen in stenine rove beetles (Coleoptera: Staphylinidae) (Schmitz 1943), but the mechanism of extension is hydraulically based and slingshot in mode, rather than rapid extension of a multi-jointed apparatus (Weinreich 1968).

#### 17.4.3.5 Paleodictyopteroidea: Mouthparts and Partitioning Plant Tissues

The Paleodictyopteroidea represent a diverse lineage that constitutes four subordinate orders—Paleodictyoptera (Fig. 17.3c), Megasecoptera (Fig. 17.3a, b), Diaphanopterodea (Fig. 17.3d), and Dicliptera. Paleodictyopteroidea represent the only superordinal lineage of insects to become extinct, ranging from the earliest Pennsylvanian to the end of the Permian (Carpenter 1992), with one family surviving into the Triassic (Béthoux et al. 2010). Paleodictyopteroids represent about half of all named insect species from the Paleozoic (Carpenter 1971) and have an immense breadth of body sizes, ranging from mosquito-sized diclipterans to paleodictyopterans with 55 cm wingspans (Wootton 1981). Paleodictyopteroids had piercing-and-sucking mouthparts (Fig. 17.3b–d), that are best exemplified by a generalized paleodictyopteroid reconstruction (Fig. 17.3d), emphasizing the external features and cross-sectional elements of the distinctive, robustly constructed beak.

The heads of paleodictyopteroids tend to be globular or spheroidal, not elongated medially or transversely, and with inconspicuous genal and gular head regions. The nonretractile beak typically is positioned hypognathously in feeding repose (Fig. 17.3b), but often is depicted as prognathous when the insect is oriented in flight (Pecharová and Prokop 2017) (Fig. 17.3a). The subsegmented clypeus is somewhat domed, revealing a cibarial pump below, and is attached to a similarly subsegmented labrum that offers an upper brace to the five mouthpart stylets, the distal-most part of which often is drawn out into an acuminate extension (Fig. 17.3d, left and right). Anatomically below the labrum are the paired mandibular stylets and below the mandibular stylets are the paired maxillary stylets (Fig. 17.3c), which are

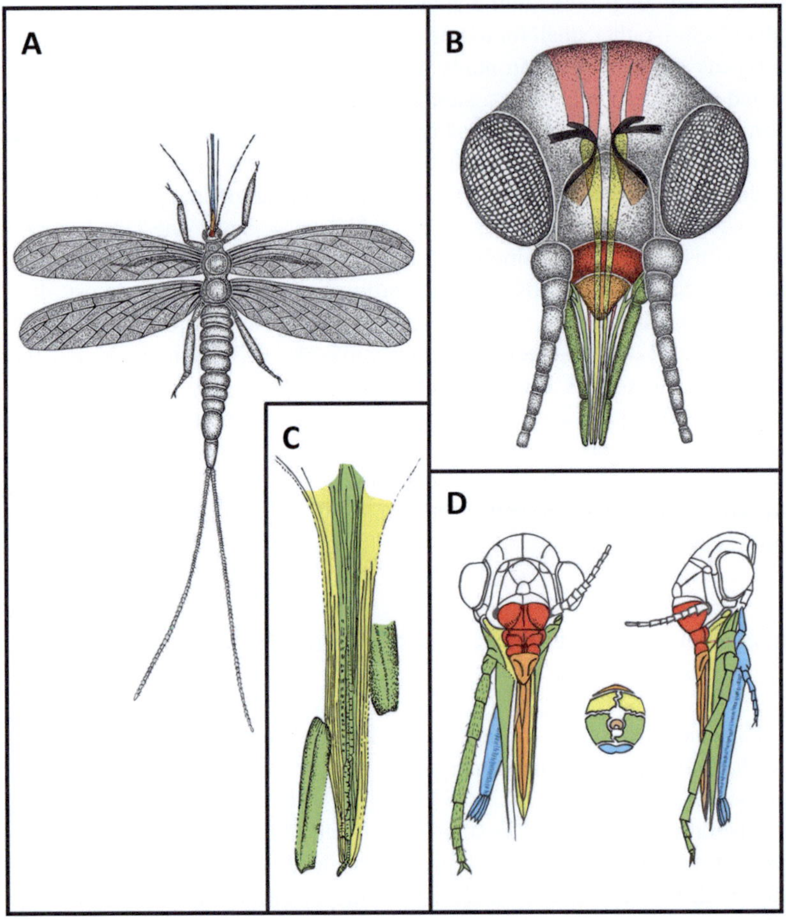


**Fig. 17.3** Initial expansion of mouthpart morphologies during the Pennsylvanian Period to early Permian Period from 323 to 272 million years ago. Part 2: Paleodictyopteroidea and mouthparts of the Robust Beak Class. (**a**) Reconstruction of *Brodioptera sinensis* Pecharová, Ren, and Prokop (2015) (Megasecoptera: Brodiopteridae), from the Early Pennsylvanian of the Czech Republic, showing the disposition of Robust Beak mouthparts during flight. Redrawn from Prokop et al. (2016), from Fig. 1 of page 2. (**b**) Mechanism for internal head musculature for mandible and lacinia stylet protraction and retraction of *Protohymen novokshonovi* Pecharová and Prokop (2017) (Megasecoptera: Protohymenidae). This reconstruction is interpreted from a composite of *P. novokshonovi*, *P. permianus* Tillyard (1924), and *P. carpenteri* (Novokshonov) (1995) (specimens MCZ 3060ab, PIN 4987/115, PIN 1700/445, and PIN 1700/3231), from the early Permian of the central Ural Mountains, Russia. Black arms are internal tentorium elements; protracting musculature is light brown; adducting musculature is pink. Redrawn from Pecharová and Prokop (2017), from Fig. 2g on page 5. (**c**) Ventral view of the enlarged beak of *Mosteropterum moravicum* Kukalová-Peck (1972) (Paleodictyoptera: Homoiopteridae), from the early Permian of the Czech Republic, displaying a beak evenly separated along a median plane, exposing a pair of mandibular and lacinial stylets on each side, with adjacent maxillary palp segments; specimen PICU 5/1972. Redrawn from Shear and Kukalová-Peck (1990), from Fig. 23 on page 1822. (**d**) Head and mouthpart morphology of *Paruralia rohdendorfi* Kukalová-Peck and Sinichenkova (1992)

protracted and retracted by pairs of muscles strategically attached to tentorial apodemes in the inner head capsule (Fig. 17.3b). In addition, there is a centrally positioned hypopharyngeal stylet. The entire stylet ensemble is supported from below by the lower brace of the highly elongated ligula of the labium. In cross section, from anterior to posterior view, the beak at mid-section is supported by the upper brace of the labrum, followed posteriorly by the two interlocking mandibular stylets and in turn followed by the two interlocking maxillary stylets. Both stylet pairs encircled the hypopharyngeal stylet and were supported by the lower brace of the labial ligula that formed a gutter to accommodate the stylet fascicle (Fig. 17.3d, center). Also supporting the beak was leg-like, maxillary palps and forwardly directed prothoracic legs that buttressed the beak during piercing (Shear and Kukalová-Peck 1990). At the center of the beak cross section are two tubular structures formed at paired stylet contacts and on opposite sides of the hypopharyngeal stylet. The more anteriorly positioned food tube was used for incoming nutritive fluids such as plant sap and was powered by the subclypeal cibarial pump with dilator muscles that provided negative pressure. By contrast, the posteriorly positioned and narrower salivary duct allowed for outgoing salivary fluids, provided positive pressure, and was presumably powered by an associated salivary pump. In insects such as piercing-and-sucking Hemiptera, the salivary pump is a piston pump, but such a pump has been difficult to discern in paleodictyopteroid compression fossils. Examples of both types of pumps are known, for example, from a mid-Cretaceous amber-preserved scorpionfly, illustrated and discussed in Lin et al. (2019).

The aspect ratios of paleodictyopteroid beaks varied considerably. Compact, short, triangular beaks about 0.6 cm long likely incurred some lateral movement, allowing for protraction-retraction as well as adduction-abduction stylet movements, such as the beak of the megasecopteran *Permohymen* (Fig. 17.3b), documented by Pecharová and Prokop (2017) and by Brauckmann and Koch (1982) for the paleodictyopteran *Heterologopsis*. Such beaks were associated with large heads and likely involved consumption of spores and pollen that have been found in the guts of diaphanopterodean nymphs with similarly small, truncate beaks (Kukalová-Peck 1987), as well as evidence from coprolites consisting entirely of *Punctatisporites* marattialean tree-fern spores and *Florinites* cordaite pollen in permineralized coal-ball deposits (Labandeira 1998, 2006a). By contrast, at the other end of the aspect-ratio divide, adult insects such as the paleodictyopteran *Eugereon boeckingi* possessed a long, gracile, narrow and probably flexible beak 3.2 cm long (Müller 1978). Similar species with small heads, comparatively smaller cibarial pumps, and equally gracile beaks included the megasecopteran *Brodioptera sinensis* (Pecharová et al. 2015; Prokop et al. 2016), estimated as 1.4 cm long, and

**Fig. 17.3** (continued) (Diaphanopterodea: Paruraliidae), as interpreted by Kukalová-Peck, with elongate mandibles acting as scissors-like fashion in conjunction with elongate lacinial, mandibular, and hypopharyngeal stylets. Redrawn from Shear and Kukalová-Peck (1990), from Fig. 23 on page 1822. Mouthpart color scheme: red, clypeus; orange, labrum; yellow, mandibles; brown, hypopharynx; green, maxilla; blue, labium; light brown, protracting muscles; and pink retracting muscles

likely were feeding on nutritious phloem (Fig. 17.3a). Bearers of such beaks targeted vascular strands in plants, seated shallowly in parenchymatous tissue, as indicated by curvilinear stylet paths through parenchymatous tissue that terminated in enlarged feeding pools embedded in phloem and xylem tissues (Scott and Taylor 1983; Labandeira and Phillips 1996a). An alternative food source for long, gracile beaked paleodictyopteroids could have been pollination drops from seed plants such as medullosans or cordaites, either through micropylar secretions or from tubular structures that exude ovular pollination drops (Labandeira 2000; Labandeira et al. 2007a). Such a food source may represent one of the earliest stages in insect pollination of seed plants (Labandeira 2000), given piercing-and-sucking damage documented *Dolerotheca* pollen organs of medullosan trees mentioned by Retallack and Dilcher (1988). Paleodictyopteroid taxa such as *Monsteropterum moravicum*, from the early Permian of the Czech Republic (Kukalová-Peck 1972), bore a 2-cm-long flaring beak of intermediate aspect ratio, size, robustness, mandibular stylets longer than maxillary stylets, and were buttressed by adjacent palps (Fig. 17.3c). This combination of characters would indicate feeding on more indurated tissues, such as calamitalean reproductive cones, tree-fern synangia, or larger seeds (Shear and Kukalová-Peck 1990) that were inaccessible to more gracile beaked species.

Paleodictyopteroid insects were herbivorous members of the piercing-and-sucking functional feeding group that exhibited considerable mouthpart variation, given the evidence from head and mouthpart structure, plant reproductive features, gut contents, coprolites, and plant damage (Scott and Taylor 1983; Labandeira 1998). The diversity of plant fluids and secretions represented considerable variation in the amounts of nutritive ingredients such as carbohydrates, lipids, and proteins as well as their concentrations ranging from dilute xylem to more concentrated phloem to viscous pollination drops (Labandeira et al. 2007a). This variation in fluid food quality was matched by paleodictyopteroid structural features such as the extent and development of cibarial pumps; aspect ratios; length and flexibility of stylet ensembles; and the largely unknown role of the hypopharyngeal stylet and its salivary pump. From analogy, feeding on xylem, a nutritionally dilute food source, may have been similar to the activities of modern, large, long-beaked cicadas (Brues 1972) for medium-sized paleodictyopteroids. By contrast, shorter beaked species likely were phloem feeders on more nutritious and concentrated food sources, similar to modern planthoppers and leafhoppers (Risebrow and Dixon 1987), or alternatively consuming megagametophytic embryonic tissues of reproductive structures, analogous to extant seed bugs feeding on the liquefied macerate from the endosperm of angiosperm seeds (Slansky and Panizzi 1987). These feeding styles have evidence in the fossil record that includes shorter (Scott and Taylor 1983) and longer (Labandeira and Phillips 1996a) stylet tracks into plant tissues, and punctures in seeds such as the medullosan *Trigonocarpus* (Scott and Taylor 1983; Jennings 1974) and the cordaite *Samaropsis* (Sharov 1973; Shcherbakov et al. 2009). Although the possibility of hematophagy has been advanced (Shear and Kukalová-Peck 1990), it appears unlikely that vertebrate blood and lymph was a food source, although there is evidence for insect predation by a large megasecopteran, based on spinose, forwardly directed, jack-knifed forelegs (Carpenter 1971).

#### 17.4.3.6 Archaeorthoptera: Mouthparts and Dietary Diversity

The systematic position of early members of orthopteroid insects currently is unsettled for some Pennsylvanian lineages, such as the Geraridae (Fig. 17.4e), which formerly were considered as ancestors of Paraneoptera (Kukalová-Peck 1990), and now are adjudged as closely affiliated with early Polyneoptera, as Archaeorthoptera (Béthoux and Briggs 2008). Other Pennsylvanian lineages were once considered members of the diverse, polyphyletic assemblage "Protorthoptera," possessing mandibulate mouthparts and generalized orthopteroid venation (Carpenter 1992; Grimaldi and Engel 2005). However, recent, detailed, systematic studies focusing on particular taxa and lineages of "Protorthoptera" have reassigned some of the constituent taxa to more robust, character-supported clades (Béthoux 2007; Béthoux and Nel 2004). Most of these taxonomic re-affiliations have been to the recently erected, monophyletic, stem-group Archaeorthoptera (Béthoux 2006, 2008, 2009; Béthoux and Briggs 2008; Gu and Béthoux 2011), but other taxa have been referred to clades such as the Dictyoptera (Béthoux and Wieland 2009; Béthoux et al. 2009) and Orthoptera (Béthoux and Nel 2002; Béthoux et al. 2004). These two latter clades diversified during the Permian and Triassic and are major components of the present insect fauna.

Two notable Middle Pennsylvanian taxa from the Euramerica paleocontinent, with distinctive prognathous, mandibulate mouthparts of the Adult Ectognathate mouthpart class, but with different mouthpart and prothoracic structures, are *Gerarus danielsi* (Archaeorthoptera: Geraridae) and *Eucaenus ovalis* (Archaeorthoptera: Eucaenidae) (Burnham 1983; Carpenter 1997). *Gerarus danielsi* (Fig. 17.4e) exhibited a prominently domed clypeus, interpreted by Kukalová-Peck and Brauckmann (1992) as supporting a hemipteroid-type cibarial pump that was inconsistent with the observations of Béthoux and Briggs (2008) who considered the taxon a member of Archaeorthoptera. Ancillary thoracic features supporting a functional-feeding-group assignment are the prominent armature of thick, ominous spines and an elongate, jutting prothorax that subtended a movable head protruding from a collar equipped by a circular rim of shorter spines (Burnham 1983). This defensive network of antipredator structures and the rather large size of the insect indicate that this species was probably an herbivore and lacked features of the prothoracic legs and mouthparts such as spinules, and raptorial mandibles, respectively, which would indicate predation. *Eucaenus ovalis* (Fig. 17.4f) is provided with a different complement of prothoracic, head, and mouthpart structures than that of *Gerarus*. These features are a considerably abbreviated prothorax; an elongate, forwardly jutting head about twice as long as wide; prominent prognathous mandibles; and conspicuous, tactile, maxillary palps consisting of four articles. The maxillary palps are highly elongate as are the somewhat longer filiform antennae (Carpenter 1997). The distance of the prothorax, head, and mouthparts from the main body and the prognathous

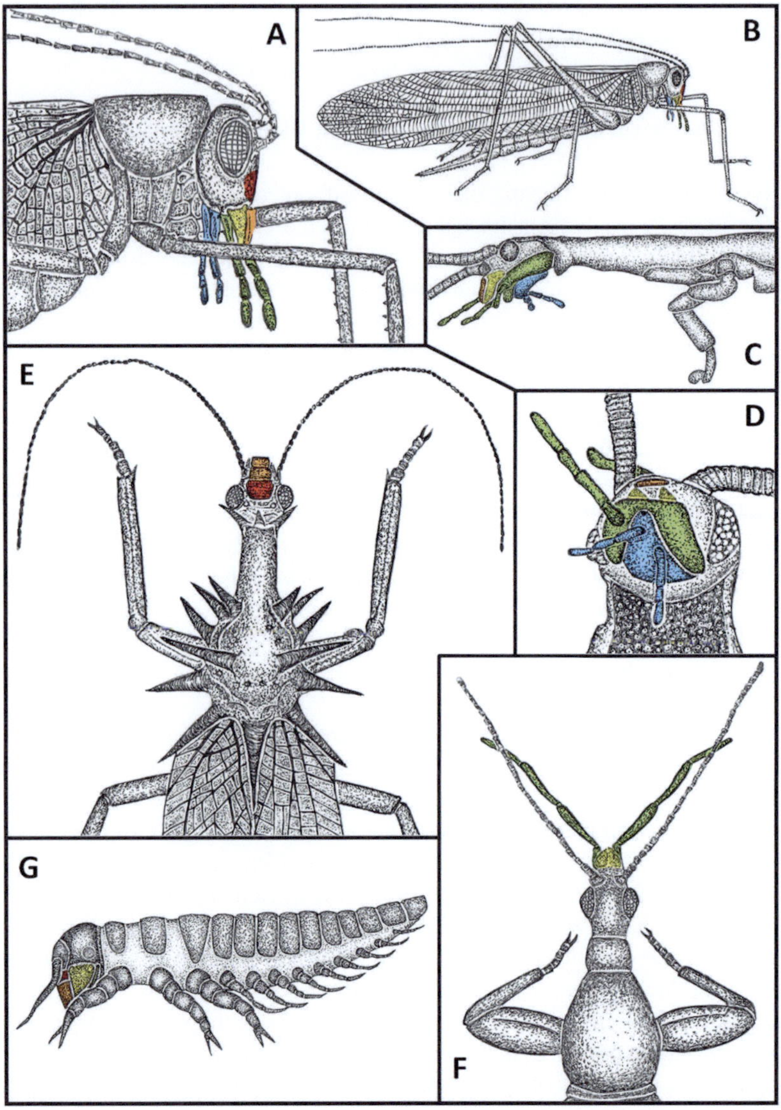


**Fig. 17.4** Initial expansion of mouthpart morphologies during the Pennsylvanian Period to early Permian Period from 323 to 272 million years ago. Part 3: Roachoids, Orthopteroids, and Holometabola representing the Adult Ectognathate (**a–f**) and Larval Ectognathate (**g**) mouthpart classes. (**a**) Enlargement of the head, mouthparts, and prothorax of *Jubilaeus beybienkoi* Sharov (1968) (Orthoptera: Tcholmanvissiidae) in right-lateral view showing the antiquity of herbivorous orthopteran mouthparts, from the early Permian of Chekarda, Central Urals, Russia; specimens PIN514/6 and 1700/1487. Redrawn from Sharov (1968) from Fig. 11a on page 28. (**b**) Habitus of *J. beybienkoi* in (**a**). Redrawn from Sharov (1968) from Fig. 11a on page 28. (**c**) Left lateral view of head, mouthparts, and prothorax of the roachoid nymph *Anebos phrixos* Garwood, Ross, Sotty, Chabard, Charbonnier, Sutton, and Withers (2012) (Blattodea: family uncertain), showing a highly flattened head and mouthparts; specimen MNHN F.SOT005630, from the Late Pennsylvanian of

mouthparts indicate a probing insect exploring its environment for small, arthropod prey items.

#### 17.4.3.7 Dictyoptera: Roachoids and Litter Feeding

The Dictyoptera encompasses the three modern, albeit paraphyletic, orders of Blattaria (cockroaches), Isoptera (termites), and Mantodea (mantises) that likely originated in the mid Mesozoic. Nevertheless, dictyopteran, roach-like fossils occur in Pennsylvanian and Permian deposits that are very difficult to distinguish taxonomically, resulting in their identification initially and colloquially as "roachoids." Although initially an informal term, the designation, roachoids, conveys a distinct meaning to paleobiologists studying Paleozoic cockroaches that their taxonomic identification is virtually impossible. The highly flattened body, lack of distinctive features other than wings, and highly variable wing venation have discouraged taxonomic progress in this apparently diverse. This lack of taxonomic resolution is emblematic in that cockroaches were the only group of fossil insects that was not included in Carpenter's (1992) two-volume taxonomic treatise on fossil Hexapoda.

However, recent advances in X-ray tomographic reconstruction of roachoid nymphs of Late Pennsylvanian age from Commentry, France, have revealed unexpected morphologic details of the head and mouthparts of one particular species that never would have been gleaned from traditional paleobiological techniques of preparation and imaging. The taxon, *Anebos phrixos* (Fig 17.4c, d), is unassigned to order or family (Garwood et al. 2012). A left lateral view shown in Fig. 17.4c shows an anteriorly prolonged prothorax ending in a rimmed collar region that, together with apparently long filiform antennae, is very reminiscent of the structure shown in *Gerarus* in Fig. 17.4e. Unlike *Gerarus*, the head of *Anebos* bears hypognathous mouthparts and unspecialized, medium-length labial palps and medium-length max-

**Fig. 17.4** (continued) Montceau-les-Mines, Massif Central, France. Redrawn from Garwood et al. (2012), from Fig. 1c on page 5. (**d**) Ventral view of the head, mouthparts, and prothorax of the roachoid specimen in (**c**), exhibiting diminutive mandibles and expansive maxillae. Redrawn from Garwood et al. (2012), from Fig. 1d on page 5. (**e**) Head, mouthparts, and prothorax of *Gerarus danielsi* Handlirsch (1906) (Archaeorthoptera: Geraridae), with prognathous mandibles, from the Middle Pennsylvanian Mazon Creek site of Illinois, USA. This is a composite based on specimen FMNH PE5276, other specimens (Burnham 1983), and subsequent emendations. Redrawn from Shear and Kukalová-Peck (1990) from Fig. 42 on page 1825. (**f**) Head, mouthparts, and prothorax of *Eucaenus ovalis* Scudder (1885) (Archaeorthoptera: Eucaenidae), with prominent, elongate, probably tactile maxillary palps; a composite from FMNH PE20790 and other specimens, from the same site as (**e**). Redrawn from Carpenter (1997) from Fig. 14a.26 on page 192. (**g**) Reconstruction of the earliest endopterygote larva, *Srokalarva berthei* J. Haug, Labandeira, Santiago-Blay, C. Haug, and Brown (2016) (Holometabola: Srokalarvidae), with chewing mouthparts, indicating phytophagy and possibly possessing a compound eye. The specimen (FMNH MCP322) is from the same site as (**e**) and (**f**). Redrawn from Haug et al. (2015) from Fig. 4b on page 6. Mouthpart color scheme: red, clypeus; orange, labrum; yellow, mandibles; brown, hypopharynx; green, maxilla; and blue, labium

illary palps unlike the long, tactile, maxillary palps with highly elongated articles occurring in *Eucaenus*. The elongate mandibles are hypognathous, simple, apparently edentulous—although the latter character could be attributable to the lack of sufficient tomographic resolution from suboptimal preservation. The paired maxillae, each of which apparently consists of a cardo, stipes, conjoined galeolacinia and prominent palp of likely four articles, is a robust feature present on each side of the median labium. The labium is an apparent unitary body that bears an unsegmented mentum region with labial palps, each with three articles. The lack of subdivision of maxillae and labium main bodies into sclerite subdivisions is likely due to a lack of tomographic resolution. In ventral view (Fig. 17.4d), almost all standard mouthpart features seen in lateral view are accounted for. If the *Anebos* prothorax is typical of most roachoid insects of the Paleozoic, the mouthparts are generalized, resemble those of the Adult Ectognathate class, and bear certain resemblances to the mouthparts of *Gerarus*. The flattened body habitus and head and mouthpart design suggest detritivory.

#### 17.4.3.8 Orthoptera: Early Appearance of Grasshopper-Like Mouthparts

Among the defining characters of Orthoptera are their saltatorial hind legs and generalized herbivore mouthparts. Lineages bearing such distinctive hind legs and other orthopteran features occur in early Permian deposits and likely represent precursor taxa to Ensifera (crickets) and Caelifera (grasshoppers) that became monophyletically distinct during the Late Permian to Middle Triassic (Grimaldi and Engel 2005). However, one early group, Oedeschiidae and closely related lineages, occurred during the later Pennsylvanian and possessed enlarged, probably saltatorial hind legs, and likely were the stem group to true Orthoptera (Grimaldi and Engel 2005). However, little is known of the head and mouthparts of the Pennsylvanian forms, but a species related to Oedeschiidae, *Jubilaeus beybienkoi*, of Tcholmanvissiidae, (Fig. 17.4b), from the early Permian of Russia, displays a well-preserved body including prothorax, head, and mouthparts, as well as the distinctive pronotal hood and saltatorial hind legs of modern Orthoptera (Sharov 1968; Béthoux and Nel 2002).

The anterior aspect of *Jubilaeus* displays a distinctive pronotal hood extending to the occipital region of the head and a head frontal region with an indistinct clypeus supporting a thin labrum (Fig. 17.4a). Although the dicondylous mandibles are robust, features of their dentition are unclear, and thus inferences cannot be made regarding diet (Gangwere 1966). The maxillae bear four- or five-segmented palps and the labium supports three-segmented palps; both pairs of palps apparently were used to manipulate food items. Nevertheless, all features of the generalized, orthopteran mouthparts indicate that *Jubilaeus* was an external foliage feeder.

#### 17.4.3.9 Acercaria: Laciniate Mouthparts Transition Between the Mandibulate and Stylate Condition

The Acercaria are a group, probably not a clade in the sense that it consists of several lineages ancestral or related to paraneopterous insects (Shear and Kukalová-Peck 1990; Prokop et al. 2017), includes the monophyletic Paraneoptera (Grimaldi and Engel 2005). The Paraneoptera frequently are referred to as "hemipteroid" insects. A fundamental aspect of understanding the early evolution of the Acercaria is documentation of the transformation of mouthpart morphology from the mandibulate to the stylate condition and therefore the shift from chewing to piercing-and-sucking mouthpart movement. The early history of Acercaria is fraught with a welter of Pennsylvanian and early Permian lineages that display various mouthpart transformations of nominally mandibulate forms in the process of becoming incipient piercing-and-sucking forms. This process typically involves establishment of a prominent, bulging cibarial pump underneath the clypeus, elongation of the mandibles, and extension of the maxillary laciniae into long, blade-like structures that are edentulous or fitted with small, often serrated teeth (Yoshizawa and Lienhard 2016; Huang et al. 2016). Groups or lineages with these head and mouthpart features, expressed as the Laciniate mouthpart class, have variously included Blattinopsodea, Cacurgodea, Caloneurodea, Geraridae, Herdiniidae, Hypoperlidae, Permopsocidae, and Synomaloptilidae. Of these, the head and mouthparts of Cacurgidae, Caloneurodea, and Synomaloptilidae were examined, recognizing that the systematic placement of many of these and related taxa have been allocated to other major clades, such as Caloneurodea, Cacurgidae and Geraridae (see above) to the Archaeorthoptera (Béthoux et al. 2004; Béthoux 2006; Béthoux and Briggs 2008).

One of the earliest lineages showing incipient transformation from Adult Ectognathate to Laciniate mouthparts is *Heterologopsis ruhrensis* (Eoblattida: Cacurgidae), from the Early Pennsylvanian of Germany (Brauckmann and Koch 1982). The prothorax, head, and mouthparts of this specimen (Fig. 17.5a) reveal a suite of characters, specifically a domed clypeus; elongation of the drawn-out, prognathous mandibles to probably reflect adduction-abduction as well as protraction-retraction movements; and robust maxillary palps that reflect the buttressing of the mandibles during feeding. A species with greater head and mouthpart data is *Paleuthygramma tenuicornis* (Caloneurodea: Paleuthygrammatidae), from the Middle Permian of Russia (Martynov 1930; Sharov 1966; Rasnitsyn 1980), sharing closely related taxa extending back into the Late Pennsylvanian. The specimen of *Paleuthygramma* (Fig. 17.5b) displays a pronounced clypeal bulge indicating a capacious cibarial pump, hypognathous mouthparts with elongate, blade-like laciniae whose inner margin bore small, serrated teeth, and leg-like maxillary palps that could support lacinial puncturing into tissue (Sharov 1966; Shear and Kukalová-Peck 1990). An equally informative species is the *Synomaloptila longipennis* (Acercaria: Synomaloptilidae), from the Early Permian of Russia (Rasnitsyn 1980; Kukalová-Peck 1990). This species (Fig. 17.5c) exhibits a bulging clypeus, prolonged and triangular mandibles adjacent to elongate laciniae that have two distal prongs reminiscent of the

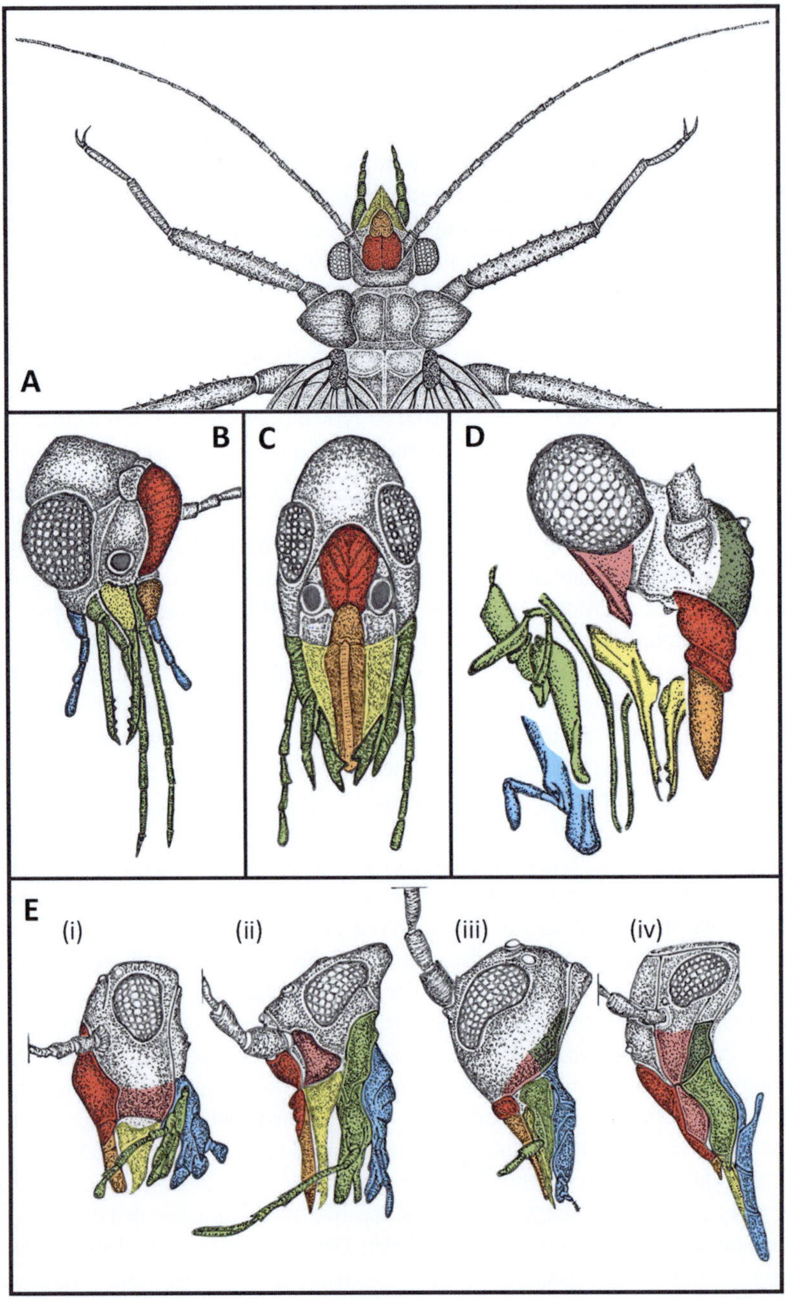


**Fig. 17.5** Initial expansion of mouthpart morphologies during the Pennsylvanian Period to early Permian Period from 323 to 272 million years ago. Part 4: The Acercaria (Hemipteroid) Complex representing the Laciniate mouthpart class. (**a**) Reconstruction of the head, mouthparts, and prothorax of *Heterologopsis ruhrensis* Brauckmann and Koch (1982) (Eoblattida: Cacurgidae). The specimen is

maxillary pick in modern Psocoptera, and long maxillary palps with supernumerary articles, similar to *Paleuthygramma* (Shear and Kukalová-Peck 1990).

The structural progression from mandibulate to piercing-and-sucking mouthparts (Fig. 17.5d) is shown in the "paleodissected" mouthparts of *Mydiognathus eviohlhoffae* (Acercaria: Archipsyllidae). *Mydiognathus* is a relict, anachronous lineage occurring in mid-Cretaceous Myanmar amber that represents a 200 million-year-earlier Late Pennsylvanian–Early Permian stage in an evolutionary transformation series. In this specimen with Laciniate mouthparts, the clypeus supports a cibarial pump; the labrum is robust and elongate, providing an upper brace; mandibles are long and blade-like with internally directed teeth; maxillary laciniae are styliform and edentulous; and the labial glossa is drawn out to serve as a fleshy gutter to accommodate the mandibles and laciniae. Instead of using a single, structurally intermediate member of the Laciniate mouthpart class to illustrate this transformation in the Acercaria, a similar head and mouthpart transformation is presented by a series of four head and mouthpart structures representing the shift from mandibulate to piercing-and-sucking taxa. This morphological series begins from the (1) Psocodea (Mortar-and-Pestle mouthparts), to (2) Permopsocida (Laciniate mouthparts), to (3) Thysanoptera (Mouthcone mouthparts), and ends in (4) Hemiptera (Segmented Beak mouthparts), each consisting of a distinctive "groundpattern" (Fig. 17.5e). This transformation series was inspired by the discovery of *Psocorrhyncha burmitica* (Fig. 17.13a) from the mid Cretaceous of Myanmar (Huang et al. 2016). The role of Laciniate mouthparts in understanding the shift from mandibulate (chewing) to stylate (piercing-and-sucking) mouthparts probably is applicable to other

**Fig. 17.5** (continued) a member of the Archaeorthoptera, (Béthoux 2008) or possibly an early representative of Acercaria (Kukalová-Peck and Brauckmann 1990), with elongated mandibles, from the Early Pennsylvanian of Hagen-Vorhalle, Ruhr Region, Germany; specimen FMW P20622. Redrawn from Kukalová-Peck and Brauckmann (1990), Fig. 7 on page 1109. (**b**) Right lateral reconstruction of the head and mouthparts of *Paleuthygramma tenuicornis* Martynov (1930) (Caloneurodea: Paleuthygrammatidae), with long, serrated laciniae and prolonged palps, from the mid Permian of the central Ural Mountains, Russia; specimen PIN 1700/1424 (Sharov 1966; Rasnitsyn 1980). Redrawn from Shear and Kukalová-Peck (1990), Fig. 40 on page 1825. (**c**) Anterior view of the head and mouthparts of *Synomaloptila longipennis* Martynov (1938) (Acercaria: Synomaloptilidae), with long, toothed laciniae and prolonged mandibles (Sharov 1966), from the early Permian of Chekarda, central Ural Mountains, Russia; specimen PIN 3353/456. Redrawn from Shear and Kukalová-Peck (1990) from Fig. 38 on page 1825. (**d**) Separated mouthpart elements of *Mydiognathus eviohlhoffae* Yoshizawa and Lienhard (2016) (Acercaria: Archipsyllidae), representing a transitional stage from mandibulate to stylate mouthparts, of the Early Cretaceous–Late Cretaceous boundary interval of Myanmar Amber; specimen SMNS Bu-174. Redrawn from Yoshizawa and Lienhard (2016) from Fig. 2a on page 233. (**e**) The hypothesis of head and mouthpart transformations in the Acercaria involving the evolution of stylate mouthparts from an ancestral mandibulate condition (Huang et al. 2016). From left to right: (i) the Psocodean condition that is also represented in Hypoperlidae; (ii) the condition in Permopsocidae; (iii) the condition in Thysanoptera, in particular the head of the adult tubuliferan *Moundthrips*; and (iv) the modern hemipteran condition. Note the changes in the clypeus into an upper postclypeus and lower anteclypeus. Mouthpart color scheme: red, clypeus; orange, labrum; yellow, mandibles; brown, hypopharynx; green, maxilla; blue, labium; pink, gena; and dark green, maxillary lobe

transformation series, such as those culminating in the Entognathous Stylate and Ectognathous Stylate mouthpart classes.

#### 17.4.3.10 Holometabola: Mouthparts and Feeding Habits of the Earliest Terrestrial Larva

Recently, the first body fossils of a holometabolous larva, *Metabolarva bella*, were described from the Late Pennsylvanian of Lower Saxony, Germany (Nel et al. 2013). However, details of their head and mouthparts were unclear. Subsequently, a better-preserved, interpretable, and earlier larva, *Srokalarva berthei*, from the Middle Pennsylvanian of Illinois, USA, was described in greater detail (Haug et al. 2015). *Metabolarva* and *Srokalarva* were externally feeding, eruciform and oligopod caterpillars, consistent with interpretations of their head structure (Fig. 17.4g). The head capsule of *Srokalarva*, the earliest known holometabolan insect, is preserved in left-lateral view, and bore multi-articled antennae of moderate length, a clypeus with surrounding sutures that articulated with the labrum, large triangular and dicondylous mandibles, maxillae, and a labium. The maxillae and labrum have downward, palp-like extensions. Evidence for a hypopharynx may be present. Many possible sutures of the cephalic capsule, such as an inverted Y-shaped frontal suture, are unclear, as is a large circular, upraised structure above the mandible articulatory line that likely was a compound eye. *Srokalarva* is the earliest documented larval holometabolan to possess Larval Ectognathate mouthparts.

Based on the eruciform larval habitus and mouthpart structure of *Srokalarva*, the most plausible interpretation for its diet is that of an externally foliage feeding herbivore (Haug et al. 2015). In the same Mazon Creek deposit in which *Srokalarva* occurs, are pinnules of *Macroneuropteris scheuchzeri*, foliage of a medullosan seed fern that occasionally contains cuspate excisions about 1 cm in chordal length along pinnular margins (Scott and Taylor 1983)—damage that could have been inflicted by *Srokalarva*. In addition to insect and plant damage contemporaneity, the mouthpart structure of *Srokalarva* is consistent with such inflicted external feeding damage. However, given the external appearance of the mandible, and the nature of the contemporaneous damage, it is unlikely that these mandibles were specialized for feeding on plant tissues of a particular degree of toughness or abrasiveness such as that demonstrated for the two, very different types of mandible construction in modern sphingid and saturniid moths (Bernays 1998).

### *17.4.4 New Mouthpart Classes and Within-Mouthpart Innovation During the Permian*

During the Permian, several new developments changed the spectrum of mouthpart classes, functional feeding activity, and the seeking of new food sources from that of the Pennsylvanian. Perhaps most important was expansion of Paraneoptera, including Psocoptera, Lophioneurida and especially Hemiptera, with their distinctive Mortar-and-Pestle, Mouthcone, and Segmented Beak mouthparts, respectively. Innovations in Hemiptera mouthparts gradually replaced paleodictyopteroid mouthparts, particularly during the early Permian. This replacement had already begun during the Pennsylvanian, when various, disparate lineages shifted from mandibulate chewing mouthparts to stylate piercing-and-sucking mouthparts, as evidenced by the emergence of Laciniate mouthpart class. Second was the modification of the mouthparts in Coleoptera and their immediate stem-group ancestors into robust structures capable of boring through wood and other indurated plant substrates. A third development was the independent, earliest origins of the Siphonate mouthpart class in small-bodied species of Neuroptera and Mecoptera, possibly indicating a Permian origin for pollination. By the end of Permian time, seven new mouthpart classes originated, and including the throughput of eleven earlier originating mouthpart classes, there were 18 mouthpart classes present. By the end of the Permian, 48.6% of all mouthpart classes had originated (Table 17.3).

#### 17.4.4.1 Hemiptera: Early Mouthpart Specialization on Fluid Plant Tissues

The earliest well-defined Hemiptera are the Archescytinidae (Grimaldi and Engel 2005), a group that occurred during the Early Permian and historically has been closely allied to the Sternorrhyncha. The Archescytinidae were present with several other early to middle Permian hemipteran lineages such as the Prosbolidae, Probolopseidae, Protopsyllidiidae, Scytinopteridae and Sojanoneuridae representing a significant species diversity of the existing fauna, particularly in the several, historically best-sampled localities from the Central Ural Mountains of Russia (Becker-Migdisova 1940, 1946, 1948a, b, 1960, 1985). Notably, the Archescytinidae had well-developed Segmented Beak mouthparts that moved in linear, up-and-down fashion (Fig. 17.6a), as in modern sternorrhynchan Hemiptera. However, this group bore a very long, structurally advanced thread-like ovipositor that was coiled like a watchspring and evidently was released to insert eggs in plant tissues. Such an analogous device did not evolve for sternorrhynchan mouthparts during the Permian, but rather occurred much later for Heteroptera, probably in the late Mesozoic, as documented in recent Aradidae by China (1931). By Middle Permian time, the diversity of Hemiptera increased immeasurably beyond that of Archescytinidae. During the Middle Permian, particularly Prosbolidae, Prosobolopseidae, Scytinopteridae (Fig. 17.6) and several other lineages offered several, new feeding strategies and associated diets that were in the process of replacing the ecologically similar piercing-and-sucking Paleodictyopteroidea.

Some of the best-preserved head and mouthpart structures of early Hemiptera originate from several Early and Middle Permian localities in Uralian Russia. A specimen (Fig. 17.6a) of *Sojanoneura kazanensis* (Hemiptera: Prosbolidae) displays a triangular shaped head with a bulging clypeus revealing a cibarial pump with a

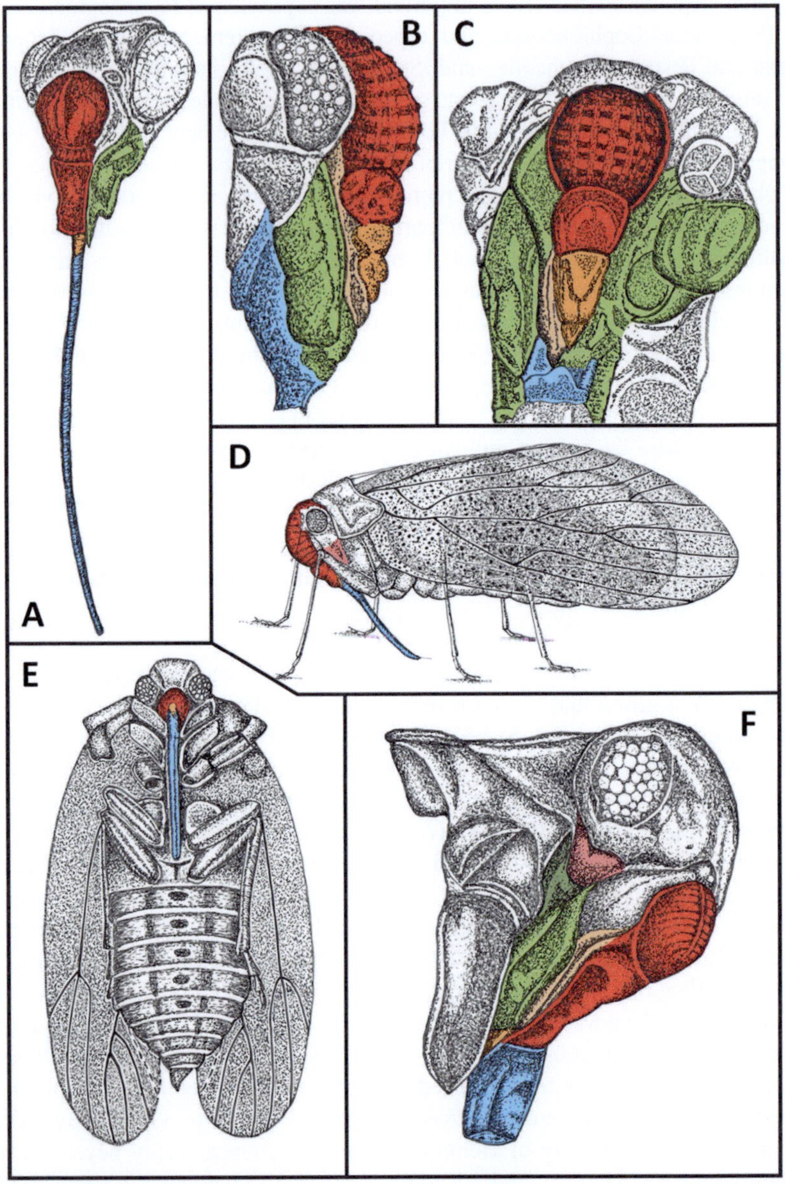


**Fig. 17.6** Specialization of Segmented Beak mouthparts in early Hemiptera from the mid Permian, 279 to 265 million years ago, along the central Ural Mountains region of Russia, based on the work of Becker-Migdisova (1940, 1946, 1948a, b, 1960). (**a**) An oblique anterior view of the head and mouthparts of *Sojanoneura kazanensis* M. Zalessky (1930) (Hemiptera: Prosbolidae), displaying a gracile elongate beak about twice the length of the head, from the mid Permian of Sojana, central Ural Mountains of Russia; specimen PIN 2124/117. Redrawn from Becker-Migdisova (1940), Fig. 42 on Page 51. (**b**) Left lateral view of the head and mouthparts of *Permocicada integra* Becker-Migdisova (1940) (Hemiptera: Prosbolidae), showing the extensive lorum lobe (in green)

median, ridge-like structure, probably an apodeme, but no underlying muscle attachment lineations (Becker-Migdisova 1961). The clypeus appears subsegmented and attached to a trapezoidally shaped labrum that likely was originally a triangular beak brace. The side of the head displays a rather inconspicuous lorum, and a prominently long, hypognathous beak that is somewhat more than twice the dorsoventral length of the head. The relatively modest cibarial pump, combined with a significantly long beak, indicates that this form was a phloem feeder. In contrast to the gracile mouthparts of *Sojanoneura*, the robust mouthparts of *Permocicada integra* (Hemiptera: Prosbolidae) indicate that it likely was a xylem feeder (Fig. 17.6b–d). A diet of nutritionally poor xylem is evidenced by the capacious size of the cibarial pump that extended to the dorsum of the head, strong cibarial dilator musculature on the clypeus surface, thick labral brace, and large protruding lorum (Becker-Migdisova 1940, 1946). Although the opisthognathous beak of *Permocicada* is comparatively short as seen in reconstruction (Fig. 17.6d), it is more massively built than that of *Sojanoneura*, and its life habits may be analogous to xylem feeders of modern Cicadidae. A reconstruction (Fig. 17.6e) of a third lineage of earlier Permian hemipterans, *Prosbolopsis* sp. (Hemiptera: Probolopseidae), shows a medium-length beak extending almost to the hind coxae (Becker-Migdisova 1960), but of unknown dietary habits. The fourth lineage, *Scytinoptera cubitalis* (Hemiptera: Scytinopteridae), is a specimen (Fig. 17.6f) of a head with a modest clypeal region showing relatively thin dilator muscle insertion lineations, and a cut-off, opisthognathous beak (Becker-Migdisova 1948a). Although most details of the beak are absent, the modest cibarium of this specimen (Fig. 17.6f), when compared to the massive cibarium with thicker muscle dilator scars of *Permocicada* (Fig. 17.6b, d), indicates a qualitatively different food source. There also is evidence in the plant damage record from the Williamson Drive locality, in north-central Texas, that scale insects were present on pteridophytes and pteridosperms and were accessing a variety of epidermal, mesophyll, and perhaps vascular tissues (Xu et al. 2018).

**Fig. 17.6** (continued) and compressor muscles of the cibarial pump at upper right, from the mid Permian of Iva Gora and Letopala of the central Ural Mountains in Russia; specimen PIN 94/906. Redrawn from Becker-Migdisova (1946), Fig. 1 on page 743. (**c**) A somewhat oblique, frontal view of the head and mouthparts of *P. integra*, also illustrated in (**b**), displaying the position of the hypopharynx and a robust labral brace, from the same locality in (**b**); specimen PIN 94/906. Redrawn from Becker-Migdisova (1946), Fig. 2 on page 743. (**d**) Whole-body reconstruction of *P. integra* in left lateral aspect, also figured in (**b**) and (**c**) from the same locality, showing a prominent, bulging cibarial pump and elongate beak that indicates xylem feeding; specimen PIN 94/106. Redrawn from Becker-Migdisova (1940) from Fig. 20 on page 20. (**e**) Reconstruction of *Prosbolopsis* sp. (Hemiptera: Probolopseidae), exhibiting stylate mouthparts extending nearly to the hind coxae, from Sheimo-Gora of the central Ural Mountains, Russia; from a composite of specimens PIN 2113/117 and 1891/117. Redrawn from Becker-Migdisova (1960), Fig. 46 on page 89. (**f**) Right-lateral view of the head, mouthparts, and prothorax of *Scytinoptera cubitalis* Becker-Migdisova (1948a) (Hemiptera: Scytinopteridae) from the mid Permian of Letopala, central Ural Mountains of Russia; specimen 1939/117. Redrawn from Becker-Migdisova (1946), Fig. 15 on page 754. Mouthpart color scheme: red, clypeus; orange, labrum; yellow, mandibles; brown, hypopharynx; green, maxilla; blue, labium; pink, gena; and dark green, epistomium

#### 17.4.4.2 Lophioneurida: The Role of *Zoropsocus* in the Development of Thrips Mouthparts

The relationship between the Lophioneurida and the Thysanoptera (thrips) has been a contentious issue. Given the explicit characters that define the Thysanoptera (Grimaldi and Engel 2005), the Lophioneurida is excluded from the Thysanoptera as a paraphyletic group. One subgroup of the Lophioneurida, most likely the lineage containing *Zoropsocus*, gave rise to the Thysanoptera during the mid Permian, and this group provided many of the fundamental features that constitute the head and Mouthcone mouthparts of Mesozoic to recent Thysanoptera. These features probably included an asymmetrical mouthcone, stylet-like mandible and laciniae, and reduction or absence of the right mandible (Nel et al. 2014). Some of the Permian lophioneurid lineages persisted into the Cretaceous, where these early, Permian clades uniquely occur with newly emerging lineages of today in deposits such as Lebanese, Álava, and Myanmar ambers (Labandeira 2014a), where preservation is sufficient to trace mouthpart details of these lineages to Permian taxa.

Four specimens of Lophioneurida (Figs. 17.7a–c and 17.8a) provide data for prothorax, head, and mouthpart structures before and after the end-Permian ecological crisis. During the Permian, one particular lineage, exemplified by *Zoropsocus tomiensis* (Lophioneurida: Zoropsocidae), possessed features such as an inflated clypeus, elongate to somewhat styliform mandibles, stubby maxillary palps, and the beginnings of asymmetrical mouthcone formation (Fig. 17.7a). These mouthpart elements were precursors to the emergence of Thysanoptera as a clade immediately before the Late Triassic (Grimaldi et al. 2004). These features are consistent with the punch-and-suck method of feeding on pollen grains by modern thrips (Kirk 1984), and evidence for this type of feeding has been found on circular punctures on noeggeranthialean spores in the Late Permian of north-central China (Wang et al. 2009). By the Middle Jurassic, Zoropsocidae, such as *Zoropsocus itschetuensis*, from southern Siberia in Russia (Fig. 17.8a), had evolved mouthparts with gracile maxillary palps, mouthcone compactness and asymmetry, and exhibited a convergence with certain modern Thysanoptera. Other Permian lineages of Lophioneuridae that survived the Late Permian ecological crisis continued to the Cretaceous, as represented by *Jantardachus perfectus* and *J. reductus* (Lophioneurida: Lophioneuridae), from Late Cretaceous Taimyr amber (Fig. 17.7b, c) of northern European Russia and eventually becoming extinct before the end of the Cretaceous. This mouthpart type likely was replaced by the emergence of Thysanoptera, some of which probably were pollinators of seed plants such as ginkgophytes (Peñalver et al. 2012). Although the features of *Jantardachus* mouthparts during the Late Cretaceous are separated from their most closely related Permian taxa by about 200 million years, the Lophioneuridae still exhibited the same fundamental structure of Mouthcone mouthparts.

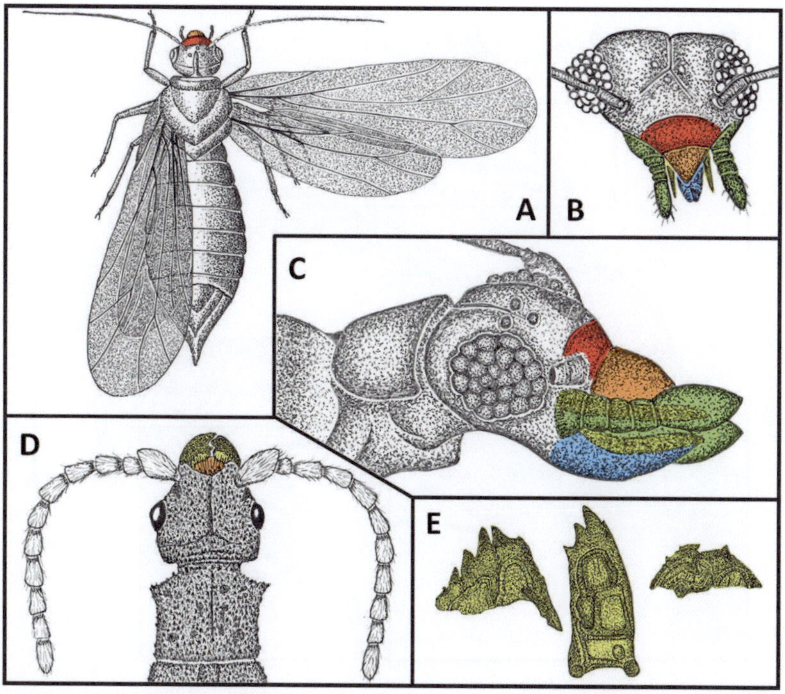


**Fig. 17.7** The establishment of generalized and specialized mouthpart types in Thysanoptera and Coleoptera during the Permian Period from 272 to 252 million years ago. (**a**) *Zoropsocus tomiensis* Becker-Migdisova (1961), in Rohdendorf (1962a) (Lophioneurida: Lophioneuridae), is a form that has some features of Thysanoptera, and likely housed Mouthcone mouthparts, from the middle Permian of Suriekova, south-central Siberia, Russia; specimen PIN 675/63. Redrawn from Becker-Migdisova (1961), Fig. 205 on page 276. (**b**) Foreshortened, anterior view of the head and Mouthcone mouthparts of *Jantardachus perfectus* Vishniakova (1981) (Lophioneurida: Lophioneuridae) from the Late Cretaceous (Campanian), from amber of the Taimyr Peninsula of northern European Russia; specimen PIN 3130/167, representing a relict lineage expressing an unplaced mouthpart type at a Permian stage of development. Redrawn from Vishniakova (1981), Fig. 47b on page 62. (**c**) Right-lateral view of the head, mouthparts, and prothorax of *Jantardachus reductus* Vishniakova (1981) (Lophioneurida: Lophioneuridae), from the same locality as (b) (PIN 3311/506), also representing a relict lineage possessing unplaced, Permian-style mouthparts that are prognathous and a precursor to the Mouthcone class mouthparts of thrips; specimen PIN 3311/506. Redrawn from Vishniakova (1981), Fig. 48b on page 62. (**d**) Dorsal view of the head, mouthparts, and prothorax of modern *Priacma serrata* Leconte (1861) (Coleoptera: Cupedidae), representing an Adult Ectognathate mouthpart type very similar to Permian Archostemata or a Coleoptera stem group with robust, prognathous mandibles and generalized maxillary and labial elements. Redrawn from Ponomarenko (1969), Fig. 60 (right) on page 111. (**e**) Three larval instar mandibles of an unidentified species of a polyphagan beetle occurring in a conifer wood boring, from the Late Permian of Ningxia, in north-central China, representing the Larval Ectognathate mouthpart class; specimens from borings YU YKLP20008, YKLP20009 and YKLP20010. Redrawn from Feng et al. (2017), Fig. 1h–j on page 2. Mouthpart color scheme: red, clypeus; orange, labrum; yellow, mandibles; brown, hypopharynx; green, maxilla; and blue, labium

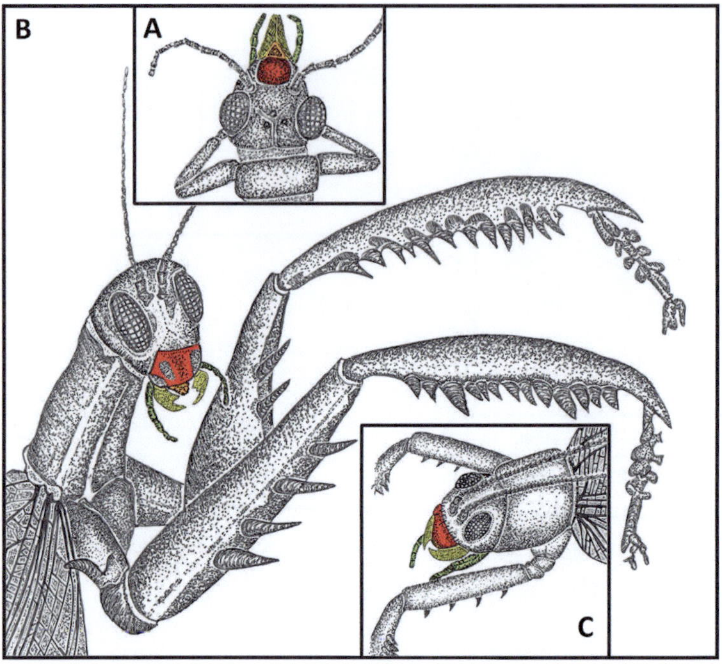


**Fig. 17.8** Lineages with distinctive mouthpart types surviving the end-Permian ecological crisis and a newly evolved, short-lived lineage occurring during the Middle to Late Triassic 252 to 199 million years ago. (**a**) Dorsal view of the head, mouthparts, and prothorax of *Zoropsocus itschetuensis* Vishniakova (1981) (Lophioneurida: Lophioneuridae), with Mouthcone mouthparts, from the Middle Jurassic of Novospassikoe, central Russia, a holdover from the Middle Permian; specimen PIN 3000/901. Redrawn from Vishniakova (1981), Fig. 37b, on page 48. (**b**) Right oblique reconstruction of the titanopteran *Gigatitan vulgaris* Sharov (1968) (Titanoptera: Gigatitanidae), a member of very large, predatory orthopteroids with Adult Ectognathate mouthparts possessing sharp, incisiform mandibles, from the Middle Triassic of Madygen, Kyrgyzstan; reconstructed from several specimens. Redrawn from Rohdendorf and Rasnitsyn (1980), Fig. 88 on page 169. (**c**) Another reconstruction, in dorsal oblique view, of the head, mouthparts, and prothorax of *Gigatitan*, the same species and same locality as (**b**). Redrawn from Sharov (1968), Fig. 50 on page 131; a composite of specimens PIN 2440/4502 and 2555/1541. Mouthpart color scheme: red, clypeus; orange, labrum; yellow, mandibles; brown, hypopharynx; green, maxilla; and blue, labium

#### 17.4.4.3 Coleoptera and the Early Development of Mouthparts for Wood Boring

Little is known of the heads and mouthparts of Permian beetle ancestors or relatives, although three sources of data are relevant. The first type of evidence is the compression fossils of Permian "Protocoleoptera" from various deposits in Central Asia and Eastern Europe (Ponomarenko 1969; Shear and Kukalová-Peck 1990). These fossils lack several features that would make them definitive Coleoptera. These features include possession of 13 antenna articles instead of 11; loosely configured, leathery, heavily veined forewings instead of compact, sclerotized elytra that lack veins; the presence of widely separated rather than adjacent fore coxae; and the presence of an obvious external ovipositor rather than one that is hidden internally (Grimaldi and Engel 2005). From these same fossils, the mouthpart evidence indicates that the prothorax consisted of a pronotal hood that had angulate extensions on each side of the head occipital region. The clypeus lacked a prominent bulge and was attached to a labrum, and the mandibles were prominent, prognathous, and structured for penetrating hard tissue. Little is known of the maxillae and labium.

Permian protocoleopteran beetles share many similarities with one of their descendant groups, such as *Priacma serrata* of the Archostemata (Fig. 17.7d), constituting the second line of evidence. *Priacma* shows prominent, jutting mandibles with asymmetrically matching incisor regions (Fig. 17.7d), generalized mouthpart morphology, and a flattened body form (Hörnschemeyer et al. 2002, 2006). A third kind of evidence is the recent discovery of fragments of beetle head capsules and mouthparts, particularly a population of mandibles from multiple larval and adult instars in coniferophyte wood borings from the Late Permian of China (Feng et al. 2017). The mandibles from the borings (Fig. 17.7e) provide rare insight into the ontogenesis of larval mandible development and wear as well as the earliest occurrence of a beetle assigned to Polyphaga. This distinctive, late Permian network of borings occurred in indurated coniferophyte bark, cambial, and wood tissues (Feng et al. 2017), and can be contrasted with earlier middle Permian tunneling in punky, partly decomposed coniferophyte wood (Naugolynkh and Ponomarenko 2010). The two occurrences suggest that mouthparts of multiple early beetle lineages were designed to bore through wood, and larvae and adults likely were feeding on wood-associated fungi (e.g., Batra and Batra 1967).

#### 17.4.4.4 Neuroptera and Mecoptera: The Antiquity of Siphonate Mouthparts

Almost all of the documentation for fossil long-proboscid mouthparts with a tubular siphon for ingesting fluid food such as gymnosperm pollination drops (Labandeira et al. 2007a) has come from the mid Mesozoic of Eurasia (Ren et al. 2009; Lin et al. 2019). Several major lineages of Neuroptera, Mecoptera, Diptera, and Lepidoptera evolved independently the siphonate proboscis approximately 13 times, resulting in about 45–50 described species among these four orders of insects (Peris et al. 2017). To that list, two small, long-proboscid insect species from the Permian of Russia add two additional examples of mouthpart convergence to the long-proboscid condition. A late Permian form was *Nedubrovia shcherbakovi* (Mecoptera: Nedubroviidae), from the late Permian of Vogolda Province in northern European Russia (Bashkuev 2011). This miniscule species possessed a long proboscis estimated to be 1 mm long, although the fossil only reveals 0.4 mm of the proboscis between the head capsule and a fracture in the rock matrix. A second species was *Tschekardithonopis ?oblitus* Vilesov 1995 (Neuroptera: Permithonidae), from the early Permian of Perm Province, in the central Ural Mountains of Russia (Labandeira 2010). This species had a complete, 1.7-mm-long proboscis with a thickened terminus reinforced with small setae. These earliest occurrences of the Siphonate mouthpart class indicate that fluids such as pollination drops (Labandeira et al. 2007a) or similar secretory enticements were offered by a variety of Permian seed plants to mosquito size insects approximately 150 million years before the recorded appearance of angiosperms in the fossil record.

## 17.5 The Mesozoic: New Modes of Mouthpart Innovation and Refinements in Targeting Food Substrates

There was considerable ecological transformation of the terrestrial and freshwater realms during the Mesozoic. After the major ecological crisis and mass extinction at the end of the Permian, insect lineages in both realms experienced a major pruning of taxa (Labandeira 2005a). Many that survived, such as the rare monuran and paleodictyopteroid lineages, were functionally extinct as "dead clades walking" (Jablonski 2005), becoming ecologically irrelevant for any contribution to mouthpart diversity. Within 15–20 million years after end-Permian event, terrestrial insect diversity exceeded that at the end of the Permian. This dramatic increase is associated with the added number of insect damage types to plant tissues (Labandeira et al. 2007a), elevated insect herbivory intensity, and terrestrial plant and insect body fossils beyond Permian levels in one heavily sampled locality in the Late Triassic of South Africa (Labandeira et al. 2018). However, such an opening did not occur globally in the freshwater realm until the Late Jurassic to Early Cretaceous in freshwater ecosystems, concomitant with major ecological changes of the Mesozoic Lacustrine Revolution (Buatois et al. 2016). Two other events during the mid Mesozoic likely were responsible for the significant increase in insect taxa and their mouthpart diversity. These episodes were the Parasitoid Revolution that represented a major expansion of hymenopteran and dipteran parasitoid diversity (Rasnitsyn 1980; Labandeira 2002b), and the Cretaceous ecological expansion of Angiosperms (Grimaldi 1999; Labandeira 2014b). It is a fair estimation that the overwhelmingly majority of mouthpart innovation in insects largely ended at the end of the mid Mesozoic.

During the Mesozoic, there was a rapid rise in the number of mouthpart classes, so much so that by the end of the era, all but two mouthpart classes had arisen, although there was the disappearance of two mouthpart classes that originated during the Pennsylvanian (Table 17.3). By the end of the Triassic, there was the net origin of seven new mouthpart classes, and including the throughput of 18 earlier originating classes, there were 25 mouthpart classes present by the end of the period, indicating that 67.6% of the total number of mouthpart classes had originated. Similarly, by the end of the Jurassic, there was the net origin of seven new mouthpart classes, and including the throughput of 25 earlier originating mouthpart classes, there were

31 mouthpart classes present by the end of the period, indicating that 88.9% of the total number of mouthpart classes had originated. Finally, by the end of the Cretaceous, there was a net origin of three mouthpart classes by the end of the period, indicating that 94.4% of the total number of mouthpart classes had originated. Approximately half of all insect mouthpart classes originated during the Paleozoic; almost the other half originated during the Mesozoic; and the Cenozoic offered very little in major insect mouthpart innovation (Table 17.3).

### *17.5.1 Reestablishment of Mouthpart Design After the Permian–Triassic Ecological Crisis*

As indicated above and in Table 17.3, about half of the mouthpart classes had originated before the Permian–Triassic mass extinction and ecological crisis. After this critical event, another approximate half of mouthpart classes originated and were Mesozoic in origin. This indicates that the restricted number of lineages that emerged from the Permian bottleneck evolved new mouthpart classes or alternatively the new mouthpart classes emerged from lineages that had originated in the early Mesozoic, soon after the Permian–Triassic event. Given the taxonomic affiliations of taxa bearing mouthpart classes during the Mesozoic, it appears that it were those early Mesozoic lineages that provided the raw material for the 16 new classes of the Mesozoic (Table 17.3).

### *17.5.2 Singular Innovations in Mouthpart Design*

Very occasionally, mouthpart structures evolve that are considered key innovations allowing the bearer of the novel mouthpart structure to exploit food resources in a new way. Three examples illustrate this phenomenon for the earlier Mesozoic. First, a largely herbivorous group of medium-sized, orthopteroid insects with generalized mouthparts and unmodified forelegs were transformed into a lineage of very large, predatory insects with specialized mouthparts and spinose, mantid-like forelegs (Sharov 1968). Second is the development of the weevil elongate rostrum as a device to penetrate plant tissues as diverse as wood, the thick parenchyma of plant stems, the interior embryonic tissues of seeds and fruits, and the novel use of the rostrum as a functional ovipositor (Anderson 1995). Third, the protractile concealed nectar extraction device is a fusion and functional co-optation of the maxillary and labial regions of parasitoid wasp mouthparts that allowed access to plant fluids such as pollination drops, floral nectar, and extrafloral secretions (Jervis 1993). While these three mouthpart-related complexes may or may not have all of the required features for a key innovation, they nevertheless do allow greater and more efficient access to food resources.

#### 17.5.2.1 Titanoptera: Orthopteroid Mouthparts and Forelegs Transitioning to Carnivory

Closely related to the Geraridae mentioned previously were the Titanoptera (Gorochov 2001), a geochronologically fleeting clade of very large insects, with wingspans up to 40 cm, from the Middle and Late Triassic of Central Asia and Australia (Sharov 1968). A common form was *Gigatitan vulgaris* (Titanoptera: Gigatitanidae) from the Middle Triassic of Kyrgyzstan (Fig. 17.8b, c), an orthopteroid insect that displayed head, mouthpart and foreleg structures for predation. The prothoracic legs were heavily spinose for impalement, and the mandibles and maxillary laciniae and galeae were equipped with sharp, incisiform teeth (Rohdendorf and Rasnitsyn 1980).

#### 17.5.2.2 Coleoptera: The Weevil Rostrum, a Key Innovation?

The Curculionoidea (weevils, bark beetles, and ambrosia beetles) are the most diverse major lineage of Coleoptera. A reason for their success may be their distinctive rostrum, the key feature of the Rhynchophorate mouthpart class, associated with modification and enlargement of certain features of the beetle head and mouthparts (Anderson 1995). The weevil rostrum, an extension of the head capsule with mandibulate mouthparts at the terminus, is characterized by a medially elongated, ventral sclerite, the gula, which allows forward extension of the mouthparts (Fig. 17.9f), as illustrated by the early primitive weevil *Distenorrhinus antennatus* (Coleoptera: Nemonychidae) from the Late Jurassic of Karatau, Kazakhstan (Arnol'di et al. 1991). This weevil possesses a protective collar of the prothorax encasing the posterior region of the head capsule, similar to the heads of many other mid-Mesozoic weevils (Davis et al. 2013). During the Mesozoic, Rhynchophorate mouthparts were modified in a variety of modes, ranging from short and blunt morphologies in wood-boring bark and ambrosia beetles to extremely long snouts in cycad weevils (Cai et al. 2018). Previously, a weevil-like rostrum had originated during the Triassic among the early beetle lineage Obrieniidae (Legalov 2015), becoming extinct before the structurally convergent curculionoid rostrum originated.

#### 17.5.2.3 Hymenoptera: The Concealed Nectar Extraction Apparatus of Wasps

Most adult parasitoid Hymenoptera are small wasps that have a combination of mouthpart elements that allow feeding on fluids such as nectar. These elements form a mouthpart mechanism consisting overwhelmingly of maxillary and labial elements known as the concealed nectar extraction apparatus that occurs for example in fig wasps (Peñalver et al. 2006). The protractile concealed nectar extraction apparatus forms the basis of the Maxillolabiate mouthpart class (Jervis et al. 1993). This functional fusion of the maxillary and labial segments provides a mechanism to extract nectar, other fluids, and pollen from present-day flowering plants (Jervis et al. 1993). Maxillolabiate mouthparts formed the primary mode of extracting fluids from fern vegetative tissues and seed plant reproductive organs during the Parasitoid Revolution of the Middle Jurassic to Early Cretaceous (Labandeira 2002b).

### 17.5.3 Processes and Patterns Involved in Mouthpart Design

Two major processes occurred during the later Mesozoic that affected the evolution of mouthpart design in the Mesozoic. The first process involved global changes in the environment that led to a major change in the biota, such as new lineages accessing new food resources in lacustrine ecosystems that became available during the Early Jurassic and culminated in the Early Cretaceous. This event was the Mesozoic Lacustrine Revolution that resulted in numerous insect lineages invading or evolving in lacustrine ecosystems that produced new mouthpart types to take advantage of new opportunities for seeking food that largely became available to herbivores and predators.

The second category or processes took place on land and largely involved evolutionary transformations of major mouthpart types within particular major insect lineages. These macroevolutionary transformations included four major, specific themes. First, there was the evolution of the early Siphonapteran proboscis from an ancestral Mecoptera ancestor into a robust structure capable of puncturing thick integument and evolving to the more delicate mouthpart features of modern fleas. Second, a broad repertoire of mouthpart innovation occurred in Diptera to exploit and access a newfound resource in terrestrial vertebrates, blood and lymph, but in more diverse functional and structural ways than that of early Siphonaptera. Third, there is the overlap, captured in several well-preserved Cretaceous amber deposits, of lineages of insects displaying anachronistic mouthpart morphologies at a Permian and Triassic stage of evolutionary development that occur in the same deposit amid new and modern lineages of insects possessing the mouthparts of today. Fourth, occurring in Cretaceous amber deposits, particularly Myanmar Amber, is the presence of bizarre mouthpart morphologies that remain unassignable to any known mouthpart class. The global process of the Mesozoic Lacustrine Revolution and the four, relevant, lineage-directed processes indicate that the Mesozoic was a time interval of considerable mouthpart innovation, much of which undoubtedly remains to be discovered.

#### 17.5.3.1 Ephemeroptera, Odonatoptera, Chresmodida, Megaloptera, Coleoptera and Diptera: Mouthparts and the Mesozoic Lacustrine Revolution

A phenomenal ecological transformation of large bodies of freshwater occurred during an interval from the Late Triassic to Early Cretaceous, known as the Mesozoic Lacustrine Revolution (MLR, Buatois et al. 2016). This event, centered principally in the vast Middle Jurassic to Early Cretaceous lakes of Eurasia, resulted in an ecological transformation that emphasized the emergence of aquatic vascular plants around lake margins, and a trophic shift from detritivorous to herbivorous invertebrate and vertebrate taxa (Miller and Labandeira 2002). There also was a more thorough reworking of muddy and sandy lake substrates by organisms, and, importantly, the emergence of a distinctive insect lake fauna in the benthos, water column, and neuston zones, particularly along lake shorefaces. The MLR is analogous to the expansion of the terrestrial insect taxa during the Pennsylvanian, except for its delay by about 175 million years.

Many Eurasian insect lineages that were involved with the MLR (Fig. 17.9a–e) bore newly evolved mouthparts, commensurate with a shift from detritivore to herbivore driven aquatic ecosystems that also favored new lineages of predators. Common aquatic herbivores were *Mesohelophorus elongatus* (Coleoptera: Hydrophilidae), an Adult Ectognathate external feeder on foliage (Fig. 17.9d), from the Early Cretaceous of Siberia (Ponomarenko, in Rasnitsyn 1990). Another herbivore was *Chironomoptera* sp. (Diptera: Chaoboridae), a filter-feeding midge larva (Fig. 17.9e), and a member of the Mouthbrush mouthpart class, from the Early Cretaceous Laiyang Formation of Shandong, China (Rasnitsyn and Quicke 2002). Predators are disproportionately represented in these ecosystems and include the dragonfly *Bellabrunetia catherinae* (Odonata: Campterophlebiidae), an aerial, adult pursuit predator of the Raptorial Ectognathate mouthpart class (Fig. 17.9a), from the Early Cretaceous Yixian Formation of Liaoning, China (Fleck and Nel 2002). Predatory naiads of other dragonfly taxa bore prehensile Labial Mask mouthparts (Fleck et al. 2002). An aquatic larval predator (Fig. 17.9b) is the beetle *Coptoclava longipoda* (Coleoptera: Coptoclavidae), bearing Larval Ectognathate mouthparts with mandibular and foreleg modifications probably for ambush predation, from the Early Cretaceous of Shandong, China (Ping 1928; Ponomarenko, in Arnol'di et al. 1991). This extinct form has head and mouthparts similar to modern predaceous diving beetles (DeMarzo 1976a, b, 1977). Another predator is *Parahygrobia natans* (Coleoptera: Parahygrobiidae), displaying prominent, incisiform, mandibular teeth of Adult Ectognathate mouthparts (Fig. 17.9c), from the Late Jurassic of Uda, in Buryat, Russia (Ponomarenko, in Arnol'di et al. 1991).

Although these five species of Odonatoptera, Coleoptera, and Diptera were major elements of the MLR, other major lineages contributed to the MLR. These MLR-associated lineages included naiads of the mayfly *Shantous lacustris* (Ephemeroptera: Hexagenitidae), from the Middle Jurassic of Inner Mongolia, China (Zhang and Kluge 2007), and the orthopteroid *Chresmoda libanica* (Chresmodida: Chresmodidae), a very large, water-strider-like insect from the

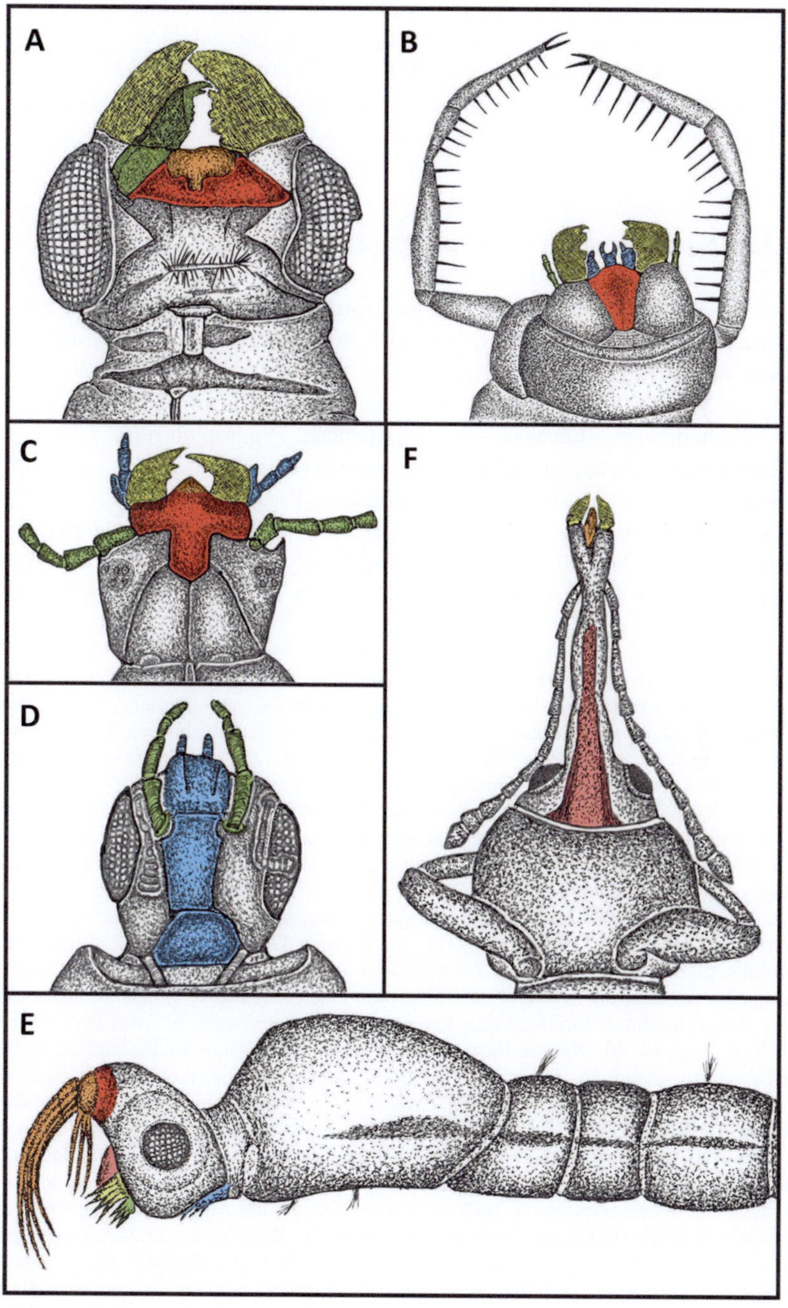


**Fig. 17.9** Mouthparts of the Adult Ectognathate (**a**, **c**, **d**) and Larval Ectognathate (**b**) and Mouthbrush (**e**) classes involved in the expansion of the Mesozoic Lacustrine Revolution from the Middle Jurassic to the Early Cretaceous, from 174 to 100 million years ago, representing the extensive development of multitrophic food webs in lake ecosystems (Buatois et al. 2016). The

Early Cretaceous of Lebanon (Nel et al. 2004; also see Martínez-Delclòs 1989). Also present was the alderfly larva *Sharasialis fusiformis* (Megaloptera: Sialidae) from the Late Jurassic Shar-Teg locality of Mongolia (Ponomarenko 2012). The last two taxa were immatures that bore formidable mandibulate mouthparts indicating obligate insectivory.

#### 17.5.3.2 Siphonaptera: Evolution of the Flea Proboscis

One of the more important mouthpart transformations in the insect fossil record is the structural evolution of flea or flea-like mouthparts from a likely mecopteran ancestor with a siphonate proboscis, such as Aneuretopsychidae (Ren et al. 2011), during the Middle Jurassic (Gao et al. 2013). During this formative interval, at least three Siphonaptera lineages—Pseudopulicidae, Saurophthiridae, and early Siphonaptera—evolved, characterized by an intervening phase of major mouthpart development (Gao et al. 2013). The earliest clade, Pseudopulicidae (Fig. 17.10a–e), consisted of forms lacking body compression, principally *Pseudopulex*, and occurred from the late Middle Jurassic to mid Early Cretaceous of China. The last known member of the Pseudopulicidae is *Tarwinia*, from the late Early Cretaceous

---

**Fig. 17.9** (continued) establishment of the Rhynchophorate mouthpart class (**f**) occurred in terrestrial ecosystems. (**a**) Dorsal view of the head, mouthparts, and prothorax of the aquatic *Bellabrunetia catherinae* Fleck and Nel (2002) (Odonata: Campterophlebiidae), displaying robust, incisiform mandibles and maxillary laciniae, and divided upper and lower compound eye regions, from the Early Cretaceous Yixian Formation of Liaoning Province, northeastern China; specimen MNHN-LP-R.55232a-b. Redrawn from Fleck and Nel (2002), Fig. 5d on page 1131. (**b**) Slightly oblique, dorsal view of the head, mouthparts, and prothorax of the aquatic *Coptoclava longipoda* Ping (1928) (Coleoptera: Coptoclavidae), bearing Larval Ectognathate mouthparts and exhibiting predatory mandibular and prothoracic leg modifications, from the Early Cretaceous Laiyang Formation, Shandong Province, China; repository unknown, specimen 2145. Redrawn from Ponomarenko, in Arnol'di et al. (1991), Fig. 13c on page 47. (**c**) Dorsal view of the head and mouthparts of the aquatic predator *Parahygrobia natans* Ponomarenko (1990) (Coleoptera: Parahygrobiidae), showing incisiform mandibles, from the Late Jurassic of Uda, in Transbaikalia, Buryat, Russia; specimen PIN 3053/423. Redrawn from Ponomarenko, in Arnol'di et al. (1991), Fig. 3b on page 24. (**d**) Ventral view of the head and mouthparts of the aquatic herbivore *Mesohelophorus elongatus* Ponomarenko (1990) (Coleoptera: Hydrophilidae), with Adult Ectognathate mouthparts, from the Early Cretaceous of Tunga, southeastern Siberia, Russia; specimen PIN 3063/735. Redrawn from Ponomarenko, in Rasnitsyn (1990), Fig. 43b on page 47. (**e**) Left lateral view of the head, mouthparts, and thorax of a reconstruction of the larva of the phantom midge *Chironomaptera* sp. (Diptera: Chaoboridae) from the same provenance as (**b**), showing filter-feeding mouthparts, such as a labral fan, of the Mouthbrush mouthpart class; specimen is a composite of fossil and related modern species. Redrawn from Rasnitsyn and Quicke (2002), Fig. 489 on page 397. (**f**) Ventral view of head, mouthparts, and prothorax of *Distenorrhinus antennatus* Arnol'di et al. (1991) (Coleoptera: Nemonychidae), bearing weevil Rhynchophorate mouthparts, with terminally positioned mouthparts and an expanded gula, from the Late Jurassic of Karatau, southern Kazakhstan; specimen PIN 2554/721. Redrawn from Arnol'di et al. (1991), Fig. 104 on page 235. Mouthpart color scheme: red, clypeus; orange, labrum; yellow, mandibles; brown, hypopharynx; green, maxilla; blue, labium; and pink, gula

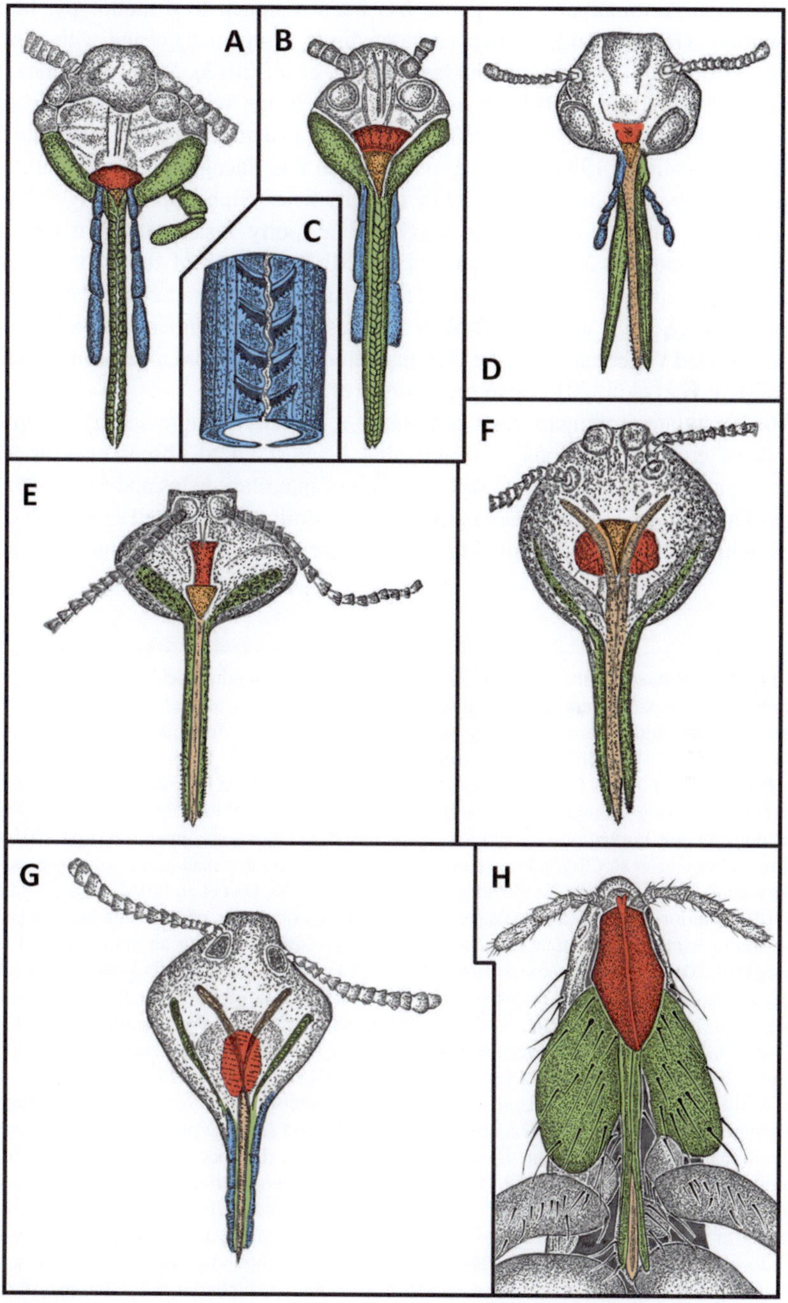


**Fig. 17.10** The mid-Mesozoic evolution of the Tristylate mouthpart class, as evidenced by mouthpart modifications of two Middle Jurassic and Early Cretaceous flea lineages, Pseudopulicidae and Saurophthiridae, spanning the interval from 170 to 120 million years ago, as well as Linognathidae of modern Siphonaptera. (**a**) Anterior view of *Pseudopulex jurassicus* Gao,

of Australia. Alternatively, this genus has been assigned to a monotypic family, Tarwiniidae (Huang 2015), which may constitute a separate lineage with a sister-group relationship to Pseudopulicidae (Huang et al. 2013). A second lineage is Saurophthiridae (Fig. 17.10f, g) and comprises forms with dorsoventrally compressed bodies such as *Saurophthirus*, although structurally similar taxa such as *Strashila* (Rasnitsyn 1992) are likely included in this lineage. The Saurophthiridae shares a sister-group relationship with modern fleas, Siphonaptera (Fig. 17.10h), possessing a laterally compressed body and a spotty fossil record in Cenozoic ambers (Beaucoumu 2003; Perrichot et al. 2012). In each of these three or four siphonapteran lineages, the evolution of the Tristylate mouthpart condition has taken a new pathway, likely circumscribed by features of the integument and pelage of its warm-blooded vertebrate hosts, be they feathered dinosaur, bird, or mammal (Huang et al. 2013; Gao et al. 2013; also see Dittmar et al. 2016).

The Tristylate mouthpart class consists of a laterally compressed head approximately of equal dimensions in profile that have small to absent eyes, a labral (epipharyngeal) and two lacinial stylets, a pair of maxillary palps and a pair of labial palps (Table 17.2). The maxillary palps often are similar in form and size to the labial palps, with the exception that the first article of the maxillary palps is modified into a

**Fig. 17.10** (continued) Shih, and Ren (2012) (Siphonaptera: Pseudopulicidae), a probable stem-group lineage to Siphonaptera, with maxillary lacinial stylets bearing inwardly directed robust teeth, from the latest Middle Jurassic Jiulongshan Formation of Inner Mongolia, northeastern China; specimen CNU-NN-2010001. Redrawn from Gao et al. (2012), Fig. 1d on page 733. (**b**) Anterior view *Pseudopulex magnus* Gao, Shih, and Ren (2012) (Siphonaptera: Pseudopulicidae), displaying protracting-retracting lacinial stylets that are braced by adjacent labial palps, from the mid Early Cretaceous Yixian Formation of Liaoning, in northeastern China; specimen CNU-ND-2010002. Redrawn from Gao et al. (2012), Fig. 2c on page 734. (**c**) An enlargement of the conjoined laciniae at midsection in (**b**), showing interlocking teeth. Redrawn from Gao et al. (2012), Fig. 2d on page 734. (**d**) Anterior view of a female specimen from a composite of part and counterpart of *Pseudopulex wangi* (Siphonaptera: Pseudopulicidae), possibly revealing an epipharyngeal stylet (Huang et al. 2013), from the same general locality as (**a**); specimen NIGP 154244b. Redrawn from Huang et al. (2012), Fig. 2b on page 202. (**e**) Anterior view of the giant flea *Tyrannopsylla beipiaoensis* Huang, Engel, Cai, and Nel (2013) (Siphonaptera: Pseudopulicidae), a female exhibiting lacinial stylets that lack teeth or serrations, from the same general locality as (**b**); specimen NIGP 154250. Redrawn from Huang et al. (2013), from Fig. 3f on page 7. (**f**) Anterior view of the flea *Saurophthirus exquisitus* Gao, Shih, Rasnitsyn, and Ren (2013) (Siphonaptera: Saurophthiridae), showing a median epipharyngeal stylet and separate protractor–retractor muscle pairs within the head capsule, from the same general locality as (**b**); specimen CNU-LL2010016CP. Redrawn from Gao et al. (2013), Fig. 2c on page 1263. (**g**) Interpretation by Gao et al. (2013) of *S. exquisitus* in (**f**), distinctly revealing a protractor muscle pair and a retractor muscle pair. Redrawn from Gao et al. (2013), Fig. 1f on page 1262. (**h**) Head, mouthparts, and prothorax of the modern groundhog flea, *Oropsylla arctomys* Baker (Hamilton 1934) (Siphonaptera: Ceratophyllidae), for comparison to mid-Mesozoic fleas in (**a**–**g**) above, showing an epipharyngeal stylet, pair of finely serrated lacinial stylets, and a large, ovoidal maxillary lever for lacinial stylet protraction. Redrawn from Photosearch Stock Photography and Stock Footage (https://www.fotosearch.com/ulv395u20110003), accessed on November of 2018. Mouthpart color scheme: red, clypeus; orange, labrum; yellow, mandibles; brown, epipharynx; green, maxilla; and blue, labium

large structure that is musculated such that it is deployed as a lever to downwardly engage the piercing and serrated lacinial stylets to penetrate host tissue below (Fig. 17.10h) (Kluge 2002). The labial palps and mentum are involved in guiding the stylets preparatory to and during host penetration (Kluge 2002). Most of these structures can be seen in anterior (frontal) views of Mesozoic and modern fleas in Fig. 17.10. In *Pseudopulex jurassicus* and *P. magnus* (Siphonaptera: Pseudopulicidae), from the latest Middle Jurassic of northeastern China (Fig. 17.10a, b), details of lacinial stylet dentition or serrations differ from that of modern Siphonaptera (Gao et al. 2012, as illustrated by an enlargement of the adjacent lacinial stylets in the inset of Fig. 17.10c). The lacinial stylets are robust and thickened, and display intermeshing, highly notched serrations that are directed inwardly (medially), rather than outward features, different from modern fleas. The differences between the two species of *Pseudopulex* involve overall size and the distribution and development of the cutting serrations along the lacinial stylets. Also from the same locality as the material above is a specimen (Fig. 17.10d) of *Pseudopulex wangi*, which has a labial palpus of four fused articles that house the fascicle of delicately serrated stylets (Huang et al. 2013).

From the Early Cretaceous of northeastern China is a specimen (Fig. 17.10e) of *Tyrannopsylla beipiaoensis* (Siphonaptera: Pseudopulicidae), which shares some features with *Pseudopulex*, but has significant differences in mouthpart construction (Huang et al. 2013). In particular, formation of the labial palps into "two half-tubes" that are sealed (Huang et al. 2012) is a distinctive modification ensheathing the robust and rather long lacinial stylets that apparently lack serrations. In contrast to the head and mouthpart morphology of Pseudopulicidae is *Saurophthirus exquisitus* (Siphonaptera: Saurophthiridae), shown as the original fossil specimen (Fig. 17.10f) and as a reconstruction (Fig. 17.10g), displaying protractor–retractor musculature (Gao et al. 2013). The reconstruction shows restored stylet protractor–retractor musculature, a labial palpal sheath composed of three articles that forms a loose cover that incompletely surrounds the stylet ensemble, and a beak that is shorter than the head length (Gao et al. 2013). The labral stylet is prominent and the lacinial stylets have outwardly directed serrations along the margin. These mouthpart features indicate significant differentiation of the Saurophthiridae from the Pseudopulicidae. Both Mesozoic siphonapteran families, however, have differences from modern-aspect Tristylate mouthparts of modern fleas, such as *Linognathoides marmotae* (Siphonaptera: Linognathidae). The most conspicuous of these elements are exceptionally long stylets and especially the substantially expanded first maxillary palp article that is used as a lever for downward engagement of the lacinial stylets (Fig. 17.10h).

#### 17.5.3.3 Diptera: Mouthpart Innovation and Hematophagy

Another prominent hematophagous group, in addition to Siphonaptera, are Diptera. Unlike Siphonaptera, in which the original mouthpart structure for hematophagy occurred once and early in the evolution of the clade, among Diptera, hematophagy originated about 20 times among several mouthpart classes (Labandeira 1990). Hematophagy occurs in 20 families of Diptera (Balashov 1999; Lukashevich and Mostovski 2003; Martins-Neto 2003), but the distribution of its origination is highly variable across lineages. For example, hematophagy originated once in the Culicomorpha that currently consists of seven blood-feeding families, once in each of the families Psychodidae (sandflies) and Tabanidae (horseflies), and several times within the Muscidae (houseflies) (Grimaldi and Engel 2005). The significant difference about hematophagy between Siphonaptera and Diptera is that in the former, it is associated with highly stereotyped Tristylate mouthparts (Fig. 17.10), whereas in the latter it is associated with a major clade of insects with the most variable mouthpart morphology of any major insect clade (Fig. 17.11).

Diptera experienced a major radiation in lineages during the Middle Jurassic to Early Cretaceous, in which mouthparts of the Hexastylate and Distylate/Tetrastylate classes evolved parasitic adaptations for hematophagy, or alternatively for predation on insects that represented a prelude toward hematophagy. Three groups played a prominent role in mouthpart development and the origin of hematophagy: Culicomorpha (Fig. 17.11a, d), early lineages of Brachycera (Fig. 17.11b, e), and Tabanomorpha within Brachycera (Fig. 17.11c). For Culicomorpha, the head and mouthparts of the phantom midge *Dixamima villosus* (Diptera: Chaoboridae), from the Late Jurassic of Kazakhstan, represented a clade with Hexastylate mouthparts (Fig. 17.11a). *Dixamima* possessed an elongate labium housing needle-like stylets for solenophagy (Rohdendorf 1962b), a mode of feeding in which stylets pierce individual subdermal capillaries for extraction of blood (Oldroyd 1964). Similarly, the biting midge *Culicoides canadensis* (Diptera: Ceratopogonidae), from the Late Cretaceous of Canada (Fig. 17.11d), probably was another solenophage with Hexastylate mouthparts (Borkent 1995). Within the Culicomorpha, the Culicidae, noted for Hexastylate mouthparts and solenophagy, were one of the latter lineages to appear, having an earliest occurrence in mid-Cretaceous Myanmar amber (Borkent and Grimaldi 2004).

An alternative mode of blood feeding is telmophagy, in which mandibles and laciniae are modified into broad blades, often with delicate sawtooth edges, for slashing through integument and flesh, resulting in a pool of blood and lymph that wells up for uptake by the insect (Oldroyd 1964). Hematophagous tabanomorphs are excellent telmophages and would include the snipe fly *Palaeoarthroteles mesozoicus* (Diptera: Rhagionidae) from the Jurassic–Cretaceous boundary interval of the Chita Region in Russia (Kovalev and Mostovski 1997). This fly possessed a massive labium housing Distylate/Tetrastylate mouthparts containing an armature of blades with delicate sawtooth edges (Fig. 17.11c). More basal or early brachyceran taxa, such as an unnamed species of woodgnat (Diptera: Anisopodoidea) from the Early Cretaceous of Siberia, Russia (Rasnitsyn 1990), likely housed Distylate/Tetrastylate mouthparts of some sort, although they were housed in a rather broad labium (Fig. 17.11b). Another early or basal brachyceran lineage was *Pachyrhyphus transbaicalicus* (Diptera: Protorhyphidae), an extinct lineage from the Late Jurassic to Early Cretaceous boundary interval of Siberia (Kovalev, in Rasnitsyn 1990), with Distylate/Tetrastylate mouthparts likely containing blade-like stylets

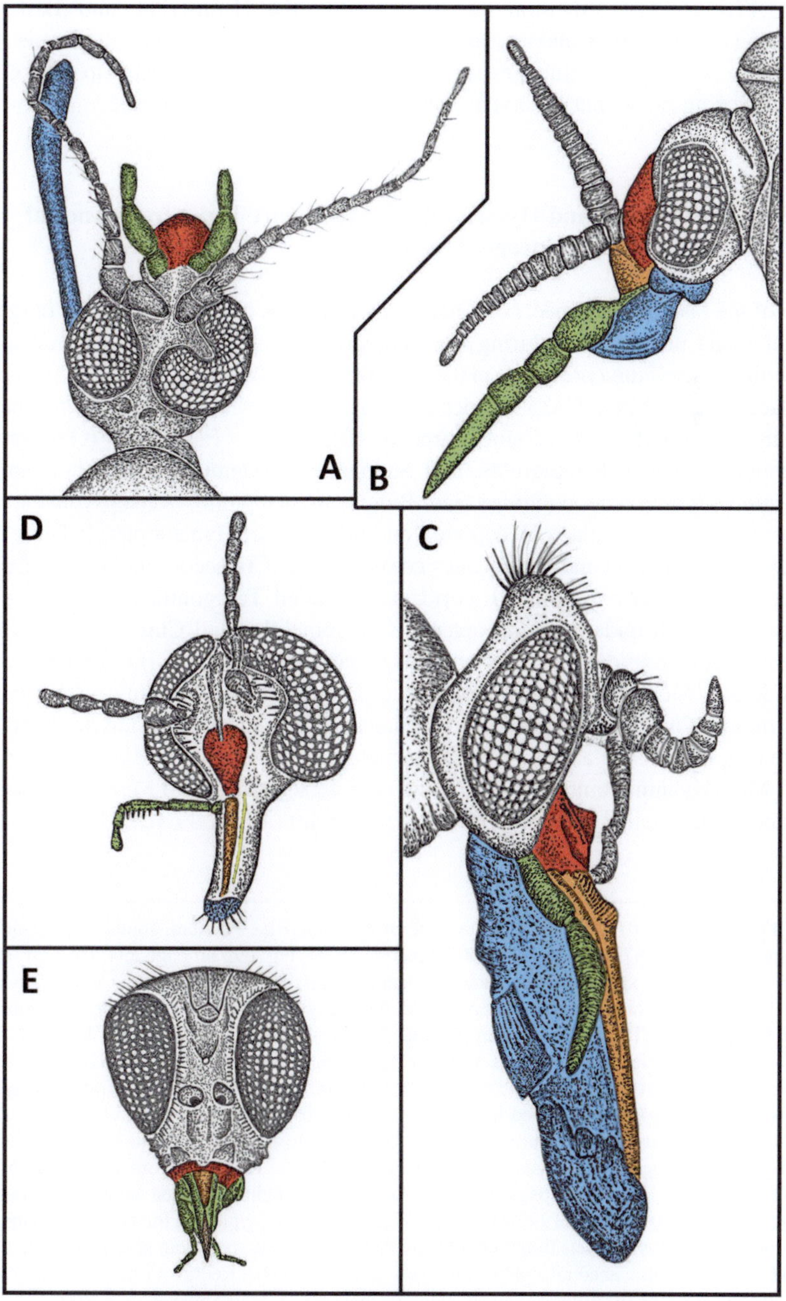


**Fig. 17.11** The expansion of dipteran Hexastylate and Distylate/Tetrastylate mouthpart classes involved in hematophagy and insectivory during the mid Mesozoic, approximately 170–90 million years ago. (**a**) Dorsal view of *Dixamima villosus* Rohdendorf (1962b) (Diptera: Chaoboridae), with Hexastylate mouthparts, based on modern analogs with very similar heads and mouthparts, from

(Fig. 17.11e). The structural variation among the Hexastylate and Distylate/Tetrastylate mouthpart classes, as well as multiple feeding styles, indicates that mid-Mesozoic dipteran lineages were broadly exploring many ways to exploit the blood-feeding niche that involved dinosaur, bird, and mammal hosts.

#### 17.5.3.4 Hemiptera and Hymenoptera: Mouthpart Types from Ancient and Modern Lineages Overlap

One of the fascinating aspects of Cretaceous ambers is that they record an outgoing insect fauna predominant during the Permian and early Mesozoic as well as an incoming insect fauna prevalent in the world today (Labandeira 2014a). A variety of Cretaceous amber insect taxa that occurred in a much earlier fern- and gymnosperm-dominated world were Caloblattinidae (Blattodea); Elcanidae (Orthoptera), Protopsyllidiidae, Schizopteridae, and Steingeliidae (Hemiptera); Lophioneuridae (Lophioneurida); Nemonychidae and Belidae (Coleoptera); Mesopsychidae and Pseudopolycentropodidae (Mecoptera); and Serphitidae (Hymenoptera). These currently extinct or relict ancient lineages coexisted with Cretaceous amber taxa from a modern, angiosperm-dominated world that included Tettigoniidae and Tetrigidae (Orthoptera); Thripidae (Thysanoptera); Chrysomelidae and Curculionidae (Coleoptera); Cecidomyiidae and Therevidae (Diptera); Gracillariidae (Lepidoptera); and Formicidae, Melittosphecidae, and Apidae (Hymenoptera). In many instances, the occurrence of ancient lineages from Permian and Triassic compression deposits now present in Cretaceous ambers such as Lebanese amber (ca. 125 Ma), Álava amber (110 Ma), Myanmar amber (99 Ma), and New Jersey amber (90 Ma), were the first instance that detailed structures such as mouthparts (Figs. 17.12a, d and 17.13a)

---

**Fig. 17.11** (continued) the early Late Jurassic of Michailkovka, Karatau, southern Kazakhstan; specimen PIN 335/167. Redrawn from Rohdendorf (1974), Fig. 69b on page 234. (**b**) Left lateral view of genus and species 1 (Diptera: Anisopodoidea), showing prominent, four-segmented maxillary palp and broad labium indicative of the Distylate/Tetrastylate mouthparts, from the Early Cretaceous of Turga, southeastern Siberia, Russia; specimen PIN 1742/686. Redrawn from Rasnitsyn (1990), Fig. 112b on page 145. (**c**) Right lateral view of a female *Palaeoarthroteles mesozoicus* Kovalev and Mostovski (1997) (Diptera: Rhagionidae), exhibiting Distylate/Tetrastylate mouthparts represented by a massive proboscis with decurved maxillary palps and a prominent, robust, styliform labrum, from early Late Jurassic–Early Cretaceous boundary interval of Daya, Chita Region of Russia; specimen PIN 3063/171. Redrawn from Lukashevich and Mostovski (2003), Fig. 2 on page 154. (**d**) An anterior, slightly oblique view of *Culicoides canadensis* (Boesel) Borkent (1995) (Diptera: Ceratopogonidae), typical of the Hexastylate mouthpart condition, with six stylets though not all are displayed in this specimen; specimen CAS 1154, from Late Cretaceous Canadian amber from Cedar Lake, Manitoba. Redrawn from Borkent (1995), Fig. 10g on page 199. (**e**) Anterior view of *Pachyrhyphus mesozoicus* Kovalev, in Rasnitsyn (1990) (Diptera: Protorhyphidae), with Distylate/Tetrastylate mouthparts, showing a pair of mandibular and apical lacinial stylets, from Unda, Late Jurassic to Earl Cretaceous of southeastern Siberia, Russia; specimen PIN 3015/295. Mouthpart color scheme: red, clypeus; orange, labrum; yellow, mandibles; brown, hypopharynx; green, maxilla; and blue, labium

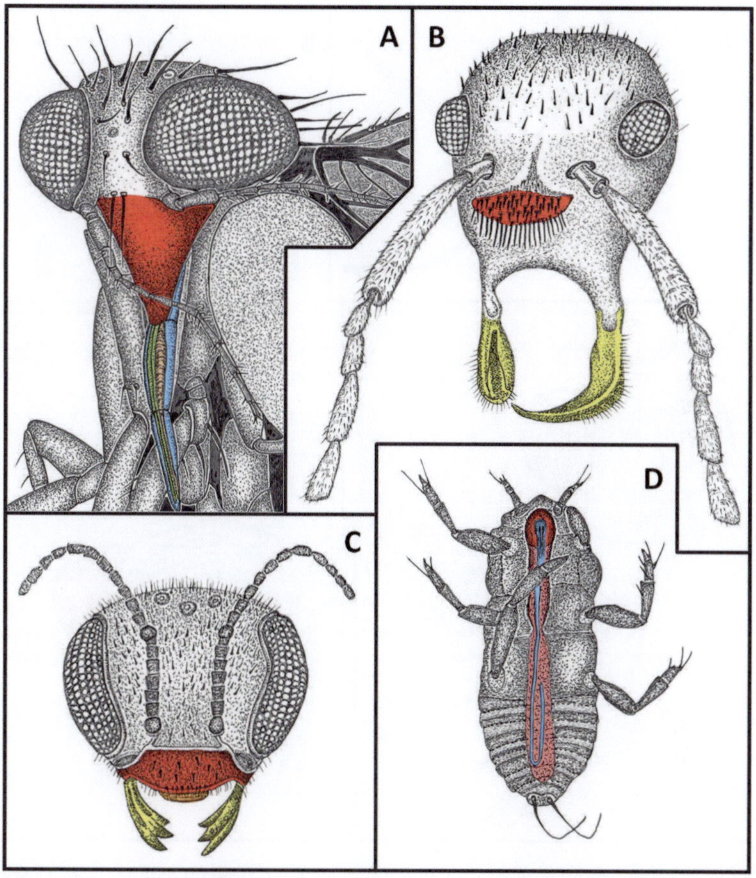


**Fig. 17.12** Insect mouthpart diversity representing a variety of mouthpart classes from Cretaceous ambers of Lebanon, Kachin State in Myanmar, and New Jersey, USA, ranging from 130 to 80 million years ago. (**a**) A slightly oblique anterior view of the relict *Postopsyllidium emilyae* Grimaldi (2003) (Hemiptera: Protopsyllidiidae), showing Segmented Beak mouthparts of a lineage prevalent approximately 180–160 million years earlier during the Middle to Late Permian, surviving to the Late Cretaceous in New Jersey amber, USA; specimen AMNH Bu137. This specimen exhibits a broad expansion of the labrum and a modest cibarial food-pump bulge and probably raptorial forelegs, indicating a predatory lifestyle. Redrawn from Grimaldi (2003), Fig. 2 on page 332. (**b**) A slightly oblique, anterior view of head and partial mouthparts of *Haidomyrmex cerberus* Dlussky (1996) (Hymenoptera: Formicidae), displaying specialized, sickle-shaped, internally scooped mandibles and a clypeus with a margin of stout bristles of the Maxillolabiate mouthpart class, from mid-Cretaceous Myanmar amber; specimen NHM 20182. (**c**) Reconstruction of the head and mouthparts of the earliest documented bee, *Melittosphex burmensis* Poinar and Danforth (2006) (Hymenoptera: Melittosphecidae), a member of the Glossate mouthpart class, and showing the distinctive tridentate mandibles and branched hairs of bees associated with pollination, from mid-Cretaceous Myanmar Amber; specimen OSU B-Hy-7. (**d**) Ventral view of the scale-insect nymphal crawler *Palaeosteingella* sp. (Hemiptera: Steingeliidae), from Early Cretaceous Lebanese Amber (Koteja and Azar 2008), revealing a posteriorly oriented folded stylet fascicle in an infrabuccal pouch of the Segmented Beak mouthpart class, extending almost to the abdomen terminus; specimen MNHN HAM-86—Cocc0848. Redrawn from Koteja and Azar (2008), Fig. 14b on page 153. Mouthpart color scheme: red, clypeus; orange, labrum; yellow, mandibles; brown, hypopharynx; green, maxilla; blue, labium; and pink, buccal pouch

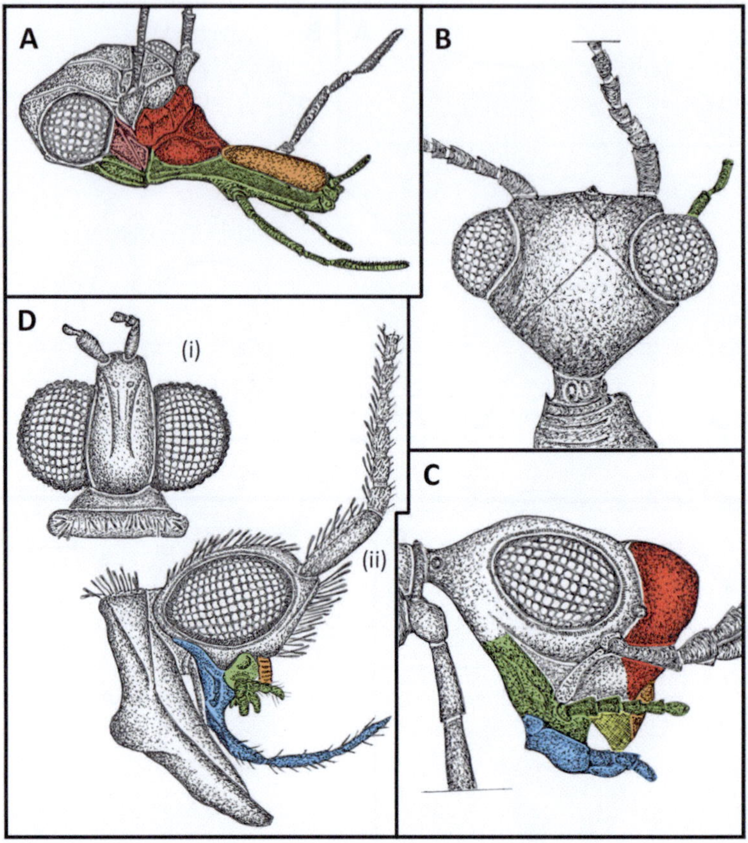


**Fig. 17.13** Three new or otherwise distinctive mouthpart types from 99 million-year-old Myanmar amber, of the mid Cretaceous, revealing the previously unknown diversity of mouthpart types present in well-preserved deposits. (**a**) Reconstruction in an oblique to right-lateral view of the head and mouthparts of *Psocorrhyncha burmitica* (Permopsocida: Archipsyllidae) Huang et al. (2016), showing a prominent, forwardly directed rostrum and mouthparts typical of the Laciniate mouthpart class predominant during the Permian and Triassic; specimen SMNS Bu-157. Redrawn from Huang et al. (2016), Fig. 2g on page 3. (**b**) Dorsal aspect of the head, mouthparts, and prothorax of *Aethiocarenus burmanicus* Poinar and Brown (2017) (Aethiocarenodea: family uncertain), exhibiting a triangular-pyramidal head shape and a miniscule neck, unassignable to mouthpart class; specimen OSU B-De-2. Redrawn from Poinar and Brown (2017), Fig. 1b on page 101. (**c**) Right-lateral view of same specimen in (**b**), showing in addition hypognathous, mandibulate mouthparts, probably representing the Adult Ectognathate mouthpart class. Redrawn from Poinar and Brown (2017), from Fig. 2b on page 102. (**d**) At bottom (ii) is a right-lateral view of the distinctive mouthparts of *Tarachochelis microlepidopterella* Mey, Wichard, Müller, and Wang (2017) (Amphiesmenoptera: Tarachochelidae), showing significantly diminished but complete mouthparts, the ensemble of which is unassignable to a mouthpart class in Labandeira (1997). At upper left in (i) is a dorsal view of the same specimen; specimen NMS G2010.20.36. Both are redrawn from Mey et al. (2017), Figs. 1.1 and 1.2 on page 252. Mouthpart color scheme: red, clypeus; orange, labrum; yellow, mandibles; brown, hypopharynx; green, maxilla; blue, labium, and pink, gena

were observed and documented in detail. Almost all of these ancient lineages, however, failed to survive the end-Cretaceous ecological crisis and extinctions (Labandeira et al. 2016) and were replaced by more familiar lineages of today or their precursors (Figs. 17.12b, c and 17.14) (Labandeira 2014b).

One holdover from the Middle to Late Permian is *Postopsyllidium rebeccae* (Hemiptera: Protopsyllidiidae) from Myanmar Amber (Fig. 17.12a), representing a relict lineage that likely was a sister group to early Permian Sternorrhyncha (Grimaldi 2003). The broader clade of Protopsyllidiidae originated in the middle Permian, underwent a radiation during the late Triassic, and suffered major extinction at the Jurassic–Cretaceous boundary, apparently with a sole representative, *Postopsyllidium*, surviving to the early Late Cretaceous (Grimaldi 2003). The abundantly setose head had a pronounced clypeal bulge, prominent eyes, and a moderately long, stout beak, tucked between the fore coxae, with stylets exposed and ensheathed by a four-segmented labium. Based on the presence of probable raptorial forelegs, this specimen with Segmented Beak mouthparts was probably predatory. Another relict was *Palaeosteingella* sp. (Hemiptera: Steingeliidae), from Early Cretaceous Lebanese Amber (Koteja and Azar 2008), which had developed an infrabuccal pouch extending almost to the abdominal terminus that housed a very long, folded stylet fascicle (Fig. 17.12d). Such mouthparts were used to access vascular tissue, particularly more nutritious phloem rather than xylem. Also occurring in Cretaceous ambers were the earliest representatives of modern bees and ants. For example, an early ant *Haidomyrmex cerberus* (Hymenoptera: Formicidae), from mid-Cretaceous Myanmar amber (Dlussky 1996), displays exceptionally specialized mouthparts of the Maxillolabiate mouthpart class, a clypeus with a comb of bristles, and edentulous, scooped mandibles (Fig. 17.12b). Likewise, the earliest bee, *Melittosphex burmensis* (Hymenoptera: Melittosphecidae), also from Myanmar amber (Poinar and Danforth 2006), bears the typical head and mouthpart features of distinctive tridentate mandibles and branched hairs of Glossate mouthparts in modern bees associated with pollination (Fig. 17.12c). A bee, *Cretatrigona priscum* (Michener and Grimaldi 1988; Engel 2000), with an indeterminate mouthpart type, was found in the latest Cretaceous of New Jersey, USA. These curious hemipteran holdovers and earliest appearances of modern hymenopterans provide a unique taxonomic composition to Myanmar amber.

#### 17.5.3.5 Enigmatic Clades and Atypical Mouthpart Diversity from Myanmar Amber

A consequence of a taxonomically very diverse fossil insect fauna is the high probability of encountering an associated, elevated diversity of mouthpart types. Such mouthpart types would include those that are entirely novel, or have unexpected or unconventional combinations of characters from known mouthpart types, or mouthparts that were once thought as extinct but reappear after a hiatus of tens of millions of years (Bai et al. 2016; Huang et al. 2016; Poinar and Brown 2017; Mey et al. 2017). The insect fauna of Myanmar amber, from the Early Cretaceous–Late

Cretaceous boundary interval (99 Ma) of Kachin State, in northern Myanmar, offers such a diversity of unusual mouthpart types, most of which remain difficult to characterize, as well as the enigmatic insect lineages that bear those mouthparts (Fig. 17.13).

One holdover lineage from the past (Fig. 17.13a) is *Psocorrhyncha burmitica* (Permopsocida: Archipsyllidae), a member of Acercaria that is at a Late Pennsylvanian to Early Permian phase of stylate beak development of approximately 200 million years earlier. This specimen (Huang et al. 2016) is a member of the Laciniate mouthpart class (Fig. 17.5) and displays a precursor morphology to the Segmented Beak mouthpart class (Fig. 17.6). Specializations include an enlarged but not bulbous clypeus; a prominent, elongate labrum; highly prolonged mandibles transitional between the mandibulate and styliform condition, but with terminal, inwardly directed teeth; a prolonged maxilla with a galeal guide for mandibular movement and nonstyliform laciniae; and a fleshy labium, possibly offering support to other projecting mouthparts. A second atypical mouthpart type (Fig. 17.13b, c) is *Aethiocarenus burmanicus* (Aethiocarenodea: Aethiocarenidae), the sole member of a newly established order (Poinar and Brown 2017), reminiscent of Zoraptera and Dermaptera (Engel and Grimaldi 2000). The head is tetragonal in shape, attached by a very narrow neck, bearing a forwardly enlarged clypeus and nominally mandibulate mouthparts, but with inconspicuous mandibles, short maxillary and labial palps, and a robust labium. Although the preservation is suboptimal, the prothorax, head, and mouthparts of this specimen are unlike the mouthparts of any known insect past or present. An equally incongruous but very different head and mouthparts (Fig. 17.13d) is *Tarachocelis microlepidopterella* (Amphiesmenopera: Tarachocelidae). This specimen has a vertically elongate pronotum; an elongate head somewhat flattened dorsoventrally; is covered dorsally and anteriorly with flat, elliptical scales; and has very large compound eyes (Mey et al. 2017). The mouthparts consist of a normal clypeus; a large and quadrangular labrum; mandibles apparently absent; a maxilla with probable, small, protruding galeal and lacinial lobes and minute, three-segmented palpi; and a labium with long, three-segmented, upwardly curving palpi. There is no evidence of a haustellum, and with the exception of the prominent labial palpi, the mouthparts are diminutive and recessed into the face of the insect. Some of these features are reminiscent of basal Lepidoptera and basal Trichoptera, although the distinctive head and mouthparts cannot be affiliated with any known insect taxon.

## 17.6 The Cenozoic: Additional Specializations in Mouthpart Structure Leading to the Modern World

The evolution of mouthpart design during the Cenozoic essentially involved modifications of existing mouthpart elements or the addition of a key feature, such as a maxillary tentacle or mandibular appendage for pollinating insects. Cenozoic mouthpart evolution involved the origination of two new mouthpart classes. The

Tubulostylate mouthpart class originated during the Paleocene Period, after which 97.1% of all mouthparts had originated. During the Neogene, Siphonostylates originated, completing the 37 mouthpart classes. The panoply of mouthpart classes that were present during the mid Cretaceous essentially set the themes for their continuation, with some modification, during the Cenozoic.

### 17.6.1 Effects of the Cretaceous–Paleogene Ecological Crisis on Insect Mouthpart Design

The effects or lack thereof of the mass extinction and ecological crisis at the Cretaceous–Paleogene boundary (66 Ma) on plant–insect interactions have been examined extensively. Plant–insect interaction studies have focused on the Western Interior of North America (Labandeira 2002a, b; Wilf et al. 2006; Donovan et al. 2014), Western Europe (Wappler et al. 2009), and Patagonia, Argentina (Donovan 2016, 2018). In these, and other studies such as those involving pollination (Labandeira et al. 2016), there has been no indication that this significant global event had an effect at the level of mouthpart class diversity (Tables 17.2 and 17.3).

### 17.6.2 Mouthpart Innovations of the Cenozoic

With the probable exception of vestigial mouthparts in Strepsiptera, the six examples of mouthpart modification from holometabolous insects presented in Fig. 17.14 likely originated within the Cenozoic Era, given what is known of the phylogenetic relationships of the insect clade causing the interaction as well as the host clade. These examples represent a diverse repertoire of the Maxillolabiate, Mouthhook, Siphonate, Tubulostylate, Siphonostylate, and Reduced Trophic mouthpart classes, indicating that mouthpart innovation can occur in virtually any mouthpart design in time and space. These examples involve a variety of relationships with their host organisms, such as herbaceous angiosperms, yuccas, figs, and mammals, including humans. The several functional feeding groups among these examples are leaf mining, pollination, parasitism, endoparasitoidism, and ectoparasitoidism. Some of these may be considered key innovations, such as mouthhooks in cyclorrhaphan larvae, the tsetse proboscis, the maxillary tentacle of yucca moths, and the mandibular appendage of fig wasps. If so, they could be equivalent to earlier key innovations such as the weevil rostrum and the concealed nectar extraction apparatus of parasitoid wasps during the Mesozoic. Interestingly, some of these innovations in mouthparts are the defining features of their respective mouthpart classes: the concealed nectar extraction apparatus for the Maxillolabiate, the weevil rostrum for the Rhynchophorate, and mouthhooks for the Mouthhook mouthpart classes.

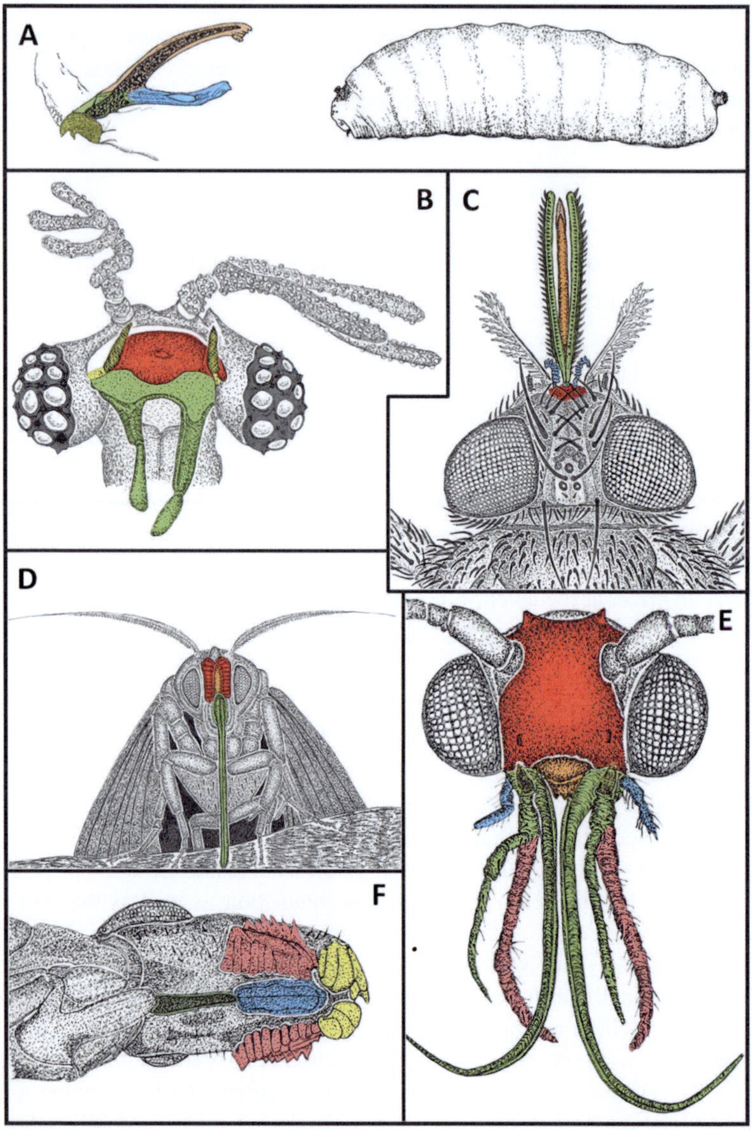


**Fig. 17.14** Four geochronologically recent, host-specialized mouthpart classes mostly from the Cretaceous to Neogene periods, 100 million years ago to the present, and two Neogene mouthpart structures involved in pollination. (**a**) The Mouthhook mouthpart class indicated by leaf-mining-fly leaf mines on a *Platanites* sycamore leaf from the early Paleogene Epoch, 64 million years ago, from Mexican Hat, Montana, USA (Winkler et al. 2010). The illustrated specimen is modern, *Phytomyza chelonei* Spencer (1969) (Diptera Agromyzidae), showing mouthhooks and cephalopharyngeal apparatus at left and the larva bearing the inconspicuous mouthparts at right. More generalized forms of this mouthpart class, however, extend to the Jurassic. Redrawn from Spencer (1992), Figs. 23 (left) and 24 (right) on page 876. (**b**) Ventral view of the head and vestigial mouthparts of the nonfeeding adult of the twisted-wing insect *Bohartilla kinzelbachi* Kathirithamby

#### 17.6.2.1 Diptera: Expansion of Mouthhook Design in Leaf-mining Fly Larvae

Often the best evidence for the presence of inconspicuous or otherwise hard-to-determine fossil mouthparts is to examine the damage that such mouthparts inflict on fossil plants (Labandeira 2007b). Indeed, this approach was used to provide indirect evidence that an externally feeding insect of the Adult Ectognathate mouthpart class was present in a Late Mississippian deposit, based on cuspate leaf-margin excisions on a seed plant in the absence of a suspect body fossil (Iannuzzi and Labandeira 2008). Because of the very distinctive, stereotyped leaf mines that leaf-mining flies (Diptera: Agromyzidae) produce on modern angiosperms, detection of the mouthhook mouthpart class in the fossil record is significantly better than that of the body-fossil record (Winkler et al. 2010). The Mouthhook mouthpart class is barely distinguishable externally on the larvae of cyclorrhaphan Diptera (Fig. 17.14a). In cyclorrhaphan larvae, the external head capsule has been reduced to an internal cephalopharyngeal apparatus and has been retracted internally into the anterior thorax (Spencer 1992). Agromyzid larvae virtually lack a fossil record (Winkler et al. 2010). The earliest agromyzid leaf mines were described from an early Paleocene (64 Ma) leaf mine from the Mexican Hat site in eastern Montana, USA, documenting, indirectly, the earliest occurrence of mouthhooks in Agromyzidae. This trace-fossil evidence considerably antedates by 20 million years the direct middle Eocene, body-fossil evidence for the presence of the Mouthhook mouthpart class (Winkler et al. 2010).

**Fig. 17.14** (continued) and Grimaldi (1993) (Strepsiptera: Bohartillidae), with highly reduced mandibles and a two-articled maxillary palp, both indicative of the Reduced Trophic mouthpart class, from the early Neogene Period of amber from the northern Dominican Republic; specimen AMNH DR-10-6. Redrawn from Kathirithamby and Grimaldi (1993), from Fig. 2 on page 34. (**c**) Reconstruction of the dorsal head, mouthparts, and prothorax of a modern female tsetse fly, *Glossina pallidipes* Austen (1903) (Diptera: Glossinidae), in Grimaldi (1992), showing a distinctive, elongate, medial, labrum above the hypopharyngeal stylet and lateral labial stylets with serrated exodont teeth of the Tubulostylate mouthpart class. Tsetse flies vector the causative parasite of sub-Saharan African sleeping sickness in humans and nagana in bovid vertebrates, whose body structure is virtually the same as *Glossina oligocenus* Scudder (1892) and *G. osborni* Cockerell (1918) from the latest Eocene Florissant Formation of Colorado, USA. Redrawn from Grimaldi (1992), Fig. 6.7 on page 187. (**d**) The distinctive, Recent piercing-and-sucking mouthparts of hematophagous *Calyptra* (*Calpe*) *eustrigata* Hampson (1926) (Lepidoptera: Erebidae), feeding on a human finger in Thailand (Bänziger 1960), typical of the Siphonostylate mouthpart class. Redrawn from Bänziger (1968), Fig. 1 on plate IX. (**e**) The Siphonate mouthparts of the yucca moth *Tegeticula yuccasella* (Riley 1872) (Lepidoptera: Prodoxidae), showing the distinctive maxillary tentacle (in pink) involved in pollination of yucca flowers. Redrawn from Davis (1967), Fig. 41 on page 138. (**f**) A slightly oblique, ventral view of the head, mouthparts, and prothorax of the fig wasp *Tetrapus delclosi* Peñalver, Engel, and Grimaldi (2006) (Hymenoptera: Agaonidae), a member of the Maxillolabiate mouthpart class, showing the distinctive mandibular appendage in pink, from the early Paleogene of the Dominican Republic; specimen AMNH DR-14-282. Redrawn from Peñalver et al. (2006), Fig. 2a on page 6. Mouthpart color scheme: red, clypeus or fronto-clypeus; orange, labrum; yellow, mandibles; brown, hypopharynx; green, maxilla; and blue, labium

#### 17.6.2.2 Strepsiptera: Atrophied Mouthparts of Adult Parasitoids

In some insect lineages, adults and males, often in combination, become nonfeeding because of their ephemeral existence and their sole role as providers of sperm to fertilize the eggs of longer-lived, conspecific females. This condition is present in several insect groups such as adult mayflies, male scale insects, rhipiphorid beetles, some parasitoid wasps, and strepsipterans (twisted wing parasitoids) (Labandeira 1990). A nonfeeding habit forms the basis of the Reduced Trophic and Nontrophic mouthpart classes (Table 17.2). Atrophied, nonfunctional mouthparts, or their complete absence, have been documented in taxa such as *Bohartilla kinzelbachi* (Strepsiptera: Bohartillidae), from early Miocene amber of the Dominican Republic (Fig. 17.14b). This mouthpart morphology has underwent stasis for about 21million years (Kathirithamby and Grimaldi 1993), and subsequently was extended to 42 million years, based on the virtually identical mouthparts of *Mengea tertiaria* (Mengeidae) occurring in older Baltic amber of Northern Europe (Pohl et al. 2010).

#### 17.6.2.3 Diptera: Tsetse Proboscis and Evolution of Sleeping Sickness

A crucial vector for the major diseases affecting mammals (nagana) and humans (sleeping sickness) in Africa is the tsetse, one of which, *Glossina pallidipes* (Diptera: Glossinidae), attacks humans (Askew 1971). The tsetse head and mouthpart structure exhibits many features designed for vectoring the diseases *Trypanosoma gambiense* and *T. brucei* in humans in sub-Saharan Africa. Disease vectoring is effected by a long, rigid proboscis. The proboscis consists of a rigid labral stylet armed with prestomal teeth for puncturing host integument and transporting blood through a food tube to the fly's esophagus (Fig. 17.14c) (Grimaldi 1992). Located below the labral stylet is a narrow, needle-like hypopharyngeal stylet for depositing anticoagulant and possibly anesthetic fluids into the wound. The third element of the proboscis is the lowermost labellar lobe that supports the two superposing stylets, which is responsible for the transmission of the *Trypanosoma* protozoans into the active wound. Notably, fossil *Glossina* bearing this unique mouthpart type of the Tubulostylate mouthpart class occurred in the late Eocene (34 Ma) of Colorado, USA (Cockerell 1918), and early Oligocene (30 Ma) of Germany (Wedmann 2000). This former widespread biogeographical distribution indicates that tsetse and the diseases it vectored may have afflicted a variety of extinct mammalian hosts before its current confinement to sub-Saharan Africa.

#### 17.6.2.4 Lepidoptera: The Piercing Siphonate Proboscis of Owlet Moths

An important modification of the lepidopteran siphon is its transformation into a stylet capable of puncturing fruit or the integument of warm vertebrates or humans. In a series of papers, Bänziger (1968, 1970, 1980) has described a clade of owlet moths, of which *Calyptra eustrigata* (Lepidoptera: Erebidae) has been most intensively studied. *Calyptra eustrigata* pierces the skin of humans and even the more indurated integument of bovids in Southeastern Asia (Fig. 17.14d). Two key features of the proboscis that allow penetration of mammalian integument are the stiffening of the siphon and erection of terminal barbs that result from an increase in hemolymph hydrostatic pressure (Bänziger 1970, 1980). Penetration is achieved functionally by antiparallel movements of the two siphonal galeae against each other, producing a sawtooth, scissors-like motion for initial cutting into integument (Bänziger 1980). The process is assisted by a capacious cibarial pump that provides suction for imbibition of blood. This siphon modification, constituting the Siphonostylate mouthpart class, also has evolved, probably independently, in a clade of fruit-piercing owlet moths in West Africa (Büttiker 1962).

#### 17.6.2.5 Lepidoptera: The Yucca Moth Maxillary Tentacle

One of the few plant–insect pollination mutualisms that is said to have close fidelity between a pollinator and its pollinated host are yuccas and their yucca moth pollinators (Davis 1967). One such relationship is between the yucca moth *Tegeticula yuccasella* (Lepidoptera: Prodoxidae) and its pollinated yucca plant host, *Yucca filamentosa* (Asparagaceae), both of which lack a relevant fossil record. A key innovation, the maxillary tentacle (Fig. 17.14e), originated in female yucca moths, *Tegeticula,* and its sister genus *Parategeticula*, approximately during the middle Eocene (44 Ma) (Pellmyr and Krenn 2002). The maxillary tentacle is a 2.4-mm-long, elongate, appendicular extension from the first maxillary palp segment that lacks homologues in other lepidopteran taxa. The singular origin of the maxillary tentacle in yucca moths renders these structures as a novel, key innovation within the Siphonate mouthpart class that enhanced the efficiency in the active collection of substantial pollen loads by yucca moths.

#### 17.6.2.6 Hymenoptera: The Mandibular Plate of Fig Wasps

Other mouthpart structures have developed in pollinating insects that facilitate the pollination of their plant hosts. One such structure is the mandibular appendage, occurring in extant female fig wasps, but also in fossils such as *Tetrapus delclosi* (Hymenoptera: Agaonidae) from early Miocene amber (21 Ma) of the Dominican Republic (Peñalver et al. 2006). This fig wasp specimen (Fig. 17.14f) is directly associated with pollen of a fossil species of *Ficus* (Moraceae), its fig host. In *Tetrapus delclosi* (Fig. 17.14f), the cylindrical shaped head is two to two-and-a-half times longer than wide and houses prognathous to slightly hypognathous mouthparts; the labiomaxillary complex appears poorly developed, even though fig wasps have essentially Maxillolabiate mouthparts. The most prominent structure is the mandible, which bears the distinctive mandibular appendage (Weiblen 2002). The mandibular appendage is an extension of the mandible that contains serrate ridges along its medial axis, as well as an adjacent, outer, lateral row of serrate-appearing denticles. The mandibular appendage of fig wasps has morphological parallels with the maxillary tentacle of yucca moths, and similarly involves greater pollination efficiency in a highly specific mutualism involving a plant host and its insect pollinator.

### *17.6.3 Zombie Ant Mouthparts, Death-Grip Plant Damage, and a Parasitoid Fungus*

In addition to distinctive fossil plant damage occasionally revealing herbivory associated with a particular mouthpart type, fossil plant damage can also reveal the life history and behavior of the insect that created the damage (Labandeira 2007b). A good example of extracting fossil life-history information from a particular type of plant damage is the case of death-grip scars inflicted on fossil leaves by an ant's erratic behavior, induced by brain control of a parasitoid fungus (Hughes et al. 2011). This system of the manipulation of ant behavior by a parasitoid fungus recently has been documented in Thailand (Andersen et al. 2009). In this system, *Camponotus leonardi* carpenter ants are infected by a fungal parasitoid, *Ophiocordyceps unilateralis* (Hypocreales: Ophiocordycipitaceae), which affects brain function and causes attachment of the ant to the abaxial surface of a major leaf vein. This attachment results in a dumbbell-shaped death-grip scar by the ant's mandibles on the understory plant's leaf with the ant's carcass still attached. During attachment, the parasitoid fungus produces a prominent sporophore with a mid-stalk fructification that eventually liberates spores to infect conspecific ants in the immediate area, thus repeating the ant–plant–parasitoid fungus life cycle.

The dumbbell-shaped death grip is very distinctive type of hole damage (Labandeira et al. 2007b and addenda), and, if fossil leaves have this damage, its presence seemingly would be detected by considerable searching. The stereotyped damage, however, was found in the voluminous fossil plant collections of the Messel Biota at the Senckenberg Institute in Frankfurt, Germany, and was assigned the unique damage type DT212 (Labandeira et al. 2007b, and addenda). The Messel Formation crops out near Darmstadt, in Hesse, Germany, and represents an exceptionally well-preserved deposit from the Middle Eocene (48 Ma) that includes microorganisms, insects, plants, and vertebrates (Dunne et al. 2014). The environmental setting at Messel is significantly different from the context of the ant–plant–parasitoid fungus from in contemporary Thailand. This shared, highly honed association between Eocene Messel and modern Thailand indicates a significant spatiotemporal continuity across 48 million years of a specialized behavior associated with the Maxillolabiate mouthpart class (Hughes et al. 2011).

## 17.7 Discussion and Summary: Major Evolutionary Developments Involving Insect Mouthparts

The fossil record of fossil insect mouthparts reveals macroevolutionary patterns and processes in two different modes. The first mode portrays major developments in insect mouthpart structure based on important ecological events or clade-specific evolutionary developments that document mouthpart access to novel food sources (Fig. 17.15). A second mode lists the geochronologic distribution of mouthpart classes based on direct and indirect fossil data (Fig. 17.16), and plots of data through time to express time intervals of elevated and flat mouthpart diversity (Fig. 17.17). Both geochronologically based assessments of mouthpart data record intervals of morphological innovation that are separated by hiatuses of relative mouthpart structural stasis.

The earliest development is the origin of insect mouthparts (Fig. 17.15, development 1), an acquisition that is synonymous with the origin of the Hexapoda probably sometime during the late Silurian but first entering the fossil record in the Early Devonian. Based on direct and indirect evidence of the existence of four mouthpart classes during the Early Devonian, some of which were specialized, evidently there already was a dietary partitioning of food resources. In particular, Collembola, with Entognathate mouthparts, likely were feeding on plant tissues or fungi (Kevan et al. 1975) and spores (Habgood et al. 2004). However, the Hexapoda fossil record soon disappears, lacking direct evidence for insect mouthpart types during a 60-million-year-long period between the Middle Devonian to the Mississippian–Pennsylvanian boundary. Indirect evidence indicates the presence of small insects feeding on live fern and seed-fern foliage with Adult Ectognathate mouthparts.

At the Mississippian–Pennsylvanian boundary, there was a major expansion of approximately 15 major insect lineages and the appearance of several mouthpart classes within a short, several million-year-long interval (Fig. 17.15, development 2). This proliferation of insect diversity during Pennsylvanian time consisted of the clades Monura, Ephemeroptera, Odonatoptera, Paleodictyopteroidea, Archaeorthoptera, Acercaria and Holometabola, among others. On land, the Larval Ectognathate, Raptorial Ectognathate, Sericterate, Laciniate, and Robust Beak mouthpart classes (Table 17.2) were used to access foods ranging from dead plant, fungal and animal plant tissues, live fungal and plant tissues, to insects from the small to the very large (Scott and Taylor 1983; Shear and Kukalová-Peck 1990; Labandeira and Phillips 1996a, b; Labandeira 2006a; Haug et al. 2015). Functional feeding groups associated with these mouthpart classes included detritivory, external feeding on live foliage, piercing and sucking, galling, pith boring, seed predation, palynivory, sporangia feeding and solid feeding on insect tissues (Shear and Kukalová-Peck 1990; Labandeira 1998). Various members of the Robust Beak mouthpart class engaged in piercing and sucking of mesophyll from *Etapteris* fern petioles and vascular tissues from *Psaronius* tree ferns (Scott and Taylor 1983; Labandeira and Phillips 1996a). Perhaps most notable was the Raptorial Ectognathate mouthpart class of huge adult Odonatoptera, likely preying and ambushing other large insects on vegetation (Shear

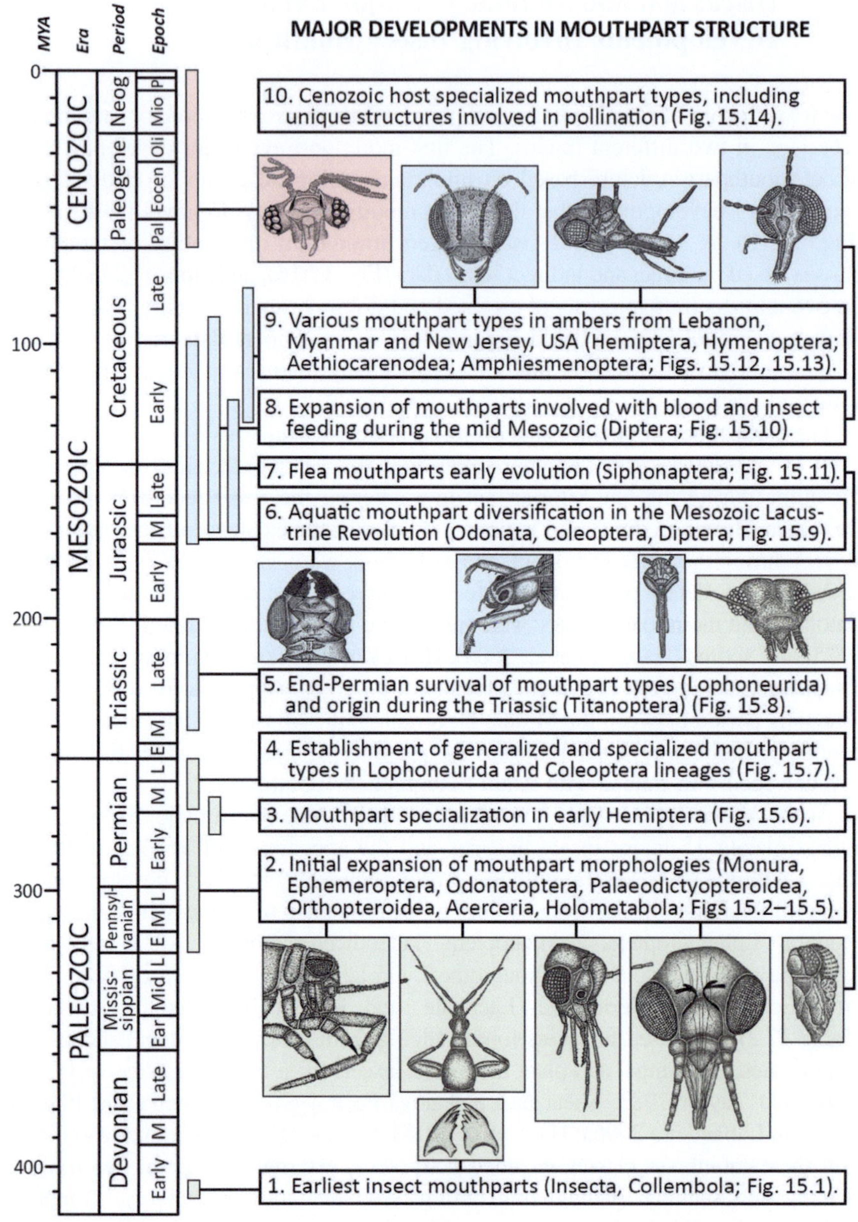


**Fig. 17.15** The geochronology of major ten developments in insect mouthpart structure based on major ecological events and clade-specific evolutionary developments in mouthpart structure that provided access to novel food sources

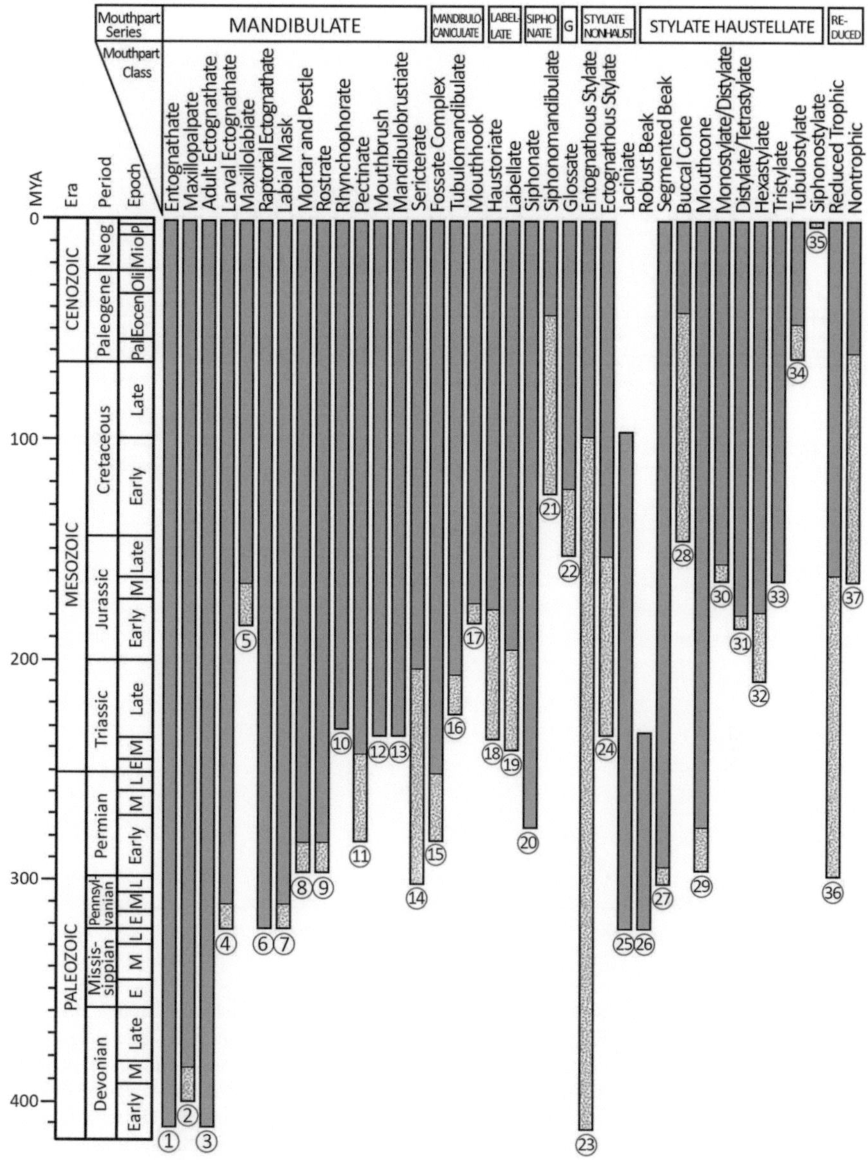


**Fig. 17.16** The geochronologic distribution of modern and fossil mouthpart classes outlined in Tables 17.2 and 17.3. The range-through data is updated from Labandeira (1990, 1997). Solid grey segments of vertical bars indicate direct evidence for the presence of a mouthpart class. Stippled blue, downward extensions of grey bars indicate all indirect evidence for a mouthpart class based on close phylogenetic relationships, the presence of a distinctive insect-damage pattern on a host organism indicative of a particular mouthpart class, or other indirect data. Numbers at the bottom of refer to mouthpart classes in Table 17.2. Abbreviations: Neog, Neogene Period; and for epochs: Pal, Paleogene; Oli, Oligocene; Mio, Miocene; P, Pliocene; E, Early; M, Middle; L, Late. Above the Pliocene Epoch, unlabeled, is the Pleistocene Epoch ending 11,000 year ago. Abbreviation for mouthpart series: G, Glossate

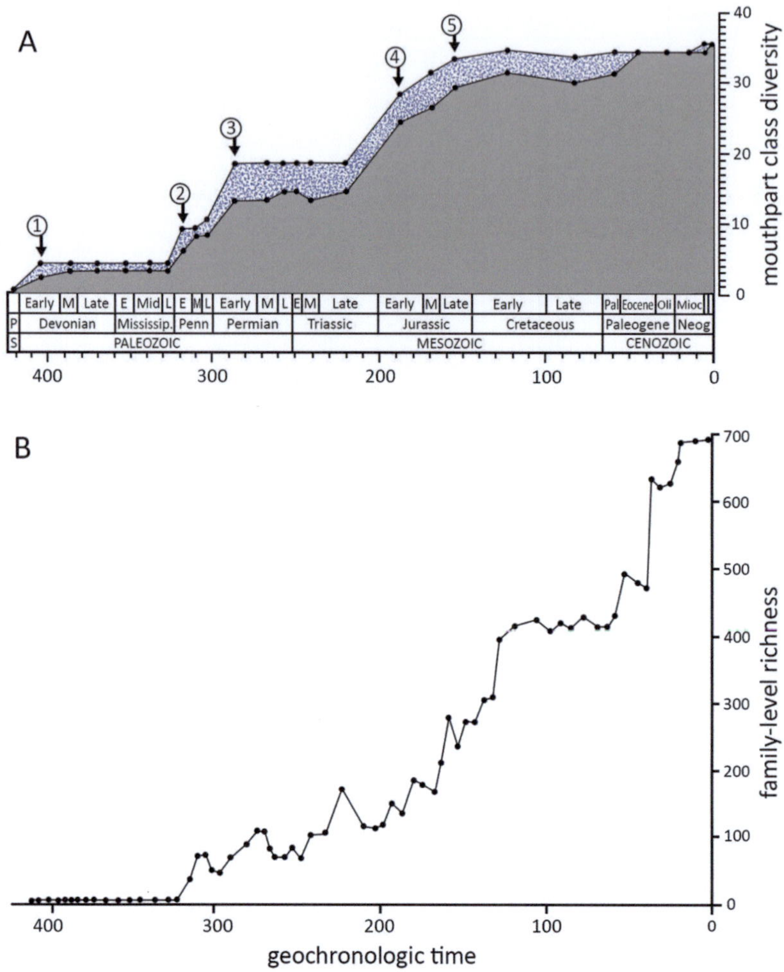


**Fig. 17.17** Updated comparison of mouthpart diversity with insect richness through time. (**a**) Updated mouthpart data from Fig. 17.16, using the convention of direct data (solid gray pattern) and indirect data (stippled blue pattern) for mouthpart class occurrence. (**b**) Insect family-level richness using range-through data (Nicholson et al. 2015), updated from Labandeira and Sepkoski (1993), redrawn from Fig. 3a of Nicholson et al. (2015). The data consist of 37 mouthpart classes instead of the 34 of Labandeira (1997), are resolved at the geochronologic level of epoch, and retain the five phases of mouthpart class diversification (numbered arrows), although details of timing have somewhat changed. See Fig. 17.16 for geologic time-scale abbreviations

and Kukalová-Peck 1990). All these diets and functional feeding groups continued to the present, with the exception of leaf mining, which originated at the latest Permian (Krassilov and Karasev 2008). The Labial Mask and probably Pectinate mouthpart classes (Table 17.3) were the first to be present in the aquatic realm (Minter et al. 2016), the former involved in predation and the latter likely involved in detritivory. The low level of mouthpart class diversity in the freshwater realm apparently continued into the Permian and Triassic and did not significantly expand until the Mesozoic Lacustrine Revolution of the mid Mesozoic.

During early Permian time, there was continued expansion of mouthpart classes (Fig. 17.15, development 2). For the Segmented Beak mouthpart class of early Hemiptera (Fig. 17.15, development 3), mid-Permian mouthpart structures indicate piercing of deep vascular tissues such as xylem, phloem, and mesophyll by long and short stylate mouthparts (Becker-Migdisova 1940, 1946), evidenced by stylet mark traces consistent with such mouthparts (Xu et al. 2018). Some late Permian lineages exhibited mouthpart specializations for partitioning of plant food resources (Fig. 17.15, development 4), which included functional feeding groups such as woodborers (Feng et al. 2017) and palynivores (Wang et al. 2009), two well-documented interactions (Fig. 17.15, development 4). Newly added mouthpart classes were the Mortar-and-Pestle, Rostrate, Fossate Complex, Siphonate, Segmented Beak, Mouthcone, and Reduced Trophic mouthpart classes, all of which occurred in the terrestrial realm. For the Mouthcone mouthpart class, damage on noeggeranthialean spores from the late Permian of China indicates that lophioneurids were engaged in the same punch-and-suck feeding technique of consuming spore protoplasts (Wang et al. 2009), as documented in modern thrips (Kirk 1984). The Larval Ectognathate class was involved in complex, subsocial tunneling in conifer wood during the late Permian, indicating the strengthening of certain mouthpart elements, particularly mandibles among larval instars (Feng et al. 2017). Siphonate mouthparts make their debut in the mid Permian (as Neuroptera), having another occurrence during the Late Permian (as Mecoptera), but were borne by very small insects (Labandeira 2010; Bashkuev 2011), suggesting they sequestered fluids from ovulate pollination drops of seed plants, or perhaps glandular secretory tissues of ferns (Labandeira et al. 2007a).

At the Permian–Triassic ecological crisis and extinction event, there was a drastic decrease in insect lineages (Labandeira 2005a). Nevertheless, all mouthpart classes survived into the Triassic (Fig. 17.15, development 5), although the Robust Beak mouthpart class soon became extinct during the Triassic and the Laciniate mouthpart class was extirpated much later during the Cretaceous. The Early Triassic was a time of depauperate biotas in the terrestrial and freshwater realms (Hochuli et al. 2010; Chen and Benton 2012). However, during the early Late Triassic, there was a pivotal biotal diversification event, as demonstrated from preliminary insect data from the Molteno Biota of the Karoo Basin in South Africa (Anderson et al. 1998). This proliferation of lineages led to a major increase in feeding damage, numbers of functional feeding groups and an elevated level of interactions, as demonstrated by the Molteno biotas in general (Labandeira 2006b) and the Aasvoëlberg 411 biota in particular (Labandeira et al. 2018). Associated with this Late Triassic surge in biotal diversity are the earliest occurrences of the Haustoriate, Labellate, Ectognathous Stylate, Distylate/Tetrastylate, and Hexastylate mouthpart classes in the terrestrial realm and the Mouthbrush and Mandibulobrustiate mouthpart classes in the freshwater realm. Most of the new mouthpart classes in the terrestrial realm involve modifications of the dipteran proboscis for feeding on insects and sponging surface fluids on plant reproductive structures, fungi, and vertebrates. This latter process was convergent with the Haustoriate mouthparts of emerging Trichoptera. Ectognathous Stylate mouthparts, a mouthpart class convergently arrived at by several lineages of beetles (Besuchet 1972; Vit 1981; Pakaluk 1987), were likely fungivores, although direct evidence does not emerge for this mouthpart class until later in the Mesozoic.

For the Jurassic, the most important process was the prolonged Mesozoic Lacustrine Revolution (Fig. 17.15, development 6), which had profound global effects on increasing the number of insect lineages and their mouthpart diversity in lake ecosystems. This process was caused by the broad, trophic shift from detritivore to herbivore driven food webs that began in the middle Jurassic and culminated during the early Cretaceous (Buatois et al. 2016). The six major, aquatic insect lineages and their mostly mouthpart classes of naiads, nymphs and larvae were Odonatoptera (Labial Mask), Ephemeroptera (Pectinate), Chresmodida (Adult Ectognathate), Coleoptera (Adult Ectognathate, Larval Ectognathate), Diptera (Mouthbrush), and Trichoptera (Mandibulobrustiate). These mouthpart classes were involved in the entire trophic gamut of lacustrine feeding. This spectrum included detritivory and herbivory of epibenthic, planktic, and neuston material by collectors, sievers and filterers herbivores of submerged aquatic vegetation; and insects preying on arthropods and small vertebrates at multiple trophic levels (Buatois et al. 2016). The second major feature that began during the middle Jurassic was the evolution from Mecoptera ancestors of flea-like and flea Tristylate mouthparts involved in hematophagy that evolved robust structures that penetrated thicker skinned mid-Mesozoic vertebrates (Huang et al. 2013; Gao et al. 2013) to more delicate mouthparts of modern Siphonaptera, probably of Late Cretaceous origin (Fig. 17.15, development 7). Apparently, early evolution of Tristylate mouthparts involved the evolution of very stout, robust lacinial stylets that had serrated teeth directed along their inward lateral margin (Gao et al. 2012, 2013), rather than the outward lateral margin of modern Siphonaptera. Inwardly directed dentition of lacinial stylets could have been a specialized feature for penetration of dinosaur or other pachydermous integument. The third major mouthpart process, also begun during the Middle Jurassic, was the diversification of dipteran Monostylate/Distylate, Distylate/Tetrastylate, Hexastylate, and probably Labellate mouthparts through various structural devices for feeding on blood (Fig. 17.15, development 8). Although paralleling in a minor way the transformation of Tristylate mouthparts involved in hematophagy, the diversity of Diptera mouthpart classes and associated feeding strategies allowed a much greater spectrum partitioning of blood-containing tissues.

The Early Cretaceous represents the end of the effects of the Mesozoic Lacustrine Revolution, transformation of the flea proboscis, and the proliferation of dipteran hematophage mouthparts. Other features of mouthpart evolution, such as broad inventory of mouthpart types in Cretaceous ambers, suggest that such amber insects may reveal a considerable amount of new mouthpart diversity that normally would not be available in adpression deposits (Fig. 17.15, development 9). Myanmar amber (99 Ma) is the best example that reveals a hidden insect fauna in three ways. First, there is anachronistic overlap of Permian and Triassic lineages of insects, often thought to have been long extinct, with the earliest occurrences of modern insect lineages (Labandeira 2014a). This temporal overlap in superbly preserved fossils provides details of head and mouthpart structure of the latest occurrences of ancient lineages and the earliest occurrences of modern lineages that are important for evolutionary studies. Second, Myanmar amber has revealed new, high-ranked insect taxa with head and mouthpart morphologies that do not occur elsewhere in the fossil or modern records and thus expand the possibilities of what is structurally possible in insect body design (Bai et al. 2016; Poinar and Brown 2017; Mey et al. 2017). Lastly, Myanmar amber has revealed associations with other organisms, such as scorpionflies and gymnosperms that are not found elsewhere in the fossil or modern record (Lin et al. 2019). Given the wealth of new insect material coming from Myanmar amber, there will be many additional surprises, undoubtedly including new mouthpart classes.

For mouthpart morphologies, the Paleogene and Neogene often represent, when fossils are present, the extension of modern clades that occur in the past that may or may not have corroborating body fossils (Fig. 17.15, development 10). In the case of the Mouthhook mouthpart class, within 2 million years after the Cretaceous–Paleogene ecological crisis and mass extinction (66 Ma), distinctive leaf-mining fly mines were found in a species of Paleocene sycamore leaf (64 Ma), and not by the presence of a diagnosable body fossil (Winkler et al. 2010). Other mouthpart classes, such as the Siphonostylate mouthpart class of Southeastern Asian erebid moths, lack a record, and their Late Pleistocene origin can be established from the distribution and history of their bovid hosts (Bänziger 1986; Zaspel et al. 2012). Still other distinctive mouthparts, such as the Tubulostylate mouthparts of tsetse from modern Subsaharan Africa, have Paleogene records, but in unexpected localities, such as Colorado, USA, and Germany, indicating substantial, Cenozoic biogeographic shifts. As for mouthparts of the deeper past, specific mouthpart elements, such as the pollination-related maxillary tentacle of yucca moths or the mandibular appendage of fig wasps, can reveal the life habits of insects from the recent past (Fig. 17.15, development 10).

An update of the geochronologic distribution of individual mouthpart classes through time is shown in Fig. 17.16. These direct and indirect, range-through data are plotted as diversity curves through time in Fig. 17.17b, based on Tables 17.2 and 17.3. Notably, the five phases of insect mouthpart expansion (Fig. 17.16, vertical arrows) are recovered in the new data (this report). Nevertheless, Phases 2 and 3 are displaced backward in geologic time, indicating an earlier timing of mouthpart innovation than previously indicated. Phase 2 commences at the beginning of the Pennsylvanian and is short lived. Phase 3 commences during the Early Permian, after a sharp rise in mouthpart class diversity, and remains flat until the Late Triassic, evidently unaffected by the Late Permian–Early Triassic ecological crisis and extinction event. Phase 4 retains its former position in the Early Jurassic. By contrast, Phase 5 is displaced back in time to the Late Jurassic, from a former Early Cretaceous position. This temporal shift is attributable to the extensive discoveries of diverse, new, insect taxa of the Middle Jurassic to Late Jurassic boundary interval, particularly from the Jiulongshan Formation of northeastern China (Ren et al. 2019). Importantly, the pattern of mouthpart morphological diversity occurs significantly earlier (convex trendline in Fig. 17.17a) than the major increase in insect taxic diversity (concave trendline in Fig. 17.17b). Apparently, the geochronologic patterns of both the richness data of family-level taxa and the diversity of mouthpart classes (Labandeira 1997) remain robust upon re-analyses.

**Acknowledgements** I am grateful to Harald Krenn for the invitation to provide this review. Jennifer Wood assembled and finalized the drawings into figures. The Smithsonian Institution Libraries provided facilities and interlibrary loan articles essential for the completion of this review. This is contribution 373 of the Evolution of Terrestrial Ecosystems Consortium at the National Museum of Natural History, in Washington, D.C.

## References

Achtelig M (1967) Über die Anatomie des Kopfes von *Raphidia flavipes* Stein und die Verwandtschaftsbeziehungen der Raphidiidae zu den Megaloptera. Zool Jb Anat 84:249–312

Acker TS (1958) The comparative morphology of *Stenorrhachus walkeri* (McLachlan) and of *Nemopterella* sp. (Neuroptera: Nemopteridae). Microentomology 23:106–114

Adler S, Theodor O (1926) The mouth parts, alimentary tract and salivary apparatus of the female in *Phlebotomus papatsii*. Ann Trop Med Parasitol 20:109–142

Andersen SB, Gerritsma S, Yusah KM, Mayntz D, Hywel-Jones NL, Billen J, Boomsma JJ, Hughes DP (2009) The life of a dead ant: the expression of an adaptive extended phenotype. Am Nat 174:424–433

Anderson RH (1995) An evolutionary perspective on diversity in Curculionoidea. Mem Entomol Soc Wash 14:103–114

Anderson LI, Trewin NH (2003) An early Devonian arthropod fauna from the Windyfield cherts, Aberdeen, Scotland. Palaeontology 46(3):467–509

Anderson JM, Anderson HM, Cruickshank ARI (1998) Late Triassic ecosystems of the Molteno/lower Elliott biome of southern Africa. Palaeontology 41:387–421

Arnol'di LV, Zherikhin VV, Nikritin LM, Ponomarenko AG (1991) Mesozoic Coleoptera. In: Vandenberg NJ (ed) Smithsonian Institution Libraries and National Science Foundation, Washington, DC, p 285 (trans: Russian)

Arora GL (1959) External morphology of the larva of *Diprion pini* (L.) (Symphyta: Hymenoptera). Res Bull Punjab Univ NS 10:55–64

Arora GL, Singh S (1962) Morphology and musculature of the head and mouth-parts of *Idiocerus atkinsoni* Leth (Jassidae, Homoptera). J Morphol 110:131–140

Asahina S (1954) A morphological study of a relic dragonfly, *Epiophlebia superstes* Selys (Odonata, Anisozygoptera). Japan Society for the Promotion of Science, Tokyo, p 153

Askew RR (1971) Parasitic insects. Elsevier, New York, p 316

Austen EE (1903) A monograph on the tsetse-flies (Genus *Glossina,* Westwood) based on the collection in the British Museum. William Clowes and Sons, London, p 319

Badonnel A (1931) Copéognathes de France (11$^{e}$ note). Sur les pieces buccales de *Scoliopsyllopsis latreillei* (Enderlein). Bull Soc Zool Fr 56:250–257

Bai M, Beutel RG, Klass K-D, Zhang W, Yang X, Wipfler B (2016) Alienoptera – a new insect order in the roach–mantodean twilight zone. Gondw Res 39:317–326

Baker GT, Ellsbury MM (1988) Morphology of the mouth parts and antenna of the larva of the clover stem borer, *Languria mozardi* Latreille (Coleoptera: Languriidae). Proc Entomol Soc Wash 91:15–21

Balachowsky A (1937) Les cochenilles de France, d'Europe, du nord de l'Afrique et du Bassin Méditerranéen. Caractères generaux des cochenilles—morphologie externe. Actual Sci Indust 526:1–67

Balashov YS (1999) Evolution of hematophagy in insects and ticks. Entomol Rev 79:943–954

Bänziger H (1960) Biological and morphological notes on the fruit piercing and eye frequenting moths. Proc X Int Kongr Entomol Wien 2:10–17

Bänziger H (1968) Preliminary observations on a skin-piercing blood-sucking moth (*Calyptra eustrigata* (Hmps.) (Lep., Noctuidae)) in Malaya. Bull Entomol Res 58:159–164

Bänziger H (1970) The piercing mechanism of the fruit-piercing moth *Calpe* (*Calyptra*) *thalictri* Bkh (Noctuidae) with reference to the skin-piercing blood-sucking moth *C. eustrigata* Hmps. Acta Trop 27:54–88

Bänziger H (1980) Skin-piercing blood-sucking moths III: feeding act and piercing mechanism of *Calyptra eustrigata* (Hmps.) (Lep., Noctuidae). Mitt Schweiz Entomol Ges 53:127–142

Bänziger H (1986) Skin-piercing blood-sucking moths IV: biological studies on adults of 4 *Calyptra* species and 2 subspecies (Lep., Noctuidae). Bull Soc Entomol Suisse 59:111–138

Barth R (1952) Estudos anatomicos e histologicos sobre a subfamilia Triatominae (Heteroptera, Reduviidae). I parte: A cabeça do *Triatoma infestans*. Mem Inst Oswaldo Cruz 5:69–196

Barth R (1953) Estudos anatomicos e histologicos sobre a subfamilia Triatominae (Heteroptera, Reduviidae). III parte: Pesquisas sobre o mecanismo da picada dos Triatominae. Mem Inst Oswaldo Cruz 51:11–72

Barth FG (1985) Insects and flowers: the biology of a partnership. Princeton University Press, Princeton, NJ, p 297

Bashkuev AS (2011) Nedubroviidae, a new family of Mecoptera: the first Paleozoic long-proboscid scorpionflies. Zootaxa 2895:47–57

Batra L, Batra ST (1967) The fungus gardens of insects. Sci Am 217:112–120

Beaucormu JC (2003) *Palaeopsylla groehni* n. sp., quatrième espèce de puce connue de l'ambre de la Baltique (Siphonaptera, Ctenophthalmidae). Bull Soc Entomol Fr 108:217–220

Bechly G, Stockar R (2011) The first Mesozoic record of the extinct apterygote insect genus *Dasyleptus* (Insecta: Archaeognatha: Monura: Dasyleptidae) from the Triassic of Monte San Giorgio (Switzerland). Palaeodiversity 4:23–37

Becker-Migdisova EE (1940) Fossil Permian cicadas of the family Prosbolidae from the Sojana River. Trans Paleontol Inst Akad Nauk SSSR 11(2):1–98 (in Russian)

Becker-Migdisova EE (1946) A contribution to knowledge of the comparative morphology of recent and Permian Homoptera. Part 1. Bull Acad Sci USSR 1946(6):741–766

Becker-Migdisova EE (1948a) Permian cicadas of family Scytinopteridae from Soyana. Trudy Paleontol Inst USSR 15(2):1–42 (in Russian)

Becker-Migdisova EE (1948b) A contribution to knowledge of the comparative morphology of recent and Permian Homoptera. Part II. Bull Acad Sci USSR 1948(1):123–142. in Russian

Becker-Migdisova EE (1960) New Permian Homoptera from the European part of the USSR. Trudy Paleontol Inst 76:1–112 (in Russian)

Becker-Migdisova EE (1961) Order Psocoptera. In: Rohdendorf BB, Becker-Migdisova EE, Martynova OM, Sharov AG (eds) Paleozoic insects of the Kuznetsk Basin, pp 271–286 (in Russian)

Becker-Migdisova EE (1985) Fossil insect Psyllomorpha. Trudy Paleontol Inst 206:1–93. in Russian

Benwitz G (1956) Der Kopf von *Corixa punctata* Ill. (*geoffroyi* Leach) (Hemiptera, Heteroptera). Zool Jb Anat 75:311–378

Bernays EA (1998) Evolution of feeding behavior in insect herbivores. Bioscience 48(1):35–44

Besuchet C (1972) Les coleoptérès aculagnathides. Rev Suisse Zool 79:99–145

Béthoux O (2006) Revision of *Cacurgus* Handlirsch, 1911, a basal Pennsylvanian Archaeorthoptera (Insecta: Neoptera). Bull Peabody Mus Nat Hist 47(1–2):29–35

Béthoux O (2007) Emptying the Paleozoic wastebasket for insects: member of a Carboniferous 'protorthopterous family' assigned to natural groups. Alavesia 1:41–48

Béthoux O (2008) Revision and phylogenetic affinities of the lobeatid species *bronsoni* Dana, 1864 and *silvatica* Laurentiaux & Laurentiaux-Vieira, 1980 (Pennsylvanian; Archaeorthoptera). Arthopod Syst Phylogeny 66(2):145–163

Béthoux O (2009) Head and leg morphology of *Elongata brongniarti*, 1893 (Late Carboniferous, Archaeorthoptera): phylogenetic and palaeoecological implications. Ann Zool 59:141–147

Béthoux O, Briggs DEG (2008) How *Gerarus* lost its head: stem-group Orthoptera and Paraneoptera revisited. Syst Entomol 33:529–547

Béthoux O, Nel P (2002) New data on Tcholmanvissiidae (Orthoptera, Permian). J Orthop Res 11(2):223–235

Béthoux O, Nel A (2004) Some Palaeozoic 'Protorthoptera' are 'ancestral' orthopteroids: major wing braces as clues to a new split among the 'Protorthoptera' (Insecta). J Syst Palaeontol 2(4):285–309

Béthoux O, Wieland F (2009) Evidence for Carboniferous origin of the order Mantodea (Insecta: Dictyoptera) gained from forewing morphology. Zool J Linnean Soc 156:79–113

Béthoux O, Nel A, Lapeyrie J (2004) The extinct order Caloneurodea (Insecta: Pterygota: Panorthoptera): wing venation, systematics and phylogenetic relationships. Ann Zool 54(2):289–318

Béthoux O, Klass KD, Schneider JW (2009) Tackling the Protoblattoidea problem: revision of *Protoblattinopsis stubblefieldi* (Dictyoptera: late Carboniferous). Eur J Entomol 106:145–152

Béthoux O, Voigt S, Schneider JW (2010) A Triassic palaeodictyopteran from Kyrgystan. Palaeodiversity 3:9–13

Bierbrodt E (1943) Der Larvenkopf von *Panorpa communis* L. und seine Verwandlung, mit besonderer Berückssichtigung des Gehirns und der Augen. Zool Jb Anat 68:49–136

Bitsch J (1963) Morphologie cephalique des machilides (Insecta, Thysanura). Ann Soc Entomol Fr NS 11:255–324

Blanke A, Machida R, Szucsich NU, Wilde F, Misof B (2015) Mandibles with two joints evolved much earlier in the history of insects: dicondyly is a synapomorphy of bristletails, silverfish and winged insects. Syst Entomol 40:357–364

Bletchly JD (1954) The mouth-parts of the dance fly, *Empis livida* L. (Diptera, Empididae). Proc Zool Soc Lond 124:317–334

Bletchly JD (1955) The mouth-parts of the down-looker fly, *Rhagio* (=*Leptis*) *scolopacea* (L.) (Diptera, Rhagionidae). Proc Zool Soc Lond 125:779–794

Blunck H (1918) Die Entwicklung des *Dytiscus marginalis* L. vom Ei bis zur Imago. Z Wiss Zool 117:1–129

Bohart RM (1943) New species of *Halictophagus* with a key to the genus in North America (Strepsiptera, Halictophagidae). Ann Entomol Soc Am 36:341–359

Bonhag PF (1951) The skeleton-muscular mechanism of the head and abdomen of the adult horsefly (Diptera: Tabanidae). Trans Am Entomol Soc 77:131–202

Borkent A (1995) Biting midges in the Cretaceous amber of North America (Diptera, Ceratopogonidae). Backhuys, Leiden, p 237

Borkent A, Grimaldi DA (2004) The earliest fossil mosquito (Diptera: Culicidae), in mid-Cretaceous Burmese amber. Ann Entomol Soc Am 97(5):882–888

Borkent A, Wood DM (1986) The first and second larval instars and the egg of *Parasimulium stonei* Peterson (Diptera: Simuliidae). Proc Entomol Soc Wash 88:287–296

Botosaneanu L (1956) Contributions à la connaissance des stades aquatiques des Trichoptérès crénobiontes: *Rhyacophila laevis* Pict., *Wormaldia triangulifera* MacLachl., *Drusus romanicus* Murg. & Bots., *Silovaripilosa* Bots. (Trichoptera). Beitr Entomol 6:590–624

Brack-Egg A (1973) Der Kopf von *Rhynchites auratus* Scop. (Curculionidae)—Eine konstruktionsmorphologische Untersuchung mit einem Beitrag zum Brutfürsorgeverhalten. Zool Jb Anat 91:500–545

Brauckmann C, Koch L (1982) Neue Insekten aus den Vorhalle-Schichten (oberes Namurium B) von Hagen-Vorhalle. Dortmund Beitr Landes Natur Mitt 16:15–26

Brauckmann C, Lutz K, Kemper M (1985) Spinnentiere (Arachnida) und Insekten aus den Vorhalle-Schichten (Namurium B, Ober-Karbon) von Hagen-Vorhalle (West-Deutschland). Geol Paläontol Westfalen 3:1–131

Briolini G (1960) Richerche su quattro specie di microlepidotteri miniatori delle fogli di melo. *Nepticula malella* Staint., *Stigmella pomella* Vaugh. (Nepticulidae); *Leucoptera scitella* Zell. (Bucculatricidae); *Lithocolletis blancardella* F. (Gracillariidae). Boll Ist Entomol Bologna Univ 24:239–269

Briolini G (1965) Ricerche morfologiche ed etologiche su *Paraleucoptera sinuella* Rtti. (Lepidoptera, Lyonetiidae). Boll Ist Entomol Bologna Univ 27:229–243

Brongniart C (1884) Sur un gigantesque Néurorthoptère, provenant des terrains houillers de Commentry (Allier). C R Hebd Seances Acad Sci 98:832–833

Brongniart C (1893) Recherches pour servir à l'histoire des insectes fossiles des temps primaires. Précédées d'une étude sur la nervation des ailes des insectes. Imprimerie Théolier, Saint-Etienne, p 493

Brown DS (1961) The morphology and functioning of the mouthparts of *Chloeon dipterum* L. and *Baetis rhodani* Pictet (Insecta, Ephemeroptera). Proc Zool Soc Lond 136:147–176

Brues CT (1972) Insect dietary. Harvard University Press, Cambridg, p 466

Buatois LA, Labandeira CC, Mángano MG, Cohen AC, Voigt S (2016) The Mesozoic Lacustrine Revolution. In: Mángano MG, Buatois LA (eds) The trace-fossil record of major evolutionary events. Top Geobiol 40:179–263

Bucher GE (1948) The anatomy of *Monodontomerus dentipes* Boh., an entomophagous chalcid. Can J Res 26D:230–281

Buckup L (1959) Der Kopf von *Myrsidea cornicis* (DeGeer) (Mallophaga—Amblycera). Zool Jb Anat 77:241–288

Burnham L (1983) Studies on upper Carboniferous insects: 1. The Geraridae (Order Protorthoptera). Psyche 90(1–2):1–57

Butt FH (1951) Feeding habits and mechanism of the Mexican bean beetle. Mem Cornell Univ Agric Exp Station 306:1–32

Büttiker W (1962) Biological and morphological notes on the fruit-piercing and eye-frequenting moths. Verh XI Int Congr Entomol 2:10–15

Cai C, Escalona HE, Li L, Yin Z, Huang D, Engel MS (2018) Beetle pollination of cycads in the Mesozoic. Curr Biol 28:2806–2812

Carpenter FM (1960) Studies on North American Carboniferous insects. 1. The Protodonata. Psyche 67:98–110

Carpenter FM (1971) Adaptations among Paleozoic insects. Proc North Am Paleontol Conv [1969] 1:1236–1251

Carpenter FM (1992) Volume 3. Superclass Hexapoda. In: Moore RC, Kaesler RL, Brosius E, Keim J, Priesner J (eds) Treatise on invertebrate paleontology. Part R. Arthropoda 4. Geological Society of America and University of Kansas: Boulder, Colorado and Lawrence, Kansas, p 655

Carpenter FM (1997) Insecta. In: Shabica CW, Hay AA (eds) Richardson's guide to the fossil fauna of Mazon Creek. Northeastern Illinois Press, Chicago, pp 184–193

Chao H-F (1953) The external morphology of the dragonfly *Onychogomphus ardens* Needham. Smithson Misc Collect 122:1–56

Chassagnard M-T, Tsacas L (1974) Morphologie de la tête et des pieces buccales imaginales de *Chrysopilus auratus* F. et de *Vermileo vermileo* DeG (Diptera: Rhagionidae). Int J Ins Morphol Embryol 3:13–32

Chaudonneret J (1986) Les pieces buccales des insectes: thème et variations. V. Bull Sci Bourgonge 39:26–45

Chaudonneret J (1990) Les pieces buccales des insectes: thème et variations. Imprimerie Berthier, Dijon, p 256

Chen Z, Benton MJ (2012) The timing and pattern of biotic recovery following the end-Permian mass extinction. Nat Geosci 5:375–383

China WE (1931) Morphological parallelism in the structure of the labium in the hemipterous genera *Coptosomoides*, gen. nov., and *Bozius* Dist. (Fam Plataspidae) in connection with mycetophagous habits. Ann Mag Nat Hist 7(10):281–286

Clancy DW (1946) The insect parasites of the Chrysopidae (Neuroptera). Univ Calif Publ Entomol 7:403–496

Cleal CJ, Shute CH, Zodrow EL (1990) A revised taxonomy for Paleozoic neuropteroid foliage. Taxon 39:486–492

Cobben RH (1978) Evolutionary trends in Heteroptera. Part II. Mouthpart-structures and feeding strategies. Meded Landbouwhoges Wagen 78-5:1–407

Cockerell TDA (1918) New species of North American fossil beetles, cockroaches and tsetse flies. Proc US Natl Mus 54:301–311

Common IFB (1973) A new family of Dacnonympha (Lepidoptera) based on three new species from southern Australia, with notes on the Agathiphagidae. J Aust Entomol Soc 12:11–23

Cope OB (1940) The morphology of *Psocus confraternus* Bankson (Psocoptera: Psocidae). Microentomology 6:71–92

Cowley DR (1978) Studies on the larvae of New Zealand Trichoptera. N Z J Zool 5:639–750

Craig DA (1977) Mouthparts and feeding behaviour of Tahitian larval Simuliidae (Diptera: Nematocera). Quaest Entomol 13:195–218

Crampton GC (1930) The head structures of the orthopteron *Stenopelmatus*—a contribution to the study of the external anatomy of *Stenopelmatus*. Pan Pac Entomol 6:97–110

Crampton GC (1942) The external morphology of the Diptera. Bull Connecticut Geol Nat Hist Surv 64:10–165

Cregan MB (1941) Generic relationships of the Dolichopodidae (Diptera) based on a study of the mouth parts. Ill Biol Mon 18:1–68

Crichton ME (1957) Zur Morphologie und Anatomie der Larve von *Oryctes nasicornis* L. Deut Entomol Z 4:228–262

Dajoz R (1976) Les coléoptérès Cerylonidae. Etude des espéces de la fauna Paléarctique. Bull Mus Natl Hist Nat 3(360):249–281

Davidson J (1914) On the mouth-parts and mechanism of suction in *Schizoneura lanigera* Hausmann. J Linn Soc 32:307–330

Davis DR (1967) A revision of the moths of the subfamily Prodoxinae (Lepidoptera: Incurvariidae). Bull US Natl Mus 255:1–170

Davis SR, Engel MS, Legalov A, Ren D (2013) Weevils of the Yixian Formation, China (Coleoptera: Curculionoidea): phylogenetic considerations and comparison with other Mesozoic faunas. J Syst Palaeontol 11(4):399–429

DeMarzo L (1976a) Studi sulle larve dei Coleotteri Ditiscidi. IV. Morfologia de tre stadi larvali di *Copelatus haemorrhoidalis* F. Entomologia 12:89–106

DeMarzo L (1976b) Studi sulle larve dei Coleotteri Ditiscidi. V. Note morfologiche sulle larve di tre specie del genere *Laccophilus* Leach. Entomologica 12:107–129

DeMarzo L (1977) Studi sulle larve dei Coleotteri Ditiscidi. VII. Considerazioni sul significato evolutivo del comportamento dei caratteri delle mandibole nelle larve di alcune specie della subf. Dytiscinae. Entomologica 13:71–84

Demoll R (1908) Die Mundteile der solitären Apiden. Z Wiss Zool 91:1–51

Dennell R (1942) The structure and function of the mouth-parts, rostrum and fore-gut of the weevil *Calandra granaria* L. Philos Trans R Soc Lond B 231:247–291

Ding Q, Labandeira CC, Ren D (2014) Biology of a leaf miner (Coleoptera) on *Liaoningocladus boii* (Coniferales) from the Early Cretaceous of northeastern China and the leaf-mining biology of possible insect culprit clades. Arthopod Syst Phylogeny 72:281–308

Dittmar K, Zhu Q, Hastriter MW, Whiting MF (2016) On the probability of dinosaur fleas. BMC Evol Biol 16(9):9. https://doi.org/10.1186/s12862-015-0568x

Dlussky GM (1996) Ants (Hymenoptera: Formicidae) from Burmese amber. Paleontol J 30:449–454

Donges J (1954) Der Kopf von *Cionus scrophulariae* L. (Curculionidae). Zool Jb Anat 74:1–76

Donovan MP, Wilf P, Labandeira CC, Johnson KR, Peppe DJ (2014) Novel insect leaf-mining after the end-Cretaceous extinction and the demise of Cretaceous leaf miners, Great Plains USA. PLoS One 9(7):e103542

Donovan MP, Iglesias A, Wilf P, Labandeira CC, Cúneo NR (2016) Rapid recovery of Patagonian plant–insect associations after the end-Cretaceous extinction. Nat Ecol Evol 1:12. https://doi.org/10.1038/s41559-016-0012

Donovan MP, Iglesias A, Wilf P, Labandeira CC, Cúneo NR (2018) Diverse plant–insect associations from the latest Cretaceous and early Paleocene of Patagonia, Argentina. Ameghiniana 55(3):303–338
Dumbleton LJ (1952) A new genus of seed-infesting micropterigid moths. Pac Sci 6:17–29
Dunne JA, Labandeira CC, Williams RJ (2014) Highly resolved early Eocene food webs show development of modern trophic structure after the end-Cretaceous extinction. Proc R Soc B 281:20133280. https://doi.org/10.1098/rspb.2013.3280
Durand J-B (1955) Contribution a l'étude des Trichoptérès. Trav Lab Zool Univ Dijon 9:1–45
Eastham LES, Eassa YEE (1955) The feeding mechanism of the butterfly *Pieris brassicae* L. Philos Trans R Soc Lond B 239:1–43
Edmunds GF Jr (1984) Ephemeroptera. In: Merritt RW, Cummins KW (eds) An introduction to the aquatic insects of North America, 2nd edn. Kendall-Hunt, Dubuque, pp 94–125
Eisenbeis G, Wichard W (1987) Atlas on the biology of soil arthropods. Springer, Berlin, p 437
Elzinga RJ, Broce AB (1986) Labellar modifications of the Muscomorpha flies (Diptera). Ann Entomol Soc Am 79:150–209
Engel MS (2000) A new interpretation of the oldest fossil bee (Hymenoptera: Apidae). Am Mus Novit 3296:1–11
Engel MS, Grimaldi DA (2000) A winged *Zorotypus* in Miocene amber from the Dominican Republic (Zoraptera: Zorotypidae), with discussion on relationships of and within the order. Acta Geol Hisp 35(1):149–164
Engel MS, Grimaldi DA (2004) New light shed on the oldest insect. Nature 427:627–630
Felix M (1962) Recherches sur l'anatomie céphalique d'une larve de *Dixa* (*Paradixa*) (Diptère Nématocère). Trav Lab Zool Fac Sci Dijon 43:1–49
Feng Z, Wang J, Rößler R, Ślipiński A, Labandeira CC (2017) Late Permian wood borings reveal an intricate network of ecological relationships. Nat Commun 8:556. https://doi.org/10.1038/s41467-017-00696-0
Fleck G, Nel A (2002) The first isophlebioid dragonfly (Odonata: Isophlebioptera: Campterophlebiidae) from the Mesozoic of China. Palaeontology 45(6):1123–1136
Fleck G, Nel A, Bechly G, Escuillié F (2002) The larvae of the Mesozoic family Aeschnidiidae and their phylogenetic implications (Insecta, Odonata, Anisoptera). Palaeontology 45(1):165–184
Foote BA (1970) The larvae of *Tanypeza longimana* (Diptera: Tanypezidae). Ann Entomol Soc Am 63:235–238
Fotius-Jaboulet MC (1964) Description de la larve de *Rhyacophila praemorsa* McL. et comparison de cette larve a *Rhyacophila septentrionis* McL. Trav Lab Zool Univ Dijon 62:1–13
François J (1959) Squelette et musculature cephaliques d'*Acerentomon propinquum* (Condé) (Diplura: Campodeidae). Zool Jb Anat 87:331–376
François F (1969) Anatomie et morphologie cephalique des protoures (Insecta Apterygota). Mem Mus Natl Hist Nat 59:1–144
Frochot B (1962) La larve de *Stenophylax permistus* McL. (Trichoptera, Limnophilidae). Trav Lab Zool Univ Dijon 42:1–16
Froehlich CG (1964) The feeding apparatus of the nymph of *Arthroplea congener* Bengtsson (Ephemeroptera). Opusc Entomol 29:188–208
Gad AM (1951) The head-capsule and mouth-parts in the Ceratopogonidae. Bull Soc Fouad I Inst Entomol 35:17–75
Gangwere SK (1966) Relationships between the mandibles, feeding behavior, and damage inflicted on plants by the feeding of certain acridids (Orthoptera). Mich Entomol 1:13–16
Gao T, Shih CK, Xu X, Wang S, Ren D (2012) Mid-Mesozoic flea-like ectoparasites of feathered or haired vertebrates. Curr Biol 22:732–735
Gao T, Shih CK, Rasnitsyn AP, Xu X, Wang S, Ren D (2013) New transitional fleas from China highlighting diversity of early Cretaceous ectoparasitic insects. Curr Biol 23:1261–1266
Gardner CF, Nielsen LT, Knight KL (1973) Morphology of the mouthparts of larval *Aedes communis* (DeGeer) (Diptera; Culicidae). Mosq Syst 5:163–182

Garwood R, Ross A, Sotty D, Chabard D, Charbonnier S, Sutton M, Withers PJ (2012) Tomographic reconstruction of neopterous Carboniferous insect nymphs. PLoS One 7(9):e45779

Gibbins EG (1938) The mouth-parts of the female in *Simulium damnosum* Théobald, with special reference to the transmission of *Onchocerca volvulus* Leuckart. Ann Trop Med Parasitol 32:9–20

Giliomee JD (1968) Morphology and relationships of the male of an *Asterolecanium* species (Homoptera: Coccoidea: Asterolecaniidae). J Entomol Soc S Afr 31:297–308

Gorochov AV (2001) The higher classification, phylogeny and evolution of the superfamily Stenopelmatoidea. In: Field LH (ed) The biology of wetas, king crickets and their allies. CABI, Wallingford, pp 3–33

Goto HE (1972) On the structure and function of the mouthparts of the soil-inhabiting collembolan *Folsomia candida*. Biol J Linn Soc 4:147–168

Gouin FJ (1958) L'appareil buccal de *Chrysochroma bipunctatum* Scop. (Dipt. Stratiom.). Proc Tenth Int Congr Entomol 1:517–519

Grandi G (1924) Studi sullo sviluppo postembrionale delle varie razze del *Bombyx mori* L. Boll Lab Zool Gen Agric Portici 27:3–40

Greene CT (1929) Characters of the larvae and pupae of certain fruit flies. J Agric Res 38:489–503

Greenslade P, Whalley PES (1986) The systematic position of *Rhyniella praecursor* Hirst & Maulik (Collembola). The earliest known hexapod. In: Dallai R (ed) Second international seminar on Apterygota. University of Siena, Italy, pp 319–323

Grimaldi D (1992) Vicariance biogeography, geographic extinctions, and the North American Oligocene tsetse flies. In: Novacek MJ, Wheeler QD (eds) Extinction and phylogeny. Columbia University Press, New York, pp 178–204

Grimaldi D (1999) The co-radiations of pollinating insects and angiosperms in the Cretaceous. Ann Mo Bot Gard 86(2):373–406

Grimaldi D (2003) First amber fossils of the extinct family Protopsyllidiidae and their phylogenetic significance among Hemiptera. Inst Syst Evol 34:329–344

Grimaldi D, Engel MS (2005) Evolution of the insects. Cambridge University Press, New York, p 755

Grimaldi D, Shmakov A, Fraser N (2004) Mesozoic thrips and early evolution of the order Thysanoptera (Insecta). J Paleontol 78(5):941–952

Grinfel'd EK (1975) Anthophily in beetles (Coleoptera) and a critical evaluation of the cantharophilous hypothesis. Entomol Rev 54:18–22

Gu JJ, Béthoux RD (2011) *Longzhua loculata* n. gen. n. sp., one of the best documented Pennsylvanian Archaeorthoptera (Insecta; Ningxia, China). J Paleontol 85:304–314

Gurney AB (1947) Notes on Dilaridae and Berothidae, with special reference to the immature stages of the Nearctic genera (Neuroptera). Psyche 54:145–169

Habgood K, Haas H, Kerp H (2004) Evidence for an early terrestrial food web: coprolites from the early Devonian Rhynie Chert. Trans R Soc Edinburgh Earth Sci 94:371–389

Hamilton WJ Jr (1934) The life history of the rufescent woodchuck, *Marmota monax rufescens Howell*. Ann Carnegie Mus 25:85–178

Hampson GF (1926) Descriptions of new genera and species of Lepidoptera Phalaenae of the subfamily Noctuinae (Noctuidae) in the British Museum (Natural History). Taylor and Francis, London, p 641

Handlirsch A (1906) Revision of American Paleozoic insects. Proc US Natl Mus 29:661–820

Handschin E (1929) Ein neuer Rüsseltypus bei einem Käfer. Biologische und morphologische Beobachtungen an *Leptopalpus rostratus* F. Z Morphol Ökol Tiere 14:513–521

Hansen HJ (1903) The mouth-parts of *Glossina* and *Stomoxys*. In: Austen EE (ed) A monograph of the Tse-Tse flies genus *Glossina*, Westwood based on the collection in the British Museum. Longmans, London, pp 105–120

Hartigan J (1988) Program 1M: cluster analysis of variables. In: Dixon WJ, Brown MV, Engelman L, Hill MA, Jennrich RIBMDP (eds) Statistical software manual, vol 2, 2nd edn. University of California Press, Berkeley, CA, pp 745–754

Haub F (1967) Der Kopf von *Pseudomenopon pilosum* (Scopoli) (Mallophaga–Amblycera). Zool Jb Anat 84:493–558
Haug C, Haug JT (2017) The presumed oldest flying insect: more likely a myriapod? PeerJ 5:e3402. https://doi.org/10.7717/peerj.3402
Haug JT, Labandeira CC, Santiago-Blay JA, Haug C, Brown S (2015) Life habits, hox genes, and affinities of a 311 million-year-old holometabolan larva. BMC Evol Biol 15:208. https://doi.org/10.1186/s12862-015-0428-8
Heddergott H (1938) Kopf und Vorderdarm von *Panorpa communis* L. Zool Jb Anat 64:229–294
Heming BS (1978) Structure and function of the mouthparts in larvae of *Haplothrips verbasci* (Osborn) (Thysanoptera, Tubulifera, Phlaeothripidae). J Morphol 156:1–37
Hepburn HR (1969) The skeleton-muscular system of Mecoptera: the head. Univ Kans Sci Bull 48:721–765
Hirsch F (1986) Die Mundwerkzeuge von *Phthirius pubis* L. (Anoplura). Zool Jb Anat 114:167–204
Hirst S, Maulik S (1926) On some arthropod remains from the Rhynie chert (Old Red Sandstone). Geol Mag 63:69–71
Hiznay PA, Krause JB (1955) The structure and musculature of the larval head and mouthparts of the horned passalus beetle, *Popilius disjunctus* Illiger. J Morphol 97:55–75
Hochuli PA, Hermann E, Vigran JO, Bucher H, Weissert H (2010) Rapid demise and recovery of plant ecosystems across the end-Permian extinction event. Glob Planet Change 74:144–155
Honomichl K (1975) Beitrag zur Morphologie des Kopfes der Imago von *Gyrinus substriatus* Stephens, 1829 (Coleoptera, Insecta). Zool Jb Anat 94:218–295
Hopkin SP (1997) Biology of the springtails (Insecta: Collembola). Oxford University Press, Oxford, p 330
Hopkin SP, Read HJ (1992) The biology of millipedes. Oxford University Press, Oxford, p 248
Hopkins GHE, Traub R (1955) The genus *Cratynius* Jordan (Siphonaptera) and its systematic position, with a description of a new species. Trans R Entomol Soc Lond 107:249–264
Hörnschemeyer T, Beutel RG, Pasop F (2002) Head structures of *Priacma serrata* Leconte inferred from X-ray tomography. J Morphol 252:298–314
Hörnschemeyer T, Goebbels J, Weidemann G, Faber C, Haase A (2006) Head structures of *Priacma serrata* Leconte (Coleoptera: Archostemata) inferred from X-ray tomography. Eur J Entomol 103:409–423
Hörnschemeyer T, Haug JT, Béthoux O, Beutel RG, Charbonnier S, Hegna TA, Koch M, Rust J, Wedmann S, Willmann R (2013) Is *Strudiella* a Devonian insect? Nature 494(7437):E3–E4
Houston TF (1987) Fossil brood cells of stenotritid bees (Hymenoptera: Apoidea) from the Pleistocene of South Australia. Trans R Soc S Aust 3:93–97
Hoyt CP (1952) The evolution of the mouth parts of adult Diptera. Microentomology 17:61–125
Huang D (2015) *Tarwinia australis* (Siphonaptera: Tarwiniidae) from the lower Cretaceous Koonwarra fossil bed: morphological revision and analysis of its evolutionary relationship. Cretac Res 52:507–515
Huang D, Engel MS, Cai C, Wu H, Nel A (2012) Diverse transitional giant fleas from the Mesozoic era of China. Nature 483:201–204
Huang D, Engel MS, Cai C, Nel A (2013) Mesozoic giant fleas from northeastern China (Siphonaptera): taxonomy and implications for palaeodiversity. Chin Sci Bull 58(14):1682–1690. https://doi.org/10.1007/s11434-013-5769-3
Huang D, Bechly G, Nel P, Engel MS, Prokop J, Azar D, Cai CY, van de Kamp T, Staniczek AH, Garrouste R, Krogmann L, dos Santos RT, Baumbach T, Ohlhoff R, Shmakov AS, Bourgoin T, Nel A (2016) New fossil insect order Permopsocida elucidates major radiation and evolution of suction feeding in hemimetabolous insects (Hexapoda: Acercaria). Sci Rep 6:23004. https://doi.org/10.1038/srep23004
Hubbard MD, Kukalová-Peck J (1980) Permian mayfly nymphs: new taxa and systematic characters. In: Flanagan JE, Marshall KE (eds) Advances in Ephemeroptera biology. Plenum, New York, pp 19–31

Hughes DP, Wappler T, Labandeira CC (2011) Ancient death-grip leaf scars reveal ant–fungal parasitism. Biol Lett 7:67–70
Iannuzzi R, Labandeira CC (2008) The oldest record of external foliage feeding and the expansion of insect folivory on land. Ann Entomol Soc Am 101(1):79–94
Ippolito R (1977) Note morfo-biologische su *Cnephasia incertana* Tr. e *Cnephasia cinareana* Chr (Lepidoptera–Tortricidae–Cnephasiini) viventi su carciofo. Entomologica 13:121–147
Irwin ME, Lyneborg L (1981) Therevidae. In: McAlpine JF, Peterson BV, Shewell GE, Teskey HJ, Vockeroth JR, Wood DM (eds) Manual of Nearctic Diptera. vol 1. Res Br Agric Can 27:513–523
Jablonski D (2005) Mass extinctions and macroevolution. Paleobiology 31:192–210
Jaboulet MC (1960) Contribution a l'étude des larves d'Haliplides. Trav Lab Zool Fac Sci Dijon 31:1–17
Jancke GD (1955) Zur Morphologie der männlichen Cocciden. Z Angew Entomol 37:265–314
Jayewickreme SH (1940) A comparative study of the larval morphology of leaf-mining Lepidoptera in Britain. Trans R Entomol Soc Lond 90:63–105
Jennings JR (1974) Lower Pennsylvanian plants of Illinois. I. A flora from the Pounds Sandstone Member of the Caseyville Formation. J. Paleontol 48:459–473
Jervis MA, Kidd NAC, Fitton MG, Huddleston T, Dawah HA (1993) Flower-visiting by hymenopteran parasitoids. J Nat Hist 27:67–105
Jobling B (1926) A comparative study of the structure of the head and mouth parts in the Hippoboscidae (Diptera Pupipara). Parasitology 18:319–349
Jobling B (1928) The structure of the head and mouth-parts in the Nycteribiidae (Diptera Pupipara). Parasitology 20:254–272
Jobling B (1929) A comparative study of the structure of the head and mouthparts in the Streblidae (Diptera Pupipara). Parasitology 21:417–445
Jobling B (1933) A revision of the head, mouth-part and salivary glands of *Glossina palpalis* Rob.-Desv. Parasitology 24:449–490
Jones T (1954) The external morphology of *Chirothrips hamatus* (Trybom) (Thysanoptera). Trans R Entomol Soc Lond 105:163–187
Kathirithamby J, Grimaldi DA (1993) Remarkable stasis in some lower Tertiary parasitoids: descriptions, new records and review of Strepsiptera in the Oligo-Miocene amber of the Dominican Republic. Entomol Scand 24:31–41
Kelsey LP (1954) The skeleto-motor mechanism of the dobson fly, *Corydalis cornutus*. Part 1. Head and prothorax. Cornell Univ Agric Exp Station Mem 334:1–51
Kevan PG, Chaloner WG, Savile DBO (1975) Interrelationships of early terrestrial arthropods and plants. Palaeontology 18(2):391–417
Kinzelbach RK (1966) Zur Kopfmorphologie der Fächerflügler (Strepsiptera, Insecta). Zool Jb Anat 84:559–684
Kirk WDJ (1984) Pollen feeding in thrips (Insecta: Thysanoptera). J Zool 204:107–117
Klemm N (1966) Die Morphologie des Kopfes von *Rhyacophila* Pict. (Trichoptera). Zool Jb Anat 83:1–51
Kluge NJ (2002) The homology of mouthparts in fleas (Insecta, Aphaniptera). Entomol Rev 82(8):1020–1026
Korboot K (1964) Comparative studies of the external and internal anatomy of three species of caddis flies (Trichoptera). Pap Univ Qld Dep Entomol 2:1–44
Korn W (1943) Die Muskulatur des Kopfes und des Thorax von *Myrmeleon europaeus* und ihre Metamorphose. Zool Jb Anat 68:273–330
Koteja J, Azar D (2008) Scale insects from the lower Cretaceous amber of Lebanon (Hemiptera: Sternorrhyncha: Coccinea). Alavesia 2:133–167
Kovalev VG, Mostovski MB (1997) A new genus of snipe flies (Diptera, Rhagionidae) from the Mesozoic of Eastern Transbaikalia. Paleontol J 31(5):523–527
Krassilov V, Karasev E (2008) First evidence of plant–arthropod interaction at the Permian–Triassic boundary in the Volga Basin, European Russia. Alavesia 2:247–252

Kristensen NP (1968a) The skeletal anatomy of the heads of adult Mnesarcheidae and Neopseustidae (Lep., Daconympha). Entomol Meddel 36:137–151

Kristensen NP (1968b) The anatomy of the head and the alimentary canal of adult Eriocraniidae (Lep., Dacnonympha). Entomol Meddel 36:239–315

Kühl G, Rust J (2009) *Devonohexapodus bocksbergensis* is a synonym of *Wingertshelicus backesi* (Euarthropoda) – no evidence for marine hexapods living in the Devonian Hunsrück Sea. Org Divers Evol 9(3):215–231

Kukalová-Peck J (1972) Unusual structures in the Paleozoic insect orders Megasecoptera and Palaeodictyoptera, with a description of a new family. Psyche 79:243–268

Kukalová-Peck J (1987) New Carboniferous Diplura, Monura and Thysanura, the hexapod ground plan, and the role of thoracic lobes in the origin of wings (Insecta). Can J Zool 65:2327–2345

Kukalová-Peck J (1990) Fossil history and the evolution of hexapod structures. In: Naumann ID, Carne PB, Lawrence JF, Nielsen ES, Spradbery JP, Taylor RW, Whitten MJ, Littlejohn MJ (eds) The insects of Australia, vol 1. Cornell University Press, Ithaca, pp 141–179

Kukalová-Peck J (2009) Carboniferous protodonatoid dragonfly nymphs and the synapomorphies of Odonatoptera and Ephemeroptera (Insecta: Palaeoptera). Palaeodiversity 2:169–198

Kukalová-Peck J, Brauckmann C (1990) Wing folding in pterygote insects, and the oldest Diaphanopterodea from the early late Carboniferous of West Germany. Can J Zool 68:1104–1111

Kukalová-Peck J, Brauckmann C (1992) Most Paleozoic Protorthoptera are ancestral hemipteroids: major wing braces as clues to a new phylogeny. Can J Zool 70:2452–2473

Kukalová-Peck J, Sinitshenkova ND (1992) The wing venation and systematics of lower Permian Diaphanopterodea from the Ural Mountains, Russia (Insecta: Paleoptera). Can J Zool 70:229–235

Labandeira CC (1990) Use of a phenetic analysis of recent hexapod mouthparts for the distribution of hexapod food resource guilds in the fossil record. Dissertation, The University of Chicago, pp xxvii + 1186. https://archive.org/details/labandeirahexapod1990chicago

Labandeira CC (1997) Insect mouthparts: ascertaining the Paleobiology of insect feeding strategies. Annu Rev Ecol Syst 28:153–193

Labandeira CC (1998) Early history of arthropod and vascular plant associations. Annu Rev Earth Planet Sci 26:329–377

Labandeira CC (1999) Insects and other hexapods. In: Singer R (ed) Encyclopedia of paleontology, vol 1. A–L. Fitzroy Dearborn, London, pp 603–624

Labandeira CC (2000) The paleobiology of pollination and its precursors. In: Gastaldo RA, DiMichele WA (eds) Phanerozoic terrestrial ecosystems. Paleontol Soc Pap 6:233–269

Labandeira CC (2001) Rise and diversification of insects. In: Briggs DEG, Crowther PR (eds) Palaeobiology II. Blackwell Science, London, pp 82–88

Labandeira CC (2002a) The history of associations between plants and animals. In: Herrera C, Pellmyr O (eds) Plant–animal interactions: an evolutionary approach. Blackwell Science, Oxford, p 26–74, 248–261

Labandeira CC (2002b) Paleobiology of predators, parasitoids and parasites: death and accommodation in the fossil record of continental invertebrates. In: Kowalewski M, Kelley PH (eds) The fossil record of predation. Paleontol Soc Pap 8:211–249

Labandeira CC (2005a) The fossil record of insect extinction: new approaches and future directions. Am Entomol 51:14–29

Labandeira CC (2005b) Invasion of the continents: cyanobacterial crusts to tree-inhabiting arthropods. Trends Ecol Evol 20(5):253–262

Labandeira CC (2006a) Silurian to Triassic plant and hexapod clades and their associations: new data, a review, and interpretations. Arthopod Syst Phylogeny 64(1):53–94

Labandeira CC (2006b) The four phases of plant–arthropod associations in deep time. Geol Acta 4:409–438

Labandeira CC (2007a) The origin of herbivory on land: initial patterns of plant tissue consumption by arthropods. Inst Sci 14:259–275

Labandeira CC (2007b) Assessing the fossil record of plant–insect associations: ichnodata versus body-fossil data. In: Bromley RG, Mángano MG, Genise JF, Melchor RN (eds) Sediment–organism interactions: a multifaceted ichnology. Soc Econ Paleontol Miner Spec Publ 88:9–26
Labandeira CC (2010) The pollination of mid-Mesozoic seed plants and the early history of long-proboscid insects. Ann Mo Bot Gard 97:469–513
Labandeira CC (2013) Deep-time patterns of tissue consumption by terrestrial arthropod herbivores. Naturwissenschaften 100:355–364
Labandeira CC (2014a) Amber. In: Laflamme M, Schiffbauer JD, Darroch SAF (eds) Reading and writing of the fossil record: preservational pathways to exceptional fossilization. Paleontol Soc Pap 20:163–216
Labandeira CC (2014b) Why did terrestrial insect diversity not increase during the angiosperm radiation? Mid-Mesozoic, plant-associated insect lineages harbor clues. In: Pontarotti P (ed) Evolutionary biology, genome evolution, speciation, coevolution and origin of life. Springer, Cham, pp 261–299
Labandeira CC, Phillips TL (1996a) Insect fluid feeding on upper Pennsylvanian tree ferns (Palaeodictyoptera, Marattiales) and the early history of the piercing-and-sucking functional feeding group. Ann Entomol Soc Am 89(2):157–183
Labandeira CC, Phillips TL (1996b) A Carboniferous petiole gall: insight into early ecologic history of the Holometabola. Proc Natl Acad Sci U S A 93:8470–8474
Labandeira CC, Phillips TL (2002) Stem borings and petiole galls from Pennsylvanian tree ferns of Illinois, USA: implications for the origin of the borer and galler functional-feeding-groups and holometabolous insects. Palaeontographica B 264:1–86
Labandeira CC, Sepkoski JJ Jr (1993) Insect diversity in the fossil record. Science 261:310–315
Labandeira CC, Beall BS, Hueber FM (1988) Early insect diversification: evidence from a lower Devonian bristletail from Quebec. Science 242:913–916
Labandeira CC, Johnson KR, Wilf P (2001) Impact of the terminal Cretaceous event on plant–insect associations. Proc Natl Acad Sci U S A 99(4):2061–2066
Labandeira CC, Johnson KR, Lang P (2002) Preliminary assessment of insect herbivory across the Cretaceous–Tertiary boundary: major extinction and minimum rebound. Geol Soc Am Spec Pap 361:297–327
Labandeira CC, Kvaček J, Mostovski MB (2007a) Pollination drops, pollen, and insect pollination of Mesozoic gymnosperms. Taxon 56:663–695
Labandeira CC, Wilf P, Johnson KR, Marsh F (2007b) Guide to insect (and other) damage types on compressed plant fossils. Version 3.0. Smithsonian Institution, Washington, DC. http://paleobiology.si.edu/dfs/InsectDamageGuide3.01
Labandeira CC, Rodriguez-Tovar FJ, Uchman A (2016) The end-Cretaceous extinction and ecosystem change. In: Mángano MG, Buatois LA (eds) The trace-fossil record of major evolutionary events. Top Geobiol 40:265–300
Labandeira CC, Anderson JM, Anderson HM (2018) Expansion of arthropod herbivory in late Triassic South Africa: the Molteno Biota, Aasvoëlberg 411 site and developmental biology of a gall. In: Tanner LH (ed) The Late Triassic world. Top Geobiol 46:623–719
Lass M (1905) Beiträge zur Kenntnis des histologisch-anatomischen Baues des weiblichen Hundflohes (*Pulex canis* Dugès s. *Pulex serraticeps* Taschenberg). Z Wiss Zool 79:73–131
Lawrence JF, Stephan K (1975) The North American Cerylonidae (Coleoptera: Clavicornia). Psyche 82:131–166
Leconte JL (1861) Classification of the Coleoptera of North America. Part 1. Smithsonian Institution, Washington, DC, p 285
Legalov AA (2015) Fossil Mesozoic and Cenozoic weevils (Coleoptera, Obrienioidea, Curculionoidea). Paleontol J 49(13):1442–1513

Lin X, Labandeira CC, Shih CK, Hotton CH, Ren D (2019) Evolutionary biology of new two-winged long-proboscid scorpionflies from mid-Cretaceous Myanmar amber. Nat Commun 10:1235. https://doi.org/10.1038/s41467-019-09236-4

Lobanov AM (1970) The morphology of last instar larvae of Helomyzidae (Diptera). Zool Zhurnal 49:1671–1675 (in Russian)

Lopez AW (1932) Morphological studies of the head and mouthparts of the mature codling-moth larva *Carpocapsa pomonella* (Linn.). Univ Calif Publ Entomol 5:19–36

Lozinski P (1908) Beitrag zur Anatomie und Histologie der Mundwerkzeuge der Myrmeleonidenlarven. Zool Anz 33:473–484

Ludwig CE (1949) Embryology and morphology of the larval head of *Calliphora erythrocephala* (Meigen). Microentomology 14:75–111

Lukashevich ED, Mostovski MB (2003) Hematophagous insects in the fossil record. Paleontol J 37(2):153–161

Mángano MG, Buatois LA, Maples CG, Lanier WP (1997) *Tonganoxichnus*, a new insect trace from the upper Carboniferous of eastern Kansas. Lethaia 30:113–125

Marshall WS (1914) On the anatomy of the dragonfly *Libellula quadrimaculata*. Trans Wis Acad Sci 17:755–790

Martínez-Delclòs X (1989) *Chresmoda aquatica* n.sp. insecto Chresmodidae del Cretácico Inferior de la Sierra del Montsec (Lleida, España). Rev Esp Paleont 4:67–74

Martins-Neto RG (2003) The fossil tabanids (Diptera Tabanidae): when they began to appreciate warm blood and when they began transmit diseases? Mem Inst Oswaldo Cruz 98(Suppl 1):29–34

Martynov AV (1930) On two new primordial insects from Kungurian localities of Perm. Trans Paleontol Inst Akad Sci SSSR 8:36–47

Martynov AV (1938) Permian fossil insects from the Arkhangelsk district. Part V [sic!]. The family Euthygrammatidae and its relatives. Trudy Paleontol Inst Akad Sci SSSR 7(3):69–80. [in Russian]

Mathur PN, Mathur KC (1961) Morphology of the head capsule and its appendages of *Orthetrum taeniolatum* Schneider (Odonata, Anisoptera, Libellulidae, Libellulinae). Proc Ind Acad Sci B 53:187–203

Matsuda R (1957) Morphology of the head of a sawfly, *Macrophya pluricincta* Norton (Hymenoptera, Tenthredinidae). J Kans Entomol Soc 30:99–109

Mayhew PJ (2007) Why are there so many insect species? Perspectives from fossils and phylogenies. Biol Rev 82:425–454

McAlpine DK (1966) Description and biology of an Australian species of Cypselosomatidae (Diptera), with a discussion of family relationships. Aust J Zool 14:673–685

Metcalf CL (1929) The mouthparts of insects. Trans Ill State Acad Sci 21:109–135

Mey W, Wichard W, Ross E, Ross A (2017) On the systematic position of a highly derived amphiesmenopteran insect from Burmese amber (Insecta, Amphiesmenoptera). Trans R Soc Edinb Earth Environ Sci 107:249–254

Michener CD (2000) The bees of the world. Johns Hopkins University Press, Baltimore, p 913

Michener CD, Grimaldi D (1988) The oldest fossil bee: apoid history, evolutionary stasis, and antiquity of social behavior. Proc Natl Acad Sci U S A 85:6424–6426

Mickoleit E (1963) Untersuchungen zur Kopfmorphologie der Thysanopteren. Zool Jb Anat 81:101–150

Miller JY (1971) The head capsule of selected Hesperioidea. J Res Lepidopterol 9:193–214

Miller MF, Labandeira CC (2002) Slow crawl across the salinity divide: delayed colonization of freshwater ecosystems by invertebrates. GSA Today 12(12):4–10

Minter NJ, Buatois LA, Mángano MG, Davies NS, Gibling MR, Labandeira C (2016) The establishment of continental ecosystems. In: Mángano MG, Buatois LA (eds) The trace-fossil record of major evolutionary events. Top Geobiol 39:205–324

Misof B, Liu S, Meusemann K, Peters RS, Donath A, Mayer C, Frandsen PB, Ware J, Flouri T, Beutel RG, Niehuis O, Petersen M, Izquierdo-Carrasco F, Wappler T, Rust J, Aberer AJ,

Aspöck U, Aspöck H, Bartel D, Blanke A, Berger S, Böhm A, Buckley TR, Calcott B, Chen J, Friedrich F, Fukui M, Fujita M, Greve C, Grove P, Gu S, Huang Y, Jermiin LS, Kawahara AY, Krogmann L, Kubiak M, Lanfear R, Letsch H, Li Y, Li Z, Li J, Lu H, Meng G, Nakagald YI, Navarrete-Heredia JL, Ott M, Ou Y, Pass G, Podsiadlowski L, Pohl H, von Reumont BM, Schütte K, Sekiya K, Shimizu S, Ślipiński A, Stamatakis A, Song W, Su X, Szucsich NU, Tan M, Tan X, Tang M, Tang J, Timelthaler G, Tomizuka S, Trautwein M, Tong X, Uchifume T, Walzl MG, Wiegmann BM, Wilbrandt J, Wipfler B, TKF W, Wu Q, Wu G, Xie Y, Yang S, Yang Q, Yeates DK, Yoshizawa K, Zhang Q, Zhang R, Zhang W, Zhang Y, Zhao J, Zhou C, Zhou L, Ziesmann T, Zou S, Li Y, Xu X, Zhang Y, Yang H, Wang J, Wang J, Kjer KM, Zhou X (2014) Phylogenomics resolves the timing and pattern of insect evolution. Science 346(6210):763–767

Morimoto K (1962) Comparative morphology and phylogeny of the superfamily Curculionoidea of Japan (Comparative morphology, phylogeny and systematics of the superfamily Curculionoidea of Japan I). J Fac Agric Kyushu Univ 11:331–373

Moritz G (1989) Die Ontogenese der Thysanoptera (Insecta) unter besonderer Berücksichtigung des Fränsenflüglers *Hercinothrips femoralis* (O.M. Reutter, 1891) (Thysanoptera, Thripidae, Panchaetothripinae). IV. Mitteilung: Imago-Kopf. Zool Jb Anat 118:273–307

Müh C (1985) Kopfmorphologie der Larven der Tanypodinae (Chironomidae, Diptera) am Beispeil von *Macropelopia nebulosi* (Meigen). Zool Jb Anat 113:331–362

Mukerji D, Sen-Sarma P (1955) Anatomy and affinity of the elephant louse, *Haematomyzus elephantis* Piaget (Insecta: Rhynchophthiraptera). Parasitology 45:5–30

Müller AH (1978) Zur Entomofauna des Permokarbon über die Morphologie, Taxonomie und Ökologie von *Eugereon boeckingi* (Palaeodictyoptera). Freib Forsch 334C:7–20

Muñiz R, Barrera A (1969) *Rhopalotria dimidiata* Chervrolat, 1878: studio morfológico del adulto y Descripción de la larva (Ins. Col. Curcul.: Oxycoryninae). Rev Soc Mex Hist Nat 30:205–222

Munscheid L (1933) Die Metamorphose des Labiums der Odonaten. Z Wiss Zool 143:201–240

Murphy HE (1922) Notes on the biology of some of our North American species of may-flies. I. The metamorphosis of may-fly mouth parts. Bull Lloyd Libr Entomol Ser 2:1–39

Naugolnykh SV, Ponomarenko AG (2010) Possible traces of feeding by beetles in coniferophyte wood from the Kazanian of the Kama River basin. Paleontol J 44:468–474

Neiswander CR (1926) On the anatomy of the head and thorax in *Ranatra* (Heteroptera). Trans Am Entomol Soc 51:311–320

Nel P, Azar D, Martínez-Delclòs X, Makhoul E (2004) A new upper Cretaceous species of *Chresmoda* from Lebanon – a latest representative of Chresmodidae (Insecta: Polyneoptera inc. sed.): first record of homeotic mutations in the fossil record of insects. Eur J Entomol 101:145–151

Nel A, Roques P, Nel P, Prokin AA, Bourgoin T, Prokop J, Szwedo J, Azar D, Desutter-Grandcolas L, Wappler T, Garrouste R, Coty D, Huang DY, Engel MS, Kirejtshuk AG (2013) The earliest known holometabolous insects. Nature 503:257–261

Nel P, Retana-Salazar AP, Azar D, Nel A, Huang DY (2014) Redefining the Thripida (Insecta: Paraneoptera). J Syst Palaeontol 12(7):865–878

Nicholson DV, Mayhew PJ, Ross AJ (2015) Changes to the fossil record of insects through fifteen years of discovery. PLoS One 19(7):e0128554

Nielson ES, Scoble MJ (1986) *Afrotheora*, a new genus on primitive Hepialidae from Africa (Lepidoptera: Hepialoidea). Entomol Scand 17:29–54

Noars R (1956) Contribution a la connaissance de la larve *d'Orechtochilus villosus* Mull. Trav Lab Zool Univ Dijon 17:1–32

Novoksonov VG (1995) New fossil insects from the Kungurian of the middle Urals. Paleontol J 29(3):76–85

Oldroyd H (1964) The natural history of flies. Norton, New York, p 324

Otanes FQ (1922) Head and mouth-parts of Mecoptera. Ann Entomol Soc Am 15:310–323

Pakaluk J (1987) Revision and phylogeny of the Neotropical genus *Hoplicnema* Matthews (Coleoptera: Corylophidae). Trans Am Entomol Soc 113:73–116

Parker SP (1982) Synopsis and classification of living organisms, vol 2. McGraw-Hill, New York, pp 1166–1232

Parsons MC (1966) Studies on the cephalic anatomy of Naucoridae (Heteroptera). Trans R Entomol Soc Lond 118:119–151

Peacock AD (1919) The structure of the mouthparts and mechanism of feeding in *Pediculus humanus*. Parasitology 11:98–117

Pecharová M, Prokop J (2017) The morphology of mouthparts, wings and genitalia of Paleozoic insect families Protohymenidae and Scytohymenidae reveals new details and supposed function. Arthopod Struct Dev 47(1):117–129

Pecharová M, Ren D, Prokop J (2015) A new palaeodictyopteroid (Megasecoptera: Brodiopteridae) from the early Pennsylvanian of northern China reveals unique morphological traits and intra-specific variability. Alcheringa 39:236–249

Pellmyr O, Krenn HW (2002) Origin of a complex key innovation in an obligate insect–plant mutualism. Proc Natl Acad Sci U S A 99(8):5498–5502

Peñalver E, Engel MS, Grimaldi DA (2006) Fig wasps in Dominican amber (Hymenoptera: Agaonidae). Am Mus Novit 3541:1–16

Peñalver E, Labandeira CC, Barrón DX, Nel P, Nel A, Tafforeau P, Soriano C (2012) Thrips pollination of Mesozoic gymnosperms. Proc Natl Acad Sci U S A 109(22):8623–8628

Peris D, Pérez-de la Fuente R, Peñalver E, Delclòs X, Barrón E, Labandeira CC (2017) False blister beetles and the expansion of gymnosperm–insect pollination modes before angiosperm dominance. Curr Biol 27(6):897–904

Perrichot V, Beaucournu JC, Velten J (2012) First extinct genus of a flea (Siphonaptera: Pulicidae) in Miocene amber from the Dominican Republic. Zootaxa 3438:54–61

Photosearch Stock Photography and Stock Footage (2018) https://www.fotosearch.com/ulv395u20110003. Accessed Nov, 2018

Picker MD (1987) An unusual species of spoon-wing lacewing (Neuroptera: Nemopteridae) from South Africa, with notes on its biology. Syst Entomol 12:239–248

Ping C (1928) Cretaceous fossil insects of China. Paleontol Sin B 13(1):1–56

Pohl H, Wipfler B, Grimaldi DA, Beckmann F, Beutel RG (2010) Reconstructing the anatomy of the 42-million year old fossil *Mengea tertiaria* (Insecta, Sterpsiptera). Naturwissenschaften 97:855–859

Poinar GO Jr, Brown AE (2017) An exotic insect *Aethiocarenus burmanicus* gen. et sp. nov. (Aethiocarenodea ord. nov., Aethiocarenidae fam. nov.) from mid-Cretaceous Myanmar amber. Cretac Res 72:100–104

Poinar GO Jr, Danforth BN (2006) A fossil bee from early Cretaceous Burmese amber. Science 314(5799):614

Poinar GO Jr, Miller JC (2002) First fossil record of endoparasitism of adult ants (Formicidae: Hymenoptera) by Braconidae (Hymenoptera). Ann Entomol Soc Am 95:41–43

Ponomarenko AG (1969) Historical development of archostematan beetles. Trudy Paleontol Inst Akad Nauk SSSR 125:1–240 (in Russian)

Ponomarenko AG (1990) Scarabaeida. In: Rasnitsyn AG (ed) Late Mesozoic insects of eastern Transbaikalia. Nauka Press, Moscow, pp 39–87

Ponomarenko AG (2012) Supposed alderfly larva (Insecta, Megaloptera) from the Jurassic locality Shar-Teg, southwestern Mongolia. Paleontol J 46(4):378–382

Popadić A, Abzhanov A, Rusch D, Kaufman TC (1998) Understanding the genetic basis of morphological evolution: the role of homeotic genes in the diversification of the arthropod bauplan. Int J Dev Biol 42:453–461

Popham EJ (1961) The functional morphology of the mouthparts of the cockroach *Periplaneta americana* L. Entomologiste 94:185–192

Popham EJ, Bevan E (1979) Functional morphology of the feeding apparatus in larval and adult *Aeshna juncea* (L.) (Anisoptera: Aeshnidae). Odonatologia 8:301–318

Principi MM (1954) Contributi allo studio dei neurotteri Italiani. XI. *Chrysopa viridana* Schn. Boll Inst Entomol Bologna Univ 20:359–376

Prokop J, Pecharová M, Ren D (2016) Hidden surface microstructures on Carboniferous insect *Brodioptera sinensis* (Megasecoptera) enlightens functional morphology and sensorial perception. Sci Rep 6:28316. https://doi.org/10.1038/srep28316

Prokop J, Pecharová M, Garrouste R, Beattie R, Chintauan-Marquier IC, Nel A (2017) Redefining the extinct orders Miomoptera and Hypoperlida as stem acercarian insects. BMC Evol Biol 17:205. https://doi.org/10.1186/s12862-017-1039-3

Puri VD (1958) Studies on the anatomy of the sugarcane moth borers. III. Comparative study of the head and its appendages. Proc Ind Acad Sci B 46:96–109

Quenneday A (1965) Contribution à la connaissance de quelques types larvaires de Sphaeriinae (Col. Hydrophilidae). Trav Lab Zool Fac Sci Dijon 65:1–56

Raj JS (1943) Mouth parts of *Lamprophorus tenebrosus* Wlk. Curr Sci 12:83–84

Ramcke J (1965) Kopf der Schweinelaus (*Haematopinus suis* L., Anoplura). Zool Jb Anat 82:547–663

Randriamamonj F (1963) Squelette et musculature cephalique de *Phthorimaea heliopa* Low. (Gelechiidae, Lepidoptera). Trav Lab Zool Univ Dijon 51:1–76

Rasnitsyn AP (1980) Origin and evolution of Hymenoptera. Trudy Paleontol Inst Akad Nauk SSSR 174:1–190. (in Russian)

Rasnitsyn AP (1990) Late Mesozoic insects of eastern Transbaikalia. Trans Paleontol Inst Akad Nauk SSSR 239:1–224. (in Russian)

Rasnitsyn AP (1992) *Strashila incredibilis*, a new enigmatic mecopteroid insect with possible siphonapteran affinities from the upper Jurassic of Siberia. Psyche 99:323–333

Rasnitsyn AP, Quicke DLJ (2002) History of insects. Kluwer, Dordrecht, p 517

Rathore YK (1961) Studies on the mouth-parts and feeding mechanism in *Dysdercus cingulatus* Fabr. (Pyrrhocoridae: Heteroptera). Ind J Entomol 23:163–185

Ren D, Labandeira CC, Santiago-Blay JA, Rasnitsyn A, Shih C, Bashkuev A, Logan MAV, Hotton CL, Dilcher D (2009) A probable pollination mode before angiosperms: Eurasian, long-proboscid scorpionflies. Science 362:840–847

Ren D, Shih CK, Labandeira CC (2011) A well-preserved aneuretopsychid from the Jehol Biota of China (Insecta: Mecoptera: Aneuretopsychidae). ZooKeys 121:17–28

Ren D, Shih CK, Gao TP, Wang YJ, Yao YZ (eds) (2019) Rhythms of insect evolution: evidence from the Jurassic and Cretaceous in northern China. Wiley Blackwell, Chichester, p 707

Rentz DCF (1980) A new family of ensiferous Orthoptera from the coastal sands of Southeast Queensland. Mem Qld Mus 20:49–63

Retallack GJ, Dilcher DL (1988) Reconstructions of selected seed ferns. Ann Mo Bot Gard 79:1010–1057

Reyne A (1927) Untersuchungen über die Mundteile der Thysanopteren. Zool Jb 49:391–500

Rieger C (1976) Skelett und Muskulatur des Kopfes und Prothorax von *Ochterus marginatus* Latreille. Zoomorphologie 83:109–191

Riek EF (1968) A new family of caddis-flies from Australia (Trichoptera: Tasimiidae). J Aust Entomol Soc 7:109–114

Riek EF, Kukalová-Peck J (1984) A new interpretation of dragonfly wing venation based upon early Carboniferous fossils from Argentina (Insecta: Odonatoidea) and basic character states inn pterygote wings. Can J Zool 62:1150–1166

Riley CV (1872) The fertilization of the yucca plant by *Pronuba yuccasella*. Can Entomol 4:182

Risebrow A, Dixon AFG (1987) Nutritional ecology of phloem-feeding insects. In: Slansky F Jr, Rodriguez JG (eds) Nutritional ecology of insects, mites, spiders, and related invertebrates. Wiley, New York, pp 421–448

Risler H (1957) Der Kopf von *Thrips physapus* L. (Thysanoptera, Terebrantia). Zool Jb Anat 76:251–302

Rivnay E (1929) Revision of the Rhipiphoridae of North and Central America (Coleoptera). Mem Am Entomol Soc 6:1–68

Robinson GG (1939) The mouthparts and their function in the female mosquito, *Anopheles maculipennis*. Parasitology 31:212–242

Rogers BT, Peterson MD, Kaufman TC (2002) The development and evolution of insect mouthparts as revealed by the expression patterns of gnathocephalic genes. Evol Dev 4(2):96–110

Rohdendorf BB (1962a) The order Psocoptera. In: Rohdendorf BB (ed) Fundamentals of paleontology, Arthropoda: Tracheata and Chelicerata, vol 9. Akademia Nauk, Moscow, pp 226–236. (in Russian)

Rohdendorf BB (1962b) The order Diptera. In: Rohdendorf BB (ed) Fundamentals of paleontology, Arthropoda: Tracheata and Chelicerata, vol 9. Akademia Nauk, Moscow, pp 444–502. (in Russian)

Rohdendorf BB (1974) The historical development of Diptera. Trans: Russian by JE Moore, I Thiele. University of Alberta Press, Edmonton

Rohdendorf BB, Rasnitsyn AP (1980) Historical development of the Class Insecta. Trans Paleont Inst Acad Nauk SSSR 175:1–258

Rothschild NC (1906) The physiological anatomy of the mouth-parts and alimentary canal of the Indian rat flea, *Pulex cheopis*, Rothschild. J Hyg 6:486–495

Rouchy R (1964) Squelette et musculature cephaliques de deux types d'imagos de lepidoptérès heteroneurus. Trav Lab Zool Univ Dijon 61:1–59

Rousset A (1936) Sur l'anatomie céphalique les larves Coniopterygidae (Neuroptérès Planipennes). C R Acad Sci Paris 242:934–936

Russell LM (1941) A classification of the scale insect genus *Asterolecanium*. U S Dept Agric Misc Publ 424:1–322

Sattler W (1958) Beiträge zur Kenntnis von Lebensweise und Körperbau der Larve und Puppe von *Hydropsyche* Pict. (Trichoptera) mit besonderer Berücksichtigung des Netzbaues. Z. Morphol Ökol Tiere 47:115–192

Schachat SR, Labandeira CC, Saltzman M, Cramer B, Payne J, Boyce CJ (2018) Phanerozoic $pO_2$ and the early evolution of terrestrial animals. Proc R Soc B 285:20172631

Schiemenz H (1957) Vergleichende funktionelle-anatomische Untersuchungen der Kopfmuskulatur von *Theobaldia* und *Eristalis* (Dipt. Culicid. und Syrphid.). Deut Entomol Z 4:268–331

Schmitt JB (1938) The feeding mechanism of adult Lepidoptera. Smithson Misc Collect 97:1–28

Schmitz G (1943) Le Labium et les structures buccopharyngiennes du genre *Stenus* Latreille. Cellule 49:291–334

Schonmann H (1981) Zur Kopfmorphologie der Ephemeriden-Larven *Siphlonurus aestivalis* Eaton und *Lepeorus goyi goyi* Peters. Zoologica 45:1–51

Schremmer F (1950a) Morphologische und funktionelle Analyse der Mundteile und des Pharynx der Larve von *Anopheles maculipennis* Meig. Oster Zool Z 2:173–222

Schremmer F (1950b) Bau und Funktion der Larvenmundteile der Dipterengattung *Dixa* Meigen. Oster Zool Z 2:379–413

Schremmer F (1952) Die Mundteile der Brachycerenlarven und der Kopfbau der Larve von *Stratiomys chamaeleon* L. Oster Zool Z 3:326–397

Scott AC, Taylor TN (1983) Plant/animal interactions during the upper Carboniferous. Bot Rev 49:259–307

Scourfield DJ (1940) The oldest known fossil insect (*Rhyniella praecursor* Hirst & Maulik)—further details from additional specimens. Proc Linn Soc Lond 152:113–131

Scudder SH (1885) Palaeodictyoptera; or the affinities and classification of Paleozoic Hexapoda. Mem Boston Soc Nat Hist 3:319–351

Scudder SH (1892) Some insects of special interest from Florissant, Colorado, and other points in the territories of Colorado and Utah. Bull US Geol Surv 93:1–25

Seghal SS (1963) The skeleton-musculature of the common Indian yellow wasp, *Polistes olivaceus* (De Geer). I. The head and its appendages. Proc Zool Soc Ind 16:153–167

Sen Gupta T, Crowson RA (1973) A review of the classification of Cerylonidae (Coleoptera, Clavicornia). Trans R Entomol Soc Lond 124:365–446

Shalaby AM (1956) On the mouth parts of the larval instars of *Anopheles quadrimaculatus* (Say) (Diptera–Anophelini). Bull Soc Entomol Egypte 40:137–174

Sharov AG (1957) Peculiar Paleozoic wingless insects of the new order Monura (Insecta, Apterygota). Dokl Akad Nauk SSSR 115:796–798 (in Russian)
Sharov AG (1966) The position of the orders Glosselytrodea and Caloneurodea in insect classification. Paleontol Zh 3:84–93. (in Russian)
Sharov AG (1968) Phylogeny of orthopteroid insects. Trudy Paleontol Inst Akad Nauk SSSR 118:1–218 (English translation from Russian, Israel Program for Scientific Translation, 1971)
Sharov AG (1973) Morphological features and way of life of Palaeodictyoptera. In: Bei-Benko GY (ed) Twenty-fourth annual lectures in memory of N.A. Kholodkovskogo. Academy of Sciences, Moscow, pp 49–63
Shcherbakov DE, Makarkin VN, Aristov DS, Vasilenko DV (2009) Permian insects from the Russky Island, South Primorye. Russ Entomol J 18:7–16
Shear WA, Kukalová-Peck J (1990) The ecology of Paleozoic terrestrial arthropods: the fossil evidence. Can J Zool 68:1807–1834
Shear WA, Bonamo PM, Grierson JD, Rolfe WDI, Smith EL, Norton RA (1984) Early land animals in North America: evidence from Devonian age arthropods from Gilboa, New York. Science 224:492–494
Short JRT (1955) The morphology of the head of *Aeshna cyanea* (Müller) (Odonata: Anisoptera). Trans R Entomol Soc London 106:197–211
Sikes EK (1930) Larvae of *Ceratophyllus wickhami* and other species of fleas. Parasitology 12:242–259
Silvestri F (1924) Contribuzioni alla conoscenza Tortricidi delle querce. Boll Lab Zool Gen Agric Portici 27:41–107
Slansky F Jr, Panizzi AR (1987) Nutritional ecology of seed-sucking insects. In: Slansky F Jr, Rodriguez JG (eds) Nutritional ecology of insects, mites, spiders, and related invertebrates. Wiley, New York, pp 283–320
Sluss TP, Foote BA (1971) Biology and immature stages of *Leucopis verticalis* (Diptera: Chamaemyiidae). Can Entomol 103:1427–1434
Snodgrass RE (1921) The mouth parts of the cicada. Proc Entomol Soc Wash 23:1–15
Snodgrass RE (1928) Morphology and evolution of the insect head and its appendages. Smithson Misc Collect 81(3):1–158
Snodgrass RE (1943) The feeding apparatus of biting and disease-carrying flies: a wartime contribution to medical entomology. Smithson Misc Collect 104:1–51
Snodgrass RE (1954) The dragonfly larva. Smithson Misc Collect 123(2):1–38
Spencer KA (1969) The Agromyzidae of Canada and Alaska. Mem Entomol Soc Can 64:1–311
Spencer KA (1992) Agromyzidae. In: McAlpine JF, Peterson BV, Shewell GE, Teskey HJ, Vockeroth JF, Wood DM (eds) Manual of Nearctic Diptera, vol 2. Res Br Agric Can 28:869–879
Srivastava RP, Bogawat JK (1969) Feeding mechanism of a fruit-sucking moth *Othreis materna* (Lepidoptera: Noctuidae). J Nat Hist 3:165–181
Stoffolano JG Jr, Yin LRS (1983) Comparative study of the mouthparts and associated sensillae of adult male and female *Tabanus nigrovittatus* (Diptera, Tabanidae). J Med Entomol 20:11–32
Stojanovich CJ (1945) The head and mouthparts of the sucking lice (Insecta: Anoplura). Microentomology 10:1–46
Strenger A (1949) Funktionsstudie des Kopfes von *Forficula auricularia*. Zool Jb Anat 70:557–575
Strenger A (1952) Die funktionelle und morphologische Bedeutung der Nähte am Insektenkopf. Zool Jb Anat 72:468–521
Strenger A (1954) Zur Kopfmorphologie der Ephemeriden-Larven. Erster Teil. *Ecdyonurus* und *Rithrogena*. Öster Zool Z 4:191–228
Sutcliffe JF, Deepan PD (1988) Anatomy and function of the mouthparts of the biting midge, *Culicoides sanguisuga* (Diptera: Ceratopogonidae). J Morphol 198:363–365
Taylor JS (1957) Notes on the proboscis in Lepidoptera. Entomol Rec 69(25–29):53–58
Taylor TN, Osborn JM (1996) The importance of fungi in shaping the paleoecosystem. Rev Palaeobot Palynol 90:249–262

Taylor TN, Klavins SD, Krings M, Taylor EL, Kerp H, Hass H (2004) Fungi from the Rhynie Chert: a view from the dark side. Trans R Soc Edinb Earth Sci 94:457–473
Theron JG (1958) Comparative studies on the morphology of male scale insects (Hemiptera: Coccoidea). Ann Univ Stellenbosch A 34:1–71
Tillyard RJ (1923) On the mouth-parts of the Micropterygoidea. Trans Entomol Soc Lond 1923:181–206
Tillyard RJ (1924) Kansas Permian insects. Part 3: the new order protohymenoptera. Am J Sci 8 (44):111–122
Tillyard RJ (1928) Some remarks on the Devonian fossil insects from the Rhynie Chert beds, Old Red Sandstone. Trans Entomol Soc Lond 76:65–71
Tower DG (1914) The mechanism of the mouth parts of the squash bug, *Anasa tristis tristis* DeGeer. Psyche 21:99–108
Tragardh I (1913) Contributions towards the comparative morphology of the trophi of the lepidopterous leaf-miners. Arkiv Zool 8:1–48
Tuxen SL (1959) The phylogenetic significance of entognathous apterygotes. Smithson Misc Collect 137:379–416
Ulrich W (1924) Die Mundwerkzeuge der Spheciden (Hym, Foss.). Beitrag zur Kenntnis der Insektenmundwerkzeuge. Z Morphol Ökol Tiere 1:539–636
Vaillant F, Delhom MM (1956) Les forms adaptives de l'appareil bucco-pharyngien chez les larves de Stratiomyiidae (Diptera). Bull Soc Hist Nat Afr Nord 47:217–250
Vasudeva S (1956) Studies on the morphology of the common lemon butterfly, *Papilio demoleus demoleus* L. J Zool Soc Ind 8:211–234
Verma SR, Tyagi MP, Gupta SP (1973) Studies on the morphology of *Belostoma indicus* Lep. et Serv. (head capsule and mouth parts). Zool Jb Anat 91:449–463
Vilesov AP (1995) Permian neuropterans (Insecta: Myrmeleontida) from the Chekarda locality in the Urals. Paleontol J 29:115–129
Vishniakova VN (1981) New Palaeozoic and Mesozoic Lophioneuridae (Thripida). In: Vishniakova VN, Dlussky GM, Pritykina LN (eds) New fossil insects from the territory of the USSR. Trans Paleontol Inst Acad Nauk SSSR 183:43–63 (in Russian)
Vit S (1977) Contribution a la connaissance des Eucinetidae (Coleoptera). Rev Suisse Zool 84:917–935
Vit S (1981) Une nouvelle espèce du genre *Tohlezkus* Vit de Taiwan (Coleoptera, Eucinetidae). Rev Suisse Zool 88:669–774
Vogel R (1915) Beitrag zur Kenntnis des Baues und der Lebensweise der Larve von *Lampyris noctiluca*. Z Wiss Zool 112:291–432
Walker EM (1933) On the anatomy of *Grylloblatta campodeiformis* Walker. 2. Comparisons of head with those of other orthopteroid insects. Ann Entomol Soc Am 24:309–344
Walker JD, Geissman JW, Bowring SA, Babcock LE (2013) The Geological Society of America time scale. Geol Soc Am Bull 125(3/4):259–272
Wang J, Labandeira CC, Zhang ZF, Bek J, Pfefferkorn HW (2009) Permian *Circulipuncturites discinisporis* Labandeira, Wang, Zhang, Bek et Pfefferkorn gen. et spec. nov. (formerly *Discinispora*) from China, an ichnotaxon of punch-and-sucking insect on Noeggeranthialean spores. Rev Palaeobot Palynol 156:277–282
Wappler T, Currano ED, Wilf P, Rust J, Labandeira CC (2009) No post-Cretaceous ecosystem depression in European forests? Rich insect-feeding damage on diverse middle Palaeocene plants, Menat, France. Proc R Soc B 276:4271–4277
Ward P, Labandeira CC, Laurin M, Berner RA (2006) Confirmation of Romer's gap as a low oxygen interval constraining the timing of initial arthropod and vertebrate terrestrialization. Proc Natl Acad Sci U S A 103(45):16818–16822
Watson MG (1955) The utilization of mandibular armature in taxonomic studies of anisopterous nymphs. Trans Entomol Soc Am 81:155–205
Weber H (1929) Kopf und Thorax von *Psylla mali* Schmidb. (Hemiptera—Homoptera): eine morphogenetische Studie. Z Morphol Ökol Tiere 14:59–165

Weber H (1938) Beiträge zur Kenntnis der Überordnung Psocoidea. 1. Die Labialdrüsen der Copeognathen. Zool Jb Anat 64:243–286
Wedmann S (2000) Die Insekten der oberoligozänen Fossillagerstätte Enspel (Westerwald, Deutschland): Systematik, Biostrationomie und Paläoökologie. Mainz Naturwiss Arch 23:1–154
Weiblen GD (2002) How to be a fig wasp. Annu Rev Entomol 47:299–330
Weinreich E (1968) Über den Klebfangapparat der Imagines von *Stenus* Latr. (Coleopt., Staphylinidae) mit einem Beitrag zur Kenntnis der Jugendstadien dieser Gattung. Z Morphol Ökol Tiere 62:162–310
Wenk P (1953) Der Kopf von *Ctenocephalus canis* (Curt.) (Aphaniptera). Zool Jb Anat 73:103–164
Wenk P (1962) Anatomie des Kopfes von *Wilhelmia equina equina* L. (Simuliidae syn. Melusinidae, Diptera). Zool Jb Anat 80:81–134
Wesché W (1902) Undescribed palpi on the proboscis of some dipterous flies, with remarks on the mouth-parts in several families. J R Micros Soc 1902:412–416
Wesché W (1904) The labial and maxillary palpi in Diptera. Trans Linn Soc Lond 9:219–230
Wesché W (1909) The mouth-parts of the Nemocera and their relations to the other families in Diptera. Corrections and additions to the paper published in 1904. J R Micros Soc 1909:1–16
Whalley PES, Jarzembowski EA (1981) A new assessment of *Rhyniella*, the earliest known insect, from the Devonian of Rhynie, Scotland. Nature 291:317
Whiting JH Jr, Black HL, Jorgensen CD (1989) A scanning electron microscopy study of the mouthparts of *Paraponera clavata* (Hymenoptera: Formicidae). Pan Pac Entomol 65:302–309
Wilf P, Labandeira CC, Johnson KR (2006) Decoupled plant and insect diversity after the end-Cretaceous extinction. Science 313:1112–1115
Williams IW (1933) The external morphology of the primitive tanyderid dipteron *Protoplasa fitchi* O.S., with notes on the other Tanyderidae. J N Y Entomol Soc 41:1–34
Willis DM (1949) The anatomy and histology of the head, gut and associated structures of *Typhlocyba ulmi*. Proc Zool Soc Lond 118:984–1001
Willmann R (2005) Reinterpretation of an alleged marine hexapod stem-group representative. Org Div Evol 5(3):199–202
Winkler IS, Labandeira CC, Wappler T, Wilf P (2010) Distinguishing Agromyzidae (Diptera) leaf mines in the fossil record: new taxa from Paleogene of North America and Germany and their evolutionary implications. J Paleontol 84:935–954
Winston ML (1979) The proboscis of the long-tongued bees: a comparative study. Univ Kans Sci Bull 51(22):631–667
Wirth WW (1952) Three new Nearctic species of *Systenus* with a description of the immature stages from tree cavities. Proc Entomol Soc Wash 54:236–244
Wolter H (1963) Vergleichende Untersuchungen zur Anatomie und Funktionsmorphologie der stechend-saugenden Mundwerkzeuge der Collembolen. Zool Jb Anat 81:27–100
Wong HR (1963) The external morphology of the adults and ultimate larval instar of the larch sawfly, *Pristiphora erichsonii* (Htg.) (Hymenoptera: Tenthredinidae). Can Entomol 95:897–921
Wootton RJ (1981) Palaeozoic insects. Annu Rev Entomol 26:319–344
Wundt H (1961) Der Kopf der Larve von *Osmylus chrysops* L. (Neuroptera, Planipennia). Zool Jb Anat 79:557–662
Xu Q, Jin J, Labandeira CC (2018) Williamson Drive: herbivory of a north-central Texas flora of latest Pennsylvanian age showing discrete component community structure, early expansion of piercing and sucking, and plant counterdefenses. Rev Palaeobot Palynol 251:28–72
Yang AS (2001) Modularity, evolvability, and adaptive radiations: a comparison of the hemi- and holometabolous insects. Evol Dev 3(2):59–72
Yoshizawa K, Lienhard C (2016) Bridging the gap between chewing and sucking in the hemipteroid insects: new insights from Cretaceous amber. Zootaxa 4079(2):229–245
Zaitland LM, Larsen JR (1984) Morphologie of the head of *Mydas clavatus* Drury (Diptera, Mydaidae). Int J Inst Morphol Embryol 13:105–136

Zalessky MD (1930) Sur deux représentants nouveaux des Paléohémiptères du Permien de la Kama et du Perebore dans le Bassin de la Péchora. Izv Akad Nauk SSSR 10:1–17

Zaspel JM, Zahiri R, Hoy MA, Janzen D, Weller SJ, Wahlberg N (2012) A molecular phylogenetic analysis of the vampire moths and their fruit-piercing relatives (Lepidoptera: Erebidae: Calpinae). Mol Phylo Evol 65:786–791

Zhang J, Kluge NJ (2007) Jurassic larvae of mayflies (Ephemeroptera) from the Daohugou Formation in inner Mongolia, China. Orient Insects 41:351–366

Zwick P (1967) Beschreibung der aquatischen Larve von *Neurorthus fallax* (Rambur) und Errichtung der neuen Planipennien-Familie Neurorthidae fam. nov. Gewass Abwass 44(45):65–86

# Index





H. W. Krenn (ed.), *Insect Mouthparts*, Zoological Monographs 5,
https://doi.org/10.1007/978-3-030-29654-4

**N**



**O**

## T



## V

MIX
Papier aus verantwortungsvollen Quellen
Paper from responsible sources
FSC® C105338

If you have any concerns about our products,
you can contact us on
**ProductSafety@springernature.com**

In case Publisher is established outside the EU,
the EU authorized representative is:
**Springer Nature Customer Service Center GmbH**
**Europaplatz 3, 69115 Heidelberg, Germany**

Printed by Libri Plureos GmbH
in Hamburg, Germany